Particle Physics and Introduction to Field Theory

Contemporary Concepts in Physics

A series edited by
Henry Primakoff,
University of Pennsylvania

Associate Editors:

Eli Burstein
University of Pennsylvania

Willis Lamb
University of Arizona

Leon Lederman
Fermi National
Accelerator Laboratory

Sir Rudolf Peierls
Oxford University

Mal Ruderman
Columbia University

Volume 1
Particle Physics and Introduction to Field Theory

Additional volumes in preparation.

ISSN: 0272-2488

Science Press (Beijing)

Particle Physics and Introduction to Field Theory

粒子物理和场论简引

T.D. Lee
Columbia University

harwood academic publishers
chur london new york

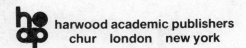

Published under license and distributed by:
Harwood Academic Publishers GmbH
Poststrasse 22
CH-7000 Chur, Switzerland

Editorial Office for the United Kingdom:
61 Grays Inn Road
London WC1X 8TL

Editorial Office for the United States of America:
Post Office Box 786
Cooper Station
New York, New York 10276

A Chinese edition was published by
Science Press, Beijing

Library of Congress Cataloging in Publication Data
Lee, T.D., 1926—
Particle physics and introduction to field theory.
(Contemporary concepts in physics; v. 1)
1. Particles (Nuclear physics) 2. Field theory (Physics)
3. Symmetry (Physics) 4. Nuclear reactions. I. Title. II. Series.
QC793.2.L43 1981 539.7'21 80-27348
ISBN 3-7186-0032-3 AACR1
ISBN3-7186-0033-1 (pbk.)

Library of Congress catalog card number 80-27348. ISSN 0272-2488. ISBN 3-7186 0032-3 (cloth); 3-7186-0033-1 (paperback).

Printed in the United States of America.

TO JEANNETTE

Contents

PREFACE TO THE SERIES

The series of volumes, Concepts in Contemporary Physics, of which the present book by T. D. Lee is the first, is addressed to the professional physicist and to the serious graduate student of physics. The subjects to be covered will include those at the forefront of current research. It is anticipated that the various volumes in the series will be rigorous and complete in their treatment, supplying the intellectual tools necessary for the appreciation of the present status of the areas under consideration and providing the framework upon which future developments may be based. An examination of Professor Lee's book reveals that these criteria are more than amply fulfilled and that his lucid, unique and profound presentation of the entire subject of elementary particle physics and quantum field theory is an excellent model for the volumes to follow in this series.

H. Primakoff

PREFACE

In the spring of 1979 I was invited by Academia Sinica to give lectures on particle physics and statistical mechanics in Beijing. Parts of the lecture notes have been published in Chinese, and the particle physics section evolved into this volume.

The intention of the course was to bring both theoretical and experimental students of physics to the forefront of this very exciting and active field. Because of the different backgrounds of those who attended my lectures, this book is self-contained. Whenever possible, I have adopted an approach that is more pragmatic than axiomatic. All derivations are done explicitly, which at times to a more sophisticated reader may appear pedantic.

Among the topics not treated are the details of renormalization in field theory, the use of dispersion techniques in particle physics and the very beautiful topological soliton solutions in gauge theories. Fortunately, some of these subjects are well covered by the well-known books of Bjorken and Drell and the recent volume by Itzykson and Zuber.

Many people have given me valuable suggestions, and I wish to thank in particular N. H. Christ and A. H. Mueller. It would not have been possible for me to complete this volume without the essential assistance of Irene Tramm.

T. D. Lee

New York
July 1981

I. INTRODUCTION TO FIELD THEORY

The main purpose of this course is to discuss particle physics. Because field theory, and especially relativistic local field theory, is the main theoretical tool for analyzing particle physics, we shall first give an introduction to this subject. As we shall see, through weak and electromagnetic processes, the local field theory has been found applicable at least to regions larger than $\sim 10^{-15}$ cm. For smaller regions, while there still does not exist sufficient experimental evidence, we shall assume that it is also valid.

To begin with, let us discuss the dimensions that will be used in this book. Let $[M]$, $[L]$ and $[T]$ represent respectively the dimensions of mass, length and time. The dimensions of other physical constants can be expressed in terms of these three basic dimensions. For example,

$$\text{velocity of light } c : \quad [c] = \frac{[L]}{[T]} ,$$

$$\hbar = \frac{1}{2\pi} \text{ Planck's constant} : \quad [\hbar] = \frac{[M][L]^2}{[T]} ,$$

$$\text{fine structure constant } \alpha = \frac{e^2}{4\pi\hbar c} : [\alpha] = [1]$$

in which e is the electric charge and $[A]$ represents the dimension of A.

The units of these three independent dimensions can be arbitrarily chosen. In our discussions we shall adopt the natural units:

$$c = \hbar = 1 .$$

Therefore, we have

$$[L] = [T] ,$$
$$[M] = [L]^{-1} ,$$
$$[e^2] = [1] \quad .$$

Any equation in physics, say $A = B$, must satisfy the requirement that the dimension of A should be the same as that of B: $[A] = [B]$. This seemingly elementary requirement can serve the useful purpose of verifying the correctness of one's equations, especially after a long calculation. In the natural units, because there remains one dimension whose unit has not been fixed, say $[L]$, one can still use such dimensional considerations for purposes of verification.

In the following, we denote a three-dimensional position vector by $\vec{r}$, and the components of a four-dimensional position vector by $x_k = \vec{r}_k$ for $k = 1, 2, 3$ and $x_4 = it$. A vector in the Hilbert space (see p. 14) is represented by a ket $|a>$, or the corresponding column matrix $\psi(a)$; its conjugate is the bra $<a|$, or $\psi(a)^\dagger$. The scalar product between two vectors $|a>$ and $|b>$ is

$$<a \mid b> = \psi(a)^\dagger \psi(b) = \sum_n \psi_n^*(a) \psi_n(b)$$

where $\dagger$ denotes the Hermitian conjugation, $*$ the complex conjugation and the sum n extends over all the components of the ψ's.

Chapter 1

MECHANICS OF A FINITE SYSTEM (REVIEW)

1.1 Classical Mechanics

Let us first consider a classical system of particles whose generalized coordinates are q_i $(i = 1,2, \cdots, N)$. For example, $N = 3n$ if we have n particles in three dimensions. Suppose the Lagrangian is

$$L = L(q_i , \dot{q}_i) \tag{1.1}$$

where $\dot{q}_i$ denotes the time derivative of q_i. The Lagrangian equation of motion is given by the variational principle

$$\delta \int_{t_1}^{t_2} L dt = 0 , \tag{1.2}$$

in which δ denotes the variation with the boundary condition $\delta q_i = 0$ at the initial and final times, t_1 and t_2. This is the well-known action principle, and it leads to Lagrange's form of the equations of motion

$$\frac{d}{dt} \frac{\partial L}{\partial \dot{q}_i} - \frac{\partial L}{\partial q_i} = 0 . \tag{1.3}$$

The generalized momentum p_i is

$$p_i \equiv \frac{\partial L}{\partial \dot{q}_i} . \tag{1.4}$$

The Hamiltonian of the system is then given by

$$H(q_i , p_i) \equiv \sum_i p_i \dot{q}_i - L . \tag{1.5}$$

The transformation relating $L(q_i, \dot{q}_i)$ and $H(q_i, p_i)$ is called the Legendre transformation, in which L is regarded as a function of q_i and $\dot{q}_i$, while H is a function of q_i and p_i.

Quite often, we shall adopt the convention that repeated indices are supposed to be summed over. Thus, (1.5) can be simply written as

$$H(q_i, p_i) = p_i \dot{q}_i - L .$$

From (1.3)-(1.5) one readily obtains Hamilton's form of the equations of motion

$$\dot{p}_i = -\frac{\partial H}{\partial q_i} , \quad \dot{q}_i = \frac{\partial H}{\partial p_i} . \tag{1.6}$$

1.2 Quantization

Next, we shall discuss the quantization of the system. Assuming that the Hamiltonian $H(q_i, p_i)$ is given for a classical system, in order to quantize this system we first regard $q_i(t)$ and $p_i(t)$ as operators which satisfy the commutation relations

$$[q_i(t), p_j(t)] = i\delta_{ij}$$

and

$$[q_i(t), q_j(t)] = [p_i(t), p_j(t)] = 0 \tag{1.7}$$

in which $[A, B] = AB - BA$ and δ_{ij} is the Kronecker symbol

$$\delta_{ij} = \begin{cases} 1 & i = j \\ 0 & i \neq j . \end{cases}$$

In passing from classical to quantum mechanics, each physical observable becomes a Hermitian operator. If we represent a Hermitian operator in its matrix form, then its matrix elements satisfy

$$A_{ij} = (A^\dagger)_{ij} \equiv (A^*)_{ji} = (A_{ji})^*$$

where $\dagger$ denotes Hermitian conjugation and $*$ complex conjugation.

Thus we have

$$q_i = q_i^\dagger \ , \quad p_i = p_i^\dagger \ , \quad L = L^\dagger \ \text{ and } \ H = H^\dagger \ . \qquad (1.8)$$

In classical mechanics the time dependence of q_i and p_j is given by Hamilton's equations (1.6). In quantum mechanics the time derivative $\dot{O}$ of any operator $O(t)$ is determined by Heisenberg's equation

$$[H, O(t)] = -i\dot{O}(t) \ . \qquad (1.9)$$

By regarding H as a polynomial of q_i and p_i, one can verify that Heisenberg's equation leads to the same Hamilton equations when $O(t) = q_i$ and p_i. [See the example below.]

In classical mechanics q_i and p_j commute. Thus there can be ambiguities in passing from the classical Hamiltonian $H(q_i, p_j)$ to its quantum-mechanical form. For example

$$H_1 = p^3 q^2 + q^2 p^3 \qquad \text{and} \qquad H_2 = 2pqpqp$$

represent identical systems in classical mechanics, but in quantum mechanics they correspond to different Hamiltonians. One may therefore ask which form one should choose. The answer is that these are two different quantum-mechanical systems, each having the same classical limit. Knowing the classical limit does not always imply a unique determination of the corresponding quantum-mechanical system. For a realistic physical system, only through direct comparison between the experimental result and the theoretical analysis can one be sure which Hamiltonian form is the correct one.

Example. The harmonic oscillator

The simplest harmonic oscillator is one with unit frequency. Its Lagrangian is

$$L = L(q, \dot{q}) = \tfrac{1}{2}(\dot{q}^2 - q^2) \ .$$

Hence

$$p = \frac{\partial L}{\partial \dot{q}} = \dot{q} \quad , \tag{1.10}$$

and therefore

$$H(q, p) = \tfrac{1}{2}(p^2 + q^2) \quad .$$

By using Hamilton's equations (1.6), one obtains

$$\dot{q} = \frac{\partial H}{\partial p} = p \quad \text{and} \quad \dot{p} = -\frac{\partial H}{\partial q} = -q \quad . \tag{1.11}$$

In classical mechanics, the commutators between these functions are all zero. They are called c. number (for commuting) functions.

To quantize the system we change all the above c. number functions to the appropriate q. number (for quantum) operators. By using Heisenberg's equation (1.9) we derive

$$\begin{aligned} -i\dot{p} &= [H, p] = \tfrac{1}{2}[q^2, p] \\ &= \tfrac{1}{2}(q(qp - pq) - (pq - qp)q) = iq \quad , \end{aligned}$$

i.e.,

$$\dot{p} = -q \quad .$$

Likewise, we find

$$-i\dot{q} = [H, q] = -ip \quad ,$$

i.e.,

$$\dot{q} = p \quad .$$

Thus, Heisenberg's equation gives the identical result as Hamilton's equations.

To analyze the eigenvalue problem we introduce

$$a \equiv \frac{1}{\sqrt{2}}(q + ip) \quad . \tag{1.12}$$

In accordance with (1.8), we have $q = q^\dagger$ and $p = p^\dagger$, and therefore the Hermitian conjugate of a is

$$a^\dagger = \frac{1}{\sqrt{2}}(q - ip) \quad . \tag{1.13}$$

We may solve q and p in terms of a and $a^\dagger$:

$$q = \frac{1}{\sqrt{2}}(a + a^\dagger) , \tag{1.14}$$

$$p = \frac{-i}{\sqrt{2}}(a - a^\dagger) . \tag{1.15}$$

Because $[p, q] = -i$, we find

$$[a, a^\dagger] = 1 . \tag{1.16}$$

Moreover,

$$a^\dagger a = \tfrac{1}{2}(q - ip)(q + ip)$$
$$= \tfrac{1}{2}\{q^2 + p^2 - i(pq - qp)\} = H - \tfrac{1}{2} ,$$

which leads to $H = a^\dagger a + \frac{1}{2}$. Let

$$N \equiv a^\dagger a , \tag{1.17}$$

then the Hamiltonian H can also be written as

$$H = N + \tfrac{1}{2} . \tag{1.18}$$

The operator N is non-negative since its expectation value over any state vector is an absolute value squared and is therefore $\geqslant 0$.

We shall now show that the eigenvector $|n>$ of H satisfies

$$H|n> = (n + \tfrac{1}{2})|n> \tag{1.19}$$

in which n can be any positive integer 0, 1, 2, ··· . Furthermore, let $|0>$ be the eigenvector with the smallest eigenvalue, i.e.

$$H|0> = \tfrac{1}{2}|0> ; \tag{1.20}$$

then the other eigenvectors can be written as

$$|n> = \frac{1}{\sqrt{n!}}(a^\dagger)^n |0> . \tag{1.21}$$

<u>Proof.</u> Let $|>$ be any eigenvector of N :

$$N|> = \ell |> \tag{1.22}$$

where ℓ is a number. Because

$$Na^\dagger = a^\dagger a a^\dagger = a^\dagger(a^\dagger a + 1) = a^\dagger(N + 1) \tag{1.23}$$

and

$$Na = a^\dagger aa = (aa^\dagger - 1)a = a(N-1) \quad , \tag{1.24}$$

it follows that $Na^\dagger | > = (\ell + 1)\, a^\dagger | >$ and $Na | > = (\ell - 1)\, a | >$. Next, replace $| >$ respectively by $a^\dagger | >$ in the first equation and by $a | >$ in the second. After repeating this process n times, we find

$$N(a^\dagger)^n \; | > \; = \; (\ell + n)\,(a^\dagger)^n \; | > \tag{1.25}$$

and

$$Na^n \; | > \; = \; (\ell - n)\, a^n \; | > \quad . \tag{1.26}$$

Now, if $\ell \neq$ integer, by choosing in (1.26) n = integer $> \ell$ we would obtain a negative eigenvalue for N ; that is impossible since, as noted before, N is non-negative. Hence ℓ = integer, and according to (1.26)

$$| 0 > \; \equiv \; a^\ell \; | >$$

satisfies

$$N \, | \, 0 > \; = \; 0 \quad . \tag{1.27}$$

Furthermore, zero must be the smallest eigenvalue of N . By setting $| 0 >$ in place of $| >$ in (1.25), we find

$$N(a^\dagger)^n \; | \, 0 > \; = \; n(a^\dagger)^n \; | \, 0 >$$

where n can be any positive integer. Thus, (1.19) - (1.21) are proved. Let us choose the normalization of the state $| 0 >$ so that $< 0 \,|\, 0 > = 1$. Then from (1.21) all other states $| n >$ are also normalized: $< n \,|\, n > = 1$.

From (1.24) and (1.27), we see that

$$Na \, | \, 0 > \; = \; - a \, | \, 0 > \quad .$$

Since N is non-negative, we must have

$$a \, | \, 0 > \; = \; 0 \quad . \tag{1.28}$$

In the coordinate representation p is $-i \dfrac{\partial}{\partial q}$. Let $\psi(q) = < q \,|\, 0 >$. Equation (1.28) becomes

$$a\psi = \frac{1}{\sqrt{2}}(q + ip)\psi = \frac{1}{\sqrt{2}}(q + \frac{\partial}{\partial q})\psi = 0$$

which determines the solution $\psi(q) \propto e^{-\frac{1}{2}q^2}$. Therefore, not only does $|0>$ exist, it is also non-degenerate. This completes the proof.

According to (1.21)

$$a^\dagger | n> = \sqrt{n+1} \; | n+1> , \tag{1.29}$$

therefore $a^\dagger | n-1> = \sqrt{n} \; | n>$, which leads to

$$a | n> = \frac{a a^\dagger}{\sqrt{n}} | n-1> = \frac{1}{\sqrt{n}} (a^\dagger a + 1) | n-1> ,$$

i.e.,

$$a | n> = \sqrt{n} \; | n-1> . \tag{1.30}$$

The operator $N = a^\dagger a$ is commonly called the occupation-number operator. Because of (1.29) and (1.30), $a^\dagger$ is called the creation operator and a the annihilation operator.

In this example of the harmonic oscillator we may consider a space whose basis vectors are the eigenvectors $|0>, |1>, \cdots$. Due to the orthonormality of these vectors we may write

$$|0> = \begin{pmatrix} 1 \\ 0 \\ 0 \\ \vdots \end{pmatrix} , \qquad |1> = \begin{pmatrix} 0 \\ 1 \\ 0 \\ \vdots \end{pmatrix} , \cdots .$$

The corresponding matrix forms of N, $a^\dagger$ and a are

$$N = \begin{pmatrix} 0 & 0 & 0 & \cdots \\ 0 & 1 & 0 & \cdots \\ 0 & 0 & 2 & \cdots \\ \vdots & \vdots & \vdots & \vdots \end{pmatrix} ,$$

$$a^\dagger = \begin{pmatrix} 0 & 0 & 0 & \cdots \\ \sqrt{1} & 0 & 0 & \cdots \\ 0 & \sqrt{2} & 0 & \cdots \\ \vdots & \vdots & \vdots & \vdots \end{pmatrix} , \quad a = \begin{pmatrix} 0 & \sqrt{1} & 0 & \cdots \\ 0 & 0 & \sqrt{2} & \cdots \\ 0 & 0 & 0 & \cdots \\ \vdots & \vdots & \vdots & \vdots \end{pmatrix} . \tag{1.31}$$

Exercise. Repeat the same steps for a harmonic oscillator whose Hamiltonian is $H = \frac{1}{2m} p^2 + \frac{1}{2} k q^2$ where m and k are constants.

1.3 Some General Theorems

Let H be a Hermitian operator in a linear space

$$\{v_a\} \quad , \qquad a = 0,1,2, \cdots \tag{1.32}$$

where the basis vectors v_a form an orthonormal set. A vector $| >$ in this space is sometimes called a state vector, or simply a state. We shall assume that the Hermitian operator H is bounded from below; i.e., for any state $| >$ the ratio $\frac{< | H | >}{< | >}$ is always larger than a fixed constant c. Its eigenvector equation can be written as $H \mid a > = E_a \mid a >$, and its eigenvalues E_a can be arranged so that (for $a = 0,1,2, \cdots$)

$$E_0 \leqslant E_1 \leqslant \cdots \leqslant E_m \leqslant E_{m+1} \leqslant \cdots \quad . \tag{1.33}$$

Since H is Hermitian, we can always choose

$$< a \mid a' > = \delta_{aa'} \tag{1.34}$$

for any eigenvectors $| a >$ and $| a' >$ of H.

Theorem 1. The minimum of $\frac{< | H | >}{< | >}$ is

(i) E_0, if $| >$ can be any state vector,

(ii) E_1, if $| >$ can be any state vector that satisfies the constraint $< 0 | > = 0$,

(iii) E_n, if $| >$ can be any state vector that satisfies the constraints $< 0 | > = < 1 | > = \cdots = < n-1 | > = 0$.

Proof. Let

$$E \equiv \frac{< | H | >}{< | >} \quad ,$$

and denote

$$\psi = | > , \quad \psi^\dagger = < | \quad .$$

To find the minimum of E we may consider the variation $\psi \to \psi + \delta\psi$. The corresponding variation in E is

$$\delta E = \frac{1}{\psi^\dagger \psi}(\delta\psi^\dagger H\psi + \psi^\dagger H \delta\psi) - \frac{\psi^\dagger H \psi}{\psi^\dagger \psi}\left(\frac{\delta\psi^\dagger \psi}{\psi^\dagger\psi} + \frac{\psi^\dagger \delta\psi}{\psi^\dagger\psi}\right)$$

$$= \frac{1}{\psi^\dagger\psi}[\delta\psi^\dagger (H-E)\psi + \psi^\dagger (H-E)\delta\psi] \ .$$

Set $f = (H-E)\psi$. If $f = 0$, then clearly $\delta E = 0$. If $f \neq 0$, we may choose $\delta\psi = \epsilon f$ where ϵ is an infinitesimal real quantity. Thus, the above equation can be written as

$$\delta E = \frac{2}{\psi^\dagger\psi}\,\epsilon f^\dagger f \ .$$

The necessary condition for E to be a minimum is $\delta E = 0$ for arbitrary $\delta\psi$. Hence $f = 0$, which implies $(H-E)\psi = 0$. Since E_0 is the smallest eigenvalue of H, (i) is then established.

In (1.32) we may choose $v_0 = |\,0>$. Let us consider the subspace $\{v_i\}$ which is spanned by all v_i with $i \geqslant 1$. By definition, all vectors orthogonal to $|\,0>$ are in this subspace. Furthermore, for any v_i $(i \geqslant 1)$, H satisfies

$$< v_0 \,|\, H \,|\, v_i > \; = \; E_0 < v_0 \,|\, v_i > \; = \; 0 \ .$$

Hence Hv_i also belongs to this subspace $\{v_i\}$. Now consider the Hamiltonian in this subspace. By following the same argument as that used in proving (i) we can establish (ii), and likewise also (iii).

We call a set of basis vectors $\{|\,a>\}$ complete if for any state vector $|\,>$, there exists a set of numerical constants $\{C_a\}$ such that

$$\lim_{m\to\infty} < R_m \,|\, R_m > \; = \; 0 \ ,$$

where

$$|\,R_m > \; \equiv \; |\,> - \sum_{a=0}^{m} C_a \,|\, a> \ .$$

Theorem 2. If a Hermitian operator H is bounded from below, but not from above (i.e., for any real constant c there exists a state vector $|\,>$ such that $\dfrac{<|\,H\,|>}{<\,|\,>}$ is larger than c), then the set of all its eigenvectors $\{|\,a>\}$ is complete.

Proof. Since H is a Hermitian operator, we can choose its eigenvectors to satisfy (1.34). The theorem is obvious if $\{|\,a>\}$ spans a finite dimensional space. We need only consider the case when $\{|\,a>\}$ spans a space of infinite dimensions.

Let us arrange the eigenvalues of H in the form (1.33). Because H is bounded from below, by replacing $H \to H +$ a constant we can set

$$E_0 > 0 \quad .$$

Let us choose $C_a = <a\,|\,>$. Consequently

$$|\,R_m> \; = \; |\,> - \sum_{a=0}^{m} C_a \,|\,a>$$

satisfies

$$<a\,|\,R_m> \; = 0 \qquad \text{when} \qquad a \leqslant m \quad .$$

From Theorem 1 we have

$$\frac{<R_m\,|\,H\,|\,R_m>}{<R_m\,|\,R_m>} \geqslant E_{m+1} \geqslant E_m \quad . \tag{1.35}$$

Because H is not bounded from above, it follows that

$$E_m \to \infty \qquad \text{when} \qquad m \to \infty \quad . \tag{1.36}$$

In addition,

$$<R_m\,|\,H\,|\,R_m> \; = (<| - \sum_a C_a^* <a\,|\,)\; H \;(\,|> - \sum_b C_b \,|\,b>)$$

$$= <|\,H\,|> - \sum_a C_a^* <a\,|\,H\,|> - \sum_b C_b <|\,H\,|\,b> + \sum_{a,b} C_a^* C_b <a\,|\,H\,|\,b>$$

where the sum extends over a and $b = 1, 2, \cdots, m$. Since $|\,a>$

and $|b>$ are eigenstates of H, this leads to

$$<R_m|H|R_m> = <|H|> - \sum_a C_a^* C_a E_a - \sum_b C_b C_b^* E_b + \sum_a C_a^* C_a E_a$$

$$= <|H|> - \sum_b C_b C_b^* E_b$$

which is $\leqslant$ the first term $<|H|>$, since the second term $-\sum_b C_b C_b^* E_b$ is always negative. Thus, by using (1.35), we find

$$<R_m|R_m> \leqslant \frac{1}{E_m} <R_m|H|R_m> \leqslant \frac{1}{E_m} <|H|> . \quad (1.37)$$

Because $<|H|>$ is independent of m and $<R_m|R_m>$ is positive, by using (1.36) we obtain

$$<R_m|R_m> \to 0 \qquad \text{when} \qquad m \to \infty .$$

This establishes Theorem 2. Therefore the set of all eigenvectors of H can be used as the complete set of basis vectors in the Hilbert space.

The following are a few examples:

(i) $H = \frac{1}{2}p^2 + V(x)$, in which $V(x)$ has a lower bound.

In the x-representation $p^2 = -\frac{d^2}{dx^2}$. Choose

$$<x|> \propto e^{-x^2/\lambda^2} ,$$

then, when $\lambda \to 0$,

$$\frac{<|p^2|>}{<1>} \to \infty .$$

Hence, H is bounded from below but not from above. The set of all its eigenvectors forms a complete set of functions.

(ii) Consider a circle in a two-dimensional space. Let $H = p_\phi^{\,2} = -\frac{d^2}{d\phi^2}$ where ϕ is the angular variable which varies from 0 to 2π.

Let $|m>$ be an eigenstate of H. In the ϕ-representation, we may denote $\psi_m(\phi) = <\phi|m>$, which is a periodic function of ϕ. The solutions of the eigenstate equation, $H\psi_m = m^2\psi_m$, are

$$\psi_m = \frac{1}{\sqrt{2\pi}} e^{im\phi} \qquad \text{where} \qquad m = 0, \pm 1, \cdots .$$

From Theorem 2 we know that, except for a set of points of zero measure, any function $f(\phi)$ can be expanded in terms of these eigenfunctions:

$$f(\phi) = \sum C_m e^{im\phi} .$$

This is the main content of the well-known Fourier theorem.

(iii) Next, we consider the surface of a unit sphere in a three-dimensional space. Let $H = -\nabla^2$ on the surface; i.e., in terms of the spherical coordinates

$$H = -\frac{1}{\sin\theta}\frac{\partial}{\partial\theta}\left(\sin\theta\frac{\partial}{\partial\theta}\right) - \frac{1}{\sin^2\theta}\frac{\partial^2}{\partial\phi^2} .$$

The eigenfunctions of H are called the spherical harmonics $Y_{\ell m}(\theta, \phi)$ which satisfy

$$H Y_{\ell m}(\theta, \phi) = \ell(\ell+1) Y_{\ell m}(\theta, \phi) , \tag{1.38}$$

where $\ell = 0, 1, 2, \cdots$. The ϕ-dependence of $Y_{\ell m}$ is

$$Y_{\ell m} \propto e^{im\phi} \tag{1.39}$$

with $m = 0, \pm 1, \pm 2, \cdots \pm \ell$. Theorem 2 tells us that, except for a set of points of zero measure, any function $f(\theta, \phi)$ can be expanded in terms of $Y_{\ell m}$:

$$f(\theta, \phi) = \sum C_{\ell m} Y_{\ell m}(\theta, \phi) .$$

Equations (1.38) and (1.39) determine $Y_{\ell m}$ up to a multiplicative factor. The usual choices are

$$Y_{0,0} = \frac{1}{\sqrt{4\pi}} , \quad Y_{1,\pm 1} = \mp\sqrt{\frac{3}{8\pi}} \sin\theta \, e^{\pm i\phi} ,$$

$$Y_{1,0} = \sqrt{\frac{3}{4\pi}} \cos\theta , \quad \text{etc.} \tag{1.40}$$

The application of Theorem 2 is quite general. The space spanned by such a complete set of vectors is called Hilbert space.

Problem 1.1 Let H be a Hermitian operator that is bounded from below. Arrange its eigenvalues $E_0, E_1, \cdots$ and the corresponding eigenvectors $|0>, |1>, \cdots$ in the order given by (1.33).

(i) Let $|b>$ be an arbitrarily chosen vector. Define F(b) to be the minimum of $\frac{<|H|>}{<|>}$ where $|>$ can be any vector that satisfies $<b|>=0$. By varying $|b>$, prove that the maximum of F(b) is E_1, the second lowest eigenvalue of H.

Hint: To find F(b), consider the vector

$$<b|1>|0> - <b|0>|1> .$$

(ii) Let $|b_1>, |b_2>, \cdots, |b_n>$ be n arbitrarily chosen linearly independent vectors. Define $F(b_1, b_2, \cdots, b_n)$ to be the minimum of $\frac{<|H|>}{<|>}$ where $|>$ can be any vector that satisfies $<b_1|>=<b_2|>= \cdots =<b_n|>=0$. Prove that the maximum of $F(b_1, b_2, \cdots, b_n)$ is E_n. (maximum-minimum principle)

Problem 1.2 (i) In the above problem, suppose a constraint C is imposed on all state vectors. Correspondingly, all eigenvalues and eigenvectors will be changed: $E_n \to E_n'$ and $|n> \to |n'>$. By applying the maximum-minimum principle, prove that $E_0 \leqslant E_0'$, $E_1 \leqslant E_1', \cdots, E_n \leqslant E_n', \cdots$.

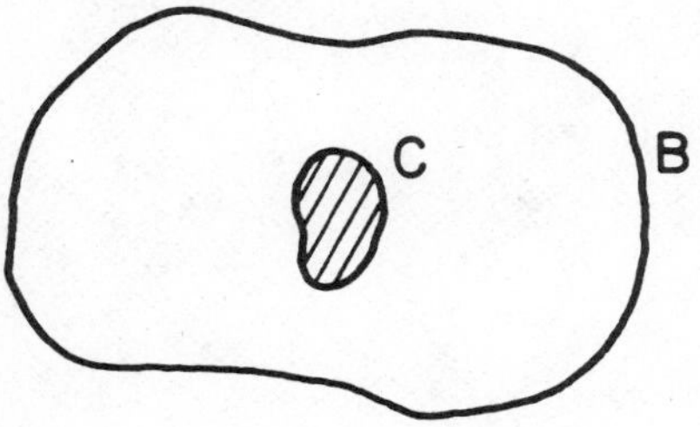

Fig. 1.1

(ii) Consider the vibration of a membrane with a fixed boundary B. The characteristic frequency ω_n is determined by $-\nabla^2\phi = \omega_n^2\phi$ where the vibrational amplitude ϕ is zero at the boundary. Arrange these frequencies in the order

$\omega_0 \leqslant \omega_1 \leqslant \omega_2 \leqslant \cdots$. Impose the constraint that ϕ is also zero inside a closed curve C within the membrane, as shown in Fig. 1.1 ; correspondingly, the characteristic frequency ω_n is changed to $\omega_n{}'$. Show that $\omega_n \leqslant \omega_n{}'$ for all n .

References

Some standard textbooks on subjects discussed in this chapter are:

R. Courant and D. Hilbert, Methods of Mathematical Physics, 2 vols. (New York, Interscience Publishers, Inc., 1962).

P. A. M. Dirac, Quantum Mechanics (Oxford, The Clarendon Press, 1958).

E. T. Whittaker, Analytical Dynamics (Cambridge, The University Press, 1960).

Chapter 2

THE SPIN-0 FIELD

2.1 General Discussion

Now we turn to the quantization of a local field theory. Let $\phi(x)$ be a local field where $x = x_\mu = (\vec{r}, it)$; i.e., $x_i = \vec{r}_i$ for $i = 1,2,3$ and $x_4 = it$. If, under the Lorentz transformation ϕ is invariant, then we call it a spin-0 field. In addition, we call ϕ a scalar field if it does not change sign under the space inversion; otherwise, a pseudoscalar field. Let us begin our discussions by considering a Hermitian field:

$$\phi(x) = \phi(\vec{r}, t) = \phi^\dagger(x) \quad .$$

We may first enclose the whole system in a finite rectangular box of size Ω , and assume ϕ to satisfy the periodic boundary condition, and then in the end let $\Omega \to \infty$. It may be emphasized that this particular procedure is no less physical than that in which Ω is set to be infinite at the beginning. For all we know our universe may well be finite, since within the present experimental degree of accuracy in particle physics it is not even remotely possible to determine its size, let alone its boundary conditions. Therefore, different routes allowing Ω to go to infinity should lead to the same theoretical result.

Let the Lagrangian density of the system be

$$\mathcal{L} = -\tfrac{1}{2}\left(\frac{\partial\phi}{\partial x_\mu}\right)^2 - V(\phi) \ , \tag{2.1}$$

in which, here as well as in later discussions, the repeated index μ is summed over from 1 to 4. Therefore, in the above expression,

$$\left(\frac{\partial\phi}{\partial x_\mu}\right)^2 = \frac{\partial\phi}{\partial x_\mu}\frac{\partial\phi}{\partial x_\mu} = (\vec{\nabla}\phi)^2 - \dot{\phi}^2 \ .$$

The Lagrangian L is given by

$$L = \int_\Omega \mathcal{L}\, d^3r = \int_\Omega \mathcal{L}(\phi(\vec{r}, t), \dot{\phi}(\vec{r}, t))\, d^3r \ , \tag{2.2}$$

which is regarded as a functional $L(\phi, \dot{\phi})$ of $\phi(\vec{r}, t)$ and its time derivative $\dot{\phi}(\vec{r}, t)$. By comparing it with (1.1), we see that ϕ corresponds to the generalized coordinate with $\dot{\phi}$ the corresponding velocity. The main difference is that while in (1.1) the index i is discrete and of finite value, in the case of a field the corresponding index is $\vec{r}$, which is continuous and has an infinite number of values.

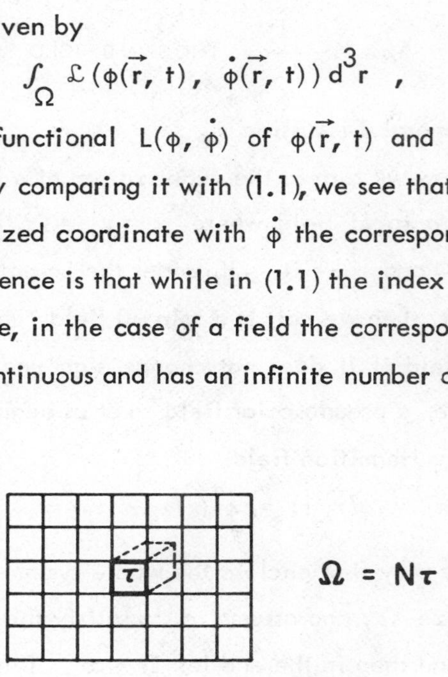

Fig. 2.1 Division of Ω into N tiny cubes, each of size τ.

For convenience, we may divide Ω into many small cubes of size τ, as in Fig. 2.1. Let the value of $\phi(\vec{r}, t)$ in each particular small cube τ be represented by $\phi(\vec{r}_i, t) \equiv \phi_i(t)$ where $\vec{r}_i$ is the coordinate of any arbitrarily chosen fixed point in this little cube. Set $q_i(t) \equiv \tau\,\phi_i(t)$. Equation (2.2) will now be written as

$$L = \frac{1}{2}\int_\Omega \dot{\phi}^2\, d^3r - \int_\Omega \left[\frac{1}{2}(\nabla\phi)^2 + V(\phi)\right] d^3r$$

$$= \frac{1}{2}\sum_i \frac{\dot{q}_i^{\,2}}{\tau} - \cdots ,$$

where the $\cdots$ denotes terms independent of $\dot{q}_i$. Therefore the corresponding generalized momentum is

$$p_i(t) = \frac{\partial L}{\partial \dot{q}_i} = \frac{\dot{q}_i}{\tau} = \dot{\phi}_i(t) \equiv \Pi(\vec{r}_i, t) \quad . \tag{2.3}$$

When $\tau \to 0$, by using (1.5) we find the Hamiltonian to be

$$H = \sum_i p_i \dot{q}_i - L = \int \left[\frac{1}{2}\Pi^2 + \frac{1}{2}(\nabla\phi)^2 + V(\phi)\right] d^3r \, . \tag{2.4}$$

According to the general rules of quantization

$$[p_i(t), \ q_j(t)] = -i\,\delta_{ij} \ ,$$

we have

$$[\Pi(\vec{r}_i, t), \ \phi(\vec{r}_j, t)] = -i\,\frac{\delta_{ij}}{\tau} \ ,$$

which leads to, when $\tau \to 0$,

$$[\Pi(\vec{r}, t), \ \phi(\vec{r}\,', t)] = -i\,\delta^3(\vec{r} - \vec{r}\,') \ , \tag{2.5}$$

where $\delta^3(\vec{r} - \vec{r}\,')$ is the three-dimensional Dirac δ-function. The definition of the δ-function is

$$\delta^3(\vec{r} - \vec{r}\,') = 0 \qquad \text{if} \qquad \vec{r} \neq \vec{r}\,'$$

and

$$\int \delta^3(\vec{r} - \vec{r}\,')\, d^3r = 1$$

in which the integration extends over any volume that includes the point $\vec{r}\,'$. In (2.5) both points $\vec{r}$ and $\vec{r}\,'$ are assumed to be inside the box Ω. Likewise, because $[q_i(t), q_j(t)] = [p_i(t), p_j(t)] = 0$,

$$[\phi(\vec{r}, t), \ \phi(\vec{r}\,', t)] = [\Pi(\vec{r}, t), \ \Pi(\vec{r}\,', t)] = 0 \ . \tag{2.6}$$

The equation of motion of the field remains given by the same Heisenberg equation, (1.9). By setting $O(t)$ to be $\phi(\vec{r}, t)$ in (1.9) we find

$$[H, \ \phi(\vec{r}, t)] = -i\,\dot{\phi}(\vec{r}, t) \ .$$

On account of (2.4)-(2.6) we see that the lefthand side is $-i\,\Pi(\vec{r}, t)$, and that leads to

$$\Pi(\vec{r}, t) = \dot{\phi}(\vec{r}, t) \quad , \tag{2.7}$$

in agreement with (2.3). Likewise, by setting $O(t) = \Pi(\vec{r}, t)$ in Heisenberg's equation (1.9) we have

$$[H, \Pi(\vec{r}, t)] = -i\,\dot{\Pi}(\vec{r}, t) \quad . \tag{2.8}$$

Now, (2.5) can also be written as

$$[\phi(\vec{r}, t), \; \Pi(\vec{r}\,', t)] = i\,\delta^3(\vec{r} - \vec{r}\,') = i\,\frac{\delta\phi(\vec{r}, t)}{\delta\phi(\vec{r}\,', t)} \quad ,$$

where, in the variational derivative of the last expression, we may regard

$$\phi(\vec{r}, t) = \int \delta^3(\vec{r} - \vec{r}\,')\,\phi(\vec{r}\,', t)\,d^3r \quad .$$

Consequently,

$$[H, \Pi(\vec{r}, t)] = i\,\frac{\delta H}{\delta\phi(\vec{r}, t)} = i(-\nabla^2\phi + \frac{dV}{d\phi}) \quad ,$$

which, together with (2.7) and (2.8), gives

$$\ddot{\phi} - \nabla^2\phi + \frac{dV}{d\phi} = 0 \quad . \tag{2.9}$$

The same equation of motion can also be derived by using the variational principle

$$\delta \int L\,dt = \delta \int \mathcal{L}\,d^4x = 0 \quad . \tag{2.10}$$

Of course, this is not an accident. The underlying reason is the same as that discussed in Chapter 1 for a finite system.

Thus we see that the quantum field theory is merely an extension of the ordinary quantum mechanics of a finite system to an infinite system.

2.2 Fourier Expansion (Free or Interacting Fields)

At any given time t the operators $\phi(\vec{r}, t)$ and $\Pi(\vec{r}, t)$ can be expanded in terms of the Fourier series:

$$\phi(\vec{r}, t) = \sum_{\vec{k}} \frac{1}{\sqrt{\Omega}} e^{i\vec{k}\cdot\vec{r}} q_{\vec{k}}(t) \tag{2.11}$$

and

$$\Pi(\vec{r}, t) = \sum_{\vec{k}} \frac{1}{\sqrt{\Omega}} e^{i\vec{k}\cdot\vec{r}} p_{-\vec{k}}(t) \tag{2.12}$$

in which $q_{\vec{k}}(t)$ and $p_{\vec{k}}(t)$ are time-dependent operators in the Hilbert space, and components of $\vec{k}$ are given by

$$k_i = \frac{2\pi \ell_i}{L_i} \ , \qquad i = 1, 2, 3 \tag{2.13}$$

with

$$\ell_i = 0, \pm 1, \pm 2, \cdots \tag{2.14}$$

and L_1, L_2, L_3 are, respectively, the lengths of the three sides of the rectangular box Ω. The validity of this expansion depends only on the completeness property of the Fourier series, discussed in the last chapter. [We may view the operators $\phi(\vec{r}, t)$ and $\Pi(\vec{r}, t)$ as matrices; each of their matrix elements is a c. number function of $\vec{r}$ and t which at any given time t can be expanded in terms of the Fourier series, and that results in the above expansions.] Since ϕ and Π are Hermitian operators, i.e., $\phi(\vec{r}, t) = \phi^\dagger(\vec{r}, t)$ and $\Pi(\vec{r}, t) = \Pi^\dagger(\vec{r}, t)$, we have

$$q_{\vec{k}}(t) = q^\dagger_{-\vec{k}}(t) \ , \qquad p_{\vec{k}}(t) = p^\dagger_{-\vec{k}}(t) \ . \tag{2.15}$$

Let us define

$$a_{\vec{k}}(t) \equiv \sqrt{\frac{\omega}{2}} \left(q_{\vec{k}} + \frac{i}{\omega} p_{-\vec{k}}\right) \ , \tag{2.16}$$

where

$$\omega \equiv \sqrt{\vec{k}^2 + m^2} \tag{2.17}$$

and m is an arbitrarily chosen real parameter. Because of (2.15), the Hermitian conjugate of $a_{\vec{k}}(t)$ is

$$a^\dagger_{\vec{k}}(t) = \sqrt{\frac{\omega}{2}} \left(q_{-\vec{k}} - \frac{i}{\omega} p_{\vec{k}}\right) \ . \tag{2.18}$$

By changing the sign of $\vec{k}$ we obtain

$$a^\dagger_{-\vec{k}}(t) = \sqrt{\frac{\omega}{2}} \left(q_{\vec{k}} - \frac{i}{\omega} p_{-\vec{k}}\right) \ . \tag{2.19}$$

From (2.16) and (2.19) we may express $q_{\vec{k}}$ and $p_{-\vec{k}}$ in terms of $a_{\vec{k}}$ and $a^{\dagger}_{-\vec{k}}$:

$$q_{\vec{k}}(t) = \frac{1}{\sqrt{2\omega}} [a_{\vec{k}}(t) + a^{\dagger}_{-\vec{k}}(t)]$$

and

$$p_{-\vec{k}}(t) = \frac{-i\omega}{\sqrt{2\omega}} [a_{\vec{k}}(t) - a^{\dagger}_{-\vec{k}}(t)] \quad .$$

By substituting these expressions into (2.11) and (2.12) we find

$$\phi(\vec{r}, t) = \sum \frac{1}{\sqrt{2\omega\Omega}} [a_{\vec{k}}(t) e^{i\vec{k}\cdot\vec{r}} + a^{\dagger}_{\vec{k}}(t) e^{-i\vec{k}\cdot\vec{r}}] \qquad (2.20)$$

and

$$\Pi(\vec{r}, t) = \sum \frac{-i\omega}{\sqrt{2\omega\Omega}} [a_{\vec{k}}(t) e^{i\vec{k}\cdot\vec{r}} - a^{\dagger}_{\vec{k}}(t) e^{-i\vec{k}\cdot\vec{r}}] \ . \qquad (2.21)$$

Next, we observe that on account of (2.13) and (2.14),

$$\Omega^{-1} \int e^{i(\vec{k}-\vec{k}')\cdot\vec{r}} d^3r = \delta_{\vec{k},\vec{k}'} \qquad (2.22)$$

which is 0 if $\vec{k} \neq \vec{k}'$ and is 1 if $\vec{k} = \vec{k}'$. Therefore, from (2.11) and (2.12) we have

$$q_{\vec{k}}(t) = \frac{1}{\sqrt{\Omega}} \int e^{-i\vec{k}\cdot\vec{r}} \phi(\vec{r}, t) d^3r$$

and

$$p_{\vec{k}}(t) = \frac{1}{\sqrt{\Omega}} \int e^{i\vec{k}\cdot\vec{r}} \Pi(\vec{r}, t) d^3r \quad ,$$

which leads to

$$[p_{\vec{k}}(t), q_{\vec{k}'}(t)] = \frac{1}{\Omega} \int e^{i\vec{k}\cdot\vec{r} - i\vec{k}'\cdot\vec{r}'} [\Pi(\vec{r},t), \phi(\vec{r}',t)] d^3r d^3r'.$$

By using (2.5) and (2.22) we find

$$[p_{\vec{k}}(t), q_{\vec{k}'}(t)] = -i\delta_{\vec{k},\vec{k}'} \quad . \qquad (2.23)$$

Likewise, by using (2.6) we have

$$[q_{\vec{k}}(t), q_{\vec{k}'}(t)] = [p_{\vec{k}}(t), p_{\vec{k}'}(t)] = 0 \quad . \qquad (2.24)$$

These expressions enable us to derive the commutators between the $a_{\vec{k}}$'s and $a^{\dagger}_{\vec{k}}$'s. From (2.16) and (2.18) it follows that

$$[a_{\vec{k}}(t),\ a^{\dagger}_{\vec{k}'}(t)] = \delta_{\vec{k},\vec{k}'} \ ,$$

$$[a_{\vec{k}}(t),\ a_{\vec{k}'}(t)] = [a^{\dagger}_{\vec{k}}(t),\ a^{\dagger}_{\vec{k}'}(t)] = 0 \ . \tag{2.25}$$

Remarks

1. We may reverse the above proof by starting from the commutation relations (2.25) between the $a_{\vec{k}}$'s and $a^{\dagger}_{\vec{k}}$'s, and then establishing the commutation relations (2.5) and (2.6) between ϕ and Π: From (2.25) and the Fourier expansions (2.20) and (2.21), we can directly derive (2.6) and

$$[\Pi(\vec{r}, t),\ \phi(\vec{r}', t)] = -i \sum \frac{1}{\Omega} e^{i\vec{k}\cdot(\vec{r}-\vec{r}')} \ . \tag{2.26}$$

Furthermore, we observe that for $\vec{r}$ and $\vec{r}'$ within the volume Ω, the function $\delta^3(\vec{r}-\vec{r}')$ may be expanded in terms of the Fourier series:

$$\delta^3(\vec{r}-\vec{r}') = \sum_{\vec{k}} C_{\vec{k}}(\vec{r}') \frac{1}{\sqrt{\Omega}} e^{i\vec{k}\cdot\vec{r}} \ .$$

Because of the orthonormality relation (2.22), we have

$$C_{\vec{k}}(\vec{r}') = \int \frac{d^3r}{\sqrt{\Omega}} e^{-i\vec{k}\cdot\vec{r}} \delta^3(\vec{r}-\vec{r}') = \frac{1}{\sqrt{\Omega}} e^{-i\vec{k}\cdot\vec{r}'} \ ,$$

and therefore

$$\delta^3(\vec{r}-\vec{r}') = \sum_{\vec{k}} \frac{1}{\Omega} e^{i\vec{k}\cdot(\vec{r}-\vec{r}')} \ . \tag{2.27}$$

Substituting (2.27) into (2.26), we obtain for $\vec{r}$ and $\vec{r}'$ within the volume Ω

$$[\Pi(\vec{r}, t),\ \phi(\vec{r}', t)] = -i\delta^3(\vec{r}-\vec{r}')$$

which is (2.5). So far, $\vec{r}$ and $\vec{r}'$ are restricted to points within Ω. In the limit $\Omega \to \infty$, the above expression becomes valid everywhere.

2. When $\Omega \to \infty$, we may replace the sum over $\vec{k}$ vectors by an integration:

$$\frac{1}{\Omega} \sum_{\vec{k}} \to \frac{1}{8\pi^3} \int d^3k \ . \tag{2.28}$$

This can be proved by using (2.13) and observing that, in accordance with (2.14), the parameter ℓ_i runs over the discrete values $\cdots, -2, -1, 0, 1, 2, \cdots$ with a spacing

$$\Delta \ell_i = 1 \quad ; \tag{2.29}$$

therefore, the corresponding variation in k_i is

$$\Delta k_i = \frac{2\pi}{L_i} \Delta \ell_i = \frac{2\pi}{L_i} \quad . \tag{2.30}$$

Because of (2.29), (2.30) and $\Omega = L_1 L_2 L_3$ we may write

$$\frac{1}{\Omega} \sum_{\vec{k}} = \frac{1}{\Omega} \sum_{\vec{k}} \Delta \ell_i \, \Delta \ell_2 \, \Delta \ell_3 = \frac{1}{8\pi^3} \sum_{\vec{k}} \Delta k_1 \, \Delta k_2 \, \Delta k_3 \, .$$

When $L_i \to \infty$, we have $\Delta k_i \to 0$. Hence the above expression leads to (2.28).

In the same limit $\Omega \to \infty$, Eq. (2.27) becomes

$$\delta^3(\vec{r} - \vec{r}\,') = \frac{1}{8\pi^3} \int e^{i\vec{k}\cdot(\vec{r}-\vec{r}\,')} \, d^3k \quad . \tag{2.31}$$

While (2.27) is valid only for $\vec{r}$ and $\vec{r}\,'$ within the volume Ω, the above formula is valid for arbitrary $\vec{r}$ and $\vec{r}\,'$.

3. We note that the validity of the Fourier expansion (2.20)-(2.21) and the commutation relations (2.25) is independent of the detailed form of the Hamiltonian. Thus in (2.4) the function $V(\phi)$ can be of arbitrary form. If $V(\phi)$ is a quadratic function of ϕ, then it is a free-field theory, otherwise not. Furthermore, the parameter m in (2.17) is as yet completely arbitrary.

This situation is analogous to the quantum mechanics of a finite system of particles, as discussed in Chapter 1. There, the choice of the generalized coordinates $q_i(t)$ and the generalized momenta $p_i(t)$ can also be made independently of the detailed form of the interaction potential between particles.

2.3 Hilbert Space (Free or Interacting Fields)

Without any loss of generality, we can write

$$V(\phi) = \tfrac{1}{2} m^2 \phi^2 + \mathcal{H}_{int}(\phi)$$

where m is the same parameter introduced in (2.17) and $\mathcal{H}_1$ is simply defined to be

$$\mathcal{H}_{int} = V(\phi) - \tfrac{1}{2} m^2 \phi^2 \quad . \tag{2.32}$$

Thus the Hamiltonian (2.4) can be written as

$$H = H_0 + H_{int} \tag{2.33}$$

where

$$H_0 = \int \mathcal{H}_0 \, d^3r \quad , \qquad H_{int} = \int \mathcal{H}_{int} \, d^3r \tag{2.34}$$

and

$$\mathcal{H}_0 = \tfrac{1}{2} (\Pi^2 + (\nabla\phi)^2 + m^2 \phi^2) \quad .$$

Through partial integration H_0 becomes

$$H_0 = \tfrac{1}{2} \int [\Pi^2 + \phi(-\nabla^2 + m^2)\phi] \, d^3r \quad . \tag{2.35}$$

By using (2.17) and (2.20) we find

$$(-\nabla^2 + m^2)\phi = \sum_{\vec{k}} \frac{\omega^2}{\sqrt{2\omega\Omega}} [a_{\vec{k}}(t) \, e^{i\vec{k}\cdot\vec{r}} + a_{\vec{k}}^\dagger(t) \, e^{-i\vec{k}\cdot\vec{r}}] \quad .$$

Upon substituting this expression and (2.21) into (2.35) we have

$$H_0 = \tfrac{1}{2} \sum_{\vec{k}} \omega (a_{\vec{k}} a_{\vec{k}}^\dagger + a_{\vec{k}}^\dagger a_{\vec{k}}) = \sum_{\vec{k}} \omega (a_{\vec{k}}^\dagger a_k + \tfrac{1}{2}) \, . \tag{2.36}$$

From the example discussed in Section 1.2 we know that the eigenvalues of

$$N_{\vec{k}} \equiv a_{\vec{k}}^\dagger a_{\vec{k}} \tag{2.37}$$

are $n_{\vec{k}} = 0, 1, 2, \cdots$. Hence the operator $\sum \omega a_{\vec{k}}^\dagger a_{\vec{k}}$ is one that is bounded from below, but not from above. The totality of its eigenvectors forms a complete set, and it spans the entire Hilbert space of this system. These eigenvectors, properly normalized, are

$$|0>, \ a_{\vec{k}}^\dagger |0>, \ a_{\vec{k}}^\dagger a_{\vec{k}'}^\dagger |0> \text{ if } \vec{k} \neq \vec{k}', \ \frac{1}{\sqrt{2}} (a_{\vec{k}}^\dagger)^2 |0>, \cdots \tag{2.38}$$

where the state $|0>$ satisfies

$$a_{\vec{k}} |0> = 0 \qquad \text{for all } \vec{k} . \tag{2.39}$$

The lowest-energy state of H_0 is $|0>$, which will therefore be called the vacuum state of H_0. Similarly, we may call $a^\dagger_{\vec{k}} |0>$ the corresponding one-particle state and $a^\dagger_{\vec{k}} a^\dagger_{\vec{k}'} |0>$ the two-particle state, etc. Because $a^\dagger_{\vec{k}}$ commutes with $a^\dagger_{\vec{k}'}$ we have

$$a^\dagger_{\vec{k}} a^\dagger_{\vec{k}'} |0> = a^\dagger_{\vec{k}'} a^\dagger_{\vec{k}} |0> . \tag{2.40}$$

Consequently, these particles automatically satisfy Bose statistics.

<u>Remarks</u>

1. As emphasized before, so far the parameter m in (2.17) can be arbitrarily chosen. Different m values give different $a_{\vec{k}}$'s and $a^\dagger_{\vec{k}}$'s in the expansion (2.20)-(2.21), and therefore different basis vectors (2.38) of the same Hilbert space. Consequently, a change in m implies a canonical transformation between the $a_{\vec{k}}$'s and the $a^\dagger_{\vec{k}}$'s.

2. If in (2.33) H_{int} is 0, i.e., the field is free, the corresponding Hamiltonian becomes

$$H = H_0 = \tfrac{1}{2} \int [\Pi^2 + (\nabla\phi)^2 + m^2 \phi^2] \, d^3r . \tag{2.41}$$

Now, set the parameter m in (2.17) to be the same one as above. We have in accordance with (2.36)

$$H = H_0 = \sum_{\vec{k}} \omega (a^\dagger_{\vec{k}} a_{\vec{k}} + \tfrac{1}{2}) . \tag{2.42}$$

By using Heisenberg's equation (1.9) and setting $O(t) = a_{\vec{k}}$ and $a^\dagger_{\vec{k}}$, we find

$$a_{\vec{k}}(t) \propto e^{-i\omega t} \quad \text{and} \quad a^\dagger_{\vec{k}}(t) \propto e^{i\omega t} , \tag{2.43}$$

which is valid for a free-field system. Since a change of the Hamiltonian from $H \to H +$ a constant does not alter the dynamics of the system, we may replace (2.42) by

$$H = H_0 = \sum_{\vec{k}} \omega a^\dagger_{\vec{k}} a_{\vec{k}} \quad . \tag{2.44}$$

Consequently, the energy of the vacuum state defined by (2.39) becomes zero.

3. If $H_{int} \neq 0$, then the time variation of $a_{\vec{k}}(t)$ and $a^\dagger_{\vec{k}}(t)$ will in general be much more complicated than (2.43). Suppose that there is no bound state in the system. Because of Lorentz invariance, the spectrum of the total Hamiltonian $H = H_0 + H_{int}$ must be given by (apart from an additive term which can be chosen to be zero)

$$\sum n_k \, \omega_{phys} \tag{2.45}$$

where $\omega_{phys} = (\vec{k}^2 + m^2_{phys})^{\frac{1}{2}}$ and $n_k = 0, 1, 2, \cdots$. By definition, m_{phys} is the observed mass of the physical particle in the system. It is convenient to choose in H_0 the parameter m to be m_{phys} ; in this case the spectrum of H_0, given by (2.44), is identical to that of H, given by (2.45). As will be discussed in Chapter 5, this choice

$$m = m_{phys} \tag{2.46}$$

in H_0 brings great convenience to making the perturbation series expansion in powers of H_{int}.

4. The discussion above can readily be extended to a system of n Hermitian fields $\phi_1, \phi_2, \cdots, \phi_n$. The corresponding Lagrangian density can be written as

$$\mathcal{L} = -\sum_{i=1}^{n} \left[\tfrac{1}{2} \left(\frac{\partial \phi_i}{\partial x_\mu}\right)^2 + \tfrac{1}{2} m_i^2 \phi_i^2 \right] - V(\phi_i) \tag{2.47}$$

where $\phi_i = \phi_i^\dagger$ and $i = 1, 2, \cdots, n$. In the case that $n = 2$ and $m_1 = m_2 = m$, we may express the above Lagrangian in terms of the complex field

$$\phi = \frac{1}{\sqrt{2}} (\phi_1 + i\phi_2) \tag{2.48}$$

and its Hermitian conjugate

$$\phi^\dagger = \frac{1}{\sqrt{2}} (\phi_1 - i\phi_2) \quad . \tag{2.49}$$

Accordingly, (2.47) becomes

$$\mathcal{L} = - \frac{\partial \phi^\dagger}{\partial x_\mu} \frac{\partial \phi}{\partial x_\mu} - m^2 \phi^\dagger \phi - V(\phi^\dagger, \phi) \quad . \tag{2.50}$$

<u>Problem 2.1</u> Show that for a free Hermitian field ϕ

$$[\phi(\vec{r}, t), \phi(\vec{r}\,', t')] = -i D(x-x')$$

where $x = (\vec{r}, it)$, $x' = (\vec{r}\,', it')$,

$$D(x-x') = (2\pi)^{-3} \int d^3k\, e^{i\vec{k}\cdot(\vec{r}-\vec{r}\,')} \omega^{-1} \sin \omega(t-t')$$

and

$$\omega = \sqrt{\vec{k}^{\,2} + m^2} \quad .$$

Furthermore, prove that $D(x)$ satisfies

(i) $\left(-\frac{\partial^2}{\partial t^2} + \nabla^2 - m^2\right) D(x) = 0$,

(ii) $D(x) = 0$ at $t = 0$

and (iii) $\dot{D}(x) = \delta^3(\vec{r})$ at $t = 0$.

<u>Reference</u>

G. Wentzel, <u>Quantum Theory of Fields</u> (New York, Interscience Publishers, Inc., 1949).

Chapter 3

THE SPIN-1/2 FIELD

3.1 Mathematical Preliminaries

We first introduce three 2 X 2 Pauli matrices

$$\tau_1 = \begin{pmatrix} 0 & 1 \\ 1 & 0 \end{pmatrix}, \quad \tau_2 = \begin{pmatrix} 0 & -i \\ i & 0 \end{pmatrix}, \quad \tau_3 = \begin{pmatrix} 1 & 0 \\ 0 & -1 \end{pmatrix} \tag{3.1}$$

which satisfy

$$\tau_i = \tau_i^\dagger ,$$

$$[\tau_i , \tau_j] = \tau_i \tau_j - \tau_j \tau_i = 2i\epsilon_{ijk}\tau_k , \tag{3.2}$$

$$\{\tau_i , \tau_j\} \equiv \tau_i \tau_j + \tau_j \tau_i = 2\delta_{ij} \tag{3.3}$$

where δ_{ij} is the Kronecker symbol used before, and

$$\epsilon_{ijk} = \begin{cases} 1, & \text{if } ijk \text{ is an even permutation of } 1, 2, 3, \\ -1, & \text{if } ijk \text{ is an odd permutation} \\ 0, & \text{otherwise.} \end{cases} \tag{3.4}$$

Quite often, we use the vector notation

$$\vec{\tau} = (\tau_1 , \tau_2 , \tau_3) .$$

Throughout this book, we denote the commutator and anticommutator between two matrices a and b by $[a, b] = ab - ba$ and $\{a, b\} = ab + ba$ respectively.

Next we introduce the definition of the direct product $A \times B$ of an $n \times n$ matrix $A = (A_{aa'})$ times an $m \times m$ matrix $B = (B_{bb'})$:

$$(A \times B)_{ab,\, a'b'} \equiv A_{aa'} B_{bb'} \qquad (3.5)$$

where the subscripts a, a' can vary from 1 to n and the subscripts b, b' vary from 1 to m. Thus the matrix $A \times B$ is of dimension $nm \times nm$. One can verify readily that if matrices A and C are of the same dimension $n \times n$ and if matrices B and D are also of the same dimension $m \times m$, then

$$(A \times B) \cdot (C \times D) = (A \cdot C) \times (B \cdot D) \ ,$$

where the dot denotes the usual matrix multiplication.

The Dirac matrices $\vec{\sigma}$ and $\vec{\rho}$ are 4×4 matrices which can be expressed as the direct product between the 2×2 Pauli matrices (3.1) and the 2×2 unit matrix I :

$$\vec{\sigma} \equiv \vec{\tau} \times I \ , \qquad \vec{\rho} \equiv I \times \vec{\tau} \ . \qquad (3.6)$$

Consequently, we have

$$\vec{\sigma} = \begin{pmatrix} \vec{\tau} & 0 \\ 0 & \vec{\tau} \end{pmatrix} , \qquad (3.7)$$

and $\vec{\rho} = (\rho_1, \rho_2, \rho_3)$ is given by

$$\rho_1 = \begin{pmatrix} 0 & I \\ I & 0 \end{pmatrix}, \quad \rho_2 = \begin{pmatrix} 0 & -iI \\ iI & 0 \end{pmatrix}, \quad \rho_3 = \begin{pmatrix} I & 0 \\ 0 & -I \end{pmatrix} . \qquad (3.8)$$

The $\vec{\sigma}$ and $\vec{\rho}$ matrices satisfy the following relations:

$$\begin{aligned} &\rho_i = \rho_i^\dagger \ , \qquad \sigma_i = \sigma_i^\dagger \ , \\ &[\rho_i, \rho_j] = 2i\,\epsilon_{ijk}\,\rho_k \ , \qquad [\sigma_i, \sigma_j] = 2i\,\epsilon_{ijk}\,\sigma_k \ , \\ &\{\rho_i, \rho_j\} = \{\sigma_i, \sigma_j\} = 2\delta_{ij} \ , \\ &[\rho_i, \sigma_j] = 0 \ . \end{aligned} \qquad (3.9)$$

Furthermore, we shall define

$$\vec{\alpha} \equiv \rho_1 \vec{\sigma} \ , \qquad \beta \equiv \rho_3 \ , \qquad (3.10)$$

$$\gamma_i = -i\beta\alpha_i = \rho_2\sigma_i \quad \text{and} \quad \gamma_4 = \beta = \rho_3 \ . \qquad (3.11)$$

These matrices satisfy

$$\alpha_i = \alpha_i^\dagger \ , \quad \beta = \beta^\dagger \ , \quad \gamma_\mu = \gamma_\mu^\dagger \ ,$$

$$[\alpha_i \, , \alpha_j] = [\gamma_i \, , \gamma_j] = 2i\,\epsilon_{ijk}\,\sigma_k \ , \tag{3.12}$$

$$\{\alpha_i \, , \alpha_j\} = 2\delta_{ij} \ , \quad \{\alpha_i \, , \beta\} = 0$$

and

$$\{\gamma_\mu \, , \gamma_\nu\} = 2\delta_{\mu\nu} \ .$$

In the above, as well as later on, all Roman subscripts i, j, k vary from 1 to 3 and all Greek subscripts μ, ν, λ vary from 1 to 4.

3.2 Free Field

The Lagrangian density of a free spin-½ field is

$$\mathcal{L}_{\text{free}} = -\psi^\dagger \gamma_4 \left(\gamma_\mu \frac{\partial}{\partial x_\mu} + m\right)\psi \tag{3.13}$$

where ψ is a 4×1 column matrix. (In the quantum theory, each of its matrix elements is a Hilbert space operator.)

If ψ were a classical field, then from the variational principle

$$\delta \int \mathcal{L}\, d^4x = 0 \ ,$$

we have

$$\left(\gamma_\mu \frac{\partial}{\partial x_\mu} + m\right)\psi = 0 \tag{3.14}$$

which, because of (3.11), can also be written as

$$(-i\vec{\alpha} \cdot \vec{\nabla} + \beta m)\,\psi = i\dot{\psi} \ . \tag{3.15}$$

We observe that Eq. (3.13) can be written as

$$\mathcal{L} = i\psi^\dagger \dot{\psi} + \cdots \tag{3.16}$$

where the $\cdots$ term does not contain $\dot{\psi}$. Let ψ_λ be the λth component of the 4×1 matrix ψ, where $\lambda = 1, 2, 3, 4$. By regarding $\psi_\lambda(\vec{r}, t)$ as a generalized coordinate, we see that its conjugate momentum is, on account of (3.16)

$$\mathcal{P}_\lambda(\vec{r}, t) = \frac{\partial \mathcal{L}}{\partial \dot{\psi}_\lambda} = i\psi_\lambda^\dagger(\vec{r}, t) \ . \tag{3.17}$$

The corresponding Hamiltonian density is

$$\mathcal{H}_{free} = \mathcal{P}_\lambda \dot{\psi}_\lambda - \mathcal{L}_{free} = \psi^\dagger \gamma_4 (\gamma_i \frac{\partial}{\partial x_i} + m)\psi \quad ,$$

which, because of (3.10)-(3.11), can also be written as

$$\mathcal{H}_{free} = \psi^\dagger \left(\frac{\vec{\alpha}\cdot\vec{\nabla}}{i} + \beta m\right)\psi \quad . \tag{3.18}$$

3.3 Quantization (Free or Interacting Fields)

We first generalize the above discussion to systems with interactions. The Lagrangian density is now given by

$$\mathcal{L} = \mathcal{L}_{free} + \mathcal{L}_{int} \tag{3.19}$$

where $\mathcal{L}_{free}$ remains given by (3.13). If the system consists of only the ψ field, then $\mathcal{L}_{int} = \mathcal{L}_{int}(\psi, \psi^\dagger)$; if there are additional fields, such as ϕ, then $\mathcal{L}_{int} = \mathcal{L}_{int}(\psi, \psi^\dagger, \phi, \dot{\phi})$. We shall assume that $\mathcal{L}_{int}$ does not contain $\partial\psi/\partial x_\mu$. Consequently the conjugate momentum of $\psi(\vec{r}, t)$ remains given by (3.17). The corresponding Hamiltonian density can now be written as

$$\mathcal{H} = \mathcal{H}_{free} + \mathcal{H}_{int} \tag{3.20}$$

where $\mathcal{H}_{free}$ is given by (3.18) and $\mathcal{H}_{int} = -\mathcal{L}_{int}(\psi, \psi^\dagger)$ if the system consists only of the ψ field. If there are additional fields such as ϕ, then $\mathcal{H}_{int} = \mathcal{H}_{int}(\psi, \psi^\dagger, \phi, \Pi)$, where Π is the conjugate momentum of ϕ.

Now we turn to the quantization problem. Following Jordan and Wigner, the quantization of a spin-$\frac{1}{2}$ field differs from that of an integer-spin field by the replacement of all equal-time commutation relations by anticommutation relations; i.e.,

$$\{\psi_\mu(\vec{r}, t), \mathcal{P}_\lambda(\vec{r}\,', t)\} = i\delta^3(\vec{r} - \vec{r}\,')\delta_{\mu\lambda} \quad ,$$

$$\{\psi_\mu(\vec{r}, t), \psi_\lambda(\vec{r}\,', t)\} = \{\mathcal{P}_\mu(\vec{r}, t), \mathcal{P}_\lambda(\vec{r}\,', t)\} = 0 \quad .$$

Because of (3.17) these relations can also be written as

$$\{\psi_\mu(\vec{r}, t), \ \psi_\lambda^\dagger(\vec{r}', t)\} = \delta^3(\vec{r} - \vec{r}')\,\delta_{\mu\lambda} \ , \tag{3.21}$$

$$\{\psi_\mu(\vec{r}, t), \ \psi_\lambda(\vec{r}', t)\} = \{\psi_\mu^\dagger(\vec{r}, t), \ \psi_\lambda^\dagger(\vec{r}', t)\} = 0 \ . \tag{3.22}$$

The equation of motion remains given by Heisenberg's equation (1.9):

$$[\int \mathcal{H} d^3r, \ O(t)] = -i\,\dot{O}(t) \ .$$

In the case of a free field, by setting $H = \int \mathcal{H}_{free}\, d^3r$ and $O(t) = \psi(\vec{r}, t)$ we have

$$[\int \mathcal{H}_{free}\, d^3r, \psi] = -i\dot{\psi} \ . \tag{3.23}$$

On account of (3.21)-(3.22), we have, for any 4×4 matrix Γ whose matrix elements are c. numbers,

$$\begin{aligned}
&[\int \psi^\dagger(\vec{r}', t)\,\Gamma\,\psi(\vec{r}', t)\, d^3r', \ \psi_\lambda(\vec{r}, t)] \\
&= \int d^3r' (\psi_\mu^\dagger(\vec{r}', t)\ \Gamma_{\mu\nu}\,\psi_\nu(\vec{r}', t)\,\psi_\lambda(\vec{r}, t) \\
&\qquad\qquad - \psi_\lambda(\vec{r}, t)\,\psi_\mu^\dagger(\vec{r}', t)\ \Gamma_{\mu\nu}\,\psi_\nu(\vec{r}', t)) \\
&= \int d^3r' (-\psi_\mu^\dagger(\vec{r}', t)\ \Gamma_{\mu\nu}\,\psi_\lambda(\vec{r}, t)\,\psi_\nu(\vec{r}', t) \\
&\qquad\qquad - \psi_\lambda(\vec{r}, t)\,\psi_\mu^\dagger(\vec{r}', t)\ \Gamma_{\mu\nu}\,\psi_\nu(\vec{r}', t)) \\
&= -\int d^3r'\,\delta_{\mu\lambda}\,\Gamma_{\mu\nu}\,\delta^3(\vec{r} - \vec{r}')\,\psi_\nu(\vec{r}', t) = (-\Gamma\psi(\vec{r}, t))_\lambda \ .
\end{aligned} \tag{3.24}$$

Consequently, for the free field, (3.23) gives the same equation of motion (3.15) for the operator ψ:

$$(-i\vec{\alpha}\cdot\vec{\nabla} + \beta m)\psi = i\dot{\psi} \ .$$

Likewise, we can show that in the case of interacting fields Heisenberg's equation also leads to the same equations of motion as those given by the variational principle $\delta \int \mathcal{L}\, d^4x = 0$.

<u>Exercise.</u> Show that

$$\begin{aligned}
&[\int \psi^\dagger(\vec{r}', t)\ \Gamma\psi(\vec{r}', t)\, d^3r' \ , \ \int \psi^\dagger(\vec{r}, t)\,\Gamma'\psi(\vec{r}, t)\, d^3r] \\
&\qquad = \int \psi^\dagger(\vec{r}, t)\,[\Gamma, \Gamma']\,\psi(\vec{r}, t)\, d^3r
\end{aligned} \tag{3.24a}$$

where Γ and Γ' are both 4×4 matrices whose matrix elements are c. numbers.

3.4 Fourier Expansion (Free or Interacting Fields)

Just as in (2.11)-(2.12), at any given time t, the operator $\psi(\vec{r}, t)$ can be expanded in terms of the Fourier series:

$$\psi(\vec{r}, t) = \sum_{\vec{p}} S_{\vec{p}}(t) \frac{e^{i\vec{p}\cdot\vec{r}}}{\sqrt{\Omega}} \quad . \tag{3.25}$$

In the above expression $S_{\vec{p}}(t)$ is, like $\psi(\vec{r}, t)$, a 4×1 matrix with its matrix elements the Hilbert-space operators. The only difference is that, unlike $\psi(\vec{r}, t)$, $S_{\vec{p}}(t)$ is independent of $\vec{r}$.

Let us regard the 4×1 column matrices as vectors in a 4-dimensional space, called spinor space. For a given $\vec{p}$ it is convenient to introduce the following set of c. number basis vectors $u_{\vec{p},s}$ and $v_{-\vec{p},s}$ in the spinor space. These vectors satisfy

$$(\vec{\alpha}\cdot\vec{p} + \beta m) \begin{cases} u_{\vec{p},s} \\ v_{-\vec{p},s} \end{cases} = E_p \begin{cases} u_{\vec{p},s} \\ -v_{-\vec{p},s} \end{cases} \tag{3.26}$$

and

$$\vec{\sigma}\cdot\hat{p} \begin{cases} u_{\vec{p},s} \\ v_{-\vec{p},s} \end{cases} = 2s \begin{cases} u_{\vec{p},s} \\ v_{-\vec{p},s} \end{cases} \tag{3.27}$$

where

$$E_p = \sqrt{\vec{p}^{\,2} + m^2} > 0 \ , \tag{3.28}$$

$$\hat{p} = \frac{\vec{p}}{|\vec{p}|}$$

and the parameter

$$s = \pm \tfrac{1}{2} \tag{3.29}$$

is called helicity, whose physical significance will be discussed in Section 3.7. We shall normalize these vectors so that

$$u^\dagger_{\vec{p},s} u_{\vec{p},s} = v^\dagger_{-\vec{p},s} v_{-\vec{p},s} = 1 \ . \tag{3.30}$$

At a given $\vec{p}$, the 4×4 matrices $(\vec{\alpha} \cdot \vec{p} + \beta m)$ and $\vec{\sigma} \cdot \hat{p}$ are both Hermitian. According to (3.26) and (3.27) the four vectors $u_{\vec{p},s}$ and $v_{-\vec{p},s}$ with $s = \pm\frac{1}{2}$ are eigenvectors of these two Hermitian matrices with different eigenvalues. Consequently, these four vectors are orthogonal to each other; because of (3.30) they form a complete orthonormal set of basis vectors in the spinor space. The $S_{\vec{p}}(t)$ in Eq. (3.25) can be expanded in terms of this set of bases:

$$S_{\vec{p}}(t) = \sum_{s=\pm\frac{1}{2}} \left(a_{\vec{p},s}(t)\, u_{\vec{p},s} + b^{\dagger}_{-\vec{p},s}(t)\, v_{-\vec{p},s}\right) \tag{3.31}$$

where the coefficients $a_{\vec{p},s}(t)$ and $b^{\dagger}_{-\vec{p},s}(t)$ are Hilbert-space operators. Combining (3.25) and (3.31) we have

$$\psi(\vec{r},t) = \frac{1}{\sqrt{\Omega}} \sum_{\vec{p},s} \left(a_{\vec{p},s}(t)\, u_{\vec{p},s}\, e^{i\vec{p}\cdot\vec{r}} + b^{\dagger}_{\vec{p},s}(t)\, v_{\vec{p},s}\, e^{-i\vec{p}\cdot\vec{r}}\right) . \tag{3.32}$$

Its Hermitian conjugate is

$$\psi^{\dagger}(\vec{r},t) = \frac{1}{\sqrt{\Omega}} \sum_{\vec{p},s} \left(a^{\dagger}_{\vec{p},s}(t)\, u^{\dagger}_{\vec{p},s}\, e^{-i\vec{p}\cdot\vec{r}} + b_{\vec{p},s}(t)\, v^{\dagger}_{\vec{p},s}\, e^{i\vec{p}\cdot\vec{r}}\right) . \tag{3.33}$$

From the anticommutation relations (3.21)–(3.22) we can readily verify

$$\begin{aligned}
\{a_{\vec{p},s}(t),\, a^{\dagger}_{\vec{p}\,',s'}(t)\} &= \{b_{\vec{p},s}(t),\, b^{\dagger}_{\vec{p}\,',s'}(t)\} = \delta_{\vec{p},\vec{p}\,'}\, \delta_{s,s'} \;, \\
\{a_{\vec{p},s}(t),\, a_{\vec{p}\,',s'}(t)\} &= \{b_{\vec{p},s}(t),\, b_{\vec{p}\,',s'}(t)\} = 0 \;, \\
\{a_{\vec{p},s}(t),\, b_{\vec{p}\,',s'}(t)\} &= \{a_{\vec{p},s}(t),\, b^{\dagger}_{\vec{p}\,',s'}(t)\} = 0 \;.
\end{aligned} \tag{3.34}$$

As noted before, any function of $\vec{r}$ can be expanded in terms of the complete set $\{ \frac{1}{\sqrt{\Omega}}\, e^{i\vec{p}\cdot\vec{r}} \}$, and any 4×1 column matrix in the spinor space can be expanded in terms of the four orthonormal basis vectors: $u_{\vec{p},\pm\frac{1}{2}}$ and $v_{-\vec{p},\pm\frac{1}{2}}$ where $\vec{p}$ is fixed. Consequently the expansion (3.32) is valid for the free, as well as the interacting, field.

For the free-field case, the Hamiltonian density is given by

$$\mathcal{H} = \mathcal{H}_{free} = \psi^{\dagger}(-i\vec{\alpha}\cdot\vec{\nabla} + \beta m)\,\psi \quad . \tag{3.35}$$

By substituting (3.32) into the above expression, we obtain

$$H = H_{free} = \int \mathcal{H}_{free}\, d^3 r = \sum_{\vec{p},s} (a^{\dagger}_{\vec{p},s}\, a_{\vec{p},s} - b_{\vec{p},s}\, b^{\dagger}_{\vec{p},s})\, E_p$$

$$= \sum_{\vec{p},s} (a^{\dagger}_{\vec{p},s}\, a_{\vec{p},s} + b^{\dagger}_{\vec{p},s}\, b_{\vec{p},s} - 1)\, E_p \ . \tag{3.36}$$

Since a change $H \to H +$ a constant does not alter the dynamics of the system, we may drop the -1 inside the parentheses in the above formula. The Hamiltonian (3.36) is then replaced by

$$H = H_{free} = \sum_{\vec{p},s} (a^{\dagger}_{\vec{p},s}\, a_{\vec{p},s} + b^{\dagger}_{\vec{p},s}\, b_{\vec{p},s})\, E_p \quad . \tag{3.37}$$

By using Heisenberg's equation (1.9) and by setting $O(t) = a_{\vec{p},s}(t)$, we have for the free-field case

$$-i\,\dot{a}_{\vec{p},s}(t) = [H, a_{\vec{p},s}(t)] = -E_p\, a_{\vec{p},s}(t) \quad ,$$

and therefore

$$a_{\vec{p},s}(t) \propto e^{-iE_p t} \quad . \tag{3.38}$$

Likewise we can derive

$$b_{\vec{p},s}(t) \propto e^{-iE_p t} \quad . \tag{3.39}$$

In the case of an interacting field, the time-dependence of $a_{\vec{p},s}(t)$ and $b_{\vec{p},s}(t)$ will in general be more complicated.

Exercise. Show that

$$\sum_s u_{\vec{p},s}\, u^{\dagger}_{\vec{p},s}\, \beta = \frac{\not{p} + m}{2p_0} \quad ,$$

$$\sum_s v_{\vec{p},s}\, v^{\dagger}_{\vec{p},s}\, \beta = \frac{\not{p} - m}{2p_0} \quad ,$$

$$\tfrac{1}{4}\,\mathrm{trace}\,(\not{A}\,\not{B}) = -(A\cdot B) \equiv A_0 B_0 - \vec{A}\cdot\vec{B} \quad ,$$

$$\tfrac{1}{4}\,\mathrm{trace}\,(\not{A}\,\not{B}\,\not{C}\,\not{D}) = (A\cdot B)(C\cdot D) - (A\cdot C)(B\cdot D) + (A\cdot D)(B\cdot C)$$

where A, B, C and D can be any c. number 4-vectors,

$\not{A} = -i\gamma_\mu A_\mu$, $A_4 = iA_0$, $\not{B} = -i\gamma_\mu B_\mu$, $B_4 = iB_0$, etc.

3.5 Hilbert Space (Free or Interacting Fields)

Just as in (2.38), the Hilbert space is spanned by the set of orthonormal basis vectors:

$$|0>,\quad a^\dagger_{\vec{p},s}|0>,\quad b^\dagger_{\vec{p},s}|0>,\quad a^\dagger_{\vec{p},s}a^\dagger_{\vec{p}',s'}|0>,$$

$$a^\dagger_{\vec{p},s}b^\dagger_{\vec{p}',s'}|0>,\quad b^\dagger_{\vec{p},s}b^\dagger_{\vec{p}',s'}|0>,\ \cdots \tag{3.40}$$

where the state vector $|0>$ satisfies

$$a_{\vec{p},s}|0> = 0 \quad \text{and} \quad b_{\vec{p},s}|0> = 0 \tag{3.41}$$

for all $\vec{p}$ and s. In order to analyze the structure of this Hilbert space, we must first discuss some elementary algebraic properties of these anticommuting operators $a_{\vec{p},s}$, $b_{\vec{p},s}$ and their Hermitian conjugates.

(i) We first discuss the case of a single mode. Let a and $a^\dagger$ satisfy the following anticommutation relations

$$\{a, a^\dagger\} = 1\ , \tag{3.42}$$

and

$$\{a, a\} = \{a^\dagger, a^\dagger\} = 0\ ;$$

the latter can also be written as

$$a^2 = (a^\dagger)^2 = 0\ . \tag{3.43}$$

Let us define

$$N \equiv a^\dagger a\ . \tag{3.44}$$

Because of (3.42)-(3.43), we find

$$N^2 = a^\dagger a\, a^\dagger a = a^\dagger(1 - a^\dagger a)\, a = N\ ,$$

which implies that the eigenvalues of N can only be 0 or 1. Assuming that N does have an eigenstate, denoted by $|0>$, with the eigenvalue 0 :

$$N\,|\,0> \; = \; 0 \quad . \tag{3.45}$$

Then it follows that

$$N a^\dagger\,|\,0> \; = \; a^\dagger a\, a^\dagger\,|\,0> \; = \; a^\dagger(1 - a^\dagger a)\,|\,0> \; = \; a^\dagger\,|\,0> \quad . \tag{3.46}$$

By designating

$$|\,1> \; \equiv \; a^\dagger\,|\,0> \quad , \tag{3.47}$$

we can write (3.46) as

$$N\,|\,1> \; = \; |\,1> \quad . \tag{3.48}$$

Thus, the existence of the eigenstate $|\,0>$ implies that of $|\,1>$. The converse is also true, since from (3.47) we can establish

$$|\,0> \; = \; a\,|\,1> \quad . \tag{3.49}$$

Therefore, both eigenstates exist. Furthermore, because of (3.43) we have

$$a^\dagger\,|\,1> \; = \; 0 \quad \text{and} \quad a\,|\,0> \; = \; 0 \quad . \tag{3.50}$$

These two eigenvectors $|\,0>$ and $|\,1>$ span a two-dimensional Hilbert space. We may represent

$$|\,0> \; = \begin{pmatrix} 1 \\ 0 \end{pmatrix} \quad \text{and} \quad |\,1> \; = \begin{pmatrix} 0 \\ 1 \end{pmatrix} \quad . \tag{3.51}$$

The operators a , $a^\dagger$ and N can be expressed in matrix form:

$$\begin{aligned} a &= \begin{pmatrix} 0 & 1 \\ 0 & 0 \end{pmatrix} \equiv \tau_+ = \tfrac{1}{2}(\tau_1 + i\,\tau_2) \quad , \\ a^\dagger &= \begin{pmatrix} 0 & 0 \\ 1 & 0 \end{pmatrix} \equiv \tau_- = \tfrac{1}{2}(\tau_1 - i\,\tau_2) \quad , \\ N &= \begin{pmatrix} 0 & 0 \\ 0 & 1 \end{pmatrix} = \tfrac{1}{2}(1 - \tau_3) \end{aligned} \tag{3.52}$$

where τ_1 , τ_2 and τ_3 are 2×2 Pauli matrices given by (3.1). As in the case of the boson field, we call N the occupation-number operator, a the annihilation operator and $a^\dagger$ the creation operator.

(ii) Next we consider the case of two modes. There are now two annihilation operators a_1 and a_2 ; their Hermitian conjugates form two creation operators. These operators satisfy

$$\{a_i\,,\ a_j^\dagger\} = \delta_{ij}$$

and $$\{a_i\,,\ a_j\} = \{a_i^\dagger\,,\ a_j^\dagger\} = 0 \tag{3.53}$$

where i and j can be 1 or 2. We may define

$$N_1 = a_1^\dagger a_1 \quad \text{and} \quad N_2 = a_2^\dagger a_2 \ . \tag{3.54}$$

Because of (3.53), N_1 commutes with N_2. By following an argument similar to that in case (i) we can show that the eigenvalue of each N_i can be 0 or 1. Thus, the eigenvalues of the set (N_1, N_2) can be (0, 0), (1, 0), (0, 1) and (1, 1). By regarding the corresponding eigenstates as the basis vectors, we form a four-dimensional Hilbert space. In this space the matrix representations of N_1 and N_2 are

$$N_1 = \begin{pmatrix} 0 & & & \\ & 1 & & \\ & & 0 & \\ & & & 1 \end{pmatrix} , \quad N_2 = \begin{pmatrix} 0 & & & \\ & 0 & & \\ & & 1 & \\ & & & 1 \end{pmatrix} . \tag{3.55}$$

The corresponding matrices for a_i and $a_i^\dagger$ may be given by the following direct products:

$$a_1 = \tau_+ \times I\ , \quad a_1^\dagger = \tau_- \times I\ ,$$

$$a_2 = \tau_3 \times \tau_+\ , \quad a_2^\dagger = \tau_3 \times \tau_- \tag{3.56}$$

where I is a 2×2 unit matrix. In explicit form, (3.56) can also be written as

$$a_1 = \begin{pmatrix} \tau_+ & 0 \\ 0 & \tau_+ \end{pmatrix} , \quad a_1^\dagger = \begin{pmatrix} \tau_- & 0 \\ 0 & \tau_- \end{pmatrix} ,$$

$$a_2 = \begin{pmatrix} 0 & \tau_3 \\ 0 & 0 \end{pmatrix} , \quad a_2^\dagger = \begin{pmatrix} 0 & 0 \\ \tau_3 & 0 \end{pmatrix} . \tag{3.57}$$

By using (3.56) one may verify directly that the anticommutation relations given by (3.53) hold. Furthermore, the matrices N_1 and

N_2 are given by (3.55).

(iii) The above considerations can be readily generalized to the case of n modes. The anticommutation relations have the same form as (3.53); i.e.,

$$\begin{aligned} \{a_i, a_j^\dagger\} &= \delta_{ij}, \\ \{a_i, a_j\} &= \{a_i^\dagger, a_j^\dagger\} = 0 \end{aligned} \qquad (3.58)$$

except that i and j can now vary from 1 to n. By induction we can generalize (3.56) to

$$\begin{aligned} a_1 &= \tau_+ \times I \times I \times \cdots \times I, \quad a_1^\dagger = \tau_- \times I \times I \times \cdots \times I, \\ a_2 &= \tau_3 \times \tau_+ \times I \times \cdots \times I, \quad a_2^\dagger = \tau_3 \times \tau_- \times I \times \cdots \times I, \\ &\cdots\cdots \\ a_n &= \tau_3 \times \tau_3 \times \cdots \times \tau_+, \quad a_n^\dagger = \tau_3 \times \tau_3 \times \cdots \times \tau_- \end{aligned} \qquad (3.59)$$

in which the expression for a_1 contains $n - 1$ factors of I, that for a_2 $n - 2$ factors of I, etc. These matrices in (3.59) can be easily seen to satisfy the anticommutation relations (3.58). Furthermore, the operator

$$N_i \equiv a_i^\dagger a_i$$

has the eigenvalue 0 or 1, where the subscript i can be $1, 2, \cdots, n$.

Remarks. In the general case, the Hilbert space is spanned by the basis vectors (3.40), in which $|0>$ is called the vacuum state of H_{free}, which is determined by (3.41). Likewise, $a^\dagger_{\vec{p},s} |0>$ is called a one-particle state, $b^\dagger_{\vec{p},s} |0>$ a one-antiparticle state, $a^\dagger_{\vec{p},s} a^\dagger_{\vec{p}',s'} |0>$ a two-particle state, etc. As we shall discuss in the next section, the subscripts $\vec{p}$ and s denote respectively the momentum and helicity of the particle (or antiparticle), where "helicity" means the component of angular momentum along the direction of $\vec{p}$.

Because of the anticommutation relations (3.34), we have

$$a^\dagger_{\vec{p},s}\, a^\dagger_{\vec{p}',s'} \mid 0 > \; = \; - a^\dagger_{\vec{p}',s'}\, a^\dagger_{\vec{p},s} \mid 0 > \tag{3.60}$$

which implies that these particles obey Fermi statistics. Thus, for example, when $\vec{p} = \vec{p}'$ and $s = s'$ the vector (3.60) is a null vector, showing that these particles do satisfy Pauli's exclusion principle.

Exercise. In the case of bosons with n modes, the commutation relations between the annihilation and creation operators are given by

$$[a_i, a_j^\dagger] = \delta_{ij}$$

and

$$[a_i, a_j] = [a_i^\dagger, a_j^\dagger] = 0 ,$$

where i and j vary from 1 to n. Show that the matrix forms of these operators can be written as the following direct products

$$a_1 = a \times I \times I \times \cdots \times I , \quad a_1^\dagger = a^\dagger \times I \times I \times \cdots \times I ,$$

$$a_2 = I \times a \times I \times \cdots \times I , \quad a_2^\dagger = I \times a^\dagger \times I \times \cdots \times I ,$$

$$\cdots\cdot$$

$$a_n = I \times I \times \cdots \times I \times a , \quad a_n^\dagger = I \times I \times \cdots \times I \times a^\dagger$$

where a and $a^\dagger$ are given by (1.31) and I is an $\infty \times \infty$ unit matrix.

3.6 Momentum and Angular Momentum Operators

Let us define the momentum operator $\vec{P}$ and the angular momentum operator $\vec{J}$ of a spin-$\frac{1}{2}$ field to be

$$\vec{P}(t) \equiv -i \int \psi^\dagger(\vec{r}, t)\, \vec{\nabla} \psi(\vec{r}, t)\, d^3r \tag{3.61}$$

and

$$\vec{J}(t) \equiv \int \psi^\dagger(\vec{r}, t)\, (\vec{\ell} + \tfrac{1}{2}\vec{\sigma})\, \psi(\vec{r}, t)\, d^3r \tag{3.62}$$

where

$$\vec{\ell} = -i \vec{r} \times \vec{\nabla} . \tag{3.63}$$

By using (3.24), we find

$$[\vec{P}(t), \psi(\vec{r}, t)] = i \vec{\nabla} \psi(\vec{r}, t) \tag{3.64}$$

and

$$[\vec{J}(t),\ \psi(\vec{r},t)] = -(-i\,\vec{r}\times\vec{\nabla} + \tfrac{1}{2}\vec{\sigma})\,\psi(\vec{r},t) \quad . \tag{3.65}$$

Thus, operating on $\psi(\vec{r}, t)$, $\vec{P}(t)$ acts as an infinitesimal displacement operator and $\vec{J}(t)$ as an infinitesimal rotation operator. [See Problem 3 at the end of Chapter 10.]

By using the exercise given in Section 3.3 we can readily verify that the components of $\vec{P}(t)$ and $\vec{J}(t)$ satisfy

$$[P_i(t),\ P_j(t)] = 0 \ , \tag{3.66}$$

$$[J_i(t),\ J_j(t)] = i\,\epsilon_{ijk}\,J_k(t) \ , \tag{3.67}$$

$$[P_1(t),\ J_1(t)] = [P_2(t),\ J_2(t)] = [P_3(t),\ J_3(t)] = 0 \ , \tag{3.68}$$

but

$$[P_i(t),\ J_j(t)] \neq 0 \quad \text{if} \quad i \neq j \ , \tag{3.69}$$

where i and j can be 1, 2 or 3, and ϵ_{ijk} is given by (3.4). The above relations merely reflect the fact that the differential operators ∇_i and ∇_j commute,

$$[\ell_i,\ \ell_j] = i\,\epsilon_{ijk}\,\ell_k \ , \qquad [\tfrac{1}{2}\sigma_i,\ \tfrac{1}{2}\sigma_j] = i\,\epsilon_{ijk}\,\tfrac{1}{2}\sigma_k$$

and while ∇_i commutes with $(\vec{r}\times\vec{\nabla})_i$, it does not commute with $(\vec{r}\times\vec{\nabla})_j$ for $j \neq i$.

If the system is a free spin-$\frac{1}{2}$ field, then the Hamiltonian is

$$H_{free} = \int \mathcal{H}_{free}\, d^3r \tag{3.70}$$

where $\mathcal{H}_{free}$ is given by (3.18). It is straightforward to derive

$$[\vec{P}(t),\ H_{free}] = 0$$

and

$$[\vec{J}(t),\ H_{free}] = 0 \ . \tag{3.71}$$

Both $\vec{P}$ and $\vec{J}$ are therefore constants of motion. In the case of interacting fields, the total momentum consists of $\vec{P}(t)$ of the spin-$\frac{1}{2}$ field, plus the momenta of other fields. Consequently, conservation of total momentum does not imply that $\vec{P}(t)$ is a constant of motion.

Similar considerations also apply to the total angular momentum.

We may express $\vec{P}$ in terms of the Fourier components of ψ. Let us substitute (3.32)-(3.33) into (3.61). By using (2.22) and the orthonormality relations between the spinors $u_{\vec{p},s}$ and $v_{\vec{p},s}$ we obtain

$$\begin{aligned}\vec{P} &= \sum_{\vec{p},s} \vec{p}\,(a^{\dagger}_{\vec{p},s}\, a_{\vec{p},s} - b_{\vec{p},s}\, b^{\dagger}_{\vec{p},s}) \\ &= \sum_{\vec{p},s} \vec{p}\,(a^{\dagger}_{\vec{p},s}\, a_{\vec{p},s} + b^{\dagger}_{\vec{p},s}\, b_{\vec{p},s} - 1) \ .\end{aligned}$$

Since $\sum_{\vec{p}} \vec{p} = 0$ due to symmetry, the above expression can be written as

$$\vec{P} = \sum_{\vec{p},s} \vec{p}\,(a^{\dagger}_{\vec{p},s}\, a_{\vec{p},s} + b^{\dagger}_{\vec{p},s}\, b_{\vec{p},s}) \ . \tag{3.72}$$

For definiteness, we may consider ψ to be the electron field, and denote in the Hilbert space (3.40)

$$| e^{-}_{\vec{p},s} > \ \equiv \ a^{\dagger}_{\vec{p},s}(t) \ | \ 0 >$$

and

$$| e^{+}_{\vec{p},s} > \ \equiv \ b^{\dagger}_{\vec{p},s}(t) \ | \ 0 > \ . \tag{3.73}$$

From (3.72), it follows that these states are eigenvectors of $\vec{P}$:

$$\vec{P} \ | \ 0 > \ = \ 0 \tag{3.74}$$

and

$$\vec{P} \ | \ e^{\pm}_{\vec{p},s} > \ = \ \vec{p} \ | \ e^{\pm}_{\vec{p},s} > \ . \tag{3.75}$$

The former can also be derived by noting that the validity of all our expressions from (3.40) on is independent of the Hamiltonian H, provided that all operators are synchronized at the same time t. Thus, we need only consider the special case $H = H_{free}$. Because of (3.71) and the fact that $| 0 >$ is the ground state of H_{free} with no degeneracy, (3.74) follows. Likewise, we have

$$\vec{J} \mid 0 > = 0 \; . \tag{3.76}$$

Next, we shall prove

$$\vec{J} \cdot \vec{P} \mid e^{\pm}_{\vec{p},s} > = |\vec{p}| \, s \mid e^{\pm}_{\vec{p},s} > \tag{3.77}$$

which, together with (3.75), means that $\vec{p}$ is the momentum of the state and the helicity $s = \pm \frac{1}{2}$ is its component of angular momentum along $\vec{p}$. To see this, we take the Hermitian conjugates of (3.64) and (3.65) and apply them onto $| 0 >$. That gives, on account of (3.74) and (3.76)

$$P_k(t) \, \psi^\dagger(\vec{r}, t) \mid 0 > = -i \nabla_k \psi^\dagger(\vec{r}, t) \mid 0 > \, ,$$

$$J_k(t) \, \psi^\dagger(\vec{r}, t) \mid 0 > = -(i \vec{r} \times \vec{\nabla} \psi^\dagger + \tfrac{1}{2} \psi^\dagger \vec{\sigma})_k \mid 0 >$$

and therefore

$$\vec{J} \cdot \vec{P} \, \psi^\dagger(\vec{r}, t) \mid 0 > = i \tfrac{1}{2} \nabla_k \psi^\dagger(\vec{r}, t) \, \sigma_k \mid 0 > . \tag{3.78}$$

Because

$$a^\dagger_{\vec{p},s}(t) = \int \frac{1}{\sqrt{\Omega}} e^{i\vec{p}\cdot\vec{r}} \, \psi^\dagger(\vec{r}, t) \, u_{\vec{p},s} \, d^3r \, , \tag{3.79}$$

we obtain

$$\vec{J} \cdot \vec{P} \, a^\dagger_{\vec{p},s} \mid 0 > = |\vec{p}| \, s \, a^\dagger_{\vec{p},s} \mid 0 > . \tag{3.80}$$

Identical considerations can be applied to $b^\dagger_{\vec{p},s} \mid 0 >$. Equation (3.77) now follows.

From (3.68) and (3.69) it follows that the component of $\vec{J}$ along $\vec{P}$ commutes with $\vec{P}$. Therefore these two operators can be diagonalized simultaneously; their eigenvalues for the states (3.73) are respectively the helicity s and the momentum $\vec{p}$. Under a space rotation, helicity is invariant because it is the scalar product of two vectors. As we shall see in Section 3.8, if the particle is of zero mass, then helicity is also invariant under a Lorentz transformation.

3.7 Phase Factor Conventions between the Spinors

In the expansion (3.31), the four orthogonal basis vectors $u_{\vec{p},s}$ and $v_{-\vec{p},s}$ in the spinor space are determined by (3.26)-(3.30) where $\vec{p}$ is a given fixed vector. Because (3.26) and (3.27) are homogeneous equations, the phase factors of these c. number spinors $u_{\vec{p},s}$ and $v_{-\vec{p},s}$ remain arbitrary. In this section we shall show that it is possible to choose their phase factors such that

$$\text{(i)} \quad v_{\vec{p},s} = \gamma_2\, u^*_{\vec{p},s}\,, \qquad u_{\vec{p},s} = \gamma_2\, v^*_{\vec{p},s}\,, \tag{3.81}$$

$$\text{(ii)} \quad \gamma_4\, u_{\vec{p},s} = u_{-\vec{p},-s}\,, \qquad \gamma_4\, v_{\vec{p},s} = -v_{-\vec{p},-s}\,, \tag{3.82}$$

and

$$\text{(iii)} \quad \sigma_2\, u^*_{\vec{p},s} = e^{i\theta_{\vec{p},s}}\, u_{-\vec{p},s}\,, \qquad \sigma_2\, v^*_{\vec{p},s} = e^{-i\theta_{-\vec{p},s}}\, v_{-\vec{p},s} \tag{3.83}$$

where

$$e^{i\theta_{\vec{p},s}} = -\,e^{i\theta_{-\vec{p},s}}\,. \tag{3.84}$$

These phase factor conventions will be adopted throughout this book.

<u>Proof.</u> We note that from (3.1)-(3.11) the matrices

$$\rho_1\,,\ \rho_3\,,\ \sigma_1\,,\ \sigma_3\,,\ \alpha_1\,,\ \alpha_3\,,\ \gamma_2$$

are real and the matrices

$$\rho_2\,,\ \sigma_2\,,\ \alpha_2\,,\ \gamma_1\,,\ \gamma_3$$

are imaginary.

(i) Because $\gamma_2 = \rho_2 \sigma_2$, we have

$$\gamma_2 \vec{\sigma}^* \gamma_2 = -\vec{\sigma} \ , \qquad \gamma_2 \vec{\alpha}^* \gamma_2 = \vec{\alpha} \ . \tag{3.85}$$

By taking the complex conjugate of (3.26) and noting that $\vec{p}$, m, β and E_p are all real, we obtain

$$(\vec{\alpha}^* \cdot \vec{p} + \beta m)\, u^*_{\vec{p},s} = E_p\, u^*_{\vec{p},s} \ .$$

Through (3.85) and $\{\gamma_2, \beta\} = 0$, the above equation becomes

$$(-\vec{\alpha} \cdot \vec{p} + \beta m)(\gamma_2 u^*_{\vec{p},s}) = -E_p\, \gamma_2 u^*_{\vec{p},s} \ . \tag{3.86}$$

Likewise the complex conjugate of (3.27) is

$$(\vec{\sigma}^* \cdot \hat{p})\, u^*_{\vec{p},s} = 2s\, u^*_{\vec{p},s} \ ,$$

which, on account of (3.85), can also be written as

$$(-\vec{\sigma} \cdot \hat{p})\, \gamma_2 u^*_{\vec{p},s} = 2s\, \gamma_2 u^*_{\vec{p},s} \ . \tag{3.87}$$

By comparing (3.86) and (3.87) with (3.26) and (3.27) we see that $\gamma_2 u^*_{\vec{p},s}$ and $v_{\vec{p},s}$ satisfy identical equations; therefore they must be proportional to each other. Because of the normalization condition (3.30), the proportionality constant must only be a phase factor, which can be chosen to be 1. Thus, we derive the first equation in (3.81):

$$v_{\vec{p},s} = \gamma_2 u^*_{\vec{p},s} \ . \tag{3.88}$$

Since $\gamma_2 = \rho_2 \sigma_2$ is a real matrix and $\gamma_2^2 = 1$, the complex conjugate of the above equation gives

$$u_{\vec{p},s} = \gamma_2 v^*_{\vec{p},s} \ .$$

(ii) Because

$$\gamma_4 \vec{\alpha} \gamma_4 = -\vec{\alpha} \ , \qquad \gamma_4 \vec{\sigma} \gamma_4 = \vec{\sigma} \ ,$$

Eq. (3.26), when multiplied by γ_4 on the left, becomes

$$(-\vec{\alpha}\cdot\vec{p}+\beta m)\,\gamma_4\, u_{\vec{p},s} = E_p\,\gamma_4\, u_{\vec{p},s} \quad . \tag{3.89}$$

Likewise, a similar multiplication onto (3.27) leads to

$$(-\vec{\sigma}\cdot\hat{p})\,\gamma_4\, u_{\vec{p},s} = -2s\,\gamma_4\, u_{\vec{p},s} \quad . \tag{3.90}$$

By comparing (3.89) and (3.90) with (3.26) and (3.27) we see that $\gamma_4\, u_{\vec{p},s}$ and $u_{-\vec{p},-s}$ satisfy identical equations; consequently they differ only in a phase factor. By choosing that phase factor to be 1, we have

$$\gamma_4\, u_{\vec{p},s} = u_{-\vec{p},-s} \quad .$$

Taking the complex conjugate and multiplying on the left by γ_2, we obtain

$$\gamma_2\,\gamma_4\, u^*_{\vec{p},s} = \gamma_2\, u^*_{-\vec{p},-s} \quad ,$$

which, because of (3.88), leads to

$$\gamma_4\, v_{\vec{p},s} = -v_{-\vec{p},-s} \quad .$$

(iii) We note that

$$\sigma_2\,\vec{\sigma}^*\,\sigma_2 = -\vec{\sigma} \quad , \quad \sigma_2\,\vec{\alpha}^*\,\sigma_2 = -\vec{\alpha} \quad .$$

By first taking the complex conjugate of (3.26) and then multiplying it on the left by σ_2, we find

$$(-\vec{\alpha}\cdot\vec{p}+\beta m)\,\sigma_2\, u^*_{\vec{p},s} = E_p\,\sigma_2\, u^*_{\vec{p},s} \quad .$$

A similar operation on (3.27) leads to

$$(-\vec{\sigma}\cdot\hat{p})\,\sigma_2\, u^*_{\vec{p},s} = 2s\,\sigma_2\, u^*_{\vec{p},s} \quad .$$

Thus $\sigma_2\, u^*_{\vec{p},s}$ and $u_{-\vec{p},s}$ satisfy identical equations, which implies

$$\sigma_2\, u^*_{\vec{p},s} \propto u_{-\vec{p},s} \quad ,$$

and therefore

$$\sigma_2\, u^*_{\vec{p},s} = e^{i\theta_{\vec{p},s}}\, u_{-\vec{p},s} \tag{3.91}$$

where the phase angle $\theta_{\vec{p},s}$ is real. Because σ_2 is an imaginary matrix, the complex conjugate of (3.91) gives

$$-\sigma_2 u_{\vec{p},s} = e^{-i\theta_{\vec{p},s}} u^*_{-\vec{p},s} , \tag{3.92}$$

which can also be written as

$$\sigma_2 u^*_{-\vec{p},s} = -e^{i\theta_{\vec{p},s}} u_{\vec{p},s} . \tag{3.93}$$

By changing the subscript $\vec{p}$ to $-\vec{p}$, we convert the above equation into

$$\sigma_2 u^*_{\vec{p},s} = -e^{i\theta_{-\vec{p},s}} u_{-\vec{p},s} . \tag{3.94}$$

A comparison between (3.91) and (3.94) gives (3.84). By using (3.81), (3.84) and (3.92), we have

$$\begin{aligned} \sigma_2 v^*_{\vec{p},s} &= \sigma_2(\gamma_2 u^*_{\vec{p},s})^* = \gamma_2 \sigma_2 u_{\vec{p},s} \\ &= -e^{-i\theta_{\vec{p},s}} \gamma_2 u^*_{-\vec{p},s} = -e^{-i\theta_{\vec{p},s}} v_{-\vec{p},s} \\ &= e^{-i\theta_{-\vec{p},s}} v_{-\vec{p},s} . \end{aligned} \tag{3.95}$$

That completes the proof of (3.81)-(3.84).

As we shall see later, the convention (3.81) is useful for discussion of particle-antiparticle conjugation, (3.82) for parity, and (3.83) for time reversal. We note that for a given helicity s, spinors with different $\vec{p}$ in the set $\{u_{\vec{p},s}\}$ are connected through rotations and Lorentz transformations. In (3.81) we fix the relative phase factors between members of the set $\{u_{\vec{p},s}\}$ and the corresponding ones in $\{v_{\vec{p},s}\}$; likewise (3.82) relates those between $\{u_{\vec{p},s}\}$ and $\{u_{-\vec{p},-s}\}$. On the other hand, (3.83) connects the relative phase between $u_{\vec{p},s}$ and $u_{-\vec{p},s}$, which are members of the same set $\{u_{\vec{p},s}\}$. Because of continuity, the phase factor $e^{i\theta_{\vec{p},s}}$ in (3.91) cannot be 1 for all $\vec{p}$;

as shown explicitly by (3.84).

3.8 Two-component Theory

Let us first consider the case of a quantized free spin-$\frac{1}{2}$ field with $m = 0$. The equation of motion (3.14) now becomes

$$\gamma_\mu \frac{\partial}{\partial x_\mu} \psi = 0 \quad . \tag{3.96}$$

It is useful to introduce

$$\gamma_5 \equiv \gamma_1 \gamma_2 \gamma_3 \gamma_4 \quad .$$

Because of (3.12) we obtain

$$\{\gamma_5, \gamma_\mu\} = 0 \quad , \tag{3.97}$$

where $\mu = 1, 2, 3, 4$, and

$$\gamma_5^{\ 2} = 1 \quad . \tag{3.98}$$

Thus, from (3.96) we also have

$$\gamma_\mu \frac{\partial}{\partial x_\mu} (\gamma_5 \psi) = 0 \quad . \tag{3.99}$$

We may decompose

$$\psi = \psi_L + \psi_R \tag{3.100}$$

where

$$\psi_L = \tfrac{1}{2}(1 + \gamma_5)\psi \quad \text{and} \quad \psi_R = \tfrac{1}{2}(1 - \gamma_5)\psi \quad . \tag{3.101}$$

From (3.96) and (3.99) it follows that ψ_L and ψ_R separately satisfy the equation of motion; i.e.

$$\gamma_\mu \frac{\partial}{\partial x_\mu} \psi_L = 0 \quad \text{and} \quad \gamma_\mu \frac{\partial}{\partial x_\mu} \psi_R = 0 \quad . \tag{3.102}$$

In the representation (3.11), $\gamma_i = \rho_2 \sigma_i$ and $\gamma_4 = \rho_3$, the matrix γ_5 is

$$\gamma_5 = -\rho_1 \quad . \tag{3.103}$$

When $m = 0$, (3.26) becomes simply

$$\vec{\alpha}\cdot\vec{p}\begin{cases}u_{\vec{p},s}\\ v_{-\vec{p},s}\end{cases} = |\vec{p}|\begin{cases}u_{\vec{p},s}\\ -v_{-\vec{p},s}\end{cases},$$

which, in the representation $\vec{\alpha} = \rho_1\vec{\sigma}$ and $\gamma_5 = -\rho_1$ can also be written as

$$\gamma_5\,\vec{\sigma}\cdot\hat{p}\begin{cases}u_{\vec{p},s}\\ v_{-\vec{p},s}\end{cases} = \begin{cases}-u_{\vec{p},s}\\ v_{-\vec{p},s}\end{cases}.$$

By comparing this expression with (3.27) we find

$$\gamma_5\,u_{\vec{p},s} = -2s\,u_{\vec{p},s}$$

and

$$\gamma_5\,v_{-\vec{p},s} = 2s\,v_{-\vec{p},s} \qquad (3.104)$$

Thus for any $\vec{p}$, $u_{\vec{p},-\frac{1}{2}}$ and $v_{\vec{p},\frac{1}{2}}$ are eigenvectors of γ_5 with an eigenvalue $+1$, while $u_{\vec{p},\frac{1}{2}}$ and $v_{\vec{p},-\frac{1}{2}}$ are those of γ_5 with an eigenvalue -1. Because, according to (3.101),

$$\gamma_5\,\psi_L = \psi_L \quad\text{and}\quad \gamma_5\,\psi_R = -\psi_R,$$

the Fourier expansion (3.32) can be separated into two parts

$$\psi_L = \frac{1}{\sqrt{\Omega}}\sum_{\vec{p}}\{a_{\vec{p},-\frac{1}{2}}\,u_{\vec{p},-\frac{1}{2}}\,e^{i\vec{p}\cdot\vec{r}} + b^{\dagger}_{\vec{p},+\frac{1}{2}}\,v_{\vec{p},+\frac{1}{2}}\,e^{-i\vec{p}\cdot\vec{r}}\}, \qquad (3.105)$$

and

$$\psi_R = \frac{1}{\sqrt{\Omega}}\sum_{\vec{p}}\{a_{\vec{p},\frac{1}{2}}\,u_{\vec{p},\frac{1}{2}}\,e^{i\vec{p}\cdot\vec{r}} + b^{\dagger}_{\vec{p},-\frac{1}{2}}\,v_{\vec{p},-\frac{1}{2}}\,e^{-i\vec{p}\cdot\vec{r}}\}. \qquad (3.106)$$

Since the equation $\psi = \psi_L + \psi_R$ is simply a restatement of (3.32), it is therefore valid in all cases, whether $m = 0$ or not. However, this decomposition is a particularly useful one in the case of $m = 0$. The free Hamiltonian density (3.35) is invariant under a phase transformation

$$\psi \to e^{i\theta}\,\psi. \qquad (3.107)$$

When $m = 0$, (3.35) becomes simply

$$\mathcal{H}_{free} = \psi^{\dagger}(-i\vec{\alpha}\cdot\vec{\nabla})\,\psi$$

which, because $[\gamma_5, \vec{\alpha}] = 0$, is also invariant under the transformation

$$\psi \rightarrow \gamma_5 \psi \quad . \tag{3.108}$$

Therefore, we may impose a supplementary condition: Either $\gamma_5 \psi = \psi$, and as a result

$$\psi = \psi_L \quad , \tag{3.109}$$

or $\gamma_5 \psi = -\psi$, so that

$$\psi = \psi_R \quad . \tag{3.110}$$

Clearly this is possible only when $m = 0$. Physically, this can also be understood in a simple way: When $m = 0$, under a Lorentz transformation the particle momentum $\vec{p} \rightarrow \vec{p}\,'$, but its helicity s remains unchanged. However, if $m \neq 0$, for any given $\vec{p}$ we can always perform a Lorentz transformation along $\vec{p}$ but with a velocity larger than $\vec{p}/E_p$, whereupon the particle-momentum direction will change sign but its spin direction will be the same. Consequently, the particle helicity s will be changed to $-s$. Such a Lorentz transformation is not possible when $m = 0$. This explains why, in accordance with (3.105) and (3.106), ψ_L can only annihilate particle states with helicity $s = -\frac{1}{2}$ (and create antiparticle states with $s = +\frac{1}{2}$) while ψ_R can only annihilate those with $s = +\frac{1}{2}$ (and create antiparticle states with $s = -\frac{1}{2}$).

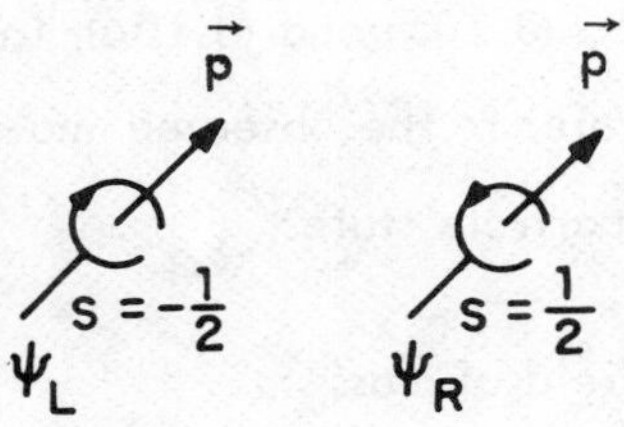

Fig. 3.1. The lefthand field ψ_L can only annihilate particle states with helicity $s = -\frac{1}{2}$, while the righthand field ψ_R can only annihilate those with helicity $s = +\frac{1}{2}$.

Next, we discuss the case with interactions. Let us assume that the interaction Hamiltonian H_{int} is, like the free Hamiltonian, also invariant under the phase transformation (3.107) and the γ_5 transformation (3.108). For example, H_{int} can be a function only of j_μ where

$$j_\mu = i\,\psi^\dagger \gamma_4\, \gamma_\mu (1 + \gamma_5)\, \psi \quad .$$

In such a case we can impose the supplementary condition (3.109), $\psi = \psi_L$, and require the physical mass of the particle m_{phys} to remain 0. [See Problem 3.2.] All particle states are of helicity $s = -\frac{1}{2}$ and all antiparticle states are of $s = +\frac{1}{2}$. Unlike the non-zero mass case where, in general, for each $\vec{p}$ there are four spinor states, $u_{\vec{p},\pm\frac{1}{2}}$ and $v_{\vec{p},\pm\frac{1}{2}}$, here there are only two.

As we shall discuss later, the two-component theory is well-suited for neutrinos.

Remarks. It is possible to transform the matrix representation of $\gamma_1, \gamma_2, \cdots, \gamma_5$ by a unitary transformation

$$\gamma_\alpha \rightarrow u\, \gamma_\alpha\, u^\dagger \quad ,$$

where u is a 4×4 unitary matrix. Equations (3.96)-(3.102) and the supplementary condition (3.109), or (3.110), are all invariant under such a transformation. The particular matrix representations (3.1)-(3.11) and (3.103) are adopted for convenience. They lead to equations (3.26), (3.27) and the expansion (3.32), (3.105) and (3.106); in these expansions the subscripts $\vec{p}$ and s refer to the observed momentum and helicity of the particle or antiparticle states.

Problem 3.1. Show that for a free spin-$\frac{1}{2}$ field of mass m

$$\{\psi_\alpha(x),\ \overline{\psi}_\beta(0)\} = i\left(\gamma_\mu \frac{\partial}{\partial x_\mu} - m\right)_{\alpha\beta} D(x)$$

where $D(x)$ is given in Problem 2.1, and $\bar{\psi} = \psi^\dagger \gamma_4$.

Problem 3.2. Let ψ be a usual quantized four-component Dirac spinor field, and ψ_L be the corresponding two-component field defined by (3.101):

$$\psi_L \equiv \tfrac{1}{2}(1 + \gamma_5)\,\psi \quad .$$

We may define a two-component theory to be one in which the Lagrangian density is a function only of ψ_L, $\psi_L^\dagger$ and possibly also other fields, but not of ψ_R and $\psi_R^\dagger$. Clearly, a two-component theory is always invariant under the γ_5 transformation: $\psi \to \gamma_5 \psi$.

(i) Show that a two-component theory which is also invariant under the phase transformation (3.107), $\psi \to e^{i\theta}\psi$, implies a zero physical mass for the particle. [Such a theory may be called the Weyl theory.]

As we shall discuss later in Chapter 10, the phase invariance is connected with the fermion-number conservation law.

(ii) Assume that the Lagrangian density of a two-component theory is given by

$$\mathcal{L} = -\psi_L^\dagger \gamma_4 \gamma_\mu \frac{\partial}{\partial x_\mu} \psi_L - \frac{m}{2}\left(\psi_L^\dagger \gamma_4 \psi_L^c + \text{h.c.}\right)$$

where

$$(\psi_L^c)_\alpha = (\gamma_2)_{\alpha\beta} (\psi_L^\dagger)_\beta \quad .$$

The Hamiltonian density is then given by

$$\mathcal{H} = \psi_L^\dagger(-i\vec{\alpha}\cdot\vec{\nabla})\,\psi_L + \frac{m}{2}\left(\psi_L^\dagger \beta\, \psi_L^c + \text{h.c.}\right)$$

and h.c. denotes the Hermitian conjugate. Calculate the energy spectrum in this theory and show that the physical mass of the particle is m.

Thus, a two-component theory that is not invariant under the phase transformation (3.107) can acquire a nonzero mass. [Such a theory may be called the Majorana theory.*]

(iii) Show that identical conclusions can be reached if ψ_L is replaced by $\psi_R = \frac{1}{2}(1-\gamma_5)\psi$.

Hint for (ii): Define the Majorana field operator

$$\psi_M = (\psi_L + \psi_L^c)/\sqrt{2} \quad ,$$

which is invariant under the particle-antiparticle conjugation; i.e.,

$$\psi_M^c = \psi_M$$

where $(\psi_M^c)_\alpha = (\gamma_2)_{\alpha\beta}(\psi_M^\dagger)_\beta$. Note that the Lagrangian density

$$\mathcal{L}_M \equiv -\psi_M^\dagger \gamma_4 (\gamma_\mu \frac{\partial}{\partial x_\mu} + m)\psi_M$$

differs from $\mathcal{L}$ only by a total derivative.

References

P. A. M. Dirac, Quantum Mechanics (Oxford, The Clarendon Press, 1958).

W. Pauli, Reviews of Modern Physics 13, 203 (1941).

G. Wentzel, Quantum Theory of Fields (New York, Interscience Publishers, Inc., 1949).

---

* See K. Case, Phys.Rev. 107, 307 (1957).

Chapter 4

THE SPIN-1 FIELD ($m \neq 0$)

4.1 Free Field

The Lagrangian density for a free spin-1 field with mass $m \neq 0$ is

$$\mathcal{L} = \mathcal{L}_{free} = -\tfrac{1}{4} F_{\mu\nu}^2 - \tfrac{1}{2} m^2 A_\mu^2 , \tag{4.1}$$

where, like x_μ, the space components of A_μ are Hermitian and the time component anti-Hermitian, i.e.

$$\vec{A} = \vec{A}^\dagger , \qquad A_4 = - A_4^\dagger ,$$

and

$$F_{\mu\nu} = \frac{\partial}{\partial x_\mu} A_\nu - \frac{\partial}{\partial x_\nu} A_\mu . \tag{4.2}$$

From the variational principle

$$\delta \int \mathcal{L} \, d^3r \, dt = 0 , \tag{4.3}$$

we find

$$\frac{\partial}{\partial x_\mu} F_{\mu\nu} - m^2 A_\nu = 0 . \tag{4.4}$$

By taking its divergence

$$\frac{\partial}{\partial x_\nu} \left(\frac{\partial}{\partial x_\mu} F_{\mu\nu} - m^2 A_\nu \right) = 0$$

and by noting that $m \neq 0$ and $F_{\mu\nu} = - F_{\nu\mu}$, we derive for the free field

$$\frac{\partial}{\partial x_\mu} A_\mu = 0 . \tag{4.5}$$

As will be discussed later, this equation is not necessarily valid when there are interactions.

To carry out the quantization, we observe that in (4.1) $\mathcal{L}$ does not contain $\dot{A}_4$. Thus we may regard A_4 as a dependent variable. From

$$F_{4j} = -i\dot{A}_j - \nabla_j A_4$$

and

$$-\tfrac{1}{4}F_{\mu\nu}^{\ 2} = \tfrac{1}{2}(\dot{A}_j - i\nabla_j A_4)^2 - \tfrac{1}{2}(\vec{\nabla}\times\vec{A})^2 ,$$

we find the conjugate momenta of A_j to be

$$\Pi_j = \frac{\partial\mathcal{L}}{\partial\dot{A}_j} = \dot{A}_j - i\nabla_j A_4 = iF_{4j} , \qquad (4.6)$$

where, as before, all Roman subscripts i, j vary from 1 to 3, all Greek subscripts μ, ν vary from 1 to 4, and all repeated indices are to be summed over. By setting $\nu = 4$ in (4.4), we have

$$A_4 = \frac{1}{m^2}\nabla_j F_{j4} = \frac{i}{m^2}\vec{\nabla}\cdot\vec{\Pi} , \qquad (4.7)$$

which, together with (4.6), gives

$$\vec{\Pi} = \dot{\vec{A}} + \frac{1}{m^2}\vec{\nabla}(\vec{\nabla}\cdot\vec{\Pi}) . \qquad (4.8)$$

In order to derive the Hamiltonian, we regard $\vec{A}$ and $\vec{\Pi}$ as independent variables, but A_4 as a function of $\vec{\Pi}$ through (4.7). From (4.8), it follows that

$$\vec{\Pi}\cdot\dot{\vec{A}} = \vec{\Pi}^{\,2} - \frac{1}{m^2}\vec{\Pi}\cdot\vec{\nabla}(\vec{\nabla}\cdot\vec{\Pi}) ,$$

and therefore

$$\int \vec{\Pi}\cdot\dot{\vec{A}}\, d^3r = \int [\vec{\Pi}^{\,2} + \frac{1}{m^2}(\vec{\nabla}\cdot\vec{\Pi})^2]\, d^3r .$$

Thus

$$H_{free} = \int (\vec{\Pi}\cdot\dot{\vec{A}} - \mathcal{L}_{free})\, d^3r$$

$$= \int \tfrac{1}{2}[\vec{\Pi}^{\,2} + (\frac{\vec{\nabla}\cdot\vec{\Pi}}{m})^2 + (\vec{\nabla}\times\vec{A})^2 + m^2\vec{A}^{\,2}]\, d^3r . \qquad (4.9)$$

According to the usual quantization rule, the equal-time commutator between $A_i(\vec{r}, t)$ and $\Pi_j(\vec{r}', t)$ is

$$[A_i(\vec{r}, t), \ \Pi_j(\vec{r}', t)] = i\delta_{ij}\, \delta^3(\vec{r} - \vec{r}') \tag{4.10}$$

while all other equal-time commutators $[A_i, A_j]$ and $[\Pi_i, \Pi_j]$ are zero.

To carry out the Fourier expansion, we introduce for any given $\vec{k}$ a set of three orthogonal unit vectors $\hat{k} = \vec{k}/|\vec{k}|$, $\hat{e}_1$ and $\hat{e}_2$:

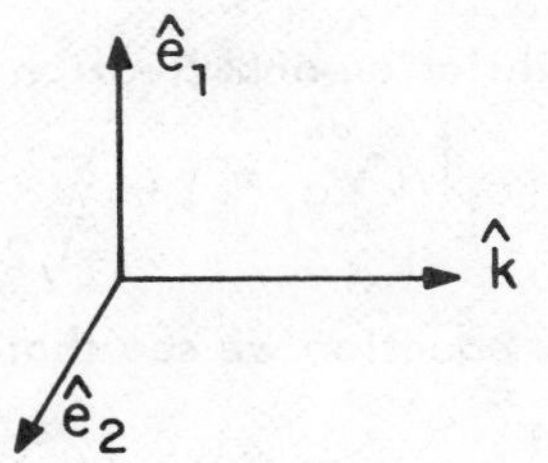

Fig. 4.1. An orthonormal set of three vectors.

At any fixed time t we may expand $\vec{A}(\vec{r}, t)$ and $\vec{\Pi}(\vec{r}, t)$ in terms of the Fourier series:

$$\vec{A}(\vec{r}, t) = \sum_{\vec{k}} \frac{1}{\sqrt{2\omega\,\Omega}} \left\{ [\hat{k}\, a_L(\vec{k})\, \frac{\omega}{m} + \sum_{T=1,2} \hat{e}_T\, a_T(\vec{k})]\, e^{i\vec{k}\cdot\vec{r}} + \text{h.c.} \right\} \tag{4.11}$$

and

$$\vec{\Pi}(\vec{r}, t) = \sum_{\vec{k}} \frac{1}{\sqrt{2\omega\,\Omega}} \left\{ -i\,[m\hat{k}\, a_L(\vec{k}) + \omega \sum_{T=1,2} \hat{e}_T\, a_T(\vec{k})]\, e^{i\vec{k}\cdot\vec{r}} + \text{h.c.} \right\} \tag{4.12}$$

in which h.c. denotes the Hermitian conjugate terms and, as before in (2.17),

$$\omega = \sqrt{\vec{k}^2 + m^2} \ . \tag{4.13}$$

The subscripts L and T indicate the longitudinal and transverse components respectively and the $a_L(\vec{k})$'s and $a_T(\vec{k})$'s are all functions of t . From (4.10) and by following similar arguments used in

Chapter 2 to derive (2.25), we obtain the following equal-time commutator between $a_i(\vec{k})$ and $a_j^\dagger(\vec{k}')$:

$$[a_i(\vec{k}), a_j^\dagger(\vec{k}')] = \delta_{ij}\,\delta_{\vec{k},\vec{k}'} \quad . \tag{4.14}$$

All other equal-time commutators are zero; i.e.

$$[a_i(\vec{k}), a_j(\vec{k}')] = 0 \quad \text{and} \quad [a_i^\dagger(\vec{k}), a_j^\dagger(\vec{k}')] = 0 \tag{4.15}$$

where i and j denote either the longitudinal component L or the transverse components T = 1 or 2 .[See the exercise on the next page.] In terms of these annihilation and creation operators, H_{free} is

$$H_{free} = \sum_{\vec{k}} \omega\,[a_L^\dagger(\vec{k})\,a_L(\vec{k}) + \sum_{T=1,2} a_T^\dagger(\vec{k})\,a_T(\vec{k}) + \tfrac{3}{2}\,] \ . \tag{4.16}$$

By using Heisenberg's equation we see that

$$a_i(\vec{k}) \propto e^{-i\omega t} \qquad \text{and} \qquad a_i^\dagger(\vec{k}) \propto e^{i\omega t} \quad . \tag{4.17}$$

From (4.7) and (4.12), it follows that

$$A_4(\vec{r}, t) = i \sum_{\vec{k}} \frac{1}{\sqrt{2\omega\,\Omega}}\ \frac{|\vec{k}|}{m}\, a_L(\vec{k})\, e^{i\vec{k}\cdot\vec{r}} + \text{h.c.} \tag{4.18}$$

By substituting (4.11), (4,17) and (4.18) into (4.5), we can verify that, for free fields, $\partial A_\mu / \partial x_\mu = 0$.

4.2 Interacting Fields

Next we consider the case that the spin-1 field A_μ has interactions with other fields. For simplicity, let us assume the Lagrangian density to be given by

$$\mathcal{L} = -\tfrac{1}{4} F_{\mu\nu}^2 - \tfrac{1}{2} m^2 A_\mu^2 - j_\mu A_\mu + \cdots , \tag{4.19}$$

where j_μ and the ··· terms depend only on other fields, <u>independent</u> of A_μ .

From the action principle (4.3), we have

$$\frac{\partial}{\partial x_\mu} F_{\mu\nu} - m^2 A_\nu = j_\nu \quad , \tag{4.20}$$

from which it follows that

$$\frac{\partial}{\partial x_\mu} A_\mu = - \frac{1}{m^2} \frac{\partial}{\partial x_\nu} j_\nu \quad . \tag{4.21}$$

Since, by assumption, the current j_ν is independent of A_μ, the question whether j_ν is conserved or not depends on the equations of motion of other fields and on the detailed form of the current. If $\partial j_\nu / \partial x_\nu = 0$, then $\partial A_\mu / \partial x_\mu = 0$; otherwise not.

Just as in the previous section, the new Lagrangian (4.19) does not contain $\dot{A}_4$; we shall regard A_4 as a dependent variable. From (4.19) we see that the conjugate momentum Π_j remains given by (4.6); i.e.

$$\Pi_j = \frac{\partial \mathcal{L}}{\partial \dot{A}_j} = i F_{4j} \quad . \tag{4.22}$$

By setting $\nu = 4$ in (4.20), we find that (4.7) should be replaced by

$$A_4 = \frac{i}{m^2} \vec{\nabla} \cdot \vec{\Pi} - \frac{1}{m^2} j_4 \quad . \tag{4.23}$$

The equal-time commutator between A_i and Π_j is still given by (4.10) and, as before, all other equal-time commutators $[A_i, A_j]$ and $[\Pi_i, \Pi_j]$ remain zero.

Exercise. In the present case of interacting fields, show that at any time t we may still expand $\vec{A}(\vec{r}, t)$ and $\vec{\Pi}(\vec{r}, t)$ in terms of the Fourier series (4.11)-(4.12). In addition, the equal-time commutation relations (4.14)-(4.15) remain valid.

From the Lagrangian density (4.19) and by regarding A_4 as a function of $\vec{\Pi}$ and other field variables through (4.23), the Hamiltonian density can be constructed in the usual way. Unlike the free-field case, the time dependences of $a_i(\vec{k})$ and $a_i^\dagger(\vec{k})$ are in general quite complicated, in contrast to (4.17).

Remarks. The presence of A_4 makes some of the discussions of a spin-1 field quite different from those of a spin-0 field. For a given momentum $\vec{k}$, a spin-1 particle with a non-zero physical mass, $m \neq 0$, has three modes: longitudinal L and transverse T = 1, 2. They correspond to, in the rest frame of the particle, the three z-component angular momentum states $j_z = 0$ and ± 1. On the other hand, because of Lorentz invariance, the spin-1 field is represented by A_μ which has four components. Consequently, one of the four components is not an independent variable.

Because in many of the above expressions the parameter m is often in the denominator, the limit $m = 0$ for a spin-1 field is more complicated than that for a spin-$\frac{1}{2}$ field. The details will be given in Chapter 6, when we discuss quantum electrodynamics.

Problem 4.1. Consider a system which consists of a spin-1 field A_μ and a spin-$\frac{1}{2}$ field ψ. Assume that the Lagrangian density is given by (4.19) in which the ... term is

$$- \psi^\dagger \gamma_4 (\gamma_\mu \frac{\partial}{\partial x_\mu} + m') \psi$$

and j_μ is either

(i) $j_\mu = i g \psi^\dagger \gamma_4 \gamma_\mu \psi$

or

(ii) $j_\mu = i g \psi^\dagger \gamma_4 \gamma_\mu \gamma_5 \psi$.

Construct the Hamiltonian for this system and work out the details of the Fourier expansion for ψ, $\vec{A}$ and A_4. Use Heisenberg's equation to obtain the equation of motion. Find out in which of the above cases $\partial A_\mu / \partial x_\mu = 0$.

Problem 4.2. Show that for a free spin-1 field of mass m

$$[A_\mu(x),\ A_\nu(0)] = -i\left(\delta_{\mu\nu} - m^{-2}\frac{\partial^2}{\partial x_\mu \partial x_\nu}\right) D(x)$$

where $D(x)$ is given in Problem 2.1.

References

W. Pauli, Reviews of Modern Physics 13, 203 (1941).

G. Wentzel, Quantum Theory of Fields (New York, Interscience Publishers, Inc., 1949).

Chapter 5

FEYNMAN DIAGRAMS

5.1 Heisenberg, Schrödinger and Interaction Representations

In quantum mechanics all experimental results can be expressed in terms of the matrix elements $\langle a | O | b \rangle$ of different operators O between various state vectors $| a \rangle$ and $| b \rangle$. There are many ways to describe the time variation of such matrix elements.

1. Heisenberg representation

In the Heisenberg representation only the physical operators vary with time. All physical state vectors are time-independent. If we denote $O_H(t)$ and $| t \rangle_H$ as the operator $O(t)$ and the state vector $| t \rangle$ in the Heisenberg representation, their equations of motion are

$$[H_H, O_H(t)] = -i \dot{O}_H(t) \tag{5.1}$$

and

$$\frac{\partial}{\partial t} | t \rangle_H = 0 , \tag{5.2}$$

where H_H is the Hamiltonian in the Heisenberg representation.

2. Schrödinger representation

In the Schrödinger representation only the physical state vectors are time-dependent; all physical operators are time-independent. Let $O_S(t)$ and $| t \rangle_S$ be the operator $O(t)$ and the state vector $| t \rangle$ in the Schrödinger representation. The equations of motion are

$$\dot{O}_S(t) = 0 \tag{5.3}$$

and

$$H_S \mid t >_S = -\frac{1}{i}\frac{\partial}{\partial t} \mid t >_S , \tag{5.4}$$

where H_S is the Hamiltonian in the Schrödinger representation.

3. Interaction representation

Let H_I be the Hamiltonian in the interaction representation, which will be decomposed into

$$H_I = (H_0)_I + (H_{int})_I . \tag{5.5}$$

We denote $O_I(t)$ and $\mid t >_I$ as the operator $O(t)$ and the state vector $\mid t >$ in the interaction representation. The equations of motion are now

$$[(H_0)_I , O_I(t)] = -i \dot{O}_I(t) , \tag{5.6}$$

and

$$(H_{int}(t))_I \mid t >_I = -\frac{1}{i}\frac{\partial}{\partial t} \mid t >_I . \tag{5.7}$$

So far, the decomposition (5.5) is quite arbitrary. If $H_0 = 0$, then the interaction representation is identical to the Schrödinger representation; if $H_{int} = 0$, then it is identical to the Heisenberg representation.

We shall now show the equivalence of these different representations. For clarity of notation, we shall write

$$H = H_S , \tag{5.8}$$

i.e., we shall omit the subscript S when the Hamiltonian is in the Schrödinger representation. Likewise, the decomposition (5.5) will be written in the Schrödinger representation as

$$H = H_0 + H_{int} ,$$

i.e., just as in (5.8),

$$H_0 = (H_0)_S \quad \text{and} \quad H_{int} = (H_{int})_S . \tag{5.9}$$

In the Schrödinger representation, the state $| t >_S$ can be readily expressed in terms of its form $| 0 >_S$ at $t = 0$:

$$| t >_S = e^{-iHt} | 0 >_S \quad . \tag{5.10}$$

The corresponding state vector $| t >_H$ in the Heisenberg representation is time-independent, and may be set to be equal to that in the Schrödinger representation at $t = 0$; i.e.

$$| t >_H = | 0 >_H = | 0 >_S \qquad \text{at all } t \ .$$

Hence, the state vectors $| t >_H$ and $| t >_S$ are related by the unitary transformation e^{iHt} :

$$| t >_H = e^{iHt} | t >_S \quad . \tag{5.11}$$

Under the same unitary transformation, the corresponding operators in these two representations are related by

$$O_H(t) = e^{iHt} \, O_S \, e^{-iHt} \quad . \tag{5.12}$$

By setting the operator $O(t)$ to be the total Hamiltonian, we see that it is unchanged when we switch from the Schrödinger representation to the Heisenberg representation. Hence, in the notation (5.8),

$$H_H = H_S = H \quad . \tag{5.13}$$

By differentiating (5.11) and (5.12), and by using (5.3)–(5.4) and (5.13) we see that

$$\begin{aligned} - i \, \dot{O}_H(t) &= - i \frac{\partial}{\partial t} (e^{iHt} \, O_S \, e^{-iHt}) \\ &= (H e^{iHt} \, O_S \, e^{-iHt} - e^{iHt} \, O_S \, e^{-iHt} H) \\ &= H \, O_H(t) - O_H(t) \, H = [H_H , O_H] \end{aligned}$$

and

$$- \frac{1}{i} \frac{\partial}{\partial t} | t >_H = e^{iHt} (- H + H) | t >_S = 0 \quad .$$

Hence the equations of motion (5.3)-(5.4) in the Schrödinger representation imply those in the Heisenberg representation. Similarly, we can verify that the converse is also correct.

The state vector $|\, t >_I$ in the interaction representation is related to that in the Schrödinger representation by

$$|\, t >_I \;=\; e^{iH_0 t}\, |\, t >_S \tag{5.14}$$

where H_0 is given by (5.9). Under the same unitary transformation $e^{iH_0 t}$, the corresponding operators in these two representations are related by

$$O_I(t) = e^{iH_0 t}\, O_S\, e^{-iH_0 t} \;. \tag{5.15}$$

By setting the operator $O(t)$ to be H_0 we see that

$$(H_0)_I \;=\; (H_0)_S \;=\; H_0 \;. \tag{5.16}$$

Hence, H_0 is unchanged when we switch from the Schrödinger representation to the interaction representation. In the following, H_0 will be called the unperturbed Hamiltonian. By differentiating (5.14) and (5.15) and by using (5.3)-(5.4) and (5.9), we find

$$\begin{aligned} - i\, O_I(t) \;&=\; - i\, \frac{\partial}{\partial t}\, (e^{iH_0 t}\, O_S\, e^{-iH_0 t}) \\ &= H_0\, e^{iH_0 t}\, O_S\, e^{-iH_0 t} - e^{iH_0 t}\, O_S\, e^{-iH_0 t}\, H_0 \\ &= [\, (H_0)_I \;,\; O_I(t)\,] \end{aligned}$$

and

$$\begin{aligned} - \frac{1}{i}\, \frac{\partial}{\partial t}\, |\, t >_I \;&=\; - \frac{1}{i}\, \frac{\partial}{\partial t}\; e^{iH_0 t}\, |\, t >_S \\ &= e^{iH_0 t} (- H_0 + H)\, |\, t >_S \\ &= e^{iH_0 t}\, H_{int}\, |\, t >_S \end{aligned}$$

$$= e^{iH_0 t} H_{int} e^{-iH_0 t} e^{iH_0 t} | t >_S$$

$$= (H_{int}(t))_I \, | t >_I \; .$$

Therefore, the equations of motion (5.3)-(5.4) in the Schrödinger representation also imply those in the interaction representation. Likewise we can show that the converse is also correct. From (5.11)-(5.12) and (5.14)-(5.15), it follows that at any time t

$$(< a \,|\, O \,|\, b >)_H = (< a \,|\, O \,|\, b >)_S = (< a \,|\, O \,|\, b >)_I$$

where O can be any operator and $| a >$, $| b >$ can be any two state vectors. That completes the proof of equivalence between these different representations. [See also (24.100).]

5.2 S - Matrix

In this section, as well as throughout the remainder of this chapter, we shall stay in the interaction representation. For simplicity, the subscript I will be omitted. Therefore (5.7) becomes

$$-\frac{1}{i} \frac{\partial}{\partial t} \, | t > = H_{int}(t) \, | t > \tag{5.17}$$

where in accordance with (5.6) $H_{int}(t)$ satisfies

$$-i \dot{H}_{int} = [H_0 , H_{int}] \; . \tag{5.18}$$

Both H_0 and H_{int} are Hermitian. Let $U(t, t_0)$ be the Green's function of (5.17). Then we have

$$| t > = U(t, t_0) \, | t_0 > \tag{5.19}$$

where $U(t, t_0)$ satisfies

$$-\frac{1}{i} \frac{\partial}{\partial t} U(t, t_0) = H_{int}(t) \, U(t, t_0) \; , \tag{5.20}$$

with the initial condition

$$U(t_0, t_0) = 1 = \text{unit matrix} . \qquad (5.21)$$

The Hermitian conjugate of (5.20) is

$$\frac{1}{i}\frac{\partial}{\partial t} U^\dagger(t, t_0) = U^\dagger(t, t_0) H_{int}(t) \quad .$$

Thus,

$$\frac{1}{i}\frac{\partial}{\partial t} (U^\dagger(t, t_0) U(t, t_0)) = 0$$

which, together with the initial condition (5.21), implies

$$U^\dagger(t, t_0) U(t, t_0) = 1 \quad , \qquad (5.22)$$

i.e., $U(t, t_0)$ is unitary.

The S-matrix is defined to be the limit

$$S \equiv \lim_{\substack{t_0 \to -\infty \\ t \to \infty}} U(t, t_0) \quad . \qquad (5.23)$$

It connects the state vector from time $= -\infty$ to $+\infty$. Any scattering problem can be described by the transformation between the initial and final state vectors; the former is given in the remote past ($t_0 = -\infty$) while the latter is in the remote future ($t = +\infty$). The scattering amplitude is therefore given by the corresponding matrix element of the S-matrix.

So far the decomposition $H = H_0 + H_{int}$ is quite arbitrary. However, in order for the double limits (5.23) to exist, there are some simple requirements.

To see this, let us consider a simple example of a single harmonic oscillator whose Hamiltonian is given by

$$H = \omega a^\dagger a$$

where a and $a^\dagger$ are the usual annihilation and creation operators. Suppose H is decomposed into the sum of

$$H_0 = \omega_0 a^\dagger a \quad ,$$

and

$$H_{int} = (\omega - \omega_0)\, a^\dagger a \quad . \tag{5.24}$$

The solution of (5.20) is

$$U(t, t_0) = e^{-i(\omega-\omega_0)(t-t_0)N} \tag{5.25}$$

where N is the occupation number operator

$$N = a^\dagger a \quad .$$

Let $| n >$ be the eigenvector of N with eigenvalue n. The diagonal matrix element $< n | U(t, t_0) | n >$ is

$$e^{-i(\omega-\omega_0)(t-t_0)n} \quad ;$$

for any $n \neq 0$; its limit when either $t_0 \to -\infty$ or $t \to \infty$ does not exist unless $\omega = \omega_0$.

There are several different approaches by which we can bypass this difficulty. One way is to modify the differential equation (5.18) or (5.20) in the asymptotic region in time. For example, one may keep H_{int} intact but assume t to have an appropriate small imaginary component, or in a completely equivalent way, keep t real but replace (5.24) by

$$H_{int}(t) = \begin{cases} (\omega-\omega_0)\, a^\dagger a & \text{for } |t| \leqslant T \\ (\omega-\omega_0 - i\epsilon)\, a^\dagger a & \text{for } |t| > T > 0 \end{cases}$$

where $\epsilon = 0+$ and T is a large constant. Thus, (5.25) is valid only when t_0 and t are both within the range $-T$ to T, but not outside; e.g. when $t_0 < -T$ and $t > T$, (5.25) is replaced by

$$U(t, t_0) = e^{-i(\omega-\omega_0)(t-t_0)N}\, e^{-\epsilon(t-t_0-2T)N}$$

which, at a fixed $\epsilon > 0$, does have a limit as $t \to \infty$ and $t_0 \to -\infty$, but this limit is not unitary.

Of course, one may try a different modification: For example, (5.24) may be replaced by

$$H_{int}(t) = \begin{cases} (\omega - \omega_0)\, a^\dagger a & \text{for } |t| \leq T \\ 0 & \text{for } |t| > T . \end{cases}$$

As above, (5.25) remains valid when t_0 and t are within the range $-T$ to T ; outside this range, for $t > T$ and $t_0 < -T$, we have

$$U(t, t_0) = U(T, -T) = e^{-i\,2(\omega - \omega_0)\,T N} .$$

Thus, the limit (5.23) exists and is now unitary; however, it depends on an artificial parameter T.

There exists still another method, and this is the one we shall adopt. There is no change in either of the differential equations (5.18) or (5.20) and t remains real. However, in this approach we require the spectrum of H_0 to be identical to that of H. As we shall see, for a large class of relativistic theories, it turns out that this seemingly stringent condition can be easily satisfied. In the above simple example, this condition leads to $\omega = \omega_0$ and therefore $H_{int} = 0$ and $U = S = 1$.

Next, let us consider a spin-0 field ϕ whose Hamiltonian density is given by

$$H = \tfrac{1}{2}(\Pi^2 + (\vec{\nabla}\phi)^2 + m_0^2\, \phi^2) + g_0^2\, \phi^4 - E_{vac} , \qquad (5.26)$$

where Π is the conjugate momentum of ϕ, and m_0, g_0 and E_{vac} are real numbers. The parameter m_0 is called the mechanical mass and g_0 the unrenormalized coupling constant. We define the vacuum state $|\, vac >$ to be the lowest-energy state of H. The parameter E_{vac} in (5.26) is determined by the condition

$$H\, |\, vac > = 0 .$$

Since in (5.26) the interaction $g_0^2\, \phi^4$ is positive, it gives rise only to repulsive forces between the spin-0 quanta. We may therefore assume that there is no bound state in the system. From relativistic invariance we know that the energy spectrum of the system must be of the form

$$E = \sum_{\vec{k}} n_{\vec{k}} \sqrt{\vec{k}^2 + m^2} \tag{5.27}$$

where m is the physical mass and $n_{\vec{k}} = 0, 1, 2, \cdots$, representing the number of physical spin-0 quanta with momentum $\vec{k}$. Consequently, we may introduce

$$H_0 = \tfrac{1}{2}(\Pi^2 + (\vec{\nabla}\phi)^2 + m^2\phi^2) - E_0 \ , \tag{5.28}$$

and E_0 is chosen so that the lowest energy state $|0>$ of H_0 also has a zero eigenvalue, i.e.

$$H_0 \,|\, 0> = 0 \ . \tag{5.29}$$

Clearly, the spectrum of H_0 is <u>identical</u> to that of H provided the parameter m in (5.28) is the physical mass.

In a general case of interacting relativistic fields, when the theory has no stable bound states (as in the above example or in quantum electrodynamics, to be discussed in the next chapter), the unperturbed Hamiltonian H_0 will be chosen to be that of free fields, but with their mass parameters set to be the physical masses of the interacting system. Consequently, the spectrum of H_0 becomes the same as that of H, which makes it possible to take the double limits (5.23) for the S-matrix.

In the case that there are stable bound states, we shall still adopt the same procedure, setting H_0 to be the free-particle Hamiltonian with the m 's the physical mass parameters. This enables us to make the formal perturbation series expansion in terms of diagrams, as will be discussed in the subsequent sections. Each stable bound state can be identified as an infinite sum of an appropriate set of diagrams. After isolating these infinite sums, we can then carry out the limit (5.23). For the moment, we shall ignore such complications.

Therefore, in our interaction representation all field operators satisfy the free equations; their time-dependences are known. For example, the Fourier transformation (2.20) can then be written as

$$\phi(x) = a(x) + a^\dagger(x) , \tag{5.30}$$

where

$$a(x) = \sum_{\vec{k}} \frac{1}{\sqrt{2\omega\Omega}} a_{\vec{k}}\, e^{i\vec{k}\cdot\vec{r} - i\omega t} , \tag{5.31}$$

$a^\dagger(x)$ is its Hermitian conjugate, as before $x = (\vec{r}, it)$ and

$$\omega = \sqrt{\vec{k}^2 + m^2} > 0$$

in which m is chosen to be the physical mass of the spin-0 field. Here, the operator $a_{\vec{k}}$ is time-independent, whereas the operator $a_{\vec{k}}(t)$ in (2.20) has a time-dependence $e^{-i\omega t}$ in the interaction representation; these two are related by $a_{\vec{k}}(t) = a_{\vec{k}}\, e^{-i\omega t}$.

Likewise, for a spin-$\frac{1}{2}$ field the Fourier expansion (3.32) in the interaction representation can be written as

$$\psi(x) = u(x) + \bar{v}(x) , \tag{5.32}$$

where

$$u(x) = \frac{1}{\sqrt{\Omega}} \sum_{\vec{p},s} a_{\vec{p},s}\, u_{\vec{p},s}\, e^{i\vec{p}\cdot\vec{r} - iE_p t} , \tag{5.33}$$

$$\bar{v}(x) = \frac{1}{\sqrt{\Omega}} \sum_{\vec{p},s} b^\dagger_{\vec{p},s}\, v_{\vec{p},s}\, e^{-i\vec{p}\cdot\vec{r} + iE_p t} , \tag{5.34}$$

and

$$E_p = \sqrt{\vec{p}^2 + m^2} > 0 ,$$

with m the physical mass of the spin-$\frac{1}{2}$ field. Here, the operators $a_{\vec{p},s}$ and $b^\dagger_{\vec{p},s}$ are again both time-independent; they are related to the corresponding operators $a_{\vec{p},s}(t)$ and $b^\dagger_{\vec{p},s}(t)$ in (3.32) by $a_{\vec{p},s}(t) = a_{\vec{p},s}\, e^{-iE_p t}$ and $b^\dagger_{\vec{p},s}(t) = b^\dagger_{\vec{p},s}\, e^{iE_p t}$. It is convenient to introduce

$$\overline{\psi} \equiv \psi^{\dagger}(x)\,\gamma_4 \equiv \bar{u}(x) + v(x) \,, \tag{5.35}$$

where

$$\bar{u}(x) = \frac{1}{\sqrt{\Omega}} \sum_{\vec{p},s} a^{\dagger}_{\vec{p},s}\, u^{\dagger}_{\vec{p},s}\, \gamma_4\, e^{-i\vec{p}\cdot\vec{r}+iE_p t} \tag{5.36}$$

and

$$v(x) = \frac{1}{\sqrt{\Omega}} \sum_{\vec{p},s} b_{\vec{p},s}\, v^{\dagger}_{\vec{p},s}\, \gamma_4\, e^{i\vec{p}\cdot\vec{r}-iE_p t} \,. \tag{5.37}$$

Similar expansions can also be written for a spin-1 field. The details will, however, be omitted here.

5.3 Time-ordered Products, Normal Products and Contractions

In (5.30), (5.32) and (5.35), $\alpha(x)$, $u(x)$ and $v(x)$ contain only annihilation operators, while $\alpha^{\dagger}(x)$, $\bar{u}(x)$ and $\bar{v}(x)$ contain only creation operators. Let $X_i(x_i)$ be any one of these operators at $x_i = (\vec{r}_i\,,\, it_i)$. We define the time-ordered product of $\prod_{i=1}^{n} X_i(x_i)$ to be

$$T(X_1(x_1)\,X_2(x_2)\cdots X_n(x_n)) \equiv \delta_p X_{p_1}(x_{p_1})\,X_{p_2}(x_{p_2})\cdots X_{p_n}(x_{p_n}) \,, \tag{5.38}$$

so that the time sequence x_{p_i} satisfies

$$t_{p_1} \geqslant t_{p_2} \geqslant t_{p_3} \geqslant \cdots \geqslant t_{p_n} \,.$$

Whenever $t_i = t_j$, the relative order of the corresponding operators X_i and X_j is the same on both sides of (5.38). The normal product of $\prod_{i=1}^{n} X_i(x_i)$ is defined to be

$$:X_1(x_1)\,X_2(x_2)\cdots X_n(x_n): \equiv \delta_p X_{p_1}(x_{p_1})\,X_{p_2}(x_{p_2})\cdots X_{p_n}(x_{p_n}) \,, \tag{5.39}$$

where the permuted sequence on the righthand side is arranged so that annihilation operators always appear to the right of creation operators. In both definitions (5.38) and (5.39), δ_p can be $+1$ or -1, depending only on the permutation of the fermion operators. If the order of fermion operators on the lefthand side is an even permutation of those on the righthand side then $\delta_p = +1$, otherwise -1. The subscripts $p_1, p_2, \cdots, p_n$ represent a permutation

$$p = \begin{pmatrix} 1 & 2 \cdots n \\ p_1 & p_2 \cdots p_n \end{pmatrix} .$$

The time-ordered product and the normal product of a sum are defined to be the sum of the corresponding products; i.e.

$$T(A+B) = T(A) + T(B)$$

and

$$:A+B: = :A: + :B: \quad . \tag{5.40}$$

The following are a few examples of such products

$$T(\psi_\alpha(1)\ \psi_\beta(2)) = \begin{cases} \psi_\alpha(1)\ \psi_\beta(2) & t_1 > t_2 \\ -\psi_\beta(2)\ \psi_\alpha(1) & t_1 < t_2 \end{cases} , \tag{5.41}$$

$$T(\phi(1)\ \phi(2)) = \begin{cases} \phi(1)\ \phi(2) & t_1 > t_2 \\ \phi(2)\ \phi(1) & t_1 < t_2 \end{cases} , \tag{5.42}$$

$$\begin{aligned} :\psi_\alpha(1)\ \bar{\psi}_\beta(2): \ &= -\bar{u}_\beta(2)\, u_\alpha(1) + u_\alpha(1)\, v_\beta(2) \\ &\quad + \bar{v}_\alpha(1)\, v_\beta(2) + \bar{v}_\alpha(1)\, \bar{u}_\beta(2) \ , \end{aligned} \tag{5.43}$$

$$:\phi(1)\ \phi(2): \ = a(1)\,a(2) + a^\dagger(2)\,a(1) + a^\dagger(1)\,a(2) + a^\dagger(1)\,a^\dagger(2) \tag{5.44}$$

where 1 and 2 stand for x_1 and x_2 respectively, and the subscripts α and β denote the spinor indices.

The Dyson-Wick contraction between $X_1(1)$ and $X_2(2)$ is defined to be

$$\underbrace{X_1(1)\ X_2(2)} \equiv T(X_1(1)\ X_2(2)) - :X_1(1)\ X_2(2): \quad . \tag{5.45}$$

As an example, we may obtain the contraction between $\phi(x)$ and $\phi(0)$ by using (5.42) and (5.44)-(5.45):

$$\underbrace{\phi(x)\,\phi(0)} = \begin{cases} a(x)\,a^\dagger(0) - a^\dagger(0)\,a(x) & t \geqslant 0 \\ a(0)\,a^\dagger(x) - a^\dagger(x)\,a(0) & t < 0 \end{cases}$$

where $x = (\vec{r}, it)$. The righthand side of the above expression can be written as commutators. We have

$$\underbrace{\phi(x)\,\phi(0)} = \begin{cases} [a(x),\ a^\dagger(0)] & t \geqslant 0 \\ [a(0),\ a^\dagger(x)] & t < 0 \end{cases} . \tag{5.46}$$

Likewise, the contraction between $\psi(x)$ and $\overline{\psi}(0)$ is

$$\underbrace{\psi_\alpha(x)\,\overline{\psi}_\beta(0)} = \begin{cases} \{u_\alpha(x),\ \bar{u}_\beta(0)\} & t \geqslant 0 \\ -\{v_\beta(0),\ \bar{v}_\alpha(x)\} & t < 0 \end{cases} . \tag{5.47}$$

We note that these contractions are all c. numbers, while the corresponding time-ordered products and normal products are operators.

Next, we discuss the four-dimensional integral representations of these contractions. By using (5.31) and the commutation relation $[a_{\vec{k}},\ a^\dagger_{\vec{k}'}] = \delta_{\vec{k},\vec{k}'}$, we obtain

$$[a(x),\ a^\dagger(0)] = \sum_{\vec{k}} \frac{1}{2\omega\Omega}\, e^{i\vec{k}\cdot\vec{r} - i\omega t} ,$$

$$[a(0),\ a^\dagger(x)] = \sum_{\vec{k}} \frac{1}{2\omega\Omega}\, e^{-i\vec{k}\cdot\vec{r} + i\omega t} .$$

Let us define the Feynman propagator $D_F(x)$ to be the contraction between $\phi(x)$ and $\phi(0)$. We have

$$D_F(x) = \underbrace{\phi(x)\,\phi(0)} = \sum_{\vec{k}} \frac{1}{2\omega\Omega}\, e^{i\vec{k}\cdot\vec{r} \mp i\omega t} \tag{5.48}$$

in which the minus sign in the exponent holds when t is > 0, and the positive sign when $t < 0$. At $t = 0$ the function $D_F(x)$ is continuous. We shall now establish the following integral representation

$$D_F(x) \equiv -\frac{i}{(2\pi)^4}\int \frac{e^{ik\cdot x}\, d^4k}{k^2 + (m - i\epsilon)^2} \quad , \tag{5.49}$$

where

$$k\cdot x = \vec{k}\cdot\vec{r} - k_0 t \quad , \qquad d^4k = d^3k\, dk_0 \quad ,$$

$$k^2 = \vec{k}^2 - k_0^2 \qquad \text{and} \qquad k_4 = ik_0 \quad .$$

The parameter ϵ is a positive real infinitesimal; i.e., $\epsilon = 0+$. To see this, we may first perform the integration from $k_0 = -\infty$ to ∞ :

$$\int \frac{e^{-ik_0 t}\, dk_0}{\vec{k}^2 - k_0^2 + m^2 - i\epsilon} = \int \frac{-e^{-ik_0 t}\, dk_0}{(k_0 + \omega - i\epsilon)(k_0 - \omega + i\epsilon)} \quad , \tag{5.50}$$

where $\omega = \sqrt{\vec{k}^2 + m^2} > 0$.

In the complex k_0-plane, the integrand of (5.50) has two singularities $k_0 = \omega - i\epsilon$ and $-\omega + i\epsilon$. When $t > 0$, we may consider the contour integral of the same integrand along the closed curve shown in Fig. 5.1(a). When the radius of the half-circle becomes

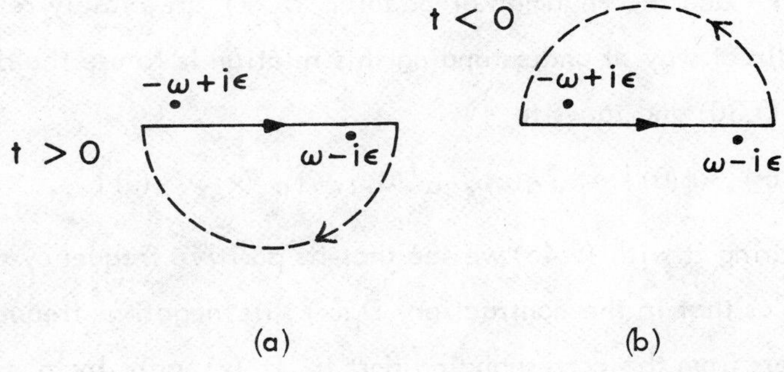

Fig. 5.1. Contours for the Feynman integral (5.46) in the complex k_0-plane.

infinite, the contour integration along this half-circle approaches 0. The integral (5.50) equals $-2\pi i$ times the residue of the integrand at the pole $k_0 = \omega - i\epsilon$. Upon substituting it into (5.49), we obtain for $t > 0$,

$$D_F(x) = \frac{1}{(2\pi)^3} \int \frac{d^3k}{2\omega} e^{i\vec{k}\cdot\vec{r} - i\omega t} \equiv D_+(x) \quad . \qquad (5.51)$$

Likewise, when t is negative we consider the contour in Fig. 5.1(b). That leads to, for $t < 0$

$$D_F(x) = \frac{1}{(2\pi)^3} \int \frac{d^3k}{2\omega} e^{i\vec{k}\cdot\vec{r} + i\omega t} \equiv D_-(x) \quad . \qquad (5.52)$$

From (2.28) and (5.51)-(5.52), we see the equivalence between (5.48) and (5.49). The $D_+(x)$ denotes the positive frequency part of $D_F(x)$ and the $D_-(x)$ the corresponding negative frequency part.

By using Problem 2.1 and (5.51)-(5.52) we find the commutator between $\phi(x)$ and $\phi(0)$ in the interaction representation to be

$$[\phi(x), \phi(0)] = -iD(x) = D_+(x) - D_-(x) \qquad (5.53)$$

where

$$D(x) \equiv \int \frac{1}{8\pi^3} d^3k \frac{\sin \omega t}{\omega} e^{i\vec{k}\cdot\vec{r}} \quad . \qquad (5.54)$$

Thus, $D(x)$ and the Feynman propagator $D_F(x)$ are closely related. A more direct way of understanding this relation is to use the decomposition (5.30) and to write

$$[\phi(x), \phi(0)] = [\alpha(x), \alpha^\dagger(0)] + [\alpha^\dagger(x), \alpha(0)] \quad . \qquad (5.55)$$

By comparing it with (5.46) we see that its positive frequency part is the same as that in the contraction $D_F(x)$; its negative frequency part differs from the corresponding part in $D_F(x)$ only by a minus sign.

Identical reasoning can be applied to the spin-$\frac{1}{2}$ field ψ. On account of (5.32) and (5.35), the anticommutator between ψ and its adjoint $\bar{\psi} \equiv \psi^\dagger \gamma_4$ in the interaction representation is

$$\{\psi_\alpha(x), \bar{\psi}_\beta(0)\} = \{u_\alpha(x), \bar{u}_\beta(0)\} + \{\bar{v}_\alpha(x), v_\beta(0)\} \quad (5.56)$$

where, as before, the subscripts α and β are spinor indices. From (5.33) we see that $\{u_\alpha(x), \bar{u}_\beta(0)\}$ consists only of positive frequency terms; i.e. terms proportional to $e^{-iE_p t}$. Likewise from (5.34), $\{\bar{v}_\alpha(x), v_\beta(0)\}$ consists only of negative frequency parts which are proportional to $e^{iE_p t}$. On the other hand, from Problem 3.1 and Eq. (5.53), we see that the same anticommutator is also given by

$$\{\psi(x), \bar{\psi}(0)\} = i(\gamma_\mu \frac{\partial}{\partial x_\mu} - m) D(x)$$

$$= -(\gamma_\mu \frac{\partial}{\partial x_\mu} - m) [D_+(x) - D_-(x)] . \quad (5.57)$$

We now equate, respectively, the positive and negative frequency parts of (5.56) and (5.57). That gives

$$\{u_\alpha(x), \bar{u}_\beta(0)\} = -(\gamma_\mu \frac{\partial}{\partial x_\mu} - m)_{\alpha\beta} D_+(x)$$

and

$$\{\bar{v}_\alpha(x), v_\beta(0)\} = (\gamma_\mu \frac{\partial}{\partial x_\mu} - m)_{\alpha\beta} D_-(x) .$$

By substituting the above expressions into (5.47) and by using (5.51)-(5.52), we derive

$$\underbrace{\psi(x) \bar{\psi}(0)} = -(\gamma_\mu \frac{\partial}{\partial x_\mu} - m) D_F(x) \quad (5.58)$$

which, because of (5.49), can also be written as

$$\underbrace{\psi(x) \bar{\psi}(0)} = i \int \frac{1}{(2\pi)^4} \frac{i\gamma_\mu k_\mu - m}{k^2 + (m - i\epsilon)^2} e^{ik\cdot x} d^4k .$$

It is convenient to introduce (as in the exercise on page 36)

$$\not{k} \equiv -i\gamma_\mu k_\mu . \quad (5.59)$$

One can readily verify that

$$(\not k - m)(\not k + m) = \not k^2 - m^2 = -k^2 - m^2$$

and

$$\frac{1}{\not k - m} = \frac{\not k + m}{(\not k - m)(\not k + m)} = \frac{-\not k - m}{k^2 + m^2} .$$

Thus, (5.58) can also be written as

$$\underbrace{\psi(x)\,\bar\psi(0)} = \int \frac{1}{(2\pi)^4} \frac{i}{\not k - (m - i\epsilon)} e^{ik\cdot x} d^4x \equiv S_F(x) . \tag{5.60}$$

5.4 Perturbation Series

We begin with (5.20):

$$-\frac{1}{i}\frac{\partial}{\partial t} U(t, t_0) = H_{int}(t)\, U(t, t_0) .$$

Let us replace in the above

$$H_{int} \to \lambda H_{int} ,$$

and write

$$U(t, t_0) = \sum_{n=0}^{\infty} \lambda^n U_n(t, t_0) . \tag{5.61}$$

Equation (5.20) then becomes

$$-\frac{1}{i}\frac{\partial}{\partial t} \sum_{n=0}^{\infty} \lambda^n U_n(t, t_0) = \lambda H_{int}(t) \sum_{n=0}^{\infty} \lambda^n U_n(t, t_0) . \tag{5.62}$$

By equating the coefficients of λ^n on both sides we have

$$-\frac{1}{i}\frac{\partial}{\partial t} U_0(t, t_0) = 0 , \tag{5.63}$$

and

$$-\frac{1}{i}\frac{\partial}{\partial t} U_n(t, t_0) = H_{int}(t)\, U_{n-1}(t, t_0) \quad \text{for } n \geqslant 1 . \tag{5.64}$$

Similarly, by substituting (5.61) into the initial condition (5.21) we find at $t = t_0$

$$U_0(t_0 , t_0) = 1 \tag{5.65}$$

and

$$U_n(t_0, t_0) = 0 \qquad \text{for } n \geqslant 1 \quad . \tag{5.66}$$

Equations (5.63) and (5.65) give

$$U_0(t, t_0) = 1 \quad . \tag{5.67}$$

The first-order perturbation term, $n = 1$, can be readily obtained by using (5.64), (5.66) and (5.67). The solution is

$$U_1(t, t_0) = -i \int_{t_0}^{t} H_{int}(t')\, dt' \quad . \tag{5.68}$$

Likewise, the solution for $n = 2$ is

$$\begin{aligned} U_2(t, t_0) &= (-i)^2 \int_{t_0}^{t} dt_1 \int_{t_0}^{t_1} dt_2\, H_{int}(t_1)\, H_{int}(t_2) \\ &= \frac{(-i)^2}{2!} \int_{t_0}^{t} dt_1 \int_{t_0}^{t} dt_2\, T(H_{int}(t_1)\, H_{int}(t_2)) \ , \end{aligned} \tag{5.69}$$

etc. By setting $\lambda = 1$ and because of (5.23), we obtain the perturbation series expansion of the S-matrix

$$\begin{aligned} S = U(\infty, -\infty) &= 1 - i \int_{-\infty}^{\infty} H_{int}(t)\, dt \\ &+ \frac{(-i)^2}{2!} \int_{-\infty}^{\infty} dt_1 \int_{-\infty}^{\infty} dt_2\, T(H_{int}(t_1)\, H_{int}(t_2)) \\ &+ \frac{(-i)^3}{3!} \int_{-\infty}^{\infty} dt_1 \int_{-\infty}^{\infty} dt_2 \int_{-\infty}^{\infty} dt_3\, T(H_{int}(t_1)\, H_{int}(t_2)\, H_{int}(t_3)) \\ &+ \cdots \ , \end{aligned} \tag{5.70}$$

where, as in (5.69), T denotes the time-ordered product. For actual computation, it is useful to convert all these time-ordered products in the series into normal products. The systematics of this conversion is given by Wick's theorem.

5.5 Wick Theorem

We first generalize the definition of the normal product (5.39) to include a c. number multiplicative factor:

$$: c\, X_1\, X_2 \cdots X_n : \; = \; c : X_1\, X_2 \cdots X_n : \tag{5.71}$$

where c is a c. number. Thus, we have

$$: \underbrace{X_1\, X_2}\, X_3 \cdots X_n : \; = \; \underbrace{X_1\, X_2} : X_3 \cdots X_n : \; ,$$

$$: \underbrace{X_1\, X_2\, X_3} \cdots X_n : \; = \; \delta_p\, \underbrace{X_1\, X_3} : X_2 \cdots X_n : \; , \quad \text{etc.} \tag{5.72}$$

in which the definition of δ_p is the same as that in (5.39); i.e., $\delta_p = -1$ if X_2 and X_3 are fermion operators, otherwise $\delta_p = +1$.

Let the space-time position of the operator $X_i (x_i)$ be, as before, $x_i = (\vec{r}_i\, , it_i)$, and $Y(y)$ be an operator like X_i but at the space-time point $y = (\vec{r}_y\, , it_y)$. We first establish a lemma.

<u>Lemma.</u> If

$$t_y \leqslant t_i \qquad \text{for all} \quad i = 1, 2, \cdots, n \; , \tag{5.73}$$

then

$$\begin{aligned} : X_1\, X_2 \cdots X_n : Y \; = \; & : X_1\, X_2 \cdots \underbrace{X_n\, Y} : \\ & + : X_1\, X_2 \cdots \underbrace{X_{n-1}\, X_n\, Y} : + \cdots \\ & + : \underbrace{X_1\, X_2 \cdots X_n\, Y} : + : X_1\, X_2 \cdots X_n\, Y : \quad . \end{aligned} \tag{5.74}$$

<u>Proof.</u> (i) In the case that Y is an annihilation operator, we have $: X_i\, Y : = X_i\, Y$. Since $T(X_i\, Y) = X_i\, Y$ on account of (5.73), it follows then that

$$\underbrace{X_i\, Y} \; = \; T(X_i\, Y) - : X_i\, Y : \; = \; 0 \; .$$

Hence, except for the last term $: X_1\, X_2 \cdots X_n\, Y :$, all other terms on the righthand side of (5.74) are zero. One easily sees that the lemma is correct.

(ii) In the case that Y is a creation operator, we shall prove the lemma in three steps:

(a) Suppose that $X_1, X_2, \cdots, X_n$ are all annihilation operators. We have

$$:X_1 X_2 \cdots X_n : Y = X_1 X_2 \cdots X_n Y \quad , \qquad (5.75)$$

and

$$:X_i Y: = \delta_i Y X_i$$

where $\delta_i = -1$ if Y and X_i are fermion operators, otherwise $\delta_i = +1$. Consequently,

$$:X_1 X_2 \cdots X_n Y: = \delta_p : Y X_1 X_2 \cdots X_n :$$

$$= \delta_p Y X_1 X_2 \cdots X_n \quad .$$

In addition, because of the hypothesis (5.73) we also have

$$T(X_i Y) = \underbrace{X_i Y} \quad ,$$

and therefore

$$X_i Y = \delta_i Y X_i + \underbrace{X_i Y} \quad ,$$

which leads to

$$X_1 \cdots X_n Y = \delta_n X_1 \cdots X_{n-1} Y X_n + X_1 \cdots X_{n-1} \underbrace{X_n Y} \quad .$$

We can permute Y and X_{n-1} in the first term on the righthand side:

$$X_1 \cdots X_n Y = \delta_n \delta_{n-1} X_1 \cdots X_{n-2} Y X_{n-1} X_n$$

$$+ : X_1 \cdots X_{n-2} \underbrace{X_{n-1} X_n Y} :$$

$$+ : X_1 \cdots X_{n-1} \underbrace{X_n Y} : \quad .$$

By repeating this process, we obtain

$$X_1 \cdots X_n Y = \delta_p Y X_1 \cdots X_n + :X_1 \cdots X_{n-1} \underbrace{X_n Y}:$$

$$+ :X_1 \cdots X_{n-2} \underbrace{X_{n-1} X_n Y} : + \cdots + : \underbrace{X_1 \cdots X_n Y} : \ .$$

Since, according to (5.75) the lefthand side of the above expression is the same as that in (5.74), the lemma is established in this case.

(b) Next, we suppose that $X_1, X_2, \cdots, X_j$ are creation operators, but $X_{j+1}, X_{j+2}, \cdots, X_n$ are annihilation operators. Because t_y is $\leqslant t_i$ and Y is a creation operator, we have

$$\underbrace{X_i Y} = 0 \qquad \text{for} \qquad i = 1, 2, \cdots, j$$

and

$$:X_1 X_2 \cdots X_n : Y = X_1 \cdots X_j X_{j+1} \cdots X_n Y$$

$$= X_1 \cdots X_j : X_{j+1} \cdots X_n : Y \ .$$

Since $X_{j+1}, \cdots, X_n$ are all annihilation operators, we can use the result of (a) and apply the lemma to the last term. This gives

$$:X_1 \cdots X_n : Y = X_1 \cdots X_j (: X_{j+1} \cdots X_n Y :$$

$$+ : X_{j+1} \cdots \underbrace{X_n Y} : + : X_{j+1} \cdots \underbrace{X_{n-1} X_n Y} :$$

$$+ \cdots + : \underbrace{X_{j+1} \cdots X_n Y} :) \ .$$

The lemma then follows since $X_1, \cdots, X_j$ are all creation operators.

(c) We now consider the general case that each X_i can be either an annihilation operator or a creation operator. From the definition of the normal product (5.39) we can permute their orders so that on the lefthand side of (5.74) the orders of X_i are arranged to have annihilation operators to the right of creation operators. By using the result of (b) we complete the proof of the lemma.

Wick's theorem deals with the conversion of a time-ordered

product into a sum of normal products.

Theorem.

$$
\begin{aligned}
T(X_1\, X_2 \cdots X_n) = {} & :X_1\, X_2 \cdots X_n: \\
& + :\underbrace{X_1\, X_2} \cdots X_n: + :\underbrace{X_1\, X_2\, X_3} \cdots X_n: \\
& + \cdots + :\underbrace{X_1\, X_2}\, \underbrace{X_3\, X_4} \cdots X_n: \\
& + :X_1^{\bullet}\, X_2^{\bullet\bullet}\, X_3^{\bullet}\, X_4^{\bullet\bullet} \cdots X_n: + \cdots \,. \qquad (5.76)
\end{aligned}
$$

Each term on the righthand side consists of a number of contractions between different pairs of the X_i's, and the righthand side is the sum total of all such terms.

Proof. When $n = 2$, the theorem holds because of the definition of contraction (5.45). Let us assume that for $n \leqslant N$, the theorem is correct. We then consider the time-ordered product when $n = N + 1$.

Among the $N + 1$ operators, $X_1, X_2, \cdots, X_{N+1}$, there must be one X_i whose time t_i is the earliest. Let that X_i be X_{N+1}, i.e. $t_{N+1} \leqslant t_j$ for all $j = 1, 2, \cdots, N$. Hence,

$$
T(X_1\, X_2 \cdots X_N\, X_{N+1}) = T(X_1\, X_2 \cdots X_N)\, X_{N+1} \,.
$$

Because of the assumption that the theorem holds for $n = N$, we may convert $T(X_1\, X_2 \cdots X_N)$ into a sum of normal products by using (5.76). The result is that $T(X_1\, X_2 \cdots X_N)\, X_{N+1}$ can be written as a sum of terms; each is of the form of the lefthand side of (5.74). By using the lemma, we establish (5.76) for $n = N + 1$. The theorem is then proved by induction.

5.6 Applications

The following example illustrates how Wick's theorem can lead to Feynman diagrams and the evaluation of the S-matrix.

Let the Hamiltonian of a spin-0 field be

$$H = H_0 + H_{int} \quad , \tag{5.77}$$

$$H_0 = \tfrac{1}{2} \int : \Pi^2 + (\vec{\nabla}\phi)^2 + m^2 \phi^2 : d^3r \quad , \tag{5.78}$$

$$H_{int} = \int : -\tfrac{1}{2}\delta m^2 \phi^2 + \frac{1}{3!} g_0 \phi^3 + \frac{1}{4!} f_0 \phi^4 : d^3r \tag{5.79}$$

where, as before, Π is the conjugate momentum, g_0 and f_0 are the unrenormalized coupling constants, m is the physical mass and m_0, defined by the following equation, is the mechanical mass

$$m_0^2 = m^2 - \delta m^2 \quad . \tag{5.80}$$

In the weak coupling, δm^2 is assumed to be of the order of g_0^2 and f_0^2. In the following, we shall stay in the interaction representation and expand $\phi(x)$ in the form of (5.30)-(5.31). Because in (5.78) H_0 is expressed in terms of the normal product, we have

$$H_0 = \sum_{\vec{k}} \omega\, a_{\vec{k}}^\dagger a_{\vec{k}} \quad , \tag{5.81}$$

where

$$\omega = \sqrt{\vec{k}^2 + m^2} \quad ,$$

just as in (2.44).

We shall now discuss the scattering process in Fig. 5.2. The initial state is

$$|\, 1, 2 > \equiv a_1^\dagger\, a_2^\dagger \;|\, 0 > \quad , \tag{5.82}$$

and the final state is

$$|\, 1', 2' > \equiv a_{1'}^\dagger\, a_{2'}^\dagger \;|\, 0 > \tag{5.83}$$

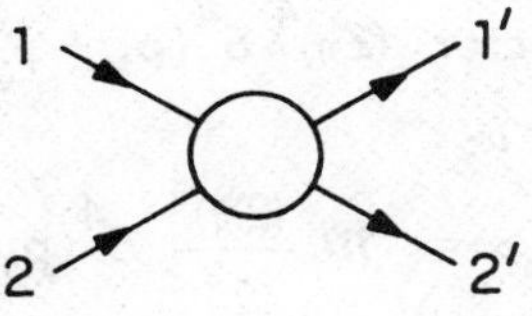

Fig. 5.2. Scattering process $1 + 2 \to 1' + 2'$.

where

$$a_i^\dagger = a_{\vec{p}_i}^\dagger \quad \text{and} \quad a_i'^\dagger = a_{\vec{p}_i'}^\dagger ,$$

$i = 1, 2$ and $\vec{p}_i$, $\vec{p}_i'$ are respectively the 3-momenta of particles i and i' . Our purpose is to compute the matrix element $\langle 1', 2' \mid S \mid 1, 2 \rangle$ where S is the S-matrix given by (5.70).

1. For simplicity, we first discuss the case

$$g_0 = 0 \ , \quad \text{but} \quad f_0 \neq 0 \ . \tag{5.84}$$

To the lowest order of f_0, we have

$$\langle 1', 2' \mid S \mid 1, 2 \rangle = \frac{-i}{4!} f_0 \int d^4x \langle 1', 2' \mid :\phi^4(x): \mid 1, 2 \rangle \ . \tag{5.85}$$

This is because δm^2 is $O(f_0^2)$; hence to $O(f_0)$, there is only $\frac{1}{4!} f_0 \phi^4$ in the expression (5.79) for H_{int} . In (5.85) we have to select from $:\phi^4(x):$ only terms proportional to $a_1'^\dagger a_2'^\dagger a_1 a_2$; there are $4! = 24$ such terms due to the different ways of selecting such factors from the four ϕ's . Thus, to $O(f_0)$ we obtain

$$\langle 1', 2' \mid S \mid 1, 2 \rangle = -i f_0 \frac{1}{4\sqrt{\omega_1 \omega_2 \omega_1' \omega_2'}} \int \frac{d^4x}{\Omega^2} e^{i(p_1 + p_2 - p_1' - p_2')\cdot x}$$

where p_1 , p_2 , p_1' and p_2' are respectively the 4-momenta of particles 1, 2, 1' and 2', the ω_i and ω_i' are the corresponding energies; as before, $x = (\vec{r}, it)$ and $d^4x = d^3r\,dt$. The 4-dimensional integration yields a factor $(2\pi)^4\,\delta^4(p_1 + p_2 - p_1' - p_2')$. It is convenient to introduce $\mathcal{M}$:

$$< 1', 2' \,|\, S \,|\, 1, 2 > \;\equiv\; \mathcal{M}\,\frac{(2\pi)^4}{\Omega^2}\,\delta^4(p_1 + p_2 - p_1' - p_2'). \quad (5.86)$$

In this simple example

$$\mathcal{M} = -if_0\,\frac{1}{4\sqrt{\omega_1\omega_2\omega_1'\omega_2'}} \quad . \quad (5.87)$$

In general, $\mathcal{M}$ can be represented by a diagram, or sum of diagrams, called Feynman diagrams. Their rules are called Feynman rules. In Fig. 5.3 we draw the diagram for the $\mathcal{M}$ given by (5.87). The four lines represent the two incoming and two outgoing particles; their intersection is called a 4-point vertex, representing the action of a $:\phi^4:$ interaction.

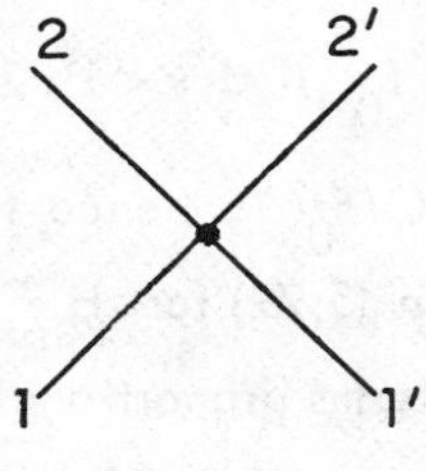

Fig. 5.3. The lowest-order Feynman diagram for $1 + 2 \rightleftarrows 1' + 2'$ when $H_{int} = :\frac{1}{4!}\,f_0\,\phi^4:$

The Feynman rule for Fig. 5.3 is: Each external line

 gives a factor $\dfrac{1}{\sqrt{2\omega}}$

where ω is the energy carried by the external line; each 4-point vertex

 gives a factor $-i\, f_0$. (5.88)

The product of all these factors gives the $\mathcal{M}$ of (5.87).

Next, we illustrate the calculation of the cross section. Since

$$\int d^4x\, e^{ipx} = (2\pi)^4\, \delta^4(p) \quad ,$$

we have

$$(2\pi)^4\, \delta^4(p) \int d^4x\;\, e^{ipx} = [\,(2\pi)^4\, \delta^4(p)\,]^2 \quad .$$

On the lefthand side, because of the δ^4-function we may set $p = 0$ in the integrand. The integral becomes

$$\int d^4x = \Omega\, T$$

where the $\vec{r}$-integration extends over the volume Ω and the time-integration over the interval T. Both Ω and T will $\to \infty$ in the end. Hence we can combine the above two expressions and write

$$[\,(2\pi)^4\, \delta^4(p)\,]^2 = (2\pi)^4\, \delta^4(p)\, \Omega\, T \tag{5.89}$$

and that leads to, for the square of (5.86),

$$|\langle 1', 2' \,|\, S \,|\, 1, 2\rangle|^2 = |\mathcal{M}|^2\, \frac{(2\pi)^4}{\Omega^4}\, \delta^4(p_1 + p_2 - p_1' - p_2')\, \Omega T\, . \tag{5.90}$$

Let us adopt the laboratory frame in which particle 2 is at rest and particle 1 is moving with velocity $\vec{v}_1$. Assuming that the cross section for $1 + 2 \to 1' + 2'$ is $d\sigma$, we consider a cylinder of volume $v_1 T\, d\sigma$, as shown in Fig. 5.4, where $v_1 = |\vec{v}_1|$. Since our initial

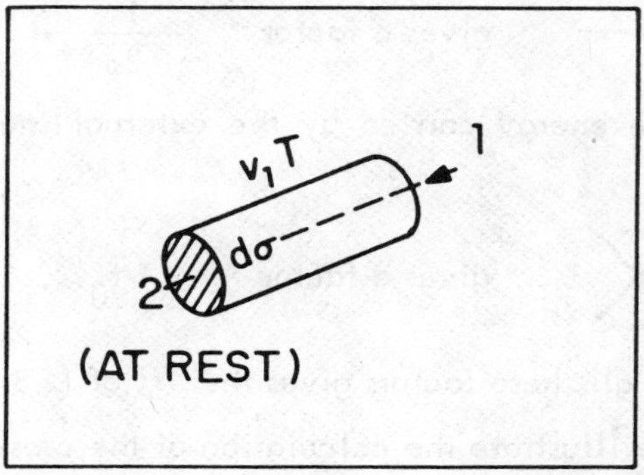

Fig. 5.4. If in the rest frame of particle 2, particle 1 is inside the cylinder $v_1 T d\sigma$, then a reaction $1 + 2 \to 1' + 2'$ will occur in time T.

state (5.82) corresponds to a state in which there are only two particles 1 and 2 in the volume Ω, the probability that particle 1 lies inside this cylinder is

$$\frac{v_1 T d\sigma}{\Omega} . \tag{5.91}$$

If particle 1 is inside this cylinder, then a reaction $1 + 2 \to 1' + 2'$ will occur in time T. Hence, this probability is equal to

$$| < 1', 2' \mid U(\tfrac{T}{2}, -\tfrac{T}{2}) \mid 1, 2 > |^2 \tag{5.92}$$

where the U-matrix is given by (5.19). In the limit $\Omega \to \infty$ and then $T \to \infty$, (5.92) becomes (5.90), and therefore upon equating it to (5.91) we have

$$d\sigma = \sum \frac{(2\pi)^4}{\Omega^2 v_1} \, |\mathcal{M}|^2 \, \delta^4(p_1 + p_2 - p_1' - p_2') , \tag{5.93}$$

where the sum extends over different 3-momenta $\vec{p}_1{}'$ and $\vec{p}_2{}'$ of the final state. Because of (2.28)

$$\sum_{\vec{p}_i} \to \frac{\Omega}{8\pi^3} \int d^3 p_i \ ,$$

we obtain from (5.93)

$$d\sigma = 2\pi \int \frac{d^3 p_1' \, d^3 p_2'}{8\pi^3 v_1} \, |\mathcal{M}|^2 \, \delta^4(p_1 + p_2 - p_1' - p_2') \quad (5.94)$$

where v_1 and $\mathcal{M}$ are evaluated in the rest frame of particle 2.

We note that the passage from the definition (5.86) of $\mathcal{M}$ to the above expression for $d\sigma$ has a general validity independent of the special form of the interaction Hamiltonian.

2. We now consider the general case $g_0 \neq 0$ and $f_0 \neq 0$. For simplicity, we shall evaluate the S-matrix element for the reaction $1 + 2 \to 1' + 2'$ only to $O(f_0)$ and $O(g_0{}^2)$. From (5.70) we find, to these orders and for $|1,2\rangle \neq |1',2'\rangle$,

$$\langle 1',2' | S | 1,2\rangle = -i f_0 \frac{1}{4!} \int \langle 1',2' | :\phi^4(x): | 1,2\rangle \, d^4x$$

$$+ \frac{(-i g_0)^2}{2!\,3!\,3!} \int \langle 1',2' | T([:\phi^3(x):][:\phi^3(y):]) | 1,2\rangle \, d^4x \, d^4y \ . \quad (5.95)$$

We first apply Wick's theorem which converts the T-product in the second integral into a sum of normal products, then retain in this sum only terms proportional to $a_{1'}^\dagger a_{2'}^\dagger a_1 a_2$, as before. There are $(3!)^2$ terms of the form

$$\frac{a_{1'}^\dagger a_{2'}^\dagger a_1 a_2}{4\Omega^2 \sqrt{\omega_1 \omega_2 \omega_1' \omega_2'}} \quad (5.96)$$

multiplied by

$$e^{i(p_1 + p_2)\cdot x - i(p_1' + p_2')\cdot y} + \text{(same terms, but interchanging } x \text{ with } y) \ ; \quad (5.97)$$

there are also $(3!)^2$ terms of the form (5.96) multiplied by

$$e^{i(p_1 - p_1')\cdot x + i(p_2 - p_2')\cdot y} + \text{(same terms, but interchanging } x \text{ with } y) \; ; \tag{5.98}$$

in addition there are $(3!)^2$ terms of the form (5.96) multiplied by

$$e^{i(p_1 - p_2')\cdot x + i(p_2 - p_1')\cdot y} + \text{(same terms, but interchanging } x \text{ with } y) \; . \tag{5.99}$$

In each case the $(3!)^2$ terms and the $x - y$ interchanging terms exactly cancel out the factor $\frac{1}{2!\,3!\,3!}$ outside the integral sign. The three classes (5.97)-(5.99) of terms lead to the three diagrams in Fig. 5.5.

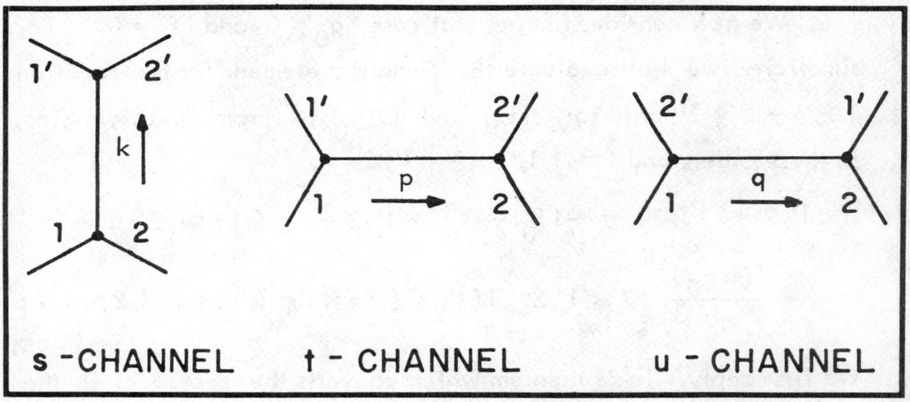

Fig. 5.5. The three $O(g_0{}^2)$ diagrams for reaction $1 + 2 \to 1' + 2'$. The arrows indicate the directions of momentum flow. *

To evaluate these diagrams we use the following Feynman rules: As in Fig. 5.3 each external line

$$\bullet\!\!\longrightarrow \quad \text{carries a factor} \quad \frac{1}{\sqrt{2\omega}} \; ; \tag{5.100}$$

each internal line

---

* Because each momentum component can be positive or negative, the arrow directions of these momenta can be arbitrarily drawn.

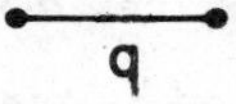

carries a factor $\dfrac{-i}{q^2 + (m - i\epsilon)^2}$ (5.101)

where q denotes its 4-momentum and $q^2 = \vec{q}^{\,2} - q_0^{\,2}$; each 3-point vertex

gives a factor $-ig_0$. (5.102)

The product of these component-factors is the contribution of the Feynman diagram to $\mathcal{M}$, where $\mathcal{M}$ is defined by (5.86).

By summing over the diagram in Fig. 5.3 and those in Fig. 5.5, we obtain

$$\mathcal{M} = \prod_i \frac{1}{\sqrt{2\omega_i}\ \sqrt{2\omega_i{}'}}$$

$$[-if_0 + (-ig_0)^2 \left(\frac{-i}{k^2 + (m - i\epsilon)^2} + \frac{-i}{p^2 + (m - i\epsilon)^2} + \frac{-i}{q^2 + (m - i\epsilon)^2} \right)] , \tag{5.103}$$

where the subscript i extends over 1 and 2, $k = p_1 + p_2 = p_1' + p_2'$, $p = p_1 - p_1' = p_2' - p_2$ and $q = p_1 - p_2' = p_1' - p_2$.

The differential cross section for the reaction $1 + 2 \rightarrow 1' + 2'$ is again given by (5.94).

The Feynman rules (5.88) and (5.100)-(5.102) can be applied to any Feynman diagram of arbitrary order. At each vertex the flow of 4-momenta is conserved. In the high-order diagrams, there will be those that contain loops. An example is given in Fig. 5.6, which is one of the $O(f_0^2)$ diagrams for the scattering process $1 + 2 \rightarrow 1' + 2'$.

Fig. 5.6. A sample loop diagram.

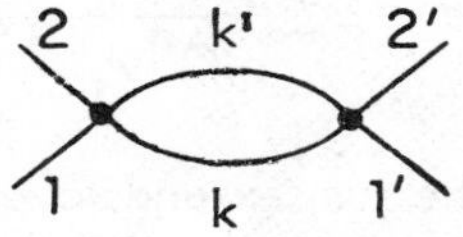

Neither Fig. 5.3 nor Fig. 5.5 contains any loop; they are called tree diagrams. In a tree diagram, the momenta carried by the internal lines are all determined by the external momenta. However, in a loop diagram each loop carries a free momentum to be integrated over. For example, in Fig. 5.6 the internal 4-momentum k is a free variable, and that gives an additional factor

$$\int \frac{d^4k}{(2\pi)^4} \quad . \tag{5.104}$$

In any loop diagram, there is such a factor for every free momentum.

5.7 Differential Cross Sections for $1 + 2 \rightarrow 1' + 2' + \cdots + n'$

1. Equation (5.94) for the differential cross section can be generalized to an n-body final state:

$$1 + 2 \rightarrow 1' + 2' + \cdots + n' \quad . \tag{5.105}$$

As in (5.86) we define $\mathcal{M}$ by

$$< 1', 2', \cdots n' \mid S \mid 1,2 > = \frac{(2\pi)^4}{\Omega^{1+(n/2)}} \delta^4 \left(\sum_{i=1}^{n} p_i' - p_1 - p_2 \right) \mathcal{M} \tag{5.106}$$

where p_i and p_i' denote respectively the 4-momenta of particles i and i'. The process (5.105) can again be expressed as a sum of Feynman diagrams. By using exactly the same Feynman rules (5.88), (5.100)-(5.102) and (5.104), we obtain the contribution of each diagram to $\mathcal{M}$. By following the same considerations given between (5.90) and (5.93), we see that

$$d\sigma = \sum \frac{(2\pi)^4}{\Omega^n v_1} |\mathcal{M}|^2 \delta^4 \left(\sum_{i=1}^{n} p_i' - p_1 - p_2 \right)$$

where the sum extends over different 3-momenta $\vec{p}_1{}', \vec{p}_2{}', \cdots, \vec{p}_n{}'$. When $\Omega \rightarrow \infty$, the above expression becomes

$$d\sigma = \frac{2\pi}{v_1}\left(\frac{1}{8\pi^3}\right)^{n-1} \int |\mathcal{M}|^2 \, \delta^4\left(\sum_{i=1}^{n} p_i' - p_1 - p_2\right) \prod_{i=1}^{n} d^3 p_i' \tag{5.107}$$

where all momenta and v_1, the velocity of particle 1, are in the rest frame of 2.

We note that the Feynman rule (5.101) for the internal line is Lorentz-invariant; so are the rules (5.88), (5.102) and (5.104) for the vertices and for the loop-momentum integration. However, the rule (5.100) for the external line is not. Thus, it is convenient to separate out the external-line factors in $\mathcal{M}$ by introducing

$$A \equiv (\sqrt{2}\,)^{n+2} \sqrt{\omega_1' \cdots \omega_n' \omega_1 \omega_2}\; \mathcal{M}\,, \tag{5.108}$$

which is invariant under Lorentz transformation. The differential cross section (5.107) can now be written as

$$d\sigma = \frac{2\pi}{v_1} (\text{phase space})_n \cdot |A|^2 \frac{1}{4\omega_1\omega_2} \tag{5.109}$$

where

$$(\text{phase space})_n \equiv \left(\frac{1}{8\pi^3}\right)^{n-1} \int \prod_{i=1}^{n} \frac{d^3 p_i'}{2\omega_i'} \, \delta^4\left(\sum_{i=1}^{n} p_i' - p_1 - p_2\right) \tag{5.110}$$

is the Lorentz-invariant n-body phase space.

Sometimes it is convenient to choose the center-of-mass system. We shall show that, in the c.m. system, the differential cross section for reaction (5.105) is given by

$$d\sigma = \frac{2\pi}{|\vec{v}_1 - \vec{v}_2|}\left(\frac{1}{8\pi^3}\right)^{n-1} \int |\mathcal{M}|^2 \, \delta^4\left(\sum_{i=1}^{n} p_i' - p_1 - p_2\right) \prod_{i=1}^{n} d^3 p_i' \tag{5.111}$$

where $\vec{v}_1$ and $\vec{v}_2$ are respectively the velocities of initial particles 1 and 2. [Actually, (5.111) is valid in any frame Σ given in Fig. 5.7.]

Let Σ_{lab} denote the reference frame in which particle 2 is at rest and particle 1 has velocity $\vec{v}_1$. Consider another frame Σ, as shown in Fig. 5.7, which is moving with a constant velocity $\vec{u} \,/\!/\, \vec{v}_1$

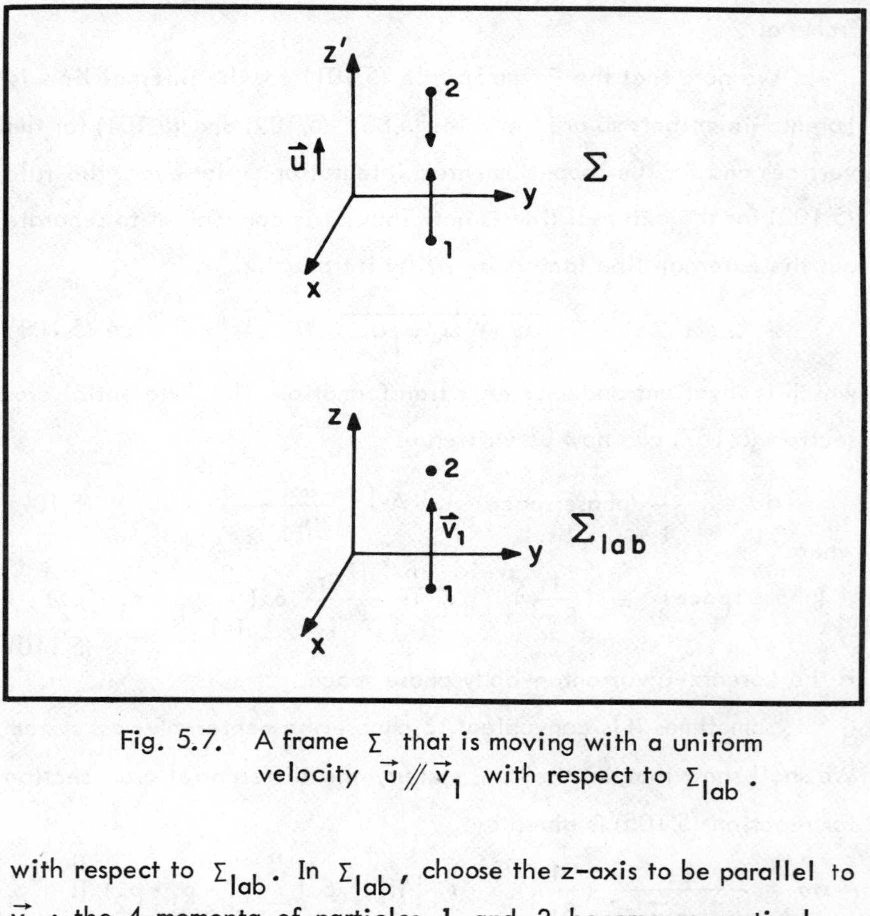

Fig. 5.7. A frame Σ that is moving with a uniform velocity $\vec{u} \,/\!/\, \vec{v}_1$ with respect to Σ_{lab}.

with respect to Σ_{lab}. In Σ_{lab}, choose the z-axis to be parallel to $\vec{v}_1$; the 4-momenta of particles 1 and 2 become respectively

$$p_1 = \omega_1(0, 0, v_1, i) \quad \text{and} \quad p_2 = m(0, 0, 0, i) \quad .$$

Let $(\omega_i)_\Sigma$ be the energy of particle i in Σ. We have, according to the Lorentz transformation

$$(\omega_1)_\Sigma = \frac{\omega_1 - u\, v_1 \omega_1}{\sqrt{1 - u^2}} \quad \text{and} \quad (\omega_2)_\Sigma = \frac{m}{\sqrt{1 - u^2}} \quad ,$$

which gives

$$(\omega_1 \omega_2)_\Sigma = \omega_1 m \frac{1 - u v_1}{1 - u^2} \quad . \tag{5.112}$$

The velocities of 1 and 2 in Σ are both parallel to the z-axis; their components are respectively

$$(v_1)_\Sigma = \frac{v_1 - u}{1 - u v_1} \quad \text{and} \quad (v_2)_\Sigma = -u \quad .$$

Hence we have

$$|\vec{v}_1 - \vec{v}_2|_\Sigma = (v_1 - v_2)_\Sigma = v_1 \frac{1 - u^2}{1 - u v_1} \quad . \tag{5.113}$$

Throughout, ω_1 and v_1 without the subscript Σ refer to the energy and the velocity of 1 in Σ_{lab}. Combining (5.112) and (5.113), we obtain

$$(\omega_1 \omega_2 |\vec{v}_1 - \vec{v}_2|)_\Sigma = \omega_1 m v_1 \quad . \tag{5.114}$$

Next, we note that in Σ_{lab}, since $\omega_2 = m$, (5.109) can be written as

$$d\sigma = \frac{2\pi}{v_1} (\text{phase space})_n \, |A|^2 \frac{1}{4\omega_1 m} \quad . \tag{5.115}$$

Since A and $(\text{phase space})_n$ are both Lorentz-invariant, by using (5.114) we derive the expression (5.111) for any such moving frame Σ, and therefore also for $\Sigma_{c.m.}$.

2. The passage from (5.107) for $d\sigma$ in Σ_{lab} to (5.111) in Σ is a general one, valid for particles of arbitrary spin. Therefore it may be useful to give an alternative proof.

In any frame Σ, let ρ_1 and ρ_2 be the number densities of initial particles 1 and 2, Ω the volume of the system and T the total time interval. It is convenient to introduce the reaction rate R, defined by

$$(\rho_1 \rho_2 R)_\Sigma \equiv (\text{total number of reactions}/\Omega T)_\Sigma \ .$$

Because ΩT is the four-dimensional volume, it is Lorentz invariant, as is the total number of reactions. Consequently, the lefthand side is also Lorentz invariant; i.e.

$$(\rho_1 \rho_2 R)_\Sigma = (\rho_1 \rho_2 R)_{\Sigma'} \tag{5.116}$$

where Σ' denotes any other frame of reference. In any frame Σ, for $\rho_1 = \rho_2 = \frac{1}{\Omega}$, the total number of reactions in ΩT is given by (5.92), except that the bra is now $\langle 1', 2', \cdots, n' |$. Hence in the limit $\Omega \to \infty$ and $T \to \infty$ we have

$$R_\Sigma = \sum \frac{\Omega}{T} |\langle 1', 2', \cdots, n' | S | 1, 2 \rangle|^2$$

where the sum extends over all final 3-momenta. Let $\mathcal{M}$ be related to the matrix element of S by (5.106). The above expression becomes

$$R_\Sigma = \sum \frac{(2\pi)^4}{\Omega^n} \delta^4\Big(\sum_{i=1}^{n} p_i' - p_1 - p_2\Big) |\mathcal{M}|^2 \ .$$

Because of (2.28), it can be written as

$$R_\Sigma = \frac{2\pi}{(8\pi^3)^{n-1}} \int \prod_{i=1}^{n} d^3 p_i' \, |\mathcal{M}|^2 \, \delta^4\Big(\sum_{i=1}^{n} p_i' - p_1 - p_2\Big) \tag{5.117}$$

in which all momenta are to be evaluated in the same system Σ. Now, in the laboratory frame we have, on account of (5.91),

$$d\sigma = \left(\frac{R}{v_1}\right)_{\Sigma_{lab}} \ , \tag{5.118}$$

which gives (5.107).

Let j_1 and j_2 be, respectively, the four-dimensional current vectors of particles 1 and 2. In Σ_{lab}, since particle 2 is at rest, they are given by

$$(j_1)_{\Sigma_{lab}} = \rho_1(\vec{v}_1, i)$$

and

$$(j_2)_{\Sigma_{lab}} = \rho_2(0, i) \quad .$$

Consider a frame Σ which is moving with a constant velocity $\vec{u} /\!/ \vec{v}_1$ with respect to Σ_{lab}. On account of Lorentz transformation, the number densities of particles 1 and 2 in Σ and in Σ_{lab} are related by

$$(\rho_1)_\Sigma = \frac{1 - uv_1}{\sqrt{1-u^2}} (\rho_1)_{\Sigma_{lab}}$$

and

$$(\rho_2)_\Sigma = \frac{1}{\sqrt{1-u^2}} (\rho_2)_{\Sigma_{lab}} \quad .$$

Thus,

$$(\rho_1 \rho_2)_\Sigma = \frac{1 - uv_1}{1 - u^2} (\rho_1 \rho_2)_{\Sigma_{lab}} \quad .$$

By using (5.113), we see that

$$(\rho_1 \rho_2 |\vec{v}_1 - \vec{v}_2|)_\Sigma = (\rho_1 \rho_2 v_1)_{\Sigma_{lab}} \quad ,$$

which together with (5.116) and (5.118) gives

$$d\sigma = \left(\frac{R}{v_1}\right)_{\Sigma_{lab}} = \left(\frac{R}{|\vec{v}_1 - \vec{v}_2|}\right)_\Sigma \quad . \tag{5.119}$$

Because of (5.117), the expression (5.111) for $d\sigma$ follows.

<u>Problem 5.1.</u> Let the total Hamiltonian of a spin-0 field

$$\phi(\vec{r}, t) = \sum_{\vec{k}} (2\omega\Omega)^{-\frac{1}{2}} (a_{\vec{k}}(t)\, e^{i\vec{k}\cdot\vec{r}} + a_{\vec{k}}^\dagger(t)\, e^{-i\vec{k}\cdot\vec{r}})$$

be

$$H = H_0 + \int J(\vec{r})\, \phi(\vec{r}, t)\, d^3r \tag{5.120}$$

where

$$H_0 = \tfrac{1}{2} \int :(\Pi^2 + (\vec{\nabla}\phi)^2 + \mu^2\phi^2): d^3r \quad , \tag{5.121}$$

$\omega = (\vec{k}^2 + \mu^2)^{\frac{1}{2}}$, $J(\vec{r}) = J(\vec{r})^*$ is a c. number function independent of time, and $\Pi(\vec{r}, t)$ is the conjugate momentum of $\phi(\vec{r}, t)$ which satisfies

$$[\Pi(\vec{r}, t), \ \phi(\vec{r}\,', t)] = -i\delta^3(\vec{r} - \vec{r}\,') \ .$$

(i) Prove that

$$S H S^\dagger = H_0 + \Omega \lambda_J$$

where S is a unitary matrix given by

$$S = \exp \sum_{\vec{k}} (-a_{\vec{k}} j_{\vec{k}}^* + a_{\vec{k}}^\dagger j_{\vec{k}}) (2\omega^3)^{-\frac{1}{2}} \ ,$$

$$j_{\vec{k}} = \int \Omega^{-\frac{1}{2}} J(\vec{r}) e^{-i\vec{k}\cdot\vec{r}} d^3r$$

and

$$\lambda_J = -\int (16\pi^3 \omega^2)^{-1} |j_{\vec{k}}|^2 d^3k \ .$$

[Note: $e^{A+B} = e^A e^B e^{-\frac{1}{2}[A, B]}$ for any two operators A and B whose commutator $[A, B]$ is a c. number.] Thus, the eigenstate $| n_{\vec{k}} >_0 \equiv \prod_k (n_{\vec{k}}!)^{-\frac{1}{2}} (a_{\vec{k}}^\dagger)^{n_{\vec{k}}} | \text{vac} >_0$ of H_0, with $n_{\vec{k}} = 0, 1, 2, \cdots$, is related to the corresponding eigenstate $| n_{\vec{k}} >$ of H by $| n_{\vec{k}} > = S^\dagger | n_{\vec{k}} >_0$, where

$$H | n_{\vec{k}} > = (\sum_{\vec{k}} n_{\vec{k}} \omega + \Omega \lambda_J) | n_{\vec{k}} > \ .$$

The new vacuum state $| \text{vac} >$ satisfies $H | \text{vac} > = \Omega \lambda_J | \text{vac} >$.

(ii) Regard $\Omega \lambda_J$ as a functional of $J(\vec{r})$. Show that the variational derivative

$$\frac{\delta \lambda_J}{\delta J(\vec{r})} \equiv \bar{\phi}(\vec{r}) = -\sum_{\vec{k}} \frac{1}{2\omega^2 \Omega^{\frac{1}{2}}} (j_{\vec{k}} e^{i\vec{k}\cdot\vec{r}} + j_{\vec{k}}^* e^{-i\vec{k}\cdot\vec{r}}) \ .$$

Furthermore, $\bar{\phi}(\vec{r})$ is the vacuum expectation of $\phi(\vec{r}, t)$:

$$\bar{\phi}(\vec{r}) = < 0 | \phi(\vec{r}, t) | 0 > \ .$$

Note that because $J(\vec{r})$ is assumed to be independent of t, so is $\bar{\phi}(\vec{r})$.

(iii) Show that

$$\mathcal{E}(\bar{\phi}) \equiv \Omega \lambda_J - \int J(\vec{r})\, \bar{\phi}(\vec{r})\, d^3 r$$

is the minimum of $< | H_0 | >$, taken among all states $| >$ under the constraint

$$< | \phi(\vec{r}, t) | > = \bar{\phi}(\vec{r}) \ .$$

Problem 5.2.   Replace (5.120) in Problem 5.1 by

$$H = H_0 + H_{int}$$

where H_0 remains given by (5.121), but

$$H_{int} = \int (J\phi + \tfrac{1}{2} m^2 \phi^2)\, d^3 r$$

where, as before, $J(\vec{r}) = J(\vec{r})^*$ is a c. number function.   Regard H_{int} as the interaction Hamiltonian.   In the interaction representation the propagator $\underset{}{\overset{p}{\rule{2cm}{0.4pt}}}$ is $-i(\vec{p}^2 - p_0^2 + \mu^2 - i\epsilon)^{-1}$ where $\epsilon = 0+$, and there are vertices $\rule{1cm}{0.4pt}\!\bullet\!\rule{1cm}{0.4pt} = -i m^2$, $\overset{k}{\rule{1cm}{0.4pt}\!\!\rightarrow\!\bullet} = -i j_k$ and $\overset{k}{\rightarrow\!\bullet} = -i j_k^*$, where the arrow indicates the flow direction of the momentum k, whose fourth component is 0 because the c. number function $J(\vec{r})$ is assumed to be independent of time. By summing over graphs, show that the full propagator is

$$D_F(p) \equiv \overset{p}{\rule{2cm}{0.4pt}} + \overset{p}{\rule{1cm}{0.4pt}}\!\!\overset{p}{\rule{1cm}{0.4pt}} + \overset{p}{\rule{1cm}{0.4pt}}\!\bullet\!\overset{p}{\rule{1cm}{0.4pt}}\!\bullet\!\overset{p}{\rule{1cm}{0.4pt}} + \cdots$$

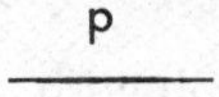

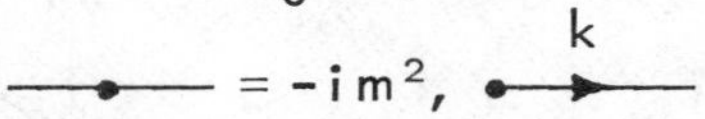

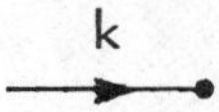

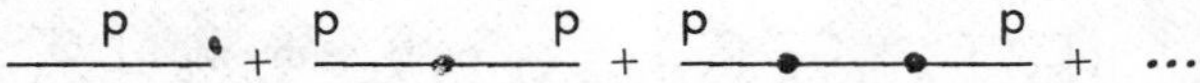

$$= -i(\vec{p}^2 - p_0^2 + \mu^2 + m^2 - i\epsilon)^{-1}$$

and the new vacuum energy is $\Omega(\Lambda_J + \Delta)$ where

$$-i\Lambda_J = \bullet\!\rule{1cm}{0.4pt}\!\bullet + \bullet\!\rule{0.7cm}{0.4pt}\!\bullet\!\rule{0.7cm}{0.4pt}\!\bullet + \bullet\!\rule{0.5cm}{0.4pt}\!\bullet\!\rule{0.5cm}{0.4pt}\!\bullet\!\rule{0.5cm}{0.4pt}\!\bullet + \cdots$$

$$= i\tfrac{1}{2} \int [8\pi^3(\vec{k}^2 + \mu^2 + m^2)]^{-1}\, |j_{\vec{k}}|^2\, d^3 k$$

and

$-i\Delta =$  $+$ 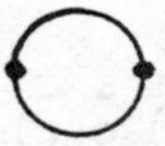 $+$  $+ \cdots$

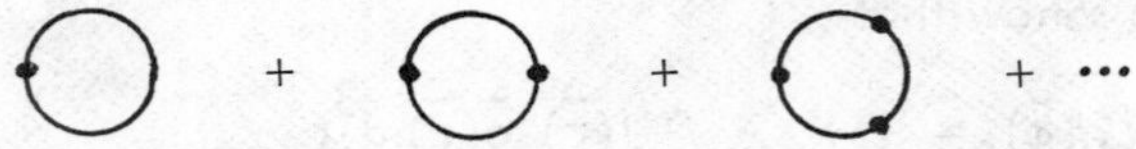

$$= -i\tfrac{1}{2} \int (8\pi^3)^{-1} \left[(\vec{k}^2 + \mu^2 + m^2)^{\frac{1}{2}} - (\vec{k}^2 + \mu^2)^{\frac{1}{2}} \right] d^3k \,.$$

Notice that the graphs  ,  ,  , etc. all have a symmetry number* factor $\frac{1}{2}$.

Problem 5.3. (i) Consider the following matrix element of the operator

$$J_\mu \equiv : i \psi_b^\dagger \gamma_4 \gamma_\mu (C_V + C_A \gamma_5) \psi_a : \tag{5.122}$$

$$< J_\mu > \equiv < b \mid J_\mu \mid a > \tag{5.123}$$

where ψ_a and ψ_b are both spin-$\frac{1}{2}$ quantized fields, C_V and C_A are constants, $\mid a >$ is the free a-particle state of helicity s_a , mass m_a and 4-momentum a_μ , and $\mid b >$ the free b-particle state of helicity s_b , mass m_b and 4-momentum b_μ .

$$T_{\mu\nu} \equiv \pm \sum_{s_a, s_b} < J_\mu >< J_\nu >^* \tag{5.124}$$

where $+$ is for $\nu \neq 4$ and $-$ is for $\nu = 4$. Show that by using the exercise on page 36 and setting the volume $\Omega = 1$,

$$a_0 b_0 T_{\mu\nu} = (\mid C_V \mid^2 + \mid C_A \mid^2)(a_\mu b_\nu + a_\nu b_\mu - \delta_{\mu\nu}\, a \cdot b)$$
$$+ (C_V^* C_A + C_A^* C_V)\, \epsilon_{\mu\nu\lambda\delta}\, a_\lambda b_\delta - m_a m_b (\mid C_V \mid^2 - \mid C_A \mid^2)\, \delta_{\mu\nu} \tag{5.125}$$

where $a_0 = (\vec{a}^2 + m_a^{\ 2})^{\frac{1}{2}}$ and $b_0 = (\vec{b}^2 + m_b^{\ 2})^{\frac{1}{2}}$, $\delta_{\mu\nu} = 1$ if $\mu = \nu$, but $= 0$ if $\mu \neq \nu$, while $\epsilon_{\mu\nu\lambda\delta} = 1$ or -1 depending on whether $\mu\nu\lambda\delta$ is an even or odd permutation of 1234 , and $= 0$ otherwise.

---

* See page 500 for the definition of symmetry number. Note that each Hermitian boson line is unarrowed. [In contrast, the line of a charged (complex) field carries an arrow pointing in the direction of the charge flow. Cf. also pages 530–32 and 90.]

(ii) Show that (5.125) remains valid, if instead of (5.123)

$$\langle J_\mu \rangle = \langle \bar{a} \mid J_\mu \mid \bar{b} \rangle$$

where $\mid \bar{a} \rangle$ is the free $\bar{a}$-particle state of 4-momentum a_μ and $\mid \bar{b} \rangle$ the free $\bar{b}$-particle state of 4-momentum b_μ.

(iii) If, instead of Eq. (5.123)

$$\langle J_\mu \rangle = \langle 0 \mid J_\mu \mid a\bar{b} \rangle \quad \text{or} \quad \langle \bar{a}b \mid J_\mu \mid 0 \rangle ,$$

then $T_{\mu\nu}$, defined by (5.124), remains given by (5.125), except for the change

$$- m_a m_b \rightarrow + m_a m_b$$

on the righthand side, where $\mid a\bar{b} \rangle$, or $\mid \bar{a}b \rangle$, is the state of a free a-particle (or $\bar{a}$-particle) of 4-momentum a_μ together with a free $\bar{b}$-particle (or b-particle) of 4-momentum b_μ.

Problem 5.4. Phenomenologically, the weak interaction Lagrangian for

$$\nu_\ell + a \rightarrow \ell^- + b \tag{5.126}$$

and

$$\bar{\nu}_\ell + b \rightarrow \ell^+ + a \tag{5.127}$$

can be written as

$$\mathcal{L} = 2^{-\frac{1}{2}} [j_\mu J_\mu + j_\mu^\dagger J_\mu^\dagger]$$

where a and b are some spin-$\frac{1}{2}$ hadrons, $\ell^\pm$ denotes the charged lepton $e^\pm$ or $\mu^\pm$,

$$j_\mu = i \psi_\ell^\dagger \gamma_4 \gamma_\mu (1 + \gamma_5) \psi_{\nu_\ell}$$

and

$$J_\mu = i \psi_b^\dagger \gamma_4 \gamma_\mu (C_V + C_A \gamma_5) \psi_a .$$

Neglect the strong interaction of a and b, assume the hadron mass $m_a = m_b = m$, and set the lepton mass $m_\ell = 0$. Show that to the lowest order in C_V and C_A, the differential cross sections for reactions (5.126) and (5.127) are respectively

$$\frac{d\sigma_\nu}{dy} = \frac{mE_\nu}{2\pi}\left[\,|C_V+C_A|^2 + |C_V-C_A|^2(1-y)^2 + (|C_A|^2-|C_V|^2)\frac{m}{E_\nu}\,y\,\right]$$

and (5.128)

$$\frac{d\sigma_{\bar\nu}}{dy} = \frac{mE_\nu}{2\pi}\left[\,|C_V-C_A|^2 + |C_V+C_A|^2(1-y)^2 + (|C_A|^2-|C_V|^2)\frac{m}{E_\nu}\,y\,\right]$$

where

$$y \equiv \frac{E_\nu - E_\ell}{E_\nu}\,,$$

E_ν and E_ℓ are, respectively, the energies of the neutrino and $\ell^\pm$ in the laboratory frame (i.e., the rest frame of the initial hadron).

Note that in any frame $y = q\cdot a/k\cdot a$ for (5.126) and $q\cdot b/k\cdot b$ for (5.127) where $q_\mu \equiv k_\mu - k'_\mu$ and k_μ, k'_μ, a_μ, b_μ are, respectively, the 4-momenta of the neutrino, $\ell^\pm$, a and b, the range of y is from 0 to 1. For further discussions see Chapters 21 and 23.

References

F. Dyson, Phys.Rev. 75, 486 (1949).

R. P. Feynman and A. R. Hibbs, Quantum Mechanics and Path Integrals (New York, McGraw-Hill, 1965).

G. C. Wick, Phys.Rev. 80, 268 (1950).

See also Chapter 19 for a discussion of the path-integration method, which is the original way Feynman invented his diagrams.

Chapter 6

QUANTUM ELECTRODYNAMICS

In quantum electrodynamics we consider the electromagnetic interaction between photons, electrons and positrons. If we wish, we may also include other charged leptons, such as $\mu^{\pm}$ and $\tau^{\pm}$. Because the photon is of spin 1 and mass 0, this also serves as an example of how to deal with the $m = 0$ limit of a vector field.

6.1 Lagrangian

Let ψ be the electron field, A_μ be the electromagnetic 4-potential and $F_{\mu\nu}$ be the electromagnetic field tensor, which is related to A_μ by

$$F_{\mu\nu} \equiv \frac{\partial}{\partial x_\mu} A_\nu - \frac{\partial}{\partial x_\nu} A_\mu \quad . \tag{6.1}$$

The Lagrangian density in quantum electrodynamics is

$$\mathcal{L} = \mathcal{L}_e + \mathcal{L}_\gamma + \mathcal{L}_{int} \quad , \tag{6.2}$$

where

$$\mathcal{L}_\gamma = -\tfrac{1}{4} F_{\mu\nu}^{\,2} = \tfrac{1}{2}(\vec{E}^2 - \vec{B}^2) \quad , \tag{6.3}$$

$$\mathcal{L}_e = -\psi^\dagger \gamma_4 (\gamma_\mu \frac{\partial}{\partial x_\mu} + m)\,\psi \tag{6.4}$$

and

$$\mathcal{L}_{int} = e\, j_\mu A_\mu + \psi^\dagger \gamma_4 \psi\, \delta m \quad . \tag{6.5}$$

In (6.3), $\vec{E}$ is the electric field and $\vec{B}$ the magnetic field, related to $A_\mu = (\vec{A}, iA_0)$ by

$$\vec{E} \equiv -\vec{\nabla} A_0 - \dot{\vec{A}} \ , \qquad \vec{B} \equiv \vec{\nabla} \times \vec{A} \ . \tag{6.6}$$

In (6.5), the electromagnetic current is

$$j_\mu \equiv i \psi^\dagger \gamma_4 \gamma_\mu \psi \ , \tag{6.7}$$

e is the unrenormalized charge and δm is defined to be the difference between the physical mass m and the mechanical mass m_0:

$$\delta m = m - m_0 \ . \tag{6.8}$$

From the variational principle (2.10), we obtain the equations of motion for A_μ and ψ:

$$\frac{\partial}{\partial x_\mu} F_{\mu\nu} = -e j_\nu \tag{6.9}$$

and

$$\gamma_\mu \left(\frac{\partial}{\partial x_\mu} - ie A_\mu \right) \psi + m_0 \psi = 0 \ . \tag{6.10}$$

6.2 Coulomb Gauge

The Lagrangian density is invariant under the gauge transformation

$$A_\mu \rightarrow A_\mu + \frac{\partial \theta}{\partial x_\mu} \tag{6.11}$$

and

$$\psi \rightarrow e^{ie\theta} \psi \ . \tag{6.12}$$

By choosing a suitable function θ, we may impose the transversality condition on $\vec{A}$:

$$\vec{A} = \vec{A}^{tr} \ , \qquad \text{i.e.,} \quad \vec{\nabla} \cdot \vec{A} = 0 \ . \tag{6.13}$$

This particular choice is called the Coulomb gauge. Since the theory is gauge invariant, any choice of gauge should lead to the same physical results. The Coulomb gauge is, however, particularly convenient for the purpose of quantization, as we shall see. Since the Lagrangian density (6.2) does not contain $\dot{A}_0$, we shall regard A_0 as a dependent variable, just as in Chapter 4. When $\nu = 4$, (6.9) becomes simply

$$\vec{\nabla} \cdot \vec{E} = e\rho \tag{6.14}$$

where $j_4 = i\rho = i\psi^\dagger\psi$. Because of (6.6) and (6.13), the above equation can be written as

$$\nabla^2 A_0 = -e\rho \quad . \tag{6.15}$$

In the Coulomb gauge, we regard A_0 as a functional of $\psi^\dagger\psi$, given by the solution of the Laplace equation

$$A_0(\vec{r}, t) \equiv \int \frac{e\rho(\vec{r}\,', t)}{4\pi\,|\vec{r} - \vec{r}\,'|}\, d^3r' \quad . \tag{6.16}$$

We may decompose the electric field in (6.6) into two terms:

$$\vec{E} = \vec{E}^{tr} + \vec{E}^{\ell} \tag{6.17}$$

in which the transverse component is

$$\vec{E}^{tr} = -\dot{\vec{A}} \tag{6.18}$$

and the longitudinal component is

$$\vec{E}^{\ell} = -\vec{\nabla} A_0 \quad . \tag{6.19}$$

Clearly, $\vec{E}^{\ell}$ is irrotational and $\vec{E}^{tr}$, because of (6.13), is divergence free. Through partial integration, the following volume integral can be reduced to 0 :

$$\int \vec{E}^{tr} \cdot \vec{E}^{\ell}\, d^3r = \int \dot{\vec{A}} \cdot \vec{\nabla} A_0\, d^3r = 0 \quad .$$

Therefore we may replace the Lagrangian density $\mathcal{L}_\gamma$ in (6.3) by

$$\mathcal{L}_\gamma = \tfrac{1}{2}[\,(\vec{E}^{tr})^2 + (\vec{E}^{\ell})^2 - \vec{B}^2\,] \quad , \tag{6.20}$$

without changing the resulting Lagrangian $L = \int \mathcal{L}\, d^3r$. The conjugate momentum Π of the electromagnetic potential $\vec{A}$ is

$$\vec{\Pi} \equiv \frac{\partial \mathcal{L}}{\partial \dot{\vec{A}}} = -\vec{E}^{tr} , \tag{6.21}$$

and the conjugate momentum of the electron field is

$$\mathcal{P} \equiv \frac{\partial \mathcal{L}}{\partial \dot{\psi}} = i\psi^{\dagger} . \tag{6.22}$$

By following the usual canonical procedure, we find the Hamiltonian density to be

$$\mathcal{H} = \mathcal{H}_{\gamma} + \mathcal{H}_{e} + \mathcal{H}_{int} , \tag{6.23}$$

where

$$\mathcal{H}_{\gamma} = \tfrac{1}{2}(\vec{E}^{tr})^2 + \tfrac{1}{2}\vec{B}^2 , \tag{6.24}$$

$$\mathcal{H}_{e} = \psi^{\dagger}(-i\vec{\alpha}\cdot\vec{\nabla} + \beta m)\psi \tag{6.25}$$

and

$$\mathcal{H}_{int} = -\psi^{\dagger}\beta\psi\cdot\delta m - \tfrac{1}{2}(\vec{E}^{\ell})^2 - e j_{\mu} A_{\mu} . \tag{6.26}$$

It is convenient to separate out from $\mathcal{H}_{int}$ a part that corresponds to the Coulomb interaction between the charge density. We define

$$\mathcal{H}_{Coul} \equiv -\tfrac{1}{2}(E^{\ell})^2 + e\rho A_0 .$$

Because of (6.15)–(6.16) and (6.19), the space integral of $\mathcal{H}_{Coul}$ is

$$\begin{aligned} H_{Coul} &= \int \mathcal{H}_{Coul}\, d^3r = \int (\tfrac{1}{2} A_0 \nabla^2 A_0 + e\rho A_0)\, d^3r \\ &= \tfrac{1}{2} \int e\rho A_0\, d^3r \\ &= \tfrac{1}{2} \int \frac{e^2 \rho(\vec{r},t)\,\rho(\vec{r}\,',t)}{4\pi\,|\vec{r}-\vec{r}\,'|}\, d^3r\, d^3r' . \end{aligned} \tag{6.27}$$

The interaction Hamiltonian is then given by

$$\begin{aligned} H_{int} &= \int \mathcal{H}_{int}\, d^3r \\ &= -\int \psi^{\dagger}\beta\psi\cdot\delta m\, d^3r + H_{Coul} - e\int \vec{j}\cdot\vec{A}\, d^3r . \end{aligned} \tag{6.28}$$

The total Hamiltonian H is given by the space integral of (6.23). In H, the generalized coordinates are $\vec{A} = \vec{A}^{tr}$ and ψ; the generalized momenta are $\vec{\Pi} = -\vec{E}^{tr}$ and $\mathcal{P} = i\psi^\dagger$.

6.3 Quantization

The quantization procedure in the Coulomb gauge can be carried out in a straightforward manner. Because of the transversality condition (6.13), the equal-time commutator between $\vec{\Pi}$ and $\vec{A}$ is

$$[\Pi_i(\vec{r}, t), \; A_j(\vec{r}\,', t)] = -i(\delta_{ij} - \nabla^{-2}\nabla_i \nabla_j)\, \delta^3(\vec{r} - \vec{r}\,') \tag{6.29}$$

where the factor $(\delta_{ij} - \nabla^{-2}\nabla_i \nabla_j)$ is to insure that the righthand side satisfies the same divergence-free constraints,

$$\vec{\nabla} \cdot \vec{A} = 0 \qquad \text{and} \qquad \vec{\nabla} \cdot \vec{\Pi} = 0 \;. \tag{6.30}$$

Between the electron field ψ and its Hermitian conjugate $\psi^\dagger$ we have the usual equal-time anticommutator

$$\{\psi(\vec{r}, t), \; \psi^\dagger(\vec{r}\,', t)\} = \delta^3(\vec{r} - \vec{r}\,') \;. \tag{6.31}$$

Likewise, the equal-time anticommutator between ψ (and that between $\psi^\dagger$) at different space-positions is 0, and so are the equal-time commutators between other pairs of field operators.

At any given time t we may expand $\vec{A}(\vec{r}, t)$ and $\vec{\Pi}(\vec{r}, t)$ in terms of the Fourier series

$$\vec{A}(\vec{r}, t) = \sum_{\vec{k}} \frac{1}{\sqrt{2\omega\Omega}} [\vec{a}_{\vec{k}}(t)\, e^{i\vec{k}\cdot\vec{r}} + \text{h.c.}] \tag{6.32}$$

and

$$\vec{\Pi}(\vec{r}, t) = -\vec{E} = \sum_{\vec{k}} \sqrt{\frac{\omega}{2\Omega}} [-i\vec{a}_{\vec{k}}(t)\, e^{i\vec{k}\cdot\vec{r}} + \text{h.c.}] \tag{6.33}$$

where $\omega = |\vec{k}|$. Because of (6.30), the $\vec{a}_{\vec{k}}$'s satisfy

$$\vec{a}_{\vec{k}}(t) \cdot \vec{k} = 0 \;. \tag{6.34}$$

The expansions (6.32) and (6.33) are valid in any representation. In the interaction representation, the $\vec{a}_{\vec{k}}$'s have the same time dependence as that in the free field. We have

$$\vec{a}_{\vec{k}}(t) \propto e^{-i\omega t} . \tag{6.35}$$

As in Chapter 4, we may introduce for any given $\vec{k}$ a set of three orthogonal unit vectors $\hat{k} = \vec{k}/|\vec{k}|$, $\hat{e}_1$ and $\hat{e}_2$:

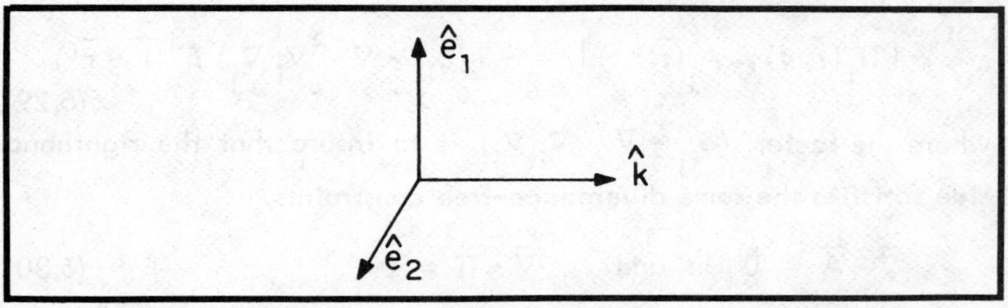

Fig. 6.1. A righthanded orthonormal set of three vectors.

It is convenient to define

$$a^{\dagger}_{\vec{k},s=\pm 1} \equiv \frac{1}{\sqrt{2}} \vec{a}^{\dagger}_{k} \cdot (\hat{e}_1 \pm i\hat{e}_2) , \tag{6.36}$$

and its Hermitian conjugate

$$a_{\vec{k},s=\pm 1} \equiv \frac{1}{\sqrt{2}} \vec{a}_{k} \cdot (\hat{e}_1 \mp i\hat{e}_2) . \tag{6.37}$$

From (6.29) and

$$[A_i(\vec{r}, t), A_j(\vec{r}', t)] = [\Pi_i(\vec{r}, t), \Pi_j(\vec{r}', t)] = 0 ,$$

one can readily verify that

$$[a_{\vec{k},s}(t), a^{\dagger}_{\vec{k}',s'}(t)] = \delta_{\vec{k},\vec{k}'} \delta_{ss'} , \tag{6.38}$$

and

$$[a_{\vec{k},s}(t), a_{\vec{k}',s'}(t)] = 0 \tag{6.39}$$

where s and s' can be +1 or −1. It can also be shown that the subscript s denotes the helicity of the photon (= its spin component along the direction of motion). Thus $a^{\dagger}_{\vec{k},s}$ is the creation operator of a photon with momentum $\vec{k}$ and helicity s ; its Hermitian conjugate $a_{\vec{k},s}$ is the corresponding annihilation operator.

The S-matrix can be derived by following the steps discussed in Chapter 5. Its perturbation series is given by (5.70) in which the interaction Hamiltonian is the integral of the normal product of $\mathcal{H}_{int}$,

$$H_{int} = \int :\mathcal{H}_{int}: d^3r \tag{6.40}$$

where $\mathcal{H}_{int}$ is given by (6.26). Likewise, the unperturbed Hamiltonian is

$$H_0 = \int :\mathcal{H}_\gamma + \mathcal{H}_e: d^3r \tag{6.41}$$

where $\mathcal{H}_\gamma$ and $\mathcal{H}_e$ are given by (6.24) and (6.25).

6.4 Photon Propagator and Relativistic Invariance

In the Coulomb gauge, while one can easily carry out the quantization procedure, the Lorentz-invariant character of QED is less obvious, but will be demonstrated using Feynman diagrams.

Let us consider the diagram shown in Fig. 6.2. The amplitude

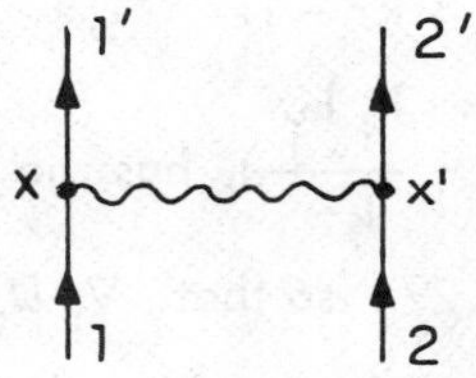

Fig. 6.2. A diagram for $1 + 2 \to 1' + 2'$.

of this diagram consists of two terms. The first is due to

$$-i\int H_{Coul}\,dt = -\frac{i}{2}\int \frac{e^2}{4\pi}\,\rho(\vec{r},t)\,\rho(\vec{r}\,',t)\,\frac{d^3r\,d^3\vec{r}\,'}{|\vec{r}-\vec{r}\,'|}\,dt$$

$$= \frac{i}{2}\int \frac{e^2}{4\pi}\,j_4(x)\,j_4(x')\,\frac{\delta(t-t')}{|\vec{r}-\vec{r}\,'|}\,d^4x\,d^4x' \,, \tag{6.42}$$

where $j_4 = i\rho$. The second term is due to

$$\frac{(-i)^2}{2!}\int T(e^2\,\vec{j}(x)\cdot\underbrace{\vec{A}(x)\,\vec{j}(x')\cdot\vec{A}(x')})\,d^4x\,d^4x' \tag{6.43}$$

where T denotes the time-ordered product. Their sum is

$$\frac{(-i)^2}{2!}\int e^2\,T(j_\mu(x)\,D_{\mu\nu}^{Coul}(x-x')\,j_\nu(x'))\,d^4x\,d^4x' \,, \tag{6.44}$$

where $D_{\mu\nu}^{Coul}(x-x')$ is the photon propagator in the Coulomb gauge:

$$= D_{\mu\nu}^{Coul}(x-x') = \begin{cases} D_{ij}^{tr}(x-x') = \underbrace{A_i(x)\,A_j(x')} & \text{if } \mu = i \neq 4 \text{ and } \nu = j \neq 4 \,, \\ \dfrac{-i\,\delta(t-t')}{4\pi\,|\vec{r}-\vec{r}\,'|} & \text{if } \mu = \nu = 4 \,, \\ 0 & \text{otherwise} \,. \end{cases} \tag{6.45}$$

Because $A_i = A_i^{tr}$, in the above expression when $\mu = i \neq 4$ and $\nu = j \neq 4$ the photon propagator is given by

$$D_{ij}^{tr}(x) = \frac{-i}{(2\pi)^4}\int \frac{d^4k}{k^2 - i\epsilon}\left(\delta_{ij} - \frac{k_i\,k_j}{\vec{k}^2}\right) e^{ik\cdot x} \,, \tag{6.46}$$

in which the factor $(\delta_{ij} - \frac{k_i k_j}{\vec{k}^2})$ has the same origin as the factor $(\delta_{ij} - \nabla^{-2}\,\nabla_i\nabla_j)$ in (6.29), so that $\nabla_i\,D_{ij}^{tr}(x) = 0$. As before,

$$\epsilon = 0+ \qquad \text{and} \qquad k^2 = \vec{k}^2 - k_0^2 \,.$$

We note that

$$\frac{1}{8\pi^3}\int \frac{1}{\vec{k}^2}\, e^{i\vec{k}\cdot\vec{r}}\, d^3k = \frac{2\pi}{8\pi^3}\int e^{ikr\cos\theta}\, dk\, d\cos\theta$$

$$= \frac{1}{4\pi^2}\int_0^\infty \frac{e^{ikr} - e^{-ikr}}{ikr}\, dk = \frac{1}{4\pi r}\ .$$

Thus, in the momentum space the photon propagator becomes

$$D^{Coul}_{\mu\nu}(k) = \begin{cases} \dfrac{-i}{k^2 - i\epsilon}\left(\delta_{ij} - \dfrac{k_i k_j}{\vec{k}^2}\right) & \text{if } \mu = i \neq 4 \text{ and } \nu = j \neq 4\ , \\ -\dfrac{i}{\vec{k}^2} & \text{if } \mu = \nu = 4\ , \\ 0 & \text{otherwise} \end{cases} \qquad (6.47)$$

where

$$D^{Coul}_{\mu\nu}(k) \equiv \int D^{Coul}_{\mu\nu}(x)\, e^{-ik\cdot x}\, d^4x\ .$$

According to (6.44) the amplitude for the Feynman diagram in Fig. 6.2 is given by

$$\frac{1}{2!}(-ie)^2 \int d^4x\, d^4x'\, [\langle 1'\,|\, j_\mu(x)\,|\, 1\rangle\, D^{Coul}_{\mu\nu}(x - x')\langle 2'\,|\, j_\mu(x')\,|\, 2\rangle$$

$$+ \text{ same terms, but interchanging } x \text{ with } x'\,]\ . \qquad (6.48)$$

Let p_i and p_i' be respectively the 4-momenta of particles i and i'. The matrix element $\langle 1'\,|\, j_\mu(x)\,|\, 1\rangle$ is proportional to $e^{i(p_1 - p_1')\cdot x}$; likewise $\langle 2'\,|\, j_\mu(x')\,|\, 2\rangle$ is proportional to $e^{i(p_2 - p_2')\cdot x'}$. Thus, we may write

$$\langle 1'\,|\, j_\mu(x)\,|\, 1\rangle \equiv e^{i(p_1 - p_1')\cdot x}\, a_\mu\ ,$$

and

$$\langle 2'\,|\, j_\mu(x')\,|\, 2\rangle \equiv e^{i(p_2 - p_2')\cdot x'}\, b_\mu \qquad (6.49)$$

where a_μ and b_μ are independent of x and x'. Due to momentum conservation, we have

$$k \equiv p_1 - p_1' = -p_2 + p_2'\ . \qquad (6.50)$$

Because of current conservation, we also have

$$\frac{\partial j_\mu(x)}{\partial x_\mu} = 0 \ ,$$

which leads to, on account of (6.49)-(6.50),

$$k_\mu a_\mu = k_\nu b_\nu = 0 \ . \tag{6.51}$$

Transforming to momentum space, we can re-label Fig. 6.2 in the following form:

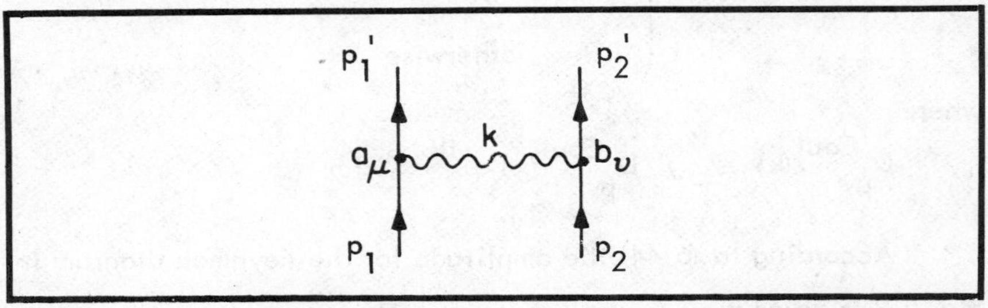

Fig. 6.3. Diagram for 1 + 2 → 1' + 2' in momentum space.

Because of (6.47), its amplitude is proportional to

$$a_\mu D^{Coul}_{\mu\nu}(k)\, b_\nu = -i\left[\frac{1}{k^2}\left(\vec{a}\cdot\vec{b} - \frac{(\vec{k}\cdot\vec{a})(\vec{k}\cdot\vec{b})}{\vec{k}^2}\right) - \frac{a_0 b_0}{\vec{k}^2}\right] \tag{6.52}$$

where $a_\mu = (\vec{a}, i a_0)$, $b_\mu = (\vec{b}, i b_0)$ and, as before, $k^2 = \vec{k}^2 - k_0^2$. For simplicity we have omitted $-i\epsilon$ in the denominator of $\frac{1}{k^2}$. Since $\vec{k}\cdot\vec{a} = k_0 a_0$ and $\vec{k}\cdot\vec{b} = k_0 b_0$, in accordance with (6.51), the amplitude (6.52) becomes

$$-i\left[\frac{\vec{a}\cdot\vec{b}}{k^2} - \frac{a_0 b_0}{\vec{k}^2}\left(\frac{k_0^2}{k^2} + 1\right)\right] = \frac{-i}{k^2}\left(\vec{a}\cdot\vec{b} - a_0 b_0\right) \ .$$

Consequently, we can replace the non-covariant Coulomb propagator

$D_{\mu\nu}^{Coul}$ by the covariant propagator $D_{\mu\nu}$:

$$a_\mu D_{\mu\nu}^{Coul}(k)\, b_\nu = a_\mu D_{\mu\nu}(k)\, b_\nu \tag{6.53}$$

where

$$D_{\mu\nu}(k) = \frac{-i}{k^2 - i\epsilon}\left[\delta_{\mu\nu} + \lambda \frac{k_\mu k_\nu}{k^2}\right] . \tag{6.54}$$

Because of (6.51), the Feynman amplitude (6.53) is independent of the parameter λ. The choice of λ is therefore arbitrary. When $\lambda = 0$, it is referred to as the Feynman gauge; when $\lambda = -1$, as the Landau gauge.

It is possible to prove that this replacement is valid in all Feynman diagrams, and thereby establish the Lorentz invariance of the theory.

6.5 Remarks

Because positronium states are all unstable, quantum electrodynamics is one example in which no stable bound state exists. Therefore, in the notation of (6.23)-(6.26) the spectra of the total Hamiltonian

$$H = \int : \mathcal{H}_\gamma + \mathcal{H}_e + \mathcal{H}_{int} : d^3r$$

and

$$H_0 = \int : \mathcal{H}_\gamma + \mathcal{H}_e : d^3r \tag{6.55}$$

are the same, provided the mass m in H_0 is the physical mass of the electron. Any eigenstate $|\, n >$ of the free Hamiltonian,

$$H_0 \,|\, n > = E_n \,|\, n > ,$$

can be written as

$$|\, n > = \Pi\, a^\dagger_{\vec{p}_i, s_i}\, b^\dagger_{\vec{p}_j, s_j}\, a^\dagger_{\vec{p}_k, s_k} \,|\, 0 > \tag{6.56}$$

where $a^\dagger_{\vec{p}_i, s_i}$, $b^\dagger_{\vec{p}_j, s_j}$ and $a^\dagger_{\vec{p}_k, s_k}$ are respectively the creation operators of the electron, positron and photon. The state $| 0 >$ is the vacuum state of H_0 ,

$$H_0 \,|\, 0 > = 0 \,. \qquad (6.57)$$

Let $U(t, t_0)$ be the solution of (5.20)-(5.21). It is useful to introduce

$$| n^{in} > \equiv U(t, -\infty) \,|\, n > \qquad (6.58)$$

and

$$| n^{f} > \equiv U^\dagger(\infty, t) \,|\, n > \,, \qquad (6.59)$$

where t can be any finite time, the superscripts in and f denote respectively the initial and final states. It can be shown * that these two states are both eigenstates of the total Hamiltonian; i.e.,

$$H \,|\, n^{in} > = E_n \,|\, n^{in} > \qquad (6.60)$$

and

$$H \,|\, n^{f} > = E_n \,|\, n^{f} > . \qquad (6.61)$$

For a multiparticle state, $| n^{in} >$ represents plane waves described by (6.56) plus outgoing waves due to the interaction. This is illustrated in Fig. 6.4, in which the free state $| n >$ consists of two particles

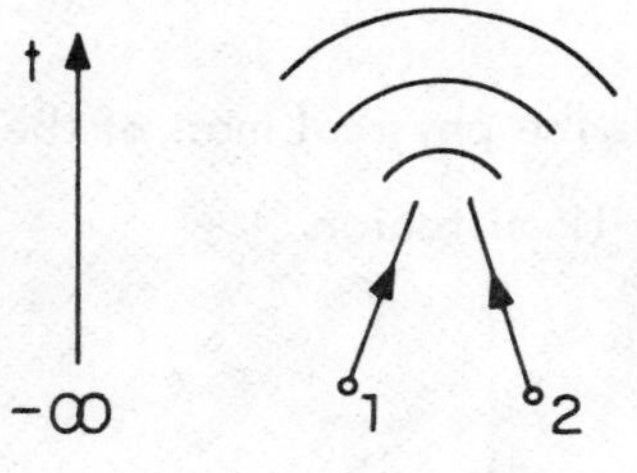

Fig. 6.4. The evolution from a state of free particles 1 and 2 at time = $-\infty$ to one with outgoing waves at time t .

---

* See, e.g., S. S. Schweber, An Introduction to Relativistic Quantum Field Theory (New York, Row, Peterson and Co., 1961), pages 320-25, and T. D. Lee and M. Nauenberg, Phys. Rev. 133, B1549 (1964), Appendix A.

1 and 2. Through the time interval from $-\infty$ to t these two particles have interacted continuously, leading to the outgoing-wave component of $U(t, -\infty)\,|\,n\rangle$. In a similar way, by considering the time evolution from t to $+\infty$, one can show that $|\,n^f\rangle$ represents the superposition of plane waves plus incoming waves. The S-matrix between two states of "free particles" $|\,n\rangle$ and $|\,n'\rangle$ is

$$\langle n'\,|\,S\,|\,n\rangle = \langle n'\,|\,U(\infty, -\infty)\,|\,n\rangle$$

where $|\,n\rangle$ and $|\,n'\rangle$ are both of the form (6.56). Because of (6.58)-(6.59), the matrix element of S can also be written as

$$\langle n'\,|\,S\,|\,n\rangle = \langle n'^{f}\,|\,n^{in}\rangle\,. \tag{6.62}$$

Since the sets $\{|n^{in}\rangle\}$ and $\{|\,n^f\rangle\}$ are each a complete orthonormal set of basis vectors in Hilbert space, the S-matrix is simply the unitary transformation between these two sets. For any scattering process

$$1 + 2 \rightarrow 1' + 2' + \cdots ,$$

the plane-wave part of $|\,n^{in}\rangle$ is associated with the initial particles $1 + 2$, and that of $|\,n^f\rangle$ with the final particles $1' + 2' + \cdots$.

Problem 6.1. Show that to order α^2 the differential and total cross sections of $e^+ + e^- \rightarrow \mu^+ + \mu^-$ are

$$d\sigma = \tfrac{1}{2}\pi\alpha^2 v(2 - v^2 \sin^2\theta)\,\frac{d\cos\theta}{4E^2}$$

and $\sigma = \frac{1}{2}\pi\alpha^2 v(1 - \frac{1}{3}v^2)/E^2$ where $2E$, v and θ are, respectively, the total energy, the muon velocity and the angle between the e^- and μ^- momenta in the center-of-mass system, $\alpha = e^2/4\pi$ and, for simplicity, the electron mass m_e is set to be 0.

Problem 6.2. Show that to order α^2 the differential cross section of $e^- + p \rightarrow e^- + p$, in the approximation that $m_e = 0$ and the strong interaction of the proton is neglected, is

$$\frac{d\sigma}{dq^2} = 4\pi \left(\frac{\alpha}{q^2}\right)^2 \left[1 - \frac{q^2}{q_{max}^2} + \frac{1}{2}\left(\frac{q^2}{2m_p E}\right)^2\right]$$

where m_p is the proton mass, E is the initial e^- energy in the rest system of the initial p (laboratory system), and q^2 is the (4-momentum transfer)^2 between e^- and p :

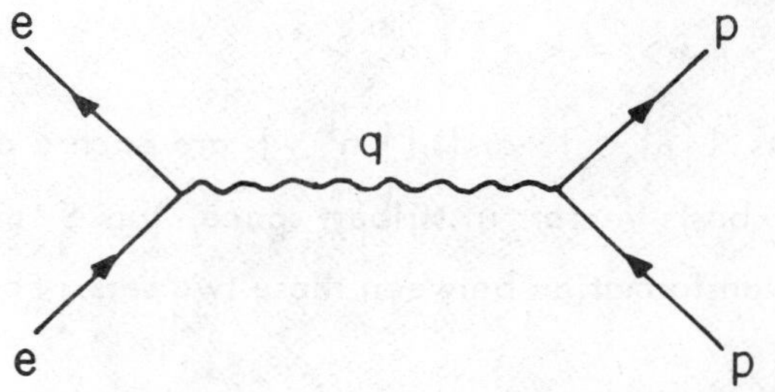

Hence, $q^2 = \vec{q}^{\,2} - q_0^{\,2}$ can vary between 0 to

$$q_{max}^2 = \frac{2m_p E}{\left(1 + \frac{m_p}{2E}\right)} .$$

References

J. Bjorken and S. D. Drell, Relativistic Quantum Mechanics (New York, McGraw-Hill, 1964).

R. Feynman, Quantum Electrodynamics (New York, W. A. Benjamin, 1962).

G. Wentzel, Quantum Theory of Fields (New York, Interscience Publishers, Inc., 1949).

Chapter 7

SOLITONS

The usual description of a bound state is in terms of the Schrödinger equation if it is nonrelativistic, or its generalization, the Bethe-Salpeter equation, in the relativistic case. Such an approach is highly successful in the case of atoms and molecules; it is also reasonably adequate with regard to nuclear structure. In these descriptions, Planck's constant plays an essential role. These bound states exist only in quantum mechanics. Indeed, in the case of the coupling between matter and the electromagnetic field, classical physics is totally inadequate to provide a stable atomic structure against radiation. It is that failure which led to the discovery of quantum mechanics in the first place. Since then, it has usually been thought that, in a relativistic field theory, in order to have stationary bound states, quantum mechanics must be crucial. As we shall see, this turns out <u>not</u> to be the case. In a nonlinear field theory, with an appropriate amount of nonlinearity, stable bound states can exist on a classical, as well as quantum mechanical, level. Such bound states are called solitons.

7.1 Early History

The earliest discussion of the subject was given by J. Scott Russell in the Report of the British Association for the Advancement of Science, published in 1845. In his own words (given below):

Report on Waves. *By* J. Scott Russell, *Esq., M.A., F.R.S. Edin., made to the Meetings in* 1842 *and* 1843.

Members of Committee { Sir John Robison*, *Sec. R.S. Edin.*
J. Scott Russell, *F.R.S. Edin.*

I believe I shall best introduce this phænomenon by describing the circumstances of my own first acquaintance with it. I was observing the motion of a boat which was rapidly drawn along a narrow channel by a pair of horses, when the boat suddenly stopped—not so the mass of water in the channel which it had put in motion; it accumulated round the prow of the vessel in a state of violent agitation, then suddenly leaving it behind, rolled forward with great velocity, assuming the form of a large solitary elevation, a rounded, smooth and well-defined heap of water, which continued its course along the channel apparently without change of form or diminution of speed. I followed it on horseback, and overtook it still rolling on at a rate of some eight or nine miles an hour, preserving its original figure some thirty feet long and a foot to a foot and a half in height. Its height gradually diminished, and after a chase of one or two miles I lost it in the windings of the channel. Such, in the month of August 1834, was my first chance interview with that singular and beautiful phænomenon which I have called the Wave of Translation, a name which it now very generally bears; which I have since found to be an important element in almost every case of fluid resistance, and ascertained to be the type of that great moving elevation of the sea, which, with the regularity of a planet, ascends our rivers and rolls along our shores.

Fig. 7.1

Scott Russell then went on to propose that the solitary object which he encountered actually represents a general class of solutions of hydrodynamics, which he first called "wave of translation", and later "solitary wave". Unlike the shock wave, which is singular at the shock front, the "solitary wave" is regular everywhere without singularity. The solitary wave is nondispersive and stable; therefore, it is different from any wave packet composed of the usual plane-wave solutions. However, Scott Russell did not succeed in convincing all his colleagues. As we can see from Fig. 7.2, taken from an 1876 paper by Lord Rayleigh, the subject of the solitary wave was still in hot dispute among various leading physicists of the time. The dispute was not settled until 1895, when Korteweg and de Vries * gave the complete analytic explanation in terms of what is now called the soliton solution of the nonlinear hydrodynamical equation—the Korteweg-de Vries equation.

Nevertheless, the question remains whether such stable, nonsingular and nondispersive solutions can occur in other domains of physics, outside hydrodynamics. This problem received a new impetus through the work done by Fermi, Pasta and Ulam ** in the early fifties. By using one of the first large electronic computers, Maniac I, they investigated the approach to equipartition of energy between 64 harmonic oscillators, coupled with some very weak nonlinear couplings. Initially, all energy lay only in one oscillator. To their great surprise, the usual idea of how the thermal equilibrium is reached turned out to be quite incorrect.

---

* D. J. Korteweg and G. de Vries, Phil. Mag. 39, 422 (1895).

** Collected Papers of Enrico Fermi, general editor E. Segré (University of Chicago Press, 1965), Vol. II, 978.

THE

LONDON, EDINBURGH, AND DUBLIN

PHILOSOPHICAL MAGAZINE

AND

JOURNAL OF SCIENCE.

[FIFTH SERIES.]

APRIL 1876.

XXXII. *On Waves.* *By* LORD RAYLEIGH, *M.A., F.R.S.**

. . .

The Solitary Wave.

This is the name given by Mr. Scott Russell to a peculiar wave described by him in the British-Association Report for 1844.

. . .

Airy, in his treatise on Tides and Waves, still probably the best authority on the subject, appears not to recognize any thing distinctive in the solitary wave.

. . .

On the other hand, Professor Stokes says*:—"It is the opinion of Mr. Russell that the solitary wave is a phenomenon *sui generis*, in no wise deriving its character from the circumstances of the generation of the wave.

Fig. 7.2

266.

STUDIES OF NON LINEAR PROBLEMS

E. FERMI, J. PASTA, and S. ULAM

Document LA-1940 (May 1955).

ABSTRACT.

A one-dimensional dynamical system of 64 particles with forces between neighbors containing nonlinear terms has been studied on the Los Alamos computer MANIAC I. The nonlinear terms considered are quadratic, cubic, and broken linear types. The results are analyzed into Fourier components and plotted as a function of time.

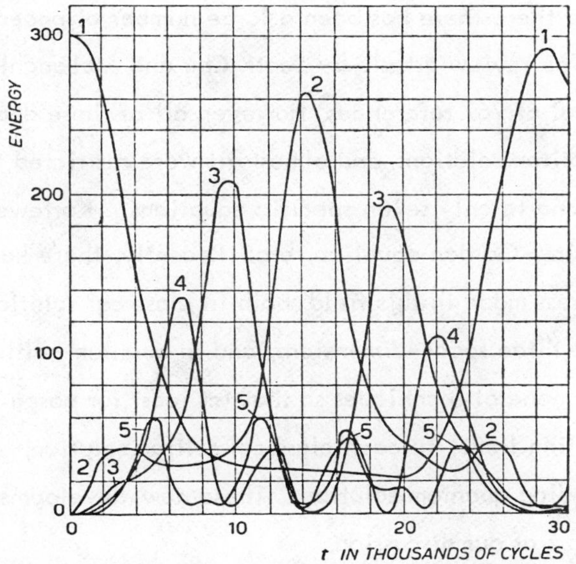

Fig. 1. – The quantity plotted is the energy (kinetic plus potential in each of the first five modes). The units for energy are arbitrary. N = 32 ; $\alpha = 1/4$; $\delta t^2 = 1/8$. The initial form of the string was a single sine wave. The higher modes never exceeded in energy 20 of our units. About 30,000 computation cycles were calculated.

Fig. 7.3

As we can see from Fig. 7.3, after some tens of thousands of cycles, the energy invariably returned nearly completely to the original mode, leaving only a few percent of the total energy to very few other oscillators. [This is not the Poincaré cycle, which requires a much longer time duration.] The development of such collective modes is a general phenomenon; it can be approximately represented by the soliton solution of the Toda lattice.* An important and general feature of the soliton solutions is that they exist even if the nonlinear coupling is extremely weak:

$$\text{weak coupling} \neq \text{weak amplitude}. \tag{7.1}$$

Since then, there has been a large number of papers on soliton solutions. The review article by Scott, Chu and McLaughlin** in 1973 listed a total of 267 references. However, all of these dealt only with classical soliton solutions, and almost all were restricted to one space-dimension and to only seven specific equations: Korteweg - de Vries equation, sine-Gordon equation, etc. Recently, there has been some major progress made in this field, both in classical solutions, extending them to three space-dimensions, and in quantum soliton solutions, developing general techniques so that (at least for boson fields in the weak-coupling limit) to each classical soliton solution, there exists a corresponding quantum solution. These new developments will be the main part of our discussion.

---

* M. Toda, Progr. Theor. Phys. Suppl. 45, 174 (1970).

** A. C. Scott, F. Y. F. Chu and D. W. Mclaughlin, Proc. IEEE 61, 1443 (1973).

7.2 Definition, Classification and Some General Remarks

Let us begin with the definition:

A classical soliton is any spatially confined and nondispersive solution of a classical field theory.

Throughout our discussions we shall be interested only in relativistic local field theories. In order to have soliton solutions, there must be nonlinear couplings; otherwise, the only solutions are plane waves. While wave packets can be formed through the superposition of plane waves, these packets are always dispersive and therefore not solitons.

The following remarks are applicable to any boson-field solitons.

1. In a general case, the Lagrangian may consist of several fields and many different couplings. It is convenient to represent the various fields collectively as ϕ, and to write the Lagrangian density $\mathcal{L}$ as

$$\mathcal{L} = -\frac{1}{2}\left(\frac{\partial\phi}{\partial x_\mu}\right)^2 - \frac{1}{g^2} V(g\phi) \tag{7.2}$$

where g is dimensionless and V has its minimum at $\phi = 0$. Without any loss of generality, this minimum value may be chosen to be zero. Thus, in a power series expansion

$$\frac{1}{g^2} V(g\phi) = \frac{1}{2} m^2 \phi^2 + O(g\phi^3) + O(g^2\phi^4) + \cdots , \tag{7.3}$$

in which the quadratic ϕ^2-term is independent of g. If there is only a single field in the theory, then m is a number. Otherwise, ϕ represents a column matrix with n components, and $\frac{1}{2}m^2\phi^2$ stands for $\frac{1}{2}\tilde{\phi}\, m^2 \phi$ where m is an $n \times n$ matrix. The equation of motion can be obtained through the variational principle (2.10); it is

$$\frac{\partial^2\phi}{\partial x_\mu^2} - \frac{1}{g} V'(g\phi) = 0 \tag{7.4}$$

where $V'(\sigma) = dV(\sigma)/d\sigma$ and $\sigma = g\phi$. Because of (7.3), the above equation becomes

$$\frac{\partial^2 \phi}{\partial x_\mu^2} = m^2\phi + O(g\phi^2) + O(g^2\phi^3) + \cdots . \tag{7.5}$$

When $g = 0$ the equation becomes linear. Therefore, g characterizes the various nonlinear couplings in the equation. As mentioned before, since soliton solutions are nondispersive, they do not exist when $g = 0$. As we shall see, all soliton solutions are singular when $g \to 0$.

2. In a classical theory, this singularity is always a simple pole. To show this, we may write

$$\phi_{\text{classical}} = \frac{1}{g}\sigma . \tag{7.6}$$

The Lagrangian density $\mathcal{L}$ then becomes

$$\mathcal{L} = \frac{1}{g^2}\mathcal{L}_\sigma \tag{7.7}$$

where

$$\mathcal{L}_\sigma = -\tfrac{1}{2}\left(\frac{\partial \sigma}{\partial x_\mu}\right)^2 - V(\sigma) \tag{7.8}$$

which is g-independent. Since the classical solution is determined by the extremity of the action integral, the g-independence of $\mathcal{L}_\sigma$ implies that the corresponding soliton solution σ is also g-independent, and that establishes (7.6). Therefore, the existence of soliton solutions does not depend on the strength of g, so long as $g \neq 0$. This is why, as noted in (7.1), even in the case of weak coupling it is not possible to neglect the soliton solutions. Unlike the plane-wave solution, the soliton solution $\to \infty$ when $g \to 0$.

3. An important and delightful feature which was discovered only relatively recently is that for any boson-field system, once the

classical soliton exists there is always a corresponding quantum soliton solution, at least in the weak coupling.

The simplest way to anticipate this is to note that the action A in a quantum theory can be set to be $\hbar^{-1}$ times the classical action; i.e.

$$A = \hbar^{-1} \int \mathcal{L}\, d^4x = (\hbar g^2)^{-1} \int \mathcal{L}_\sigma\, d^4x \tag{7.9}$$

where $\mathcal{L}$ and $\mathcal{L}_\sigma$ are related by (7.7). In the quantum theory, one considers all paths leading from an initial to a final configuration. [See Fig. 7.4.] Each path carries an amplitude proportional to e^{iA},

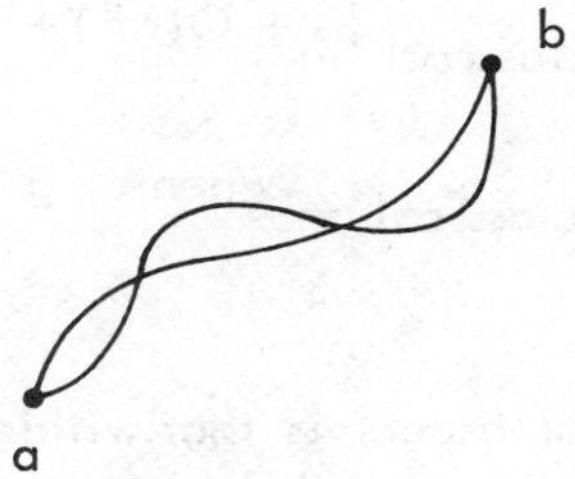

Fig. 7.4. In quantum theory each path from a to b carries an amplitude proportional to e^{iA} where A is the action integral. [See Chapter 19.]

and the superposition of all these amplitudes is the state vector. When $\hbar$ approaches 0, the only important path is the one with a stationary phase, $\delta A = 0$, and that leads to the classical description. The formal expansion in terms of $\hbar$ gives the familiar W. K. B. approximation. Because of (7.9), we expect g^2 to play the same role as $\hbar$:

$$g^2 \sim \hbar \ . \tag{7.10}$$

Therefore, an expansion in g^2 is equivalent to that in $\hbar$; the leading term must be the same as the classical limit. The details of how to carry out such a quantum expansion will be given in Section 7.6. Here, we only note that (7.10) explains why in the weak coupling the existence of a classical soliton implies a corresponding quantum solution. For example, on account of (7.6)-(7.7), the energy of a classiical soliton is of the form

$$E_{\text{classical}} = O(g^{-2}) \quad . \tag{7.11}$$

In a perturbation expansion, the energy of the corresponding quantum solution becomes then

$$E_{\text{quantum}} = E_{\text{classical}}\,[1 + O(g^2) + O(g^4) + \cdots] \quad . \tag{7.12}$$

Thus,

$$E_{\text{quantum}} \to E_{\text{classical}} \quad \text{when} \quad g \to 0 \quad , \tag{7.13}$$

at least formally.

Another pleasant aspect is that within the conventional class of renormalizable theories, all radiative corrections $O(g^2)$, $O(g^4)$, $\cdots$ in (7.12) are expected to be automatically finite for the soliton solutions. This is closely tied to the fact that the classical soliton solution is regular everywhere. At very high frequencies, the scattering amplitude of an incident plane wave by the soliton must be negligibly small; hence, the existence of soliton solutions should not alter the high-energy behavior of the theory. For the renormalizable theories, because all radiative corrections are finite, the limit (7.13) is valid in a real sense, and thereby connects the classical to the quantum solution.

Because of the uncertainty principle, a quantum soliton cannot be confined in space all the time. The definition of a quantum soliton is tied to that of the corresponding classical solution through (7.12)-(7.13).

If we restrict ourselves to renormalizable relativistic local field theories, then all soliton solutions can be classified into two general types (the details of which will be given in the next few sections).

(1) Topological solitons. The necessary condition is that there should be degenerate vacuum states so that the boundary condition at infinity for a soliton state is topologically different from that of the physical vacuum state. Some typical examples of the topological soliton solutions are those of the sine-Gordon equation* in one space-dimension, the vortex solution of Nielsen and Olesen** in two space-dimensions, and the magnetic monopole solution of 't Hooft and Polyakov*** in three space-dimensions.

(2) Nontopological solitons. The boundary condition at infinity for a nontopological soliton is the same as that for the vacuum state. Thus, there is no need of the degenerate vacuum states. The necessary condition for the existence of nontopological solitons is that there should be an additive conservation law. The nontopological soliton solutions can also exist in any space dimension ****, as we shall discuss in the subsequent sections.

---

* See Problem 7.1 for its definition.

** H. B. Nielsen and P. Olesen, Nucl. Phys. B61, 45 (1973).

*** G. 't Hooft, Nucl. Phys. B79, 276 (1974); A. M. Polyakov, JETP Lett. 20, 194 (1974).

**** R. Friedberg, T. D. Lee and A. Sirlin, Phys. Rev. D13, 2739 (1976), Nucl. Phys. B115, 1, 32 (1976).

7.3 One-space-dimensional Examples

For simplicity, we first consider soliton solutions in one space-dimension (plus the time-dimension). In view of (7.10)-(7.13), we need only examine the classical system. The quantum solution can then be derived by the perturbation series, as will be shown in Section 7.6.

1. Topological soliton. Let ϕ be a Hermitian field. In accordance with (7.2), the Lagrangian density can be written as

$$\mathcal{L} = -\tfrac{1}{2}\left(\frac{\partial \phi}{\partial x_\mu}\right)^2 - \frac{1}{g^2} V(g\phi) \quad , \qquad (7.14)$$

where $x_\mu = (x, it)$. Through the substitution

$$\phi = \frac{1}{g} \sigma ,$$

$\mathcal{L}$ becomes

$$\mathcal{L} = \frac{1}{g^2} \mathcal{L}_\sigma \qquad (7.15)$$

where

$$\mathcal{L}_\sigma = -\tfrac{1}{2}\left(\frac{\partial \sigma}{\partial x_\mu}\right)^2 - V(\sigma) \quad . \qquad (7.16)$$

Since the necessary condition for the topological soliton is the existence of degenerate vacuum, there must be more than one minimum of $V(\sigma)$. Without any loss of generality, we may choose the minimum of V to be 0. Therefore, as shown in Fig. 7.5, V is ≥ 0 and it has more than one minimum, say $V(\sigma) = 0$, at $\sigma = a, b, \cdots$. In terms of σ, the equation of motion is

$$\frac{\partial^2 \sigma}{\partial x_\mu^2} - \frac{dV}{d\sigma} = 0 \qquad (7.17)$$

whose time-independent solution is determined by

$$\frac{d^2 \sigma}{dx^2} - \frac{dV}{d\sigma} = 0 \quad .$$

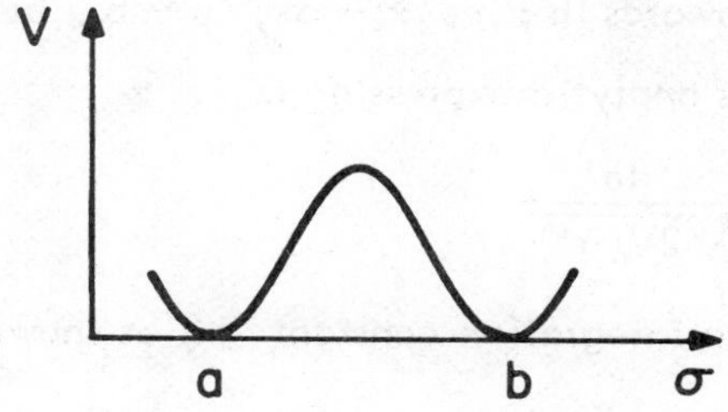

Fig. 7.5. The schematic drawing of $V(\sigma)$ for a theory that has topological soliton solutions.

We may multiply it by $d\sigma/dx$ and then integrate. This leads to

$$\frac{1}{2}\left(\frac{d\sigma}{dx}\right)^2 - V(\sigma) = \text{constant} \quad . \tag{7.18}$$

There exists a simple mechanical analog: We may consider a point particle with σ as its "position" and x its "time", moving in a "potential" $-V(\sigma)$, as shown in Fig. 7.6. The above equation is then simply the energy conservation law in the analog problem. In this analog problem, let us set at "time" $x = -\infty$ the "position" of the particle σ at a. We may start the motion by pushing the

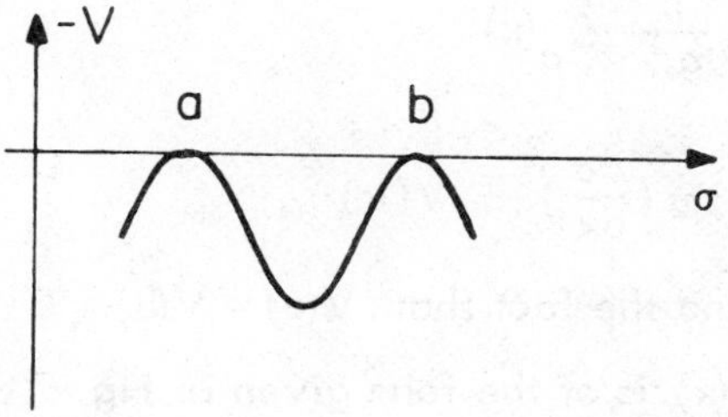

Fig. 7.6. The "potential" $-V$ in the mechanical analog problem.

particle very gently towards the right. As the "time" x increases, σ moves from a towards b ; as $x \to \infty$, $\sigma \to b$, because of energy conservation. Its analytic expression is

$$x = \int^{\sigma} \frac{d\sigma'}{\sqrt{2V(\sigma')}} \tag{7.19}$$

which contains an integration constant ξ, as shown in Fig. 7.7.

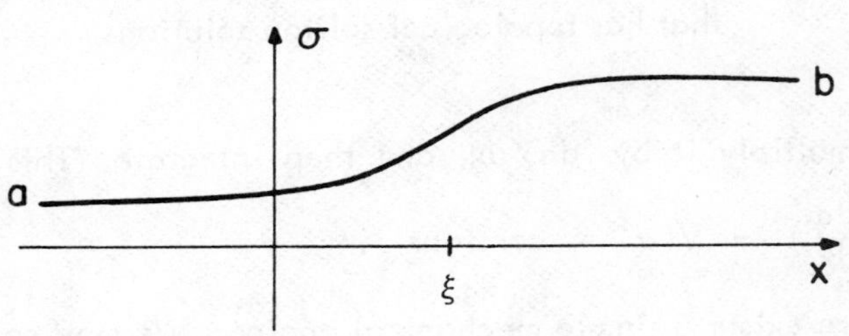

Fig. 7.7. A topological soliton solution.

Returning now to the original field-theory problem, because of (7.14)-(7.16), the energy density for a time-independent solution can be written as

$$\mathcal{E}(x) = \frac{1}{g^2}\, \mathcal{E}_\sigma(x)$$

where

$$\mathcal{E}_\sigma(x) = \tfrac{1}{2}\left(\frac{d\sigma}{dx}\right)^2 + V(\sigma) \quad . \tag{7.20}$$

From Fig. 7.7 and the fact that $V(a) = V(b) = 0$ we find that the energy density $\mathcal{E}(x)$ is of the form given in Fig. 7.8, which is confined in space at all times. Because the boundary conditions of the field $\phi = \sigma/g$ at $x = \pm\infty$ are different, it is called the topological

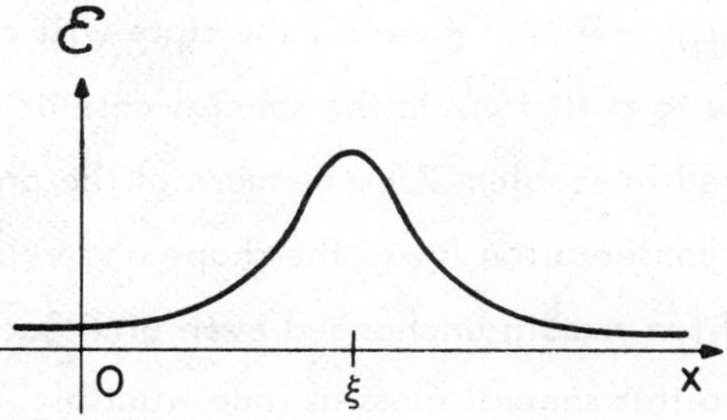

Fig. 7.8. A schematic drawing of energy density for the soliton given by Fig. 7.7.

soliton solution. Its stability is insured by the boundary conditions at infinity.

In the mechanical-analog problem, we may derive another solution by setting at "time" $x = -\infty$, the "position" σ at b. As x increases, σ moves from b to a. If we call the solution given by Fig. 7.7 the soliton, then this new solution is the anti-soliton. Both have the same energy. Thus, the concept of particle-antiparticle conjugation already exists on the classical level.

Because of Lorentz invariance, if $\sigma(x)$ is a solution of (7.17), then

$$\sigma(\gamma x - \gamma v t)$$

must also satisfy the same field equation, where $\gamma = (1 - v^2)^{-\frac{1}{2}}$. Consequently, we also have the solution for a moving soliton, or anti-soliton.

To describe the scattering between a soliton and an anti-soliton (or between two solitons or anti-solitons) we may consider the initial condition that at $t = -\infty$, one of them is, e.g., at $x = -\infty$

moving with velocity $v > 0$, while the other is at $x = +\infty$ moving with velocity $-v$. In general, the state will change in the course of time due to collision. In the special case of the sine-Gordon equation (defined in Problem 7.1), because of the presence of an infinite number of conservation laws, the shape and velocity of each soliton or anti-soliton remain unchanged even after such a head-on collision. We refer to this special class as indestructible solitons. Indestructible solitons exist only in one space-dimension. If one requires relativistic invariance, then it exists only for the sine-Gordon equation.

2. Nontopological soliton

To construct nontopological solitons, one does not need degenerate vacuum. However, as we shall see, because of the requirement of an additive conservation law, there must at least be a complex field.

Again, we shall first consider the case of one space-dimension. Let ϕ be a complex field. In accordance with our general form (7.2), the Lagrangian density is assumed to be

$$\mathcal{L} = -\frac{\partial \phi^\dagger}{\partial x_\mu}\frac{\partial \phi}{\partial x_\mu} - \frac{1}{g^2}\, U(g^2 \phi^\dagger \phi) \tag{7.21}$$

where $\dagger$ denotes the Hermitian conjugate. By using the variation principle (2.10), one finds the equation of motion to be

$$\frac{\partial^2 \phi}{\partial x_\mu^{\,2}} - \phi \frac{d}{d(g^2\phi^\dagger\phi)}\, U(g^2 \phi^\dagger \phi) = 0 \tag{7.22}$$

where $x_\mu = (x, it)$, as before. We shall assume $U(\phi^\dagger \phi)$ has a single minimum at $\phi = 0$. Furthermore, the minimum value of U is 0. Hence, just as in (7.3),

$$\frac{1}{g^2}\, U \rightarrow m^2 \phi^\dagger \phi \qquad \text{as} \qquad \phi \rightarrow 0 \tag{7.23}$$

where m is the mass of the usual plane wave solution. [See (7.35)-(7.36) below.]

The Lagrangian density (7.21) is invariant under the phase transformation

$$\phi \to e^{-i\theta} \phi \ . \tag{7.24}$$

Hence, as can be verified directly, the current

$$j_\mu = i\phi^\dagger \frac{\partial \mathcal{L}}{\partial\left(\frac{\partial \phi^\dagger}{\partial x_\mu}\right)} - i \frac{\partial \mathcal{L}}{\partial\left(\frac{\partial \phi}{\partial x_\mu}\right)} \phi \tag{7.25}$$

satisfies

$$\frac{\partial j_\mu}{\partial x_\mu} = 0 \ . \tag{7.26}$$

On account of (7.21), in our case the current j_μ is given by

$$j_\mu = i \frac{\partial \phi^\dagger}{\partial x_\mu} \phi - i\phi^\dagger \frac{\partial \phi}{\partial x_\mu} \ . \tag{7.27}$$

The particle density ρ is given by the time-component of j_μ multiplied by $-i$. From (7.27), one finds

$$\rho = i(\phi^\dagger \dot{\phi} - \dot{\phi}^\dagger \phi) \ . \tag{7.28}$$

Its space integral is the particle number N,

$$N = \int \rho \, dx \ . \tag{7.29}$$

Because of (7.26) N is conserved; i.e.

$$\dot{N} = 0 \ . \tag{7.30}$$

From (7.28)-(7.29) one sees that for $N \neq 0$, ϕ must vary with time. It is not difficult to show that the lowest-energy classical solution should be of the form

$$\phi = \frac{1}{g} \sigma(x) \, e^{-i\omega t} \tag{7.31}$$

where $\sigma(x)$ is real. In terms of σ, (7.22) becomes

$$\frac{d^2\sigma}{dx^2} + \omega^2 \sigma - \sigma \frac{d}{d\sigma^2} U(\sigma^2) = 0 \ , \tag{7.32}$$

which, after being multiplied by $d\sigma/dx$, can be integrated. The result is

$$\frac{1}{2} \left(\frac{d\sigma}{dx}\right)^2 - V(\sigma) = \text{constant} \tag{7.33}$$

where

$$V(\sigma) = \tfrac{1}{2} U(\sigma^2) - \tfrac{1}{2} \omega^2 \sigma^2 \ . \tag{7.34}$$

In Fig. 7.9, an example of the function U is plotted against σ ; as mentioned before, U has a single minimum at $\sigma = 0$.

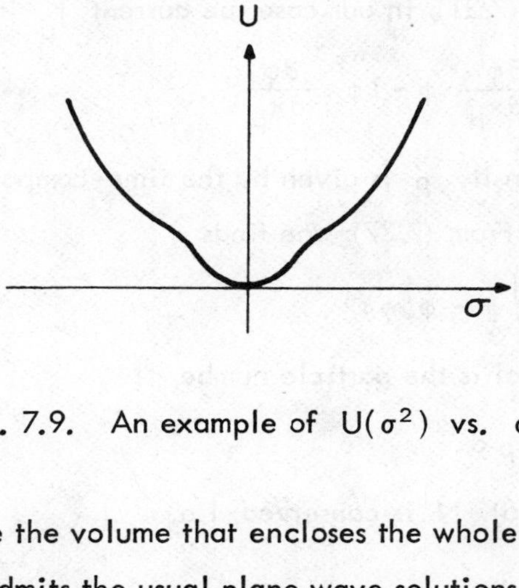

Fig. 7.9. An example of $U(\sigma^2)$ vs. σ .

Let Ω be the volume that encloses the whole system. When $\Omega \to \infty$, (7.22) admits the usual plane-wave solutions

$$\phi = \sqrt{\frac{N}{2\omega\Omega}} \ e^{i(kx - \omega t)} \tag{7.35}$$

where

$$\omega = \sqrt{\vec{k}^2 + m^2} \ . \tag{7.36}$$

This is because in this limit the amplitude ϕ becomes infinitesimal;

therefore, on account of (7.23), (7.22) reduces to

$$\frac{\partial^2 \phi}{\partial x_\mu^{\ 2}} - m^2 \phi = 0 \qquad (7.37)$$

and (7.35) is the solution. The soliton solution differs from the plane wave solution, since at finite x its amplitude does not become infinitesimal as $\Omega \to \infty$. Furthermore, when $x \to \pm\infty$, the soliton amplitude approaches zero exponentially; therefore $\omega^2 < m^2$. Hence we may regard these two types of solution as analytical continuations of each other:

$$\omega^2 > m^2 \quad \text{for the plane-wave solution,}$$

$$\text{and} \quad \omega^2 < m^2 \quad \text{for the soliton solution.} \qquad (7.38)$$

This relation is valid in any space-dimension.

We shall now show that in order to have the nontopological soliton solutions, the function $V = \frac{1}{2}(U - \omega^2 \sigma^2)$, defined by (7.34), must be of the form given by Fig. 7.10, at least when $\omega^2 = m_-^{\ 2}$. More specifically, the condition

$$U(\sigma^2) - \omega^2 \sigma^2 = 0 \qquad (7.39)$$

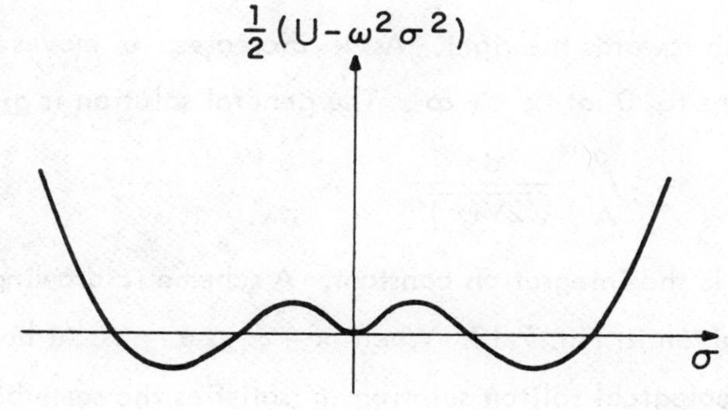

Fig. 7.10. A schematic drawing of $V = \frac{1}{2}(U - \omega^2 \sigma^2)$.

has, for $\omega^2 \leqq m^2$, besides the solution $\sigma = 0$ also some other $\sigma \neq 0$ solutions. Assuming that this is indeed the case, just as in the previous example of the topological soliton, we may consider the mechanical analog in which there is a point particle at a "position" σ and a "time" x, moving in a potential $-V = -\frac{1}{2}[U - \omega^2\sigma^2]$, shown in Fig. 7.11. At the "time" $x = -\infty$, we may set the particle at the

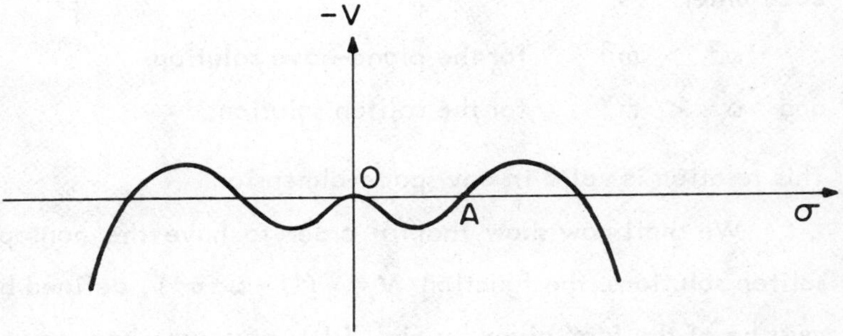

Fig. 7.11. The "potential" $-V$ in the mechanical analog problem.

"position" $\sigma = 0$. Again, we may start the motion by an extremely gentle push towards the right. As x increases, σ moves to A and then returns to 0 at $x = +\infty$. The general solution is given by

$$x - \xi = \int_A^\sigma \frac{d\sigma}{\sqrt{2V(\sigma)}} \qquad (7.40)$$

where ξ is the integration constant. A schematic drawing of the solution is given in Fig. 7.12. When $x = \xi$, $\sigma = A$. At both infinities, the nontopological soliton solution σ satisfies the same boundary condition:

$$\sigma \to 0 \quad \text{when} \quad x \to \infty \quad \text{or} \quad -\infty \ .$$

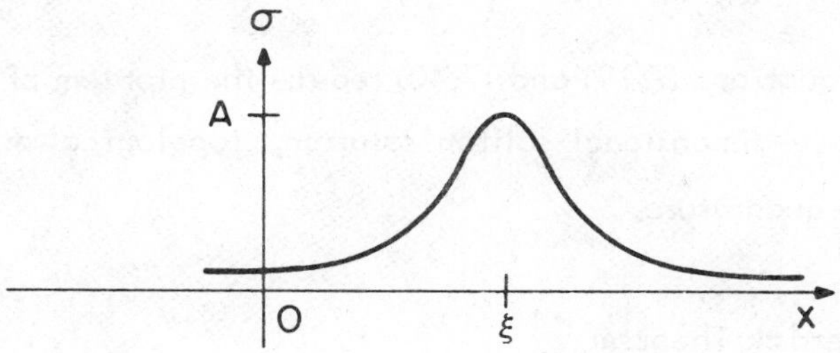

Fig. 7.12. A nontopological soliton solution.

We note that when $\sigma \to 0$, on account of (7.23), the function $V = \frac{1}{2}(U - \omega^2\sigma^2)$ becomes

$$V \to \tfrac{1}{2}(m^2 - \omega^2)\,\sigma^2 + O(\sigma^4) \quad . \tag{7.41}$$

Thus, in order that in Fig. 7.10 the curve $V(\sigma)$ should be concave upward at the origin $\sigma = 0$, we must have

$$\omega^2 < m^2 \quad ,$$

which confirms (7.38). Furthermore, condition (7.39) can be most easily satisfied if the σ^4-term is < 0, which corresponds to attraction between the fields.

As an explicit example, we may refer to Problem 7.2 in which

$$\frac{1}{g^2}\,U = \frac{m^2\phi^\dagger\phi}{1+\epsilon^2}\,[(1 - g^2\phi^\dagger\phi)^2 + \epsilon^2] \quad .$$

The solution is

$$\phi = \frac{1}{g}\left[\frac{a}{1+\sqrt{1-a}\,\cosh y}\right]^{\frac{1}{2}} e^{-i\omega t} \tag{7.42}$$

where

$$a = (1+\epsilon^2)\,\frac{1}{m^2}\,(m^2 - \omega^2)$$

and

$$y = 2\sqrt{m^2 - \omega^2} \quad (x - \xi) \quad .$$

Equations (7.19) and (7.40) reduce the problem of finding any one-space-dimensional soliton solution, topological or nontopological, to quadrature.

7.4 Derrick Theorem

A theorem due to G. H. Derrick* imposes severe restrictions on the types of soliton solutions that can exist when the space-dimension (excluding the time-dimension) is $D > 1$. Let us consider a classical system consisting only of scalar fields $\phi_1, \cdots, \phi_N$, whose Lagrangian density is

$$\mathcal{L} = \tfrac{1}{2} \sum_a [\, \dot{\phi}_a^2 - (\vec{\nabla} \phi_a)^2 \,] - U(\phi_a)$$

where $a = 1, 2, \cdots, N$, $\vec{\nabla}$ is the D-dimensional gradient vector and U is assumed to be $\geqslant 0$, with its minimum value given by

$$\min U(\phi_a) = 0 \quad . \tag{7.43}$$

Thus, the ground state (i.e., the vacuum) is of zero energy. There may, however, exist more than one such ground state.

<u>Theorem.</u> For $D \geqq 2$, the only time-independent solutions of finite energy are the ground states; i.e., $\phi_a =$ constant everywhere for which $U(\phi_a) = 0$.

<u>Proof.</u> From the Lagrangian density, it follows that the Hamiltonian density is

$$\mathcal{H} = \tfrac{1}{2} \sum_a [\, \Pi_a^2 + (\vec{\nabla} \phi_a)^2 \,] + U(\phi_a)$$

---

* G. H. Derrick, J.Math.Phys. 5, 1252 (1964).

where $\Pi_a = \dot{\phi}_a$ is the conjugate momentum of ϕ_a. Let $\phi_a(\vec{r})$ be a time-independent solution of the theory; hence $\phi_a(\vec{r})$ satisfies the field equation

$$\nabla^2 \phi_a - \frac{dU}{d\phi_a} = 0 \quad .$$

The corresponding total energy $\int \mathcal{H} d^3 r$ is

$$E(1) = T_1 + V_1$$

with

$$T_1 \equiv \tfrac{1}{2} \sum_a \int (\vec{\nabla} \phi_a(\vec{r}))^2 \, d^D r$$

and

$$V_1 \equiv \int U(\phi_a(\vec{r})) \, d^D r \quad .$$

We may construct a new function $\phi_a^{\lambda}(\vec{r})$, defined by

$$\phi_a^{\lambda}(\vec{r}) \equiv \phi_a(\lambda \vec{r}) \quad .$$

If everywhere the field distribution is set to be this new function $\phi_a^{\lambda}(\vec{r})$, since it is also time-independent the corresponding total energy is now given by

$$E(\lambda) = T_\lambda + V_\lambda$$

where

$$T_\lambda = \tfrac{1}{2} \sum_a \int (\vec{\nabla} \phi_a^{\lambda}(\vec{r}))^2 \, d^D r$$

and

$$V_\lambda \equiv \int U(\phi_a^{\lambda}(\vec{r})) \, d^D r \quad .$$

Because $\phi_a^{\lambda}(\vec{r})$ is obtained from $\phi_a(\vec{r})$ through the scale transformation

$$\vec{r} \rightarrow \lambda \vec{r} \quad ,$$

one sees that T_λ and V_λ are related to their values at $\lambda = 1$ by

$$T_\lambda = \lambda^{2-D} \, T_1$$

and

$$V_\lambda = \lambda^{-D} V_1 \quad .$$

Hence, the derivative of $E(\lambda)$ with respect to $\ln \lambda$ is

$$\lambda \frac{dE(\lambda)}{d\lambda} = (2 - D)\, T_\lambda - D\, V_\lambda \quad .$$

When $\lambda = 1$, $\phi_a^{\lambda}(\vec{r})$ becomes $\phi_a(\vec{r})$, which is a solution of the field equation, and therefore its total energy must be stationary against any small variations. Consequently $dE(\lambda)/d\lambda$ is zero at $\lambda = 1$, from which we obtain

$$(2 - D)\, T_1 - D\, V_1 = 0 \quad . \tag{7.44}$$

Now, for $D \geqq 2$, both $(2 - D)\, T_1$ and $- D\, V_1$ are $\leqq 0$. If D is > 2, (7.44) is clearly impossible unless $T_1 = V_1 = 0$. For $D = 2$, (7.44) implies $V_1 = 0$, which means that $U(\phi_a) = 0$ everywhere and $E(1)$ equals T_1. Suppose that $U(\phi_a) = 0$ when $\phi_a =$ either constant: c_a or c_a'. If $\phi_a = c_a$ in one region of space and $\phi_a = c_a'$ in a neighboring region, then $\vec{\nabla}\phi_a$ across the boundary of these two regions contains a δ function and the integral of $(\vec{\nabla}\phi_a)^2$ gives ∞. Therefore, the only finite-energy solution is $\phi_a =$ same constant everywhere. This then completes the proof.

Thus, when D is > 1, in order to have soliton solutions, we must either include fields of nonzero spin, or consider time-dependent but nondispersive solutions. The former leads to the gauge-field topological solitons, such as the aforementioned vortex solution of Nielsen and Olesen in two dimensions and the magnetic monopole solution of 't Hooft and Polyakov in three dimensions. The latter is represented by the multidimensional nontopological solitons whose properties will be discussed in the next section.

7.5 Solitons vs. Plane Waves

As already noted in (7.35), any nonlinear relativistic field equation always admits plane wave solutions of the form

$$\sqrt{\frac{N}{2\omega\Omega}}\; e^{i\vec{k}\cdot\vec{r} - i\omega t} \tag{7.45}$$

where $\Omega \to \infty$ is the volume of the system and $\omega = \sqrt{\vec{k}^2 + m^2}$.

For the topological soliton, its stability is insured against decay into plane waves because of the different boundary conditions satisfied by these two different types of solutions. For the nontopological soliton, its stability depends on which type of solution is of the lowest energy. In the following, this question will be analysed in some detail.

Let us consider a nonlinear theory in which, say, the particle number N is conserved; furthermore, we assume that the theory has nontopological soliton solutions. For the plane-wave solution (7.45), the energy is linear in N. Hence

$$E\,(\text{plane wave}) = N\omega \quad . \tag{7.46}$$

Since

$$\omega\,(\text{plane wave}) \geqq m \quad ,$$

we have

$$E\,(\text{plane wave}) \geqq Nm \quad . \tag{7.47}$$

For the soliton solutions, the energy E is a nonlinear function of N, and as noted in (7.38) the corresponding ω is $< m$.

1. <u>One space-dimension</u>

As we shall prove, in one space-dimension, if the nontopological solution exists, then its lowest energy E is always lower than Nm:

$$E\,(\text{soliton}) < Nm \quad , \tag{7.48}$$

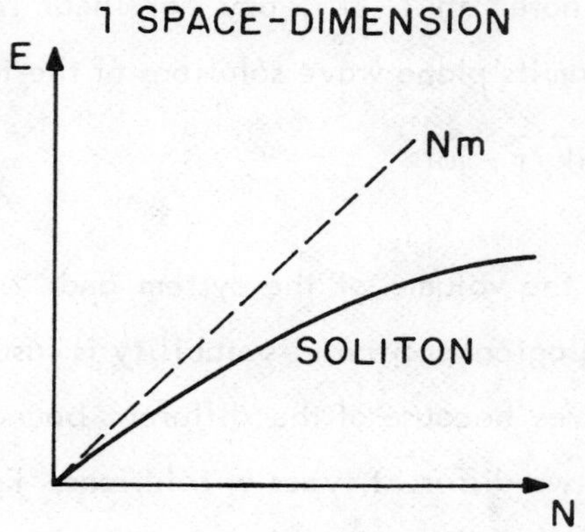

Fig. 7.13. At a given N, the energy of a one-space-dimensional soliton is always lower than that of a plane-wave solution.

which is illustrated in Fig. 7.13. To show (7.48) we first note that N and the phase angle $\theta \equiv \omega t$ of the complex field ϕ are conjugate variables. From Hamilton's equations we have

$$\dot{N} = -\frac{\partial H}{\partial \theta} \tag{7.49}$$

and

$$\dot{\theta} = \frac{\partial H}{\partial N} \;. \tag{7.50}$$

Due to the invariance under the phase transformation (7.24), the Hamiltonian H is independent of θ. Therefore (7.49) gives (7.30); i.e., N is conserved. Since $\omega = \dot{\theta}$ and the value of H is the energy E, (7.50) may be written as

$$\omega = \frac{dE}{dN} \;. \tag{7.51}$$

Now, according to (7.38)

$$\omega\,(\text{soliton}) < m \;, \tag{7.52}$$

(7.51) implies

$$\frac{d}{dN}\, E\,(\text{soliton}) = \omega < m \;. \tag{7.53}$$

Next, we shall show that as $\omega \to m-$ and at large $|x|$, the amplitude of the nontopological soliton is of order

$$|\phi| \sim \frac{\sqrt{m^2 - \omega^2}}{g}\, e^{-\sqrt{m^2-\omega^2}\,|x|} \;. \tag{7.54}$$

As can be seen from (7.42), this asymptotic behavior holds at least for the example cited. That this is true in general follows from the fact that when $|x|$ is large, $\phi = \frac{\sigma}{g}$ must be small. From (7.41), we have

$$V \to \tfrac{1}{2}(m^2 - \omega^2)\,\sigma^2 + O(\sigma^4) \;, \quad \text{as} \quad \sigma \to 0 \;.$$

In (7.54), the exponential factor arises from equating $-\frac{d^2\sigma}{dx^2}$ with $\frac{dV}{d\sigma}$; the multiplicative factor $\frac{\sqrt{m^2-\omega^2}}{g}$ comes from the approximate equipartition between the quadratic and the quartic terms in $V(\sigma)$; i.e.

$$(m^2 - \omega^2)\,\sigma^2 \sim \sigma^4 \;. \tag{7.55}$$

Thus, (7.54) is valid in any space-dimension. In one space-dimension, as $\omega \to m-$

$$N \to 2m \int |\phi|^2\, dx \sim \frac{1}{g^2}\sqrt{m^2 - \omega^2} \to 0 \;, \tag{7.56}$$

since $|\phi|^2$ carries a factor $\frac{(m^2-\omega^2)}{g^2}$ and the x-integration gives another factor $\frac{1}{\sqrt{m^2-\omega^2}}$. In Figure 7.13 the dashed line represents the lowest-energy state $\omega = m$ of the plane-wave solution. Since the nontopological soliton solution is the analytic continuation of the plane-wave solution to the region $\omega < m$, the curve E (soliton) vs. N should be connected to the straight line E (plane wave) = Nm, when $\omega \to m-$. From (7.56), this connection occurs at $N = 0$. Hence (7.53) leads to, for the one-space-dimensional nontopological soliton,

$$E(\text{soliton}) = \int_0^N \omega \, dN < m \int_0^N dN = mN \,, \qquad (7.57)$$

which establishes (7.48) for any N and g.

2. Two space-dimensions

Generalization of the nontopological soliton solutions to space-dimensions > 1 has been given in the literature. A detailed discussion lies outside the scope of this book. As in the one-space-dimensional case, all one needs is some nonlinear interaction which gives rise to attraction between the fields (or field quanta); this would be the case if the interaction is mediated by a scalar field. By using variational considerations, it is not difficult to show that

$$E(\text{soliton}) < Nm \,, \qquad \text{as} \quad N \to \infty \,. \qquad (7.58)$$

In either the two- or three-space-dimensional cases, by following the same argument which led to (7.54) we see that, as $\omega \to m-$ and at large radial distance r, the amplitude of the nontopological soliton is of order

$$|\phi| \sim \frac{\sqrt{m^2 - \omega^2}}{g} \, e^{-\sqrt{m^2-\omega^2}\, r} \,. \qquad (7.59)$$

Thus, in two space-dimensions, as $\omega \to m-$

$$N \to 2m \int |\phi|^2 d^2 r = N_c \sim O(1) \,. \qquad (7.60)$$

This is because the factor $(m^2-\omega^2)$ in $|\phi|^2$ is exactly cancelled by a corresponding factor $(m^2-\omega^2)^{-1}$ in the d^2r integration. The result is given in Fig. 7.14. There exists a critical number N_c. The nontopological soliton solution exists only for $N > N_c$; in that region, the lowest-energy solution is always the soliton, not the plane wave.

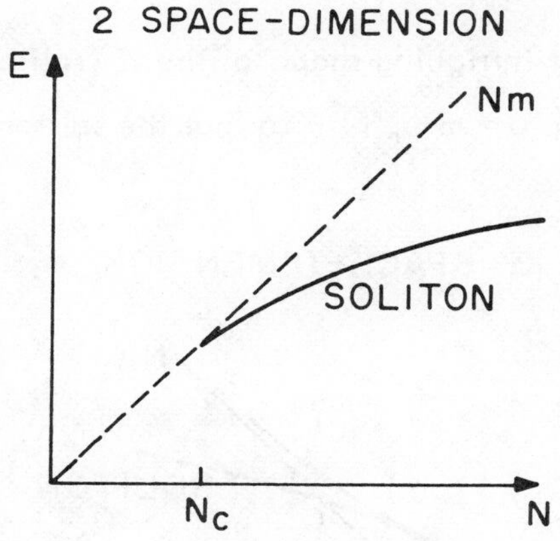

Fig. 7.14. The solid curve is E (soliton) vs. N for a two-space-dimensional nontopological soliton.

3. Three space-dimensions

As $\omega \to m-$ and when r is large, the asymptotic behavior of a nontopological soliton is still given by (7.59); the corresponding value of N is

$$N \to 2m \int |\phi|^2 d^3r \ .$$

Because of the exponential factor in (7.59), the d^3r integration gives a factor $\sim (m^2 - \omega^2)^{-3/2}$. Hence, as $\omega \to m-$,

$$N \sim O\left(\frac{1}{\sqrt{m^2 - \omega^2}}\right) \to \infty \ . \qquad (7.61)$$

It can also be shown* that when ω is sufficiently small, N has to

* R. Friedberg, T. D. Lee and A. Sirlin, Phys.Rev. D13, 2739 (1976), Nucl.Phys. B115, 1, 32 (1976).

increase with decreasing ω, and in that way we can approach the limit given by (7.58). This, together with

$$\frac{d}{dN} \; E\,(\text{soliton}) \; = \; \omega \; > \; 0 \tag{7.62}$$

produces the rather intriguing shape of the $E\,(\text{soliton})$ curve shown in Fig. 7.15. When $\omega = m-$, $N \to \infty$ but the soliton energy is $Nm+$.

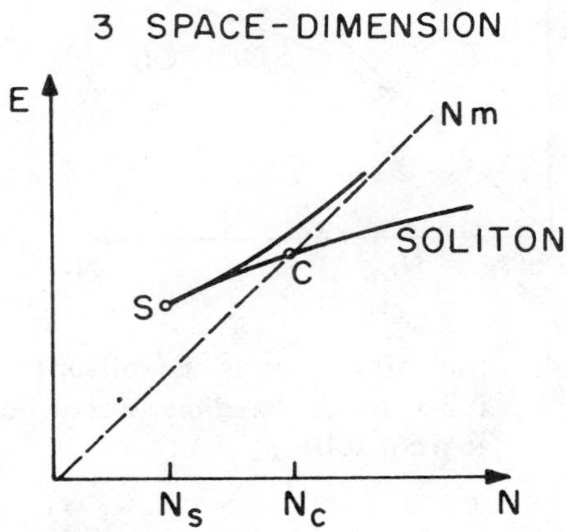

Fig. 7.15. The solid curve is $E\,(\text{soliton})$ vs. N for a three-space-dimensional non-topological soliton.

As ω decreases from $m-$, N also decreases until it reaches N_s; throughout this interval the soliton energy is $> Nm$. As ω decreases further, N then starts to increase; it crosses the line $E = Nm$ at $N = N_c$. If we decrease ω still further, then N keeps on increasing with $E\,(\text{soliton})$ always $< Nm$, until $N \to \infty$.

There now exists besides a critical point C also a spike S, with $N_c > N_s$. For $N < N_s$, there is no soliton solution. For $N_s < N < N_c$, the lowest-energy solution is the plane wave; for

$N > N_c$, the lowest-energy solution is always the soliton. In Fig. 7.15, along the lower branch of the soliton curve, for $N > N_c$, the soliton solution is absolutely stable, being the lowest-energy solution; for $N_c > N > N_s$, it can be shown that the soliton solution is stable against infinitesimal perturbations, even though it is not of the lowest energy. Along the upper branch, the soliton solution is always unstable. The numerical values of N_c and N_s depend on the parameters in the theory. It is not difficult to give examples in which N_c, and therefore also N_s, are < 1.

In a quantum theory N takes on only integer values $0, 1, 2, \cdots$. Thus, if $N_c < 1$, and if the classical results are good approximations of the quantum solutions, then the lowest-energy state of the system abruptly changes its character from the vacuum state $(N = 0)$ to any $N \neq 0$ state. It is important to note that, when N is small, the classical soliton description of an N-body bound state is quite different from the usual description in terms of solutions of the Bethe-Salpeter equation (say, under the ladder approximation). For example, in the soliton description there is no sharp difference between the $N = 1$ and the $N = 2$ states; both are "blob-like". In the Bethe-Salpeter description, the $N = 1$ state is "point-like" while the $N = 2$ state is a "blob" formed by the wave function of the two "point-like" particles.

Remarks. In a nonlinear system, it is not difficult to conceive of situations in which the ground state can change its character quite suddenly with only a small variation in its parameters. The above three-space-dimensional case is one such example; when N varies from $< N_c$ to $> N_c$, the lowest-energy solution changes from plane-wave to soliton. Similar examples can also be found in simple mechanical problems.

As one such example, let us consider a single point particle moving in a central potential. The position vector of the particle is $\vec{r}$ and its conjugate momentum is $\vec{p}$. The Hamiltonian is assumed to be

$$\tfrac{1}{2}\vec{p}^{\,2} + \tfrac{1}{2}r^2[(1-gr)^2+\Delta^2] \tag{7.63}$$

where $r = |\vec{r}|$, and g and Δ are real parameters. The angular momentum $\vec{\ell} \equiv \vec{r} \times \vec{p}$ is conserved. At a fixed value of $\ell = |\vec{\ell}|$, (7.63) becomes

$$\tfrac{1}{2}p_r^{\,2} + V_\ell(r)$$

where p_r is the radial momentum and

$$V_\ell(r) = \tfrac{1}{2}\left(\frac{\ell}{r}\right)^2 + \tfrac{1}{2}r^2[(1-gr)^2+\Delta^2] \quad .$$

Here ℓ plays the same role as the conserved quantity N in our previous discussions. For ℓ and Δ both not too large and $g > 0$, $V_\ell(r)$ has two local minima, say at $r = r_1$ and r_2 with $r_1 < r_2$. As shown in Fig. 7.16(a), when $\ell = 0$ one sees that $r_1 = 0$. Hence,

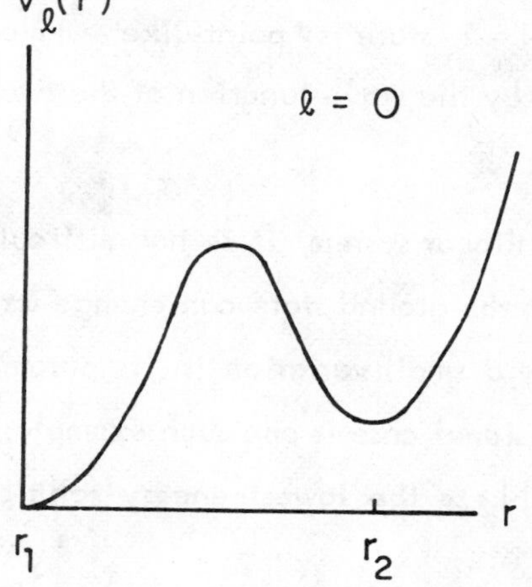

Fig. 7.16(a)

r_1 denotes the absolute minimum when ℓ is small. It is easy to show that there exists a critical value ℓ_c. For $\ell > \ell_c$, the absolute minimum of $V_\ell(r)$ changes from r_1 to r_2, as shown in Fig. 7.16(b). Now, in a quantum theory, ℓ takes on only integer values.

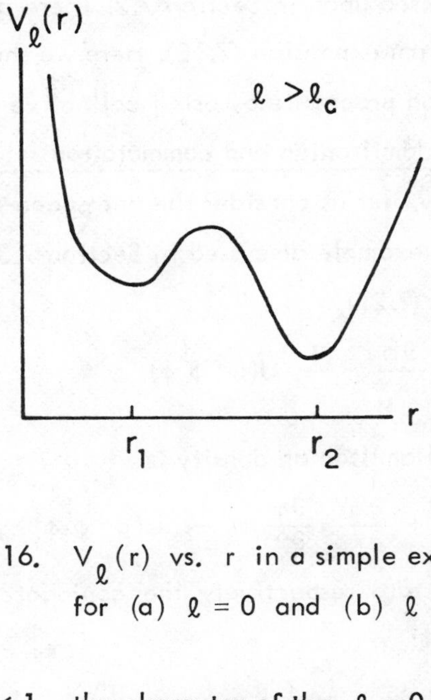

Fig. 7.16(b)

Fig. 7.16. $V_\ell(r)$ vs. r in a simple example for (a) $\ell = 0$ and (b) $\ell > \ell_c$.

Thus, if ℓ_c is < 1, the character of the $\ell = 0$ state can be drastically different from all $\ell \neq 0$ states. This is quite analogous to our field-theoretical problem, in which depending on the parameters, the vacuum state $(N = 0)$ may also be significantly different from all $N \neq 0$ states, provided that N_c is < 1.

Notice that in this example, if Δ is sufficiently small, one can always arrange to have $\ell_c < 1$, at any given g. A small g only means that the nonlinear solution is very far from the origin.

If N_c is small, then the soliton description of, say, the two-body bound state is really quite different from the usual Bethe-Salpeter description (with, e.g., the ladder approximation).

7.6 Quantization

We now return to the important question of quantization, which was briefly commented upon in Section 7.2. There are many ways* to carry out the quantum expansion (7.12). Here we shall follow the canonical quantization procedure by using collective coordinates.**

1. Lagrangian, Hamiltonian and commutation relations

For simplicity, let us consider the one-space-dimensional non-topological soliton example discussed in Section 7.3. The Lagrangian density is given by (7.21)

$$\mathcal{L} = -\frac{\partial\phi^\dagger}{\partial x_\mu}\frac{\partial\phi}{\partial x_\mu} - \frac{1}{g^2}U(g^2\phi^\dagger\phi) .$$

The corresponding Hamiltonian density is

$$\mathcal{H} = \Pi^\dagger\Pi + \frac{\partial\phi^\dagger}{\partial x}\frac{\partial\phi}{\partial x} + \frac{1}{g^2}U(g^2\phi^\dagger\phi) , \qquad (7.64)$$

where Π and $\Pi^\dagger$ are respectively the conjugate momenta of ϕ and $\phi^\dagger$, given by

$$\Pi = \frac{\partial\mathcal{L}}{\partial\dot{\phi}} = \dot{\phi}^\dagger \quad \text{and} \quad \Pi^\dagger = \frac{\partial\mathcal{L}}{\partial\dot{\phi}^\dagger} = \dot{\phi} . \qquad (7.65)$$

The classical soliton solution can be written as

$$\phi_{cl} = \frac{1}{g}\sigma(x-\xi)e^{-i\theta} \qquad (7.66)$$

where σ is given by (7.40), ξ is the integration constant and $\theta = \omega t$

---

* See the proceedings of the Conference on Extended Systems in Field Theory, Physics Reports 23C (1976) for detailed references.

** N. H. Christ and T. D. Lee, Phys. Rev. D12, 1606 (1975).

+ constant. The classical solution is degenerate under a constant variation in either ξ or θ ; i.e., ϕ_{cl} remains a solution with the same energy under the variation

$$\phi_{cl} \to \phi_{cl} + \delta\phi \tag{7.67}$$

where

$$\delta\phi = -\frac{1}{g} e^{-i\theta} \left[\frac{d\sigma}{dx} \delta\xi + i\sigma\,\delta\theta \right] . \tag{7.68}$$

The variations $\xi \to \xi + \delta\xi$ represent a space translation and $\theta \to \theta + \delta\theta$ corresponds to a phase change in ϕ ; their conjugate momenta are respectively the total momentum P and the particle number N of the system. The invariance under (7.67) - (7.68) implies that both P and N are conserved. Classically, ξ and θ commute with P and N ; therefore, when the system has a definite momentum P and a definite particle number N, both ξ and θ can vary arbitrarily, resulting in the aforementioned degeneracy of the classical solution. As we shall see, quantum mechanically this degeneracy is lifted by treating ξ and θ as collective coordinates; their conjugate momenta become the derivative operators

$$P = \frac{1}{i} \frac{\partial}{\partial\xi} \qquad \text{and} \qquad N = \frac{1}{i} \frac{\partial}{\partial\theta} . \tag{7.69}$$

We shall first follow the standard procedure given in Chapter 2 for carrying out the quantization. The commutation relations are given by

$$[\Pi(x,t),\ \phi(x',t)] = [\Pi^\dagger(x,t),\ \phi^\dagger(x',t)] = -i\,\delta(x-x') \tag{7.70}$$

and all other equal-time commutators between ϕ, $\phi^\dagger$, Π and $\Pi^\dagger$ are 0 . The dynamics of the system is then determined by requiring Heisenberg's equation

$$[H, O(t)] = -i\,\dot{O}(t)$$

to hold for any operator $O(t)$.

As in (2.48), we may decompose the complex field into its Hermitian components:

$$\phi = \frac{1}{\sqrt{2}} (\phi_1 + i\phi_2) \quad \text{and} \quad \phi^\dagger = \frac{1}{\sqrt{2}} (\phi_1 - i\phi_2) \qquad (7.71)$$

where ϕ_1 and ϕ_2 are Hermitian fields. Their Fourier series can be written as

$$\phi_\alpha(x, t) = \sum_k \sqrt{\frac{2}{L}} \, [x_{\alpha,k}(t) \cos kx + y_{\alpha,k}(t) \sin kx] \qquad (7.72)$$

where $\alpha = 1$ or 2, L is the volume (i.e., length) of the system and k is given by

$$k = \frac{2n\pi}{L} \qquad \text{with} \qquad n = 0, 1, 2, \cdots . \qquad (7.73)$$

The Hermiticity of ϕ_α implies that $x_{\alpha,k}$ and $y_{\alpha,k}$ are also Hermitian; i.e.,

$$x_{\alpha,k} = x^\dagger_{\alpha,k} \quad \text{and} \quad y_{\alpha,k} = y^\dagger_{\alpha,k} . \qquad (7.74)$$

The conjugate momenta Π and $\Pi^\dagger$ can be decomposed in a similar manner:

$$\Pi = \frac{1}{\sqrt{2}} (\Pi_1 - i\Pi_2) \quad \text{and} \quad \Pi^\dagger = \frac{1}{\sqrt{2}} (\Pi_1 + i\Pi_2) . \qquad (7.75)$$

The Fourier components of Π_1 and Π_2 can be written in terms of the differential operators of $x_{\alpha,k}$ and $y_{\alpha,k}$. We have

$$\Pi_\alpha = -i \sum_k \sqrt{\frac{2}{L}} \, [\cos kx \frac{\partial}{\partial x_{\alpha,k}} + \sin kx \frac{\partial}{\partial y_{\alpha,k}}] . \qquad (7.76)$$

Because of (7.72) and (7.76) we have

$$[\Pi_\alpha(x, t), \phi_\beta(x', t)] = -i\delta(x - x') \, \delta_{\alpha\beta}$$

and (7.77)

$$[\phi_\alpha(x,t),\ \phi_\beta(x',t)] = [\Pi_\alpha(x,t),\ \Pi_\beta(x',t)] = 0$$

where α and β can be 1 or 2, which in turn lead to the original commutation relation (7.70).

In terms of these Fourier components, because

$$\int \dot{\phi}^\dagger \dot{\phi}\, dx = \sum_{\alpha,k} \tfrac{1}{2}(\dot{x}_{\alpha,k}^{\,2} + \dot{y}_{\alpha,k}^{\,2}) \ , \tag{7.78}$$

the Lagrangian density can be written as

$$\int \mathcal{L}\, dx = \sum_{\alpha,k} \tfrac{1}{2}(\dot{x}_{\alpha,k}^{\,2} + \dot{y}_{\alpha,k}^{\,2}) - V(x_{\alpha,k},\ y_{\alpha,k}) \tag{7.79}$$

where

$$V = \int \left[\frac{\partial \phi^\dagger}{dx} \frac{\partial \phi}{\partial x} + \frac{1}{g^2} U(g^2 \phi^\dagger \phi) \right] dx \ . \tag{7.80}$$

Correspondingly, the Hamiltonian $H = \int \mathcal{H}\, dx$ is given by

$$H = -\tfrac{1}{2} \sum_{\alpha,k} \left(\frac{\partial^2}{\partial x_{\alpha,k}^{\,2}} + \frac{\partial^2}{\partial y_{\alpha,k}^{\,2}} \right) + V(x_{\alpha,k},\ y_{\alpha,k}) \ . \tag{7.81}$$

While the Fourier series is a convenient expansion for analysing the plane wave, it is particularly ill-adapted for the soliton solution. We will therefore make a change of variables from the Fourier components to a new set.

2. Collective coordinates

The new set consists of two collective coordinates $\xi(t)$ and $\theta(t)$, introduced before, and a number of vibrational coordinates $q_{\pm,n}(t)$ with $n = 2, 3, \cdots$. If one wishes, one may label $q_{+,1}(t) = \xi(t)$ and $q_{-,1}(t) = \theta(t)$ so that the new set is $\{q_{\pm,n}(t)\}$. Let us choose two independent sets of complete, real and orthonormal c. number functions $\{\psi_{+,n}(x)\}$ and $\{\psi_{-,n}(x)\}$ with the conditions

$$\psi_{+,1}(x) = \text{constant times } \frac{d\sigma}{dx} \tag{7.82}$$

and

$$\psi_{-,1}(x) = \text{constant times } \sigma \quad . \tag{7.83}$$

Hence for $n = 2, 3, \cdots$, we have

$$\int \frac{d\sigma}{dx} \psi_{+,n} dx = \int \sigma \psi_{-,n} dx = 0 \quad . \tag{7.84}$$

Except for the above conditions, these two sets of c. number functions can be arbitrarily chosen. We now expand the quantum field operator $\phi(x, t)$ as follows:

$$\phi(x, t) = \{ \frac{1}{g} \sigma(x-\xi) + \sum_{n=2}^{\infty} \frac{1}{\sqrt{2}} [q_{+,n}(t) \psi_{+,n}(x-\xi) + i q_{-,n} \psi_{-,n}(x-\xi)] \} e^{-i\theta} . \tag{7.85}$$

These new coordinates $\xi(t)$, $\theta(t)$ and $q_{\pm,n}(t)$ are all Hermitian.

To understand the two conditions given in (7.84), let us expand ϕ around the soliton solution $\frac{1}{g}\sigma(x-\xi) e^{-i\theta}$ and regard $\delta\xi$, $\delta\theta$ and $q_{\pm,n}$ with $n \geqslant 2$ all as small. By using (7.68) and (7.85) we see that

$$\delta\phi = [-\frac{1}{g}(\frac{d\sigma}{dx} \delta\xi + i\sigma\,\delta\theta) + \sum_{n=2}^{\infty} 2^{-\frac{1}{2}}(q_{+,n}\psi_{+,n} + i q_{-,n}\psi_{-,n})] e^{-i\theta} . \tag{7.86}$$

Equation (7.84) insures that motions due to the variations in the collective variables ξ and θ will not be mistaken for the vibrational modes.

By equating (7.85) with the Fourier expansion (7.71) - (7.72), we obtain the coordinate transformation from

$$x_{\alpha,k}, y_{\alpha,k} \rightarrow q = \begin{pmatrix} \xi \\ \theta \\ q_{+,2} \\ q_{-,2} \\ \vdots \end{pmatrix} . \tag{7.87}$$

This enables us to express the differential operators $\frac{\partial}{\partial x_{a,k}}$ and $\frac{\partial}{\partial y_{a,k}}$ as linear functions of $\frac{\partial}{\partial \xi}$, $\frac{\partial}{\partial \theta}$, $\frac{\partial}{\partial q_{+,2}}$, $\frac{\partial}{\partial q_{-,2}}$, $\cdots$. By substituting these linear relations into (7.81), we can obtain the Hamiltonian H in terms of these new variables. The easiest way to derive the explicit form is to first consider the classical problem. The time derivative of (7.85) gives $\dot{\phi}$ as a linear function of $\dot{\xi}$, $\dot{\theta}$, $\dot{q}_{\pm,2}$, $\cdots$; i.e.

$$\dot{\phi} = \sum_{\lambda,n} \frac{\partial \phi}{\partial q_{\lambda,n}} \dot{q}_{\lambda,n} \tag{7.88}$$

where

$$n = 1, 2, 3, \cdots, \qquad \lambda = +, - \quad ,$$

and for notational convenience, we have set

$$q_{+,1} = \xi \quad \text{and} \quad q_{-,1} = \theta \quad . \tag{7.89}$$

Thus, (7.78) can also be written as

$$\int \dot{\phi}^\dagger \dot{\phi}\, dx = \tfrac{1}{2} \tilde{\dot{q}}\, \mathcal{M} \dot{q} \tag{7.90}$$

where $\dot{q}$ is the time derivative of the column matrix in (7.87), and

$$\mathcal{M} = \mathcal{M}(q) \tag{7.91}$$

is a real, symmetric $\infty \times \infty$ matrix whose matrix elements are

$$\mathcal{M}_{\lambda,n;\lambda',n'}(q) = \int \left(\frac{\partial \phi^\dagger}{\partial q_{\lambda,n}} \frac{\partial \phi}{\partial q_{\lambda',n'}} + \frac{\partial \phi^\dagger}{\partial q_{\lambda',n'}} \frac{\partial \phi}{\partial q_{\lambda,n}} \right) dx \quad . \tag{7.92}$$

By differentiating (7.90) with respect to $\dot{q}_{\lambda,n}$, we see that the conjugate momentum $p_{\lambda,n}$ of the generalized coordinate $q_{\lambda,n}$ is the $(\lambda, n)^{th}$ component of the column matrix

$$p = \mathcal{M} \dot{q} \quad . \tag{7.93}$$

Classically, the Hamiltonian is

$$H_{cl} = \tfrac{1}{2} \tilde{p}_{cl}\, \mathcal{M}^{-1} p_{cl} + V(q) \tag{7.94}$$

where $\mathcal{M}^{-1}$ is the inverse of $\mathcal{M}$, and the function V(q) can be obtained by substituting (7.85) into (7.80). Quantum mechanically, because the transformation (7.87) is a point-transformation, the generalized Laplace operator in (7.81) satisfies (see Problem 7.3)

$$\sum_{\alpha,k} \left(\frac{\partial^2}{\partial x_{\alpha,k}^2} + \frac{\partial^2}{\partial y_{\alpha,k}^2} \right) = \frac{1}{\mathcal{J}} \tilde{p}\, \mathcal{M}^{-1} \mathcal{J} p \tag{7.95}$$

where the matrix element of the column matrix p is now the differential operator

$$p_{\lambda,n} = -i \frac{\partial}{\partial q_{\lambda,n}} , \tag{7.96}$$

$\mathcal{M}^{-1}$ remains the same inverse matrix in (7.94) and

$$\mathcal{J} = \sqrt{\det \mathcal{M}(q)} . \tag{7.97}$$

On account of (7.69) and (7.89), the first two components of p are the total momentum P and the particle number N of the system; i.e.

$$p_{+,1} = P = -i \frac{\partial}{\partial \xi} \quad \text{and} \quad p_{-,1} = N = -i \frac{\partial}{\partial \theta} . \tag{7.98}$$

The quantum Hamiltonian (7.81) is equal to the Hermitian operator

$$H = \frac{1}{2\mathcal{J}} \tilde{p}\, \mathcal{M}^{-1} \mathcal{J} p + V(q) . \tag{7.99}$$

Because the classical soliton solution (7.66) is invariant under the variation (7.67)-(7.68) in the two collective coordinates ξ and θ, the functions V(q), $\mathcal{J}$(q) and $\mathcal{M}$(q) all do not depend on ξ and θ. Thus, P and N both commute with H :

$$[P, H] = [N, H] = 0 , \tag{7.100}$$

which means that they are both constants of motion. The eigenvalue of P is continuous, and is the total momentum of the system. Since θ is a cyclic variable, N has only discrete integer eigenvalues

$\cdots, -2, -1, 0, 1, 2, \cdots$, positive for particles and negative for antiparticles. The energy of the system is determined, as usual, by the Schrödinger equation

$$H \,|\,\rangle = E\,|\,\rangle \quad . \tag{7.101}$$

3. Perturbation expansion

To show how the eigenstate of H can be solved by the perturbation series in g, we assume that the vibrational coordinates $q_{\pm,2}$, $q_{\pm,3}$, $\cdots$ and their time derivatives are all $O(1)$. Thus, (7.85) gives

$$\dot{\phi} = -\frac{1}{g}\left(\frac{d\sigma}{dx}\,\dot{\xi} + i\,\sigma\dot{\theta}\right) e^{-i\theta} + O(1) \quad . \tag{7.102}$$

Hence (7.90) is $O(g^{-2})$, and it is given by

$$\int \dot{\phi}^{\dagger}\dot{\phi}\,dx = \frac{M}{2}\,\dot{\xi}^2 + \frac{I}{2}\,\dot{\theta}^2 + O(g^{-1}) \tag{7.103}$$

where

$$M = \frac{2}{g^2}\int \left(\frac{d\sigma}{dx}\right)^2 dx \quad \text{and} \quad I = \frac{2}{g^2}\int \sigma^2\,dx \quad . \tag{7.104}$$

By substituting (7.85) into (7.80), we find $V(q)$ is also $O(g^{-2})$:

$$V(q) = \frac{1}{g^2}\int \left[\left(\frac{d\sigma}{dx}\right)^2 + U(\sigma^2)\right] dx + O(g^{-1}) \quad . \tag{7.105}$$

To simplify our analysis, let us assume the soliton is at rest; i.e.,

$$P\,|\,\rangle = 0 \quad . \tag{7.106}$$

By following the steps from (7.90)-(7.99) and by using (7.105)-(7.106), we see that

$$H = \frac{N^2}{2I} + \frac{1}{g^2}\int \left[\left(\frac{d\sigma}{dx}\right)^2 + U(\sigma^2)\right] dx + O(g^{-1}) . \tag{7.107}$$

Let us define ω by setting the eigenvalue of N to be

$$N = I\omega \quad . \tag{7.108}$$

Because N is also the derivative of (7.103) with respect to $\dot{\theta}$, we see that $N = I\dot{\theta}$ and therefore ω is the same parameter that characterizes the classical solution σ. The only restriction in the quantum solution is that now N must be an integer. Consequently, on account of (7.104) and (7.108), (7.107) can be written as

$$H = \frac{1}{g^2} \int \left[\left(\frac{d\sigma}{dx}\right)^2 + U(\sigma^2) + \omega^2 \sigma^2 \right] dx + O(g^{-1}) \ . \tag{7.109}$$

By using the classical soliton solution (7.31), we find that its energy is of exactly the same form

$$E_{cl} = \frac{1}{g^2} \int \left[\left(\frac{d\sigma}{dx}\right)^2 + U(\sigma^2) + \omega^2 \sigma^2 \right] dx \ ; \tag{7.110}$$

therefore, the quantum solution becomes

$$H \,|\,\rangle = [\, E_{cl} + O(1) \,] \,|\,\rangle \ . \tag{7.111}$$

In (7.109), there seems to be an $O(g^{-1})$ correction term. However, because σ is the solution of the classical equation, it is not difficult to show that the $O(g^{-1})$ term is in fact zero, which leads to (7.111).

Thus, when g is small the existence of a classical soliton solution insures that of a quantum solution; furthermore, when $g \to 0$, the quantum soliton mass is given by the same classical curve $E_{cl}(N)$, except that N must be integers. As noted in (7.12) - (7.13), these properties have a general validity not restricted to the particular type of one-space-dimensional examples discussed in this section.

Remarks. In this introduction to field theory we have covered the basic quantization procedures and the general method of evaluating the S-matrix. The perturbation series around the plane-wave solutions will be useful to describe leptons and photons which have no strong interactions; as we shall see, there exist compelling reasons to regard all known hadrons as bound states of quarks. The quarks seem to be permanently confined in space and that is why free quarks are **not** seen. At distances ~ the hadron radius, the dynamics of these quark composites (i.e., hadrons) can be best described in terms of the soliton solution. However, at very small distances inside the hadron, because of the asymptotic freedom property of quantum chromodynamics, the usual perturbation expansion around plane-wave solutions again becomes approximately valid. These topics will be analysed later on.

Our discussions of particle physics in the subsequent chapters will consist of two main parts: symmetry and interactions. In symmetry, the analysis will be entirely phenomenological, and yet rigorous conclusions can be derived without any detailed knowledge of the dynamics. In the chapters dealing with interactions, specific assumptions will be made about the Hamiltonian, which enable us to make more detailed calculations.

We shall now go on to particle physics.

Problem 7.1. Let the Lagrangian density of a one-space-dimensional Hermitian field σ be

$$\mathcal{L} = -\tfrac{1}{2}\left(\frac{\partial\sigma}{\partial x_\mu}\right)^2 - V(\sigma)$$

where $x_\mu = (x, it)$. Show that the classical soliton solution for

(i) $V(\sigma) = \frac{1}{8}(1-\sigma^2)^2$ is $\sigma = \tanh[\frac{1}{2}(x-\xi)]$

and (ii) $V(\sigma) = 1 - \cos\sigma$ is $\sigma = 4\tan^{-1} e^{x-\xi}$, with ξ as the integration constant.

The field equation of (ii) is called the sine-Gordon equation.

Problem 7.2. Let the Lagrangian density of a one-space-dimensional complex field σ be

$$\mathcal{L} = -\frac{\partial\sigma^\dagger}{\partial x_\mu}\frac{\partial\sigma}{\partial x_\mu} - U(\sigma^\dagger\sigma) \quad .$$

Show that the classical soliton solution for

$$U = \frac{\sigma^\dagger \sigma}{1+\epsilon^2}\left[(1-\sigma^\dagger\sigma)^2+\epsilon^2\right]$$

is

$$\sigma = \left(\frac{a}{1+\sqrt{1-a}\,\cosh y}\right)^{\frac{1}{2}} e^{-i\omega t} \qquad (7.112)$$

where $a = (1+\epsilon^2)(1-\omega^2)$ and $y = 2\sqrt{1-\omega^2}\,(x-\xi)$.

In both problems, for simplicity we set the scale so that $g = 1$ and $m = 1$.

Problem 7.3. In the transformation $x_i \rightarrow q_j = q_j(x_i)$ where the subscripts can vary from 1 to n, we define M to be the $n \times n$ matrix $M = (M_{ij}) \equiv \left(\frac{\partial x_k}{\partial q_i}\frac{\partial x_k}{\partial q_j}\right)$, $M^{-1} = (M_{ij}^{-1})$ its inverse and $|M| = \mathcal{g}^2$ its determinant. Show that

$$\text{(i)}\quad \frac{1}{|M|}\frac{\partial |M|}{\partial q_k} = M_{ji}^{-1}\frac{\partial M_{ij}}{\partial q_k} = -2\,\frac{\partial x_i}{\partial q_k}\frac{\partial}{\partial q_j}\left(\frac{\partial q_j}{\partial x_i}\right)$$

and

$$\text{(ii)}\quad \frac{\partial^2}{\partial x_i\,\partial x_i} = \frac{1}{\mathcal{g}}\frac{\partial}{\partial q_i}\left(M_{ij}^{-1}\,\mathcal{g}\,\frac{\partial}{\partial q_j}\right) . \qquad (7.113)$$

References

For quantization and especially topological solitons, there are some excellent review articles:

S. Coleman, in New Phenomena in Subnuclear Physics, Part A, ed. A. Zichichi (New York and London, Plenum Press, 1977), 297.

L. D. Faddeev and V. E. Korepin, Physics Reports C42, No. 1 (1978).

R. Jackiw, Revs. Mod. Phys. 49, 681 (1977).

II. PARTICLE PHYSICS

Chapter 8

ORDER-OF-MAGNITUDE ESTIMATIONS

From a phenomenological point of view, there are four distinct interactions between particles, given in Table 1.

Table 1

Interaction	Phenomenological Coupling Constant	Quanta of the Mediating Field
Strong	~ 1	Color gluon
Electromagnetic	$\alpha \cong \frac{1}{137}$	Photon
Weak	$G m_p^2 \sim 10^{-5}$	Intermediate boson
Gravitational	$\mathcal{G} m_p^2 \sim 6 \times 10^{-39}$	Graviton

The four phenomenological classes of interactions. α is the fine structure constant, G is the Fermi constant for beta-decay, $\mathcal{G}$ is Newton's gravitational constant and m_p is the mass of the proton; all these constants are in the natural units. Among the quanta of the mediating fields, only the photon has been detected experimentally.

Throughout our discussions we will concentrate only on the strong, weak and electromagnetic interactions. Except for the quanta of the mediating fields, all particles can be classified into two types.

Those that have strong interactions are called hadrons, all others are leptons. For example, p, n, Λ, π are all hadrons; e, μ, τ and various neutrinos are leptons. (Since the mass of τ is about 2 GeV, the word 'lepton', which means light particle, is something of a misnomer.) The detailed properties of these particles are given in a table at the end of this book.

In this chapter we shall illustrate how to estimate the order of magnitude of physical quantities, such as various particle sizes and cross sections. These estimations will employ very few input parameters, listed below:

the fine structure constant $\alpha \cong \frac{1}{137}$,

the Fermi constant $G \cong 10^{-5}/m_p^2$,

the electron mass $m_e \cong 0.51 \text{ MeV} \cong (4\times 10^{-11} \text{ cm})^{-1}$,

the proton mass $m_p \cong 1800\, m_e$, (8.1)

and

the pion mass $m_\pi \cong \frac{1}{7} m_p$.

The art of order-of-magnitude estimations is based on

(i) simple physical considerations,

and (ii) dimensional analysis,

as will be illustrated by the following examples.

8.1 Radius of the Hydrogen Atom

The hydrogen radius r is determined by the orbit of the electron. Therefore, the momentum of the electron is $p \sim \frac{1}{r}$ and its kinetic energy is $\sim \frac{p^2}{2m_e}$. The electrostatic energy should be $\sim -\frac{\alpha}{r}$. The energy E can then be estimated to be

$$E \sim \frac{1}{2m_e}\left(\frac{1}{r}\right)^2 - \frac{\alpha}{r} \quad . \tag{8.2}$$

Its minimum is determined by

$$\frac{dE}{dr} = 0 \quad ,$$

which leads to

$$r = \frac{1}{m_e \alpha} \quad . \tag{8.3}$$

By using the values of α and m_e in (8.1) we see that

$$r \cong 5 \times 10^{-9} \text{ cm} \tag{8.4}$$

which is the Bohr radius, now derived without solving any differential equations.

Remarks. In quantum electrodynamics there are three important lengths, differing from each other by powers of α :

Bohr radius	$\frac{1}{m_e \alpha}$,	
Compton wavelength of e	$\frac{1}{m_e}$,	(8.5)
classical radius of e	$\frac{\alpha}{m_e}$.	

8.2 Hadron Size

According to Table 1, the strong interaction coupling constant is ~ 1 instead of α . Therefore a glance at (8.5) tells us, for the hadrons, there is only one length, which can be taken as the Compton wavelength of the particle. Since the pion is the lowest-mass hadron, its Compton wavelength is therefore the largest. Because of strong interactions, pion clouds must exist in other hadrons such as p ,

$n, \Lambda, \cdots$. All these hadrons, including the pion, should be of a size $\sim \frac{1}{m_\pi}$ which according to (8.1) is $\sim 10^{-13}$ cm = 1 fermi.

Therefore, one expects the charge radius r_p of the proton to be of the same order. Experimentally, one finds

$$r_p \cong .81 \text{ fermi} , \tag{8.6}$$

consistent with the above simple estimation.

8.3 High-energy pp, πp and Kp Total Cross Sections

Because pp has strong interactions, we expect that at high energy the total cross section of a pp collision will be

$$\sigma_{pp} \sim \pi r_p^2 . \tag{8.7}$$

Since r_p is $\sim 10^{-13}$ cm, we estimate

$$\sigma_{pp} \sim 3 \times 10^{-26} \text{ cm}^2 \cong 30 \text{ mb} \tag{8.8}$$

where 1 mb = 10^{-3} b and 1 b = 1 barn = 10^{-24} cm^2.

As we shall see later in the discussion of the quark-parton model, a nucleon (proton or neutron) is made of three quarks while a meson (π or K) is made of two. Furthermore, at high energy we can treat these quarks as free particles, at least so far as total cross sections are concerned. Thus we expect

$$\frac{\sigma_{\pi p}}{\sigma_{pp}} \cong \frac{\sigma_{Kp}}{\sigma_{pp}} \cong \frac{2}{3}$$

which is independent of the charge of π and K. By using (8.8), we can estimate, for the total cross sections,

$$\begin{aligned} \sigma_{\pi^+ p} &\cong \sigma_{\pi^- p} \cong 20 \text{ mb} , \\ \sigma_{K^+ p} &\cong \sigma_{K^- p} \cong 20 \text{ mb} . \end{aligned} \tag{8.9}$$

Likewise, we expect

$$\sigma_{np} \cong \sigma_{pp} \cong \sigma_{\bar{p}p} \cong \sigma_{\bar{n}p} \quad . \tag{8.10}$$

The present high-energy experimental values are

$$\sigma_{pp} \cong \sigma_{np} \cong \sigma_{\bar{p}p} \cong \sigma_{\bar{n}p} \cong 45 \text{ mb} \quad ,$$

$$\sigma_{\pi^{\pm}p} \cong 25 \text{ mb}$$

and

$$\sigma_{K^{\pm}p} \cong 20 \text{ mb} \quad ,$$

all consistent with the above estimations.

8.4 $e^+ + e^- \rightarrow \mu^+ + \mu^-$

We shall estimate the total cross section for this reaction at high energy. Since $e^{\pm}$ and $\mu^{\pm}$ are leptons, the strongest interaction between them is the electromagnetic. The lowest order diagram is given in Fig. 8.1 where the wavy line represents the virtual photon.

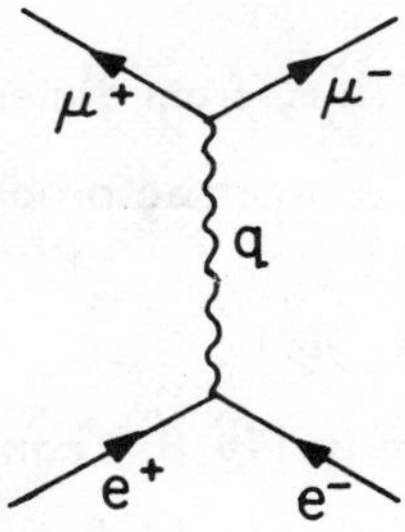

Fig. 8.1 Feynman diagram for $e^+ + e^- \rightarrow \mu^+ + \mu^-$.

Let σ be the total cross section. In Fig. 8.1 each vertex carries a factor proportional to the electric charge; the Feynman amplitude is therefore proportional to the fine structure constant α . So far as the total cross section is concerned, the only Lorentz-invariant variable in this problem is the square of the 4-momentum q carried by the virtual photon. It is useful to visualize the process in the center-of-mass system (which is also the laboratory system if this is a standard colliding beam experiment), and to denote s as the square of the center-of-mass energy $E_{c.m.}$. In our case, we have

$$s \equiv E_{c.m.}^2 = -q^2 = q_0^2 - \vec{q}^{\,2} . \tag{8.11}$$

The total cross section must therefore be of the form

$$\sigma = \alpha^2 f(s, m_e, m_\mu) \tag{8.12}$$

where m_e and m_μ are respectively the masses of e and μ, and the function f is to be determined. When $E_{c.m.}$ is much greater than either lepton mass, we may set $m_e = m_\mu = 0$ as an approximation. Now, α is dimensionless, and in the natural units the dimensions of σ and s are

$$[\sigma] = [L^2] \quad \text{and} \quad [s] = [L^{-2}] , \tag{8.13}$$

where L denotes length. Thus, from dimensional analysis the function f in (8.12) for large s must be proportional to s^{-1}; i.e.

$$\sigma \sim \frac{\alpha^2}{s} . \tag{8.14}$$

The Feynman diagram in Fig. 8.1 can be readily evaluated, and the complete answer for σ in the high-energy limit is

$$\sigma = \frac{4}{3} \pi \frac{\alpha^2}{s} . \tag{8.15}$$

[See Problem 6.1.] The point is that without any computation it is possible to give an estimation of the cross section, albeit without the factor $\frac{4}{3}\pi$.

From (8.14), one can readily estimate, e.g., that when $s \cong (1\ \text{GeV})^2$

$$\sigma(e^+e^- \to \mu^+\mu^-) \sim 4 \times 10^{-32}\ \text{cm}^2 = .04\ \mu\text{b} \quad , \tag{8.16}$$

where $1\ \mu\text{b} = 1$ microbarn $= 10^{-6}$ b .

8.5 $\nu + N \to \cdots$

Let $\sigma(\nu N)$ be the total cross section of this reaction summed over all final states. The initial ν can be either the μ-neutrino ν_μ or the e-neutrino ν_e, and N is the nucleon. As before, s denotes the square of the center-of-mass energy. Since this is a weak process, its amplitude should be proportional to the Fermi constant G. Hence, $\sigma(\nu N)$ must be of the form

$$\sigma(\nu N) = G^2 f(s, m_N) \tag{8.17}$$

where m_N is the nucleon mass and the function f is to be determined. For $s \gg m_N^2$, we can, as in the preceding example, set $m_N = 0$ as an approximation. Since the dimension of G is

$$[G] = [L^2] \quad , \tag{8.18}$$

by using (8.13) we see that in (8.17) the function f must be proportional to s ; therefore,

$$\sigma(\nu N) \sim G^2 s \quad . \tag{8.19}$$

The laboratory system in a typical high-energy neutrino experiment is one in which the initial nucleon is at rest. Let E_ν and $\vec{p}_\nu$ be respectively the energy and the momentum of the neutrino in the laboratory system. We have

$$s = E_{c.m.}^2 = (E_\nu + m_N)^2 - \vec{p}_\nu^{\,2} = m_N(2E_\nu + m_N) \cong 2m_N E_\nu \ . \tag{8.20}$$

Thus, (8.19) can also be written in terms of E_ν :

$$\sigma(\nu N) \sim G^2 m_N E_\nu \quad . \tag{8.21}$$

By using (8.1), we find at high energy

$$\sigma(\nu N) \sim 4 \times 10^{-38} \left(\frac{E_\nu}{m_N}\right) \text{cm}^2 \quad . \tag{8.22}$$

The experimental result is

$$\sigma(\nu N) \cong 0.6 \times 10^{-38} \left(\frac{E_\nu}{m_N}\right) \text{cm}^2 \quad . \tag{8.23}$$

Again, we can estimate the order of magnitude of this reaction without any computation. [Notice the huge difference between (8.8), (8.16) and (8.22) at a comparable energy.]

8.6 Compton Scattering

The lowest-order Feynman diagrams for Compton scattering

$$\gamma + e^{\pm} \rightarrow \gamma + e^{\pm} \tag{8.24}$$

are given in Fig. 8.2. By following the same reasoning which led to (8.12), we expect the total cross section σ_{Comp} of this reaction to

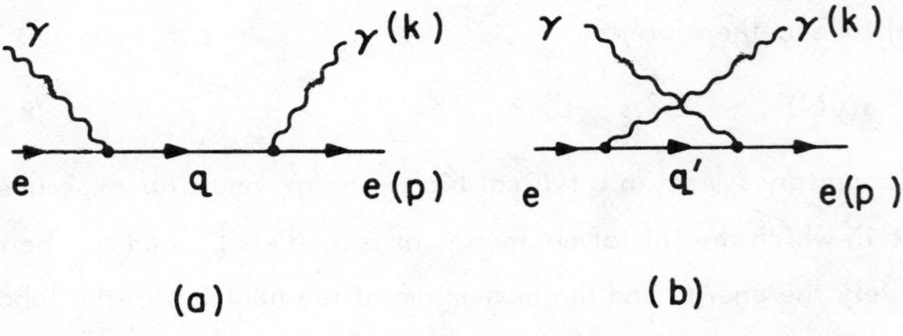

Fig. 8.2. Lowest-order diagrams for Compton scattering in which p and k denote the 4-momenta of the final e and γ .

be of the form

$$\sigma_{Comp} = \alpha^2 f(s, m_e) \tag{8.25}$$

where, as before, s is the center-of-mass energy squared, and the function f is to be determined.

At the nonrelativistic limit, $s \to m_e^2$ and therefore f is a function which depends only on m_e. From dimensional considerations, we expect

$$\sigma_{Comp} \sim \frac{\alpha^2}{m_e^2} \qquad \text{N. R.} \tag{8.26}$$

where N.R. denotes the nonrelativistic limit. When s is $\gg m_e^2$, we can neglect m_e in (8.25), and that leads to, through dimensional analysis,

$$\sigma_{Comp} \sim \frac{\alpha^2}{s} \qquad \text{E. R.} \tag{8.27}$$

where E.R. denotes the extreme relativistic limit.

An accurate calculation of the Feynman diagrams in Fig. 8.2 gives *

$$\sigma \cong \begin{cases} \frac{8}{3}\pi\left(\frac{\alpha}{m_e}\right)^2 & \text{N. R.} \\ 2\pi \frac{\alpha^2}{s} \ln \frac{s}{m_e^2} & \text{E. R.} \end{cases} \tag{8.28}$$

Our estimation (8.26) differs from the accurate nonrelativistic formula (called the Thomson limit) only by a factor $\frac{8\pi}{3}$. However, in the extreme relativistic case, the estimation (8.27) misses an s-dependent factor $2\pi \ln \frac{s}{m_e^2}$. While this is a slowly varying function of s, the existence of such log terms has a general underlying reason, as we shall explain.

---

* See W. Heitler, The Quantum Theory of Radiation (Oxford, Oxford University Press, 1944).

8.7 Mass Singularity and High-energy Behavior

We first give the technical reason for the $\ln \frac{s}{m_e^2}$ factor. Let us consider diagram (a) in Fig. 8.2. The virtual electron carries a 4-momentum

$$q = p + k \quad , \tag{8.29}$$

where p and k are, respectively, the final 4-momenta of e and γ. The components of p and k may be written as $p = (\vec{p}, ip_0)$ and $k = (\vec{k}, ik_0)$. Since the final e and γ are both on the mass shell, we have

$$p^2 + m_e^2 = 0 \qquad \text{and} \qquad k^2 = 0 \quad ,$$

i.e.

$$p_0 = \sqrt{\vec{p}^{\,2} + m_e^2} \qquad \text{and} \qquad k_0 = |\vec{k}| \quad . \tag{8.30}$$

For definiteness, we consider the laboratory frame. In the E.R. limit, p_0 is $\gg m_e$, and therefore

$$p_0 \cong |\vec{p}| + \frac{m_e^2}{2p_0} \quad . \tag{8.31}$$

In diagram (a) the electron propagator carries a denominator

$$q^2 + m_e^2 = (p+k)^2 + m_e^2 \tag{8.32}$$

which, because of (8.30), is equal to

$$2p \cdot k = 2\vec{p} \cdot \vec{k} - 2p_0 k_0 \quad .$$

In the E.R. limit and for the nearly forward scattering case, we have

$$\begin{aligned} q^2 + m_e^2 &\cong -m_e^2 \frac{k_0}{p_0} - 2|\vec{p}|\, k_0 (1 - \cos\theta) \\ &\cong -\frac{k_0}{p_0} (m_e^2 + p_0^2 \theta^2) \end{aligned} \tag{8.33}$$

where $\theta \ll 1$ is the angle between $\vec{p}$ and $\vec{k}$. Diagram (a) gives a contribution σ_a to the cross section that is inversely proportional to

$(q^2 + m_e^2)^2$. For the region under consideration we expect the deviation of σ_a from the simple estimate (8.27) to be

$$\frac{\sigma_a}{(\alpha^2/s)} \sim \int \frac{\theta^2 2\pi \sin\theta \, d\theta}{(m_e^2 + p_0^2 \theta^2)^2} \cdot p_0^4$$

$$\sim 2\pi \ln\left(\frac{p_0}{m_e}\right)^2 \sim 2\pi \ln\left(\frac{s}{m_e^2}\right) \qquad (8.34)$$

where $2\pi \sin\theta \, d\theta$ is the solid angle, p_0^4 is there to make the whole expression dimensionless, and the θ^2 factor in the numerator is due to γ_5 invariance, as will be explained below. The new estimation (8.34) is in good agreement with the exact limit given by the Klein-Nishina formula * given in (8.28).

The γ_5-transformation

$$\psi \to \gamma_5 \psi , \qquad (8.35)$$

has been discussed in Section 3.8 in connection with the two component theory. For the electron field ψ, since $m_e \neq 0$, the mass term violates the γ_5 invariance. However, the electromagnetic interaction is invariant under the γ_5 transformation, as can be readily verified by substituting (8.35) into (6.7). In the E.R. limit, m_e can be neglected; therefore, by following the discussions given in Section 3.8 one sees that the helicity of the electron is unchanged through its electromagnetic interaction. Since the helicity of the photon is either +1 or −1, a helicity-conserving electron cannot emit or absorb a photon in the exact forward direction, as can be easily seen through angular-momentum conservation along the direction of motion. Thus, for zero m_e when the angle θ between $\vec{p}$ and $\vec{k}$ is 0, the

* O. Klein and Y. Nishina, Zeit.f.Phys. 52, 853 (1929).

Feynman amplitude must also be 0 . Consequently, the matrix element for diagram (a) in Fig. 8.2 carries a factor proportional to θ, and that explains the θ^2 term in (8.34). [For $m_e \neq 0$ but $\theta = 0$, the amplitude is $\propto m_e$; terms proportional to m_e do not lead to the mass singularity.]

From (8.34) one observes that in the E.R. limit the cross section has a logarithmic singularity (called mass singularity) when the electron mass $m_e \to 0$. The origin of the mass singularity is associated with the fact that (8.33) becomes 0 when $m_e = \theta = 0$; i.e., the relevant propagator becomes infinite, which in turn means the virtual particle is approaching its mass shell. There is a simple, but general, reason for this: If we have a number of zero-mass particles moving in the same direction, their total energy $E = \sum E_a$ and the magnitude of their total 3-momentum $\vec{p} = \sum \vec{p}_a$ become equal

$$\sum_a E_a = | \sum \vec{p}_a | \tag{8.36}$$

where the subscript a denotes the ath particle. Thus, when $m_e \to 0$, conservation of energy follows from the conservation of momentum whenever the momenta of e and γ are all parallel; i.e., in

$$e \leftrightarrows e + \gamma$$

all particles can be on-mass-shell in this special limit.

A general theorem can be established which states that, for any process

$$i \to f , \tag{8.37}$$

although the Feynman diagrams have mass singularities, the square of the S-matrix element

$$\sum_{\{i\},\{f\}} | \langle f | S | i \rangle |^2 , \tag{8.38}$$

summed over the sets $\{i\}$ and $\{f\}$ which respectively contain all states that are degenerate (within an energy width $\epsilon \neq 0$, which can be arbitrarily chosen) with i and f, does not. Expression (8.38) depends on the width ϵ but has a finite limit when $m_e \to 0$ to every order in the perturbation expansion. This theorem is valid not only in quantum electrodynamics, but also in other field theories such as quantum chromodynamics. The proof is quite elementary, and is given in Chapter 23.

8.8 e^+e^- Pair Production by High-energy Photons

Let us consider the reaction

$$\gamma + Z \to e^+ + e^- + Z \tag{8.39}$$

where Z denotes a heavy nucleus of charge Z (in units of positron charge). The lowest-order diagrams are given in Fig. 8.3 in which the wavy lines represent real or virtual photons. In order to establish the cross section of (8.39) we first consider the reaction of pair production by two photons

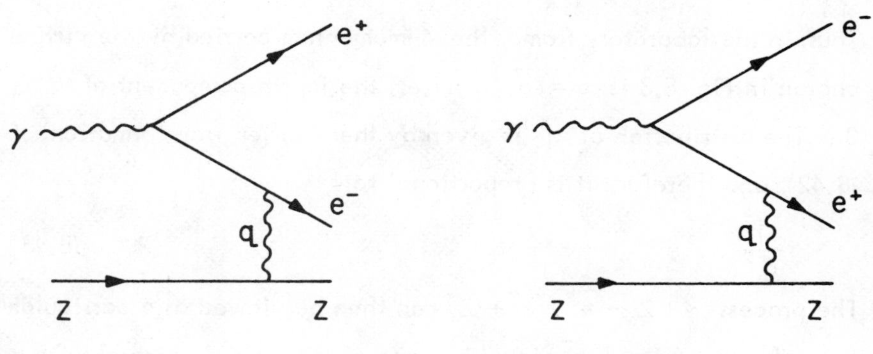

Fig. 8.3. Lowest-order Feynman diagrams for (8.39).

$$\gamma + \gamma \rightarrow e^+ + e^- \tag{8.40}$$

whose diagram is given in Fig. 8.4. By following the same argument

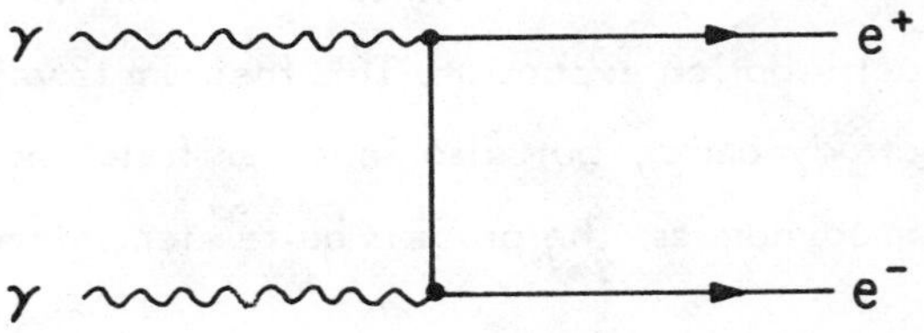

Fig. 8.4. Lowest-order diagram for $\gamma + \gamma \rightarrow e^+ + e^-$.

which leads to (8.27), we may estimate at high energy

$$\sigma(\gamma\gamma \rightarrow e^+ e^-) \sim \frac{\alpha^2}{s} \ . \tag{8.41}$$

Next we compare the difference between reactions (8.39) and (8.40). For simplicity the nucleus is assumed to be extremely heavy, at rest in the laboratory system. The electrostatic potential generated by the nucleus at distance r is

$$\frac{Ze}{r} \ . \tag{8.42}$$

Thus, in the laboratory frame, the 4-momentum carried by the virtual photon in Fig. 8.3 is $q = (\vec{q}, 0)$; i.e., the fourth component of q is 0. The distribution of $\vec{q}$ is given by the Fourier transformation of (8.42), and therefore it is proportional to

$$\frac{1}{\vec{q}^{\,2}} \ . \tag{8.43}$$

The process $\gamma + Z \rightarrow e^+ + e^- + Z$ can then be viewed as a particular case of $\gamma + \gamma \rightarrow e^+ + e^-$ in which one of the γ's is virtual with a momentum distribution given by (8.43) in the laboratory frame.

Accordingly, in reaction (8.39) the 3-momenta $\vec{p}_+$ and $\vec{p}_-$ of e^+ and e^- are independent; therefore, the final-state phase space differs from that of (8.40) by an additional factor

$$\frac{d^3\vec{p}}{p_0} \sim \vec{p}^{\,2} \sim s \tag{8.44}$$

where $p_0 = (\vec{p}^{\,2} + m_e^{\,2})^{\frac{1}{2}}$ and $\vec{p}$ can be either $\vec{p}_+$ or $\vec{p}_-$. By multiplying (8.44) and (8.41) we see that the s-factors cancel, and therefore at high energy the cross section σ_{pair} for reaction (8.39) is approximately independent of s. In Fig. 8.3, each Feynman diagram has three vertices, one is proportional to Ze and the other two to e. Thus, σ_{pair} is of the form

$$\sigma_{pair} \sim Z^2\alpha^3 f(m_e)$$

where the function f can be determined through dimensional analysis. We then obtain the estimation

$$\sigma_{pair} \sim \frac{Z^2\alpha^3}{m_e^{\,2}} \ . \tag{8.45}$$

So far, we have only made the crudest estimation, ignoring completely the possibility of mass singularity. By following arguments similar to that given between (8.29) and (8.34), we can derive a better high-energy estimation

$$\sigma_{pair} \sim \frac{Z^2\alpha^3}{m_e^{\,2}} \ln \frac{E_\gamma}{m_e} \tag{8.46}$$

where E_γ is the laboratory energy of the initial γ in reaction (8.39). The exact limit given by the Bethe-Heitler formula* is

$$\sigma_{pair} = \frac{28}{9}\,\frac{Z^2\alpha^3}{m_e^2} \ln \frac{E_\gamma}{m_e} \ , \tag{8.47}$$

---

* H. Bethe and W. Heitler, Proc.Roy.Soc. 146, 83 (1934). See also Heitler, The Quantum Theory of Radiation, loc. cit.

consistent with the above estimation.

Equation (8.47) is valid for a point nucleus without screening. If one has complete screening, then the log factor in (8.47) should be replaced by a constant $\cong \ln(183\, Z^{-\frac{1}{3}})$. Unlike σ_{Comp}, σ_{pair} does not decrease with increasing energy. This has an important experimental consequence; it means that in matter a high-energy photon has a finite, nonzero, almost energy-independent mean free path, which is, e.g., about 460 meters in air (1 atm. pressure and 0° C), 13 cm in aluminum and 7 mm in lead.

IIA. PARTICLE PHYSICS: SYMMETRY

Chapter 9

GENERAL DISCUSSION

Since the beginning of physics, symmetry considerations have provided us an extremely powerful and useful tool in our effort to understand nature. Gradually they have become the backbone of our theoretical formulation of physical laws. In this chapter we shall review these symmetry operations and examine their foundation. Such an examination is useful, especially in view of the various asymmetries that have been discovered during the past quarter century.

There are four main groups of symmetries, or broken symmetries, that are found to be of importance in physics:

1. Permutation symmetry: Bose-Einstein and Fermi-Dirac statistics.
2. Continuous space-time symmetries, such as translation, rotation, acceleration, etc.
3. Discrete symmetries, such as space inversion, time reversal, particle-antiparticle conjugation, etc.
4. Unitary symmetries, which include

    U_1 - symmetries such as those related to conservation of charge, baryon number, lepton number, etc.,

    SU_2 (isospin) - symmetry,

    SU_3 (color) - symmetry,

and SU_n (flavor) - symmetry.

Among these, the first two groups, together with some of the U_1-symmetries and perhaps the SU_3 (color)-symmetry in the last group, are believed to be exact. All the rest seem to be broken.

9.1 Non-observables, Symmetry Transformations and Conservation Laws

The root of all symmetry principles lies in the assumption that it is impossible to observe certain basic quantities; these will be called "non-observables". Let us illustrate the relation between non-observables, symmetry transformations and conservation laws by a simple example. Consider the interaction energy V between two particles at positions $\vec{r}_1$ and $\vec{r}_2$. The assumption that the absolute position is a non-observable means that we can arbitrarily choose the origin O from which these position vectors are drawn; the interaction energy should be independent of O. In other words, V is invariant under an arbitrary space translation, changing O to O';

$$\vec{r}_1 \to \vec{r}_1 + \vec{d} \quad \text{and} \quad \vec{r}_2 \to \vec{r}_2 + \vec{d} \,, \tag{9.1}$$

as shown in Fig. 9.1. Consequently, V is a function only of the

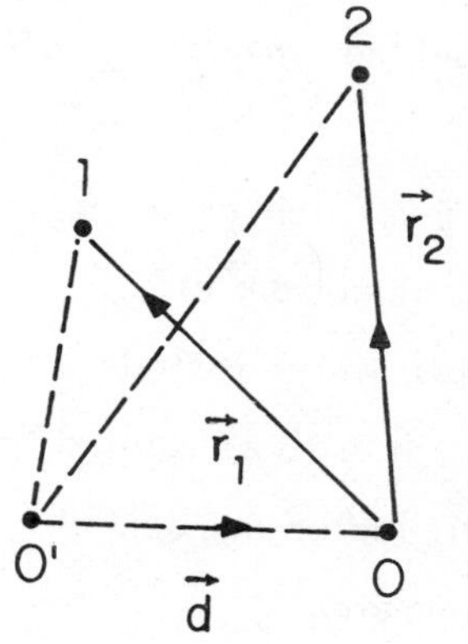

Fig. 9.1. The interaction energy V between particles 1 and 2 is invariant under a change of origin O → O'.

relative distance $\vec{r}_1 - \vec{r}_2$,

$$V = V(\vec{r}_1 - \vec{r}_2) \quad . \tag{9.2}$$

From this, we deduce that the total momentum of this system of two particles must be conserved, since its rate of change is equal to the force

$$-(\vec{\nabla}_1 + \vec{\nabla}_2)\, V$$

which, on account of (9.2), is zero.

This example illustrates the interdependence among three aspects of a symmetry principle: the physical assumption of a non-observable, the implied invariance under the connected mathematical transformation and the physical consequences of a conservation law or selection rule. In an entirely similar way, we may assume the absolute time to be a non-observable. The physical law must then be invariant under a time-translation

$$t \rightarrow t + \tau$$

which results in the conservation of energy. By assuming the absolute spatial direction to be a non-observable, we derive rotation invariance and obtain the conservation law of angular momentum. By assuming that absolute (uniform) velocity is not an observable, we derive the symmetry requirement of Lorentz invariance, and with it the conservation laws connected with the six generators of the Lorentz group. Similarly, the foundation of general relativity rests on the assumption that it is impossible to distinguish the difference between an acceleration and a suitably arranged gravitational field.

The following table summarizes these three fundamental aspects for some of the symmetry principles used in physics.

Non-observables	Symmetry Transformations	Conservation Laws or Selection Rules
difference between identical particles	permutation	B.-E. or F.-D. statistics
absolute spatial position	space translation $\vec{r} \to \vec{r} + \vec{\Delta}$	momentum
absolute time	time translation $t \to t + \tau$	energy
absolute spatial direction	rotation $\hat{r} \to \hat{r}'$	angular momentum
absolute velocity	Lorentz transformation	generators of the Lorentz group
absolute right (or absolute left)	$\vec{r} \to -\vec{r}$	parity
absolute sign of electric charge	$e \to -e$ (or $\psi \to e^{i\phi} \psi^{\dagger}$)	charge conjugation (or particle anti-particle conjugation)
relative phase between states of different charge Q	$\psi \to e^{iQ\theta} \psi$	charge
relative phase between states of different baryon number N	$\psi \to e^{iN\theta} \psi$	baryon number
relative phase between states of different lepton number L	$\psi \to e^{iL\theta} \psi$	lepton number
difference between different coherent mixture of p and n states	$\begin{pmatrix} p \\ n \end{pmatrix} \to u \begin{pmatrix} p \\ n \end{pmatrix}$	isospin

Table 9.1.

9.2 Asymmetries and Observables

Since the validity of all symmetry principles rests on the theoretical hypothesis of non-observables, the violation of symmetry arises whenever what was thought to be a non-observable turns out to be actually an observable. In a sense, the discovery of "violations" is not that surprising. It is not difficult to imagine that some of the "non-observables" may indeed be fundamental, but some may simply be due to the limitations of our present ability to measure things. As we improve our experimental techniques, our domain of observation also enlarges. It should not be completely unexpected that we may succeed in detecting one of those supposed "non-observables" at some time and therein lies the root of symmetry breaking.

The notable examples of such discoveries are the asymmetry of physical laws under the right-left mirror transformation, the particle-antiparticle conjugation and the change in the direction of time flow, past to future and future to past. It turns out that all these supposed non-observables can actually be observed. Let me illustrate the relation between "asymmetries and observables" by first considering the example of right-left asymmetry, commonly known as parity nonconservation.

Of course it is well known that even in daily life, right and left are distinct from each other. Our hearts, for example, are usually not on the right side. The word "right" also means correct, while the word "sinister" in its Latin root means left. In English, one says "right-left", but in Chinese 左右 : 左 (left) usually precedes 右 (right). However, such asymmetry in daily life is attributed to either the accidental asymmetry of our environment or initial conditions. Before the discovery of parity nonconservation in 1957, it was taken for granted that the laws of nature are symmetric under a right-left transformation.

One may wonder why, in spite of the clear difference between right and left in daily life, before 1956 practically all physicists could believe in right-left symmetry in physical laws.

Let us consider two cars which are made exactly alike, except that one is the mirror image of the other, as shown in Fig. 9.2. Car a

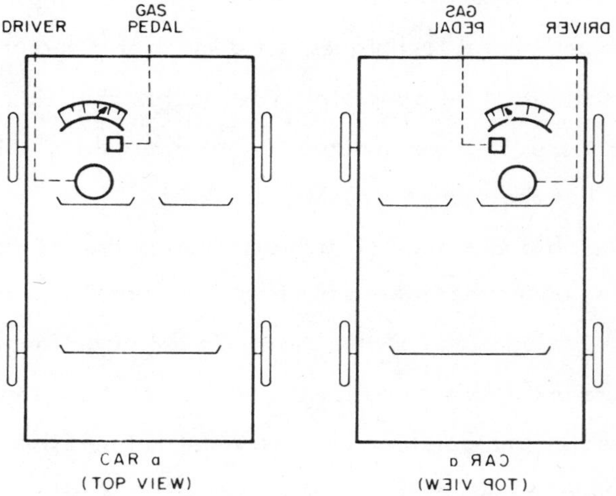

Fig. 9.2. Top view of two cars that are mirror images of each other.

has a driver on the left front seat and the gas pedal to his right, while ɒ has the driver on the right front seat with the gas pedal on his left. Both cars are filled with the same amount of gasoline. (For the sake of discussion, we may ignore the fact that the gasoline molecule is not exactly mirror-symmetric.) Now, suppose the driver of Car a starts the car by turning the ignition key clockwise and stepping on the gas pedal with his right foot, causing the car to move forward at a certain speed, say 20 mph. The other does exactly the same thing, except that he interchanges right with left; i.e., he turns the ignition key counterclockwise and steps on the gas pedal with his left foot, but keeps the pedal at the same degree of inclination. What will the motion of Car ɒ be? The reader is encouraged to make a guess.

Probably your common sense will say that both cars should move forward at exactly the same speed. If so, you are just like the pre-1956 physicists. It would seem reasonable that two arrangements, identical except that one is the mirror image of the other, should behave in exactly the same way in all respects (except of course for the original right-left difference). This is precisely what was discovered to be untrue. The possibility of right-left asymmetry in natural laws was first suggested theoretically* in 1956 in connection with the $\theta-\tau$ puzzle (see Chapter 15). It was discovered experimentally within a few months in β-decay** and in π- and μ-decays***. In principle, in the above example of two cars one may install, say, Co^{60} β-decay as part of the ignition mechanism. It will then be possible to construct two cars that are exact mirror images of each other, but may nevertheless move in a completely different way; Car a may move forward at a certain speed while Car ɒ may move at a totally different speed, or may even move backwards. That is the essence of right-left asymmetry, or parity-nonconservation.

---

* T. D. Lee and C. N. Yang, Phys.Rev. 104, 254 (1956).

** C. S. Wu, E. Ambler, R. W. Hayward, D. D. Hoppes and R. P. Hudson, Phys.Rev. 105, 1413 (1957).

*** R. L. Garwin, L. M. Lederman and M. Weinrich, Phys.Rev. 105, 1415 (1957); V. L. Telegdi and A. M. Friedman, Phys.Rev. 105, 1681 (1957).

As we shall discuss, the discoveries made in 1957 established not only right-left asymmetry, but also the asymmetry between the positive and negative signs of electric charge. In the standard nomenclature, right-left asymmetry is referred to as P violation, or parity nonconservation. The asymmetry between opposite signs of electric charge is called C violation, or charge conjugation violation, or sometimes particle-antiparticle asymmetry.

At the same time as the possibility of P and C violations was suggested, questions of possible asymmetries under time reversal and under CP were also raised.* Actual experimental confirmation did not come until quite a few years later.**

Since non-observables imply symmetry, these discoveries of asymmetry must imply observables. One may ask, what are the observables that have been discovered in connection with these symmetry-breaking phenomena? We recall that in our daily lives the sign of electric charge is merely a convention. The electron is considered to be negatively charged because we happened to assign the proton a positive charge, and the converse is also true. But now, with the discovery of asymmetry, is it possible to give an absolute definition? Can we find some absolute difference between the positive and negative signs of electricity, or between left and right?

As an illustration, let us consider the example of two imaginary, advanced civilizations A and B (see Fig. 9.3). These two civilizations are assumed to be completely separate from each other; nevertheless they manage to communicate, but only through neutral

---

* T. D. Lee, R. Oehme and C. N. Yang, Phys. Rev. 106, 340 (1957).

** J. H. Christenson, J. W. Cronin, V. L. Fitch and R. Turlay, Phys. Rev. Lett. 13, 138 (1964).

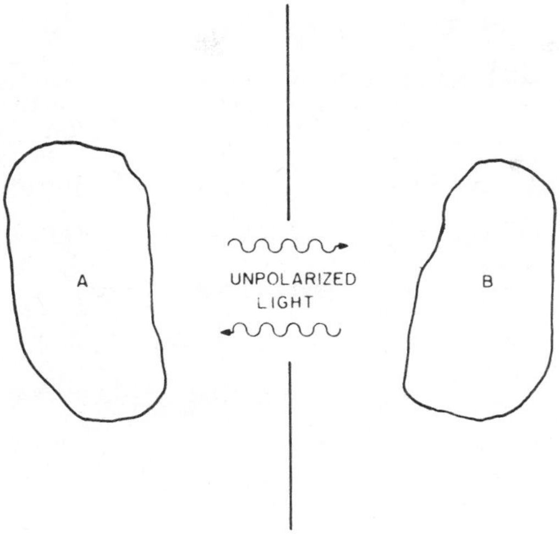

Fig. 9.3. Two imaginary civilizations communicate with each other through neutral unpolarized messages.

unpolarized messages, such as unpolarized light. After years of such communication these two civilizations may decide to increase their contact. Being very advanced, they realize that they must first agree on (i) the sign of electric charge,

and (ii) the definition of a righthand screw.

The first is important in order to establish whether the proton in civilization A corresponds to the proton or the antiproton in civilization B. If the protons in A are the same as those in B, then a closer contact is possible. Otherwise it might lead to annihilation. The definition of a righthand screw is important if these two civilizations decide to have even closer contact, such as trading machinery. The academic problem that concerns us is whether it is possible to transmit both pieces of information by using only neutral and unpolarized messages. Without these discoveries of P, C and CP asymmetries, this would not be feasible. Now, assuming that these two civilizations are as advanced as ours, such an agreement can in principle be achieved.

First, both civilizations should establish high-energy physics laboratories which can produce the long-lived neutral kaon K_L^o. By analyzing the semileptonic three-body decay modes of K_L^o under a magnetic field, they can easily separate the decay $K_L^o \to e^- + \pi^+ + \bar{\nu}$ from $K_L^o \to e^+ + \pi^- + \nu$. They would discover that although the parent particle K_L^o is neutral and spherically symmetric, nevertheless these two decay rates are different

$$\frac{\text{rate } (K_L^o \to e^+ + \pi^- + \nu)}{\text{rate } (K_L^o \to e^- + \pi^+ + \bar{\nu})} = 1.00648 \pm 0.00035. \qquad (9.3)$$

This is indeed remarkable since it means that by rate counting one can differentiate e^- from e^+. Thus, there is an absolute difference between the opposite signs of electric charge. Now, each civilization only needs to examine the faster decay mode in (9.3), and compare the charge of the final e with that of its "proton." If both civilizations have the same relative sign, then it means that they are made of the same matter.

Next, we come to the second task: the definition of a right-hand screw. This can be done by measuring the spin and momentum direction of the neutrino or antineutrino in π-decay: $\pi^+ \to e^+ + \nu$ and $\pi^- \to e^- + \bar{\nu}$. Although $\pi^\pm$ is spherically symmetric, in its decay every ν defines a lefthand screw, while every $\bar{\nu}$ a righthand screw; i.e., the helicity of ν is always $-\frac{1}{2}$ and that of $\bar{\nu}$, $+\frac{1}{2}$, as shown in Fig. 9.4. The neutrino and antineutrino can therefore be described* by the two-component theory, discussed in Section 3.8.

---

* T. D. Lee and C. N. Yang, Phys. Rev. 105, 1671 (1957); L. Landau, Nucl. Phys. 3, 127 (1957); A. Salam, Nuovo Cimento 5, 299 (1957).

Consequently, we see that through neutral and unpolarized messages these two civilizations can indeed give an absolute meaning to + and - signs of charge as well as to right and left.

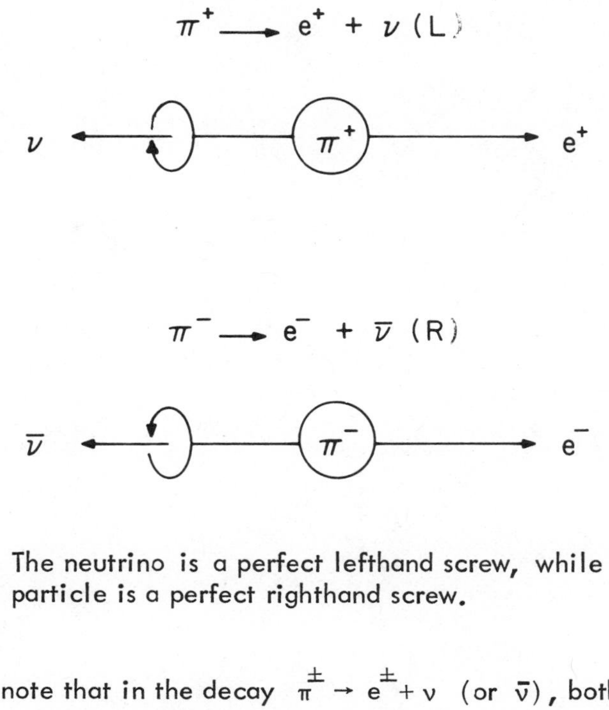

Fig. 9.4. The neutrino is a perfect lefthand screw, while its antiparticle is a perfect righthand screw.

We note that in the decay $\pi^{\pm} \to e^{\pm} + \nu$ (or $\bar{\nu}$), both C and P symmetries are violated; but if we interchange + with - , and also right with left, then it might seem that symmetry could be regained (called CP symmetry). However, CP symmetry is also violated in the K_L^0-decay, because there is an absolute rate difference between the final states $e^+\pi^-\nu$ and $e^-\pi^+\bar{\nu}$. As we shall see later in Chapter 15, from CP asymmetry and observed amplitudes of various K-decays we can deduce the asymmetry with respect to time-reversal T. At present, it appears that physical laws are not symmetrical with respect to C, P, T, CP, PT and TC. Nevertheless, all indications are that the joint action of CPT (i.e., particle ⇆ antiparticle, right ⇆ left and past ⇆ future) remains a good symmetry.

Chapter 10

U_1 SYMMETRY AND P, C INVARIANCE

10.1 QED as an Example

The use of some of these symmetry operations in a quantum theory can best be illustrated by considering quantum electrodynamics (QED). In this section we discuss the conservation of lepton number and the invariance under parity P and particle-antiparticle conjugation C. As in Chapter 6, we adopt the Coulomb gauge. Hence the independent generalized coordinates are the transverse field $\vec{A} = \vec{A}^{tr}$ and the electron field ψ. Their conjugate momenta $\vec{\Pi}$ and $\mathcal{P}$ are given by (6.21)-(6.22). In accordance with (6.30), we have

$$\vec{\nabla} \cdot \vec{A} = 0 \qquad \text{and} \qquad \vec{\nabla} \cdot \vec{\Pi} = 0 \ .$$

Theorem. The QED Hamiltonian H, given by (6.55), is invariant under the following unitary transformations:

1. $$e^{iL\theta} H e^{-iL\theta} = H \tag{10.1}$$

in which θ is any real number and L is the lepton number operator defined by the normal product

$$L \equiv \int :\psi^\dagger(x)\, \psi(x): d^3r \tag{10.2}$$

where, as before, $x_\mu = (\vec{r}, it)$.

2. $$C H C^\dagger = H \tag{10.3}$$

where C is the particle-antiparticle conjugation operator which satisfies

$$C\vec{A}(x)\,C^{\dagger} = -\vec{A}(x)\ , \qquad C\vec{\Pi}(x)\,C^{\dagger} = -\vec{\Pi}(x) \tag{10.4}$$

and

$$C\psi(x)\,C^{\dagger} = \eta_c\,\psi^c(x) \tag{10.5}$$

with η_c as a constant phase factor, $|\eta_c| = 1$, and the components of ψ^c given by, in the notation of Chapter 3,

$$\psi_{\alpha}^{\;c}(x) \equiv (\gamma_2)_{\alpha\lambda}\,\psi_{\lambda}^{\;\dagger}(x)\ . \tag{10.6}$$

3. $$P\,H\,P^{\dagger} = H \tag{10.7}$$

where P is the space inversion (i.e., parity) operator which satisfies

$$P\vec{A}(\vec{r},t)\,P^{\dagger} = -\vec{A}(-\vec{r},t)\ , \quad P\vec{\Pi}(\vec{r},t)\,P^{\dagger} = -\vec{\Pi}(-\vec{r},t)\ , \tag{10.8}$$

$$P\psi(\vec{r},t)\,P^{\dagger} = \eta_p\,\gamma_4\,\psi(-\vec{r},t)\ , \tag{10.9}$$

with η_p as a constant phase factor $|\eta_p| = 1$.

As we shall see, the operators $e^{iL\theta}$, C and P are all unitary. Furthermore, because of (10.1), (10.3) and (10.7), they are all t-independent.

Proof. 1. We first consider the commutator between L and $\psi(x)$, which on account of (3.24) and (10.2), is given by

$$[L, \psi(x)] = -\psi(x)\ . \tag{10.10}$$

Next, we introduce

$$\psi_{\theta}(x) \equiv e^{iL\theta}\,\psi(x)\,e^{-iL\theta} \tag{10.11}$$

whose derivative is

$$\frac{\partial\psi_{\theta}(x)}{\partial\theta} = e^{iL\theta}\,i[L,\psi(x)]\,e^{-iL\theta} = -i\psi_{\theta}(x)\ . \tag{10.12}$$

Equation (10.12) can be readily integrated. We find

$$\psi_\theta(x) = e^{-i\theta}\, \psi_{\theta=0}(x) = e^{-i\theta}\, \psi(x) \quad ,$$

which leads to

$$e^{iL\theta}\, \psi(x)\, e^{-iL\theta} = e^{-i\theta}\, \psi(x) \quad . \tag{10.13}$$

From the definition (10.2) we see that L is a Hermitian and therefore $e^{iL\theta}$ is unitary. Consequently, the Hermitian conjugate of (10.13) is

$$e^{iL\theta}\, \psi^\dagger(x)\, e^{-iL\theta} = e^{i\theta}\, \psi^\dagger(x) \quad . \tag{10.14}$$

Let Γ be any 4×4 matrices, such as γ_μ, $i\gamma_4\gamma_\mu$, ..., introduced in Chapter 3. It then follows that

$$\begin{aligned} e^{iL\theta}\, \psi^\dagger \Gamma \psi\, e^{-iL\theta} &= e^{iL\theta}\, \psi^\dagger\, e^{-iL\theta}\, \Gamma\, e^{iL\theta}\, \psi\, e^{-iL\theta} \\ &= \psi^\dagger e^{i\theta}\, \Gamma\, e^{-i\theta}\, \psi = \psi^\dagger \Gamma \psi \end{aligned} \tag{10.15}$$

in which we have used the fact that the matrix element of Γ is a c. number and therefore commutes with the Hilbert space operator L. In QED the electromagnetic current is $j_\mu = i : \psi^\dagger \gamma_4 \gamma_\mu \psi :$. Hence,

$$e^{iL\theta}\, j_\mu(x)\, e^{-iL\theta} = j_\mu(x) \quad . \tag{10.16}$$

Furthermore, since L depends only on the fermion field, we have

$$e^{iL\theta}\, \vec{A}(x)\, e^{-iL\theta} = \vec{A}(x) \quad \text{and} \quad e^{iL\theta}\, \vec{\Pi}(x)\, e^{-iL\theta} = \vec{\Pi}(x) \quad . \tag{10.17}$$

Because $j_4 = i\rho$, and $A_0(\vec{r}, t)$ is given by (6.16), we also have

$$e^{iL\theta}\, A_0(x)\, e^{-iL\theta} = A_0(x) \quad , \tag{10.18}$$

and therefore

$$e^{iL\theta}\, j_\mu A_\mu\, e^{-iL\theta} = j_\mu A_\mu \quad .$$

Equation (10.1) now follows. By differentiating (10.1) and setting $\theta = 0$, we obtain

$$[L, H] = 0 \quad , \tag{10.19}$$

which implies L is independent of time, i.e., L is conserved.

From the Fourier expansion (3.32),

$$\psi(x) = \sum_{\vec{p},s} \frac{1}{\sqrt{\Omega}} \left(a_{\vec{p},s}\, u_{\vec{p},s}\, e^{i\vec{p}\cdot\vec{r}} + b^{\dagger}_{\vec{p},s}\, v_{\vec{p},s}\, e^{-i\vec{p}\cdot\vec{r}} \right), \qquad (10.20)$$

we have

$$e^{iL\theta}\,\psi(x)\,e^{-iL\theta} = \sum_{\vec{p},s} \frac{1}{\sqrt{\Omega}} \left(e^{iL\theta}\, a_{\vec{p},s}\, e^{-iL\theta}\, u_{\vec{p},s}\, e^{i\vec{p}\cdot\vec{r}} + e^{iL\theta}\, b^{\dagger}_{\vec{p},s}\, e^{-iL\theta}\, v_{\vec{p},s}\, e^{-i\vec{p}\cdot\vec{r}} \right) \qquad (10.21)$$

which, together with (10.13), leads to

$$e^{iL\theta}\, b^{\dagger}_{\vec{p},s}\, e^{-iL\theta} = e^{-i\theta}\, b^{\dagger}_{\vec{p},s} \qquad (10.22)$$

and

$$e^{iL\theta}\, a_{\vec{p},s}\, e^{-iL\theta} = e^{-i\theta}\, a_{\vec{p},s} \;,$$

whose Hermitian conjugate is

$$e^{iL\theta}\, a^{\dagger}_{\vec{p},s}\, e^{-iL\theta} = e^{i\theta}\, a^{\dagger}_{\vec{p},s} \;. \qquad (10.23)$$

By using (3.32)-(3.33) we can readily verify that

$$L = \sum_{\vec{p},s} \left(a^{\dagger}_{\vec{p},s}\, a_{\vec{p},s} - b^{\dagger}_{\vec{p},s}\, b_{\vec{p},s} \right) \;. \qquad (10.24)$$

As in (3.40), the Hilbert space of QED is spanned by the set of basis vectors

$$|\,0>,\quad a^{\dagger}_{\vec{p},s}\,|\,0>,\quad b^{\dagger}_{\vec{p},s}\,|\,0>,\quad \alpha^{\dagger}_{\vec{p},s}\,|\,0>,\ \cdots$$

$$|\,n_-e^-, n_+e^+, n_\gamma \gamma> = \prod_{i=1}^{n_-} \prod_{j=1}^{n_+} \prod_{k=1}^{n_\gamma} a^{\dagger}_{\vec{p}_i,s_i}\, b^{\dagger}_{\vec{p}_j,s_j}\, \alpha^{\dagger}_{\vec{p}_k,s_k}\,|\,0>, \qquad (10.25)$$

$\cdots$, where $\alpha^{\dagger}_{\vec{p},s}$ is the photon creation operator given by (6.36). Because of (3.41) and (10.24),

$$L\,|\,0> = 0 \;. \qquad (10.26)$$

Furthermore, by following the discussions given in Section 3.5, we see that

$$L\,|n_-e^-, n_+e^+, n_\gamma\gamma> \;=\; (n_- - n_+)\,|n_-e^-, n_+e^+, n_\gamma\gamma> \;, \tag{10.27}$$

and therefore

$$e^{iL\theta}\,|n_-e^-, n_+e^+, n_\gamma\gamma> \;=\; e^{i(n_- - n_+)\theta}\,|n_-e^-, n_+e^+, n_\gamma\gamma>, \tag{10.28}$$

which can also be established directly by using (10.22)-(10.23); then by differentiating (10.28) with respect to θ and setting $\theta = 0$, we arrive back at (10.27).

2. Next, we consider the operator C defined by (10.4)-(10.5). From the Fourier expansion it follows that

$$C\psi(x)C^\dagger = \sum_{\vec{p},s} \frac{1}{\sqrt{\Omega}} \left(C a_{\vec{p},s} C^\dagger u_{\vec{p},s} e^{i\vec{p}\cdot\vec{r}} + C b^\dagger_{\vec{p},s} C^\dagger v_{\vec{p},s} e^{-i\vec{p}\cdot\vec{r}}\right), \tag{10.29}$$

in which we have used the property that $C^\dagger$ is a Hilbert space operator and therefore commutes with $e^{i\vec{p}\cdot\vec{r}}$ and the c. number spinors $u_{\vec{p},s}$ and $v_{\vec{p},s}$. In terms of the same expansion, (10.6) can be written as

$$\psi^c_\alpha(x) = (\gamma_2)_{\alpha\lambda}\,\psi^\dagger_\lambda(x) = \sum_{\vec{p},s} \frac{1}{\sqrt{\Omega}}$$

$$\cdot\left[a_{\vec{p},s}(\gamma_2)_{\alpha\lambda}(u^*_{\vec{p},s})_\lambda\, e^{-i\vec{p}\cdot\vec{r}} + b_{\vec{p},s}(\gamma_2)_{\alpha\lambda}(v^*_{\vec{p},s})_\lambda\, e^{i\vec{p}\cdot\vec{r}}\right]$$

which, because of (3.81), leads to

$$\psi^c(x) = \sum_{\vec{p},s} \frac{1}{\sqrt{\Omega}} \left(b_{\vec{p},s} u_{\vec{p},s} e^{i\vec{p}\cdot\vec{r}} + a^\dagger_{\vec{p},s} v_{\vec{p},s} e^{-i\vec{p}\cdot\vec{r}}\right). \tag{10.30}$$

From (10.5) and by equating (10.29) with η_c times (10.30), we obtain

$$C\, a_{\vec{p},s}\, C^\dagger = \eta_c\, b_{\vec{p},s} \;, \tag{10.31}$$

$$C\, b^\dagger_{\vec{p},s}\, C^\dagger = \eta_c\, a^\dagger_{\vec{p},s} \;. \tag{10.32}$$

The Hermitan conjugate of (10.31) is

$$C\, a^{\dagger}_{\vec{p},s}\, C^{\dagger} = \eta_c^{*}\, b^{\dagger}_{\vec{p},s} \quad . \tag{10.33}$$

Likewise, by using (10.4) and the Fourier expansion (6.32)-(6.33) and (6.36), we can derive for the photon creation operator

$$C\, a^{\dagger}_{\vec{p},s}\, C^{\dagger} = -\, a^{\dagger}_{\vec{p},s} \quad . \tag{10.34}$$

That there indeed exists a unitary operator C which satisfies (10.32)-(10.34) is indicated by the formula given in Problem 10.2. Here we shall give a simpler and more direct proof. Let us consider a linear transformation in the Hilbert space spanned by (10.25):

$$|0> \;\rightarrow\; |0> , \quad a^{\dagger}_{\vec{p},s}\, |0> \;\rightarrow\; \eta_c^{*}\, b^{\dagger}_{\vec{p},s}\, |0> ,$$

$$b^{\dagger}_{\vec{p},s}\, |0> \;\rightarrow\; \eta_c\, a^{\dagger}_{\vec{p},s}\, |0> , \quad a^{\dagger}_{\vec{p},s}\, |0> \;\rightarrow\; -\, a^{\dagger}_{\vec{p},s}\, |0> , \cdots$$

$$|n_- e^-, n_+ e^+, n_\gamma \gamma> \;\rightarrow\; \eta_c^{*\,n_-}\, \eta_c^{\,n_+} (-)^{n_\gamma}\, |n_- e^+, n_+ e^-, n_\gamma \gamma> . \tag{10.35}$$

The orthonormality of the complete set of basis vectors is clearly unchanged. Hence the transformation operator is unitary. Furthermore, one can easily see that it satisfies (10.32)-(10.34). Therefore, C exists and

$$C^{\dagger} C = 1 \quad . \tag{10.36}$$

In addition, in (10.35) we have fixed the overall phase factor of C by requiring $|0> \rightarrow |0>$; i.e.,

$$C\, |0> = |0> \quad . \tag{10.37}$$

Next, we shall establish the invariance of H under C. Since the free Hamiltonian H_0 of (6.55) can be written as

$$H_0 = \sum_{\vec{p},s=\pm\frac{1}{2}} E_p \left(a^{\dagger}_{\vec{p},s}\, a_{\vec{p},s} + b^{\dagger}_{\vec{p},s}\, b_{\vec{p},s} \right) + \sum_{\vec{k},s=\pm 1} \omega\, a^{\dagger}_{\vec{k},s}\, a_{\vec{k},s} \, , \tag{10.38}$$

where $\omega = |\vec{k}|$, and $E_p = \sqrt{\vec{p}^{\,2} + m^2}$ with m the physical electron mass, from (10.32)-(10.34) one sees readily

$$C H_0 C^\dagger = H_0 \quad . \tag{10.39}$$

In accordance with our convention (3.7)-(3.11), $\gamma_2 = \rho_2 \sigma_2$ is real and symmetric. Hence the Hermitian conjugate of

$$C \psi_\alpha C^\dagger = \eta_c (\gamma_2)_{\alpha\lambda} \psi_\lambda^\dagger \tag{10.40}$$

is

$$C \psi_\alpha^\dagger C^\dagger = \eta_c^* (\gamma_2)_{\alpha\lambda} \psi_\lambda = \eta_c^* \psi_\lambda (\gamma_2)_{\lambda\alpha} \quad . \tag{10.41}$$

The electromagnetic current operator is $j_\mu = i : \psi^\dagger \gamma_4 \gamma_\lambda \psi :$. By using (10.40)-(10.41), we find

$$\begin{aligned} C j_\mu C^\dagger &= i : C \psi_\alpha^\dagger C^\dagger (\gamma_4 \gamma_\mu)_{\alpha\beta} \, C \psi_\beta C^\dagger : \\ &= i : \psi_\lambda (\gamma_2)_{\lambda\alpha} (\gamma_4 \gamma_\mu)_{\alpha\beta} (\gamma_2)_{\beta\lambda'} \psi_{\lambda'}^\dagger : \\ &= -i : \psi_{\lambda'}^\dagger (\gamma_2 \gamma_4 \gamma_\mu \gamma_2)_{\lambda\lambda'} \psi_\lambda : \end{aligned} \tag{10.42}$$

where the minus sign is due to the exchange of the order of the fermion operators ψ_λ and $\psi_{\lambda'}^\dagger$, but with their anticommutator absent because of the normal product. Since γ_2, γ_4 are symmetric and γ_1, γ_3 antisymmetric, the transpose of $\gamma_2 \gamma_4 \gamma_\mu \gamma_2$ is given by

$$\widetilde{\gamma_2 \gamma_4 \gamma_\mu \gamma_2} = \gamma_2 \tilde{\gamma}_\mu \gamma_4 \gamma_2 = \gamma_4 \gamma_\mu \quad . \tag{10.43}$$

Substituting (10.43) into (10.42), we derive

$$C j_\mu C^\dagger = - j_\mu \quad . \tag{10.44}$$

Since $j_4 = i\rho$, it follows that

$$C \rho C^\dagger = -\rho \quad , \tag{10.45}$$

which, together with (6.16), yields

$$C A_0 C^\dagger = - A_0 \; . \tag{10.46}$$

Combining (10.4) with (10.46), we can write

$$C A_\mu C^\dagger = - A_\mu \; , \tag{10.47}$$

and that gives

$$C j_\mu A_\mu C^\dagger = j_\mu A_\mu$$

which establishes the particle-antiparticle conjugation symmetry of the electromagnetic interaction.

3. Lastly, we examine the parity operator P, defined by (10.8)-(10.9). Upon substituting the Fourier expansion (10.20) into (10.9), we find that its lefthand side becomes

$$P\psi(\vec{r}, t)\, P^\dagger = \sum_{\vec{p},s} \left(P a_{\vec{p},s} P^\dagger u_{\vec{p},s}\, e^{i\vec{p}\cdot\vec{r}} + P b^\dagger_{\vec{p},s} P^\dagger v_{\vec{p},s}\, e^{-i\vec{p}\cdot\vec{r}} \right), \tag{10.48}$$

and its righthand side, on account of (3.82), can be written as

$$\begin{aligned} \eta_p \gamma_4 \psi(-\vec{r}, t) &= \sum_{\vec{p},s} \eta_p \left(a_{\vec{p},s} \gamma_4 u_{\vec{p},s}\, e^{-i\vec{p}\cdot\vec{r}} + b^\dagger_{\vec{p},s} \gamma_4 v_{\vec{p},s}\, e^{i\vec{p}\cdot\vec{r}} \right) \\ &= \sum_{\vec{p},s} \eta_p \left(a_{\vec{p},s} u_{-\vec{p},-s}\, e^{-i\vec{p}\cdot\vec{r}} - b^\dagger_{\vec{p},s} v_{-\vec{p},-s}\, e^{i\vec{p}\cdot\vec{r}} \right) \\ &= \sum_{\vec{p},s} \eta_p \left(a_{-\vec{p},-s} u_{\vec{p},s}\, e^{i\vec{p}\cdot\vec{r}} - b^\dagger_{-\vec{p},-s} v_{\vec{p},s}\, e^{-i\vec{p}\cdot\vec{r}} \right). \end{aligned} \tag{10.49}$$

Hence, by equating (10.48) with (10.49), we obtain

$$P b^\dagger_{\vec{p},s} P^\dagger = - \eta_p b^\dagger_{-\vec{p},-s} \tag{10.50}$$

and

$$P a_{\vec{p},s} P^\dagger = \eta_p a_{-\vec{p},-s} \; , \tag{10.51}$$

whose Hermitian conjugate is

$$P a^\dagger_{\vec{p},s} P^\dagger = \eta_p^* a^\dagger_{-\vec{p},-s} \; . \tag{10.52}$$

Likewise, by using (10.8) and the Fourier expansion (6.32)-(6.33), we can derive for the photon creation operator

$$P \vec{a}^{\dagger}_{\vec{p}} P^{\dagger} = - \vec{a}^{\dagger}_{-\vec{p}} \tag{10.53}$$

where, as in (6.34),

$$\vec{p} \cdot \vec{a}^{\dagger}_{\vec{p}} = 0 \quad . \tag{10.54}$$

For photons of a definite helicity, we may use the creation operator defined by (6.36). Equation (10.53) takes on the form

$$P \, a^{\dagger}_{\vec{p},s} P^{\dagger} = - a^{\dagger}_{-\vec{p},-s} \quad . \tag{10.55}$$

Next, we consider a linear transformation in the Hilbert space spanned by (10.25):

$$| 0 > \rightarrow | 0 > , \quad a^{\dagger}_{\vec{p},s} | 0 > \rightarrow \eta_P^* \, a^{\dagger}_{-\vec{p},-s} | 0 > ,$$

$$b^{\dagger}_{\vec{p},s} | 0 > \rightarrow - \eta_P \, b^{\dagger}_{-\vec{p},-s} | 0 >, \quad a^{\dagger}_{\vec{p},s} | 0 > \rightarrow - a^{\dagger}_{-\vec{p},-s} | 0 > , \cdots \tag{10.56}$$

which, as in (10.35), preserves the orthonormality of the complete set of basis vectors, and is therefore unitary. One can readily see that the transformation operator satisfies (10.51)-(10.53). Therefore P exists and

$$P^{\dagger} P = 1 \quad ; \tag{10.57}$$

furthermore, because of the first expression in (10.56), we have fixed the overall phase factor of P by setting

$$P \, | 0 > = | 0 > \quad . \tag{10.58}$$

From (10.38), one observes that the free Hamiltonian is invariant under P ; i.e.

$$P \, H_0 \, P^{\dagger} = H_0 \quad . \tag{10.59}$$

As in (10.42), the transformation of the electromagnetic current operator j_μ under P is given by

$$P\, j_\mu(\vec{r}, t)\, P^\dagger = i : P\psi^\dagger(\vec{r}, t)\, P^\dagger\, \gamma_4\, \gamma_\mu\, P\,\psi(\vec{r}, t)\, P^\dagger :$$

$$= i : \psi^\dagger(-\vec{r}, t)\, \gamma_4\, \gamma_4\, \gamma_\mu\, \gamma_4\, \psi(-\vec{r}, t) :$$

$$= \begin{cases} -\, j_i(-\vec{r}, t) & \mu = i \neq 4 \\ +\, j_4(-\vec{r}, t) & \mu = 4 \quad , \end{cases} \tag{10.60}$$

which, together with (10.8), leads to

$$P\, \vec{j}(\vec{r}, t) \cdot \vec{A}(\vec{r}, t)\, P^\dagger = \vec{j}(-\vec{r}, t) \cdot \vec{A}(-\vec{r}, t) \quad . \tag{10.61}$$

Since $j_4 = i\rho$ and $\nabla^2 A_0 = -\, e\rho$, we have

$$P\, \rho(\vec{r}, t)\, A_0(\vec{r}, t)\, P^\dagger = \rho(-\vec{r}, t)\, A_0(-\vec{r}, t) \quad . \tag{10.62}$$

Furthermore, because of (10.9) and its Hermitian conjugate, we obtain

$$P\, \psi^\dagger(\vec{r}, t)\, \beta\, \psi(\vec{r}, t)\, P^\dagger = \psi^\dagger(-\vec{r}, t)\, \beta\, \psi(-\vec{r}, t) \quad . \tag{10.63}$$

Combining (10.61)-(10.63) with (6.28), we see that

$$P\, H_{int}\, P^\dagger = H_{int} \quad . \tag{10.64}$$

Equations (10.59) and (10.64) establish the invariance of H under P.

Remarks. Under the space inversion, the momentum $\vec{p}$ of a particle changes sign, but its spin $\vec{\sigma}$ does not, and therefore its helicity $s = \vec{\sigma} \cdot \hat{p} \to -s$, in accordance with (10.50)-(10.55). For a nonrelativistic particle, its spin is decoupled from its momentum; we may replace (3.27) by

$$\vec{\sigma} \cdot \hat{z} \begin{cases} u_{\vec{p}, \sigma_z} \\ v_{-\vec{p}, \sigma_z} \end{cases} = \sigma_z \begin{cases} u_{\vec{p}, \sigma_z} \\ -\, v_{-\vec{p}, \sigma_z} \end{cases} \tag{10.65}$$

where $\hat{z}$ is a unit vector along the z-axis, which can be arbitrarily chosen, and $\frac{1}{2}\sigma_z = \pm\frac{1}{2}$ is the spin of the particle along $\hat{z}$.

Correspondingly, (3.31) becomes

$$S_{\vec{p}}(t) = \sum_{\sigma_z=\pm 1} (a_{\vec{p},\sigma_z}(t)\, u_{\vec{p},\sigma_z} + b^{\dagger}_{-\vec{p},\sigma_z}(t)\, v_{-\vec{p},\sigma_z}) . \tag{10.66}$$

From (10.52)-(10.53), one can readily derive

$$\begin{aligned} P\, a^{\dagger}_{\vec{p},\sigma_z}\, P^{\dagger} &= \eta_p^*\, a^{\dagger}_{-\vec{p},\sigma_z} , \\ P\, b^{\dagger}_{\vec{p},\sigma_z}\, P^{\dagger} &= -\eta_p\, b^{\dagger}_{-\vec{p},\sigma_z} , \end{aligned} \tag{10.67}$$

which shows explicitly that spin, being an angular momentum, is a pseudovector and therefore does not change sign under P .

10.2 Applications

1. Furry theorem

Let us consider a state of n photons

$$| n\gamma > \equiv \prod_{i=1}^{n} a^{\dagger}_{\vec{p}_i,s_i} | 0 > . \tag{10.68}$$

From (10.34) and (10.37) we have

$$C | n\gamma > = (-1)^n | n\gamma > . \tag{10.69}$$

Since the QED Hamiltonian is C - invariant, the $U(t, t_0)$ matrix defined by (5.20)-(5.21) satisfies

$$C\, U(t, t_0)\, C^{\dagger} = U(t, t_0) . \tag{10.70}$$

In the limits $t \to \infty$ and $t_0 \to -\infty$, $U(t, t_0)$ becomes the S-matrix. Hence,

$$C\, S\, C^{\dagger} = S . \tag{10.71}$$

We may consider the matrix element

$$< n'\gamma | S | n\gamma > , \tag{10.72}$$

which, because of (10.69) and (10.71), is equal to

$$\langle n'\gamma | C^{\dagger} S C | n\gamma \rangle = (-1)^{n+n'} \langle n'\gamma | S | n\gamma \rangle \quad , \tag{10.73}$$

which must be zero if $n + n'$ is an odd number. Therefore, we establish the theorem

$$\text{even numbers of } \gamma \not\rightleftarrows \text{ odd numbers of } \gamma \quad , \tag{10.74}$$

valid to all orders of the electromagnetic interaction.

The above selection rule can be easily extended to include the strong interaction, provided that the strong-interaction Hamiltonian H_{st} is also assumed to be C-invariant; i.e.

$$C H_{st} C^{\dagger} = H_{st} \quad . \tag{10.75}$$

Selection rule (10.74) is then valid to all orders of the strong, as well as the electromagnetic, interaction.

A glance at the Table of Particle Properties at the end of this book tells one that the dominant decay mode of π^o is

$$\pi^o \rightarrow 2\gamma \quad . \tag{10.76}$$

From assumption (10.75) and Eq. (10.69), we conclude π^o is of $C = +1$, and therefore

$$\pi^o \not\rightarrow \text{ odd number of } \gamma \quad . \tag{10.77}$$

The experimental upper bound of the branching ratio of $\pi^o \rightarrow 3\gamma$ is $< 5 \times 10^{-6}$, which confirms the above selection rule and in turn gives support to the correctness of our assumption that H_{st} is C-invariant.

Remarks. In the case of QED, C-invariance is established explicitly; in the case of the strong interaction, it is assumed. In either event, the deduction of selection rules, such as (10.74) and (10.77), from C-invariance can be made independently of the details of the Hamiltonian.

The close correlation between the general theoretical assumptions and their rigorous experimental consequences makes the study of symmetry particularly rewarding.

2. Positronium states

It is customary to label the positronium e^+e^- states as ${}^{2S+1}L_J$ where the quantum number L is the orbital angular momentum, J the total angular momentum and S the total spin, with S = 0 being the singlet and S = 1 the triplet. We first establish that such a state is of

$$P = -(-1)^L \qquad \text{and} \qquad C = (-1)^{L+S} \quad . \tag{10.78}$$

Proof. Because the electromagnetic interaction is both C- and P-invariant, we can adiabatically reduce the electromagnetic coupling constant e^2 without altering the C and P quantum numbers of the positronium state ${}^{2S+1}L_J$. In the limit of very small e^2, the non-relativistic description becomes sufficient, and therefore we can neglect the coupling between spin and orbital angular momentum as well as the presence of virtual photons. The state vector is then given by

$$| {}^{2S+1}L_J > = \sum_{m,\sigma_z,\sigma_z'} \int d^3p\, Y_{LM}(\hat{p})\, \chi_m(\sigma_z, \sigma_z')\, C_{Mm}(p) \cdot a^\dagger_{\vec{p},\sigma_z}\, b^\dagger_{-\vec{p},\sigma_z'} | 0 > \tag{10.79}$$

where, as in (10.65)-(10.66), subscripts σ_z and σ_z' can be ± 1; they denote the spin components of e^- and e^+ along the z-axis, with the spin function $\chi_m(\sigma_z, \sigma_z')$ antisymmetric for the singlet, but symmetric for the triplet,

$$\chi_m(\sigma_z, \sigma_z') = (-1)^{S+1} \chi_m(\sigma_z', \sigma_z) \quad . \tag{10.80}$$

Quite often in the literature one writes for the triplet state, $\chi_m = \uparrow_- \uparrow_+$

for $m = 1$, $(\uparrow_- \downarrow_+ + \downarrow_- \uparrow_+)/\sqrt{2}$ for $m = 0$, and $\downarrow_- \downarrow_+$ for $m = -1$; for the singlet state one has only $m = 0$ and the corresponding $\chi_m = (\uparrow_- \downarrow_+ - \downarrow_- \uparrow_+)/\sqrt{2}$, where the subscript $-$ is for e^- and $+$ for e^+. In (10.79) the momenta of e^- and e^+ are $\vec{p}$ and $-\vec{p}$ respectively, $\hat{p} = \vec{p}/p$ with $p = |\vec{p}|$, Y_{LM} denotes the spherical harmonics given by (1.38)-(1.40) which satisfies

$$Y_{LM}(-\hat{p}) = (-1)^L Y_{LM}(\hat{p}) \quad . \tag{10.81}$$

The z-component of the total angular momentum of the state is $J_z = M + m$; $C_{Mm}(p)$ denotes the appropriate Clebsch-Gordon coefficient multiplied by the radial function, which is independent of the direction of $\vec{p}$.

We now apply the Hilbert space operator P onto the state vector (10.79):

$$P\,|{}^{2S+1}L_J\rangle = \sum_{m,\sigma_z,\sigma_z'} \int d^3p\, Y_{LM}(\hat{p})\, \chi_m(\sigma_z, \sigma_z')\, C_{Mm}(p) \cdot P\, a^\dagger_{\vec{p},\sigma_z}\, b^\dagger_{-\vec{p},\sigma_z'}\, |0\rangle \quad . \tag{10.82}$$

Because of (10.58) and (10.67),

$$\begin{aligned} P\, a^\dagger_{\vec{p},\sigma_z}\, b^\dagger_{-\vec{p},\sigma_z'}\, |0\rangle &= P\, a^\dagger_{\vec{p},\sigma_z}\, P^\dagger P\, b^\dagger_{-\vec{p},\sigma_z'}\, P^\dagger P\, |0\rangle \\ &= -\, a^\dagger_{-\vec{p},\sigma_z}\, b^\dagger_{\vec{p},\sigma_z'}\, |0\rangle \end{aligned} \tag{10.83}$$

which, together with (10.82), gives

$$P\,|{}^{2S+1}L_J\rangle = -\sum_{m,\sigma_z,\sigma_z'} \int d^3p\, Y_{LM}(\hat{p})\, \chi_m(\sigma_z, \sigma_z')\, C_{Mm}(p) \cdot a^\dagger_{-\vec{p},\sigma_z}\, b^\dagger_{\vec{p},\sigma_z'}\, |0\rangle \quad . \tag{10.84}$$

By re-labeling $-\vec{p}$ as $\vec{p}$ and by using (10.81), we can rewrite (10.84):

$$P\,|{}^{2S+1}L_J> = -\sum_{m,\sigma_z,\sigma_z'} \int d^3p\, Y_{LM}(-\hat{p})\, \chi_m(\sigma_z,\sigma_z')\, C_{Mm}(p) \cdot a^\dagger_{\vec{p},\sigma_z}\, b^\dagger_{-\vec{p},\sigma_z'}\,|0>$$

$$= -(-1)^L\,|{}^{2S+1}L_J> , \qquad (10.85)$$

which gives the first equation in (10.78).

By using (10.32)-(10.33) and (10.37), we have

$$C\,a^\dagger_{\vec{p},\sigma_z}\, b^\dagger_{-\vec{p},\sigma_z'}\,|0> = C\,a^\dagger_{\vec{p},\sigma_z}\, C^\dagger C\, b^\dagger_{-\vec{p},\sigma_z'}\, C^\dagger C\,|0>$$

$$= b^\dagger_{\vec{p},\sigma_z}\, a^\dagger_{-\vec{p},\sigma_z'}\,|0>$$

$$= -\,a^\dagger_{-\vec{p},\sigma_z'}\, b^\dagger_{\vec{p},\sigma_z}\,|0> . \qquad (10.86)$$

Hence, the application of C to the state vector (10.79) gives

$$C\,|{}^{2S+1}L_J> = \sum_{m,\sigma_z,\sigma_z'} \int d^3p\, Y_{LM}(\hat{p})\, \chi_m(\sigma_z,\sigma_z')\, C_{Mm}(p) \cdot C\,a^\dagger_{\vec{p},\sigma_z}\, b^\dagger_{-\vec{p},\sigma_z'}\,|0>$$

$$= -\sum_{m,\sigma_z,\sigma_z'} \int d^3p\, Y_{LM}(\hat{p})\, \chi_m(\sigma_z,\sigma_z')\, C_{Mm}(p) \cdot a^\dagger_{-\vec{p},\sigma_z'}\, b^\dagger_{\vec{p},\sigma_z}\,|0>$$

$$= -\sum_{m,\sigma_z,\sigma_z'} \int d^3p\, Y_{LM}(-\hat{p})\, \chi_m(\sigma_z',\sigma_z)\, C_{Mm}(p) \cdot a^\dagger_{\vec{p},\sigma_z}\, b^\dagger_{-\vec{p},\sigma_z'}\,|0> \qquad (10.87)$$

which, because of (10.80) and (10.81), is

$$(-1)^{L+S}\,|{}^{2S+1}L_J>$$

and that establishes the second equation in (10.78).

Combining (10.78) with (10.69), we derive the selection rules:

states with even $L+S \not\to$ odd γ , (10.88)

which applies to, e.g., ${}^1S_0$, ${}^3P_0$, ${}^3P_1$, ${}^3P_2$, ... states;

states with odd $L+S \not\to$ even γ , (10.89)

which applies to, e.g., ${}^3S_1$, ${}^1P_1$, ... states.

Remarks. There is a simple method for remembering the two equations in (10.78). In accordance with (10.67), the relative parity between e^+ and e^- is -1 ; under space inversion, spin is unchanged, $\vec{p} \to -\vec{p}$ and therefore $Y_{LM} \to (-1)^L Y_{LM}$. Together they give $P = -(-1)^L$.

Under C, we exchange e^+ and e^-; this gives a factor $(-1)^L$ due to the orbital angular momentum, a factor $(-1)^{S+1}$ due to the spin part and another -1 due to Fermi statistics. Together they give $C = (-1)^L (-1)^{S+1} (-1) = (-1)^{L+S}$.

3. Decay of a spin-0 particle $\to 2\gamma$

Let us consider the rest frame of the spin-0 particle. In its 2γ final state, the momenta of these two photons are equal; each γ carries a polarization direction which can be defined to be that of its electric field. We shall now establish that:

for a scalar particle $(P=+1)$, the polarization directions of these two γ's are parallel, (10.90)

for a pseudoscalar particle $(P=-1)$, the two polarization directions are perpendicular. (10.91)

Proof. Any 2γ state in its center-of-mass system can be written as

$$| 2\gamma > \; = \int d^3p \, \chi_{ij}(\vec{p}) \, a_i^\dagger(\vec{p}) \, a_j^\dagger(-\vec{p}) \; | \; 0 > \quad , \qquad (10.92)$$

where $a_i^\dagger(\vec{p})$ denotes the ith component of the photon creation operator $\vec{a}^{\dagger}_{\vec{p}}$ in (6.32)-(6.33). From (6.34) it follows that

$$p_i \, a_i^{\dagger}(\vec{p}) = 0 \tag{10.93}$$

where, as before, all repeated indices are to be summed over. For (10.92) to be of spin 0, the function $\chi_{ij}(\vec{p})$ must be a tensor of second rank under space rotation. Since it depends only on a single vector $\vec{p}$, the most general form of $\chi_{ij}(\vec{p})$ is

$$\chi_{ij}(\vec{p}) = A\delta_{ij} + B\,\epsilon_{ijk}\,p_k + C\,p_i\,p_j \quad , \tag{10.94}$$

where δ_{ij} is the Kronecker symbol and

$$\epsilon_{ijk} = \begin{cases} +1 & \text{if } ijk \text{ is an even permutation of } 1, 2, 3, \\ -1 & \text{if } ijk \text{ is an odd permutation of } 1, 2, 3, \\ 0 & \text{otherwise.} \end{cases} \tag{10.95}$$

A, B, C are functions of $|\vec{p}|$. Because of (10.93), the term $C\,p_i\,p_j$ in (10.94) makes no contribution to the $|2\gamma>$ state of (10.92); $\chi_{ij}(\vec{p})$ can be simplified to

$$\chi_{ij}(\vec{p}) = A\,\delta_{ij} + B\,\epsilon_{ijk}\,p_k \quad . \tag{10.96}$$

On account of (10.53), we have in our present notation

$$P\,a_i^{\dagger}(\vec{p})\,P^{\dagger} = -\,a_i^{\dagger}(-\vec{p}) \quad . \tag{10.97}$$

Hence,

$$\begin{aligned} P\,|\,2\gamma> &= \int d^3p\,\chi_{ij}(\vec{p})\,P\,a_i^{\dagger}(\vec{p})\,P^{\dagger}\,P\,a_j^{\dagger}(-\vec{p})\,P^{\dagger}\,P\,|\,0> \\ &= \int d^3p\,\chi_{ij}(\vec{p})\,a_i^{\dagger}(-\vec{p})\,a_j^{\dagger}(\vec{p})\,|\,0> \\ &= \int d^3p\,\chi_{ij}(-\vec{p})\,a_i^{\dagger}(\vec{p})\,a_j^{\dagger}(-\vec{p})\,|\,0> \quad , \end{aligned} \tag{10.98}$$

which means that for states with $P = \pm 1$,

$$\chi_{ij}(-\vec{p}) = \pm\,\chi_{ij}(\vec{p}) \quad .$$

Thus we obtain

$$\chi_{ij}(\vec{p}) = \begin{cases} A\,\delta_{ij} & P = 1 \;, \\ B\,\epsilon_{ijk}\,p_k & P = -1 \;. \end{cases} \tag{10.99}$$

According to the Fourier expansion (6.33), the vector direction of $\vec{a}^\dagger$ is the corresponding direction of $\vec{E}$. Equation (10.99) shows that the polarizations of two photons are parallel when $P = +1$, and perpendicular when $P = -1$, which establishes (10.90)-(10.91).

Assuming that π^o is of spin 0, we may use the final polarization directions of the two photons in

$$\pi^o \to 2\gamma$$

to determine the parity of π^o. In the decay

$$\pi^o \to \gamma_a + \gamma_b \to (e_a^+ + e_a^-) + (e_b^+ + e_b^-) \;,$$

the plane determined by the pair $e_a^+ e_a^-$ contains the electric field vector of the parent photon γ_a (and an identical relation between $e_b^+ e_b^-$ and γ_b). Via the relative orientations of the two planes, determined by $e_a^+ e_a^-$ and $e_b^+ e_b^-$, Chinovsky and Steinberger* were able to establish that π^o is a pseudoscalar; i.e., $P_{\pi^o} = -1$. Again, this conclusion is valid so long as both the strong and electromagnetic interactions are P-invariant.

Remarks. The same results (10.90)-(10.91) can also be derived phenomenologically. We may represent the parent meson by a spin-0 field ϕ, and consider an effective Lagrangian density $\mathcal{L}_{eff}$ between ϕ and the electromagnetic fields of the 2γ, which will be denoted by a and b respectively. From Lorentz-invariance, we find that if ϕ is a scalar, then

---

* W. Chinovsky and J. Steinberger, Phys. Rev. 95, 1561 (1954). See also the references cited therein.

$$\mathcal{L}_{eff} \propto \phi F_{\mu\nu}(a) F_{\mu\nu}(b) \propto \phi[\vec{E}(a)\cdot\vec{E}(b) - \vec{B}(a)\cdot\vec{B}(b)] \quad , \qquad (10.100)$$

and if ϕ is a pseudoscalar,

$$\mathcal{L}_{eff} \propto \phi\, \epsilon_{\mu\nu\lambda\delta} F_{\mu\nu}(a) F_{\lambda\delta}(b) \propto \phi[\vec{E}(a)\cdot\vec{B}(b) + \vec{B}(a)\cdot\vec{E}(b)] \qquad (10.101)$$

where

$$\epsilon_{\mu\nu\lambda\delta} = \begin{cases} +1 & \text{if } \mu\nu\lambda\delta \text{ is an even permutation of } 1,2,3,4, \\ -1 & \text{if } \mu\nu\lambda\delta \text{ is an odd permutation of } 1,2,3,4, \\ 0 & \text{otherwise.} \end{cases} \qquad (10.102)$$

It is clear that (10.100), (10.101) give (10.90), (10.91) respectively.

4. Spin-1 particle $\not\to 2\gamma$

Let us try to construct a 2γ state with a total angular momentum $J = 1$ in the center-of-mass system. Such a state must transform like a vector under space rotation; like the component v_i of a vector $\vec{v}$, it must carry a similar index $i = 1,2,3$ making a total of three $J = 1$ states. We may therefore write

$$|\, 2\gamma >_i \;=\; \int d^3p\, \chi_{ijk}(\vec{p})\, a_j^\dagger(\vec{p})\, a_k^\dagger(-\vec{p})\, |\, 0 > \qquad (10.103)$$

where $\chi_{ijk}(\vec{p})$ is a tensor of third rank under space rotation. Since χ_{ijk} depends only on the single vector $\vec{p}$, its most general form is given by

$$\begin{aligned} \chi_{ijk}(\vec{p}) = \; & A\,\epsilon_{ijk} + B p_i \delta_{jk} + C p_j \delta_{ik} + D p_k \delta_{ij} \\ & + B' p_i \epsilon_{jk\ell} p_\ell + C' p_j \epsilon_{ki\ell} p_\ell + D' p_k \epsilon_{ij\ell} p_\ell \\ & + E p_i p_j p_k \end{aligned} \qquad (10.104)$$

where A, B, $\cdots$, D', E are functions only of $|\vec{p}|$. Because $p_i a_i^\dagger(\vec{p}) = 0$, in the above expression the C, D, C', D and E terms do not contribute to the 2γ states of (10.103). The above expression can be simplified to

$$\chi_{ijk}(\vec{p}) = A\,\epsilon_{ijk} + B p_i \delta_{jk} + B' p_i \epsilon_{jk\ell} p_\ell \tag{10.105}$$

which satisfies

$$\chi_{ijk}(\vec{p}) = -\chi_{ikj}(-\vec{p}) \quad . \tag{10.106}$$

On the other hand, because the photons obey boson statistics, we may commute the two creation operators in (10.103):

$$|\,2\gamma>_i = \int d^3p\, \chi_{ijk}(\vec{p})\, a_k^\dagger(-\vec{p})\, a_j^\dagger(\vec{p})\,|\,0>$$

which, after a re-labeling of $\vec{p}$ as $-\vec{p}$ and j, k as k, j, becomes

$$|\,2\gamma>_i = \int d^3p\, \chi_{ikj}(-\vec{p})\, a_j^\dagger(\vec{p})\, a_k^\dagger(-\vec{p})\,|\,0> \quad . \tag{10.107}$$

By adding (10.103) to (10.107) and by using (10.106), we obtain

$$\begin{aligned} |\,2\gamma>_i &= \int d^3p\, \tfrac{1}{2}[\chi_{ijk}(\vec{p}) + \chi_{ikj}(-\vec{p})]\, a_j^\dagger(\vec{p})\, a_k^\dagger(-\vec{p})\,|\,0> \\ &= 0 \quad . \end{aligned} \tag{10.108}$$

That means it is not possible to put 2γ in a total angular momentum $J = 1$ state. Hence, a spin-1 particle $\not\to 2\gamma$. In the decay of positronium, all states with $J = 1$ must obey the selection rule

$$^{2S+1}L_{J=1} \not\to 2\gamma \quad . \tag{10.109}$$

However, such states can decay into 3γ,

$$^{2S+1}L_{J=1} \to 3\gamma \quad . \tag{10.110}$$

Likewise, from $\pi^0 \to 2\gamma$ one knows that the spin of $\pi^0 \neq 1$. Later on, in Chapter 13, we shall show that the pion spin is indeed 0.

10.3 General Discussion

Let us consider a general system whose Hamiltonian

$$H = H_0 + H_{int}$$

is invariant under a unitary transformation $\mathscr{S}$; i.e.,

$$\mathscr{S} H_0 \mathscr{S}^\dagger = H_0 \quad , \quad \mathscr{S} H_{int} \mathscr{S}^\dagger = H_{int} \tag{10.111}$$

and

$$\mathscr{S}^\dagger \mathscr{S} = 1 \quad . \tag{10.112}$$

For example, $\mathscr{S}$ can be $e^{iL\theta}$, or C, or P. From (10.111)-(10.112), it follows that

$$\mathscr{S} H_0 = H_0 \mathscr{S} \quad \text{and} \quad \mathscr{S} H_{int} = H_{int} \mathscr{S} \tag{10.113}$$

which, on account of (5.1) and (5.6), implies that $\mathscr{S}$ is independent of t in either the Heisenberg or the interaction representation. Furthermore, in the interaction representation the $U(t, t_0)$ matrix, defined by (5.20)-(5.21), commutes with $\mathscr{S}$,

$$[\mathscr{S}, U(t, t_0)] = 0 \quad . \tag{10.114}$$

We may denote the eigenstate of the free Hamiltonian H_0 by $|n\rangle_{free}$,

$$H_0 \, |n\rangle_{free} = E_n \, |n\rangle_{free} \quad . \tag{10.115}$$

In accordance with (6.58) and (6.60), there exists a corresponding eigenstate $|n\rangle_{phys}$ of the total Hamiltonian, related to $|n\rangle_{free}$ by

$$|n\rangle_{phys} \equiv U(t, -\infty) \, |n\rangle_{free} \quad . \tag{10.116}$$

From (10.113), the matrices H_0 and $\mathscr{S}$ can be diagonalized simultaneously. Thus, in (10.115) the state $|n\rangle_{free}$ can be chosen to be

$$\mathscr{S} \, |n\rangle_{free} = s \, |n\rangle_{free}$$

where s is the eigenvalue. Because of (10.114) and (10.116), we have

$$\mathscr{S} \, |n\rangle_{phys} = U(t, -\infty) \, \mathscr{S} \, |n\rangle_{free} = s \, |n\rangle_{phys} \quad .$$

Hence, independently of the details of H_{int}, the invariance assumption

(10.111) implies that the transformation properties of the physical state $| n >_{phys}$ are completely determined by those of the corresponding free state $| n >_{free}$, and that greatly simplifies the mathematical analysis.

10.4 Baryon Number and Lepton Number

To each physical single-particle state we assign the following eigenvalue to the baryon-number operator N :

$N = 1$ for a single-baryon state, e.g., $p, n, \Lambda, \cdots$,

$N = -1$ for a single-antibaryon state, e.g., $\bar{p}, \bar{n}, \bar{\Lambda}, \cdots$,

$N = 0$ for all other single-particle states, e.g., $e^{\pm}, \pi^{\pm}, \pi^{0}, \gamma, \cdots$.

(10.117)

For a multiparticle state, the baryon number is given by the corresponding algebraic sum, $\sum N_i$. Hadrons with $N = 0$ are called mesons, otherwise baryons or antibaryons.

By definition, N is a Hermitian operator since its eigenvalues are all real numbers (in fact, integers). Let θ be a real number. The operator

$$U = e^{iN\theta} \tag{10.118}$$

is unitary. The assumption that the total Hamiltonian H is invariant under U :

$$U H U^{\dagger} = H \tag{10.119}$$

insures that

$$[N, H] = 0 \quad , \tag{10.120}$$

as can be readily verified by differentiating (10.119) with respect to θ . Thus, N is conserved. The converse is also true; i.e. (10.120) implies (10.119). From (10.118), we have

$$U\,|\,e> \;=\; |\,e> \;, \qquad U\,|\,p> \;=\; e^{i\theta}\,|\,p> \;,$$

$$U\,|\,n> \;=\; e^{i\theta}\,|\,n> \;, \qquad U\,|\,2p> \;=\; e^{2i\theta}\,|\,2p> \;, \;\ldots\;.$$

If N is conserved, then since p is the lowest mass state with $N = 1$, it must be stable; e.g.

$$p \not\rightarrow e^+ + \gamma \qquad \text{and} \qquad p \not\rightarrow \mu^+ + \gamma \;.$$

Thus, the stability of the proton is tied to the conservation of baryon number.

Likewise, we can assign the e-lepton number L_e and the μ-lepton number L_μ:

	μ^-	μ^+	ν_μ	$\bar{\nu}_\mu$	e^-	e^+	ν_e	$\bar{\nu}_e$	other
L_μ	+1	-1	+1	-1	0	0	0	0	0
L_e	0	0	0	0	+1	-1	+1	-1	0 .

The lepton number of a multiparticle state is again the algebraic sum of the lepton numbers of its constituents. Hence, conservation of L_μ and L_e implies that, e.g. in the decay of π^+, depending on the charged lepton in the final state, the neutrino can be either ν_μ or ν_e:

$$\pi^+ \rightarrow \begin{cases} \mu^+ + \nu_\mu & (10.121) \\ e^+ + \nu_e & (10.122) \end{cases} \;.$$

Likewise, in the decay of its antiparticle π^- we have

$$\pi^- \rightarrow \begin{cases} \mu^- + \bar{\nu}_\mu & (10.123) \\ e^- + \bar{\nu}_e & (10.124) \end{cases} \;.$$

If one wishes, one may regard (10.121)-(10.124) as the definition of these neutrinos and antineutrinos. Because the strong interaction can cause virtual transitions $p \rightleftarrows n + \pi^+$ and $n \rightleftarrows p + \pi^-$, in μ^--capture and in β-decay, we must have the same kinds of neutrinos and antineutrinos,

$$\mu^- + p \to n + \nu_\mu \quad \text{and} \quad n \to p + e^- + \bar{\nu}_e \ . \tag{10.125}$$

Similarly, in μ-decay, we have

$$\mu^- \to e^- + \nu_\mu + \bar{\nu}_e \quad \text{and} \quad \mu^+ \to e^+ + \bar{\nu}_\mu + \nu_e \ .$$

The conservation of L_μ and L_e leads then to the following selection rules:

$$\mu^- \not\to e^- + \gamma \ , \tag{10.126}$$

$$\nu_\mu + n \not\to e^- + p \ , \tag{10.127}$$

$$Z \not\to (Z+2) + e^- + e^- \ . \tag{10.128}$$

Both (10.126) and (10.127) support only the conservation of the difference $L_\mu - L_e$; the absence of neutrino-less double β-decay (10.128) gives direct evidence for the conservation of L_e. Together they provide the experimental proof of these two lepton number conservations. [See Section 21.1 for a discussion of L_τ.]

Historically, the two-neutrino hypothesis, $\nu_\mu \neq \nu_e$, was introduced to "explain" the extremely small upper bound of the branching ratio for the $\mu \to e + \gamma$ decay,

$$\frac{\text{rate}\ (\mu^- \to e^- + \gamma)}{\text{rate}\ (\mu^- \to e^- + \nu_\mu + \bar{\nu}_e)} < 10^{-8} \ ,$$

and the high-energy neutrino experiment was suggested,* in part

---

* T.D. Lee and C.N. Yang, Phys. Rev. Lett. 4, 307 (1960); M. Schwartz, ibid. 306.

because of the search for ways to substantiate this hypothesis. By comparing the rate for reaction (10.127) with the allowed process $\nu_\mu + n \to \mu^- + p$, Lederman, Schwartz, Steinberger and their collaborators * were able to establish the validity of the two-neutrino hypothesis in 1962 by using high-energy neutrinos; this in turn helped to shape the present massive style of doing high-energy experimental physics.

In addition to the conservation of baryon number N and lepton numbers L_μ and L_e, we also have the familiar conservation of electric charge Q. Unlike the parity P and the particle-antiparticle conjugation C, which are multiplicatively conserved, N, L_μ, L_e and Q are all additively conserved. The unit of electric charge is a measurable quantity, as evidenced by the well-known fine-structure constant $\alpha \cong \frac{1}{137}$. In contrast, the units of N, L_μ and L_e are arbitrarily chosen. This is because while Q gives rise to the Coulomb field, so far as we know neither N nor L_μ, L_e are the sources of physical fields. This disparity is perhaps a very deep one.** It has led people to speculate that probably conservation of electric charge is truly exact, while the conservations of N, L_μ and L_e are only approximate. Hence, very slow transitions such as $p \rightleftarrows e^+ + \pi^0$, $\nu_\mu \rightleftarrows \nu_e$, etc. may be allowed. [The present lifetime limit*** of the proton is $> 2 \times 10^{30}$ years.]

---

* G. Danby, J.-M. Gaillard, K. Goulianos, L. M. Lederman, N. Mistry, M. Schwartz and J. Steinberger, Phys.Rev.Lett. 9, 36 (1962).

** T. D. Lee and C. N. Yang, Phys.Rev. 98, 1501 (1955).

*** K. Landé et al., to be published in the proceedings of the Neutrino '80 conference, Erice.

Problem 10.1. Discuss the experimental foundation of the conservation of N, Q, L_μ and L_e.

Problem 10.2. Prove that in QED, if we define

$$C \equiv \exp[i\pi \sum (\alpha^\dagger_{\vec{p},s}\alpha_{\vec{p},s} + a^\dagger_{\vec{p},s} a_{\vec{p},s})]$$

$$\cdot \exp[\frac{\pi}{2} \sum (b^\dagger_{\vec{p},s} a_{\vec{p},s} - a^\dagger_{\vec{p},s} b_{\vec{p},s})] ,$$

where $\alpha_{\vec{p},s}$, $a_{\vec{p},s}$ and $b_{\vec{p},s}$ are respectively the annihilation operators of γ, e^- and e^+ with momentum $\vec{p}$ and helicity s, then C satisfies

$$C\,\alpha_{\vec{p},s}\,C^\dagger = -\alpha_{\vec{p},s} ,$$

$$C\,a_{\vec{p},s}\,C^\dagger = b_{\vec{p},s} ,$$

$$C\,b_{\vec{p},s}\,C^\dagger = a_{\vec{p},s} ,$$

$$C^\dagger C = C^2 = 1 ,$$

and therefore C is the charge conjugation operator.

Hint: Define $M_\theta \equiv \exp[\theta \sum (b^\dagger_{\vec{p},s} a_{\vec{p},s} - a^\dagger_{\vec{p},s} b_{\vec{p},s})]$.

Through differentiation, show that

$$M_\theta\, a_{\vec{p},s}\, M_\theta^\dagger = \cos\theta\, a_{\vec{p},s} + \sin\theta\, b_{\vec{p},s} ,$$

$$M_\theta\, b_{\vec{p},s}\, M_\theta^\dagger = -\sin\theta\, a_{\vec{p},s} + \cos\theta\, b_{\vec{p},s} .$$

Problem 10.3. Any three-dimensional rotation $r_i \to r_i' = u_{ij} r_j$ can be represented by a two-dimensional rotation $\vec{\theta}$ where the magnitude of $\vec{\theta}$ is the angle of rotation and its direction the axis of rotation (with the convention that when $\vec{\theta} \to 0$, $\vec{r}' = \vec{r} + \vec{\theta} \times \vec{r}$).

(i) Show that, for a spin-$\frac{1}{2}$ field, the rotational operator $\exp(i\vec{J}\cdot\vec{\theta})$, where $\vec{J}$ is the angular momentum operator given by (3.62), satisfies

$$\exp(-i\vec{J}\cdot\vec{\theta})\ \psi(\vec{r},t)\ \exp(i\vec{J}\cdot\vec{\theta}) = e^{i\frac{1}{2}\vec{\sigma}\cdot\vec{\theta}}\ \psi(\vec{r}\,',t) \quad .$$

(ii) Verify that the QED Hamiltonian in the Coulomb gauge is invariant under the unitary transformation R defined by

$$R\,\psi(\vec{r},t)\,R^\dagger = e^{i\frac{1}{2}\vec{\sigma}\cdot\vec{\theta}}\ \psi(\vec{r}\,',t) \quad ,$$

$$R\,A_i(\vec{r},t)\,R^\dagger = u_{ij}\ A_j(\vec{r}\,',t)$$

and

$$R\,\Pi_i(\vec{r},t)\,R^\dagger = u_{ij}\ \Pi_j(\vec{r}\,',t) \quad .$$

Give explicitly the transformation properties of the annihilation and creation operators in the momentum space under R.

Note that if we write $R = \exp(-i\,\vec{\mathcal{J}}\cdot\vec{\theta})$, then $\vec{\mathcal{J}}$ is the total angular momentum operator.

<u>Problem 10.4.</u> Consider a Lorentz transformation $x'_\mu = u_{\mu\nu}\,x_\nu$ in the (z,t) plane where $u_{11} = u_{22} = 1$, $u_{33} = (1-v^2)^{-\frac{1}{2}}$, etc. In the interaction representation, the corresponding transformation on the field operators is given by

$$L\,\psi(x)\,L^{-1} = \exp\left(\frac{i}{2}\,\theta\,\gamma_3\,\gamma_4\right)\psi(x')$$

and

$$L\,A_\mu(x)\,L^{-1} = u_{\mu\nu}\ A_\nu(x')$$

where

$$\cosh\theta = (1-v^2)^{-\frac{1}{2}} \quad .$$

Show that:

(i) L is unitary, although the 4×4 matrix $\exp\left(\frac{i}{2}\,\theta\,\gamma_3\,\gamma_4\right)$ is not,

(ii) the interaction Lagrangian density $e j_\mu A_\mu$ is explicitly Lorentz-invariant, where $j_\mu = i \psi^\dagger \gamma_4 \gamma_\mu \psi$, and

(iii) the operator $\psi^c(x)$ transforms in the same way as $\psi(x)$ under an arbitrary Lorentz transformation where

$$\psi_\alpha^c(x) \equiv (\gamma_2)_{\alpha\beta} \psi_\beta^\dagger(x) \quad .$$

Remark: The use of the interaction representation is by no means necessary; it is only for convenience.

Chapter 11

ISOTOPIC SPIN AND G PARITY

11.1 Isospin

The concept of isospin was introduced in the early '30's by Heisenberg to describe the approximate charge-independent nature of the strong interaction between protons and neutrons. From a phenomenological point of view we can adopt the usage of the proton field ψ_p and the neutron field ψ_n. Together they can be represented by a column matrix

$$\psi(x) \equiv \begin{pmatrix} \psi_p(x) \\ \psi_n(x) \end{pmatrix} . \tag{11.1}$$

Each of the fields in turn is a quantized Dirac spinor operator whose indices are suppressed here, for clarity. Although, as we shall discuss (in Part II B on interactions), neither the proton nor the neutron is elementary; so far as their symmetry properties are concerned it suffices to represent them by phenomenological field operators.

1. U_2 symmetry

Let us consider the following linear transformation

$$\psi \rightarrow \psi' = u\,\psi \tag{11.2}$$

where u is a 2×2 matrix. In order to preserve the anti-commutation relation

$$\{\psi_i(\vec{r},t),\ \psi_j(\vec{r}',t)\} = \delta_{ij}\ \delta^3(\vec{r}-\vec{r}')$$

where i and j can be p or n, the matrix u must be unitary. The group spanned by all such u's is the U_2 group. As we shall see shortly, corresponding to each u there exists a unitary operator U in the Hilbert space such that

$$U\,\psi(x)\,U^\dagger = u\,\psi(x)\ . \tag{11.3}$$

We may express ψ_p and ψ_n in terms of the Fourier expansion (3.32):

$$\psi_p = \frac{1}{\sqrt{\Omega}} \sum_{\vec{k},s} (a_{\vec{k},s}(p)\ u_{\vec{k},s}\ e^{i\vec{k}\cdot\vec{r}} + b^\dagger_{\vec{k},s}(p)\ v_{\vec{k},s}\ e^{-i\vec{k}\cdot\vec{r}})\ ,$$

$$\psi_n = \frac{1}{\sqrt{\Omega}} \sum_{\vec{k},s} (a_{\vec{k},s}(n)\ u_{\vec{k},s}\ e^{i\vec{k}\cdot\vec{r}} + b^\dagger_{\vec{k},s}(n)\ v_{\vec{k},s}\ e^{-i\vec{k}\cdot\vec{r}})\ . \tag{11.4}$$

Substituting these expressions into (11.3), we obtain

$$U\begin{pmatrix} a_{\vec{k},s}(p) \\ a_{\vec{k},s}(n) \end{pmatrix} U^\dagger = u \begin{pmatrix} a_{\vec{k},s}(p) \\ a_{\vec{k},s}(n) \end{pmatrix} \tag{11.5}$$

and

$$U\begin{pmatrix} b^\dagger_{\vec{k},s}(p) \\ b^\dagger_{\vec{k},s}(n) \end{pmatrix} U^\dagger = u \begin{pmatrix} b^\dagger_{\vec{k},s}(p) \\ b^\dagger_{\vec{k},s}(n) \end{pmatrix}\ . \tag{11.6}$$

By writing u explicitly as

$$u = \begin{pmatrix} u_{11} & u_{12} \\ u_{21} & u_{22} \end{pmatrix} \tag{11.7}$$

and by calling $U a_{\vec{k},s}(p)\, U^\dagger$ and $U\, a_{\vec{k},s}(n)\, U^\dagger$, respectively, $a'_{\vec{k},s}(p)$ and $a'_{\vec{k},s}(n)$, we have from (11.5)

$$a'_{\vec{k},s}(p) = u_{11}\ a_{\vec{k},s}(p) + u_{12}\ a_{\vec{k},s}(n)$$

and

$$a'_{\vec{k},s}(n) = u_{21}\ a_{\vec{k},s}(p) + u_{22}\ a_{\vec{k},s}(n)\ .$$

Their Hermitian conjugates are

$$a'^{\dagger}_{\vec{k},s}(p) = u^*_{11}\, a^{\dagger}_{\vec{k},s}(p) + u^*_{12}\, a^{\dagger}_{\vec{k},s}(n) ,$$
$$a'^{\dagger}_{\vec{k},s}(n) = u^*_{21}\, a^{\dagger}_{\vec{k},s}(p) + u^*_{22}\, a^{\dagger}_{\vec{k},s}(n) \tag{11.8}$$

or simply

$$U \begin{pmatrix} a^{\dagger}_{\vec{k},s}(p) \\ a^{\dagger}_{\vec{k},s}(n) \end{pmatrix} U^{\dagger} = u^* \begin{pmatrix} a^{\dagger}_{\vec{k},s}(p) \\ a^{\dagger}_{\vec{k},s}(n) \end{pmatrix} . \tag{11.9}$$

Likewise, by calling $U b^{\dagger}_{\vec{k},s}(p) U^{\dagger}$ and $U b^{\dagger}_{\vec{k},s}(n) U^{\dagger}$, $b'^{\dagger}_{\vec{k},s}(p)$ and $b'^{\dagger}_{\vec{k},s}(n)$, we have

$$b'^{\dagger}_{\vec{k},s}(p) = u_{11}\, b^{\dagger}_{\vec{k},s}(p) + u_{12}\, b^{\dagger}_{\vec{k},s}(n) ,$$
$$b'^{\dagger}_{\vec{k},s}(n) = u_{21}\, b^{\dagger}_{\vec{k},s}(p) + u_{22}\, b^{\dagger}_{\vec{k},s}(n) . \tag{11.10}$$

We shall now show that for each u there indeed exists a unitary operator U in the Hilbert space which satisfies (11.3). Let us choose the basis vectors in the Hilbert space to be

$$|0>, \quad a^{\dagger}_{\vec{k},s}(p)\,|0>, \quad b^{\dagger}_{\vec{k},s}(p)\,|0>, \quad a^{\dagger}_{\vec{k},s}(n)\,|0>, \quad b^{\dagger}_{\vec{k},s}(n)\,|0>, \cdots$$

$$|N_p, N_{\bar{p}}, N_n, N_{\bar{n}}> = \prod_{i=1}^{N_p} \prod_{j=1}^{N_{\bar{p}}} \prod_{\ell=1}^{N_n} \prod_{m=1}^{N_{\bar{n}}} \cdot a^{\dagger}_{\vec{k}_i s_i}(p)\, b^{\dagger}_{\vec{k}_j s_j}(p)\, a^{\dagger}_{\vec{k}_\ell s_\ell}(n)\, b^{\dagger}_{\vec{k}_m s_m}(n)\,|0>, \tag{11.11}$$

etc., where $|0>$ satisfies

$$a_{\vec{k},s}(p)\,|0> = a_{\vec{k},s}(n)\,|0> = b_{\vec{k},s}(p)\,|0> = b_{\vec{k},s}(n)\,|0> = 0 . \tag{11.12}$$

The mapping

$$|0> \to |0> , \qquad a^{\dagger}_{\vec{k},s}(p)\,|0> \to a'^{\dagger}_{\vec{k},s}(p)\,|0> ,$$

$$b^{\dagger}_{\vec{k},s}(p)\,|\,0> \rightarrow b'^{\dagger}_{\vec{k},s}(p)\,|\,0>, \quad a^{\dagger}_{\vec{k},s}(n)\,|\,0> \rightarrow a'^{\dagger}_{\vec{k},s}(n)\,|\,0> ,$$

$$b^{\dagger}_{\vec{k},s}(n)\,|\,0> \rightarrow b'^{\dagger}_{\vec{k},s}(n)\,|\,0>, \; \cdots ,$$

$$|\,N_p, N_{\bar{p}}, N_n, N_{\bar{n}} > \rightarrow \prod_{i=1}^{N_p} \prod_{j=1}^{N_{\bar{p}}} \prod_{\ell=1}^{N_n} \prod_{m=1}^{N_{\bar{n}}} \cdot a'^{\dagger}_{\vec{k}_i s_i}(p)\, b'^{\dagger}_{\vec{k}_j s_j}(p)\, a'^{\dagger}_{\vec{k}_\ell s_\ell}(n)\, b'^{\dagger}_{\vec{k}_m s_m}(n)\,|\,0> \tag{11.13}$$

etc., clearly preserves the orthonormality relations between these vectors, and therefore it is a unitary transformation. Furthermore, the unitary transformation matrix satisfies (11.6) and (11.9) and consequently also (11.3). This establishes that U exists and is unitary. In addition, on account of the first expression in (11.13),

$$U\,|\,0> \; = \; |\,0> \; . \tag{11.14}$$

2. Isospin transformations

Let us separate from U the phase factor (10.118), related to the baryon conservation. We write

$$U = e^{iN\theta}\, S \; , \tag{11.15}$$

with the corresponding u as

$$u = e^{-i\theta}\, s \; ;$$

therefore (11.3) can be converted to

$$S\, \psi(x)\, S^{\dagger} = s\, \psi(x) \; , \tag{11.16}$$

where θ is chosen such that

$$\det\,|\,s\,| = 1 \; . \tag{11.17}$$

The group $\{s\}$, spanned by all such 2×2 unitary matrices with unit determinant, is the isospin$-SU_2$ group; on account of (11.16), so is the group $\{S\}$. The properties of $\{s\}$ are exactly the same as those of the usual three-dimensional rotational group (with spinors), except for the replacement of the ordinary space by the isospin space.

The strong interaction Hamiltonian H_{st} is assumed to be invariant under this SU_2 transformation

$$S H_{st} S^{\dagger} = H_{st} \quad , \tag{11.18}$$

which will be referred to as the isospin transformation. Since p and n have different electromagnetic interactions, clearly the electromagnetic interaction violates the isospin invariance. We may decompose H_{st} into

$$H_{st} = H_{free} + H_{int} \quad . \tag{11.19}$$

Both H_{free} and H_{int} are isospin-symmetric, with the basis vectors (11.11) as eigenstates of H_{free}. In the approximation that the electromagnetic and weak interactions are neglected, the physical vacuum state $|\,vac>$ as well as the physical nucleon or antinucleon states are all eigenvectors of H_{st} ; on account of (6.58) and (6.60) these states are given by

$$\begin{aligned}
|\,vac> &= U(0,-\infty)\,|\,0> \;, \\
|\,p> &\equiv U(0,-\infty)\; a^{\dagger}_{\vec{k},s}(p)\,|\,0> \;, \\
|\,n> &\equiv U(0,-\infty)\; a^{\dagger}_{\vec{k},s}(n)\,|\,0> \;, \\
|\,\bar{p}> &\equiv U(0,-\infty)\; b^{\dagger}_{\vec{k},s}(p)\,|\,0> \;, \\
|\,\bar{n}> &\equiv U(0,-\infty)\; b^{\dagger}_{\vec{k},s}(n)\,|\,0> \;.
\end{aligned} \tag{11.20}$$

Equations (11.14) and (11.15) imply

$$S\,|\,0> = |\,0> \quad . \tag{11.21}$$

Because of (11.18), S commutes with $U(0,-\infty)$. Hence

$$S \mid vac> \; = \; U(0, -\infty)\; S \mid 0 > \; = \; \mid vac> \quad . \tag{11.22}$$

Likewise, from (11.6), (11.9), (11.15) and (11.16) we have

$$S \begin{pmatrix} a^{\dagger}_{\vec{k},s}(n) \\ a^{\dagger}_{\vec{k},s}(n) \end{pmatrix} S^{\dagger} = s^{*} \begin{pmatrix} a^{\dagger}_{\vec{k},s}(p) \\ a^{\dagger}_{\vec{k},s}(n) \end{pmatrix}$$

and

$$S \begin{pmatrix} b^{\dagger}_{\vec{k},s}(p) \\ b^{\dagger}_{\vec{k},s}(n) \end{pmatrix} S^{\dagger} = s \begin{pmatrix} b^{\dagger}_{\vec{k},s}(p) \\ b^{\dagger}_{\vec{k},s}(n) \end{pmatrix} \tag{11.23}$$

which, together with (11.20), gives

$$S \binom{p}{n} = s^{*} \binom{p}{n} \quad \text{and} \quad S \binom{\bar{p}}{\bar{n}} = s \binom{\bar{p}}{\bar{n}} \tag{11.24}$$

where

$$\binom{p}{n} \equiv \begin{pmatrix} \mid p > \\ \mid n > \end{pmatrix} \quad \text{and} \quad \binom{\bar{p}}{\bar{n}} \equiv \begin{pmatrix} \mid \bar{p} > \\ \mid \bar{n} > \end{pmatrix} \quad . \tag{11.25}$$

Let us consider an infinitesimal isospin transformation; i.e.

$$s = 1 + \tfrac{1}{2} i \vec{\tau} \cdot \vec{\epsilon} \tag{11.26}$$

where the components of $\vec{\tau}$ are the Pauli matrices given by (3.1) and $\vec{\epsilon}$ is an infinitesimal vector. Correspondingly, the Hilbert space transformation S must also be infinitesimally close to the unit matrix; we may therefore write

$$S = 1 - i \vec{I} \cdot \vec{\epsilon} \tag{11.27}$$

which can also be regarded as the definition of $\vec{I}$, and will be referred to as the <u>isospin operator</u>. Because s is unitary and the Pauli matrices are all Hermitian,

$$s s^{\dagger} = 1 + \tfrac{1}{2} i \vec{\tau} \cdot (\vec{\epsilon} - \vec{\epsilon}^{\,*}) + O(\epsilon^{2}) = 1 \quad ,$$

and therefore

$$\vec{\epsilon} = \vec{\epsilon}^{\,*} \quad . \tag{11.28}$$

The unitarity of S then implies

$$S^\dagger S = 1 + i\vec{\epsilon}\cdot(\vec{I}^\dagger - \vec{I}) + O(\epsilon^2) = 1 \tag{11.29}$$

which leads to the Hermiticity of the isospin operator

$$\vec{I} = \vec{I}^\dagger \quad . \tag{11.30}$$

By substituting (11.27) into (11.18), we find

$$S\,H_{st}\,S^\dagger = H_{st} - i\vec{\epsilon}\cdot(\vec{I}\,H_{st} - H_{st}\,\vec{I}) + O(\epsilon^2) = H_{st} \;,$$

which gives

$$[\vec{I}, H_{st}] = 0 \quad . \tag{11.31}$$

A similar substitution of (11.27) into (11.16) results in

$$S\,\psi\,S^\dagger = \psi - i\vec{\epsilon}\cdot(\vec{I}\,\psi - \psi\,\vec{I}) + O(\epsilon^2) = (1 + \tfrac{1}{2}\,i\vec{\tau}\cdot\vec{\epsilon}\,)\,\psi \quad ,$$

and therefore

$$[\,\psi(x), \vec{I}\,] = \tfrac{1}{2}\vec{\tau}\,\psi(x) \quad . \tag{11.32}$$

We shall now show that for a finite isospin transformation

$$s = e^{i\frac{1}{2}\vec{\tau}\cdot\vec{\theta}} \quad , \tag{11.33}$$

the corresponding Hilbert space transformation matrix is

$$S = e^{-i\vec{I}\cdot\vec{\theta}} \quad . \tag{11.34}$$

<u>Proof.</u> Let

$$\theta \equiv |\vec{\theta}\,| \;, \quad \hat{\theta} \equiv \frac{\vec{\theta}}{\theta} \;, \quad I_o \equiv \vec{I}\cdot\hat{\theta} \;, \quad \tau_o \equiv \vec{\tau}\cdot\hat{\theta}$$

and

$$\psi_\theta(x) \equiv e^{-i I_o \theta}\,\psi(x)\,e^{i I_o \theta} \quad . \tag{11.35}$$

The derivative of $\psi_\theta(x)$ is, on account of (11.32) and (11.35),

$$\frac{d}{d\theta}\,\psi_\theta(x) = -i\,e^{-i I_o \theta}\,[\,I_o, \psi(x)\,]\,e^{i I_o \theta} = \tfrac{1}{2}\,i\tau_o\,\psi_\theta(x) \;,$$

which, upon integration gives $\psi_\theta(x) = e^{i\frac{1}{2}\vec{\tau}\cdot\vec{\theta}}\,\psi(x)$; consequently

$$e^{-i\vec{I}\cdot\vec{\theta}}\,\psi(x)\,e^{i\vec{I}\cdot\vec{\theta}} = e^{i\frac{1}{2}\vec{\tau}\cdot\vec{\theta}}\,\psi(x) \quad . \tag{11.36}$$

That completes the proof.

If the system consists only of nucleons and antinucleons, then by using (3.24) and (11.32) we can easily verify that the isospin operator $\vec{I}$ is given by

$$\vec{I} = \vec{I}_N = \tfrac{1}{2}\int \psi^\dagger(x)\,\vec{\tau}\,\psi(x)\,d^3r \tag{11.37}$$

where, as before, $x_\mu = (\vec{r}, it)$. Because of (11.31), $\vec{I}$ is t-independent and therefore the choice of t in the integrand of (11.37) is arbitrary. If the system contains other particles such as mesons, $\cdots$, then $\vec{I} = \vec{I}_N + \vec{I}_{Mesons} + \cdots$ where $\vec{I}_N$ is given by (11.37) and $\vec{I}_{Mesons} + \cdots$ commutes with the nucleon field $\psi(x)$.

Remarks. Since H_{st} is isospin-invariant, it can be diagonalized simultaneously with $\vec{I}^2$ and I_3 . Let $I(I+1)$ be the eigenvalue of the operator $\vec{I}^2$. Thus, each hadron carries the quantum numbers I and I_3 . For a given I, I_3 can vary from $-I$ to I, making a total of $2I+1$ states. Under isospin rotations, the quantum number I is preserved; however, these $2I+1$ states of different I_3 transform among each other, and therefore they are degenerate with respect to the strong interaction. This mass degeneracy of hadrons is lifted by the electromagnetic and weak interactions; both violate the isospin symmetry.

Exercise. Show that $[I_i , I_j] = i\,\epsilon_{ijk}\,I_k$ where ϵ_{ijk} is given by (3.4).

11.2 G Parity

1. Nucleon-antinucleon system

The proton p and the neutron n form an isospin doublet; e.g. we may regard p as ↑ and n as ↓, corresponding to $I_3 = \frac{1}{2}$ and $-\frac{1}{2}$ respectively. Therefore, apart from a phase factor, $\bar{p}$ should behave like ↓ and $\bar{n}$ like ↑. As will be shown, the precise relations are: for any transformation (11.24) of the nucleon states

$$S\binom{p}{n} = s^*\binom{p}{n} \quad , \tag{11.38}$$

we have an identical transformation of the antinucleon states

$$S\binom{\bar{n}}{-\bar{p}} = s^*\binom{\bar{n}}{-\bar{p}} \quad . \tag{11.39}$$

Proof. For an arbitrary isospin transformation, we can write, in accordance with (11.33),

$$s = e^{i\frac{1}{2}\vec{\tau}\cdot\vec{\theta}} = \sum_{n=0}^{\infty} \frac{1}{n!} (i\tfrac{1}{2}\vec{\tau}\cdot\vec{\theta})^n \quad . \tag{11.40}$$

In view of the fact that for any integer n

$$(\vec{\tau}\cdot\vec{\theta})^{2n} = \theta^{2n} \quad \text{and} \quad (\vec{\tau}\cdot\vec{\theta})^{2n+1} = \theta^{2n}\,\vec{\tau}\cdot\vec{\theta} \quad ,$$

where $\theta = |\vec{\theta}|$, (11.40) becomes

$$s = \cos\frac{\theta}{2} + i\,\vec{\tau}\cdot\hat{\theta}\,\sin\frac{\theta}{2} \tag{11.41}$$

with $\hat{\theta} = \vec{\theta}/\theta$. From (3.1) we see that

$$\tau_2\vec{\tau} = -\vec{\tau}^*\tau_2 \quad ,$$

and therefore

$$\tau_2 s = s^*\tau_2 \quad . \tag{11.42}$$

Multiplying the second equation in (11.24) by τ_2 on the left, we have

$$\tau_2 S\binom{\bar{p}}{\bar{n}} = \tau_2 s\binom{\bar{p}}{\bar{n}} \quad ,$$

which, together with (11.42), gives

$$S\tau_2\begin{pmatrix}\bar{p}\\ \bar{n}\end{pmatrix} = s^*\tau_2\begin{pmatrix}\bar{p}\\ \bar{n}\end{pmatrix} ,$$

and that establishes (11.39).

The same result can also be expressed in terms of the field operators. From (11.16), we have

$$S\begin{pmatrix}\psi_p\\ \psi_n\end{pmatrix}S^\dagger = s\begin{pmatrix}\psi_p\\ \psi_n\end{pmatrix} , \tag{11.43}$$

which leads to

$$S\begin{pmatrix}\psi_p^\dagger\\ \psi_n^\dagger\end{pmatrix}S^\dagger = s^*\begin{pmatrix}\psi_p^\dagger\\ \psi_n^\dagger\end{pmatrix} .$$

Multiplying the above equation by τ_2 on the left and using (11.42), we derive

$$S\begin{pmatrix}\psi_n^c\\ -\psi_p^c\end{pmatrix}S^\dagger = s\begin{pmatrix}\psi_n^c\\ -\psi_p^c\end{pmatrix} \tag{11.44}$$

where

$$(\psi_p^c)_\alpha = (\gamma_2)_{\alpha\beta}\,(\psi_p^\dagger)_\beta \quad \text{and} \quad (\psi_n^c)_\alpha = (\gamma_2)_{\alpha\beta}\,(\psi_n^\dagger)_\beta .$$

Remarks. If we adopt the notation that $\uparrow$ and $\downarrow$ stand for the $I_3 = \frac{1}{2}$ and $-\frac{1}{2}$ states of an isospin doublet, then from (11.38)-(11.39) the nucleon doublet can be written as

$$p = \uparrow , \qquad n = \downarrow , \tag{11.45}$$

while the antinucleon doublet is

$$\bar{n} = \uparrow \quad \text{and} \quad -\bar{p} = \downarrow . \tag{11.46}$$

2. The quantum number G

Whenever several conservation laws operate for the same system it is often possible to obtain new quantum numbers and new selection rules, as will be illustrated by the interplay between the isospin symmetry and the C invariance of the strong interaction. Under the particle-antiparticle conjugation C we have

$$\binom{p}{n} \rightarrow \binom{\bar{p}}{\bar{n}} \ .$$

Take any axis $\perp$ to the third axis in the isospin space. A 180° rotation along that axis would transform, apart from some important phase factors,

$$\uparrow \rightarrow \downarrow \quad \text{and} \quad \downarrow \rightarrow \uparrow$$

where, as before, $\uparrow$ can be p or $\bar{n}$ and $\downarrow$ can be n or $\bar{p}$. Now if we can pick the "right" axis and be in full control of the phase factors so that the above transformation becomes precisely

$$\binom{\bar{p}}{\bar{n}} \rightarrow \binom{\bar{n}}{-\bar{p}} \ ,$$

then, because of (11.38)-(11.39), the chain action of C together with this appropriate 180° isospin rotation would leave the isospin properties of the states invariant, which in turn can lead to a new quantum number, as we shall show.

It is clear that we must be careful about the phase factors. Let us adopt the convention that the nucleon fields ψ_p and ψ_n transform in the same way under C ; i.e.,

$$C \psi_p C^\dagger = \psi_p^c \qquad \text{and} \qquad C \psi_n C^\dagger = \psi_n^c \tag{11.47}$$

or, in terms of the state vector

$$C \mid p > \; = \; \mid \bar{p} > \qquad \text{and} \qquad C \mid n > \; = \; \mid \bar{n} > \ .$$

We define the G parity operator to be

$$G = C \cdot e^{i \pi I_2} \ , \tag{11.48}$$

where, on account of (11.36),

$$e^{i\pi I_2}\,\psi\, e^{-i\pi I_2} = s\,\psi\ , \tag{11.49}$$

with

$$s = e^{-i\frac{1}{2}\pi\tau_2} = -i\tau_2 = \begin{pmatrix} 0 & -1 \\ 1 & 0 \end{pmatrix} = s^*\ . \tag{11.50}$$

From (11.38)-(11.39), we also have

$$e^{i\pi I_2}\begin{pmatrix} p \\ n \end{pmatrix} = \begin{pmatrix} -n \\ p \end{pmatrix}$$

and

$$e^{i\pi I_2}\begin{pmatrix} \bar{n} \\ -\bar{p} \end{pmatrix} = \begin{pmatrix} \bar{p} \\ \bar{n} \end{pmatrix}\ . \tag{11.51}$$

<u>Theorem.</u> $[\,G\,,\,\vec{I}\,] = 0\ .$ (11.52)

<u>Proof.</u> Let us first consider the single nucleon or antinucleon state. By applying C onto the lefthand sides of (11.51) and (11.52), we find

$$G\begin{pmatrix} p \\ n \end{pmatrix} = -\begin{pmatrix} \bar{n} \\ -\bar{p} \end{pmatrix}$$

and

$$G\begin{pmatrix} \bar{n} \\ -\bar{p} \end{pmatrix} = \begin{pmatrix} p \\ n \end{pmatrix}\ . \tag{11.53}$$

Thus, for any isospin transformation S, we may combine (11.38)-(11.39) with (11.53), and derive

$$SG\begin{pmatrix} p \\ n \end{pmatrix} = -s^*\begin{pmatrix} \bar{n} \\ -\bar{p} \end{pmatrix} = GS\begin{pmatrix} p \\ n \end{pmatrix}$$

and

$$SG\begin{pmatrix} \bar{n} \\ -\bar{p} \end{pmatrix} = s^*\begin{pmatrix} p \\ n \end{pmatrix} = GS\begin{pmatrix} \bar{n} \\ -\bar{p} \end{pmatrix}\ . \tag{11.54}$$

The same considerations can also be applied to the field operators. From (11.43)-(11.44) and (11.48)-(11.50), it follows that the equivalent form of (11.53) is

$$G\begin{pmatrix} \psi_p \\ \psi_n \end{pmatrix} G^\dagger = -\begin{pmatrix} \psi_n^c \\ -\psi_p^c \end{pmatrix}$$

and

$$G\begin{pmatrix} \psi_n^c \\ -\psi_p^c \end{pmatrix} G^\dagger = \begin{pmatrix} \psi_p \\ \psi_n \end{pmatrix}\ ; \tag{11.55}$$

and that of (11.54) is

$$SG\begin{pmatrix}\psi_p\\ \psi_n\end{pmatrix}G^\dagger S^\dagger = GS\begin{pmatrix}\psi_p\\ \psi_n\end{pmatrix}S^\dagger G^\dagger$$

and

$$SG\begin{pmatrix}\psi_n^c\\ -\psi_p^c\end{pmatrix}G^\dagger S^\dagger = GS\begin{pmatrix}\psi_n^c\\ -\psi_p^c\end{pmatrix}S^\dagger G^\dagger \quad . \tag{11.56}$$

Next, we consider the arbitrary multi-nucleon antinucleon state $| > = |N_p, N_{\bar{p}}, N_n, N_{\bar{n}} >$ appearing in (11.11); from (11.56) we see that

$$SG\,| > \; = \; GS\,| > \quad .$$

Consequently, if the system consists only of nucleons and antinucleons, then

$$[S, G] = 0 \quad . \tag{11.57}$$

To extend the system to include hadrons other than nucleons and antinucleons, we observe that starting from the initial state which consists only of nucleons and antinucleons (as in any realistic high-energy experiment), it is possible to reach all known hadrons through the strong interaction. Because the strong interaction is assumed to be isospin- and C-invariant, the fact that (11.57) is valid for the initial state means that it must also be valid for the final state. Since (11.57) holds for any $S = e^{i\vec{I}\cdot\vec{\theta}}$, the theorem is established.

From (11.53) or (11.55), we see that

$$G^2\begin{pmatrix}p\\ n\end{pmatrix} = -\begin{pmatrix}p\\ n\end{pmatrix} \quad , \qquad G^2\begin{pmatrix}\bar{n}\\ -\bar{p}\end{pmatrix} = -\begin{pmatrix}\bar{n}\\ -\bar{p}\end{pmatrix}$$

and their generalization

$$G^2\,| > = \begin{cases} -\,| > & \text{if the state is of } I = \frac{1}{2}, \frac{3}{2}, \cdots \\ \;\;\,| > & \text{if the state is of } I = 0, 1, \cdots . \end{cases}$$

Thus, we can write

$$G^2 = (-1)^{2I} \quad . \tag{11.58}$$

Remarks. Under C, the third component of isospin changes sign; e.g., p is of $I_3 = \frac{1}{2}$, while $\bar{p}$ is of $I_3 = -\frac{1}{2}$. Consequently, C does not commute with $\vec{I}$. From C and $\exp(i\pi I_2)$, we derive G. The importance of G lies in its commutativity with $\vec{I}$. Since H_{st} is both isospin- and C-invariant, it is also G-invariant. This enables us to consider states which are eigenvectors of H_{st}, I, I_3 and G, thereby resulting in some rather useful selection rules, as will be discussed in the next section.

11.3 Applications to Mesons and Baryons

In this section we again consider only the strong interaction, and ignore the isospin-violating effects of the electromagnetic and weak interactions. Our discussions will be mainly on the reasoning that determined the quantum numbers of various hadrons. For convenience, we shall sometimes refer to the three components I_1, I_2 and I_3 of the isospin vector also as I_x, I_y and I_z respectively.

1. Pion

Since the virtual transition

$$\pi \rightleftarrows N\bar{N} \tag{11.59}$$

with N = p or n, can occur via the strong interaction, the pion-isospin I_π must be the same as that of $N\bar{N}$; i.e., 0 or 1. The fact that $\pi^\pm$ and π^o are of approximately the same mass, making a total of $2I+1 = 3$-fold degeneracy, means that

$$I_\pi = 1 \quad . \tag{11.60}$$

Transition (11.59) then leads to $\pi^+ \rightleftarrows p\bar{n}$ having $I_z = 1$; likewise,

π^o has $I_z = 0$, and π^- $I_z = -1$. Under an isospin rotation, the three pion states transform like a vector $\vec{r} = (x, y, z)$ in the isospin space with

$$| \pi^{\pm} > \sim \mp 2^{-\frac{1}{2}}(x \pm iy) \quad \text{and} \quad | \pi^o > \sim z \; , \tag{11.61}$$

where the sign convention is chosen in accordance with the spherical harmonics $Y_{1,\pm 1}$ and $Y_{1,0}$ given by (1.40). A 180° rotation along the y-axis in the isospin space changes z to $-z$, and therefore

$$e^{i\pi I_y} | \pi^o > = - | \pi^o > \; . \tag{11.62}$$

From (10.76), $\pi^o \rightarrow 2\gamma$, we know that

$$C | \pi^o > = | \pi^o > \; . \tag{11.63}$$

Thus under $G = Ce^{i\pi I_y}$, the state $| \pi^o >$ must change sign; i.e.,

$$G | \pi^o > = - | \pi^o > \; . \tag{11.64}$$

Because $| \pi^o >$ can be transformed into an arbitrary coherent mixture of the charged pion states through isospin rotations, the commutation relation $[G, \vec{I}] = 0$ implies that the G-quantum number of $\pi^{\pm}$ must be the same as that of π^o. From (11.64), it follows that

$$G | \pi > = - | \pi > \tag{11.65}$$

where π can be either $\pi^{\pm}$ or π^o; thereby we derive the selection rule that under the strong interaction

$$\text{even number of pions} \not\rightleftarrows \text{odd number of pions,} \tag{11.66}$$

which is valid independently of the pion charge.

The assignment (11.63) relies on the 2γ decay of π^o. Is it possible to determine the particle-antiparticle conjugation of π^o without relying on the electromagnetic interaction? The answer is "yes", as will be shown by the following exercise.

Exercise. Assuming that the pion is a pseudoscalar, the strong-interaction transition $\pi \leftrightarrows N\bar{N}$ can be described by the phenomenological Lagrangian density

$$\mathcal{L}_{int} = i\, g_{\pi N}\, \psi^{\dagger}\, \gamma_4\, \gamma_5\, \vec{\tau}\, \psi \cdot \vec{\phi} \tag{11.67}$$

where $g_{\pi N}$ is the π-nucleon coupling constant, ψ the nucleon field given by (11.1) and $\vec{\phi}$ the isovector pion field.

(i) Show that, because of (11.47), in order for $\mathcal{L}_{int}$ to be C-invariant, the z-component of $\vec{\phi}$, which represents π^{o}, must obey

$$\phi_z = C\, \phi_z\, C^{\dagger} \quad .$$

Hence, π^{o} is of $C = +1$, which is now determined by the strong interaction alone.

(ii) Work out separately the transformations of $\vec{\phi}$ under C, $\exp(i\pi I_y)$ and G.

2. Vector mesons

Let us consider the $e^{+}e^{-}$ colliding-beam experiment:

$$e^{+} + e^{-} \rightarrow \text{virtual } \gamma \rightarrow \left\{ \begin{array}{ll} \rho^{o} & (770) \\ \omega^{o} & (780) \\ \phi^{o} & (1020) \\ J/\psi & (3100) \\ \psi' & (3700) \\ \Upsilon & (9500) \end{array} \right. \tag{11.68}$$

where the numbers refer to the masses of the vector mesons in units of MeV.

Because $\rho^{\pm}$ and ρ^{o} are of approximately the same mass, the ρ-mesons form an isospin triplet; therefore, the isospin of ρ is

$$I_{\rho} = 1 \quad .$$

All the other mesons in the final states of (11.68), such as ω^o, ϕ^o, J/ψ, $\cdots$ do not have charged states of approximately the same masses. Consequently, they are all isosinglets: $I_\omega = I_\phi = \dots = 0$.

Next, we want to show that in reaction (11.68) all the final mesons are of

$$C = -1 \quad , \qquad P = -1 \qquad \text{and} \qquad \text{spin} = 1 \ . \tag{11.69}$$

Proof. Reaction (11.68) proceeds via the intermediate state of a virtual γ, whose transformation properties are the same as those of the photon field A_μ. Since $C A_\mu C^\dagger = -A_\mu$, a virtual γ must be of $C = -1$, so also must the final meson. However, the spin-parity of a virtual γ can be $1-$ or $0+$, since the space-component of A_μ is a vector while its time-component is a scalar. [For example, the well-known Coulomb excitations in atomic and nuclear physics are all via the time-like component of A_μ.] From this we conclude that the final meson can only be either $1-$ or $0+$. Next we shall show that $0+$ is impossible.

From the Feynman diagram in Fig. 11.1, we see that the amplitude of (11.68) is proportional to the following matrix element of the hadronic electromagnetic current operator $J_\mu(x)$:

$$< \text{meson} \mid J_\mu(x) \mid \text{vac} > \tag{11.70}$$

where $J_\mu(x)$ satisfies the current conservation law

$$\frac{\partial J_\mu}{\partial x_\mu} = 0 \ . \tag{11.71}$$

Let q_μ be the 4-momentum of the final meson and P_μ the 4-momentum operator of the system, with $P_4 = iH$ where H is the Hamiltonian operator. The relativistic generalization of Heisenberg's

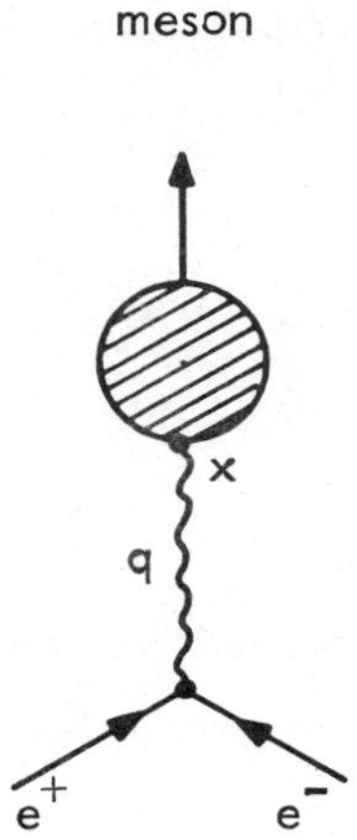

Fig. 11.1. Diagram for $e^+ + e^- \rightarrow$ virtual $\gamma(q) \rightarrow$ meson.

equation (1.9) is

$$[P_\mu , O(x)] = i \frac{\partial O}{\partial x_\mu} \tag{11.72}$$

for any local operator $O(x)$. By taking the matrix element of (11.72) between $| \text{vac} >$ and $< \text{meson} |$, we find

$$< \text{meson} | P_\mu O(x) - O(x) P_\mu | \text{vac} > = i \frac{\partial}{\partial x_\mu} < \text{meson} | O(x) | \text{vac} > . \tag{11.73}$$

Since

$$P_\mu | \text{vac} > = 0 \qquad \text{and} \qquad P_\mu | \text{meson} > = q_\mu | \text{meson} > ,$$

(11.73) becomes

$$\frac{\partial}{\partial x_\mu} < \text{meson} | O(x) | \text{vac} > = -i\, q_\mu < \text{meson} | O(x) | \text{vac} > ,$$

which relates the matrix element of $O(x)$ to its value at $x = 0$:

$$< \text{meson} | O(x) | \text{vac} > = e^{-iq_\mu x_\mu} < \text{meson} | O(0) | \text{vac} > . \tag{11.74}$$

Now, if the final meson is of spin-0, then since the matrix

element $<\text{meson} \mid J_\mu(0) \mid \text{vac}>$ depends only on a single 4-vector q_μ and because it is itself a 4-vector, we can write

$$< \text{meson} \mid J_\mu(0) \mid \text{vac} > = c q_\mu$$

where c is a constant. By using (11.74), we find

$$< \text{meson} \mid J_\mu(x) \mid \text{vac} > = c q_\mu \, e^{-i q_\mu x_\mu}$$

which, together with (11.71), gives

$$0 = \frac{\partial}{\partial x_\mu} < \text{meson} \mid J_\mu(x) \mid \text{vac} > = -i c q_\mu^2 \, e^{-i q_\mu x_\mu} \quad .$$

Because $-q_\mu^2 = (\text{c.m. energy})^2 \neq 0$, it follows that $c = 0$, i.e.,

$$< \text{meson} \mid J_\mu(x) \mid \text{vac} > = 0 \qquad \text{if} \qquad \text{spin} = 0 \; . \tag{11.75}$$

Hence the spin of the final meson cannot be 0, and (11.69) is then established. From C conservation, we conclude that all such mesons obey the selection rule

$$C = -1 \text{ meson} \not\rightarrow \text{ any number of } \pi^o \; . \tag{11.76}$$

Because the π^o's are bosons, we have

$$1^- \text{ meson} \not\rightarrow 2\pi^o \; . \tag{11.77}$$

Since the ρ^o meson is of $I = 1$, $I_z = 0$ and $C = -1$, therefore G of ρ^o is $+1$. The commutation relation (11.52) implies that $\rho^\pm$ is also of $G = +1$. Hence, G conservation requires that independently of the charge

$$\rho \not\rightarrow \text{ odd number of pions.} \tag{11.78}$$

Likewise, since ω^o, ϕ^o, J/ψ, $\cdots$ are all of $I = 0$, and therefore $G = -1$, they obey the selection rule

$$\omega,\ \phi,\ J/\psi,\ \cdots \not\rightarrow \text{ even number of pions.} \tag{11.79}$$

As we mentioned earlier, isospin symmetry and G symmetry are violated by the electromagnetic interaction, and C symmetry by the weak interaction. Hence, the violation amplitude of selection rules (11.78)-(11.79) is of the order of the fine-structure constant, while that of (11.76) is due to the weak interaction and therefore of a much smaller magnitude. So far as we know, selection rule (11.77) is exact.

Exercise. If $| A >$ and $| B >$ are eigenstates of the 4-momentum operator P_μ with eigenvalues a_μ and b_μ, show that the x-dependence of the matrix element $< B | O(x) | A >$ is completely known:

$$< B | O(x) | A > = < B | O(0) | A > e^{i(a_\mu - b_\mu) x_\mu} \quad (11.80)$$

for any local operator $O(x)$.

3. Λ and kaon

There is only one particle which is degenerate with Λ^o, and that is its antiparticle $\bar{\Lambda}^o$. Consequently, the isospin degeneracy $2 I_\Lambda + 1$ must be $\leqslant 2$, which means that $I_\Lambda = 0$ or $\frac{1}{2}$.

Consider now the strong reaction

$$\pi^- + p \rightarrow \Lambda^o + K^o \quad . \quad (11.81)$$

If I_Λ were $\frac{1}{2}$, then under an isospin rotation, say $\exp(i\pi I_y)$, Λ^o would transform into its isospin partner $\bar{\Lambda}^o$. Because of the decay

$$\Lambda^o \rightarrow \pi + N \quad , \quad (11.82)$$

Λ^o has a baryon number $= +1$; likewise since

$$\bar{\Lambda}^o \rightarrow \pi + \bar{N} \quad ,$$

the antiparticle $\bar{\Lambda}^o$ has a baryon number $= -1$. When we apply the isospin rotation $\exp(i\pi I_y)$ onto (11.81), we find the lefthand side remains a mixture of pion and nucleon; however, if $I_\Lambda = \frac{1}{2}$, the right-hand side would be a mixture of $\bar{\Lambda}^o$ and kaon, and that would violate baryon conservation. Hence we conclude that

$$I_\Lambda = 0 \ . \tag{11.83}$$

Because the isospin of π is 1 and that of the nucleon is 1/2, the total isospin of (11.81) can be either 1/2 or 3/2 which, together with (11.83), gives the kaon isospin $I_K = 1/2$ or $3/2$. Let Q_K, Q_π, Q_Λ and Q_N be the charges of K, π, Λ and N. The charges of π and N are related to their I_z's by

$$Q_\pi = I_z \quad \text{and} \quad Q_N = I_z + \tfrac{1}{2} \ .$$

Since $Q_\Lambda = I_\Lambda = 0$, reaction (11.81) implies that

$$Q_K = Q_{\pi+N} = I_z + \tfrac{1}{2}$$

where I_z refers to that of the kaon. Consequently, if $I_K = 3/2$, the kaon state that has $I_z = 3/2$ must be of charge $3/2 + 1/2 = 2$. Because there is no doubly-charged kaon, I_K cannot be 3/2. Thus, although there are four degenerate kaon states: K^+, K^-, K^o and $\overline{K}^o$, we conclude that

$$I_K = \tfrac{1}{2} \ . \tag{11.84}$$

The four kaon states form two isospin doublets

$$\begin{pmatrix} K^+ \\ K^o \end{pmatrix} \quad \text{and} \quad \begin{pmatrix} \overline{K}^o \\ -K^- \end{pmatrix} \ , \tag{11.85}$$

which transform under the isospin rotation like

$$\begin{pmatrix} p \\ n \end{pmatrix} \quad \text{and} \quad \begin{pmatrix} \bar{n} \\ -\bar{p} \end{pmatrix}$$

respectively.

4. Meson and baryon octets

Reasoning identical to that used in determining the isospin of Λ, (11.83), leads to the conclusion that no baryon can belong to the same isospin multiplet as its antiparticle. From $\pi^- + p \to \Sigma^- + K^+$ we conclude Σ has a baryon number $+1$, and from the known triplet degeneracy Σ^+, Σ^0 and Σ^- we deduce

$$I_\Sigma = 1 \quad . \tag{11.86}$$

Likewise, the allowed reaction $K^- + p \to \Xi^- + K^+$ implies that the baryon number of Ξ is 1; from the doublet degeneracy, Ξ^0 and Ξ^-, we deduce

$$I_\Xi = \tfrac{1}{2} \quad . \tag{11.87}$$

Similarly, from $K^- + p \to \Lambda^0 + \eta^0$, it follows that the baryon number of η^0 is 0. Since η^0 has no degeneracy, we conclude

$$I_\eta = 0 \quad . \tag{11.88}$$

It is useful to define hypercharge Y and strangeness* S:

$$Y \equiv 2(Q - I_z) \quad \text{and} \quad S \equiv Y - N \tag{11.89}$$

where Q is the electric charge and N the baryon number. Any interaction that conserves Q, I_z and N also conserves Y and S. In Table 11.1 we list the I, I_z, Y and S of the low-lying baryon and meson octets (the word octet refers to the SU_3 representation to be discussed in the next chapter). The baryons are all spin-$\frac{1}{2}$ with even relative parity, while the mesons are all pseudoscalars. We note that p, n and the pions are all of $S = 0$, hence non-strange particles, while Σ, Λ and the kaons have $S = \pm 1$ and therefore are called strange particles.

From Table 11.1 we see that particles with different I_z, but

---

* M. Gell-Mann, Phys.Rev. 93, 933 (1953); T. Nakano and K. Nishijima, Progr.Theoret.Phys. 10, 581 (1953).

I_z: 1	$\frac{1}{2}$	0	$-\frac{1}{2}$	-1	I	Y	S
	p^+		n^0		$\frac{1}{2}$	1	0
Σ^+		Σ^0		Σ^-	1	0	-1
		Λ^0			0	0	-1
	Ξ^0		Ξ^-		$\frac{1}{2}$	-1	-2
	K^+		K^0		$\frac{1}{2}$	1	1
π^+		π^0		π^-	1	0	0
		η^0			0	0	0
	$\overline{K}^0$		K^-		$\frac{1}{2}$	-1	-1

Table 11.1. Quantum numbers of the pseudoscalar meson and the spin-$\frac{1}{2}$ baryon octets.

the same isospin multiplet, have the same Y, S and N, as expected. Consequently, these quantum numbers commute with $\vec{I}$; i.e.

$$[Y, \vec{I}] = [S, \vec{I}] = [N, \vec{I}] = 0 \quad . \tag{11.90}$$

On the other hand, since $Q = I_z + Y/2$, we have the commutation relations

$$[Q, I_x] = [I_z, I_x] = i I_y ,$$

$$[Q, I_y] = [I_z, I_y] = -i I_x \tag{11.91}$$

and

$$[Q, I_z] = 0 \quad .$$

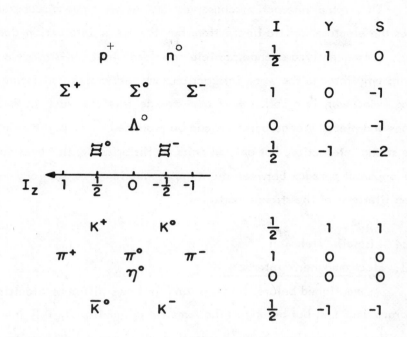

The strong interaction conserves S. As we shall discuss, so does the electromagnetic interaction, but the weak interaction does not. Consequently, a strange particle can decay into non-strange particles only through the weak interaction, and that makes its lifetime long. However, in a collision of non-strange particles such as nucleons and pions, strange particles can be produced copiously through the strong interaction, but only in pairs.* Historically, this explains the apparent paradox between the large production cross sections and long lifetimes of the strange particles.

11.4 Isospin Violation

1. Electromagnetic interaction

As mentioned before, since p and n have different electric charges and I_z, but belong to the same isospin multiplet, this implies that the electromagnetic interaction H_γ cannot be invariant under any isospin rotation that changes I_z; i.e.

$$[H_\gamma, I_x] \neq 0 \quad \text{and} \quad [H_\gamma, I_y] \neq 0 . \tag{11.92}$$

In transitions such as

$$p \rightleftarrows p+\gamma \quad \text{and} \quad n \rightleftarrows n+\gamma ,$$

the I_z of the nucleon is unchanged. In other words, these reactions conserve I_z. We will now make the assumption that in all electromagnetic processes I_z is conserved; i.e.

$$[H_\gamma, I_z] = 0 . \tag{11.93}$$

The electromagnetic interaction of a particle of charge e can

---

* A. Pais, Phys.Rev. 86, 663 (1952); Y. Nambu, K. Nishijima and Y. Yamaguchi, Progr.Theoret.Phys. 6, 615, 619 (1951).

be obtained most simply by replacing in its free Lagrangian

$$\frac{\partial}{\partial x_\mu} \rightarrow \frac{\partial}{\partial x_\mu} - ieA_\mu \quad . \tag{11.94}$$

The interaction obtained this way automatically satisfies the gauge invariance, and is called the "minimal electromagnetic interaction"; the corresponding electromagnetic current J_μ is called the minimal current. Under the isospin rotation, the minimal current, generated by (11.94), transforms in the same way as the electric-charge operator $Q = -i \int J_4 d^3r$. Because $Q = I_z + Y/2$, which consists of an iso-vector part, I_z, and an isoscalar part, $Y/2$, so does J_μ. We may decompose the minimal electromagnetic current into two terms

$$J_\mu = (J_\mu)_0 + (J_\mu)_1 \tag{11.95}$$

where the subscript 0 or 1 indicates whether the term is of $I = 0$ or 1 ; both are of $I_z = 0$.

In any transition

$a \rightarrow b + \gamma$

where a and b refer to hadron states, we may define the vector of isospin change to be

$$\Delta\vec{I} = \vec{I}_a - \vec{I}_b \quad . \tag{11.96}$$

Hence, under the assumption of minimal electromagnetic interaction and on account of (11.95), we have the following selection rules:

$$\Delta I_z = \Delta Y = \Delta S = 0 \tag{11.97}$$

and

$$|\Delta\vec{I}| = 0 \text{ or } 1 \tag{11.98}$$

where the $\Delta\vec{I} = 0$ transition is due to the isoscalar current $(J_\mu)_0$, while the $|\Delta\vec{I}| = 1$ transition is due to the isovector current $(J_\mu)_1$.

To be sure, gauge invariance alone does not imply minimal interaction. We can always add non-minimal terms to the interaction Lagrangian which depend on $F_{\mu\nu} = \partial A_\nu / \partial x_\mu - \partial A_\mu / \partial x_\nu$, instead of A_μ itself; e.g., an anomalous magnetic-moment term

$$i \psi^\dagger \gamma_4 \gamma_\mu \gamma_\nu \psi' F_{\mu\nu} \quad . \tag{11.99}$$

Such a term is always gauge-invariant; it may or may not be I_z-conserving. If $\psi = \psi_n$ and $\psi' = \psi_\Lambda$, then we could have

$$\Lambda \rightleftarrows n + \gamma \tag{11.100}$$

which violates I_z. Our assumption is that these non-minimal terms do not exist. Experimentally, reaction (11.100) has never been observed, giving support to the assumption of minimal electromagnetic interaction. On the other hand, as a further support of our selection rules,

$$\Sigma^o \rightleftarrows \Lambda + \gamma \tag{11.101}$$

does occur, and is the dominant decay mode of Σ^o. Unlike (11.100), this is an allowed decay since it satisfies $|\Delta \vec{I}| = 1$ and $\Delta I_z = 0$.

<u>Remarks.</u> We note that on account of (11.91) the total charge operator $Q = -i \int J_4 d^3 r$ commutes with the operator $\vec{I}^2$,

$$[Q, \vec{I}^2] = 0 \quad .$$

Thus, if a matrix element

$$\langle b | Q | a \rangle \neq 0 \quad ,$$

then besides the selection rules (11.96) and (11.97) we must have

$$I_a = I_b$$

where I_a and I_b denote, respectively, the magnitudes of the total isospin of the hadron states a and b. However, the electromagnetic current operator $J_\mu(x)$ does not commute with $\vec{I}^{\,2}$. This is why $\Sigma^o \leftrightarrows \Lambda + \gamma$ is an allowed transition, in which $I_\Sigma = 1$ but $I_\Lambda = 0$.

2. Weak interaction

Isospin is violated in both the nonleptonic weak processes, such as

$$\Lambda^o \to p + \pi^- \ , \qquad K^\pm \to \pi^\pm + \pi^o \qquad (11.102)$$

and semileptonic weak processes, such as

$$n \to p + e^- + \bar{\nu}_e \ , \qquad K^\pm \to \pi^o + e^\pm + \nu_e \ \ (\text{or } \bar{\nu}_e) \ . \qquad (11.103)$$

In the former we have

$$|\,\Delta I_z\,| = \tfrac{1}{2} \ , \qquad (11.104)$$

while in the latter

$$|\,\Delta I_z\,| = 1 \ \text{ or } \ \tfrac{1}{2} \ , \qquad (11.105)$$

where the change refers only to the ΔI_z of the hadrons. We now make the minimal-violation hypothesis: In all weak processes, the change of $\vec{I}$ of the hadrons satisfies

$$|\,\Delta\vec{I}\,| \leqslant 1 \ . \qquad (11.106)$$

Let us see how to apply this rule and why it is called "minimal violation."

(i) Λ-decay

In the decay

$$\Lambda^o \to \begin{cases} p + \pi^- \\ n + \pi^o \end{cases} ,$$

the initial state has $I_\Lambda = 0$, while the final-state isospin $I_{N\pi}$ can in principle be $\frac{1}{2}$ or $\frac{3}{2}$. Rule (11.106) states that $I_{N\pi}$ must assume the smaller value, $\frac{1}{2}$; i.e.,

$$|\,\Delta\vec{I}\,| = \tfrac{1}{2} \ . \qquad (11.107)$$

As we shall see in the following exercise, such a final state $|\frac{1}{2}>$ can be written as

$$|\tfrac{1}{2}> = \sqrt{\tfrac{1}{3}}\ |n\pi^{o}> - \sqrt{\tfrac{2}{3}}\ |p\pi^{-}> . \qquad (11.108)$$

Thus (11.106) leads to the relative decay rates, apart from the electromagnetic correction,

$$\frac{\text{rate } (\Lambda^{o} \to p + \pi^{-})}{\text{rate } (\Lambda^{o} \to n + \pi^{o})} \cong 2 \qquad (11.109)$$

which is consistent with the observed branching ratios of $\Lambda^{o} \to p + \pi^{-}$ and $n + \pi^{o}$ being $(64.2 \pm .5)\%$ and $(35.8 \pm .5)\%$ respectively.

<u>Exercise.</u> In the isospin space we may represent the states p , n and π by

$$|p> = \binom{1}{0} , \qquad |n> = \binom{0}{1} ,$$
$$|\pi^{\pm}> = Y_{1,\pm 1} \quad \text{and} \quad |\pi^{o}> = Y_{1,0} \qquad (11.110)$$

where $Y_{\ell,m}$ refers to the spherical harmonics given by (1.40). The state (11.108) is of $I = \frac{1}{2}$ and $I_z = -\frac{1}{2}$. Show that in the representation of (11.110), it can be written as

$$-\vec{\tau} \cdot \hat{r} \binom{0}{1} \qquad (11.111)$$

where $\hat{r} = (\sin\theta \cos\phi, \sin\theta \sin\phi, \cos\theta)$ in the isospin space. By using the above expression, verify the Clebsch - Gordon coefficients $\sqrt{\frac{2}{3}}$ and $\sqrt{\frac{1}{3}}$ in (11.108).

(ii) $K_{\pi 2}$-decay

In $K^{\pm} \to \pi^{\pm} + \pi^{o}$, K^{o} (or $\bar{K}^{o}$) $\to \pi^{+} + \pi^{-}$ or $2\pi^{o}$, since the spin of K is 0, the final two pions must be in an s-state. Bose statistics then requires that the 2π isospin $I_{2\pi} \neq 1$. Thus, $I_{2\pi}$ can be 0 or 2. Rule (11.106) states that $I_{2\pi}$ must assume the smaller value

$$I_{2\pi} = 0 \ .$$

Because $I_K = \frac{1}{2}$, we have again

$$| \Delta \vec{I} \, | = \tfrac{1}{2} \ . \tag{11.112}$$

As in the above exercise, we may represent each of the two final pions by a unit vector, $\hat{r}_a$ and $\hat{r}_b$ in the isospin space, with the subscripts referring to, say, their different momenta. The $I_{2\pi} = 0$ state can then be written as being proportional to the scalar product $\hat{r}_a \cdot \hat{r}_b$; i.e.

$$| \pi_a^+ \pi_b^- > + \, | \pi_a^- \pi_b^+ > + \, | \pi_a^o \pi_b^o > \ .$$

Hence we find, apart from the electromagnetic correction,

$$\frac{\text{rate } (K^o \rightarrow \pi^+ + \pi^-)}{\text{rate } (K^o \rightarrow 2\pi^o)} \cong 2 \tag{11.113}$$

and the same ratio if K^o is replaced by $\bar{K}^o$. Since $\pi^\pm \pi^o$ has $I_z = \pm 1$; it cannot be in an $I_{2\pi} = 0$ state and rule (11.106) implies

$$K^\pm \not\rightarrow \pi^\pm + \pi^o \ . \tag{11.114}$$

The experimental results are

$$\text{branching ratio } (K_S^o \rightarrow \pi^+ + \pi^-) = 68.61 \pm .24\% \ ,$$

$$\text{branching ratio } (K_S^o \rightarrow 2\pi^o) = 31.39 \pm .24\%$$

and

$$\frac{\text{rate } (K^\pm \rightarrow \pi^\pm + \pi^o)}{\text{rate } (K_S^o \rightarrow 2\pi)} = (1.53 \pm 0.07) \times 10^{-3}$$

in agreement with (11.113) and (11.114). [The relation between K^o, $\bar{K}^o$ and $K_S^o \cong (K^o + \bar{K}^o)/\sqrt{2}$ will be discussed in Chapter 15.]

(iii) $K_{\ell 3}$ decay

In the decays

$$K^{\pm} \rightarrow \pi^{0} + \ell^{\pm} + \nu_{\ell} \quad (\text{or } \bar{\nu}_{\ell})$$

and

$$K^{0} \ (\text{or } \bar{K}^{0}) \rightarrow \pi^{\mp} + \ell^{\pm} + \nu_{\ell} \ (\text{or } \bar{\nu}_{\ell}) \quad ,$$

where $\ell = e$ or μ, the initial isospin is $I_K = |\vec{I}_K| = \frac{1}{2}$ while the final isospin is $I_{\pi} = |\vec{I}_{\pi}| = 1$. By using the notation of (11.96), we have, according to the usual vector-addition rule of angular momenta,

$$|\Delta\vec{I}| = |\vec{I}_{\pi} - \vec{I}_{K}| = \tfrac{1}{2} \text{ or } \tfrac{3}{2} \quad . \tag{11.115}$$

[Note that $|\Delta\vec{I}| \neq \Delta|\vec{I}|$, since $\Delta|\vec{I}| = |\vec{I}_{\pi}| - |\vec{I}_{K}| = 1 - \frac{1}{2} = \frac{1}{2}$ always.]

Rule (11.106) now states that between the two possible values of (11.115), we must have, as in (11.107) and (11.112),

$$|\Delta\vec{I}| = \tfrac{1}{2} \tag{11.116}$$

only. Since the I_z of K^0 and π^+ are respectively $-\frac{1}{2}$ and $+1$, the above selection rule implies

$$K^{0} \not\rightarrow \pi^{+} + \ell^{-} + \bar{\nu}_{\ell} \quad . \tag{11.117}$$

Likewise,

$$\bar{K}^{0} \not\rightarrow \pi^{-} + \ell^{+} + \nu_{\ell} \quad . \tag{11.118}$$

By using (11.116) and by following arguments similar to those that led from (11.110)-(11.111) in the exercise to (11.108), we can derive

$$\frac{\text{rate } (K^{0} \rightarrow \pi^{-} + \ell^{+} + \nu_{\ell})}{\text{rate } (K^{+} \rightarrow \pi^{0} + \ell^{+} + \nu_{\ell})} \cong 2$$

and (11.119)

$$\frac{\text{rate } (\bar{K}^{o} \to \pi^{+} + \ell^{-} + \bar{\nu}_{\ell})}{\text{rate } (K^{-} \to \pi^{o} + \ell^{-} + \bar{\nu}_{\ell})} \cong 2 \quad .$$

In terms of $K_L^o \cong (K^o - \bar{K}^o)/\sqrt{2}$, which will be discussed in Chapter 15, these results become

$$\frac{\sum \text{rate } (K_L^{o} \to \pi^{\pm} + \ell^{\mp} + \nu_{\ell} \text{ (or } \bar{\nu}_{\ell}))}{\text{rate } (K^{\pm} \to \pi^{o} + \ell^{\pm} + \nu_{\ell} \text{ (or } \bar{\nu}_{\ell}))} \cong 2 \qquad (11.120)$$

where $\sum$ denotes the sum over + and - . The experimental values are

$$\sum \text{rate } (K_L^{o} \to \pi^{\pm} + \ell^{\mp} + \nu_{\ell} \text{ (or } \bar{\nu}_{\ell})) = \begin{cases} 5.21 \pm .10 \times 10^6/\text{sec} \\ \qquad \text{for } \ell = \mu \\ 7.49 \pm .11 \times 10^6/\text{sec} \\ \qquad \text{for } \ell = e \end{cases} \qquad (11.121)$$

and

$$\text{rate } (K^{\pm} \to \pi^{o} + \ell^{\pm} + \nu_{\ell} \text{ (or } \bar{\nu}_{\ell})) = \begin{cases} 2.58 \pm .07 \times 10^6/\text{sec} \\ \qquad \text{for } \ell = \mu \\ 3.90 \pm .04 \times 10^6/\text{sec} \\ \qquad \text{for } \ell = e , \end{cases} \qquad (11.122)$$

which are in good agreement with (11.120). We note that in the decays of the strange particles, the selection rule (11.106) always reduces to $|\Delta\vec{I}| = \frac{1}{2}$.

Problem 11.1. The pion-nucleon scattering at a laboratory π energy ~300 MeV is dominated by the resonance $\Delta(1232)$ which is of isospin $\frac{3}{2}$:

$$\pi + N \to \Delta(1232) \to \pi + N \quad .$$

Show that the ratios of the differential cross sections of

$\pi^+ + p \to \pi^+ + p$, $\pi^- + p \to \pi^0 + n$ and $\pi^- + p \to \pi^- + p$ are 9 : 2 : 1 .

Problem 11.2. Prove the selection rules listed in Tables 11.2 and 11.3 following. Discuss the nature of these rules; to what degree is each of them correct?

State:	$^1S_0$	$^3S_1$	$^1P_1$	$^3P_0$	$^3P_1$	$^3P_2$
Spin parity	0−	1−	1+	0+	1+	2+
G	−	+	+	−	−	−
$\pi^- + \pi^0$	x		x	x	x	x
$2\pi^- + \pi^+$		x	x	x		
$\pi^- + 2\pi^0$		x	x	x		
$2\pi^- + \pi^+ + \pi^0$	x			x	x	x
$\pi^- + 3\pi^0$	x			x	x	x
$3\pi^- + 2\pi^+$		x	x			
$2\pi^- + \pi^+ + 2\pi^0$		x	x			
$\pi^- + 4\pi^0$		x	x			

(x means forbidden due to $\vec{I}$, or C , or other selection rules.)

Table 11.2. Selection rules for $p + \bar{n} \to m\pi$.

State:	$^1S_0$		$^3S_1$		$^1P_1$		$^3P_0$		$^3P_1$		$^3P_2$	
Spin parity	0-		1-		1+		0+		1+		2+	
C	+		-		-		+		+		+	
I	0	1	0	1	0	1	0	1	0	1	0	1
G	+	-	-	+	-	+	+	-	+	-	+	-
$2\pi^0$	×	×	×	×	×	×		×	×	×		×
$\pi^+ + \pi^-$	×	×	×		×	×		×	×	×		×
$3\pi^0$	×		×	×	×	×	×	×	×		×	
$\pi^+ + \pi^- + \pi^0$	×			×		×	×	×	×		×	
$4\pi^0$		×	×	×	×	×		×		×		×
$\pi^+ + \pi^- + 2\pi^0$		×	×		×			×		×		×
$2\pi^+ + 2\pi^-$		×	×		×			×		×		×
$5\pi^0$	×		×	×	×	×	×		×		×	
$\pi^+ + \pi^- + 3\pi^0$	×			×		×	×		×		×	
$2\pi^+ + 2\pi^- + \pi^0$	×			×		×	×		×		×	

(× means forbidden due to $\vec{I}$, or C, or other selection rules.)

Table 11.3. Selection rules for $\bar{p} + p \rightarrow m\pi$ or $\bar{n} + n \rightarrow m\pi$.

These two tables are taken from T. D. Lee and C. N. Yang, Nuovo Cimento 3, 749 (1956). G parity was introduced in that paper.

For earlier and related work, see

K. Nishijima, Prog.Theoret.Phys. 6, 614, 1027 (1951).

A. Pais and R. Jost, Phys.Rev. 87, 871 (1952).

L. Michel, Nuovo Cimento 10, 319 (1953).

D. Amati and B. Vitale, Nuovo Cimento 2, 719 (1955).

Chapter 12

SU_3 SYMMETRY

That the strong interaction may have a much wider internal symmetry than the U_2 group was first considered by Sakata *, who explored the possibility of SU_3 symmetry generated by the unitary transformations between p, n and Λ. However, Sakata's approach encountered serious difficulties, since the Λ-nucleon force turns out to be quite different from the nucleon-nucleon force. Major progress was made by Y. Ohnuki ** in 1960 who avoided the dynamical difficulties of the Sakata model; instead he put the emphasis on the kinematics of SU_3. The observed hadrons are regarded as composites of a triplet of "baryon" fields, called X_1, X_2 and X_3 by Ohnuki. These fields have the same quantum numbers as p, n and Λ, but their quanta differ from the physical baryons because of some unspecified dynamical bound-state mechanism. By examining various representations of the SU_3 group, Ohnuki was able to identify the physical pions and kaons as members of an SU_3 octet, thereby predicting a new pseudoscalar meson, which was later discovered and is now called η^o.

---

* S. Sakata, Progr.Theoret.Phys. 16, 686 (1956).

** Y. Ohnuki, Proceedings of the International High-energy Conference, CERN (1960), p. 843.

Soon after, Gell-Mann and Ne'eman* took the decisive step of identifying the physical baryons p, n, Λ, Σ and Ξ, also, as members of an SU_3-octet. At that time the basic triplet was regarded more as a mathematical device for the construction of the octet (called the eightfold way) and the decuplet representations, which can then be directly applied to the observed hadrons. These applications of SU_3-symmetry led to great success in bringing order to the complex problems of spectroscopy, dynamics and decay rates of hadrons.

We now know that all hadrons can be viewed as composites of quarks**. In this chapter, we consider only the three low-lying quark fields: up, down and strange, which are sometimes referred to as different "flavors". Each will be represented by an element of the column matrix

$$q = \begin{pmatrix} q^1 \\ q^2 \\ q^3 \end{pmatrix} . \qquad (12.1)$$

The strong interaction is assumed to be approximately invariant under the transformation

$$q \to vq \qquad (12.2)$$

where v is unitary. The detailed form of the strong interaction will be considered later when we discuss quantum chromodynamics in Chapter 18. Here we examine only the consequences of the symmetry assumption.

---

* M. Gell-Mann, Phys.Rev. 125, 1067 (1962); Caltech Report CTSL-20 (1961); Y. Ne'eman, Nucl.Phys. 26, 222 (1961).

** G. Zweig, CERN report (unpublished); M. Gell-Mann, Phys.Lett. 8, 214 (1964).

12.1 Mathematical Preliminary

Just as in (11.17), we can separate out the overall phase of v, and consider the SU_3 group $\{v\}$ spanned by all 3×3 unitary matrices with unit determinant; i.e.,

$$v\, v^{\dagger} = 1 \qquad \text{and} \qquad \det v = 1 \quad . \tag{12.3}$$

For convenience, the matrix v will be written as

$$v = (v_b{}^a) = \begin{pmatrix} v_1{}^1 & v_2{}^1 & v_3{}^1 \\ v_1{}^2 & v_2{}^2 & v_3{}^2 \\ v_1{}^3 & v_2{}^3 & v_3{}^3 \end{pmatrix} \quad ; \tag{12.4}$$

the matrix elements of its Hermitian conjugate $v^{\dagger}$ are given by

$$v_a^{\dagger b} = (v_b{}^a)^* \quad . \tag{12.5}$$

Thus, the transformation (12.2) can be written as

$$q^a \rightarrow v_b{}^a\, q^b \quad . \tag{12.6}$$

1. Tensors

Definition: $T_{a_1 \cdots a_m}^{i_1 \cdots i_n}$ is a component of a tensor T of rank (n, m) if its transformation law is

$$T_{a_1 \cdots a_m}^{i_1 \cdots i_n} \rightarrow v_{j_1}{}^{i_1} \cdots v_{j_n}{}^{i_n}\; v^{\dagger b_1}_{a_1} \cdots v^{\dagger b_m}_{a_m}\; T_{b_1 \cdots b_m}^{j_1 \cdots j_n} \quad . \tag{12.7}$$

Throughout this section, unless otherwise indicated, all indices vary from 1 to 3. Thus, the above tensor has 3^{n+m} components.

From this definition, by adding or subtracting the corresponding components of two tensors of the same rank, we form a new tensor also of the same rank. By multiplying the components of a tensor A of rank (n, m) with those of a tensor B of rank (q, p), we can construct a new tensor of rank $(n+q\ ,\ m+p)$ whose components are

$$A_{a_1 \cdots a_m}^{i_1 \cdots i_n} \; B_{b_1 \cdots b_p}^{j_i \cdots j_q} \quad . \tag{12.8}$$

Isotropic tensors: *

$$\text{(i)} \quad \delta_a^{\ i} = \begin{cases} 1 & \text{if } i = a\ , \\ 0 & \text{otherwise}\ , \end{cases} \tag{12.9}$$

are components of a tensor of rank $(1\ ,\ 1)$.

<u>Proof.</u> Because $v\,v^{\dagger} = 1$, under the transformation

$$\delta_a^{\ i} \to v_j^{\ i}\, v_{\ a}^{\dagger b}\, \delta_b^{\ j} = \delta_a^{\ i} \quad .$$

$$\text{(ii)} \quad \epsilon^{ijk} = \begin{cases} 1 & \text{if } ijk \text{ is an even permutation of } 1,2,3, \\ -1 & \text{if } ijk \text{ is an odd permutation of } 1,2,3, \\ 0 & \text{otherwise}, \end{cases} \tag{12.10}$$

are components of a tensor of rank $(3\ ,\ 0)$.

<u>Proof.</u> Because $\det v = 1$, we have under the transformation

$$\epsilon^{ijk} \to v_a^{\ i}\, v_b^{\ j}\, v_c^{\ k}\, \epsilon^{abc} = \epsilon^{ijk} \quad .$$

(iii) Likewise, because $\det v^{\dagger} = 1$, ϵ_{ijk} given by (3.4) are components of a tensor of rank $(0\ ,\ 3)$.

Contraction: From a tensor T of rank (n, m), we can form new tensors of rank $(n-1,\ m-1)$, $(n-2,\ m+1)$ and $(n+1,\ m-2)$ by constructing respectively the following products and summing over repeated indices:

$$T_{j_1 \cdots j_m}^{i_1 \cdots i_n} \; \delta_{i_a}^{\ j_b} \quad , \tag{12.11}$$

---

* Tensors whose components are unaltered under the transformations are called isotropic.

$$T^{i_1 \cdots i_n}_{j_1 \cdots j_m} \quad \epsilon_{j_{m+1} i_a i_{a'}} \tag{12.12}$$

and

$$T^{i_1 \cdots i_n}_{j_1 \cdots j_m} \quad \epsilon^{i_{n+1} j_b j_{b'}} \quad , \tag{12.13}$$

where a and a' are two different integers between 1 and n, and likewise b and b' are different integers between 1 and m, assuming that n and m are both $\geqslant 1$ in (12.11), $n \geqslant 2$ in (12.12) and $m \geqslant 2$ in (12.13).

Definition: A tensor T of rank (n, m) is called reducible if through contraction a new nonzero tensor T' of rank (n', m') can be formed with

$$n' + m' < n + m \quad ;$$

otherwise T is irreducible.

Because of (12.12)-(12.13), an irreducible tensor $T^{i_1 \cdots i_n}_{j_1 \cdots j_m}$ must be symmetric with respect to any pair $(i_a, i_{a'})$ or $(j_b, j_{b'})$. Furthermore, in view of (12.11), T must also satisfy the trace condition

$$T^{i_1 \cdots i_n}_{j_1 \cdots j_m} \, \delta^{j_1}_{i_1} = 0 \quad . \tag{12.14}$$

Because of the symmetry and trace conditions, the components of an irreducible tensor may not all be independent.

2. Representations

Consider a tensor of rank (n, m). Let the number of its linearly-independent components be d, known as the dimension. From such a tensor, we can select d linearly-independent components $\phi_1, \phi_2, \cdots, \phi_d$ and write

$$\phi = \begin{pmatrix} \phi_1 \\ \phi_2 \\ \vdots \\ \phi_d \end{pmatrix} \quad .$$

Under (12.2), $q \rightarrow vq$, the transformation (12.7) can be written in terms of ϕ as

$$\phi \rightarrow V \cdot \phi \tag{12.15}$$

where V is a $d \times d$ matrix. Because the V's clearly satisfy the same algebra as the v's, we regard $\{V\}$ as forming a representation, denoted by ⓓ, of the SU_3 group $\{v\}$. Furthermore, the representation is called <u>irreducible</u> if the tensor is; otherwise it is reducible.

The following is a list of the low-ranking irreducible tensors:

Irreducible tensor	Rank	Representation
1	(0, 0)	①
T^i	(1, 0)	③
T_a	(0, 1)	$\bar{③}$
T_j^i	(1, 1)	⑧
T^{ij}	(2, 0)	⑥
T_{ab}	(0, 2)	$\bar{⑥}$
T^{ijk}	(3, 0)	⑩
T_{abc}	(0, 3)	$\overline{⑩}$
T_{ab}^{ij}	(2, 2)	㉗

TABLE 12.1

where the numbers in the last column denote the dimensions. We note that both T^i and T_a are of dimension three. However, because of the difference in their transformation properties, these form two different representations, called ③ and $\bar{3}$. [For further analysis, see the discussions given on page 261.] Due to the symmetry condition, $T^{ij} = T^{ji}$ and $T_{ab} = T_{ba}$, both representations are of dimension six, but each is different and hence they are labelled ⑥ and $\bar{6}$. Because of the constraint $T^{i}_{\cdot i} = 0$, the irreducible tensor T^{i}_{a} is of dimension $3^2 - 1 = 8$. For a symmetric tensor T^{ijk}, there are three independent components of the type T^{111}, six of the type T^{112} and one of the type T^{123}, making a total of ten and forming the representation ⑩ . Similarly, the irreducible tensor T_{abc} is also of dimension ten, but labelled $\overline{10}$ because it forms a representation different from that given by T^{ijk} . In like manner, it is straightforward to find that the irreducible tensor T^{ij}_{ab} is of dimension (27) .

3. Decomposition of ⑧ x ⑧

Let $A^{i}_{\cdot j}$ and $B^{i}_{\cdot j}$ be both of the ⑧ representation. Hence,

$$A^{i}_{\cdot i} = B^{i}_{\cdot i} = 0 \ . \tag{12.16}$$

Consider the product ⑧ x ⑧ :

$$A^{i}_{a} B^{j}_{b} \ , \tag{12.17}$$

which is a reducible tensor of rank (2, 2). Our task is to form linear functions of its $8^2 = 64$ components so that they become irreducible.

(i) The sum

$$S = A^{i}_{a} B^{a}_{\cdot i} \tag{12.18}$$

is clearly an irreducible tensor of rank (0, 0).

(ii) The tensor

$$F_a^{\ i} = A_j^{\ i} B_a^{\ j} - B_j^{\ i} A_a^{\ j} \tag{12.19}$$

is traceless and therefore forms an irreducible 8-dimensional representation. Likewise

$$D_a^{\ i} = A_j^{\ i} B_a^{\ j} + B_j^{\ i} A_a^{\ j} - \tfrac{2}{3} \delta_j^{\ i} S \tag{12.20}$$

forms another irreducible 8-dimensional representation. Since $F_a^{\ i}$ is antisymmetric in A and B, while $D_a^{\ i}$ is symmetric, these are two different functions, although both are ⓼ .

(iii) The symmetric tensor

$$T^{ijk} = A_a^{\ i} B_b^{\ j} \epsilon^{abk} + \text{all terms formed by permuting } ijk$$

gives an irreducible representation ⑽ . Similarly, by interchanging superscripts and subscripts, we can form a representation $\overline{⑽}$ in terms of the symmetric tensor

$$\overline{T}_{abc} = A_a^{\ i} B_b^{\ j} \epsilon_{ijc} + \text{all terms formed by permuting } abc .$$

(iv) By symmetrizing (12.17) with respect to (i, j) and (a, b), we can first form

$$R_{ab}^{\ ij} = A_a^{\ i} B_b^{\ j} + A_a^{\ j} B_b^{\ i} + A_b^{\ i} B_a^{\ j} + A_b^{\ j} B_a^{\ i} ,$$

which satisfies, on account of (12.18) and (12.20),

$$R_{ab}^{\ ib} = D_a^{\ i} + \tfrac{2}{3} \delta_a^{\ i} S .$$

We then construct the irreducible tensor

$$I_{ab}^{\ ij} = R_{ab}^{\ ij} - \frac{1}{5} (\delta_a^{\ i} D_b^{\ j} + \delta_a^{\ j} D_b^{\ i} + \delta_b^{\ i} D_a^{\ j} + \delta_b^{\ j} D_a^{\ i}) - \frac{1}{6} (\delta_a^{\ i} \delta_b^{\ j} + \delta_a^{\ j} \delta_b^{\ i}) S ,$$

which forms an irreducible representation ㉗ .

Putting together (i)-(iv), we can write

$$\textcircled{8} \times \textcircled{8} = \textcircled{1} + \textcircled{8} + \textcircled{8} + \textcircled{10} + \textcircled{\overline{10}} + \textcircled{27} .$$

It is not difficult to extend the above considerations to representations of higher dimensions.

4. Some further properties

Consider an infinitesimal SU_3 transformation which contains, because of $\det v = 1$, $3^2 - 1 = 8$ independent real infinitesimal quantities $\epsilon_1, \epsilon_2, \cdots, \epsilon_8$:

$$v = 1 + \tfrac{1}{2} i \epsilon_\ell \lambda_\ell \tag{12.21}$$

where the λ_ℓ 's are eight 3×3 matrices and 1 denotes the unit matrix. Since

$$v v^\dagger = 1 + \tfrac{1}{2} i \epsilon_\ell (\lambda_\ell - \lambda_\ell^\dagger) + O(\epsilon^2) = 1 ,$$

we have

$$\lambda_\ell = \lambda_\ell^\dagger .$$

These eight Hermitian matrices $\lambda_1, \lambda_2, \cdots, \lambda_8$ are called the generators of the SU_3 group; they play the same role as the three Pauli matrices for the SU_2 group. It is customary to write

$$\begin{aligned}
\lambda_1 &= \begin{pmatrix} 0 & 1 & 0 \\ 1 & 0 & 0 \\ 0 & 0 & 0 \end{pmatrix} , & \lambda_2 &= \begin{pmatrix} 0 & -i & 0 \\ i & 0 & 0 \\ 0 & 0 & 0 \end{pmatrix} , \\
\lambda_3 &= \begin{pmatrix} 1 & 0 & 0 \\ 0 & -1 & 0 \\ 0 & 0 & 0 \end{pmatrix} , & \lambda_4 &= \begin{pmatrix} 0 & 0 & 1 \\ 0 & 0 & 0 \\ 1 & 0 & 0 \end{pmatrix} , \\
\lambda_5 &= \begin{pmatrix} 0 & 0 & -i \\ 0 & 0 & 0 \\ i & 0 & 0 \end{pmatrix} , & \lambda_6 &= \begin{pmatrix} 0 & 0 & 0 \\ 0 & 0 & 1 \\ 0 & 1 & 0 \end{pmatrix} , \\
\lambda_7 &= \begin{pmatrix} 0 & 0 & 0 \\ 0 & 0 & -i \\ 0 & i & 0 \end{pmatrix} , & \lambda_8 &= \frac{1}{\sqrt{3}} \begin{pmatrix} 1 & 0 & 0 \\ 0 & 1 & 0 \\ 0 & 0 & -2 \end{pmatrix} .
\end{aligned} \tag{12.22}$$

The trace, commutator and anticommutator of two λ matrices are

$$\begin{aligned} \mathrm{tr}\,(\lambda_\ell\, \lambda_m) &= 2\delta_{\ell m} \;, \\ [\lambda_\ell\, , \lambda_m] &= 2i\, f_{\ell mn}\, \lambda_n \\ \text{and} \quad \{\lambda_\ell\, , \lambda_m\} &= \frac{4}{3}\, \delta_{\ell m} + 2d_{\ell mn}\, \lambda_n \end{aligned} \tag{12.23}$$

where $f_{\ell mn}$ is completely antisymmetric in its indices while $d_{\ell mn}$ is completely symmetric. The nonzero elements of $f_{\ell mn}$ and $d_{\ell mn}$ are as follows:

ℓmn	$f_{\ell mn}$	ℓmn	$d_{\ell mn}$
123	1	118	$\sqrt{\frac{1}{3}}$
147	$\frac{1}{2}$	146	$\frac{1}{2}$
156	$-\frac{1}{2}$	157	$\frac{1}{2}$
246	$\frac{1}{2}$	228	$\sqrt{\frac{1}{3}}$
257	$\frac{1}{2}$	247	$-\frac{1}{2}$
345	$\frac{1}{2}$	256	$\frac{1}{2}$
367	$-\frac{1}{2}$	338	$\sqrt{\frac{1}{3}}$
458	$\frac{1}{2}\sqrt{3}$	344	$\frac{1}{2}$
678	$\frac{1}{2}\sqrt{3}$	355	$\frac{1}{2}$
		366	$-\frac{1}{2}$
		377	$-\frac{1}{2}$
		448	$-1/(2\sqrt{3})$
		558	$-1/(2\sqrt{3})$
		668	$-1/(2\sqrt{3})$
		778	$-1/(2\sqrt{3})$
		888	$-\sqrt{\frac{1}{3}}$.

(12.24)

Let $\{v\}$ and $\{\bar{v}\}$ be two 3×3 irreducible representations of the SU_3 group. If for every $\bar{v}$ there is a v such that

$$\bar{v} = v_o \, v \, v_o^\dagger \tag{12.25}$$

where v_o is a fixed element of $\{v\}$, then all the $\bar{v}$'s also belong to $\{v\}$; therefore these two representations are regarded as the same, otherwise not. In Table 12.1 the representation associated with the irreducible tensor T^i is $\{v\}$, while that associated with T_a is $\{v^*\}$. We denote these two representations as ③ and $\textcircled{\bar{3}}$ because, as we shall now prove, they are not the same.

Proof. If we were to assume that they are, (12.25) would become

$$v^* = v_o \, v \, v_o^\dagger \quad . \tag{12.26}$$

Consider now an infinitesimal v of the form (12.21), whose complex conjugation is

$$v^* = 1 - \tfrac{1}{2} i \, \lambda_\ell^* \, \epsilon_\ell \quad .$$

Upon substituting this expression, together with (12.21), into (12.26) we find

$$\lambda_\ell^* = - v_o \, \lambda_\ell \, v_o^\dagger \quad . \tag{12.27}$$

From (12.22) we see that

$$\lambda_\ell^* = \pm \lambda_\ell \tag{12.28}$$

where the $\pm$ sign $\equiv \eta_\ell$ is given by

$$\eta_\ell = \begin{cases} + & \text{for } \ell = 1, 3, 4, 6, 8 \\ - & \text{for } \ell = 2, 5, 7 \end{cases} \tag{12.29}$$

and therefore (12.27) becomes

$$- \eta_\ell \, \lambda_\ell = v_o \, \lambda_\ell \, v_o^\dagger \tag{12.30}$$

in which the repeated index is not summed over, but can be 1 or 2, $\cdots$, or 8 . By multiplying the last equation in (12.23) on the left by v_o and on the right by $v_o^\dagger$, we see that it is invariant under the change $\lambda_\ell \to v_o \lambda_\ell v_o^\dagger$; thus on account of (12.30) we can derive

$$-\eta_\ell \eta_m \eta_n d_{\ell mn} = d_{\ell mn} \qquad (12.31)$$

where as in (12.30) the repeated indices refer to fixed numbers, not summed over. By using (12.24) and (12.29), we find that this is wrong for every nonzero $d_{\ell mn}$. Therefore, the representations $\{v\}$ and $\{v^*\}$ are different.

<u>Exercise.</u> Establish first the Jacobi identity

$$[\lambda_\ell, [\lambda_m, \lambda_n]] + [\lambda_m, [\lambda_n, \lambda_\ell]] + [\lambda_n, [\lambda_\ell, \lambda_m]] = 0$$

and then derive

$$f_{\ell\ell' k} f_{mn\ell'} + f_{m\ell' k} f_{n\ell\ell'} + f_{n\ell' k} f_{\ell m\ell'} = 0 \; .$$

5. <u>Excursion to other groups</u>

(i) SU_2 group

In the case of the SU_2 group $\{s\}$, on account of (11.42) we have for every s

$$s^* = s_o s s_o^\dagger \qquad (12.32)$$

where s_o is a fixed element given by

$$s_o = e^{i\frac{1}{2}\pi\tau_2} = i\tau_2 \; .$$

Consequently, representations $\{s\}$ and $\{s^*\}$ are the same (sometimes referred to as equivalent); hence, there exists only one two-dimensional representation of the SU_2 group.

The notion of tensors and representation can be applied equally well to the SU_2 group. We call $T_{a_1 \cdots a_n}$ the component of a tensor of rank n if its transformation law is

$$T_{a_1 a_2 \cdots a_n} \rightarrow s_{a_1 b_1} s_{a_2 b_2} \cdots s_{a_n b_n} T_{b_1 b_2 \cdots b_n} \tag{12.33}$$

where a_i and b_j can be 1 or 2. Because of (12.32), the rank is now characterized by only one number. Since det s = 1, one sees that

$$\epsilon_{ab} \equiv \begin{cases} 1 & \text{for } a = 1,\ b = 2, \\ -1 & \text{for } a = 2,\ b = 1, \\ 0 & \text{otherwise} \end{cases} \tag{12.34}$$

is an isotropic tensor of rank 2.

Similarly to (12.12)-(12.13), if a tensor T is not completely symmetric in its indices, say $T_{a_1 a_2 a_3 \cdots a_n} \neq T_{a_2 a_1 a_3 \cdots a_n}$, then by constructing

$$\epsilon_{a_1 a_2} T_{a_1 a_2 \cdots a_n}$$

we can form a nonzero tensor of lower rank n - 2. Such a tensor T is called reducible; otherwise it is irreducible. Again each tensor gives a representation of the group, and the representation is irreducible if the tensor is. A tensor of rank n has 2^n components which may not all be independent. The dimension d is defined to be the number of its linearly-independent components.

Exercise. Prove the following list of irreducible tensors and representations of the SU$_2$ group:

	Irreducible tensor	Rank	Dimension of representation
	1	0	1
	T_a	1	2
symmetric	T_{ab}	2	3
completely symmetric	T_{abc}	3	4
		. . . . .	
completely symmetric	$T_{a_1 a_2 \cdots a_n}$	n	n + 1 .

This classification of irreducible representations is identical to the usual one in terms of angular-momentum states, provided $n = 2j$ where j is the total angular-momentum quantum number.

(ii) SO_3 group

If we limit ourselves to the angular-momentum j = integer states, the corresponding group becomes SO_3, which comprises all 3×3 real, orthogonal matrices u; i.e.,

$$u = u^*, \quad u\tilde{u} = 1 \quad \text{and} \quad \det u = 1 \quad . \tag{12.35}$$

Each u can be viewed as a three-dimensional rotation of the position vector $\vec{r} = (x_1, x_2, x_3)$:

$$x_i \to u_{ij} x_j$$

where the subscripts i and j vary from 1 to 3. The component of a tensor T of rank n now transforms according to

$$T_{i_1 i_2 \cdots i_n} \to u_{i_1 j_1} u_{i_2 j_2} \cdots u_{i_n j_n} T_{j_1 j_2 \cdots j_n} \quad .$$

Because $u\tilde{u} = 1$ and $\det u = 1$, one sees that δ_{ij} and ϵ_{ijk} respectively form tensors of rank 2 and 3, where δ_{ij} is a Kronecker symbol and ϵ_{ijk} is defined by (3.4). A tensor is called reducible if it is either not totally symmetric with respect to its indices, say $T_{i_1 i_2 i_3 \cdots i_n} \neq T_{i_2 i_1 i_3 \cdots i_n}$, or it has nonzero trace. In the former we can form a nonzero tensor of rank $n - 1$ by constructing

$$T_{i_1 i_2 i_3 \cdots i_n} \epsilon_{i_1 i_2 j} \quad ;$$

while in the latter we can form a tensor of rank $n-2$ through the contraction

$$T_{i_1 i_2 \cdots i_n} \delta_{i_a i_b}$$

where a and b can be any two different integers between 1 and n . The notions, previously discussed, of dimension and representation and its reducibility can be straightforwardly applied to the present case.

Exercise. Establish the following list of irreducible tensors and representations of the SO_3 group:

Irreducible tensor	Rank	Dimension of representation
1	0	1
T_i	1	3
T_{ij}	2	5
T_{ijk}	3	7
	. . . . .	
$T_{i_1 \cdots i_\ell}$	ℓ	$2\ell + 1$.

Show that the irreducible representations can be chosen to be precisely the familiar spherical harmonics $Y_{\ell m}$ discussed in (1.38)-(1.40).

12.2 Hadron States and Their Flavor and Color Symmetries

It is convenient to denote the q^i's of (12.1) as the specific up, down and strange quarks written respectively as

$$q^1 = u \ , \quad q^2 = d \quad \text{and} \quad q^3 = s \ . \tag{12.36}$$

The approximate invariance under the transformation $q \to vq$, given by (12.2), will be referred to as the SU_3 flavor symmetry. The dynamic properties of these quarks will be examined later. Here we only use their transformation properties. The q^i's form the irreducible representation ③ ; their antiparticles $q_1 = \bar{u}$, $q_2 = \bar{d}$ and $q_3 = \bar{s}$ form the representation $\textcircled{\bar{3}}$. The isospin SU_2 is a subgroup of SU_3 with $u = \uparrow$, $d = \downarrow$ forming an isospin doublet and s an isospin singlet.

In 3×3 matrix form, each 2×2 isospin transformation matrix s corresponds to

$$v = \begin{pmatrix} s & 0 \\ 0 & 1 \end{pmatrix} \tag{12.37}$$

where 0 stands for either a 2×1 or a 1×2 null matrix.

1. Pseudoscalar octet

The eight pseudoscalar mesons listed in Table 11.1 on page 239 are regarded as forming an SU_3 octet $M_j^{\;i}$ with

$$M_i^{\;i} = 0 \;, \tag{12.38}$$

which has the same transformation properties as the quark-antiquark system:

$$M_j^{\;i} \sim q^i q_j - \tfrac{1}{3} \delta_j^{\;i} q^k q_k \;. \tag{12.39}$$

By comparing the isospin properties of the pseudoscalar mesons with those of the quarks, we find

$$\begin{aligned}
&\pi^+ \sim u\bar{d} \;, \qquad \pi^- \sim d\bar{u} \;, \\
&\pi^o \sim \frac{1}{\sqrt{2}} (u\bar{u} - d\bar{d}) \;, \\
&K^+ \sim u\bar{s} \;, \qquad K^- \sim s\bar{u} \;, \\
&K^o \sim d\bar{s} \;, \qquad \bar{K}^o \sim s\bar{d} \;, \\
&\eta^o \sim \frac{1}{\sqrt{6}} (2s\bar{s} - u\bar{u} - d\bar{d}) \;.
\end{aligned} \tag{12.40}$$

Here the $\sim$ indicates that its two sides have the same SU_3 transformation properties. Just as in (11.45)-(11.46), the isospin transformation properties can be exhibited by writing

$$u = \uparrow \;, \qquad d = \downarrow \tag{12.41}$$

for the quark isodoublet, and

$$\bar{d} = \uparrow \text{ and } -\bar{u} = \downarrow \tag{12.42}$$

for the antiquark states. Thus, the first expression in (12.40) gives

$$\pi^+ \sim \uparrow\uparrow$$

where the first $\uparrow$ on the left refers to the quark and the second to the antiquark. Likewise $\pi^- \sim -\downarrow\downarrow$ and $\pi^o \sim -\frac{1}{\sqrt{2}}(\uparrow\downarrow + \downarrow\uparrow)$ which are the familiar $I_z = -1$ and 0 states of an isospin triplet. Notice that in π^o the relative sign between $\uparrow\downarrow$ and $\downarrow\uparrow$ is $+$, while according to (12.40) that between $u\bar{u}$ and $d\bar{d}$ in π^o is $-$, which leads to, on account of (12.39),

$$\pi^o = \frac{1}{\sqrt{2}}(M_1^{\ 1} - M_2^{\ 2}) \quad . \tag{12.43}$$

Because of the trace condition (12.38), from the three diagonal matrix elements $M_1^{\ 1}$, $M_2^{\ 2}$ and $M_3^{\ 3}$ there are only two linearly independent components. One is π^o and the other is η^o. Apart from an arbitrary overall phase factor, by using (12.39) one sees that the expression for η^o is uniquely determined by its orthogonality relation with π^o and its normalization condition. The resulting expression is given by the last formula in (12.40), which may also be written as

$$\eta^o = \frac{1}{\sqrt{6}}(-M_1^{\ 1} - M_2^{\ 2} + 2\,M_3^{\ 3}) \quad . \tag{12.44}$$

To express $M_1^{\ 1}$, $M_2^{\ 2}$ and $M_3^{\ 3}$ in terms of π^o and η^o, we may temporarily refrain from using (12.38), but formally regard

$$M_i^{\ i} = M_1^{\ 1} + M_2^{\ 2} + M_3^{\ 3} \tag{12.45}$$

as an entity. Solving (12.43)-(12.45) for $M_1^{\ 1}$, we find

$$M_1^{\ 1} = \frac{\pi^o}{\sqrt{2}} - \frac{\eta^o}{\sqrt{6}} + \tfrac{1}{3} M_i^{\ i}$$

which, on account of (12.38), gives

$$M_1^{\ 1} = \frac{\pi^o}{\sqrt{2}} - \frac{\eta^o}{\sqrt{6}} \quad . \tag{12.46}$$

Likewise, we can solve $M_2^{\ 2}$ and $M_3^{\ 3}$. Thus, from (12.40) it follows that the pseudoscalar octet matrix is

$$M = (M_j^{\ i}) = \begin{pmatrix} \frac{\pi^o}{\sqrt{2}} - \frac{\eta^o}{\sqrt{6}} & \pi^+ & K^+ \\ \pi^- & -\frac{\pi^o}{\sqrt{2}} - \frac{\eta^o}{\sqrt{6}} & K^o \\ K^- & \bar{K}^o & \frac{2\eta^o}{\sqrt{6}} \end{pmatrix} . \tag{12.47}$$

2. Baryon spin-$\frac{1}{2}$ octet and spin-$\frac{3}{2}$ decuplet

The eight spin-$\frac{1}{2}$ baryons listed in Table 11.1 are also regarded as forming an SU_3 octet $B_j^{\ i}$. By matching I, I_z and Y, we replace $\pi \to \Sigma$, $\eta^o \to \Lambda^o$, $K^+ \to p$, $K^o \to n$, $K^- \to \Xi^-$ and $\bar{K}^o \to \Xi^o$; the meson octet becomes the baryon octet and (12.47) becomes

$$B = (B_j^{\ i}) = \begin{pmatrix} \frac{\Sigma^o}{\sqrt{2}} - \frac{\Lambda^o}{\sqrt{6}} & \Sigma^+ & p \\ \Sigma^- & -\frac{\Sigma^o}{\sqrt{2}} - \frac{\Lambda^o}{\sqrt{6}} & n \\ \Xi^- & \Xi^o & \frac{2\Lambda^o}{\sqrt{6}} \end{pmatrix} . \tag{12.48}$$

In the Table of Particle Properties, the low-lying spin-$\frac{3}{2}$ baryon resonances are $\Delta(1232)$, $\Sigma^*(1385)$, $\Xi^*(1530)$ and $\Omega^-(1672)$. These will be identified as forming an SU_3 decuplet D^{ijk}. The assignments can be done best in terms of the quark model.

Because of the half-integer nature of the baryons, in a quark model they must be composites of an odd number of quarks and antiquarks. The simplest way is to regard them as bound states of three quarks, each a spin-$\frac{1}{2}$ fermion. If we exhibit only the SU_3-flavor indices, as in (12.39), then we may write

$$B_j^{\ i} \sim q^i q^a q^b \epsilon_{abj} - \tfrac{1}{3}\, \delta_j^{\ i} q^k q^a q^b \epsilon_{abk}$$

and (12.49)

$$D^{ijk} \sim \text{symmetrized } q^i q^j q^k \; .$$

The assignments of the decuplet D^{ijk}, in the notation of (12.36), are

$$\Omega^- = D^{333} \sim sss \;,$$

$$\Xi^{*0} \sim \sqrt{\tfrac{1}{3}}\,(uss + sus + ssu) \;,$$

$$\Xi^{*-} \sim \sqrt{\tfrac{1}{3}}\,(dss + sds + ssd) \;,$$

$$\Sigma^{*+} \sim \sqrt{\tfrac{1}{3}}\,(suu + usu + suu) \;,$$

$$\Sigma^{*0} \sim (sud + sdu + usd + dsu + uds + dus)/\sqrt{6} \;,$$

$$\Sigma^{*-} \sim \sqrt{\tfrac{1}{3}}\,(sdd + dsd + sdd) \;, \qquad (12.50)$$

$$\Delta^{++} \sim uuu \;,$$

$$\Delta^{+} \sim \sqrt{\tfrac{1}{3}}\,(duu + udu + uud) \;,$$

$$\Delta^{0} \sim \sqrt{\tfrac{1}{3}}\,(udd + dud + ddu) \;,$$

$$\Delta^{-} \sim ddd \;,$$

in which the different orders of quarks are regarded as having some different kinematic attributes. For the moment, if one wishes, one may simply choose these different attributes to be, say, different momenta $\vec{p}, \vec{p}\,', \vec{p}\,''$ with $\vec{p} + \vec{p}\,' + \vec{p}\,'' = 0$. Hence, e.g., in Ξ^{*0} one may view the first term uss as referring to u having $\vec{p}$, the middle s having $\vec{p}\,'$ and the final s having $\vec{p}\,''$; the two other terms sus and ssu are generated by permuting the SU_3 indices, but keeping $\vec{p}, \vec{p}\,', \vec{p}\,''$ fixed. This, then, explains the $\sqrt{\frac{1}{3}}$ factor in Ξ^{*0} so that it is normalized in the same way as Ω^-. [For further details, see (12.51) and (12.54)-(12.56) below.] The actual dynamics of these bound states will be analyzed later in Chapter 20. Here we are interested in their purely kinematic aspects. From (12.50) we see that the charges of u, d and s must be $\frac{2}{3}$, $-\frac{1}{3}$ and $-\frac{1}{3}$ respectively. Their spins are $\frac{1}{2}$, and hence they are fermions.

As we shall show later, it is a reasonable approximation to assume further that each quark is in the same lowest-energy s-orbit of a common potential with no mutual interaction. This leads to an immediate problem: Consider, for example, the spin-$\frac{3}{2}$ baryon $\Omega^- \sim sss$. After we sum over the different quark momenta, because of the s-orbit assumption, the total orbital wave function is completely symmetric. So is the total spin wave function because all three $\frac{1}{2}$-quark spins are lined up to form a total $\frac{3}{2}$-spin. The Fermi statistics of quarks would make it impossible for them to be the component of a completely symmetric tensor D^{ijk} unless we assume that the quarks have some other degrees of freedom. For this reason, we shall assume that the quarks do have another degree of freedom*, called <u>color.</u> Each quark, u or d or s ···, has three varieties u(c) or d(c) or s(c) ···, with c = 1, 2, 3. [Sometimes, for more colorful designations one might choose, e.g., c = red, white and blue.] For example, in the second expression of (12.49), the baryon state with total spin = $\frac{3}{2}$ and its z-component also = $\frac{3}{2}$, assumes then the form

$$D^{ijk} \sim q_\uparrow^{\,i}(c)\, q_\uparrow^{\,j}(c')\, q_\uparrow^{\,k}(c'')\, \epsilon_{cc'c''} \tag{12.51}$$

where c, c', c" are the color indices, $\epsilon_{cc'c''}$ is the totally antisymmetric tensor given by (3.4), the subscripts ↑ and ↓ denote

---

* O. W. Greenberg, Phys. Rev. Lett. 13, 598 (1964).

the spin-up and -down configurations of the quark, and $q_\uparrow^{\,i}(c)$ represents the corresponding anticommuting quark-field operator. Consequently, the interchange of i and j on the righthand side of (12.51) results in two minus signs: one from the anticommutation of the fermion operators and the other from $\epsilon_{cc'c''}$. Hence, the symmetry of D^{ijk} is insured.

Now, let us examine other hadron states with this new degree of freedom included. Consider, e.g., (12.40); instead of $\pi^+ \sim u\bar{d}$, we might write $u(c)\,\bar{d}(c')$ which could lead to $3^2 = 9$ possible meson states because of different values of c and c'. In order to avoid this difficulty, we postulate that all interactions satisfy exact SU_3-color symmetry and all observed hadron states are color singlets. Hence, π^+ is represented by

$$\pi^+ \sim u(c)\ \bar{d}(c) \tag{12.52}$$

with the repeated color index c summed over from 1 to 3 so that the state becomes a color singlet, as in (12.51). The same rule applies to all other meson and baryon states; hence (12.39) and (12.49) become

$$\begin{aligned} M_j^{\,i} &\sim q^i(c)\,q_j(c) - \tfrac{1}{3}\,\delta_j^{\,i}\,q^k(c)\,q_k(c)\ , \\ B_j^{\,i} &\sim [\,q^i(c)\,q^a(c')\,q^b(c'')\,\epsilon_{abj} \\ &\qquad - \tfrac{1}{3}\,\delta_j^{\,i}\,q^k(c)\,q^a(c')\,q^b(c'')\,\epsilon_{abk}\,]\,\epsilon_{cc'c''} \\ \text{and}\quad D^{ijk} &\sim q^i(c)\,q^j(c')\,q^k(c'')\,\epsilon_{cc'c''} \end{aligned} \tag{12.53}$$

where, for clarity, the spin-dependence is not exhibited.

As noted before, the indices that differentiate the u, d and s quarks are referred to as flavor indices. For example, D^{ijk} has ten physically different flavor configurations corresponding to different choices of superscript:

$$
\begin{aligned}
\Omega^- &= D^{333} \, , \\
\Xi^{*o} &= \sqrt{\tfrac{1}{3}} \, (D^{133} + D^{313} + D^{331}) \, , \\
\Xi^{*-} &= \sqrt{\tfrac{1}{3}} \, (D^{233} + D^{323} + D^{332}) \, , \\
\Sigma^{*+} &= \sqrt{\tfrac{1}{3}} \, (D^{311} + D^{131} + D^{113}) \, , \\
\Sigma^{*o} &= (D^{312} + D^{321} + D^{132} + D^{231} + D^{123} + D^{213}) / \sqrt{6} \, , \\
\Sigma^{*-} &= \sqrt{\tfrac{1}{3}} \, (D^{322} + D^{232} + D^{223}) \, , \qquad (12.54) \\
\Delta^{++} &= D^{111} \, , \\
\Delta^{+} &= \sqrt{\tfrac{1}{3}} \, (D^{211} + D^{121} + D^{112}) \, , \\
\Delta^{o} &= \sqrt{\tfrac{1}{3}} \, (D^{122} + D^{212} + D^{221}) \, , \\
\Delta^{-} &= D^{222} \, ,
\end{aligned}
$$

where, for the z - component spin = $\frac{3}{2}$ states,

$$
\begin{aligned}
D^{333} &\sim s_\uparrow(c) \, s_\uparrow(c') \, s_\uparrow(c'') \, \epsilon_{cc'c''} \, , \\
D^{133} &\sim u_\uparrow(c) \, s_\uparrow(c') \, s_\uparrow(c'') \, \epsilon_{cc'c''} \, , \qquad (12.55) \\
D^{313} &\sim s_\uparrow(c) \, u_\uparrow(c') \, s_\uparrow(c'') \, \epsilon_{cc'c''} \, ,
\end{aligned}
$$

etc., in accordance with the notations of (12.36) and (12.51). Through space rotations, decuplet baryons with different spin components can be readily generated. Because D^{ijk} is symmetric with respect to i, j and k, we have from (12.54)

$$
\begin{aligned}
D^{133} &= D^{313} = D^{331} = \sqrt{\tfrac{1}{3}} \;\; \Xi^{*o} \, , \\
D^{311} &= D^{131} = D^{113} = \sqrt{\tfrac{1}{3}} \;\; \Sigma^{*+} \, , \qquad (12.56) \\
D^{211} &= D^{121} = D^{112} = \sqrt{\tfrac{1}{3}} \;\; \Delta^{+}
\end{aligned}
$$

and similar expressions for Ξ^{*-}, Σ^{*o}, Σ^{*-} and Δ^{o}.

Remarks. For the lowest-energy baryon states, the three quarks are all in the same s-state. Hence, the total angular momentum equals the total spin, which can be $\frac{1}{2}$ or $\frac{3}{2}$; the former gives the spin-$\frac{1}{2}$ octet and the latter the spin-$\frac{3}{2}$ decuplet. For a similarly constructed meson system of the s-state quark-antiquark pair, the total spin can be 0 or 1; the former corresponds to the pseudoscalar octet plus $\eta'(958)$, and the latter to the vector nonet: $\rho(770)$, $\omega(783)$, $K^*(892)$ and $\phi(1020)$. Further discussions will be given in Chapter 20.

12.3 Mass Formulas

Since π, K and η differ in their masses by a few hundred MeV, as do some of the different members of the baryon octet or decuplet, the strong interaction can at best be only approximately SU_3-flavor symmetric. In contrast, as mentioned before, the SU_3-color symmetry is assumed to be exact. In this section we shall deal only with SU_3-flavor transformations; hence the word "flavor" will be omitted for notational convenience.

Let us decompose

$$H_{st} = H_{sym} + H_{asym} \tag{12.57}$$

where H_{sym} is SU_3-invariant, while H_{asym} is not.

1. H_{asym} and the spurion formulation

We may envisage the expression of H_{asym} in terms of the quark field q^i given by (12.1). In order to conserve baryon number, the simplest form would be a linear function of the quadratic expression $q^{i\dagger} q^j$; i.e., H_{asym} transforms like an irreducible tensor of rank (1, 1). Conservation of charge and isospin then requires H_{asym} to transform as

$$H_{asym} \sim S_3^{\ 3} \tag{12.58}$$

which is the $i = j = 3$ member of an octet $S_j^{\ i}$. The precise form of $S_j^{\ i}$ is immaterial, since in this section we are interested only in the SU_3 transformation properties of H_{asym}. [As a concrete example, we may assume $S_j^{\ i}$ to be given by

$$S_j^{\ i} = q_j \beta q^i - \tfrac{1}{3} \delta_j^{\ i} q_k \beta q^k$$

where β is the Dirac matrix given by (3.10), and q_j is the Hermitian conjugate of the quark field q^j.] The energy E of a hadron h is then given by the diagonal matrix element

$$E_h = < h \mid H_{st} \mid h > . \tag{12.59}$$

In the following, we shall assume (12.57); furthermore, H_{asym} is supposed to be much weaker than H_{sym}, and therefore it can be regarded as a perturbation. Neglecting second-order effects of H_{asym} in (12.59), we need the state vector $\mid h >$ only to the accuracy of the zeroth order; i.e., $\mid h >$ satisfies

$$H_{sym} \mid h > = \gamma \mid h > \tag{12.60}$$

where γ is the eigenvalue. Hence, (12.59) becomes

$$E_h = \gamma + < h \mid H_{asym} \mid h > + O(H_{asym}^{\ 2}) . \tag{12.61}$$

To incorporate the SU_3 transformation property (12.58) of H_{asym} into the energy calculation, we introduce the "spurion" formulation of G. Wentzel.* Let us examine a typical matrix element $< h' \mid H_{asym} \mid h >$. Both $\mid h >$ and $\mid h' >$ are eigenvectors of H_{sym}

---

* G. Wentzel, in High Energy Nuclear Physics (Proceedings of the Rochester Conference), ed. J. Ballam et al. (New York, Interscience Publishers, 1956), VIII 15.

and therefore belong to some irreducible representations of SU_3. We then consider a hypothetical SU_3-conserving transition

$$h \rightarrow h' + S_j^i \tag{12.62}$$

where S_j^i denotes the spurion, which transforms as an irreducible SU_3 octet representation, and carries zero 4-momentum, zero charge and even parity; in addition, we are interested only in the final state $i = j = 3$ of the spurion. It is clear that so far as the SU_3 transformations are concerned, the matrix element $< h' \mid H_{asym} \mid h >$, with H_{asym} given by (12.58), has the same properties as that of the hypothetical SU_3-conserving transition amplitude for (12.61). Emitting a spurion with SU_3 symmetry conserved is identical to having an appropriate SU_3-violating amplitude, but without the spurion. In the following we shall see how to derive various mass formulas* by using the spurion.

2. Octet mass formulas

We first discuss the matrix element (12.59) for the baryon octet $h = B_j^i$, given by (12.48). In terms of the spurion formulation, the relevant transition (12.62) becomes

$$B_j^i \rightarrow B_b^a + S_3^3 \quad .$$

Phenomenologically we may regard H_{asym} as the transition Hamiltonian given by an SU_3-conserving sum of the products of the three octets: the baryon field $B = (B_b^a)$, its Hermitian conjugate $\overline{B} = (\overline{B}_a^b)$ and the spurion field $S = (S_j^i)$. By using (12.19) and (12.20), we see that among such products there can be only two invariants:

---

* M. Gell-Mann, Phys. Rev. 125, 1067 (1962). S. Okubo, Progr. Theor. Phys. 27, 949 (1962).

$$X_i^{\;j} S_j^{\;i} \qquad \text{and} \qquad Y_i^{\;j} S_j^{\;i} \tag{12.63}$$

where

$$X_i^{\;j} = \bar{B}_a^{\;j} B_i^{\;a}$$

and (12.64)

$$Y_i^{\;j} = \bar{B}_i^{\;a} B_a^{\;j} .$$

The H_{asym} can be written as

$$H_{asym} = (\alpha X_i^{\;j} + \beta Y_i^{\;j}) S_j^{\;i} \tag{12.65}$$

where α and β are constants and, as we shall see, the spurion amplitude can be taken to be

$$S_j^{\;i} = \begin{cases} 1 & \text{if } i = j = 3 , \\ 0 & \text{otherwise.} \end{cases} \tag{12.66}$$

For simplicity, we do not separate out the trace of $X_j^{\;i}$, $Y_j^{\;i}$ and $S_j^{\;i}$, since this would only result in a redefinition of H_{asym} and H_{sym} in the decomposition (12.57). We note that in the above expression $S_j^{\;i} = 0$ except for $i = j = 3$, which expresses the hypothesis (12.58) in the spurion language; the value $S_3^{\;3} = 1$ can be chosen without any loss of generality because of the constants α and β in (12.65). Neglecting $O(H_{asym}^{\;2})$ and combining (12.65)–(12.66) with (12.57) and (12.61), we can write for the baryon octet

$$H_{st} = \alpha X_3^{\;3} + \beta Y_3^{\;3} + \gamma \tag{12.67}$$

where γ, which is the same constant for different members of the octet, is given by (12.60). By taking the Hermitian conjugate of the baryon octet field (12.48), we have (using a bar instead of $\dagger$ for notational clarity)

$$\bar{B} = (\bar{B}_i^{\;j}) = \begin{pmatrix} \frac{\bar{\Sigma}^o}{\sqrt{2}} - \frac{\bar{\Lambda}^o}{\sqrt{6}} & \bar{\Sigma}^- & \bar{\Xi}^- \\ \bar{\Sigma}^+ & -\frac{\bar{\Sigma}^o}{\sqrt{2}} - \frac{\bar{\Lambda}^o}{\sqrt{6}} & \bar{\Xi}^o \\ \bar{p} & \bar{n} & \frac{2\bar{\Lambda}^o}{\sqrt{6}} \end{pmatrix} . \tag{12.68}$$

Therefore, on account of (12.64),

$$X_3^{\,3} = \bar{p}p + \bar{n}n + \tfrac{2}{3}\,\bar{\Lambda}^{\circ}\Lambda^{\circ}$$

and (12.69)

$$Y_3^{\,3} = \bar{\Xi}^-\,\Xi^- + \bar{\Xi}^{\circ}\,\Xi^{\circ} + \tfrac{2}{3}\,\bar{\Lambda}^{\circ}\Lambda^{\circ} \;.$$

The diagonal matrix elements (12.59) for different members of the baryon octet can now be derived by using (12.67) and (12.69):

$$E_\Lambda = \tfrac{2}{3}(\alpha+\beta)+\gamma\;, \qquad E_\Sigma = \gamma\;,$$

$$E_\Xi = \beta + \gamma \quad \text{and} \quad E_N = \alpha + \gamma$$

where N stands for p or n. We can eliminate α, β and γ among these four equations. The result is

$$3E_\Lambda - 2(E_N + E_\Xi) + E_\Sigma = 0\;. \tag{12.70}$$

Identical considerations can be applied to the pseudoscalar octet. Through the replacements $\Lambda \to \eta$, $N \to K$, $\Xi \to \bar{K}$ and $\Sigma \to \pi$, we derive from (12.70)

$$3E_\eta - 4E_K + E_\pi = 0\;. \tag{12.71}$$

Both formulas (12.70)–(12.71) are valid only to the first order in H_{asym}.

In the evaluation of $< h \mid H_{st} \mid h >$, we may assume the hadron state to be one with momentum $\vec{k}$. Hence the energy E_h is related to $\vec{k}$ and the hadron mass m_h by

$$E_h = \sqrt{\vec{k}^{\,2} + m_h^{\,2}} \quad ; \tag{12.72}$$

its variation is

$$\delta E_h = \frac{\delta(m_h^{\,2})}{2E_h}\;. \tag{12.73}$$

By substituting (12.73) into (12.70) and (12.71), we obtain the mass formulas, which agree quite well with the experimental data:

$$3m_\Lambda^{\,2} - 2(m_N^{\,2} + m_\Xi^{\,2}) + m_\Sigma^{\,2} = 0 \tag{12.74}$$

and

$$3m_\eta^{\,2} - 4m_K^{\,2} + m_\pi^{\,2} = 0 \;; \tag{12.75}$$

in this derivation we employ a reference system in which the hadron energy E_h is much bigger than the mass difference between different octet members, and therefore (12.73) holds. The resulting formulas (12.74)-(12.75) are, of course, independent of the particular reference frame. For the baryon octet we may choose the rest frame, $\vec{k}=0$, because δm_h is $\ll m_h$; hence (12.73) reduces to $\delta E_h = \delta m_h$ and (12.74) takes on the linear form

$$3m_\Lambda - 2(m_N + m_\Xi) + m_\Sigma = 0 \;. \tag{12.76}$$

For the meson octet, since m_π is quite a bit smaller than the mass difference $m_K - m_\pi$, it is not possible to linearize (12.75).

Remarks. The two invariants in (12.63), which are formed of the three octets $\overline{B}$, B and S, can be understood by considering first the ⑧ × ⑧ multiplication of the two octets $\overline{B}$ and B. Because

$$⑧ \times ⑧ = ① + ⑧ + ⑧ + ⑩ + \overline{⑩} + ㉗ \;,$$

the product consists of two octets whose components can be derived by using (12.64), and each of which can in turn be combined with $S_j^{\,i}$ to form an invariant. This accounts for the two independent constants α and β in (12.65).

3. Decuplet mass formula

From the product of the symmetric decuplet field D^{ijk} and its Hermitian conjugate $\overline{D}_{ijk} = (D^{ijk})^\dagger$, we can form only a single octet

$$Z_j^{\,i} = \overline{D}_{jab}\, D^{iab} - \tfrac{1}{3}\,\delta_j^{\,i}\,\overline{D}_{abc}\, D^{abc} \;. \tag{12.77}$$

By following the same reasoning that led to (12.67), we can write H_{st}

for the baryon decuplet as

$$H_{st} = \alpha + \beta Z_3^3 \tag{12.78}$$

where α and β are constants. Because of (12.54)–(12.56), the above expression becomes

$$H_{st} = \alpha + \beta\,[\tfrac{2}{3}\,\overline{\Omega}\Omega + \tfrac{1}{3}\,(\overline{\Xi}^{*o}\,\Xi^{*o} + \overline{\Xi}^{*-}\,\Xi^{*-}) - \tfrac{1}{3}\,(\overline{\Delta}^{++}\Delta^{++} + \overline{\Delta}^{+}\Delta^{+} + \overline{\Delta}^{o}\Delta^{o} + \overline{\Delta}^{-}\Delta^{-})\,]\ , \tag{12.79}$$

which leads to $E_\Omega = \alpha + \frac{2}{3}\beta$, $E_{\Xi^*} = \alpha + \frac{1}{3}\beta$, $E_{\Sigma^*} = \alpha$ and $E_\Delta = \alpha - \frac{1}{3}\beta$. Just as in the passage from (12.70) to (12.76), we can arrive at the mass formula

$$m_\Omega - m_{\Xi^*} = m_{\Xi^*} - m_{\Sigma^*} = m_{\Sigma^*} - m_\Delta\ . \tag{12.80}$$

The reader can easily verify that this mass formula is in good agreement with the experimental values of $\Delta(1232)$, $\Sigma^*(1385)$, $\Xi^*(1530)$ and $\Omega(1672)$ given in the Table of Particle Properties.

Remarks. Considering the fact that m_π is much smaller than the mass differences $m_K - m_\pi$ and $m_\eta - m_\pi$, it seems quite strange that H_{asym} can be treated as a perturbation. A plausible explanation will be given later in Chapter 20.

Problem 12.1. Let $|\,B_i^{\,a}>$ denote the physical spin$-\frac{1}{2}$ baryon octet state and $<B_a^{\,i}\,|$ be its Hermitian conjugate.

(i) Prove that if $S_b^{\,j}(x)$ is a scalar (i.e., spin-0, parity +) local SU_3-octet operator, then its matrix element between $|\,B_c^{\,k}>$ of 4-momentum k and $|\,B_i'^{\,a}>$ of 4-momentum k+q in the limit $q\to 0$

is given by

$$< B'^{\,i}_{a} \mid S_b^{\,j}(x) \mid B_c^{\,k} > \; = \; (D_s\, d^{ijk}_{abc} + F_s\, f^{ijk}_{abc})\; u'^{\dagger} \gamma_4\, u \qquad (12.81)$$

where u and u' are spinor solutions of (3.26) that have the same spin-momentum configurations as the initial and final physical baryon states, γ_4 is the Dirac matrix given by (3.11), D_s and F_s are constants, and the tensors

$$d^{ijk}_{abc} = \tfrac{4}{9}\,\delta_a^{\,i}\delta_b^{\,j}\delta_c^{\,k} + \delta_b^{\,i}\delta_c^{\,j}\delta_a^{\,k} + \delta_c^{\,i}\delta_a^{\,j}\delta_b^{\,k} - \tfrac{2}{3}\,[\,\delta_b^{\,i}\delta_a^{\,j}\delta_c^{\,k} + \delta_a^{\,i}\delta_c^{\,j}\delta_b^{\,k} + \delta_c^{\,i}\delta_b^{\,j}\delta_a^{\,k}\,] \qquad (12.82)$$

and

$$f^{ijk}_{abc} = \delta_b^{\,i}\delta_c^{\,j}\delta_a^{\,k} - \delta_c^{\,i}\delta_a^{\,j}\delta_b^{\,k} \;. \qquad (12.83)$$

(ii) Show that d^{ijk}_{abc} is symmetric under the interchange of either (i, a) and (j, b), or (i, a) and (k, c), or (j, b) and (k, c), while f^{ijk}_{abc} is antisymmetric. Both tensors satisfy the trace-less condition

$$d^{ijk}_{ibc} = f^{ijk}_{ibc} = 0 \;.$$

(iii) From (12.48) and (12.68), one sees that $| \, p > \, = \, | \, B_3^{\,1} >$ and $< p \,| = \, < B_1^{\,3} |$. Set $j = b = 3$ and show that

$$< p \mid S_3^{\,3} \mid p > \; = \; D_s\, d^{331}_{133} + F_s\, f^{331}_{133} = \tfrac{1}{3}\, D_s + F_s \;. \qquad (12.84)$$

Likewise, prove that

$$< \Lambda \mid S_3^{\,3} \mid \Lambda > \; = \; \tfrac{2}{3}\, D_s \;, \qquad < \Sigma \mid S_3^{\,3} \mid \Sigma > \; = \; -\tfrac{2}{3}\, D_s \;,$$
$$< \Xi \mid S_3^{\,3} \mid \Xi > \; = \; \tfrac{1}{3}\, D_s - F_s \;, \qquad (12.85)$$

and therefore

$$2 < p \mid S_3^{\,3} \mid p > + \; 2 < \Xi \mid S_3^{\,3} \mid \Xi > \; = \; 3 < \Lambda \mid S_3^{\,3} \mid \Lambda > + < \Sigma \mid S_3^{\,3} \mid \Sigma >$$

which is the octet mass formula (12.70).

Problem 12.2. Let $J_\lambda(x)_b^{\,k}$ be either the vector or the axial-vector hadron SU_3-octet current operator. Using the notation of the previous problem, show that, depending on whether $J_\lambda(x)_b^{\,j}$ is a vector or axial-vector current,

$$\lim_{q=0} \langle B_a'^{\,i} \,|\, J_\lambda(x)_b^{\,j} \,|\, B_c^{\,k} \rangle = i(D\, d_{abc}^{ijk} + F\, f_{abc}^{ijk}) \cdot \begin{cases} u'^\dagger \gamma_4 \gamma_\lambda u & \text{vector} \\ u'^\dagger \gamma_4 \gamma_\lambda \gamma_5 u & \text{axial-vector} \end{cases} \tag{12.86}$$

where D and F are constants. Thus, if h and h' denote various hadrons, the above expression can also be written as

$$\lim_{q=0} \langle h' \,|\, J_\lambda(x)_b^{\,j} \,|\, h \rangle \equiv i \langle h' \,|\, \mathcal{g}_b^{\,j} \,|\, h \rangle \cdot \begin{cases} u'^\dagger \gamma_4 \gamma_\lambda u & \text{vector} \\ u'^\dagger \gamma_4 \gamma_\lambda \gamma_5 u & \text{axial-vector} \end{cases} . \tag{12.87}$$

Prove the following table of the reduced matrix elements $\langle h' \,|\, \mathcal{g}_b^{\,j} \,|\, h \rangle$ for $b = 2$, $j = 1$ and $b = 3$, $j = 1$. [These currents are important for the Cabibbo theory of the weak interaction.]

(The table appears on the next page.)

$h \to h'$	$\langle h' \mid \mathcal{J}_2^1 \mid h \rangle$	$\langle h' \mid \mathcal{J}_3^1 \mid h \rangle$
$n \to p$	$D - F$	0
$\Sigma^- \to \Lambda^o$	$-\sqrt{\frac{2}{3}}\ D$	0
$\Lambda^o \to \Sigma^+$	$-\sqrt{\frac{2}{3}}\ D$	0
$\Sigma^- \to \Sigma^o$	$-\sqrt{2}\ F$	0
$\Xi^- \to \Xi^o$	$D + F$	0
$\Lambda^o \to p$	0	$(D - 3F)/\sqrt{6}$
$\Sigma^- \to n$	0	$D + F$
$\Xi^- \to \Lambda^o$	0	$(D + 3F)/\sqrt{6}$
$\Xi^- \to \Sigma^o$	0	$(D - F)/\sqrt{2}$
$\Xi^o \to \Sigma^+$	0	$D - F$.

Reference

Y. Ne'eman and M. Gell-Mann, eds., The Eightfold Way (New York, Benjamin, 1964).

Chapter 13

TIME REVERSAL

Consider the classical system of particles discussed at the beginning of Chapter 1. Let the generalized coordinates be q_i where $i = 1, 2, \cdots, N$, which may be written as a column matrix q. Time-reversal (T) invariance for such a system simply means that the laws of physics remain unchanged under

$$q \to q \qquad \text{and} \qquad t \to -t \ ;$$

hence, if $q(t)$ is a solution of the dynamical equation, so is $q(-t)$.

Suppose that we are shown a film of the motion of a classical system of particles. With T invariance, the time-reversed sequence also represents a possible solution of the dynamical equation. For example, the film of a drop of water falling from a faucet, hitting the sink and making a splash would, when shown backwards, start from the gathering together of the splash to re-form the drop which then jumps back into the faucet. T invariance says that such an improbable sequence can actually happen. Of course, to make it occur we must reverse simultaneously all the molecular velocities involved in the splash. This is difficult and therefore, for practical purposes, does not happen. Hence if the T-invariant classical system consists of a large number of particles, although the time-reversed sequence is always possible, it is in general improbable. It is only in this probabilistic sense that we can differentiate for a macroscopic system any time-ordered sequence of events from its time-reversed sequence, and thereby determine the direction of our macroscopic time.

It is well-known that time-reversible dynamical equations of classical mechanics can produce time-irreversible thermodynamics in the macroscopic world. As an illustration, we may think of traveling by train between New York and Princeton. In analogy to microscopic reversibility, let us assume that for each train going from New York to Princeton, there is one in the opposite direction. Imagine now that there are no signs in the railroad station. The probability that a man starting from Princeton will end in New York is 1 if he watches the train's direction; otherwise it is $\frac{1}{2}$. But once he reaches New York, if there are also no signs in Penn Station the chance that he will succeed in taking a Princeton-bound train is extremely small; the person may end up in Boston, and go from there to Chicago, $\cdots$, resulting in a macroscopic irreversibility.

In classical physics there is a major difference between a macroscopic system and a microscopic system. If the system contains only a small number of particles, then provided that T invariance holds, it is not possible in a statistical sense to differentiate between time-ordered and time-reversed sequences, both being similarly probable. As we shall see in Section 13.2, this last statement has to be modified in quantum mechanics.

We now turn to the question of time reversal in a quantum theory.

13.1 Time Reversal in the Schrödinger Representation

We start from the Schrödinger equation, (5.4),

$$H \mid t > = - \frac{1}{i} \frac{\partial}{\partial t} \mid t > \qquad (13.1)$$

which holds for any state vector $|\, t >$.

Definition: H is called T-invariant if and only if there exists a unitary operator U_T which satisfies

$$U_T H^* U_T^\dagger = H \tag{13.2}$$

where, as before, * denotes the complex conjugation. From this definition, the following theorem can readily be established.

Theorem. If H is T-invariant, then $U_T \,|\, t >^*$ satisfies the time-reversed Schrödinger equation.

Proof. In (13.1), $|\, t >$ is a column matrix and H a square matrix. Let $|\, t >^*$ and H^* be respectively their complex conjugations. From the Schrödinger equation, it follows that

$$H^* \,|\, t >^* = \frac{1}{i} \frac{\partial}{\partial t} \,|\, t >^* \;.$$

Multiplying the above formula on the left by U_T and using

$$U_T^\dagger\, U_T = 1 \;, \tag{13.3}$$

we have

$$U_T H^* U_T^\dagger U_T \,|\, t >^* = \frac{1}{i} \frac{\partial}{\partial t} U_T \,|\, t >^*$$

which, because of (13.2), can also be written as

$$H\, U_T \,|\, t >^* = -\frac{1}{i} \frac{\partial}{\partial(-t)} U_T \,|\, t >^* \;. \tag{13.4}$$

Hence $U_T \,|\, t >^*$ satisfies the time-reversed Schrödinger equation.

Thus, in quantum mechanics as well as in classical mechanics, T invariance means that from any solution of the dynamical equation one can obtain another solution which satisfies the same equation, but with $t \to -t$.

Definition: We define T to be the operator that satisfies *

---

* E. P. Wigner, Gött. Nach. Math. Naturw. Kl., p. 546 (1932).

$$T\,|\,> \;=\; U_T\,|\,>^* \tag{13.5}$$

for any state vector $|\,>$ and

$$T\,O\,T^{-1} \;=\; U_T\,O^*\,U_T^{\dagger} \tag{13.6}$$

for any linear operator O. [In matrix notation, $|\,>$ can be any column matrix and O any square matrix.]

We note that for $O =$ the unit matrix 1, (13.6) becomes

$$T\,T^{-1} \;=\; 1 \;. \tag{13.7}$$

If $|\,>$ is a linear superposition of two state vectors $|\,1>$ and $|\,2>$,

$$|\,> \;=\; c_1\,|\,1> + \; c_2\,|\,2>$$

where c_1 and c_2 are constants, then according to (13.5)

$$T\,|\,> \;=\; c_1^*\,T\,|\,1> + \; c_2^*\,T\,|\,2> \;. \tag{13.8}$$

Hence, T is <u>not</u> a linear operator. Sometimes in the literature, T is referred to as "antiunitary", which carries no more meaning than what is given above.

13.2 Improbability of Constructing Time-reversed Quantum Solutions even for a Microscopic System

In classical mechanics, as we said at the beginning of this chapter, in contrast to macroscopic systems, the time-reversed motion of a T - invariant microscopic system is not only possible but can occur with similar probability. This situation is, however, drastically altered in a quantum theory.

We may give the simple example of μ-decay to illustrate the situation. Consider a μ^- at rest with its spin $\vec{S}_\mu$ polarized in, say, the up direction, as shown in Fig. 13.1. In its decay

$$\mu^- \;\rightarrow\; e^-(L) + \bar{\nu}_e(R) + \nu_\mu(L) \;, \tag{13.9}$$

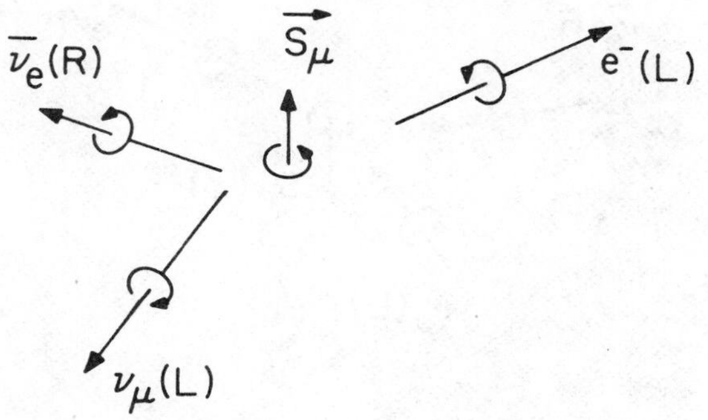

Fig. 13.1. The decay of $\mu^- \to e^- + \bar{\nu}_e + \nu_\mu$.

the e^-, $\bar{\nu}_e$ and ν_μ are emitted with helicities $-\frac{1}{2}$, $\frac{1}{2}$ and $-\frac{1}{2}$ respectively, which are denoted by the letters L, R and L indicating their helicity. [For simplicity, we approximate the mass of the electron m_e as 0.] Let us examine the specific process in which the final momenta of e^-, $\bar{\nu}_e$ and ν_μ in (13.9) are respectively $\vec{p}_e$, $\vec{p}_{\bar{\nu}}$ and $\vec{p}_\nu$; its time-reversed sequence is

$$e^-(L) + \bar{\nu}_e(R) + \nu_\mu(L) \to \mu^- , \tag{13.10}$$

which would start from the initial state of e^-, $\bar{\nu}_e$ and ν_μ with momenta $-\vec{p}_e$, $-\vec{p}_{\bar{\nu}}$ and $-\vec{p}_\nu$. Since both the directions of spin and momentum are reversed, the helicity of each particle is unchanged, as shown in Fig. 13.2. Assuming T invariance, this leads to a final state with μ^- at rest, but the question is what should be its spin direction $\vec{S}_\mu{}'$. The reader is encouraged to pause for a few moments to consider. If the system were a classical one, then T invariance would require the final μ spin $\vec{S}_\mu{}'$ to lie in the opposite direction from $\vec{S}_\mu$ in (13.9), i.e., downwards, as indicated in Fig. 13.2, but accompanied by a question mark.

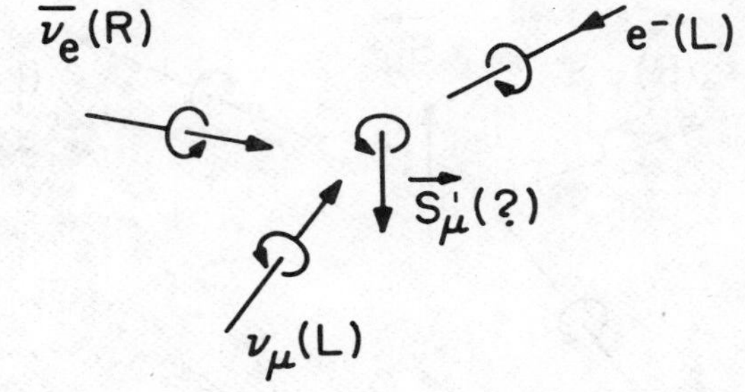

Fig. 13.2. The reversed reaction $e^- + \bar{\nu}_e + \nu_\mu \rightarrow \mu^-$, with the momenta and spins of e^-, $\bar{\nu}_e$ and ν_μ the opposite of those in Fig. 13.1.

However, this is in general not the case for the quantum system, as can most easily be seen in the special case when the momenta of $\bar{\nu}_e$ and ν_μ are parallel, as shown by Figs. 13.3 and 13.4. In the reverse reaction (13.10), since all initial momenta are assumed to be along the same line, the total angular momentum along that line must be conserved; this necessitates the final μ spin $\vec{S}'_\mu$ being the same

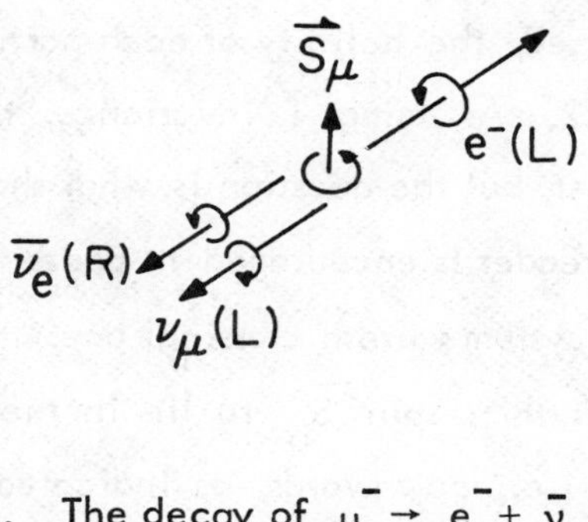

Fig. 13.3. The decay of $\mu^- \rightarrow e^- + \bar{\nu}_e + \nu_\mu$ with the two final neutrino momenta parallel.

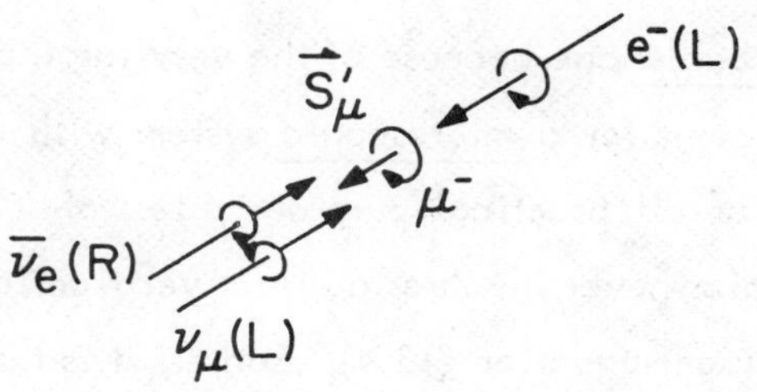

Fig. 13.4. Reaction $e^- + \bar{\nu}_e + \nu_\mu \rightarrow \mu^-$ with the two initial neutrino momenta parallel.

as the initial electron spin. Thus $\vec{S}'_\mu$ must be parallel to the momentum of the electron, which can lie in a totally different direction from $-\vec{S}_\mu$.

The reason is that in quantum mechanics the final state $|e\bar{\nu}_e\nu_\mu>$ of μ-decay, (13.9), consists of coherent outgoing spherical waves. Its time-reversed state

$$T\,|\,e^-\bar{\nu}_e\nu_\mu>$$

should be composed of coherent incoming spherical waves; i.e., we must simultaneously reverse the momentum and spin of these three leptons in all possible directions and retain the appropriate phase relationships between their wave amplitudes; only then would the final μ-spin $\vec{S}'_\mu$ in (13.10) be $-\vec{S}_\mu$. Clearly, this is extremely difficult.

The root of the problem lies in the fact that the Schrödinger wave function of any quantum system has more complexity than any wave equation in classical macroscopic physics. Only in the case of a single stable particle is the quantum-mechanical time-reversed state a simple one, since it can also be obtained by a 180°

rotation that reverses the momentum. In general, the quantum-mechanical time-reversed state $T\,|\,\rangle = U_T\,|\,\rangle^*$ is always a complicated and improbable one because of the very large degrees of freedom involved, even for a microscopic system with T invariance. Consequently, for all practical purposes a test of T invariance in quantum mechanics never involves a direct verification of the time-reversed Schrödinger equation (13.4). Rather, it is through the confirmation or violation of either the reciprocity relations or certain phase-angle equalities, as we shall discuss. But before that, we must first familiarize ourselves with the properties of the antiunitary T operator.

13.3 Properties of the T Operator

1. QED as an example

Let H be the QED Hamiltonian. As in Chapter 6, we adopt the Coulomb gauge; hence the independent generalized coordinates are the transverse electromagnetic field $\vec{A} = \vec{A}^{tr}$ and the electron field ψ. In this section, we shall stay within the Schrödinger representation. Thus, both $\vec{A}$ and ψ are independent of time t. We may write

$$\psi(\vec{r}, t) = \psi(\vec{r}, 0) \quad , \qquad \vec{A}(\vec{r}, t) = \vec{A}(\vec{r}, 0)$$

and likewise for the conjugate momentum $\vec{\Pi}$ of the electromagnetic field,

$$\vec{\Pi}(\vec{r}, t) = \vec{\Pi}(\vec{r}, 0) \quad .$$

Both $\vec{A}$ and $\vec{\Pi}$ satisfy the transversality condition

$$\vec{\nabla} \cdot \vec{A} = 0 \qquad \text{and} \qquad \vec{\nabla} \cdot \vec{\Pi} = 0 \quad ,$$

in accordance with (6.30).

Theorem. H, given by (6.55), is T invariant, where T satisfies

$$T\,\psi(\vec{r},0)\,T^{-1} = U_T\,\psi^*(\vec{r},0)\,U_T^\dagger = \eta_t\,\sigma_2\,\psi(\vec{r},0) \quad , \qquad (13.11)$$

$$T\,\vec{A}(\vec{r},0)\,T^{-1} = U_T\,\vec{A}^*(\vec{r},0)\,U_T^\dagger = -\vec{A}(\vec{r},0) \quad , \qquad (13.12)$$

$$T\,\vec{\Pi}(\vec{r},0)\,T^{-1} = U_T\,\vec{\Pi}(\vec{r},0)^*\,U_T^\dagger = \vec{\Pi}(\vec{r},0) \qquad (13.13)$$

with η_t as a constant phase factor, $|\eta_t| = 1$, and σ_2 is given by (3.6).

Proof. We shall first assume the existence of the operator T, defined by (13.11)-(13.13), so that we may use it to establish the time-reversal invariance of H. Afterwards, we shall come back and prove that such a T operator indeed exists.

Since, according to (6.6) and (6.21),

$$\vec{B} = \vec{\nabla} \times \vec{A} \qquad \text{and} \qquad \vec{E}^{tr} = -\vec{\Pi} \quad ,$$

the free electromagnetic Hamiltonian density (6.24),

$$\mathcal{H}_\gamma = \tfrac{1}{2}(\vec{E}^{tr})^2 + \tfrac{1}{2}\vec{B}^2 \quad ,$$

clearly satisfies the T- invariant condition

$$T\,\mathcal{H}_\gamma\,T^{-1} = \mathcal{H}_\gamma \quad ,$$

because of (13.12)-(13.13). Under T, the free-electron Hamiltonian density (6.25),

$$\mathcal{H}_e = \psi^\dagger(-i\,\vec{\alpha}\cdot\vec{\nabla} + \beta m)\,\psi \quad ,$$

becomes

$$T\,\mathcal{H}_e\,T^{-1} = U_T\,\mathcal{H}_e^*\,U_T^\dagger = U_T\,\psi^{*\dagger}(i\vec{\alpha}^*\cdot\vec{\nabla} + \beta^* m)\,\psi^* U_T^\dagger$$

$$= U_T\,\psi^{*\dagger}\,U_T^\dagger(i\,\vec{\alpha}^*\cdot\vec{\nabla} + \beta^* m)\,U_T\psi^* U_T^\dagger$$

which, on account of (13.11) and its Hermitian conjugate

$$T\,\psi^\dagger(\vec{r},0)\,T^{-1} = U_T\,[\psi^\dagger(\vec{r},0)]^*\,U_T^\dagger = \eta_t^*\,\psi^\dagger(\vec{r},0)\,\sigma_2 \quad , \qquad (13.14)$$

is equal to

$$\psi^{\dagger}\,\sigma_2\,(i\,\vec{\alpha}^{*}\cdot\vec{\nabla}+\beta^{*}m)\,\sigma_2\,\psi \quad . \tag{13.15}$$

We recall that in (3.10), $\vec{\alpha}=\rho_1\vec{\sigma}$ and $\beta=\rho_3$. Consequently,

$$\sigma_2\,\vec{\alpha}^{*}\,\sigma_2 \;=\; -\,\vec{\alpha} \qquad \text{and} \qquad \sigma_2\,\beta^{*}\,\sigma_2 \;=\; \beta \quad , \tag{13.16}$$

by means of which (13.15) becomes $\mathcal{H}_e$, and therefore

$$T\,\mathcal{H}_e\,T^{-1} \;=\; \mathcal{H}_e \quad .$$

The electromagnetic current operator is

$$j_\mu \;=\; i:\psi^{\dagger}\,\gamma_4\,\gamma_\mu\,\psi: \quad . \tag{13.17}$$

By using

$$i\,\gamma_4\,\gamma_\mu \;=\; \begin{cases} \rho_1\,\sigma_j \;=\; \alpha_j & \text{for}\quad \mu=j\neq 4 \\ i & \text{for}\quad \mu=4 \end{cases} \tag{13.18}$$

and (13.16), we have

$$\sigma_2(i\,\gamma_4\,\gamma_\mu)^{*}\,\sigma_2 \;=\; -\,i\,\gamma_4\,\gamma_\mu \quad , \tag{13.19}$$

from which it follows that

$$\begin{aligned} T\,i\,\psi^{\dagger}\,\gamma_4\,\gamma_\mu\,\psi\,T^{-1} \;&=\; T\,\psi^{\dagger}\,T^{-1}\,(i\,\gamma_4\,\gamma_\mu)^{*}\,T\,\psi\,T^{-1} \\ &=\; \psi^{\dagger}\,\sigma_2(i\,\gamma_4\,\gamma_\mu)^{*}\,\sigma_2\,\psi \;=\; -\,i\,\psi^{\dagger}\,\gamma_4\,\gamma_\mu\,\psi \quad . \end{aligned}$$

By combining the above equation with (13.17), we derive

$$T\,j_\mu(\vec{r},\,0)\,T^{-1} \;=\; -\,j_\mu(\vec{r},\,0) \quad . \tag{13.20}$$

This formula has a direct interpretation: because $j_\mu=(\vec{j}\,,i\,\rho)$, under time reversal the vector $\vec{j}$ reverses its direction and becomes $-\vec{j}$; the charge density ρ is unchanged,

$$T\,\rho(\vec{r},\,0)\,T^{-1} \;=\; \rho(\vec{r},\,0) \quad .$$

But because $i\rightarrow -i$, that makes $j_4=i\,\rho\rightarrow -j_4$. In the Coulomb gauge, A_0 is derived from ρ via (6.16). The T invariance of ρ implies

$$T\,A_0(\vec{r},0)\,T^{-1} = A_0(\vec{r},0)$$

which, together with (13.12), leads to

$$T\,A_\mu(\vec{r},0)\,T^{-1} = -\,A_\mu(\vec{r},0) \tag{13.21}$$

where $A_\mu = (\vec{A}, i\,A_0)$. By using (13.20)-(13.21) we establish the T invariance of the interaction Hamiltonian density (6.26), and consequently also of H ; i.e.,

$$T\,H\,T^{-1} = H\ . \tag{13.22}$$

From (13.11), we note that

$$T^2\psi\,T^{-2} = T\,(\eta_t\,\sigma_2\,\psi)\,T^{-1} = U_T(\eta_t\,\sigma_2\,\psi)^*\,U_T^\dagger$$

$$= \eta_t^*\,\sigma_2^*\,U_T\,\psi^*\,U_T = \eta_t^*\,\sigma_2^*\,\eta_t\,\sigma_2\,\psi\ .$$

Because $\sigma_2^* = -\sigma_2$, we have $\sigma_2^*\sigma_2 = -1$, which converts the above expression into

$$T^2\,\psi\,T^{-2} = -\psi\ , \tag{13.23}$$

i.e., the time reversal of the time-reversed ψ is $-\psi$.

We note further that under the phase transformation

$$\psi \to \psi' = e^{i\theta}\,\psi$$

(13.11) becomes

$$T\,\psi\,T^{-1} \to T\,\psi'\,T^{-1} = e^{-i\theta}\,T\,\psi\,T^{-1} = e^{-i\theta}\,\eta_t\,\sigma_2\,\psi = e^{-2i\theta}\,\eta_t\,\psi'\,.$$

Hence, if we define η_t' by

$$T\,\psi'\,T^{-1} \equiv \eta_t'\,\sigma_2\,\psi'\ ,$$

then

$$\eta_t' = e^{-2i\theta}\,\eta_t\ .$$

Since QED satisfies lepton-number conservation, the phase of ψ can be arbitrarily chosen; therefore we can without any loss of generality select one so that the resulting phase factor $\eta_t = 1$, and consequently (13.11) becomes

$$T\psi(\vec{r}, 0)\, T^{-1} = \sigma_2\, \psi(\vec{r}, 0) \quad , \tag{13.24}$$

or through multiplication on the left by σ_2,

$$\sigma_2\, T\, \psi(\vec{r}, 0)\, T^{-1} = \psi(\vec{r}, 0) \quad . \tag{13.24a}$$

It remains for us to establish that T, defined by (13.11)-(13.13), exists. By substituting the Fourier expansion (3.32) at $t = 0$

$$\psi(\vec{r}, 0) = \sum_{\vec{p},s} \frac{1}{\sqrt{\Omega}} \left(a_{\vec{p},s}\, u_{\vec{p},s}\, e^{i\vec{p}\cdot\vec{r}} + b^\dagger_{\vec{p},s}\, v_{\vec{p},s}\, e^{-i\vec{p}\cdot\vec{r}}\right) \tag{13.25}$$

into (13.24a), we have on the lefthand side

$$\sigma_2\, T\, \psi(\vec{r}, 0)\, T^{-1} = \sigma_2 \sum_{\vec{p},s} \frac{1}{\sqrt{\Omega}} \Big(T\, a_{\vec{p},s}\, T^{-1}\, u^*_{\vec{p},s}\, e^{-i\vec{p}\cdot\vec{r}} + T\, b^\dagger_{\vec{p},s}\, T^{-1}\, v^*_{\vec{p},s}\, e^{i\vec{p}\cdot\vec{r}}\Big) ,$$

which, because of (3.83), can be written as

$$\sum_{\vec{p},s} \frac{1}{\sqrt{\Omega}} \Big(T\, a_{\vec{p},s}\, T^{-1}\, e^{i\theta_{\vec{p},s}}\, u_{-\vec{p},s}\, e^{-i\vec{p}\cdot\vec{r}} + T\, b^\dagger_{\vec{p},s}\, T^{-1}\, e^{-i\theta_{-\vec{p},s}}\, v_{-\vec{p},s}\Big) ,$$

while the righthand side is simply (13.25). By equating the coefficients of $u_{\vec{p},s}\, e^{i\vec{p}\cdot\vec{r}}$ and $v_{\vec{p},s}\, e^{-i\vec{p}\cdot\vec{r}}$ we derive

$$\begin{aligned} T\, a_{\vec{p},s}\, T^{-1} &= e^{-i\theta_{\vec{p},s}}\, a_{-\vec{p},s} \quad , \\ T\, b^\dagger_{\vec{p},s}\, T^{-1} &= e^{i\theta_{-\vec{p},s}}\, b^\dagger_{-\vec{p},s} \quad , \end{aligned} \tag{13.26}$$

and through Hermitian conjugation,

$$\begin{aligned} T\, a^\dagger_{\vec{p},s}\, T^{-1} &= e^{i\theta_{\vec{p},s}}\, a^\dagger_{-\vec{p},s} \quad , \\ T\, b_{\vec{p},s}\, T^{-1} &= e^{-i\theta_{-\vec{p},s}}\, b_{-\vec{p},s} \quad . \end{aligned} \tag{13.27}$$

By using (3.84) and remembering the antiunitary nature of the T operator, we see that

$$T^2 a^\dagger_{\vec{p},s} T^{-2} = e^{-i\theta_{\vec{p},s} + i\theta_{-\vec{p},s}} a^\dagger_{\vec{p},s} = - a^\dagger_{\vec{p},s}$$

and

$$T^2 b^\dagger_{\vec{p},s} T^{-2} = e^{-i\theta_{\vec{p},s} + i\theta_{-\vec{p},s}} b^\dagger_{\vec{p},s} = - b^\dagger_{\vec{p},s} \qquad (13.28)$$

in conformity with (13.23). Next we substitute the Fourier expansion (6.32)-(6.33) at $t = 0$,

$$\vec{A}(\vec{r}, 0) = \sum_{\vec{k}} \frac{1}{\sqrt{2\omega\Omega}} [\vec{a}_{\vec{k}}\, e^{i\vec{k}\cdot\vec{r}} + h.c]$$

and

$$\vec{\Pi}(\vec{r}, 0) = \sum_{\vec{k}} \sqrt{\frac{\omega}{2\Omega}} [-i\, \vec{a}_{\vec{k}}\, e^{i\vec{k}\cdot\vec{r}} + h.c.]$$

into (13.12)-(13.13); we have

$$T \vec{A}(\vec{r}, 0) T^{-1} = \sum_{\vec{k}} \frac{1}{\sqrt{2\omega\Omega}} [T \vec{a}_{\vec{k}} T^{-1} e^{-i\vec{k}\cdot\vec{r}} + h.c.]$$

$$= - \vec{A}(\vec{r}, 0)$$

and

$$T \vec{\Pi}(\vec{r}, 0) T^{-1} = \sum_{\vec{k}} \sqrt{\frac{\omega}{2\Omega}} [i\, T \vec{a}_{\vec{k}} T^{-1} e^{-i\vec{k}\cdot\vec{r}} + h.c.]$$

$$= \vec{\Pi}(\vec{r}, 0) ,$$

which yield

$$T \vec{a}_{\vec{k}} T^{-1} = - \vec{a}_{-\vec{k}}$$

and

$$T \vec{a}^\dagger_{\vec{k}} T^{-1} = - \vec{a}^\dagger_{\vec{k}} \ . \qquad (13.29)$$

The transversality condition $\vec{\nabla} \cdot \vec{A} = \vec{\nabla} \cdot \vec{\Pi} = 0$ implies, as before,

$$\vec{a}_{\vec{k}} \cdot \vec{k} = \vec{a}^\dagger_{\vec{k}} \cdot \vec{k} = 0 \ .$$

If we now make use of (6.36)

$$\vec{a}^\dagger_{\vec{k},s=\pm 1} = \frac{1}{\sqrt{2}}\, \vec{a}^\dagger_{\vec{k}} \cdot [\hat{e}_1(\vec{k}) \pm i\, \hat{e}_2(\vec{k})]$$

where, as in Figure 6.1, $\hat{e}_1(\vec{k})$, $\hat{e}_2(\vec{k})$ and $\hat{k} = \vec{k}/|\vec{k}|$ form the standard righthanded set of orthonormal vectors, then under T we have

$$T \vec{a}^{\dagger}_{\vec{k},s=\pm 1} T^{-1} = -\frac{1}{\sqrt{2}} \vec{a}^{\dagger}_{-\vec{k}} [\hat{e}_1(\vec{k}) \mp i \hat{e}_2(\vec{k})] . \qquad (13.30)$$

Because $\hat{e}_1(\vec{k})$, $-\hat{e}_2(\vec{k})$ and $-\hat{k}$ also form a righthanded set of orthonormal vectors, we may set

$$\hat{e}_1(-\hat{k}) = \hat{e}_1(\hat{k}) \quad , \quad \hat{e}_2(-\hat{k}) = -\hat{e}_2(\vec{k}) \quad ,$$

which converts (13.30) into

$$T a^{\dagger}_{\vec{k},s} T^{-1} = -a^{\dagger}_{-\vec{k},s} \quad , \qquad (13.31)$$

with its Hermitian conjugate given by

$$T a_{\vec{k},s} T^{-1} = -a_{-\vec{k},s} \quad .$$

Let us consider the Hilbert space basis vectors (3.40), which may be represented by real column matrices:

$$|0> = \begin{pmatrix} 1 \\ 0 \\ 0 \\ 0 \\ \vdots \end{pmatrix} \quad , \quad a^{\dagger}_{\vec{p},s} |0> = \begin{pmatrix} 0 \\ 1 \\ 0 \\ 0 \\ \vdots \end{pmatrix} \quad ,$$

$$b^{\dagger}_{\vec{p},s} |0> = \begin{pmatrix} 0 \\ 0 \\ 1 \\ 0 \\ \vdots \end{pmatrix} \quad , \quad \cdots \quad . \qquad (13.32)$$

As in (10.35) and (10.56) we introduce the linear transformation onto the set of basis vectors (10.25) in the Hilbert space:

$$|0> \to |0> \quad , \quad a^{\dagger}_{\vec{p},s} |0> \to e^{i\theta_{\vec{p},s}} a^{\dagger}_{-\vec{p},s} |0> \quad ,$$

$$b^{\dagger}_{\vec{p},s}\,|\,0> \;\to\; -e^{i\theta_{\vec{p},s}}\, b^{\dagger}_{-\vec{p},s}\,|\,0>\,, \qquad a^{\dagger}_{\vec{k},s}\,|\,0> \;\to\; -a^{\dagger}_{-\vec{k},s}\,|\,0>\,,$$

and in general

$$|\,n_-e^-,\, n_+e^+,\, n_\gamma\gamma> \;\to\; (-1)^{n_+ + n_\gamma} \prod_{i=1}^{n_-} \prod_{j=1}^{n_+} \prod_{k=1}^{n_\gamma}$$

$$\cdot\; e^{i\theta_{\vec{p}_i,s_i}}\, a^{\dagger}_{-\vec{p}_i,s_i}\; e^{i\theta_{\vec{p}_j,s_j}}\, b^{\dagger}_{-\vec{p}_j,s_j}\; a^{\dagger}_{-\vec{p}_k,s_k}\,|\,0>\,. \tag{13.33}$$

Clearly, the orthonormality of this complete set of basis vectors is unchanged. Hence, the transformation operator is unitary, called U_T. We may now construct an antiunitary T from this U_T and the complex conjugation, in accordance with (13.5) and (13.6). Because these basis vectors are all real, as shown in (13.32), one can easily see that the T operator thus constructed satisfies (13.26), (13.27) and (13.31) with

$$T\,|\,0> \;=\; |\,0> \;. \tag{13.34}$$

This establishes the existence of U_T and T; it also completes the proof of the theorem.

<u>Remarks.</u> The above discussions can be easily extended to include other fields. For example, we may choose ψ to be the field of any charged spin-$\frac{1}{2}$ particle, such as muon, quark, proton, $\cdots$. Because the particle interacts electromagnetically, the assumption of T invariance requires its field ψ to satisfy (13.11) and its Hilbert space vectors to transform similarly to (13.33). For neutral particles, such as n, Σ^o and Λ^o, their fields must also satisfy (13.11), as can be inferred from either the assumption of T invariance of the strong interaction, or from that of the electromagnetic interaction through their magnetic moments. The same conclusion also applies to neutrinos

if we assume T invariance for the weak interaction, an assumption that will come under critical scrutiny in our later discussions.

2. Time reversal and angular momentum

Let us first consider a system of spin-$\frac{1}{2}$ particles. The angular momentum operator $\vec{J}$ is given by (3.62):

$$\vec{J} = \int \psi^\dagger (\vec{\ell} + \tfrac{1}{2}\vec{\sigma})\, \psi\, d^3r$$

where $\vec{\ell} = -i\vec{r} \times \vec{\nabla}$. From (13.11) and (13.14) it follows that

$$T\vec{J}T^{-1} = \int \psi^\dagger \sigma_2 (\vec{\ell}^* + \tfrac{1}{2}\vec{\sigma}^*)\, \sigma_2\, \psi\, d^3r$$

which, because $\vec{\ell}^* = -\vec{\ell}$ and $\sigma_2\, \vec{\sigma}^*\, \sigma_2 = -\vec{\sigma}$, can be written as

$$T\vec{J}T^{-1} = -\vec{J}\ . \tag{13.35}$$

Since all physical states can be produced by using spin-$\frac{1}{2}$ particles, the above relation is valid for all systems, provided T invariance holds.

Next, we consider any T-invariant system with H, $\vec{J}$ and J_z as the Hamiltonian, the total angular momentum operator and its z-component. Let $|j, m>$ be the eigenstate which satisfies

$$H\,|j,m> = E\,|j,m>\ ,$$

$$\vec{J}^2\,|j,m> = j(j+1)\,|j,m> \tag{13.36}$$

and

$$J_z\,|j,m> = m\,|j,m>$$

where m can vary from $-j, -j+1, \cdots, j$.

(i) We first discuss the simple case that, besides the $2j+1$ rotational degeneracy, the state $|j, m>$ does not have any other degeneracy. Applying T onto the last equation in (13.36), we obtain

$$TJ_zT^{-1}T\,|j,m> = mT\,|j,m>\ ,$$

which, because of (13.35), can be written as

$$J_z T \mid j, m > \; = \; - m T \mid j, m > \quad .$$

Likewise, from the two top equations of (13.36) we find $T \mid j, m >$ to be also an eigenstate of H and $\vec{J}^2$ with the eigenvalues E and $j(j+1)$. Hence, by our assumption of no other degeneracy, $T \mid j, m >$ must be proportional to $\mid j, -m >$; i.e.,

$$T \mid j, m > \; = \; U_T \mid j, m >^* \; = \; e^{i\theta(j,m)} \mid j, -m > \quad . \tag{13.37}$$

Let us construct the $(2j+1)$ – dimensional space whose basis vectors are $\mid j, j >$, $\mid j, j-1 >$, ..., $\mid j, -j >$. We write

$$\mid j, j > \; = \begin{pmatrix} 1 \\ 0 \\ 0 \\ \vdots \\ 0 \end{pmatrix} , \qquad \mid j, j-1 > \; = \begin{pmatrix} 0 \\ 1 \\ 0 \\ \vdots \\ 0 \end{pmatrix} ,$$

$$\cdots \; , \qquad \mid j, -j > \; = \begin{pmatrix} 0 \\ 0 \\ \vdots \\ 0 \\ 1 \end{pmatrix} . \tag{13.38}$$

In this space, the Hermitian operators J_x, J_y, J_z and the unitary operator U_T are all $(2j+1) \times (2j+1)$ matrices. As we shall see, it is possible to choose

$$\begin{aligned} &J_x \, , \; J_z && \text{real,} \\ &J_y && \text{imaginary,} \\ &U_T = e^{i\pi J_y} && \text{also real,} \end{aligned} \tag{13.39}$$

and, because the basis vectors $\mid j, m >^* = \mid j, m >$ in accordance with (13.38),

$$T\,|\,j\,,m\,> \;=\; U_T\,|\,j\,,m\,>^* \;=\; U_T\,|\,j\,,m\,> \;=\; e^{i\pi J_y}\,|\,j\,,m\,>$$
$$= \;(-1)^{j+m}\,|\,j\,,-m\,> \qquad (13.40)$$

which fixes the phase factor in (13.37).

<u>Proof.</u> In the case $j = \frac{1}{2}$, we have $\vec{J} = \frac{1}{2}\vec{\sigma}$ whose standard representation fulfills the condition J_x, J_z real and J_y imaginary, as can be seen by taking its components to be $\sigma_x = \tau_1$, $\sigma_y = \tau_2$ and $\sigma_z = \tau_3$ of (3.1). In order that (13.35) should hold, the 2×2 unitary matrix U_T must satisfy

$$U_T\,\vec{\sigma}^*\,U_T^\dagger \;=\; -\,\vec{\sigma}\quad .$$

By using (11.41), we see that U_T must be of the form

$$U_T \;=\; e^{i\alpha}\;e^{i\frac{1}{2}\pi\sigma_y}$$

where $e^{i\alpha}$ is an arbitrary factor; our convention is simply to choose $e^{i\alpha} = 1$, which makes in this two-dimensional space

$$U_T \;=\; e^{i\frac{1}{2}\pi\sigma_y} \quad \text{real},$$

and therefore (13.40) is valid.

By considering the direct product of two such $j = \frac{1}{2}$ two-dimensional spinor spaces, we obtain a four-dimensional space, which can be decomposed into a three-dimensional space for $j = 1$ and a remaining one for $j = 0$. Since J_x, J_y and J_z are additive and U_T multiplicative, it is easy to see that our choice of $\vec{J}$ and U_T matrices for $j = \frac{1}{2}$ leads to (13.39) and (13.40) for $j = 1$ as well. In the same way, by examining the direct product of N such $j = \frac{1}{2}$ spinor spaces, we can show that the same choice also gives (13.39) and (13.40) for a general j.

We note further that from (13.40)

$$T^2 \;=\; (-1)^{2j} \qquad (13.41)$$

in agreement with the minus sign in (13.23).

(ii) For discrete states, e.g. a single particle at rest, sometimes H may have other degeneracies, such as the exact one between p and $\bar{p}$, or the approximate one between p and n. In the former, the additional degeneracy can be separated out by requiring $|j, m\rangle$ also to be states with a definite baryon number; in the latter, the degeneracy can be removed by including the electromagnetic interaction in H. Consequently, in either case, (13.39) and (13.40) hold.

(iii) For free particles in a continuum state, the degeneracy can be removed by first specifying the species of particles, which reduces the problem to that of a single free particle, since the multi-free-particle wave function is a product of single-particle wave functions. Next, we may put the whole system in a large spherical box of radius R. For R finite, states of different angular momenta have different energies; thereby the degeneracy is removed. Consequently, (13.39) holds. Such states will be labeled as

$$|j, m; \text{ free}\rangle$$

and they satisfy (13.40), i.e.,

$$T|j, m; \text{ free}\rangle = (-1)^{j+m}|j, -m; \text{ free}\rangle . \qquad (13.42)$$

Since both (13.39) and (13.40) are valid for any finite R, they are also valid in the limit $R \to \infty$.

For free-particle states that are eigenstates of linear momentum, but not angular momentum, the transformation under T is given by (13.33).

(iv) For interacting particles in a continuum state with a definite angular momentum J, we may consider either (6.58), the "initial" state,

$$|(j, m)^{in}\rangle = U(0, -\infty)|j, m; \text{ free}\rangle \qquad (13.43)$$

or (6.59), the "final" state

$$| (j, m)^f > \; = \; U^\dagger(\infty, 0) \; | j, m ; \text{free} > \tag{13.44}$$

where for convenience we have set $t = 0$ in (6.58)-(6.59). As we shall prove later in (13.75), T invariance implies

$$T \, U(0, -\infty) \, T^{-1} \; = \; U(0, \infty)$$

which, because of

$$U(0, \infty) \; U(\infty, 0) \; = \; U(0, 0) \; = \; 1$$

and

$$U(\infty, 0)^\dagger \; U(\infty, 0) \; = \; 1 \quad ,$$

can also be written as

$$T \, U(0, -\infty) \, T^{-1} \; = \; U^\dagger(\infty, 0) \quad . \tag{13.45}$$

Hence, by using (13.42)-(13.45), we have

$$T \; | (j, m)^{in} > \; = \; (-1)^{j+m} \; | (j, -m)^f >$$

and

$$T \; | (j, m)^f > \; = \; (-1)^{j+m} \; | (j, -m)^{in} > \quad . \tag{13.46}$$

Hence, under time reversal the roles of initial and final states are interchanged.

<u>Exercise.</u> Show that in the $(2j + 1)$-dimensional space (13.38), the non-vanishing matrix elements of J_x, J_y and J_z can be written as

$$< j, m \pm 1 \; | \; J_x \; | \; j, m > \; = \; \tfrac{1}{2} \, [\, (j \mp m) \, (j \pm m + 1) \,]^{\frac{1}{2}} \quad ,$$

$$< j, m \pm 1 \; | \; J_y \; | \; j, m > \; = \; \mp \tfrac{1}{2} i \, [\, (j \mp m) \, (j \pm m + 1) \,]^{\frac{1}{2}} \tag{13.47}$$

and

$$< j, m \; | \; J_z \; | \; j, m > \; = \; m \quad .$$

13.4 Time Reversal in Different Representations

1. Heisenberg representation

The Heisenberg representation operator O_H is related to that in the Schrödinger representation by

$$O_H(t) = e^{iHt} O_S e^{-iHt} ,$$

and therefore it satisfies Heisenberg's equation

$$[H, O_H(t)] = -i \frac{\partial}{\partial t} O_H(t) \tag{13.48}$$

in accordance with (5.1) and (5.12). Under the T operation, $O_H(t)$ becomes

$$\begin{aligned} T O_H(t) T^{-1} &= U_T e^{-iH^*t} O_S^* e^{iH^*t} U_T^\dagger \\ &= U_T e^{-iH^*t} U_T^\dagger U_T O_S^* U_T^\dagger U_T e^{iH^*t} U_T^\dagger . \end{aligned}$$

If the theory is T-invariant, then because of (13.2) and (13.6) we can re-write the above expression as

$$T O_H(t) T^{-1} = e^{-iHt} T O_S T^{-1} e^{iHt} \tag{13.49}$$

which satisfies

$$[H, T O_H(t) T^{-1}] = -i \frac{\partial}{\partial(-t)} T O_H(t) T^{-1} , \tag{13.50}$$

as can be readily verified by differentiating (13.49) directly. Thus, $T O_H(t) T^{-1}$ satisfies the time-reversed Heisenberg equation.

Let us now continue our discussion of QED, but carry it out in the Heisenberg representation. In the case $O = \psi$, we have

$$\psi_H(\vec{r}, t) = e^{iHt} \psi(\vec{r}, 0) e^{-iHt} \tag{13.51}$$

where, at $t = 0$,

$$\psi_H(\vec{r}, 0) = \psi(\vec{r}, 0) = \psi_S(\vec{r}, 0) .$$

Equation (13.49) gives

$$T\,\psi_H(\vec{r},\,t)\,T^{-1} \;=\; e^{-iHt}\,T\,\psi(\vec{r},\,0)\,T^{-1}\,e^{iHt} \quad , \tag{13.52}$$

which, on account of (13.11) and (13.51), leads to

$$\begin{aligned} T\,\psi_H(\vec{r},\,t)\,T^{-1} \;&=\; e^{-iHt}\;\eta_t\,\sigma_2\,\psi(\vec{r},\,0)\,e^{iHt} \\ &=\; \eta_t\,\sigma_2\,\psi_H(\vec{r},\,-t) \quad . \end{aligned}$$

We observe that if the subscript H is omitted, the above equation takes on the form

$$T\,\psi(\vec{r},\,t)\,T^{-1} \;=\; \eta_t\,\sigma_2\,\psi(\vec{r},\,-t) \tag{13.53}$$

which is now also valid in the Schrödinger representation, since in that representation $\psi(\vec{r},\,t) = \psi(\vec{r},\,0)$ and therefore (13.53) becomes (13.11). Because T is antiunitary, as in (13.24), if we wish we may set $\eta_t = 1$ without any loss of generality.

Likewise, from (13.13), (13.20) and (13.21) we have

$$\begin{aligned} T\,A_\mu(\vec{r},\,t)\,T^{-1} \;&=\; -\,A_\mu(\vec{r},\,-t) \quad , \\ T\,j_\mu(\vec{r},\,t)\,T^{-1} \;&=\; -\,j_\mu(\vec{r},\,-t) \\ \text{and} \quad T\,\vec{\pi}(\vec{r},\,t)\,T^{-1} \;&=\; \vec{\pi}(\vec{r},\,-t) \quad . \end{aligned} \tag{13.54}$$

Furthermore, (13.26), (13.27) and (13.29) can now be written as

$$\begin{aligned} T\,a_{\vec{p},s}(t)\,T^{-1} &= e^{-i\theta_{\vec{p},s}}\,a_{-\vec{p},s}(-t)\,, \quad & T\,a^{\dagger}_{\vec{p},s}(t)\,T^{-1} &= e^{i\theta_{\vec{p},s}}\,a^{\dagger}_{-\vec{p},s}(-t)\,, \\ T\,b_{\vec{p},s}(t)\,T^{-1} &= e^{-i\theta_{-\vec{p},s}}\,b_{-\vec{p},s}(-t)\,, \quad & T\,b^{\dagger}_{\vec{p},s}(t)\,T^{-1} &= e^{i\theta_{-\vec{p},s}}\,b^{\dagger}_{-\vec{p},s}(-t)\,, \\ T\,a_{\vec{k},s}(t)\,T^{-1} &= -a_{-\vec{k},s}(-t)\,, \quad & T\,a^{\dagger}_{\vec{k},s}(t)\,T^{-1} &= -a^{\dagger}_{-\vec{k},s}(-t)\;. \end{aligned} \tag{13.55}$$

Both (13.54) and (13.55) are valid in the Heisenberg and Schrödinger representations.

2. <u>Interaction representation</u>

Let us decompose the total Hamiltonian into the sum of two

terms:

$$H = H_0 + H_{int} , \tag{13.56}$$

which will be written as

$$H_S = (H_0)_S + (H_{int})_S , \tag{13.57}$$

in the Schrödinger representation, and

$$H_I = (H_0)_I + (H_{int})_I \tag{13.58}$$

in the interaction representation, as in (5.5) and (5.9). The state vector $| t >_I$ and the operator $O_I(t)$ are related to those in the Schrödinger representation by (5.14) and (5.15)

$$| t >_I = e^{i H_0 t} | t >_S \tag{13.59}$$

and

$$O_I(t) = e^{i H_0 t} O_S e^{-i H_0 t} \tag{13.60}$$

where, as noted in (5.16),

$$H_0 = (H_0)_I = (H_0)_S . \tag{13.61}$$

From (13.59)-(13.60) it follows that the equations of motion (5.6)-(5.7) hold; i.e.,

$$(H_{int}(t))_I | t >_I = i \frac{\partial}{\partial t} | t >_I \tag{13.62}$$

and

$$[(H_0)_I , O_I(t)] = - i \dot{O}_I(t) . \tag{13.63}$$

We shall assume that both H_S and H_0 satisfy the T-invariance condition (13.2):

$$T H_S T^{-1} = U_T H_S^* U_T^\dagger = H_S \tag{13.64}$$

and

$$T H_0 T^{-1} = U_T H_0^* U_T^\dagger = H_0 . \tag{13.65}$$

Under the T operation, $| t >_I$ and $O_I(t)$ become respectively

$$T \mid t >_I = U_T \, e^{-i H_0^* t} \mid t >_S^* = U_T \, e^{-i H_0^* t} \, U_T^\dagger \, U_T \mid t >_S^* \tag{13.66}$$

and

$$T \, O_I(t) \, T^{-1} = U_T \, e^{-i H_0^* t} \, O_S^* \, e^{i H_0^* t} \, U_T^\dagger$$
$$= U_T \, e^{-i H_0^* t} \, U_T^\dagger \, U_T \, O_S^* \, U_T^\dagger \, U_T \, e^{i H_0^* t} \, U_T^\dagger \, .$$

By using (13.65) one sees that these two expressions can be written as

$$T \mid t >_I = e^{-i H_0 t} \, T \mid t >_S \tag{13.67}$$

and

$$T \, O_I(t) \, T^{-1} = e^{-i H_0 t} \, T \, O_S \, T^{-1} \, e^{i H_0 t} \, . \tag{13.68}$$

By direct differentiation, one can show that $T \mid t >_I$ and $T \, O_I(t) \, T^{-1}$ satisfy the same equations of motion (13.62)-(13.63), except for the replacement of $t \to -t$; i.e., they satisfy the time-reversed equations of motion.

Exercise. Show that in the case of QED, (13.53)-(13.55) are also valid in the interaction representation.

13.5 T invariance of the S-matrix

Theorem. If T invariance holds, then the S-matrix, defined by (5.23), satisfies

$$T \, S \, T^{-1} = S^\dagger \, . \tag{13.69}$$

Proof. According to (5.20)-(5.21), in the interaction representation the $U(t, t_0)$-matrix satisfies

$$-\frac{1}{i} \frac{\partial}{\partial t} U(t, t_0) = H_{int}(t) \, U(t, t_0) \tag{13.70}$$

with the boundary condition

$$U(t_0 \, , t_0) = 1 \tag{13.71}$$

where $H_{int}(t) = (H_{int}(t))_I$ of (13.58), but with the subscript I omitted for simplicity. Under T, (13.70)-(13.71) becomes

$$T[-\frac{1}{i}\frac{\partial}{\partial t}U(t,t_0)]T^{-1} = T H_{int}(t) T^{-1} T U(t,t_0) T^{-1} \tag{13.72}$$

and

$$T U(t_0, t_0) T^{-1} = 1 \quad . \tag{13.73}$$

Assuming time-reversal invariance, we have, by using (13.64)-(13.65),

$$T (H_{int})_S T^{-1} = (H_{int})_S \quad .$$

Substituting this expression into (13.68) for $O = H_{int}$, we obtain

$$T H_{int}(t) T^{-1} = H_{int}(-t) \quad . \tag{13.74}$$

Hence, (13.72) becomes

$$-\frac{1}{i}\frac{\partial}{\partial(-t)} T U(t,t_0) T^{-1} = H_{int}(-t) T U(t,t_0) T^{-1} \quad ,$$

from which it follows that

$$T U(t,t_0) T^{-1} = U(-t,-t_0) \quad , \tag{13.75}$$

since both sides satisfy the same first-order differential equation in t, and both = unit matrix when $t = t_0$. Because the S-matrix is given by

$$S = U(\infty, -\infty) \quad , \tag{13.76}$$

(13.75) implies

$$T S T^{-1} = U(-\infty, \infty) \quad . \tag{13.77}$$

By using

$$U(t,t') U(t',t_0) = U(t,t_0) \tag{13.78}$$

and the unitarity of $U(t, t_0)$, we see that

$$U(-\infty,\infty) = U(\infty,-\infty)^{-1} = U(\infty,-\infty)^{\dagger} \quad .$$

Consequently (13.77) becomes

$$T\,S\,T^{-1} = S^{\dagger}$$

which establishes the theorem.

13.6 Reciprocity

1. Reciprocity relations

Consider now the reaction

$$a + b + \cdots \rightleftarrows a' + b' + \cdots \tag{13.79}$$

in which each of the particles carries a definite momentum and helicity, labeled as $\vec{p}_i$ and s_i on the lefthand side $(i = a, b, \cdots)$ and $\vec{p}_j{}'$ and $s_j{}'$ on the right $(j = a', b', \cdots)$. Let us denote the free-particle state vectors for the left- and righthand sides respectively as $|\,\vec{p}_i\,, s_i >$ and $|\,\vec{p}_j{}'\,, s_j' >$. From the above theorem, we can establish the reciprocity relation

$$|< \vec{p}_j{}', s_j{}'\,|\,S\,|\,\vec{p}_i, s_i >| = |< -\vec{p}_i, s_i\,|\,S\,|\,-\vec{p}_j{}', s_j{}' >| \quad . \tag{13.80}$$

Proof. Under time reversal, a free particle reverses its momentum and spin vectors; therefore it retains its helicity. Consequently, we have

$$T\,|\,\vec{p}_i\,, s_i > = e^{i\theta}\,|\,-\vec{p}_i\,, s_i >$$

and

$$T\,|\,\vec{p}_j{}'\,, s_i{}' > = e^{i\theta'}\,|\,-\vec{p}_j{}'\,, s_j{}' > \tag{13.81}$$

where $e^{i\theta}$ and $e^{i\theta'}$ are phase factors. Let us take the complex conjugation of the matrix element $< \vec{p}_j{}', s_j{}'\,|\,S\,|\,\vec{p}_i, s_i >$:

$$< \vec{p}_j{}', s_j{}'\,|\,S\,|\,\vec{p}_i, s_i >^* = < \vec{p}_j{}', s_j{}'\,|\,T^{-1}\,T\,S\,T^{-1}\,T\,|\,\vec{p}_i\,, s_i >$$

which, because of (13.69) and (13.81), is

$$e^{i\theta - i\theta'} < -\vec{p}_j{}', s_j{}'\,|\,S^{\dagger}\,|\,-\vec{p}_i\,, s_i > = e^{i\theta - i\theta'} < -\vec{p}_i\,, s_i\,|\,S\,|\,-\vec{p}_j{}', s_j{}' >^*,$$

and that gives the reciprocity relation (13.80).

We note that if we start from the initial state $a+b+\cdots$ consisting of only plane waves, the final state in (13.79) would be $a'+b'+\cdots$ with outgoing spherical waves, whose time-reversed wave function should be $a'+b'+\cdots$ with incoming spherical waves. As noted before, such a time-reversed function is nearly impossible to construct physically. However, the reciprocity relation equates the transition amplitude for reactions in which both initial and final states consist of only plane waves and therefore can be directly checked experimentally.

2. Two-body reactions

Most of the tests of reciprocity relations are for two-body reactions

$$a + b \rightarrow a' + b' \tag{13.82}$$

and

$$a' + b' \rightarrow a + b \quad . \tag{13.82a}$$

Let j_i, k_i, E_i and v_i be respectively the spin, momentum, energy and velocity of particle i with

$$i = a, b, a' \text{ and } b' \quad .$$

It is convenient to stay in the c.m. system, and denote

$$\vec{k} = \vec{k}_a = -\vec{k}_b \; , \qquad \vec{k}' = \vec{k}_{a'} = -\vec{k}_{b'} \tag{13.83}$$

where in spherical coordinates

$$\vec{k} = (k, \theta, \phi) \quad \text{and} \quad \vec{k}' = (k', \theta', \phi')$$

with θ, θ' the polar angles and ϕ, ϕ' the azimuthal angles. We shall now show that if in (13.82) and (13.82a) the initial particles are not polarized and the final polarizations are not measured, then

$$\frac{d\sigma(a+b \rightarrow a'+b')}{d\sigma(a'+b' \rightarrow a+b)} = \frac{k'^2 (2j_{a'}+1)(2j_{b'}+1) \sin\theta' \, d\theta' \, d\phi'}{k^2 (2j_a+1)(2j_b+1) \sin\theta \, d\theta \, d\phi} \quad . \tag{13.84}$$

Proof. In reaction (13.82), we may denote the initial and final states as $|\vec{k}\, s_a\, s_b>$ and $|\vec{k}'\, s_{a'}\, s_{b'}>$ where s_i is the helicity of the particle i. From the reciprocity relation (13.80), we have

$$|<\vec{k}'\, s_{a'}\, s_{b'}\,|\,S\,|\,\vec{k}\, s_a\, s_b>| = |<-\vec{k}\, s_a\, s_b\,|\,S\,|\,-\vec{k}'\, s_{a'}\, s_{b'}>| \quad . \tag{13.85}$$

Under a 180° rotation along the direction $\vec{k}\times\vec{k}'$, the vector $\vec{k}$ is changed into $-\vec{k}$, and $\vec{k}'$ into $-\vec{k}'$. Hence, (13.85) becomes

$$|<\vec{k}'\, s_{a'}\, s_{b'}\,|\,S\,|\,\vec{k}\, s_a\, s_b>| = |<\vec{k}\, s_a\, s_b\,|\,S\,|\,\vec{k}'\, s_{a'}\, s_{b'}>| \quad . \tag{13.86}$$

On the left side of (13.84), the differential cross section for $a + b \to a' + b'$ can be written as, because of (5.111),

$$d\sigma = \frac{k'^2\, d\cos\theta'\, d\phi'}{4\pi^2(2j_a+1)(2j_b+1)\,|\vec{v}_a - \vec{v}_b|}\;\frac{dk'}{dE}\cdot\sum_{s_i}|<\vec{k}'\, s_{a'}\, s_{b'}\,|\,\mathcal{M}\,|\,\vec{k}\, s_a\, s_b>|^2 \tag{13.87}$$

where

$$E = E_{a'} + E_{b'} = E_a + E_b \tag{13.88}$$

is the total energy of the system, the sum extends over all four helicities $s_a, \cdots, s_{b'}$ and $\mathcal{M}$ is related to the S-matrix by (5.106). From (13.86), it follows that $\mathcal{M}$ satisfies

$$|<\vec{k}'\, s_{a'}\, s_{b'}\,|\,\mathcal{M}\,|\,\vec{k}\, s_a s_b>| = |<\vec{k}\, s_a\, s_b\,|\,\mathcal{M}\,|\,\vec{k}'\, s_{a'}\, s_{b'}>| \,. \tag{13.89}$$

Since for any particle i, $E_i\, dE_i = k_i\, dk_i$, we have, on account of (13.88),

$$\frac{dE}{dk'} = \frac{dE_{a'}}{dk'} + \frac{dE_{b'}}{dk'} = v_{a'} + v_{b'} = |\vec{v}_{a'} - \vec{v}_{b'}|$$

in the c.m. system. Substituting it into (13.87), we obtain

$$d\sigma(a+b \to a'+b') = \frac{\sum_{s_i} |\langle \vec{k}'\, s_{a'}\, s_{b'} | \mathcal{M} | \vec{k}\, s_a\, s_b \rangle|^2\, k'^2\, d\cos\theta'\, d\phi'}{4\pi^2 (2j_a+1)(2j_b+1)\, |\vec{v}_a - \vec{v}_b| \cdot |\vec{v}_{a'} - \vec{v}_{b'}|} .$$

Equation (13.84) now follows from (13.89), the above expression and a similar one for $d\sigma(a'+b' \to a+b)$.

If we integrate over all angles, then from (13.84) we have the following relations for the total cross sections:

$$\frac{\sigma(a+b \to a'+b')}{\sigma(a'+b' \to a+b)} = \frac{(2j_{a'}+1)(2j_{b'}+1)\,k'^2}{(2j_a+1)(2j_b+1)\,k^2} \cdot \begin{cases} \frac{1}{2} & \text{if } a \neq b, \text{ but } a' = b', \\ 2 & \text{if } a = b, \text{ but } a' \neq b', \\ 1 & \text{otherwise} \end{cases} \qquad (13.90)$$

where if $a \neq b$ but $a' = b'$, then because of symmetry due to identical particles the solid angle $\int d\cos\theta'\, d\phi'$ is only 2π, while $\int d\cos\theta\, d\phi$ is 4π; that accounts for the factor $\frac{1}{2}$. Likewise, there is a factor 2 if $a = b$ but $a' \neq b'$.

3. Pion spin

For reactions $\pi^+ + d \to p + p$ and $p + p \to \pi^+ + d$, (13.90) takes on the form

$$\frac{\sigma(\pi^+ + d \to p + p)}{\sigma(p + p \to \pi^+ + d)} = \frac{2}{3(2j_\pi+1)} \left(\frac{k_{pp}}{k_{\pi d}}\right)^2 \qquad (13.91)$$

where $k_{pp} = k'$, $k_{\pi d} = k$, and the numerical factor on the righthand side is due to $j_d = 1$, $j_p = \frac{1}{2}$ and, because $a' = b' = p$, there is an additional factor $\frac{1}{2}$. From this relation, the pion spin was first determined* to be 0.

---

* R. Durbin, H. Loar and J. Steinberger, Phys. Rev. 83, 646 (1951). See also R. Marshak, Phys. Rev. 82, 313 (1951).

4. Remarks

Reciprocity relations have been tested for several strong processes. Among these are *

$$Mg^{24} + d \rightleftarrows Mg^{25} + p$$

and

$$Mg^{24} + \alpha \rightleftarrows Al^{27} + p \qquad (13.92)$$

which give support for the T invariance of strong interaction, to an accuracy $\sim \frac{1}{2}\%$.

For QED, T invariance can be directly established by using the known Lagrangian. For weak processes, reciprocity relations are difficult to use. Instead, T invariance is tested by phase-angle measurements, as we shall discuss. In contrast to reciprocity relations, which are valid to all orders of the interaction provided T invariance holds, the phase-angle measurements as tests of T invariance are accurate only to the lowest order of the relevant interaction.

13.7 Phase-angle Relations

1. β decay

Consider the β decay of hadron a to b,

$$a \to b + e^{\mp} + \bar{\nu}_e \;\; (\text{or } \nu_e) \qquad (13.93)$$

where, e.g., we may have $a = n$ and $b = p$, or $a = \Sigma^{\pm}$ and $b = \Sigma^{0}$, etc. The phenomenological interaction Hamiltonian is given by the Fermi theory (modified to include parity violation):

$$H_\beta = \int [\, J_\mu(x)\; j_\mu(x) + \text{h.c.}\,]\; d^3r \qquad (13.94)$$

* W. G. Weitkamp, D. W. Storm, D. C. Shreve, W. J. Braithwaite and D. Bodansky, Phys.Rev. 165, 1233 (1968); W. von Witsch, A. Richter and P. von Brentano, Phys.Rev.Lett. 19, 524 (1967).

in which the lepton current j_μ is known explicitly,

$$j_\mu = i\,\psi_e^\dagger \gamma_4 \gamma_\mu (1+\gamma_5)\,\psi_\nu$$

with ψ_e and ψ_ν as the fields of e^- and ν_e. The hadron current operator J_μ will be analysed in our later chapters on interactions. Here, we are only interested in its matrix elements between states $|a>$ and $|b>$. From (11.80), it follows that

$$<b\,|\,J_\mu(x)\,|\,a> \;=\; <b\,|\,J_\mu(0)\,|\,a>\, e^{i(a_\mu - b_\mu)x_\mu} \qquad (13.95)$$

where a_μ and b_μ are the 4-momenta of $|a>$ and $|b>$. In most β-decay examples, the 4-momentum transfer $q_\mu \equiv (a-b)_\mu$ is extremely small. Hence, $<b\,|\,J_\mu(0)\,|\,a>$ depends only on the c. number Dirac spinors u_a and u_b of the initial and final hadron states; from Lorentz-invariance, its most general form is given by

$$<b\,|\,J_\mu(0)\,|\,a> \;=\; i\,u_b^\dagger \gamma_4 \gamma_\mu (G_V - G_A\gamma_5)\,u_a \qquad (13.96)$$

where G_V and G_A are constants. Phenomenologically, in the limit $q_\mu \to 0$, we may replace (13.96) by the operator

$$J_\mu = i\,\psi_b^\dagger \gamma_4 \gamma_\mu (G_V - G_A\gamma_5)\,\psi_a \qquad (13.97)$$

where ψ_a and ψ_b are the field operators of a and b.

Theorem 1. G_V/G_A is real, if T invariance holds.

Proof. With T invariance, we have (13.53). Hence,

$$T\,\psi_i(\vec{r},t)\,T^{-1} = \eta_i\,\sigma_2\,\psi_i(\vec{r},-t)$$

where $i = a, b, e$ or ν. Thus, by using (13.97), we find

$$T\,J_\mu(\vec{r},t)\,T^{-1} = \eta_b^*\,\eta_a\,\psi_b^\dagger\,\sigma_2 (i\gamma_4\gamma_\mu)^*\,(G_V^* - G_A^*\gamma_5^*)\,\sigma_2\,\psi_a$$

which, on account of (13.19) and $\gamma_5 = -\rho_1 =$ real, can also be written as

$$- i \eta_b^* \eta_a \psi_b^\dagger \gamma_4 \gamma_\mu (G_V^* - G_A^* \gamma_5) \psi_a ,$$

at $\vec{r}$ and $-t$. Likewise, we have

$$T j_\mu(\vec{r}, t) T^{-1} = - \eta_e^* \eta_\nu j_\mu(\vec{r}, -t) . \tag{13.98}$$

Thus, in order for H_β to be T-invariant, we must have G_V/G_A real, which establishes the theorem.

As can be seen from the Table of Particle Properties at the end of this book, for β-decay of neutron decay the phase δ, defined by

$$\frac{G_V}{G_A} = \left| \frac{G_V}{G_A} \right| e^{i\delta} ,$$

is determined experimentally to be

$$\delta = (180.20 \pm .19)^\circ \tag{13.99}$$

in agreement with the requirement of T invariance.

2. <u>Λ^o decay</u>

According to (11.107)-(11.108), in the Λ^o-decay

$$\Lambda^o \to \pi + N , \tag{13.100}$$

the final πN system is mainly in the isospin $I = \frac{1}{2}$ state. In the rest frame of Λ^o, the state of Λ^o is completely characterized by its spin wave function

$$u_\Lambda = \binom{a}{b}$$

where a and b are respectively the amplitudes of spin ↑ and spin ↓. Likewise, in the rest frame of N, the nucleon spin wave function u_N determines the physical state of N. Because of the superposition principle, the relation between these two spinors must be a linear one; i.e.,

$$u_\Lambda = M u_N$$

where M is a 2×2 matrix function. Since (13.100) is a two-body decay, besides spin, the kinematics is uniquely specified by a single unit vector

$\hat{k} \equiv$ direction of the nucleon momentum in the rest frame of Λ^o ;

so is, therefore, the functional dependence of M. From rotational invariance, we see that M must be of the form

$$M = A_s + A_p \vec{\sigma} \cdot \hat{k} \tag{13.101}$$

where σ is the Pauli spin matrix, A_s and A_p are constants. Because A_s refers to the spherically-symmetric part, it is called the s-wave amplitude; similarly, we call A_p the p-wave amplitude, since in (13.101) it is multiplied by a linear function of $\hat{k}$. Let ϕ be the relative phase between A_p and A_s, defined by

$$\frac{A_p}{A_s} = e^{i\phi} \left| \frac{A_p}{A_s} \right| . \tag{13.102}$$

Theorem 2. If T invariance holds, then

$$\phi = \delta_p - \delta_s \quad \text{or} \quad \delta_p - \delta_s + \pi \tag{13.103}$$

where δ_p and δ_s are, respectively, the p-wave and s-wave πN-scattering phase shifts (due to the strong interaction) in the $I = \frac{1}{2}$, $j = \frac{1}{2}$ state.

Proof. Since the Λ^o-decay is a weak process and the πN system has strong interactions, the T-invariance assumption applies to both strong and weak interactions. To first order of the weak-interaction Hamiltonian H_{wk}, but to all orders of the strong interaction H_{st}, the amplitude A_ℓ is given by

$$A_\ell = < (\pi N)_\ell \text{ ; free } | U(\infty, 0) H_{wk} | \Lambda^o >_m \qquad (13.104)$$

where $| (\pi N)_\ell$; free > is the free πN state, but with isospin $I = \frac{1}{2}$, orbital angular momentum ℓ, total angular momentum $j = \frac{1}{2}$ and its z-component m. For A_s, we have $\ell = 0$, and for A_p, $\ell = 1$. Clearly, because of rotational invariance, the matrix element (13.104) is independent of m. The U-matrix is determined only by the strong interaction; i.e., it satisfies (5.20) with H_{int} referring to the strong interaction. From (6.59), we see that the physical final πN state is given by

$$U(\infty, 0)^\dagger | (\pi N)_\ell \text{ ; free } > . \qquad (13.105)$$

Let $| \Lambda^o >$ denote the physical Λ^o-state. Both (13.105) and $| \Lambda^o >$ are eigenstates of the strong-interaction Hamiltonian. According to the perturbation formula, the matrix element of H_{wk} between these two states is the first-order transition amplitude, and that gives (13.104). By taking the complex conjugation of (13.104) and using (13.5)-(13.6) and $U_T^\dagger U_T = 1$, we find

$$A_\ell^* = < \pi N_\ell \text{ ; free } | T^{-1} T U(\infty, 0) T^{-1} T H_{wk} T^{-1} T | \Lambda^o >_m .$$

Because of (13.40), (13.42), (13.75) and the assumption of T invariance, the above expression can be written as

$$A_\ell^* = < \pi N_\ell \text{ ; free } | U(-\infty, 0) H_{wk} | \Lambda^o >_{-m} . \qquad (13.106)$$

We may use (13.78) and the unitarity of the U-matrix to obtain

$$\begin{aligned} U(-\infty, 0) &= U(0, -\infty)^{-1} = U(0, -\infty)^\dagger \\ &= [U(0, \infty) U(\infty, -\infty)]^\dagger \\ &= U(\infty, -\infty)^\dagger U(\infty, 0) = S^\dagger U(\infty, 0) \end{aligned} \qquad . (13.107)$$

By substituting this expression into (13.106) and using the completeness theorem, we derive

$$A_\ell^* = \langle \pi N_\ell \ ; \text{free} \mid S^\dagger U(\infty, 0)\ H_{wk} \mid \Lambda^\circ \rangle_{-m}$$
$$= \langle \pi N_\ell \ ; \text{free} \mid S^\dagger \mid n \rangle \langle n \mid U(\infty, 0)\ H_{wk} \mid \Lambda^\circ \rangle_{-m} \tag{13.108}$$

where the repeated index n denotes the sum over the complete set of all free-particle states. Here the S-matrix refers to that of the strong interaction. Thus,

$$\langle \pi N_\ell \ ; \text{free} \mid S^\dagger \mid n \rangle = \begin{cases} 0 & \text{when } \mid n \rangle \neq \mid \pi N_\ell \ ; \text{free} \rangle \\ e^{-i\,2\delta_\ell} & \text{otherwise.} \end{cases} \tag{13.109}$$

This is because the S-matrix connects $\mid \pi N_\ell \ ; \text{free} \rangle$ only to hadron states of the same quantum number and the same energy. At the energy = Λ° mass, there is only one such state, namely $\mid \pi N_\ell \ ; \text{free} \rangle$ itself. Since S is a unitary matrix, if in any row it has only one non-vanishing matrix element, then that element must be a pure phase factor $e^{2i\delta_\ell}$ where δ_ℓ is the usual phase shift. Equation (13.109) then follows. By combining (13.108) with (13.109), we have

$$A_\ell^* = e^{-2i\delta_\ell}\ A_\ell \ ,$$

which leads to Theorem 2.

The phase ϕ in (13.102) can be determined directly by measuring the density matrix of N (Problem 13.1); its experimental value, from the Table of Particle Properties, is

$$\phi = -6.5 \pm 3.5^\circ \ .$$

This is to be compared with the strong-interaction phase-shift measurement *

$$\delta_p - \delta_s = -6.5 \pm 1.3^\circ \ .$$

---

* S. W. Barnes, H. Winick, K. Miyake and K. Kinsey, Phys.Rev. 117, 238 (1960).

Remarks. As we shall discuss in Chapter 15, time-reversal asymmetry has been detected in the decay of the $K^0 - \bar{K}^0$ system. The amount of T violation observed is, however, extremely small. Its real significance is still not fully understood.

Problem 13.1. In π - N scattering,

$$\pi + N \rightarrow \pi + N \ ,$$

let $\vec{k}$ and $\vec{k}'$ be the initial and final momenta of N in the c.m. system.

(i) Assuming symmetry under both space inversion P and time reversal T, show that for an initially unpolarized nucleon, the final nucleon can be polarized, but with its spin direction $\vec{\sigma}$ perpendicular to the plane containing $\vec{k}$ and $\vec{k}'$.

(ii) Under P, we have $\vec{k} \rightarrow -\vec{k}$, $\vec{k}' \rightarrow -\vec{k}'$ and $\vec{\sigma} \rightarrow \vec{\sigma}$; this is why P symmetry makes it impossible to observe pseudoscalars such as $\vec{\sigma} \cdot \vec{k}$ and $\vec{\sigma} \cdot \vec{k}'$. Under T, we expect $\vec{k} \rightarrow -\vec{k}$, $\vec{k}' \rightarrow -\vec{k}'$ and $\vec{\sigma} \rightarrow -\vec{\sigma}$. Why is it that T symmetry does not prevent one from observing the apparently T-odd term $\vec{\sigma} \cdot (\vec{k} \times \vec{k}')$?

Problem 13.2.

(i) Show that in the decay of a completely polarized Λ^0 the spin-density matrix D of the final nucleon measured in its rest system is given by

$$D = \tfrac{1}{2}(1 + \alpha \cos \theta)(1 + \hat{S}_N \cdot \vec{\sigma})$$

where

$$\hat{S}_N = (1 + \alpha \cos \theta)^{-1} [(\alpha + \cos \theta)\, \hat{k} + \beta (\hat{k} \times \hat{S}_\Lambda) + \gamma(\hat{k} \times \hat{S}_\Lambda) \times \hat{k}] \ ,$$

$$\alpha = (|A_s|^2 + |A_p|^2)^{-1}\, 2 \mathrm{Re}(A_s^* A_p) \ ,$$

$$\beta = -(|A_s|^2 + |A_p|^2)^{-1}\, 2\,\mathrm{Im}(A_s^* A_p) \quad ,$$

$$\gamma = (|A_s|^2 + |A_p|^2)^{-1}(|A_s|^2 - |A_p|^2) \quad ,$$

$\hat{S}_\Lambda$ is the initial Λ-spin direction, $\hat{k}$ is the momentum direction of the nucleon, both measured in the rest system of Λ^o, and $\theta = \angle(\hat{S}_\Lambda, \hat{k})$. Note that $\alpha^2 + \beta^2 + \gamma^2 = 1$ and $|\hat{S}_N|^2 = 1$.

(ii) If the Λ^o is not completely polarized, but with its spin-density matrix D_Λ given by

$$D_\Lambda = \tfrac{1}{2}(1 + \vec{S}_\Lambda \cdot \vec{\sigma}) \quad ,$$

show that the angular distribution of N in the rest system of Λ^o is given by

$$\mathrm{trace}\ D_N = 1 + \alpha\, |\vec{S}_\Lambda| \cos\theta \quad .$$

Problem 13.3. Consider the decay of a spin-J particle X to a nucleon and a pion

$$X \rightarrow N + \pi \quad .$$

Let θ be the angle between the nucleon momentum and any fixed direction, measured in the rest system of X. Prove that

(i) $|<\cos\theta>_{Av.}| \leqq (2J+2)^{-1}$,

(ii) if the decay distribution is known to be a linear function in $\cos\theta$, then

$$|<\cos\theta>_{Av.}| \leqq (6J)^{-1} \quad .$$

[See Phys. Rev. 109, 1755 (1958). The spin of many hyperons, such as Λ, Σ, etc., is determined by using these inequalities.]

Chapter 14

CPT INVARIANCE

At present, there is good evidence that in our universe each of the three discrete symmetries C, P and T by itself is only approximately valid. The same applies to any of the bilinear products CP, PT, TC, CT, etc. However, as far as we know, the triple product CPT (or its permutations PTC, TCP, ⋯) does respresent an exact symmetry. Hence, we regain symmetry when we interchange particles with antiparticles, right with left and past with future. That nature should favor such combined symmetry operations fits harmoniously with our present theoretical forumulation. As we shall see, in the framework of a local field theory, Lorentz invariance and the usual spin-statistics requirement automatically lead to CPT-invariance.

14.1 CPT Theorem *

Let us consider a local field theory in which there are N_j spin-j fields, denoted by

$$\begin{aligned} &\text{spin-0}: \quad \phi_1(x),\ \phi_2(x),\ \cdots,\ \phi_{N_0}(x)\ , \\ &\text{spin-}\tfrac{1}{2}: \quad \psi_1(x),\ \psi_2(x),\ \cdots,\ \psi_{N_{\frac{1}{2}}}(x)\ , \end{aligned} \tag{14.1}$$

* W. Pauli, "Exclusion Principle, Lorentz Group and Reflection of Space-time and Charge", in Niels Bohr and the Development of Physics, ed. W. Pauli, L. Rosenfeld and V. Weisskopf (New York, McGraw-Hill, 1955). See references to earlier works of Lüders and Schwinger mentioned therein.

$$\text{spin}-1: \quad (A_1(x))_\mu, \ (A_2(x))_\mu, \ \cdots, \ (A_{N_1}(x))_\mu, \ \cdots .$$

So far as the transformation properties under the Lorentz group and the discrete symmetries C, P, T are concerned, a general integer spin-j field can always be represented by a jth-rank symmetric tensor function of x:

$$T_{\mu_1 \cdots \mu_j}(x) \ . \tag{14.2}$$

Likewise a general half-integer spin-j field can be represented by

$$S_{\mu_1 \cdots \mu_{j-\frac{1}{2}};\alpha}(x) \sim T_{\mu_1 \cdots \mu_{j-\frac{1}{2}}}(x)\, \psi_\alpha(x) \ , \tag{14.3}$$

i.e., it transforms like the direct product of a $(j-\frac{1}{2})^{th}$-rank symmetric tensor $T_{\mu_1 \cdots \mu_{j-\frac{1}{2}}}$ times a Dirac spinor ψ_α. All these fields satisfy the usual spin-statistics relation; integer-spin fields obey Bose statistics, and half-integer-spin fields Fermi statistics. The Lagrangian density $\mathcal{L}(x)$ is assumed to be of the form

$$\mathcal{L}(x) = \text{sum of normal products of } \left(\frac{\partial}{\partial x_\lambda}, \ \phi_a, \ \phi_a^\dagger, \ \psi_b, \ \psi_b^\dagger, \ (A_c)_\mu, \ (A_c^\dagger)_\mu, \ \cdots \right) \tag{14.4}$$

in which all fields are taken at the same space-time point x, with $x_\mu = (\vec{r}, it)$, as usual.

The operator $\mathcal{G}$ is defined to be

$$\mathcal{G} \equiv CPT \ . \tag{14.5}$$

[All the following discussions apply equally well to other permutations PTC, or TCP, etc.] From (13.53), we have for any spin-$\frac{1}{2}$ field

$$T\,\psi(\vec{r}, t)\,T^{-1} = \eta_t\,\sigma_2\,\psi(\vec{r}, -t) \ ,$$

which under P , given by (10.9), transforms as

$$P\,T\,\psi(\vec{r}, t)\,T^{-1}\,P^{-1} = \eta_t\,\sigma_2\,P\,\psi(\vec{r}, -t)\,P^{-1}$$
$$= \eta_t\,\eta_p\,\sigma_2\,\gamma_4\,\psi(-\vec{r}, -t) \quad .$$

By applying C of (10.5) to the above expression, we obtain

$$\mathcal{G}\,\psi(x)\,\mathcal{G}^{-1} = C\,P\,T\,\psi(x)\,T^{-1}\,P^{-1}\,C^{-1} = \eta_t\,\eta_p\,\eta_c\,\sigma_2\,\gamma_4\,\psi^c(-x) \; .$$

Since according to (10.6)

$$\psi^c_\alpha = (\gamma_2)_{\alpha\beta}\,\psi^\dagger_\beta \quad , \tag{14.6}$$

and on account of (3.11) and (3.103), $\sigma_2\,\gamma_4\,\gamma_2 = \sigma_2\,\rho_3\,\rho_2\,\sigma_2 = -\,i\,\rho_1 = i\,\gamma_5$, it follows that

$$\mathcal{G}\,\psi(x)_\alpha\,\mathcal{G}^{-1} = \eta\,(i\,\gamma_5)_{\alpha\beta}\,\psi^\dagger_\beta(-x) \tag{14.7}$$

where η is the product of phase factors $\eta_t\,\eta_p\,\eta_c$.

<u>Theorem.</u> Any Lorentz-invariant $\mathcal{L}(x)$ satisfies

$$\mathcal{G}\,\mathcal{L}(x)\,\mathcal{G}^{-1} = \mathcal{L}^\dagger(-x) \quad ,$$

provided we choose

for all $a = 1, 2, \cdots, N_0$, $\quad \mathcal{G}\,\phi_a(x)\,\mathcal{G}^{-1} = \phi_a^\dagger(-x)$, (14.8)

for all $b = 1, 2, \cdots, N_{\frac{1}{2}}$, $\quad \mathcal{G}\,(\psi_b(x))_\alpha\,\mathcal{G}^{-1} = i\,(\gamma_5)_{\alpha\beta}\,(\psi_b^\dagger(-x))_\beta$, (14.9)

for all $c = 1, 2, \cdots, N_1$, $\quad \mathcal{G}\,(A_c(x))_\mu\,\mathcal{G}^{-1} = -(A_c^\dagger(-x))_\mu$, (14.10)

for all general integer-spin j fields (14.2)

$$\mathcal{G}\,T_{\mu_1 \cdots \mu_j}(x)\,\mathcal{G}^{-1} = (-1)^j\,T^\dagger_{\mu_1 \cdots \mu_j}(-x) \quad , \tag{14.11}$$

and for all general half-integer-spin j fields (14.3)

$$\mathcal{G}\,S_{\mu_1 \cdots \mu_{j-\frac{1}{2}}\,;\,\alpha}(x)\,\mathcal{G}^{-1} = (-1)^j\,(i\,\gamma_5)_{\alpha\beta}\,S^\dagger_{\mu_1 \cdots \mu_{j-\frac{1}{2}}\,;\,\beta}(-x) \, . \tag{14.12}$$

Proof. Let us first consider the normal product of two arbitrary spin-$\frac{1}{2}$ fields $\psi_b^\dagger(x)$ and $\psi_{b'}(x)$

$$O(x) \equiv : g\, \psi_b^\dagger(x)\, \Gamma\, \psi_{b'}(x) : \tag{14.13}$$

where Γ is a 4×4 matrix. Since any 4×4 matrix has 16 matrix elements, it can always be decomposed into a linear function of the following 16 matrices, distributed among five groups: scalar (S), pseudoscalar (P), vector (V), axial vector (A) and tensor (T)

$$\begin{array}{cccccc} \Gamma = & \gamma_4 & i\gamma_4\gamma_5 & i\gamma_4\gamma_\mu & i\gamma_4\gamma_\mu\gamma_5 & \gamma_4\sigma_{\mu\nu} \\ & S & P & V & A & T \end{array} \tag{14.14}$$

where

$$\sigma_{\mu\nu} = -\tfrac{1}{2} i\,(\gamma_\mu\gamma_\nu - \gamma_\nu\gamma_\mu) . \tag{14.15}$$

In accordance with our metric $x_\mu = (\vec{r}, it)$, we have chosen the Γ-matrix to be Hermitian for S, P, the spatial components of V, A and both the spatial components and the (4, 4) component of T; otherwise Γ is anti-Hermitian, i.e. on account of (3.11)

$$\gamma_4 , \quad i\gamma_4\gamma_5 , \quad i\gamma_4\gamma_j , \quad i\gamma_4\gamma_j\gamma_5 , \quad \gamma_4\sigma_{ij} \quad \text{and} \quad \gamma_4\sigma_{44}$$

are Hermitian, while the rest of the expressions in (14.14) are anti-Hermitian. In addition, Γ satisfies

$$\begin{aligned} &\Gamma\gamma_5 = -\gamma_5\Gamma && \text{for S, P, T} \\ \text{and}\quad &\Gamma\gamma_5 = \gamma_5\Gamma && \text{for V, A,} \end{aligned} \tag{14.16}$$

whose Hermitian conjugates are

$$\begin{aligned} &\Gamma^\dagger\gamma_5 = -\gamma_5\Gamma^\dagger && \text{for S, P, T} \\ \text{and}\quad &\Gamma^\dagger\gamma_5 = \gamma_5\Gamma^\dagger && \text{for V, A .} \end{aligned} \tag{14.17}$$

The requirement that ψ_b should satisfy (14.9) also implies, for its

Hermitian conjugate,

$$\mathcal{G}\,(\psi_b^\dagger(x))_\alpha\,\mathcal{G}^{-1} = -i\,(\gamma_5)_{\alpha\beta}\,(\psi_b(-x))_\beta \tag{14.18}$$

in which we have noted that $\gamma_5{}^* = \gamma_5$. Consequently, using (14.9), (14.18) and applying $\mathcal{G}$ to (14.13), we obtain

$$\mathcal{G}\,O(x)\,\mathcal{G}^{-1} = :g^*[\psi_b(-x)]_\alpha\,(-i\,\gamma_5\,\Gamma^*\,i\,\gamma_5)_{\alpha\beta}\,[\psi_{b'}^\dagger(-x)]_\beta:$$

which, after we interchange the order of ψ_b and $\psi_{b'}^\dagger$, and use the property $\tilde{\gamma}_5 = \gamma_5$, becomes

$$-:g^*\,\psi_{b'}^\dagger(-x)\;\gamma_5\,\Gamma^\dagger\,\gamma_5\,\psi_b(-x):$$

where the minus sign appears because of the Fermi statistics, and the anticommutator has been dropped because of the normal product. Substituting (14.17) into the above expression, we derive

$$\mathcal{G}\,O(x)\,\mathcal{G}^{-1} = \begin{cases} O^\dagger(-x) & \text{for } S,\,P,\,T \\ -O^\dagger(-x) & \text{for } V,\,A\,. \end{cases} \tag{14.19}$$

It is important to note that from any spin-$\frac{1}{2}$ field ψ_b we may define its antiparticle field to be the $\psi_b{}^c$ of (14.6). Under $\mathcal{G}$, $\psi_b{}^c$ transforms as

$$\mathcal{G}\,(\psi_b{}^c(x))_\alpha\,\mathcal{G}^{-1} = (\gamma_2)^*_{\alpha\beta}\,\mathcal{G}\,(\psi_b^\dagger(x))_\beta\,\mathcal{G}^{-1}\,.$$

Because γ_2 is real and anticommutes with γ_5, we can, by using (14.18), rewrite the above expression as

$$-i\,(\gamma_2\,\gamma_5)_{\alpha\beta}\,(\psi_b(-x))_\beta = i\,(\gamma_5\,\gamma_2)_{\alpha\beta}\,(\psi_b(-x))_\beta\,,$$

which leads to

$$\mathcal{G}\,(\psi_b{}^c(x))_\alpha\,\mathcal{G}^{-1} = i\,(\gamma_5)_{\alpha\beta}\,(\psi_b{}^{c\dagger}(-x))_\beta\,. \tag{14.20}$$

From Problem 10.4 (iii), we note that under a continuous Lorentz transformation $\psi_b^{\ c}$ transforms in the same way as ψ_b ; by comparing (14.9) with (14.20) we see that they also have the same transformation under the discrete symmetry CPT .

Next, we observe that the c. number x_μ satisfies

$$\mathcal{G}\, x_\mu\, \mathcal{G}^{-1} = x_\mu^{\ *} = -(-x_\mu^{\ *}) \quad ; \tag{14.21}$$

consequently the derivative operator also satisfies

$$\mathcal{G} \frac{\partial}{\partial x_\mu} \mathcal{G}^{-1} = -\left(\frac{\partial}{\partial(-x_\mu)}\right)^* . \tag{14.22}$$

Under a continuous Lorentz transformation, x_μ , $\frac{\partial}{\partial x_\mu}$, V and A in (14.14) all transform as 1st-rank tensors; from (14.19), (14.21) and (14.22) we see that they also have the same transformation under CPT.

Let us now consider a local N^{th} - rank tensor function $F_{\mu_1 \cdot\cdot\, \mu_N}(x)$, which is a polynomial, of fermion fields ψ_1 , ψ_2 , ... and the derivative operator $\frac{\partial}{\partial x_\lambda}$:

$$F_{\mu_1 \cdot\cdot\, \mu_N}(x) = \text{normal product of } \left(\frac{\partial}{\partial x_\lambda} , \ \psi_b , \ \psi_{b'}^{\ \dagger}\right) \tag{14.23}$$

with all fields taken at the same space-time point x .

<u>Lemma.</u>

$$\mathcal{G}\, F_{\mu_1 \cdot\cdot\, \mu_N}(x)\, \mathcal{G}^{-1} = (-1)^N\, F^{\dagger}_{\mu_1 \cdot\cdot\, \mu_N}(-x) \quad . \tag{14.24}$$

To prove the lemma, let us denote the number of ψ_b in (14.23) as n, and that of $\psi_{b'}^{\ \dagger}$ as n' .

(i) n = n' . Because (14.23) is a normal product, we can permute the order of fermion fields. Since n = n' , each $\psi_{b'}^{\ \dagger}$ can be paired with a ψ_b to form bilinear products (14.13), which can then be grouped into tensors of 0th - rank (S or P) , of 1st - rank (V or A) and of 2nd - rank (T) , as in (14.14). From (14.19) and (14.22), we see that the transformation rule (14.24) applies to each of these tensors, as well as to $\frac{\partial}{\partial x_\mu}$ which is a tensor of 1st - rank; therefore (14.24) also applies to their product.

(ii) $n < n'$. In this case we can simply write the excess number of $\psi_{b'}^{\dagger}$ as $\psi_{b'}^{c}$, and treat these $\psi_{b'}^{c}$ as other ψ_b so that with this rewriting the total numbers of ψ_b and $\psi_{b'}^{\dagger}$ are the same. This case then reduces to (i).

(iii) $n > n'$. By converting the excess number of ψ_b into $\psi_b^{c\dagger}$, we can also reduce this case to (i). The lemma is then proved.

By comparing (14.11) and the lemma, we see that if a product of boson fields $T_{\mu_1 \cdots \mu_j}$ and a normal product $F_{\mu_1 \cdots \mu_N}$ of fermion fields have identical transformations under the continuous Lorentz group (i.e., they are tensors of the same rank), then they have the same transformation under CPT. From (14.12) we see that this statement can be extended to include fermion fields of arbitrary spin, and therefore (14.24) applies to any normal products of boson and fermion fields. The Lagrangian density $\mathcal{L}$ is one such product, and it is a tensor of 0th - rank. Therefore, by setting $N = 0$ in (14.24), we have

$$\mathcal{G}\,\mathcal{L}(x)\,\mathcal{G}^{-1} = \mathcal{L}^{\dagger}(-x) \qquad (14.25)$$

which establishes the CPT theorem.

In a quantum theory, the Lagrangian density $\mathcal{L}$ is a Hermitian operator and the action is the 4-dimensional volume integral $\int \mathcal{L} d^4x$. The above theorem insures the CPT invariance of the action integral and therefore also of the theory.

<u>Remarks.</u> (i) If we compare (14.7) with (14.9), we see that the phase η in (14.7) has been set to be 1 in (14.9) for all spin-$\frac{1}{2}$ fields in our proof of the CPT theorem. We may ask if it is possible to make a different phase choice in the proof.

For simplicity, let us consider a system consisting only of spin-$\frac{1}{2}$ fields. Suppose that the CPT theorem can also be proved with (14.9) replaced by

$$\text{for all } b = 1, 2, \cdots, N_{\frac{1}{2}} : \quad \mathcal{G}\,(\psi_b(x))_\alpha\, \mathcal{G}^{-1} = i\,\eta_b (\gamma_5)_{\alpha\beta} (\psi_b^\dagger(-x))_\beta \,. \tag{14.26}$$

Which is then the correct CPT operator, $\mathcal{G}$ of (14.9) or $\mathcal{G}$ of (14.26)?

We note that the quotient of these two $\mathcal{G}$ operators

$$S \equiv \mathcal{G}(14.26) / \mathcal{G}(14.9) \quad ,$$

satisfies

$$S\,\psi_b(x)\,S^{-1} = \eta_b\,\psi_b(x) \quad . \tag{14.27}$$

From the CPT theorem we know definitely that the theory is invariant under the $\mathcal{G}$ of (14.9); by our supposition it is also invariant under the $\mathcal{G}$ of (14.26). Consequently, it must be invariant under S. By looking at (14.27), we see that the S symmetry has nothing to do with CPT ; it is an unrelated internal symmetry of the theory.

For any theory, let

$$\{S\} \tag{14.28}$$

be the group of exact internal symmetries that are unrelated to C, P, T, and any continuous space-time transformations. Clearly, if $\mathcal{G}$ is a CPT operator, an equally good candidate would be $\mathcal{G}S$, or $S\mathcal{G}$, or $S\,\mathcal{G}\,S'$, where S and S' belong to the group (14.28). It makes no difference which one is called the CPT operator.

The CPT theorem states that the $\mathcal{G}$ defined by (14.8) - (14.12) is always a good symmetry operator, independently of the internal symmetry group of the system.

(ii) From (14.9) and (14.18), it follows that for any spin-$\frac{1}{2}$ field $\psi(x)$

$$\mathcal{G}^2\,\psi(x)_\alpha\,\mathcal{G}^{-2} = \mathcal{G}(i\,\gamma_5)_{\alpha\beta}\,\psi^\dagger(-x)_\beta\,\mathcal{G}^{-1}$$

$$= -i\,(\gamma_5)_{\alpha\beta}\,\mathcal{G}\,\psi^\dagger(-x)_\beta\,\mathcal{G}^{-1} = -\psi(x)_\alpha \quad .$$

Similarly, from (14.11) and (14.12) we find that for any integer-spin j field

$$\mathcal{G}^2\,T_{\mu_1 \cdots \mu_j}(x)\,\mathcal{G}^{-2} = T_{\mu_1 \cdots \mu_j}(x) \quad ,$$

while for any half-integer-spin j field

$$\mathcal{G}\,S_{\mu_1 \cdots \mu_{j-\frac{1}{2}}\,;\,\alpha}(x)\,\mathcal{G}^{-2} = -\,S_{\mu_1 \cdots \mu_{j-\frac{1}{2}}\,;\,\alpha}(x) \quad .$$

Together they imply that

$$\mathcal{G}^2 = (-1)^{2j} \quad . \tag{14.29}$$

(iii) The validity of the CPT theorem follows only from these assumptions: invariance under the continuous Lorentz group of transformations and the spin-statistics relation, plus the use of a local field theory. It is independent of whether each of the operations C, P and T is a good symmetry or not.

(iv) Consider a particle, say the proton p, at rest. Let the state $|\,p>_m$ be one with its z - component angular momentum $= m$. Apart from a multiplicative phase factor, under C the state becomes $|\,\bar{p}>_m$, under P it remains itself, but under T, m is changed into $-m$, as given by (13.40). Therefore

$$\mathcal{G}\,|\,p>_m = e^{i\theta}\,|\,\bar{p}>_{-m} \quad . \tag{14.30}$$

<u>Exercise.</u> Show that if CPT invariance holds, then the $U(t, t_0)$ - matrix defined by (5.20)-(5.21) satisfies

$$\mathcal{G}\,U(t, t_0)\,\mathcal{G}^{-1} = U(-t, -t_0) \quad . \tag{14.31}$$

Hence, under $\mathcal{G}$ the S-matrix transforms as

$$\mathcal{G} S \mathcal{G}^{-1} = S^{\dagger} . \qquad (14.32)$$

14.2 Applications

1. <u>Mass equality between particles and antiparticles</u>

For definiteness, we may consider the proton and antiproton masses. Let $|p>_m$ be the state of a proton at rest with its z-component angular momentum = m, as before. Since the mass of the proton is given by the expectation value,

$$(\text{mass})_p = <p|H|p>_m \qquad (14.33)$$

where H is the total Hamiltonian, it is clearly real and independent of m. Hence, (14.33) equals its complex conjugation. We have

$$(\text{mass})_p = <p|H|p>_m^* = <p|\mathcal{G}^{-1}\mathcal{G} H \mathcal{G}^{-1}\mathcal{G}|p>_m .$$

Because of (14.30) and $\mathcal{G} H \mathcal{G}^{-1} = H$, the above expression can also be written as

$$(\text{mass})_p = <\bar{p}|H|\bar{p}>_{-m} = (\text{mass})_{\bar{p}} . \qquad (14.34)$$

Likewise, we can prove the mass equality between e^+ and e^-, μ^+ and μ^-, etc. Similarly, we can establish

$$<K^o|H|K^o> = <\bar{K}^o|H|\bar{K}^o> . \qquad (14.35)$$

As we shall discuss in the next chapter, this identity holds to an accuracy, in units of $(\text{mass})_K$,

$$\sim 7 \times 10^{-15} \qquad (14.36)$$

which gives one of the best experimental supports of CPT invariance.

2. Opposite electromagnetic properties between particles and antiparticles

Let $j_\mu(x)$ be the electromagnetic current operator. Since it is a 4-vector, we know from (14.19), or (14.24),

$$\mathcal{G}\, j_\mu(x)\, \mathcal{G}^{-1} = -j_\mu^\dagger(-x) \quad . \tag{14.37}$$

We may write $j_\mu = (\vec{j}, i\rho)$ where $\vec{j}$ and ρ are Hermitian; (14.37) becomes then

$$\mathcal{G}\, \vec{j}(x)\, \mathcal{G}^{-1} = -\vec{j}(-x) \tag{14.38}$$

and

$$\mathcal{G}\, i\rho(x)\, \mathcal{G}^{-1} = -i\, \mathcal{G}\, \rho(x)\, \mathcal{G}^{-1} = i\rho(-x)$$

or

$$\mathcal{G}\rho(x)\, \mathcal{G}^{-1} = -\rho(-x) \quad . \tag{14.39}$$

The electric charge of a particle, say p, is given by

$$Q_p = \langle p \,|\, \int \rho d^3r \,|\, p \rangle_m \quad ,$$

which is clearly independent of the z-component of angular momentum m. Since it is real, we may write

$$\begin{aligned} Q_p &= \langle p \,|\, \int \rho\, d^3r \,|\, p \rangle_m^* \\ &= \langle p \,|\, \mathcal{G}^{-1}\mathcal{G} \int \rho\, d^3r\, \mathcal{G}^{-1}\mathcal{G} \,|\, p \rangle_m \\ &= \langle \bar{p} \,|\, -\int \rho\, d^3r \,|\, \bar{p} \rangle_{-m} = -Q_{\bar{p}} \quad , \end{aligned} \tag{14.40}$$

where in deriving the last line we have used (14.30) and (14.39). Thus the electric charges of p and $\bar{p}$ are equal in magnitude but opposite in sign. Similarly, by using (14.38) we can show, for the same spin configuration, that the magnetic moments of p and $\bar{p}$ are also equal in magnitude but opposite in sign. In the same way we can extend these considerations to all other particles and antiparticles and to the equality of electromagnetic form factors between them.

3. Lifetime equality between particles and antiparticles

Consider the decay of particle a and its antiparticle $\bar{a}$ through either the weak- or electromagnetic-interaction Hamiltonian H_{int} :

$$a \to b \qquad \text{and} \qquad \bar{a} \to \bar{b} \tag{14.41}$$

where b and $\bar{b}$ are continuum states. In these decays the initial states $|\, a>_m$ and $|\, \bar{a}>_m$ are eigenstates of the strong-interaction Hamiltonian H_{st} , and m refers to the z-component of angular momentum, as before. To the lowest order of H_{int} , the lifetimes of a and $\bar{a}$ are given by the perturbation formula

$$\tau_a^{-1} = 2\pi \sum_b \delta(E_b - E_a)\ |< b_{free}\,|\, U(\infty, 0)\, H_{int}\,|\, a>_m|^2$$

and

$$\tau_{\bar{a}}^{-1} = 2\pi \sum_{\bar{b}} \delta(E_{\bar{b}} - E_{\bar{a}})\ |< \bar{b}_{free}\,|\, U(\infty, 0)\, H_{int}\,|\, \bar{a}>_m|^2 \quad . \tag{14.42}$$

In deriving these, we have used (6.59) for the final states with the U-matrix determined by H_{st} only. By using (14.31), we can convert the above expression for τ_a into

$$\tau_a^{-1} = 2\pi \sum_b \delta(E_b - E_a)\ |< b_{free}\,|\, \mathcal{G}^{-1}\mathcal{G}\, U(\infty, 0)\, \mathcal{G}^{-1}\mathcal{G} \cdot H_{int}\, \mathcal{G}^{-1}\mathcal{G}\ |a>_m|^2$$

$$= 2\pi \sum_b \delta(E_b - E_a)\ |< \bar{b}_{free}\,|\, U(-\infty, 0)\, H_{int}\,|\, \bar{a}>_{-m}|^2 .$$

Clearly the lifetime is independent of the subscript $\pm$ m , which we shall now drop. Because of (13.107) and $E_b = E_{\bar{b}}$, $E_a = E_{\bar{a}}$, the above expression can now also be written as

$$\tau_a^{-1} = 2\pi \sum_{\bar{b}} \delta(E_{\bar{b}} - E_{\bar{a}})\ |< \bar{b}_{free}\,|\, S^\dagger U(\infty, 0)\, H_{int}\,|\, \bar{a}>|^2$$

$$= 2\pi \sum_{\bar{b}} \delta(E_{\bar{b}} - E_{\bar{a}})\ |\, \sum_{\bar{b}'} < \bar{b}_{free}\,|\, S^\dagger |\, \bar{b}'_{free}> \cdot < \bar{b}'_{free}\,|\, U(\infty, 0)\, H_{int}\,|\, \bar{a}>|^2 \tag{14.43}$$

in which the absolute-value-squared term equals

$$\sum_{\bar{b}', \bar{b}''} < \bar{b}_{free} | S^{\dagger} | \bar{b}'_{free} > < \bar{b}_{free} | S^{\dagger} | \bar{b}''_{free} >^*$$
$$< \bar{b}'_{free} | U(\infty, 0) H_{int} | \bar{a} > < \bar{b}''_{free} | U(\infty, 0) H_{int} | \bar{a} >^* . \tag{14.44}$$

Since $S^{\dagger} S = 1$ and the S-matrix has only nonvanishing matrix elements between states of equal energy, we have

$$\sum_{\bar{b}} \delta(E_{\bar{b}} - E_{\bar{a}}) < \bar{b}_{free} | S^{\dagger} | \bar{b}'_{free} > < \bar{b}_{free} | S^{\dagger} | \bar{b}''_{free} >^*$$
$$= \delta(E_{\bar{b}'} - E_{\bar{a}}) \, \delta_{\bar{b}', \bar{b}''} . \tag{14.45}$$

Substituting (14.44) into (14.43) and using (14.45), we derive

$$\tau_a^{-1} = 2\pi \sum_{\bar{b}'} \delta(E_{\bar{b}'} - E_{\bar{a}}) \, |< \bar{b}'_{free} | U(\infty, 0) H_{int} | \bar{a} >|^2$$
$$= \tau_{\bar{a}}^{-1} ,$$

which establishes the lifetime equality between a and $\bar{a}$ to the lowest order of the decay Hamiltonian H_{int}, but to all orders of H_{st}. On the other hand, the mass equality (14.34) and the opposite-charge relation (14.40) are exact to all orders of the total Hamiltonian.

We note that the validity of (14.45) depends on the summation over all $\bar{b}$. Hence, the branching ratio of a particle decaying into a particular channel can be different from that of its antiparticle.

Experimentally, the lifetime equality between a^+ and a^- has been verified to the accuracy of

$$\frac{\tau_{a^+} - \tau_{a^-}}{\tau_{a^+} + \tau_{a^-}} = \begin{cases} 0 \pm .05\% & \text{for* } a = \mu \\ .0275 \pm .0355\% & \text{for** } a = \pi \\ .045 \pm .039\% & \text{for*** } a = K . \end{cases}$$

* S. L. Meyer, E. W. Anderson, E. Bleser, L. M. Lederman, J. L. Rosen, J. Rothberg and I.-T. Wang, Phys. Rev. 132, 2693 (1963).

It is important to note that the CPT operation used in quantum field theory is by no means an obvious one. The CPT theorem implies that when we change particles to antiparticles, keeping their momenta fixed and helicity reversed, all matrix elements of the interaction should become their appropriate complex conjugates. Tests such as lifetime and mass equalities between particles and antiparticles do not verify the full extent of CPT invariance; they involve no helicity reversal, nor are they sensitive to complex conjugation. Detailed tests of CPT symmetry can be made by studying the $K - \bar{K}$ complex, as we shall discuss in the next chapter.

Problem 14.1. Show that in the decay (14.41), if the final channels b and $\bar{b}$ consist only of free particles without mutual interactions, then for any such fixed channels the branching ratios $a \to b$ and $\bar{a} \to \bar{b}$ are equal, to the lowest order of the decay Hamiltonian.

** D. S. Ayres, A. M. Cormack, A. J. Greenberg, R. W. Kenney, D. O. Caldwell, V. B. Elings, W. P. Hesse and R. J. Morrison, Phys. Rev. D3, 1051 (1971).

*** F. Lobkowicz, A. C. Melissinos, Y. Nagashima, S. Tewksbury, H. von Briesen, Jr. and J. D. Fox, Phys. Rev. 185, 1676 (1969).

Chapter 15

K - $\overline{K}$ SYSTEM

The quartet of kaons, two charged and two neutral, forms one of the most intriguing systems in nature. The experimental determination of the mass difference between K_L^o and K_S^o,

$$m_L - m_S = (3.521 \pm .015) \times 10^{-6} \text{ eV} , \qquad (15.1)$$

reaches an accuracy which is unmatched in particle physics and rivals any precision measurement in physics. It is fortunate that this small mass difference is about half the decay width of K_S^o. This makes it possible to observe the rich variety of intricate interference phenomena in the production and decay of these two close-lying states. In turn, such observations enable us to test the linear superposition principle of quantum mechanics, the interplay of different conservation laws and the validity of various symmetry principles.

Historically, the 2π and 3π decay modes of the charged kaon led to the hypothesis of nonconservation of parity. Later the observation of the $\pi^+\pi^-$ decay mode of K_L^o established the violation of CP-invariance. These details we shall now study.

15.1 Dalitz Plot

In 1953, Dalitz * pointed out that the kaon spin can be

---

* R. H. Dalitz, Phil.Mag. 44, 1068 (1953); Phys.Rev. 94, 1046 (1954).

determined through the 3π decay modes

$$K^{\pm} \rightarrow \begin{cases} \pi^{\pm} + \pi^{\pm} + \pi^{\mp} \\ \pi^{o} + \pi^{o} + \pi^{\pm} \end{cases}$$

and (15.2)

$$K^{o} \rightarrow \begin{cases} \pi^{+} + \pi^{-} + \pi^{o} \\ 3\pi^{o} . \end{cases}$$

Let $\vec{k}_i$ be the momentum of the ith pion ($i = 1$ or 2 or 3) in the rest frame of the kaon, $k_i = |\vec{k}_i|$ and

$$\omega_i = \sqrt{\vec{k}_i^{\,2} + m_\pi^{\,2}}$$

be the corresponding pion energy, with m_π its mass. Momentum conservation requires these three $\vec{k}_i$ to lie on the same plane, called the decay plane, as shown in Fig. 15.1. Energy conservation gives

$$\omega_1 + \omega_2 + \omega_3 = m_K$$

where m_K is the kaon mass. By following the same reasoning that led to (5.109)-(5.110), we find the decay probability to be

$$\text{probability} \propto |A|^2 \cdot (\text{phase space}) \tag{15.3}$$

in which A is the decay matrix element and

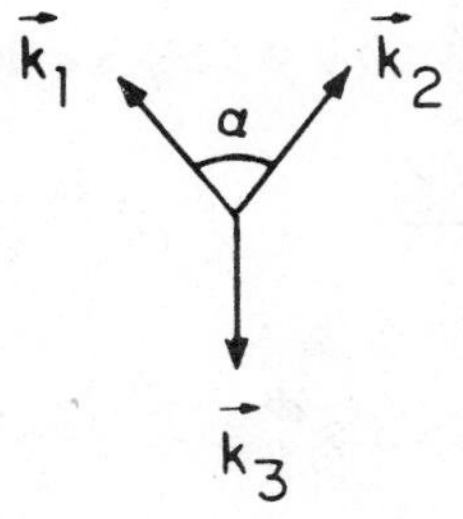

Fig. 15.1. Decay plane of $K \rightarrow 3\pi$, with $\vec{k}_i$ the momentum of the ith pion in the rest frame of K.

$$\text{phase space} \propto \prod_i \frac{d^3k_i}{\omega_i} \cdot \delta^4(\pi_1 + \pi_2 + \pi_3 - K) \tag{15.4}$$

where π_i and K refer to the 4-momenta

$$(\pi_i)_\mu = (\vec{k}_i, i\omega_i) \quad \text{and} \quad K_\mu = (0, im_K) \ .$$

1. Phase space

In this decay, each $\vec{k}_i$ has three components, giving a total of nine. Because of the overall energy-momentum conservation, there are 9 - 4 = 5 variables. If we stay in the rest frame of the kaon and ignore the production kinematics, then the decay probability (15.3) is independent of the orientation of the axis normal to the decay plane and the rotation along that axis; the former has two degrees of freedom and the latter one. Thus, altogether there are only 5 - 2 - 1 = 2 independent variables in the problem.

In Fig. 15.1 the angle between $\vec{k}_1$ and $\vec{k}_2$ is α. The phase space (15.4) can then be written as

$$d^2\rho \propto k_1{}^2 k_2{}^2 \, dk_1 \, dk_2 \, d\cos\alpha \ \delta\Big(\sum_i \omega_i - m_K\Big) \prod_i \omega_i^{-1} \ . \tag{15.5}$$

Because $\omega_3 = (k_1{}^2 + k_2{}^2 + 2k_1 k_2 \cos\alpha + m_\pi{}^2)^{\frac{1}{2}}$, we have

$$\frac{d\omega_3}{d\cos\alpha} = \frac{k_1 k_2}{\omega_3} \ .$$

Substituting this expression into (15.5), we derive

$$d^2\rho \propto \frac{k_1 k_2 dk_1 dk_2}{\omega_1 \omega_2} \ ,$$

or

$$d^2\rho \propto d\omega_1 \, d\omega_2 = d\omega_2 \, d\omega_3 = d\omega_3 \, d\omega_1 \ . \tag{15.6}$$

It is convenient to introduce the "kinetic" energy of the ith pion as

$$T_i = \omega_i - m_\pi \ , \tag{15.7}$$

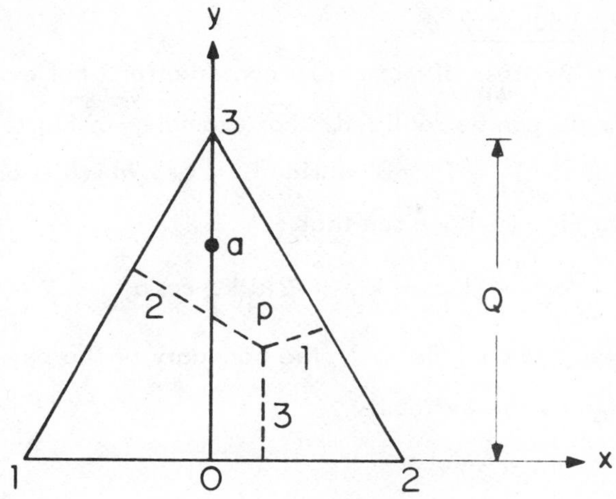

Fig. 15.2. Each $K \to 3\pi$ gives a point p inside this triangle.

and to define the Q-value of the decay as

$$Q = m_K - 3m_\pi = \sum_i T_i \quad . \tag{15.8}$$

The kinematics of each decay is thus completely determined by any two of these variables, say T_1 and T_2. A simple geometrical representation is to draw an equilateral triangle whose height is Q, as shown in Fig. 15.2. Let p be a point inside the triangle. Call its distances to the three sides T_1, T_2 and T_3; then the energy conservation (15.8) is satisfied automatically. It is easy to see that the area-element in this representation is $2\,dT_1\,dT_2/\sqrt{3}$, which together with (15.6) implies that

$$d^2\rho \propto d\,(\text{area}) \quad . \tag{15.9}$$

Hence, each decay (15.2) gives a point inside this triangle, with the phase-space volume proportional to the area. The accumulation of these points is called the Dalitz plot.

2. <u>Boundary</u>

Because of momentum conservation, not every point inside the triangle can be realized. For example, in Fig. 15.2 vertex 3 corresponds to $T_1 = T_2 = 0$ while $T_3 = Q$, which is obviously unphysical. From Fig. 15.1 we see that

$$k_3^2 = k_1^2 + k_2^2 + 2k_1 k_2 \cos\alpha \ . \tag{15.10}$$

Since $|\cos\alpha|$ is $\leqslant 1$, the boundary of the physical region is determined by the extremity

$$\cos\alpha = \pm 1 \ , \tag{15.11}$$

i.e., when the 3-momenta are collinear, and therefore

$$k_3 = \begin{cases} k_1 + k_2 & \text{for } \xleftarrow{\ 3\ } \quad \begin{matrix} \longrightarrow 1 \\ \longrightarrow 2 \end{matrix} \\ k_1 - k_2 & \text{for } \xleftarrow{\ 1\ } \quad \begin{matrix} \longrightarrow 2 \\ \longrightarrow 3 \end{matrix} \\ k_2 - k_1 & \text{for } \xleftarrow{\ 2\ } \quad \begin{matrix} \longrightarrow 3 \\ \longrightarrow 1 \end{matrix} \end{cases} \ . \tag{15.12}$$

(i) Nonrelativistic case

From (15.10) we have

$$\cos\alpha = \frac{k_3^2 - k_1^2 - k_2^2}{2k_1 k_2} \ .$$

For a nonrelativistic pion T_i is $k_i^2/2m_\pi$. Hence, the constraint $\cos^2\alpha \leqslant 1$ can be written as

$$4T_1 T_2 \geqslant (T_3 - T_1 - T_2)^2 \ ,$$

and the boundary can be expressed by a single equation

$$4T_1 T_2 = (T_3 - T_1 - T_2)^2 \ . \tag{15.13}$$

Using the coordinates x and y indicated in Fig. 15.2, we see that each T_i is a linear function of x and y, and therefore (15.13) is a quadratic function of x and y. Hence it describes a conic section, which on account of boundedness must be an ellipse. The symmetry with respect to the three pions requires the elliptical boundary to be invariant under a 120° rotation; thus it must be a circle. Because point 0 in Fig. 15.2 corresponds to $T_1 = T_2 = \frac{1}{2}Q$, and is physical, the boundary must be an inscribed circle, as shown in Fig. 15.3(a).

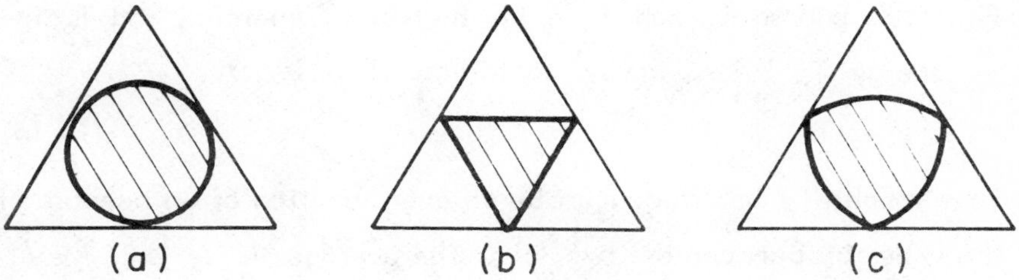

Fig. 15.3. The shaded areas correspond to the physical region of $K \to 3\pi$ in (a) the nonrelativistic limit, (b) the extreme relativistic limit and (c) the general case.

(ii) Extreme relativistic case

In this limit we can neglect m_π, and therefore $T_i = \omega_i = k_i$. The boundary (15.12) can be expressed in terms of three equations:

$$T_1 = T_2 + T_3 ,$$

$$T_2 = T_3 + T_1 \tag{15.14}$$

and

$$T_3 = T_1 + T_2 .$$

Using again the coordinates x and y of Fig. 15.2, we see that each is a straight line. The result is the shaded triangle of Fig. 15.3 (b).

(iii) General case

In general, the physical region occupies an area, resembling a shield, between the two limiting cases above, as shown in Fig. 15.3(c). The analytical form is given in Problem 15.1.

3. Spin determination

As a concrete example, let us consider the decay

$$K^+ \to \pi^+ + \pi^+ + \pi^- \quad . \tag{15.15}$$

Each decay will be represented by a point p in the Dalitz plot. In Fig. 15.2, choose T_1 and T_2 to be the two π^+ energies, and T_3 the π^- energy T_- ; i.e., the y coordinate of p is

$$y = T_- \quad . \tag{15.16}$$

The probability amplitude must be an even function of x because of the symmetry between the two π^+. The average T_i is $\cong 27$ MeV, which means that the pions are essentially nonrelativistic. Let us denote

$\vec{q}$ = relative momentum of $2\pi^+$ in the rest system of the two pions,

and

$\vec{k}$ = π^- momentum in the rest system of K^+.

Because of the low pion energy, we expect that the decay matrix element A in (15.3) can be expanded as a polynomial in $\vec{q}$ and $\vec{k}$. In this decay, each of these two momenta carries an orbital angular momentum, denoted by $\vec{\ell}_q$ and $\vec{\ell}_k$ respectively. Let the parity and the total angular momentum of the three-pion system be P and $\vec{j}$. Clearly,

$$\vec{j} = \vec{\ell}_q + \vec{\ell}_k \tag{15.17}$$

and because pions are pseudoscalars

$$P = (-1)^{\ell_q + \ell_k + 1} \tag{15.18}$$

where ℓ_q and ℓ_k denote the quantum numbers of $\vec{\ell}_q$ and $\vec{\ell}_k$. Due to the Bose statistics of $2\pi^+$, only even ℓ_q states are involved. Let j be the quantum number of $\vec{j}$.

We shall now analyze the distribution in the Dalitz plot for different jP assignments.

(i) $jP = 0+$

For $j = 0$ we must have $\ell_q = \ell_k$; hence both must be even. From (15.18) it follows that P is odd, which makes 0+ impossible.

(ii) $jP = 0-$

In this case $\ell_q = \ell_k$ = even. The simplest choice is $\ell_q = \ell_k = 0$ with the unnormalized amplitude A a constant which may be set to be unity. In general, A is of the form

$$A = F(y, x^2) = 1 + \lambda_1 y + \lambda_2 x^2 + \cdots \tag{15.19}$$

where $\lambda_1, \lambda_2, \cdots$ are constants.

(iii) $jP = 1+$

Since $j = 1$, the 3π wave function should transform as a vector. The simplest choice is $\ell_k = 1$ and $\ell_q = 0$; hence A depends linearly on $\vec{k}$. The general form is

$$A = \vec{k}\, F(y, x^2) \quad . \tag{15.20}$$

That means the decay distribution must be 0 when $\vec{k} = 0$, i.e., point O in Fig. 15.2.

(iv) $jP = 1-$

In this case, A should transform like a vector and the simplest choice is $\ell_q = \ell_k = 2$, with A depending quadratically on $\vec{q}$ and on $\vec{k}$. The most general form of A is

$$A = (\vec{q} \cdot \vec{k})\, (\vec{q} \times \vec{k})\, F(y, x^2)$$

which means the decay distribution should be 0 when $\vec{k} = 0$ and

$\vec{q} = 0$. The former refers to point O and the latter to point a (i.e., $T_1 = T_2$ and $T_3 \cong \frac{2}{3} Q$) in Fig. 15.2.

It is straightforward to see that for any $j \neq 0$, the decay amplitude must be zero either when $\vec{q} = 0$, or $\vec{k} = 0$, or both. By accumulating 3π decay events, it is possible to determine $|A|^2$ from the experimental distribution of the Dalitz plot. The result is

$$|A|^2 = 1 + g \frac{2m_K}{m_\pi^2} (\tfrac{1}{3} Q - T_-) \tag{15.21}$$

where $g = -0.215 \pm .004$, according to the Table of Particle Properties at the back of the book. Hence $|A|^2$ is approximately a constant with a slight linear dependence on y. The fact that A has no zero establishes

$$j = 0 .$$

Thus the 3π final state is $0-$. On the other hand, for $j = 0$, any 2π final state must be of even parity.

15.2 History

As early as 1947, the same year in which the π meson was discovered *, Rochester and Butler ** observed two forked tracks "of a very striking character" in their cloud-chamber investigation of cosmic rays. Both events, one charged and one neutral, were found to be the decay of unknown particles possessing masses $\approx 1{,}000$ times the electron's. This remarkable discovery, of the kaon, gave us our first glimpse into the rich world of strange particles.

---

* C. M. G. Lattes, G. P. S. Occhialini and C. F. Powell, Nature 160, 453 (1947).

** G. D. Rochester and C. C. Butler, Nature 160, 855 (1947). See also L. Leprince-Ringuet and M. L'héritier, J. Phys. Radium 7, 66, 69 (1946).

Their variety increased rapidly. Through the study of these particles, the concept of strangeness was introduced by Gell-Mann and Nishijima in the early fifties. These new developments also brought us the puzzling behavior of the θ and τ mesons.

At that time, what are now called the 2π and 3π decay modes of the kaon were referred to as

$$\theta^+ \rightarrow \pi^+ + \pi^0 ,$$
$$\theta^0 \rightarrow \pi^0 + \pi^0 , \quad \text{or} \quad \pi^+ + \pi^- ,$$
$$\tau^+ \rightarrow \pi^+ + \pi^- + \pi^0 ,$$

etc. The spin parity of θ^+ is clearly $0+$, $1-$, $2+$, $\cdots$. For θ^0, because of Bose statistics, it must be $0+$, $2+$, $4+$, $\cdots$. As we have mentioned, Dalitz was able to identify, in 1954, the spin parity of τ^+ as $0-$. Although both θ^+ and τ^+ were known to be of comparable mass, there was as yet nothing extraordinary. However, by 1955, very accurate lifetime measurements became available and were found to be, within a few percent, the same for these two mesons. In the meantime, mass determinations also became more precise, to within 10–20 MeV. We had then the paradox of two particles of approximately the same mass and same lifetime, but with very different decay modes and totally unequal phase-space volumes. This was the famous $\theta - \tau$ puzzle.

Because parity conservation was taken for granted, the first attempt to explain the puzzle was naturally via a conventional route. *

---

* It may be of interest to note how this was considered possible. As it turned out, by taking parity conservation on faith and accepting the preferred spin-parity assignments, $0-$ for τ and $0+$ for θ, one was led to assume θ to be of a much shorter lifetime than τ because of phase-space considerations. If, in addition, θ were lighter than τ by approximately 10 MeV, then the $0 - 0$ forbidden transition

$$\tau^+ \longrightarrow \theta^+ + 2\gamma$$

would have a rate comparable to the 3π decay mode of τ^+. The key to this explanation is that the lifetime measurements of both θ and τ were determined, at that time, by the attenuation of the 2π and 3π final states along the beam, which would consist of only the τ's because the θ's, due to their supposed short lifetime, would have decayed immediately after production. Hence in the cascade process (continued on next page)

However, when in early 1956 the first giant bubble-chamber experiment of Alvarez's * group failed to support such an explanation, it became clear that the solution must lie in something much deeper. Perhaps parity is not conserved, and $\theta-\tau$ are the same particle. But the immediate reaction was "So what?". If parity nonconservation occurs only in $\theta-\tau$, then the clue becomes the sole phenomenon, and one can never know for sure whether the explanation is correct. Thus it was necessary to investigate whether parity nonconservation could also occur in other weak processes.

In those days parity-selection rules were used extensively in all processes—strong, electromagnetic and weak. Because these selection rules gave results in excellent agreement with experiment, at first it seemed extremely difficult to proceed. Progress became possible only when one realized that, unlike the situation for the strong and electromagnetic interactions, not a single weak-interaction experiment designed up to that time had any relevance at all to the question of parity violation. In order to observe such violations directly, one must measure a physical pseudoscalar from an otherwise right-left symmetrical arrangement. Once this was understood, a large number of experiments were proposed ** which could serve to test the parity law.

$$\tau^+ \longrightarrow \theta^+ + 2\gamma \qquad \text{and} \qquad \tau^+ \longrightarrow 3\pi$$
$$\quad\ \ \llcorner\!\!\longrightarrow 2\pi$$

the attenuation of both 3π and 2π would be dominated by the much longer τ lifetime, and thereby give the apparent lifetime equality. This explanation was called the cascade mechanism (Phys.Rev. 100, 932 (1955)). The approximate mass equality could then be incorporated into a parity-exchange symmetry.

---

* L. Alvarez, in High Energy Nuclear Physics (Proceedings of the Sixth Annual Rochester Conference) (New York, Interscience Publishers, 1956).

** T. D. Lee and C. N. Yang, Phys.Rev. 104, 254 (1956).

With parity conservation in doubt, it was then natural to extend the questioning to other discrete symmetries. In the same year, 1956, L. M. Lederman* and his collaborators observed the long-lived neutral kaon (called θ_2^o then). This was the last great experiment that used the cloud chamber; at that time, it was presented as a proof of C conservation, based on the theoretical analysis by Gell-Mann and Pais**. A closer examination revealed that this discovery actually had no bearing on C conservation. In fact, so far as the weak interaction was concerned, just as in parity symmetry, there existed no experimental evidence for C, or T, or CP, or CT, or PT conservation. Experiments that could test all these symmetries were also proposed, with the entire K_L^o and K_S^o structure analyzed only under the assumption of CPT invariance.*** However, while the violations of P, C, CT and PT can be nearly maximal, as in the two-component theory of the neutrino[t], it was already realized then that the amount of CP and T violation in K decay must be rather small due to the unitarity limit.

Soon after these theoretical analyses were made, Wu, Ambler, Hayward, Hoppes and Hudson[tt] gave, in January 1957, the first demonstration of P and C asymmetries in β decay, which was quickly followed by the observation of the same asymmetry in π and μ

---

* K. Lande, E. T. Booth, J. Impeduglia, L. M. Lederman and W. Chinovsky, Phys. Rev. 103, 1901 (1956).

** M. Gell-Mann and A. Pais, Phys. Rev. 97, 1387 (1955).

*** T. D. Lee, R. Oehme and C. N. Yang, Phys. Rev. 106, 340 (1957).

t T. D. Lee and C. N. Yang, Phys. Rev. 105, 1671 (1957); L. Landau, Nucl. Phys. 3, 127 (1957); A. Salam, Nuovo Cimento 5, 299 (1957).

tt C. S. Wu, E. Ambler, R. W. Hayward, D. D. Hoppes and R. P. Hudson, Phys. Rev. 105, 1413 (1957).

decays by Garwin, Lederman and Weinrich * and by Friedman and Telegdi ** . The experimental discovery of CP violation came, however, much later, in 1964 by Christenson, Cronin, Fitch and Turlay ***, through the observation that the long-lived kaon can decay into $\pi^+\pi^-$.

We note that if CP were an exact symmetry, from K^o and $\bar{K}^o$ it would be possible to define two states, K_1^o and K_2^o, of opposite CP:

$$|K_1^o\rangle \equiv 2^{-\frac{1}{2}}(|K^o\rangle + |\bar{K}^o\rangle)$$

and

$$|K_2^o\rangle \equiv 2^{-\frac{1}{2}}(|K^o\rangle - |\bar{K}^o\rangle) \tag{15.22}$$

where

$$|\bar{K}^o\rangle \equiv CP\,|K^o\rangle \quad . \tag{15.23}$$

Consequently,

$$CP\,|K_1^o\rangle = |K_1^o\rangle$$

and

$$CP\,|K_2^o\rangle = -|K_2^o\rangle \quad . \tag{15.24}$$

Because $\pi^+\pi^-$ and $2\pi^o$ in a zero-angular-momentum state is of CP $=+1$, the assumption of CP invariance requires that only $K_1^o \to 2\pi$. Instead, K_2^o can decay into 3π, $\pi e \nu$, $\cdots$. Since the 2π phase-space factor is much larger than that for 3π, $\pi e\nu$, $\cdots$, it is expected that K_1^o should have a much shorter lifetime than K_2^o.

The discovery made by Christenson et al. showed that the long-lived kaon (hereafter referred to as K_L^o) can also decay into 2π; thereby they established CP nonconservation. The amplitude of CP violation observed can be characterized by the parameter

---

* R. L. Garwin, L. M. Lederman and M. Weinrich, Phys. Rev. 105, 1415 (1957).

** J. J. Friedman and V. L. Telegdi, Phys.Rev. 105, 1681 (1957).

*** J. H. Christenson, J. W. Cronin, V. L. Fitch and R. Turlay, Phys. Rev.Lett. 13, 138 (1964).

$$\eta_{+-} \equiv \frac{\text{amplitude } (K_L^o \to \pi^+ \pi^-)}{\text{amplitude } (K_S^o \to \pi^+ \pi^-)} \tag{15.25}$$

where K_S^o refers to the short-lived kaon. The actual value observed is

$$\eta_{+-} = (2.274 \pm .022) \times 10^{-3} \quad . \tag{15.26}$$

In spite of its smallness, this number signifies the violation of another cherished symmetry principle. It also shows that the physical long- and short-lived kaons cannot be the CP eigenstates K_1^o and K_2^o of (15.22). We now turn to the $K_L^o - K_S^o$ complex, following essentially the theoretical analysis given in the 1957 paper.

15.3 General Discussion of the Neutral Kaon Complex

1. Mass and decay matrices

Let us consider a coherent beam of K^o and $\overline{K}^o$, whose amplitude at its proper time τ is described by the wave function

$$\psi(\tau) = a_1(\tau) \,|\, K^o > + \; a_2(\tau) \,|\, \overline{K}^o >$$

or simply as

$$\psi(\tau) = \begin{pmatrix} a_1(\tau) \\ a_2(\tau) \end{pmatrix} \tag{15.27}$$

where the states K^o and $\overline{K}^o$ are related by

$$|\, \overline{K}^o > = e^{iS\theta} \; CPT \,|\, K^o > \tag{15.28}$$

with S as the strangeness quantum number and θ an arbitrary phase angle. Through our analysis we assume that the strong and the electromagnetic interactions, H_{st} and H_γ, do conserve CPT, leaving open the question of whether the weak interaction H_{wk} is CPT invariant. Here the state K^o is defined to be the eigenstate of $H_{st} + H_\gamma$ with strangeness $S = +1$. Since $H_{st} + H_\gamma$ commutes with CPT, the state $\bar{K}^o$, defined by (15.28), must also be an eigenstate of $H_{st} + H_\gamma$, but with $S = -1$. Thus we have

$$(H_{st} + H_\gamma) \mid K^o > = m_K \mid K^o >$$

and (15.29)

$$(H_{st} + H_\gamma) \mid \bar{K}^o > = m_K \mid \bar{K}^o > \quad .$$

The weak interaction H_{wk} connects K^o and $\bar{K}^o$ with other continuum states such as 2π, 3π, $\pi e \nu$, etc. This causes the various decay modes of K^o and $\bar{K}^o$, and removes their degeneracy. Just as in the standard line-width problem * of any atomic transition, we may eliminate the continuum amplitudes in the time-dependent Schrödinger equation. The time derivative $d\psi/d\tau$ of the wave function (15.27) can then be expressed in terms of $\psi(\tau)$ itself. By the superposition principle, such an expression clearly must be linear. The most general linear operator that connects $\psi(\tau)$ and $d\psi/d\tau$ is a 2×2 matrix, which can always be written as a sum of two Hermitian matrices. Hence we may write

$$i \frac{d}{d\tau} \psi(\tau) = (M - i\Gamma)\, \psi(\tau) \tag{15.30}$$

where Γ and M are both 2×2 Hermitian matrices,

$$\Gamma = \Gamma^\dagger \quad \text{and} \quad M = M^\dagger \quad . \tag{15.31}$$

In terms of H_{wk} and the total Hamiltonian $H = H_{st} + H_\gamma + H_{wk}$, we have the usual perturbation formula

$$(M - i\Gamma)_{\alpha\beta} = < \alpha \mid H \mid \beta > + \sum_n \frac{< \alpha \mid H_{wk} \mid n > < n \mid H_{wk} \mid \beta >}{m_K - (m_n - i\epsilon)}$$

$$+ O(H_{wk}^3) \tag{15.32}$$

---

* V. F. Weisskopf and E. P. Wigner, Z. Physik 63, 54 (1930), 65, 18 (1930).

where $\epsilon = 0+$, α and β can be K^o or $\overline{K}^o$ and $|n>$ any eigenstate of $H_{st} + H_\gamma$ with eigenvalue m_n, but with

$$n \neq K^o \text{ or } \overline{K}^o \quad . \tag{15.33}$$

Substituting the identity

$$\frac{1}{m_K - (m_n - i\epsilon)} = \mathcal{P} \frac{1}{m_K - m_n} - i\pi\delta(m_n - m_K) , \tag{15.34}$$

where $\mathcal{P}$ denotes the principal value, into (15.32), we have

$$M_{\alpha\beta} = <\alpha|H|\beta> + \sum_n \mathcal{P} \frac{<\alpha|H_{wk}|n><n|H_{wk}|\beta>}{m_K - m_n} \tag{15.35}$$

and

$$\Gamma_{\alpha\beta} = \pi \sum_n <\alpha|H_{wk}|n><n|H_{wk}|\beta> \delta(m_K - m_n) \quad . \tag{15.36}$$

In these expressions, as well as in the following, we shall neglect all $O(H_{wk}^3)$ terms.

<u>Exercise.</u> Show that for the integral

$$\int_{-a}^{a} \frac{f(z)}{z \pm i\epsilon} dz = \int_{-a}^{a} \mathcal{P} \frac{f(z)}{z} dz \mp i\pi f(0)$$

where ϵ is a positive infinitesimal quantity and the integration extends along the real axis of z from $-a$ to $a > 0$, with $f(z)$ a regular function on that interval.

<u>Theorem 1.</u> (i) Γ is a positive matrix, i.e.

$$\Gamma_{11} \geqslant 0 , \quad \Gamma_{22} \geqslant 0 \quad \text{and} \quad \det \Gamma \geqslant 0 \quad . \tag{15.37}$$

(ii) If CPT invariance holds, then independently of T symmetry

$$M_{11} = M_{22} \quad \text{and} \quad \Gamma_{11} = \Gamma_{22} \quad . \tag{15.38}$$

(iii) If T invariance holds, then independently of CPT symmetry

$$\frac{\Gamma_{12}^*}{\Gamma_{12}} = \frac{M_{12}^*}{M_{12}} . \tag{15.39}$$

Furthermore, by adjusting the arbitrary phase angle θ in (15.28), we can always choose Γ_{12} to be real; therefore if T invariance holds,

$$\Gamma = \Gamma^* = \widetilde{\Gamma} \quad \text{and} \quad M = M^* = \widetilde{M} . \tag{15.40}$$

In the above expressions, $\sim$ indicates transpose, the subscript 1 is for K^o and 2 for $\bar{K}^o$.

<u>Proof.</u> From (15.30) it follows that

$$-\frac{d}{d\tau} |\psi|^2 = 2\psi^\dagger \Gamma \psi ,$$

which must be positive since, as τ increases, the $K - \bar{K}$ beam can only decay. Hence,

$$\psi^\dagger \Gamma \psi \geqslant 0$$

for arbitrary ψ, and that gives (15.37).

To establish $M_{11} = M_{22}$, we note that from (15.35) - (15.36)

$$\Gamma_{11} = \pi \sum_n \delta(m_n - m_K) < K^o | H_{wk} | n >< n | H_{wk} | K^o > \tag{15.41}$$

and

$$M_{11} = < K^o | H | K^o > + \sum_n \mathcal{P} \frac{< K^o | H_{wk} | n >< n | H_{wk} | K^o >}{m_K - m_n} . \tag{15.42}$$

Because both are real, from (15.41) we have

$$\Gamma_{11} = \Gamma_{11}^* = \pi \sum_n \delta(m_n - m_K) < K^o | \mathcal{J}^{-1} \mathcal{J} H_{wk} \mathcal{J}^{-1} \mathcal{J} | n >$$
$$\cdot < n | \mathcal{J}^{-1} \mathcal{J} H_{wk} \mathcal{J}^{-1} \mathcal{J} | K^o > ,$$

where $\mathcal{J}$ = CPT . If one assumes CPT invariance, then $\mathcal{J} H_{wk} \mathcal{J}^{-1} = H_{wk}$. By using (15.28), we see that the above expression gives $\Gamma_{11} = \Gamma_{22}$. Likewise, from (15.42) and CPT invariance, it follows that $M_{11} = M_{22}$. From (15.36), we find

$$\Gamma_{12}^* = \pi \sum_n \delta(m_n - m_K) < K^o | T^{-1} T H_{wk} T^{-1} T | n > \cdot < n | T^{-1} T H_{wk} T^{-1} T | \overline{K}^o > . \tag{15.43}$$

If T invariance holds, then $T H_{wk} T^{-1} = H_{wk}$ and, since $T(H_{st} + H_\gamma) T^{-1} = H_{st} + H_\gamma$,

$$T | K^o > = e^{i\alpha} | K^o > ,$$

$$T | \overline{K}^o > = e^{i\bar{\alpha}} | \overline{K}^o > ,$$

where α and $\bar{\alpha}$ are real phase angles. Similarly, we can choose $T | n > \propto | n >$. Hence (15.43) can be written as

$$\Gamma_{12}^* = e^{i(\bar{\alpha} - \alpha)} \Gamma_{12} .$$

Likewise, from (15.35) and the assumption of T invariance, we have

$$M_{12}^* = e^{i(\bar{\alpha} - \alpha)} M_{12} ;$$

together they lead to (15.39). That completes the proof of the theorem.

It is clear that (15.38) depends only on CPT, not T, while (15.39) depends only on T, not CPT.

2. Eigenvalues

Because $M - i\Gamma$ is a 2×2 matrix, it has two eigenvectors, called $| L >$ and $| S >$, which satisfy

$$(M - i\Gamma) | j > = (m_j - \tfrac{1}{2} i \gamma_j) | j > , \tag{15.44}$$

where $j = L$ or S. The real parts of the eigenvalues are the masses, m_L and m_S; the imaginary parts are the half-widths, related to their lifetimes τ_j by $\tau_L = \gamma_L^{-1}$ and $\tau_S = \gamma_S^{-1}$. The subscript S denotes the one with shorter lifetime, and L the longer. From the Table of

Particle Properties, one sees that the experimental values are

$$\tau_S = (0.8923 \pm .0022) \times 10^{-10} \text{ sec.,}$$

$$\tau_L = (5.183 \pm .040) \times 10^{-8} \text{ sec.}$$

and

$$\Delta m \equiv m_L - m_S = (0.5349 \pm .0022) \times 10^{10}\ \hbar \text{ sec.}^{-1} \cong 3.5 \times 10^{-6} \text{ eV} \cong \tfrac{1}{2}\gamma_S \tag{15.45}$$

which when compared with the kaon mass m_K, gives

$$\frac{\Delta m}{m_K} \sim 10^{-14} \quad . \tag{15.46}$$

In a phenomenological analysis, the ratio of a typical matrix element of H_{wk} vs. that of H_{st} is $G m_p^{\,2}/4\pi$, where G is the Fermi constant given in (8.1). For a strangeness change $\Delta S = \pm 1$ matrix element, this should be further multiplied by the Cabibbo-angle factor, as we shall discuss in Chapter 21. This leads to

$$\left[(H_{wk})_{\Delta S = \pm 1} / H_{st} \right] \sim 10^{-7} \quad . \tag{15.47}$$

Comparing this with (15.46), we obtain

$$\frac{\Delta m}{m_K} \sim \left(\frac{H_{wk}}{H_{st}} \right)^2 \quad , \tag{15.48}$$

which means that, to the first order in H_{wk}, Δm should be 0. From (15.35) we see that the mass matrix, neglecting $O(H_{wk}^{\ 2})$, is simply $< \alpha \,|\, H \,|\, \beta >$; since it is proportional to the unit matrix we must have

$$< K^o \,|\, H \,|\, K^o > \;=\; < \overline{K}^o \,|\, H \,|\, \overline{K}^o >$$

and

$$< K^o \,|\, H \,|\, \overline{K}^o > \;=\; 0 \quad .$$

The first supports CPT invariance, at least with respect to the diagonal matrix elements, while the second implies the selection rule

$$\Delta S \neq \pm 2 \quad . \tag{15.49}$$

Because $H = H_{st} + H_\gamma + H_{wk}$, in which $H_{st} + H_\gamma$ is known to conserve strangeness S, (15.49) emphasizes the selection rule for H_{wk}.

Returning now to the eigenvectors $|j>$ given by (15.44), we may write, as in (15.27),

$$|j> = a_1(j)\,|K^o> + a_2(j)\,|\bar{K}^o> = \begin{pmatrix} a_1(j) \\ a_2(j) \end{pmatrix}. \qquad (15.50)$$

The matrix elements of M and Γ between these two eigenstates can be expressed in terms of $M_{\alpha\beta}$ and $\Gamma_{\alpha\beta}$, given by (15.35)-(15.36) as follows

$$<j\,|\,M\,|\,j'> = \sum_{\alpha\beta} a^*_\alpha(j)\,a_\beta(j')\,M_{\alpha\beta}$$

and (15.51)

$$<j\,|\,\Gamma\,|\,j'> = \sum_{\alpha\beta} a^*_\alpha(j)\,a_\beta(j')\,\Gamma_{\alpha\beta}\ .$$

Their diagonal matrix elements give m_j and $\frac{1}{2}\gamma_j$ respectively:

$$m_j = <j\,|\,M\,|\,j>$$

and (15.52)

$$\tfrac{1}{2}\gamma_j = <j\,|\,\Gamma\,|\,j>\ .$$

Because of (15.35)-(15.36) we have, neglecting $O(H^3_{wk})$,

$$m_j = <j|\,H\,|j> + \sum_n \mathcal{P}\,\frac{<j\,|H_{wk}|\,n><n\,|H_{wk}|\,j>}{m_K - m_n}$$

and (15.53)

$$\gamma_j = 2\pi \sum_n <j\,|H_{wk}|\,n><n\,|H_{wk}|\,j>\ \delta(m_n - m_K)$$

in which we sum over all eigenstates $|n>$ of $H_{st} + H_\gamma$, but with $n \neq K^o$ or $\bar{K}^o$, as in (15.33). It is sometimes convenient to group the states $|n>$ according to channels c:

$$c = \pi^+\pi^-\ ,\quad 2\pi^o\ ,\quad \pi^+ e^- \bar{\nu}_e\ ,\ \ldots$$

so that for different states n_c in each channel their matrix elements

$$<n_c\,|\,H_{wk}\,|\,j> = <c\,|\,H_{wk}\,|\,j> \qquad (15.54)$$

depend only on c , but not on individual n_c. Thus we may decompose the total width into a sum over different channels

$$\gamma_j = \sum_c \gamma_j(c) \tag{15.55}$$

where

$$\gamma_j(c) = 2\pi \rho_c \, | < c \, | \, H_{wk} \, | \, j > |^2 \tag{15.56}$$

and ρ_c is the density of states in channel c ,

$$\rho_c = \sum_{n_c} \delta(m_{n_c} - m_K) \quad . \tag{15.57}$$

To each fixed channel, a useful parameter is

$$\eta_c \equiv \frac{< c \, | \, H_{wk} \, | \, L >}{< c \, | \, H_{wk} \, | \, S >} \quad . \tag{15.58}$$

For $c = \pi^+ \pi^-$, η_c is the CP-violating parameter η_{+-} of (15.26). By using (15.56), we have

$$\eta_c = \left[\frac{\gamma_L(c)}{\gamma_S(c)} \right]^{\frac{1}{2}} e^{i \phi_c} \tag{15.59}$$

where ϕ_c is a phase angle. From (15.52) and (15.55) it follows that the diagonal matrix elements of Γ can be written as

$$< S \, | \, \Gamma \, | \, S > = \tfrac{1}{2} \sum_c \gamma_S(c)$$

and

$$< L \, | \, \Gamma \, | \, L > = \tfrac{1}{2} \sum_c \gamma_L(c) \quad . \tag{15.60}$$

Similarly from (15.36) and (15.51), we obtain

$$\begin{aligned} < S | \Gamma | L > &= \pi \sum_n < S | H_{wk} | n >< n | H_{wk} | L > \delta(m_n - m_K) \\ &= \pi \sum_c \rho_c < S | H_{wk} | c >< c | H_{wk} | L > \quad . \end{aligned} \tag{15.61}$$

As before, $\sum_c$ denotes the sum over all channels, and $\sum_n$ the sum over all eigenstates n of $H_{st} + H_\gamma$, but with $n \neq K^o$ or $\overline{K}^o$. By using the definitions $\gamma_j(c)$ and ϕ_c, we can rewrite (15.61) as

$$< S \mid \Gamma \mid L > \; = \; \tfrac{1}{2} \sum_c \, [\gamma_S(c)\, \gamma_L(c)]^{\frac{1}{2}} \; e^{i\,\phi_c} \; , \qquad (15.62)$$

a formula which will be of some use later on.

3. K_S^o and K_L^o

In this section we give the explicit form of the eigenvectors $| S >$ and $| L >$ in the notation of (15.50).

Theorem 2. (i) If CPT invariance holds, then independently of T invariance,

$$| S > \; = \; \begin{pmatrix} 1 + \epsilon \\ 1 - \epsilon \end{pmatrix} \frac{1}{\sqrt{2(1 + |\epsilon|^2)}}$$

and (15.63)

$$| L > \; = \; \begin{pmatrix} 1 + \epsilon \\ -(1 - \epsilon) \end{pmatrix} \frac{1}{\sqrt{2(1 + |\epsilon|^2)}}$$

where ϵ is a complex number.

(ii) If T invariance holds, then independently of CPT invariance we may write

$$| S > \; = \; \begin{pmatrix} 1 + \delta \\ 1 - \delta \end{pmatrix} \frac{1}{\sqrt{2(1 + |\delta|^2)}}$$

and (15.64)

$$| L > \; = \; \begin{pmatrix} 1 - \delta \\ -(1 + \delta) \end{pmatrix} \frac{1}{\sqrt{2(1 + |\delta|^2)}}$$

where δ is also a complex number.

(iii) Let us define

$$\xi \equiv \; < S \mid L > \; , \qquad (15.65)$$

then

$$\xi \text{ is } \begin{cases} \text{real} & \text{if CPT holds,} \\ \text{imaginary} & \text{if T holds.} \end{cases} \qquad (15.66)$$

Proof. In (15.44) the 2×2 matrix $M - i\Gamma$ consists of four complex matrix elements. Hence, we can always express it in terms of a complex number D and three complex components E_1, E_2, E_3 of a vector $\vec{E}$:

$$M - i\Gamma = D + \vec{E} \cdot \vec{\tau} \tag{15.67}$$

where $\vec{\tau}$ are the Pauli matrices given by (3.1). To prove (i) we start from (15.38) of Theorem 1, which states that CPT invariance implies $E_3 = 0$. Let $E_1 = E \cos\alpha$ and $E_2 = E \sin\alpha$. Equation (15.67) becomes then

$$M - i\Gamma = D + E \begin{pmatrix} 0 & e^{-i\alpha} \\ e^{i\alpha} & 0 \end{pmatrix} \tag{15.68}$$

in which D, E and α are all complex. We see readily that the eigenvectors are proportional to

$$\begin{pmatrix} 1 \\ e^{i\alpha} \end{pmatrix} \quad \text{and} \quad \begin{pmatrix} 1 \\ -e^{i\alpha} \end{pmatrix}, \tag{15.69}$$

and the corresponding eigenvalues are

$$D + E \quad \text{and} \quad D - E . \tag{15.70}$$

Introducing ϵ by

$$\frac{1-\epsilon}{1+\epsilon} \equiv e^{i\alpha}, \tag{15.71}$$

we can rewrite (15.69) as

$$\begin{pmatrix} 1+\epsilon \\ 1-\epsilon \end{pmatrix} \quad \text{and} \quad \begin{pmatrix} 1+\epsilon \\ -(1-\epsilon) \end{pmatrix},$$

which when multiplied by the normalization constant gives (15.63).

To establish (ii), it is convenient to denote $|j\rangle$ as a 2×1 column matrix ψ_j and $m_j - i\frac{1}{2}\gamma_j = \lambda_j$, where j can be L or S as before. Equation (15.44) becomes

$$(M - i\Gamma)\, \psi_S = \lambda_S\, \psi_S$$

and (15.72)

$$(M - i\Gamma)\,\psi_L = \lambda_L \psi_L \quad .$$

From (15.40) of Theorem 1, we know that T invariance leads to $\widetilde{\Gamma} = \Gamma$, $\widetilde{M} = M$, and therefore

$$\widetilde{\psi}_L (M - i\Gamma)\,\psi_S = \widetilde{\psi}_S (M - i\Gamma)\,\psi_L \quad .$$

By substituting (15.72) into the above, we obtain

$$\lambda_S \widetilde{\psi}_L \psi_S = \lambda_L \widetilde{\psi}_S \psi_L \quad .$$

Since $\widetilde{\psi}_S \psi_L = \widetilde{\psi}_L \psi_S$ and $\lambda_S \neq \lambda_L$, we see that

$$\widetilde{\psi}_L \psi_S = 0 \quad . \qquad (15.73)$$

The 2×1 column vector ψ_S can always be written as

$$\psi_S \propto \begin{pmatrix} 1 + \delta \\ 1 - \delta \end{pmatrix} \quad .$$

Because of (15.73), we must have

$$\psi_L \propto \begin{pmatrix} 1 - \delta \\ -(1 + \delta) \end{pmatrix} \quad .$$

These two expressions, when multiplied by the normalization constant, give (15.64).

To establish (iii) we note that, under the assumption of CPT invariance, (15.63) implies

$$\xi = \langle S \,|\, L \rangle = [\,2(1 + |\,\epsilon\,|^2)\,]^{-1} [\,(1+\epsilon)^* (1+\epsilon) - (1-\epsilon)^* (1-\epsilon)\,]$$

$$= (1 + |\,\epsilon\,|^2)^{-1}\,(\epsilon + \epsilon^*) \quad , \qquad (15.74)$$

which is real. Similarly, if T invariance holds, from (15.64) it follows that

$$\xi = [\,2(1 + |\,\delta\,|^2)\,]^{-1} [\,(1+\delta)^* (1-\delta) - (1-\delta)^* (1+\delta)\,]$$

$$= (1 + |\,\delta\,|^2)^{-1}\,(\delta^* - \delta) \qquad (15.75)$$

which is imaginary. Hence we establish (15.66) and the proof of Theorem 2 is completed.

Remarks. If both CPT and T were exact symmetries, then from Theorem 2 we see that ϵ and δ should be 0, and $|S>$ and $|L>$ reduce to the CP eigenstates $|K_1^o>$ and $|K_2^o>$ given by (15.22). Since CP conservation requires $\xi = 0$, ξ is a useful parameter for measuring CP violation.

Theorem 3.

$$|\xi|^2 \leqq \frac{[\sum_c \sqrt{\gamma_S(c)\,\gamma_L(c)}\,]^2}{(m_S - m_L)^2 + \frac{1}{4}(\gamma_S + \gamma_L)^2} \tag{15.76}$$

$$\leqq \frac{\gamma_S\,\gamma_L}{(m_S - m_L)^2 + \frac{1}{4}(\gamma_S + \gamma_L)^2} \,. \tag{15.77}$$

Proof. Multiplying (15.44) by i, we have

$$(\Gamma + iM)\,|j> \;=\; (\tfrac{1}{2}\gamma_j + i m_j)\,|j> \,, \tag{15.78}$$

from which it follows that

$$<S|\,\Gamma + iM\,|L> \;=\; (\tfrac{1}{2}\gamma_L + i m_L)\,<S|L> \tag{15.79}$$

and

$$<L|\,\Gamma + iM\,|S> \;=\; (\tfrac{1}{2}\gamma_S + i m_S)\,<L|S> \,.$$

The Hermitian conjugate of the latter is

$$<S|\,\Gamma - iM\,|L> \;=\; (\tfrac{1}{2}\gamma_S - i m_S)\,<S|L> \,. \tag{15.80}$$

Adding (15.79) and (15.80) and using $\xi = <S|L>$, we obtain

$$2<S|\,\Gamma\,|L> \;=\; [\tfrac{1}{2}(\gamma_L + \gamma_S) + i(m_L - m_S)]\,\xi \,, \tag{15.81}$$

which, together with (15.62), leads to

$$\xi = \frac{\sum_c [\gamma_S(c)\,\gamma_L(c)]^{\frac{1}{2}}\, e^{i\phi_c}}{\frac{1}{2}(\gamma_L + \gamma_S) + i(m_L - m_S)} \tag{15.82}$$

and, therefore, inequality (15.76). Because $\gamma_j = \sum_c \gamma_j(c)$, inequality (15.77) also follows, which completes the proof of Theorem 3.

Remarks. By using the various experimental values of $\gamma_S(c)$ and $\gamma_L(c)$ listed in the Table of Particle Properties we see that (15.76) implies $|\xi| \lesssim 6 \times 10^{-3}$, and (15.77) implies $|\xi| \lesssim 6 \times 10^{-2}$. The proof of Theorem 3 is independent of symmetry assumptions of either CPT or T; it uses only the unitarity property of quantum mechanics. Therefore these inequalities are referred to in the literature as unitarity limits. Their existence was already realized in 1956 when the validity of various discrete space-time symmetries was first questioned. Because of the unitarity limit, it was anticipated then that unlike P and C asymmetries, the experimental proof of CP violation would be much more difficult.

15.4 Interference Phenomena

As before, τ denotes the time in the rest frame of the K meson. When $\tau = 0$, let the state be

$$b_S \mid S > + \; b_L \mid L > \tag{15.83}$$

where b_S and b_L are constant parameters, determined by the production mechanism. For example, in the reaction $\pi^- + p \rightarrow \Lambda^o + K^o$, the initial state corresponds to

$$b_S \cong b_L \cong 2^{-\frac{1}{2}} \; .$$

From (15.30) and (15.44), we know that at a later time τ the state (15.83) is

$$|\tau> = b_S \, e^{-(i m_S + \frac{1}{2}\gamma_S)\tau} \mid S > + \, b_L \, e^{-(i m_L + \frac{1}{2}\gamma_L)\tau} \mid L > . \tag{15.84}$$

Hence at time τ the decay amplitude of this kaon state into channel c

is
$$\langle c \,|\, H_{wk} \,|\, \tau \rangle = b_S\, e^{-(im_S + \frac{1}{2}\gamma_S)\tau} \langle c \,|\, H_{wk} \,|\, S \rangle + b_L\, e^{-(im_L + \frac{1}{2}\gamma_L)\tau} \langle c \,|\, H_{wk} \,|\, L \rangle \; .$$

The corresponding decay rate is

$$2\pi \rho_c \,|\langle c \,|\, H_{wk} \,|\, \tau \rangle|^2 = |b_S|^2 \gamma_S(c)\; e^{-\gamma_S \tau} + |b_L|^2 \gamma_L(c)\; e^{-\gamma_L \tau} + e^{-\frac{1}{2}(\gamma_S + \gamma_L)\tau} \sqrt{\gamma_S(c)\, \gamma_L(c)}\; [b_S^* b_L\, e^{i(m_S - m_L)\tau + i\phi_c} + \text{h.c.}] \; , \tag{15.85}$$

in which ρ_c, $\gamma_S(c)$, $\gamma_L(c)$ and ϕ_c are given by (15.56) - (15.57) and (15.59). Because $\gamma_L \ll \gamma_S$ and $\Delta m = m_L - m_S \cong \frac{1}{2}\gamma_S$, as shown by (15.45), the three terms on the righthand side of (15.85) have very different time - dependences and can be measured separately. The levels $|\, L \rangle$ and $|\, S \rangle$ act like two tuning forks with frequencies differing only slightly, which are coupled into a rich variety of "sound cavities" in analogy with the production and decay channels of the kaon. These have led to some of the most refined experiments in particle physics. In turn, the quantitative verification of (15.85) also gives strong support to the linear superposition principle of quantum mechanics.

By measuring the exponents of (15.85), we can determine γ_S, γ_L and Δm ; by choosing different decay channels c , we can measure various $\gamma_S(c)$, $\gamma_L(c)$ and the phase ϕ_c. From (15.59), η_c can then be determined. For $c = \pi^+\pi^-$ and $2\pi^o$, the corresponding η_c are $\eta_{+-} = |\eta_{+-}|\; e^{i\phi_{+-}}$ and $\eta_{oo} = |\eta_{oo}|\; e^{i\phi_{oo}}$. From the Table of Particle Properties we see that

$$|\eta_{+-}| = (2.274 \pm .022) \times 10^{-3}, \quad \phi_{+-} = (44.6 \pm 1.2)^o ,$$
$$|\eta_{oo}| = (2.33 \pm .08) \times 10^{-3} \text{ and } \phi_{oo} = (54 \pm 5)^o \; . \tag{15.86}$$

Because η_{+-} and η_{oo} are $\neq 0$, K_L^o can decay into 2π. Since one of the dominant decay modes of K_L^o is known to be 3π, and as determined from the Dalitz plot the final pions are predominantly in the s-state, we have for either $\pi^+\pi^-\pi^o$ or $3\pi^o$,

$$(CP)_{3\pi} = (CP)_{2\pi}\,(CP)_{\pi^o} .$$

Because the CP of a single π^o is -1, this implies $(CP)_{3\pi} \neq (CP)_{2\pi}$. However, K_L^o has a definite lifetime and a definite mass but no degeneracy (note that $m_S \neq m_L$). The fact that its final state is not of a definite CP proves CP violation. The amplitude of the violation can be characterized by η_{+-} and η_{oo}; both are very small, only $\sim 10^{-3}$.

15.5 T Violation

If we assume CPT invariance, then CP nonconservation implies T violation. In this section we would like to show that the same conclusion can be established without assuming CPT invariance.

By using (15.59) and (15.82), we have

$$[\tfrac{1}{2}(\gamma_L + \gamma_S) + i\,(m_L - m_S)]\;\xi = \sum_c \gamma_S(c)\;\eta_c$$

$$= \gamma_S(\pi^+\pi^-)\;\eta_{+-} + \gamma_S(\pi^o\pi^o)\;\eta_{oo} + \cdots . \quad (15.87)$$

From the Table of Particle Properties we see that the dominant decay modes of K_S^o are $\pi^+\pi^-$ and $\pi^o\pi^o$. The only other decay mode that has been observed for K_S^o is $\pi^+\pi^-\gamma$, but with a branching ratio $(1.85 \pm 0.10) \times 10^{-3}$; the corresponding decay $K_L^o \to \pi^+\pi^-\gamma$ has a branching ratio $(6.0 \pm 2.0) \times 10^{-5}$. All other decay modes of K_S^o, such as 3π, $\gamma\gamma$, etc. have not been seen so far and have very small upper bounds on their branching ratios. Consequently the righthand side of (15.87) is dominated by its first two terms, both of which, according to (15.86), have phase angles $\cong 45\text{-}55^\circ$. On the other hand, because $(m_L - m_S) \cong \frac{1}{2}\gamma_S >> \frac{1}{2}\gamma_L$, the bracketed expression on the lefthand side of (15.87) also has a phase $\cong 45^\circ$. Hence we conclude that ξ is real, at least to a good approximation. However, if T invariance were correct, then according to (15.66) of Theorem 2 (iii), independently of CPT invariance ξ should be imaginary; that contradicts our conclusion, and therefore establishes T violation.

We note that, in accordance with the top line in (15.66), ξ real gives additional support for CPT invariance.

15.6 Analysis with the Assumption of CPT Invariance

1. State vectors

Assuming CPT symmetry, the state vectors $|S>$ and $|L>$ are given by (15.63). Their scalar product is ξ, which according to (15.74) is

$$\xi = <S|L> = \frac{2}{1+|\epsilon|^2}\,\mathrm{Re}\,\epsilon \;. \tag{15.88}$$

As noted before, if CP were conserved, then $\xi = <S|L> = 0$, and that would be consistent with ϵ purely imaginary, though not necessarily 0. The nonconservation of CP makes it possible for ϵ to acquire a real part. From (15.68) and (15.71), we have

$$\left(\frac{1-\epsilon}{1+\epsilon}\right)^2 = e^{2i\alpha} = \frac{(M-i\Gamma)_{21}}{(M-i\Gamma)_{12}} \;. \tag{15.89}$$

By using (15.35)-(15.36) we see that α is directly related to the arbitrary phase angle θ in (15.28), which may be written as

$$|\bar{K}^o> = e^{iS\theta}\,\mathcal{G}\,|K^o> \tag{15.90}$$

where $\mathcal{J}$ = CPT. Due to the antiunitary nature of $\mathcal{J}$, (15.90) leads to
$$| K^o > = \mathcal{J} e^{i\theta} | \overline{K}^o > = e^{-i\theta} \mathcal{J} | \overline{K}^o > .$$

We shall now discuss a convenient convention for fixing the angle θ. From the Table of Particle Properties we note that the branching ratios of K_S^o to $\pi^+\pi^-$ and $\pi^o\pi^o$ are very close to $\frac{2}{3}$ and $\frac{1}{3}$. As mentioned on p. 245, this implies that to a good approximation the final isospin in the decay $K_S^o \to 2\pi$ is

$$I_{2\pi} = 0 . \tag{15.91}$$

The angle θ will be determined by requiring* the ratio

$$R \equiv \frac{< (2\pi)_{I=0} | H_{wk} | K^o >}{< (2\pi)_{I=0} | H_{wk} | \overline{K}^o >} \quad \text{real and positive.} \tag{15.92}$$

From (15.90) and under the assumption of CPT invariance, the complex conjugate of R is

$$R^* = \frac{< (2\pi)_{I=0} | \mathcal{J}^{-1} \mathcal{J} H_{wk} \mathcal{J}^{-1} \mathcal{J} | K^o >}{< (2\pi)_{I=0} | \mathcal{J}^{-1} \mathcal{J} H_{wk} \mathcal{J}^{-1} \mathcal{J} | \overline{K}^o >}$$

$$= \frac{< (2\pi)_{I=0} | H_{wk} e^{i\theta} | \overline{K}^o >}{< (2\pi)_{I=0} | H_{wk} e^{i\theta} | K^o >} = R^{-1} .$$

The requirement (15.92) gives then

$$\frac{< (2\pi)_{I=0} | H_{wk} | K^o >}{< (2\pi)_{I=0} | H_{wk} | \overline{K}^o >} = 1 . \tag{15.93}$$

What is the advantage of this particular way of fixing θ ? As noted

---

* T. T. Wu and C. N. Yang, Phys. Rev. Lett. 13, 180 (1964).

before, if the angle θ in (15.90) is arbitrary, so would be the real part of α in (15.89); this is why even if CP were conserved, ϵ need not be 0 provided it is purely imaginary. However, with the convention (15.92), we have from (15.63) and (15.93)

$$<(2\pi)_{I=0} | H_{wk} | L> = \frac{2\epsilon}{\sqrt{2(1+|\epsilon|^2)}} <(2\pi)_{I=0} | H_{wk} | K^o> . \tag{15.94}$$

Thus, the present choice of θ implies that if CP were conserved, since K_L^o could not decay into 2π, ϵ should be 0; consequently ϵ also becomes a parameter characterizing CP violation.

Theorem 4. The phase angle of ϵ is $\cong 45^o$ or 225°.

Proof. For convenience let us first assume ϵ to be of a magnitude similar to other CP-violating parameters η_{+-}, η_{oo} and ξ; all these numbers are very small, $\sim 10^{-3}$. [This assumption will be removed later on.] From (15.89) it follows that

$$\frac{1-\epsilon}{1+\epsilon} = e^{i\alpha}$$

which implies that for $|\epsilon| \ll 1$, α should also be small. The above expression reduces to

$$2\epsilon \cong -i\alpha \quad . \tag{15.95}$$

By using (15.68), we have

$$(M-i\Gamma)_{12} - (M-i\Gamma)_{21} = -i\,2E\sin\alpha \cong -i\,2E\alpha \quad . \tag{15.96}$$

According to (15.70), we know the eigenvalues of $M-i\Gamma$ are $D \pm E$; their difference is $\pm 2E$, which together with (15.44) gives

$$\pm 2E = m_L - m_S + \tfrac{1}{2} i (\gamma_S - \gamma_L) \quad .$$

Substituting this expression into (15.96) and combining it with (15.95), we obtain

$$\pm\,\epsilon = \frac{\Gamma_{12} - \Gamma_{21} + i(M_{12} - M_{21})}{\gamma_S - \gamma_L - 2i(m_L - m_S)} \quad . \tag{15.97}$$

Now, M is Hermitian and therefore

$$i(M_{12} - M_{21}) \quad \text{is real.} \tag{15.98}$$

From (15.36) and (15.57) it follows that

$$\Gamma_{12} = \pi \sum_c \rho_c < K^o | H_{wk} | c >< c | H_{wk} | \bar{K}^o > \quad . \tag{15.99}$$

In this sum the most important channel is $c = (2\pi)_{I=0}$. From the Table of Particle Properties at the back of the book we see that the sum over all other channels, $c \neq (2\pi)_{I=0}$,

$$\pi \sum_c \rho_c |< c | H_{wk} | K^o >|^2 \lesssim 10^{-4} |\gamma_S - \gamma_L - 2i(m_L - m_S)| \; .$$

Hence, we need only consider the single channel $c = (2\pi)_{I=0}$ in (15.99); due to (15.93) it follows then, to a good approximation,

$$\Gamma_{12} \cong \Gamma_{21} \quad \text{real.} \tag{15.100}$$

On the righthand side of (15.97) the numerator is real because of (15.98) and (15.100); on account of $2(m_L - m_S) \cong \gamma_S >> \gamma_L$, we see that

$$\text{phase of } \epsilon \cong 45^o \text{ or } 225^o \quad . \tag{15.101}$$

By substituting this result into (15.88), we obtain

$$|\xi| \cong \frac{\sqrt{2}}{1 + |\epsilon|^2} |\epsilon| \quad .$$

From Theorem 3 on pages 358-9, we know that ξ is a small parameter which has an upper bound, due to unitarity, $\leqslant 6 \times 10^{-3}$. Consequently,

$$|\epsilon| < 4.3 \times 10^{-3} \quad . \tag{15.102}$$

This then substantiates our supposition, used at the beginning of the proof, that ϵ is a small number. The proof of Theorem 4 is then completed.

Because $|\epsilon| \ll 1$, we may neglect $O(\epsilon^2)$ terms. Hence (15.63) and (15.88) can be written as

$$|S> \cong \frac{1}{\sqrt{2}} \begin{pmatrix} 1+\epsilon \\ 1-\epsilon \end{pmatrix} \quad , \quad |L> \cong \frac{1}{\sqrt{2}} \begin{pmatrix} 1+\epsilon \\ -(1-\epsilon) \end{pmatrix} \tag{15.103}$$

and

$$\xi = <S|L> \cong 2 \,\mathrm{Re}\, \epsilon \quad . \tag{15.104}$$

In the next section we shall discuss how to determine ϵ and ξ experimentally.

2. K_S^o or $K_L^o \rightarrow \pi^{\mp} + \ell^{\pm} + \nu_\ell$ or $\bar{\nu}_\ell$

It is convenient to adopt the rest frame of the neutral kaon. Let $\vec{p}_\pi$, $\vec{p}_\ell$ and $\vec{p}_\nu$ be the momenta of $\pi^{\mp}$, $\ell^{\pm}$ and ν_ℓ (or $\bar{\nu}_\ell$) in that frame, where $\ell = e$ or μ. Consequently $\vec{p}_\pi + \vec{p}_\ell + \vec{p}_\nu = 0$. We define, at fixed momenta, the first-order weak-interaction decay amplitude of $K^o \rightarrow \pi^- + \ell_s^+ + \nu_\ell$ to be f, with the subscript $s = \pm\frac{1}{2}$ denoting the lepton helicity:

$$f \equiv <\pi^- \ell_s^+ \nu_\ell | H_{wk} | K^o> \quad . \tag{15.105}$$

Under $\mathcal{J} = CPT$, $\pi^{\pm} \rightarrow \pi^{\mp}$, $\ell_s^{\pm} \rightarrow \ell_{-s}^{\mp}$, $\nu_\ell\ (\bar{\nu}_\ell) \rightarrow \bar{\nu}_\ell\ (\nu_\ell)$ but the momenta of all particles are unchanged; i.e.,

$$\mathcal{J}|\pi^- \ell_s^+ \nu_\ell> \ \propto\ |\pi^+ \ell_{-s}^- \bar{\nu}_\ell> \quad .$$

Since the relative phase between $|\pi^- \ell_s^+ \nu_\ell>$ and $|\pi^+ \ell_{-s}^- \bar{\nu}_\ell>$ is not yet fixed, we may write the above expression as

$$\mathcal{J}|\pi^- \ell_s^+ \nu_\ell> \ =\ e^{i\theta} |\pi^+ \ell_{-s}^- \bar{\nu}_\ell> \tag{15.106}$$

where θ is chosen to be the same angle used in (15.90). Consequently, from (15.105) and the CPT invariance assumption, we have

$$f^* = \langle \pi^+ \ell^-_{-s} \bar{\nu}_\ell \mid H_{wk} \mid \overline{K}^o \rangle \quad . \tag{15.107}$$

Likewise, we may define the corresponding decay amplitude of $K^o \rightarrow \pi^+ + \ell^-_{-s} + \bar{\nu}_\ell$ to be g :

$$g = \langle \pi^+ \ell^-_{-s} \bar{\nu}_\ell \mid H_{wk} \mid K^o \rangle \quad . \tag{15.108}$$

By using (15.90) and (15.106), we obtain

$$g^* = \langle \pi^- \ell^+_s \nu_\ell \mid H_{wk} \mid \overline{K}^o \rangle \quad . \tag{15.109}$$

In these reactions we may define Q and S to be the charge and strangeness of only the hadrons, and

$$\begin{aligned} \Delta Q &= Q_{final} - Q_{initial} \ , \\ \Delta S &= S_{final} - S_{initial} \end{aligned} \tag{15.110}$$

where the subscripts refer to the final and initial states. Hence f, f*, g and g* represent the amplitudes of reactions that satisfy

$$\begin{aligned} f &: \ \Delta S = \Delta Q = -1 \ , \\ f^* &: \ \Delta S = \Delta Q = 1 \ , \\ g &: \ \Delta S = -\Delta Q = -1 \\ \text{and} \quad g^* &: \ \Delta S = -\Delta Q = 1 \ . \end{aligned} \tag{15.111}$$

As will be discussed in Chapter 21, there are good reasons to believe that all first-order weak processes should satisfy

$$\Delta S = \Delta Q \quad . \tag{15.112}$$

At present we shall, however, proceed with our analysis without this requirement. Define

$$x = g^*/f \quad . \tag{15.113}$$

Since f and f* satisfy the selection rule $\Delta S = \Delta Q$, but g and g* do not, the parameter x is useful for the experimental determination of how effective this selection rule really is.

The parameters x and ξ can be measured through the interference phenomena discussed in Section 15.4. Let us assume a $K-\bar{K}$ beam whose state vector $|\tau>$ is given by (15.84) at its proper time τ. For definiteness, we may consider the channels $c = \pi^- \ell_s^+ \nu_\ell$ and $\pi^+ \ell_{-s}^- \bar{\nu}_\ell$. From (15.85), we see that the decay rate of $|\tau>$ into these channels can be decomposed into three different time-dependent terms. Hence

$$\gamma_S(c) = 2\pi \rho_c |<c| H_{wk} |S>|^2 ,$$

$$\gamma_L(c) = 2\pi \rho_c |<c| H_{wk} |L>|^2$$

and the coefficient of the interference term

$$\sqrt{\gamma_S(c)\,\gamma_L(c)}\; e^{i\phi_c} = 2\pi \rho_c <c| H_{wk} |S>^* <c| H_{wk} |L> \tag{15.114}$$

can all be measured experimentally. By using (15.103), (15.105) and (15.107) - (15.109), we find

$$<\pi^- \ell_s^+ \nu_\ell | H_{wk} | L> = 1/\sqrt{2}\,[(1+\epsilon) f - (1-\epsilon) g^*] ,$$

$$<\pi^+ \ell_{-s}^- \bar{\nu}_\ell | H_{wk} | L> = 1/\sqrt{2}\,[(1+\epsilon) g - (1-\epsilon) f^*] ,$$

$$<\pi^- \ell_s^+ \nu_\ell | H_{wk} | S> = 1/\sqrt{2}\,[(1+\epsilon) f + (1-\epsilon) g^*]$$

and

$$<\pi^+ \ell_{-s}^- \bar{\nu}_\ell | H_{wk} | S> = 1/\sqrt{2}\,[(1+\epsilon) g + (1-\epsilon) f^*] . \tag{15.115}$$

These expressions together with (15.113) lead to

$$\frac{\gamma_L(\pi^- \ell_s^+ \nu_\ell)}{\gamma_L(\pi^+ \ell_{-s}^- \bar{\nu}_\ell)} = \left| \frac{(1+\epsilon) f - (1-\epsilon) g^*}{(1+\epsilon) g - (1-\epsilon) f^*} \right|^2$$

$$= \left| \frac{1+\epsilon - (1-\epsilon) x}{(1+\epsilon) x^* - (1-\epsilon)} \right|^2 = \left| \frac{1-x+\epsilon(1+x)}{x^*-1+\epsilon(x^*+1)} \right|^2$$

$$= \left| \frac{1-x}{x^*-1} \right|^2 \left[1 + (\epsilon + \epsilon^*) \left(\frac{1+x}{1-x} + \frac{1+x^*}{1-x^*} \right) + O(\epsilon^2) \right]$$

$$= 1 + 4 \,\mathrm{Re}\, \epsilon \, \frac{1 - |x|^2}{|1-x|^2} + O(\epsilon^2) \quad . \qquad (15.116)$$

In a similar way we can arrive at

$$\frac{\gamma_S(\pi^- \ell_s^+ \nu_\ell)}{\gamma_S(\pi^+ \ell_{-s}^- \bar{\nu}_\ell)} = 1 + 4 \,\mathrm{Re}\, \epsilon \, \frac{1-|x|^2}{|1+x|^2} \quad . \qquad (15.117)$$

From (15.114), we see that through the interference term it is possible to measure

$$< \pi^- \ell_s^+ \nu_\ell | H_{wk} | S >^* < \pi^- \ell_s^+ \nu_\ell | H_{wk} | L >$$

$$\propto \left(1 + \frac{1-\epsilon}{1+\epsilon} x\right)^* \left(1 - \frac{1-\epsilon}{1+\epsilon} x\right)$$

$$= [1 + x^* - x - |x|^2 + O(\epsilon)] \qquad (15.118)$$

and

$$< \pi^+ \ell_{-s}^- \bar{\nu}_\ell | H_{wk} | S >^* < \pi^+ \ell_{-s}^- \bar{\nu}_\ell | H_{wk} | L >$$

$$\propto \left(1 + \frac{1+\epsilon}{1-\epsilon} x^*\right)^* \left(1 - \frac{1+\epsilon}{1-\epsilon} x^*\right)$$

$$= [1 + x - x^* - |x|^2 + O(\epsilon)] \; . \qquad (15.119)$$

Therefore (15.118) and (15.119) depend sensitively on the imaginary part of x, while (15.116)-(15.117) can be used to determine the real parts of ϵ and x. The experimental results, as listed in the Table of Particle Properties, are

$$\mathrm{Re}\, x = 0.009 \pm 0.020 \quad , \quad \mathrm{Im}\, x = -0.004 \pm 0.026 \quad , \qquad (15.120)$$

$$\mathrm{Re}\, \epsilon = (1.621 \pm .088) \times 10^{-3} \qquad (15.121)$$

and therefore

$$\xi = 2 \,\mathrm{Re}\, \epsilon = (3.242 \pm .176) \times 10^{-3} \quad . \qquad (15.122)$$

From (15.120) we see that x is consistent with 0, which supports the $\Delta S = \Delta Q$ rule. By setting $x = 0$, we note that the ratios (15.116) and (15.117) become independent of the lepton helicity; hence we have

$$\frac{\text{rate}\ (K_L^o \to \pi^- \ell^+ \nu_\ell)}{\text{rate}\ (K_L^o \to \pi^+ \ell^- \bar{\nu}_\ell)} \cong 1 + 4\ \text{Re}\ \epsilon$$

and (15.123)

$$\frac{\text{rate}\ (K_S^o \to \pi^- \ell^+ \nu_\ell)}{\text{rate}\ (K_S^o \to \pi^+ \ell^- \bar{\nu}_\ell)} \cong 1 + 4\ \text{Re}\ \epsilon \quad .$$

As already emphasized in Chapter 9, Eq. (9.3), the fact that these two ratios are different from 1 gives a most direct evidence for the violation of C and CP. By comparing (15.86) with (15.101) and (15.121) we find $\eta_{+-} \cong \eta_{oo} \cong \epsilon$. If it turns out that these three parameters are in fact identical,

$$\eta_{+-} = \eta_{oo} = \epsilon \quad , \tag{15.124}$$

then the above relations together with $\xi = 2\ \text{Re}\ \epsilon$ indicate that all these CP-nonconserving parameters can be represented by one number, ϵ.

Since 1964 many high-energy physics laboratories throughout the world have devoted substantial effort to experiments related to CP nonconservation. It is rather remarkable that the final result of all these investigations may be summarized by this single small parameter, ϵ.

15.7 Complementarity of Symmetry Violations

Consider the case of parity nonconservation. We all know that there exist many experimental proofs which establish right - left asymmetry in the laws of physics. In our discussions so far, we have regarded the strong and electromagnetic interactions as P-conserving, but the weak as P-violating. In this section, we shall raise the question of whether such an attribution is really fundamental, or merely a matter of convenience or convention.

To begin with, let us analyze a logical point that has been glossed over in most literature on symmetry violation. It is not uncommon to find the discussion of parity nonconservation begun by first introducing a parity operator P and a Hamiltonian H, then verifying their commutator

$$[H, P] \neq 0 \quad , \tag{15.125}$$

and from that reaching the conclusion that parity is not conserved. It is important to realize that there is a self-inconsistency in this process. To see this, let us recall that in the quantum field theory P is a unitary operator in the Hilbert space. By definition, this operator P should represent the space-inversion transformation in the coordinate space,

$$P : \vec{r} \rightarrow -\vec{r} \quad , \qquad t \rightarrow t \quad .$$

According to Heisenberg's equation (1.9), the time-translation operator is

$$e^{-iH\tau} : \vec{r} \rightarrow \vec{r} \quad , \qquad t \rightarrow t + \tau \quad .$$

Mathematically we may define the Poincaré group as consisting of all space-time translations, Lorentz transformations, and space inversion and time reversal. These transformations are purely geometrical; their existence is independent of dynamics. It is obvious that in the Poincaré group the space-inversion element, $\vec{r} \rightarrow -\vec{r}$, commutes with the time-translation element, $t \rightarrow t + \tau$. On the other hand, if the corresponding operators in the Hilbert space satisfy $[H, P] \neq 0$, that would mean the time-translation operator $e^{-iH\tau}$ and the alleged space-inversion operator P fail to obey the multiplication law of the coordinate transformations that they are supposed to represent. Thus, the operator P cannot be the Hilbert-space representation of space inversion in the first place. The fact that parity is not conserved implies the impossibility of finding a unitary operator P in the Hilbert space which can represent the geometrical space-inversion transformation; i.e., the operator P is not defined.

It is, however, possible to give a definition of P by making an approximation on H. We may replace the total Hamiltonian by, say,

$$H \cong H_{st} + H_{\gamma} \tag{15.126}$$

or

$$H \cong H_{wk} \quad , \tag{15.127}$$

so that within such an approximation, a parity operator P may be defined and it commutes with the approximate Hamiltonian. Different choices of dynamical approximation may then lead to different parity operators.

To illustrate the interaction-dependent nature of the parity operator, let us consider the $\theta - \tau$ puzzle again

$$K^+ \rightarrow \begin{cases} \pi^+ \pi^o & (\theta\text{-mode}) \\ \pi^+ \pi^o \pi^o & (\tau\text{-mode}) \quad . \end{cases} \tag{15.128}$$

Since from the Dalitz plot we know that the three pions in the τ-mode are predominantly in the s-states, the parities of the final modes θ and τ satisfy,

$$\text{parity } (\tau) = \text{parity } (\theta) \cdot \text{parity } (\pi^o) \quad . \tag{15.129}$$

From the strong πN interaction, it has been determined that the pion is a pseudoscalar; the same conclusion is also reached by studying the electromagnetic decay of π^o,

$$\pi^o \rightarrow 2\gamma \rightarrow 2e^+ + 2e^- \quad .$$

Thus, by using $H_{st} + H_\gamma$ we find

$$\text{parity } (\pi^o) = -1 \quad . \tag{15.130}$$

Consequently, (15.129) and (15.130) imply the usual conclusion

$$\text{parity } (\tau) = - \text{ parity } (\theta) \quad .$$

Since K^+ decay is a weak process, this means that H_{wk} does not conserve parity; here the parity is defined by $H_{st} + H_\gamma$.

On the other hand, let us imagine that the strong and the electromagnetic interactions could both be turned off. We may then try to determine the parity through weak processes. The requirement that the weak interaction be parity-conserving would imply, in the case of K^+ decay,

$$\text{parity } (\tau) = \text{ parity } (\theta) \quad ,$$

and therefore

$$\text{parity } (\pi^o) = +1 \quad .$$

Now, by switching back the strong and electromagnetic interactions, we would conclude that the strong p-state interaction in π - N scattering and the photon polarizations in the electromagnetic decay of π^o violate parity conservation, where the parity is now determined by H_{wk}. Likewise, without the strong and electromagnetic interactions, the fact that in the weak $\pi \rightarrow \mu\nu$ decay there are two kinds of neutrino, one righthanded and the other lefthanded would, by itself, be totally consistent with right-left symmetry.

In reality, it is of course not possible to turn off any interaction, be it $H_{st} + H_\gamma$ or H_{wk}. However, because the matrix elements of $H_{st} + H_{\gamma}$ are typically much larger than those of H_{wk}, it is more convenient to adopt the approximation $H \cong H_{st} + H_{\gamma}$. Only then do we regard H_{wk} as P-violating. A similar implication also applies to our discussions of CP and T violations given in this chapter.

15.8 Phenomenological Analysis of the CP - nonconserving Interaction

It is always possible to decompose the total Hamiltonian H into two parts

$$H = H_+ + H_- \tag{15.131}$$

where

$$H_{\pm} = \tfrac{1}{2}[H \pm CP\,H\,P^{\dagger}C^{\dagger}] \quad . \tag{15.132}$$

Hence under CP, H_+ is even and H_- odd. In accordance with the above discussions, it is convenient to define the CP operator by using the strong and electromagnetic interactions.

At present all our known phenomena are consistent with the view that the CP-violating H_- are much weaker than the usual CP-conserving part of the weak interaction. In a phenomenological analysis, it is immaterial whether H_- should or should not be an integral part of H_{wk}. In the following, for convenience, we shall separate H_- from H_{wk}; hence, H_{wk} refers only to the CP-conserving part. There are two main classes of possibilities concerning the strength of H_-:

1. Milliweak

This is the obvious case. One assumes that typically

$$\frac{H_-}{H_{wk}} \sim |\epsilon| \sim 10^{-3} \tag{15.133}$$

where ϵ is the CP - violating parameter given before. By using (15.47), we have

$$\frac{H_-}{H_{st}} \sim 10^{-7} \mid \epsilon \mid \sim 10^{-10} \quad . \tag{15.134}$$

Because of (15.46)-(15.48), we conclude that

$$< K^o \mid H_- \mid \overline{K}^o > = 0 \quad ; \tag{15.135}$$

otherwise, the diagonalization of the mass matrix (15.32) would lead to a mass difference $\Delta m = m_L - m_S \sim H_- \sim 10^{-10} H_{st} \sim 10^{-10} m_K$, which disagrees with (15.46).

In order to insure (15.135), the simplest way is to assume that H_-, like H_{wk}, satisfies the selection rule $\Delta S \neq \pm 2$.

2. Superweak

In this case *, one assumes that

$$\frac{H_-}{H_{wk}} \sim 10^{-7} \mid \epsilon \mid \sim 10^{-10} \tag{15.136}$$

and, instead of (15.135),

$$< K^o \mid H_- \mid \overline{K}^o > \neq 0 \quad ; \tag{15.137}$$

i.e., $\Delta S = \pm 2$ is allowed for H_-. [The usual H_{wk}, of course, continues to satisfy the $\Delta S \neq \pm 2$ rule.] Since according to (15.47), for $\Delta S = \pm 1$, $(H_{wk}/H_{st}) \sim 10^{-7}$, we have in the superweak case

$$\frac{H_-}{H_{st}} \sim 10^{-14} \mid \epsilon \mid \quad . \tag{15.138}$$

Because of (15.137), the CP-violating parameter $\mid \epsilon \mid \sim 10^{-3}$ enters into the state vectors $\mid L >$ and $\mid S >$.

---

* L. Wolfenstein, Phys.Rev.Lett. 13, 562 (1964); T. D. Lee and L. Wolfenstein, Phys.Rev. 138, B1490 (1965).

The reason for these two disparate possibilities is because in the neutral kaon decay, while the decay matrix element is first-order in H_{wk}, the state vectors K_L^o and K_S^o depend on the second-order weak interaction. Hence, H_- can be either $\sim \epsilon H_{wk}$ or $\sim \epsilon H_{wk}^2/H_{st}$; the former is milliweak and the latter superweak.

For the neutral kaon, the dominant decay matrix element is that of 2π in the $I = 0$ channel; because it consists of a single final state, the decay matrix Γ is essentially real, as shown in (15.100). Furthermore, except for the 2π channel, none of the other decay matrix elements have been measured to an accuracy beyond 10^{-3}. This explains why in the neutral kaon decay, all known CP-violating phenomena can be characterized by a single parameter ϵ.

If H_- is superweak, then any CP-violating transition amplitude should be $\sim 10^{-10}$ times the corresponding CP-conserving weak amplitude; to detect such small amplitudes would be nearly impossible at the present. On the other hand, if H_- is milliweak, then one only needs to measure some appropriate decay amplitudes of K, or Λ^o, or $\cdots$, to the accuracy $\sim 10^{-3}$, which is somewhat more hopeful. A promising possibility is to measure the electric dipole moment $e d_n$ of, say, the neutron. In the milliweak case, according to (15.136), since $H_- \sim 10^{-10} H_{st}$ we expect typically

$$d_n \sim 10^{-10} m_n^{-1} \sim 10^{-24} \text{ cm} .$$

If it is superweak, then

$$d_n \sim 10^{-17} m_N^{-1} \sim 10^{-31} \text{ cm} .$$

The present experimental limit is*

---

* W. B. Dress, P. D. Miller, J. M. Pendlebury, P. Perrin and N. F. Ramsay, Phys. Rev. D15, 9 (1977).

$$d_n < 3 \times 10^{-24} \text{ cm} ,$$

consistent with either possibility.

Problem 15.1. In $K \to 3\pi$ decay,

(i) Show that the boundary of the Dalitz plot is given by

$$r \leqq r_o(\theta)$$

where

$$r_o^2 = (1+\epsilon)^{-1} (1 - \epsilon r_o^3 \cos 3\theta) ,$$

$$\epsilon = (2m_K - Q)^{-2}(2Q\, m_K) .$$

(ii) Prove that for different charged states, $\pi^{\pm}\pi^{\pm}\pi^{\mp}$, $\pi^{\pm}\pi^{o}\pi^{o}$, $\pi^{o}\pi^{o}\pi^{o}$ and $\pi^{+}\pi^{-}\pi^{o}$, to first order in Δm_{π}, the ratio of the respective phase space is given by the ratio of the corresponding values of

$$\Omega = (1+\epsilon)^{-1} Q^2 .$$

(iii) Assuming $|\Delta\vec{I}| = \frac{1}{2}$ rule, show that the ratios of (width$/\Omega$) for $K^+ \to \pi^+\pi^+\pi^-$, $K^+ \to \pi^o\pi^o\pi^+$, $K_L^o \to \pi^+\pi^-\pi^o$ and $K_L^o \to 3\pi^o$ are $4:1:2:3$, and the corresponding ratios* of the slope λ are $1:-2:-2:0$.

(iv) Compare these theoretical predictions with the present experimental results.

[See the review article by T. D. Lee and C. S. Wu, Ann.Rev.Nucl.Sci. 16, 471 (1966) and the references mentioned therein.]

---

* S. Weinberg, Phys.Rev.Lett. 4, 87 (1960).

Chapter 16

VACUUM AS THE SOURCE OF ASYMMETRY

16.1 What Is Vacuum?

In the last century, in order to understand how the electromagnetic force, and later the electromagnetic wave, could be transmitted in space, the vacuum was viewed as a medium called aether. In his note 3075 on experimental research Faraday wrote *

> For my own part, considering the relation of a vacuum to the magnetic force and the general character of magnetic phenomena external to the magnet, I am more inclined to the notion that in the transmission of the force there is such an action, external to the magnet, than that the effects are merely attraction and repulsion at a distance. Such an action may be a function of the aether; for it is not at all unlikely that, if there be an aether, it should have other uses than simply the conveyance of radiations.

However, since at that time the nonrelativistic Newtonian mechanics was the only one available, the vacuum was thought to provide an absolute frame which could be distinguished from other moving frames by measuring the velocity of light. As is well-known, this led to the downfall of aether and the rise of relativity.

We know now that vacuum is Lorentz-invariant, which means that just by our running around and changing the reference system we

---

* Michael Faraday, Experimental Researches in Electricity (London, R. and J. E. Taylor, 1839-55).

are not going to alter the vacuum. But Lorentz invariance does not embody all physical characteristics. We may still ask: What is this vacuum state?

In the modern treatment, we define the vacuum as the lowest-energy state of the system. It has zero 4-momentum. In most quantum-field theories, the vacuum is used only to enable us to perform the mathematical construct of a Hilbert space. From the vacuum state we build the one-particle state, then the two-particle state, $\cdots$; hopefully, the resulting Hilbert space will eventually resemble our universe. From this approach, different vacuum state means different Hilbert space, and therefore different universe.

From Dirac's hole theory we know that the vacuum, although Lorentz-invariant, is actually quite complicated. In general, we may expect the vacuum to be as complex as any spin-0 field $\phi(x)$ at the zero 4-momentum limit:

$$\text{vacuum} \sim \phi \qquad \text{at} \qquad \text{4-momentum } k_\mu = 0 \,. \tag{16.1}$$

Like a spin-0 field, it is conceivable that the vacuum state may carry quantum numbers such as isospin $\vec{I}$, parity P, strangeness S, etc. In this context we may ask: Could the vacuum be regarded as a physical medium? If under suitable conditions the properties of the vacuum, like those of any medium, can be altered physically, then the answer would be affirmative. Otherwise it might degenerate into semantics. The analysis given below will be based primarily on two of the most remarkable phenomena in modern physics:

(i) missing symmetry,

and (ii) quark confinement.

The former will be discussed now and the latter in the next chapter.

16.2 Missing Symmetry

If we add up the symmetry quantum numbers such as $\vec{I}$, S, P, C, ⋯, of all matter, we find these numbers to be constantly changing

$$\frac{d}{dt}\begin{Bmatrix} \vec{I} \\ S \\ P \\ C \\ CP \\ \vdots \end{Bmatrix}_{\text{matter}} \neq 0 \ . \tag{16.2}$$

Aesthetically, this may appear disturbing. Why should nature abandon perfect symmetry? Physically, this also seems mysterious. What happens to these missing quantum numbers? Where do they go? Can it be that matter alone does not form a closed system? If we also include the vacuum, then perhaps symmetry may be restored

$$\frac{d}{dt}\begin{Bmatrix} \vec{I} \\ S \\ P \\ C \\ CP \\ \vdots \end{Bmatrix}_{\text{matter + vacuum}} = 0 \ . \tag{16.3}$$

As a bookkeeping device, this is clearly possible. It also forms the basic idea underlying the important topic of spontaneous symmetry breaking, developed by Y. Nambu and others.* In such a scheme one often assumes that there exists some phenomenological spin-0 field ϕ which can carry the missing quantum number and whose vacuum expectation value is not zero:

$$\phi_{vac} \equiv \langle vac \,|\, \phi \,|\, vac \rangle \neq 0 \ . \tag{16.4}$$

---

* For a history of this subject, see Y. Nambu, Fields and Quanta <u>1</u>, 33 (1970).

Consequently, the observed asymmetry can be attributed entirely to the state vector of our universe, not to the physical law. [Examples will be given later.] On the other hand, unless we have other links connecting matter with vacuum, how can we be sure that this idea is right, and not merely a tautology?

A way out of this dilemma is to realize that in (16.1) the restriction $k_\mu = 0$ for the vacuum state is only a mathematical idealization. After all, very likely the universe does have a finite radius, and k_μ is therefore never strictly zero. So far as the microscopic system of particle physics is concerned, there is little difference between $k_\mu = 0$ and k_μ nearly 0; the latter corresponds to a state that varies only very slowly over a large space-time extension. This means that if the idea expressed by (16.4) is correct, then under suitable conditions, we must be able to produce excitations, or domain structures, in the vacuum. In such an excited state, there exists a volume Ω whose size is $\gg$ the relevant microscopic dimension; inside Ω we have the expectation value $\langle \phi(x) \rangle \neq \phi_{vac}$, but outside Ω $\langle \phi(x) \rangle = \phi_{vac}$. The symmetry properties inside Ω can then be different from those outside.

16.3 Vacuum Excitation

How can we produce such a change in $\langle \phi(x) \rangle$? The problem is analogous to the formation of domain structures in a ferromagnet. We may draw the analog:

$$\langle \phi(x) \rangle \longleftrightarrow \text{magnetic spin} \quad ,$$

$$J = \text{matter source} \longleftrightarrow \text{magnetic field}$$

as shown in Fig. 16.1.

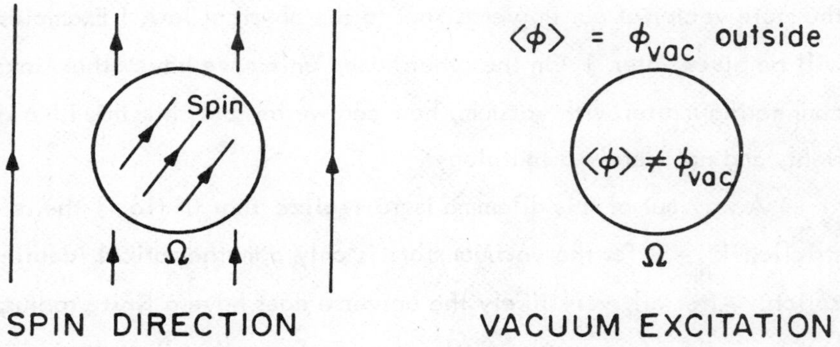

Fig. 16.1. Domain structures in a ferromagnet vs. in the vacuum.

In the case of a very large ferromagnet, because the spins interact linearly with the magnetic field, a domain structure can be created by applying an external magnetic field over a large volume. Furthermore, after domains are created, we may remove the external field; depending on the long-range forces, the surface energy and other factors, such a domain structure may persist even after the external magnetic field is removed. Similarly, by applying over a large volume any matter source J which has a linear interaction with $\phi(x)$, we may hope to create * a domain structure in $< \phi(x) >$. Depending on the dynamical theory, such domains may also remain as physical realities, even after the matter source J is removed.

As an illustration, we may consider a local scalar field theory. The Lagrangian density is

$$\mathcal{L}_\phi = -\tfrac{1}{2} \left(\frac{\partial \phi}{\partial x_\mu}\right) - U(\phi) \quad , \tag{16.5}$$

---

* T. D. Lee and G. C. Wick, Phys.Rev. D9, 2291 (1974).

where the absolute minimum of U is at $\phi = \phi_{vac}$ and with $U(\phi_{vac}) = 0$. Since we are interested in the long-wavelength limit of the field, the scalar field $\phi(x)$ is used only as a phenomenological description. The details of its microscopic structure do not concern us.

Let us now introduce an external source $J(x)$. The simplest example is to assume J to be constant inside a large volume Ω, but zero outside. For a sufficiently large Ω, we may neglect the surface energy. The energy of the system becomes

$$[\,U(\phi) + J\phi\,]\,\Omega \quad . \tag{16.6}$$

Its minimum determines the expectation value of ϕ inside Ω. The graphs in Fig. 16.2 illustrate how the new expectation value

$$\bar{\phi} = \langle \phi(x) \rangle$$

can be changed under the influence of J.

If the missing symmetry is due to $\phi_{vac} \neq 0$, then by changing $\bar{\phi}$ we may alter the symmetry properties inside Ω dynamically. Of course, to do a realistic experiment to reclaim our missing symmetry is not easy. But it is the prerogative of the theorist to contemplate such a situation.

16.4 CP Nonconservation and Spontaneous Symmetry Breaking

We discuss here one of the simplest examples that illustrates the phenomenon of spontaneous symmetry breaking. Our purpose is to give a theory * in which

(i) the Lagrangian is invariant under CP and T,

but (ii) its S-matrix violates CP and T symmetry.

---

* T. D. Lee, Physics Reports 9C, No. 2 (1974).

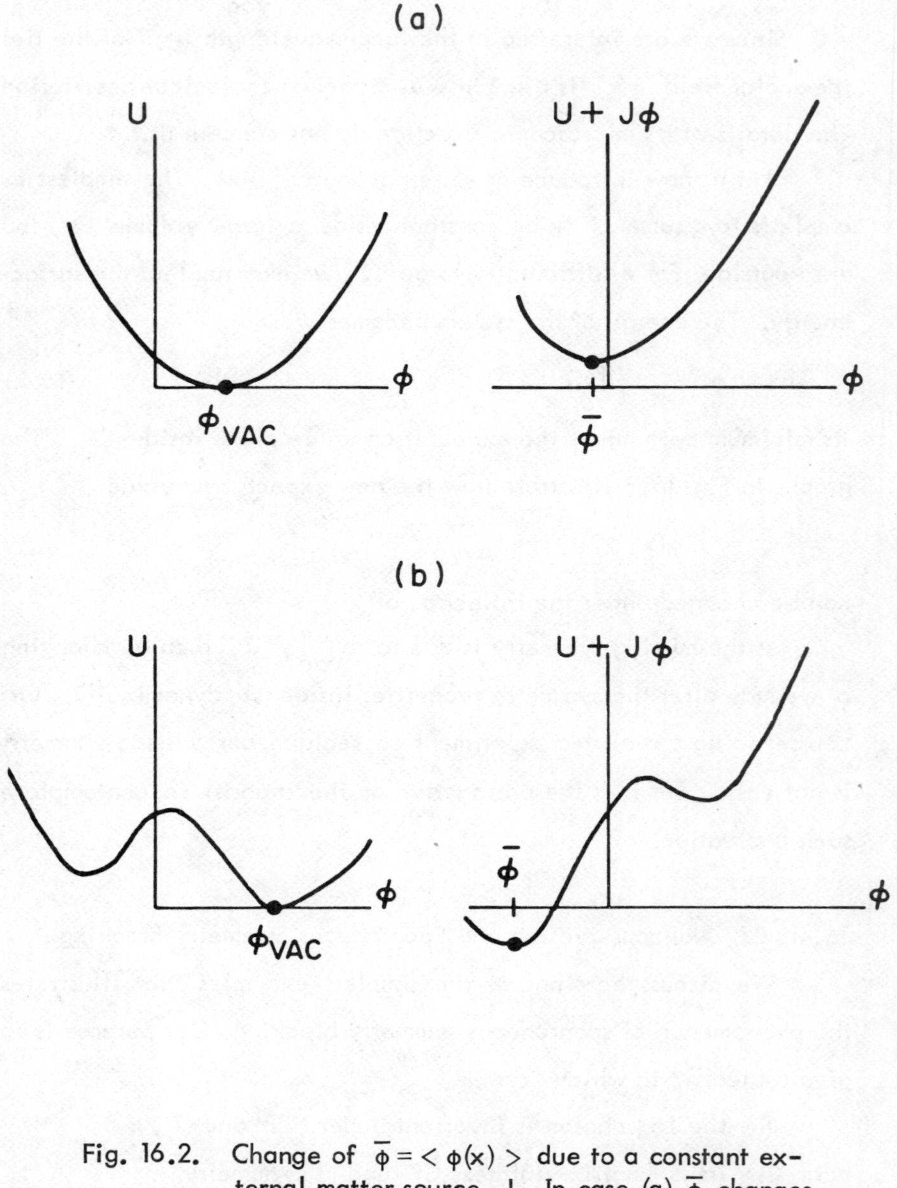

Fig. 16.2. Change of $\bar{\phi} = \langle \phi(x) \rangle$ due to a constant external matter source J. In case (a) $\bar{\phi}$ changes continuously with J. In case (b), as J increases, there is a critical value at which $\bar{\phi}$ makes a sudden jump.

Let us assume the system consists of a spin-$\frac{1}{2}$ Dirac field ψ and a spin-0 Hermitian field ϕ. The Lagrangian density is

$$\mathcal{L} = -\tfrac{1}{2}\left(\frac{\partial\phi}{\partial x_\mu}\right)^2 - U(\phi) - \psi^\dagger \gamma_4\left(\gamma_\mu \frac{\partial}{\partial x_\mu} + m\right)\psi - ig\psi^\dagger \gamma_4 \gamma_5 \psi \phi \ , \tag{16.7}$$

where

$$U(\phi) = \frac{1}{8}\kappa^2(\phi^2 - \rho^2)^2 \ . \tag{16.8}$$

From Hermiticity, the parameters m, g, ρ and κ must be real. It can readily be verified that $\mathcal{L}$ is invariant under T, C and P where

$$T\phi(\vec{r}, t)T^{-1} = -\phi(\vec{r}, -t) \ , \tag{16.9}$$

$$C\phi(\vec{r}, t)C^{-1} = \phi(\vec{r}, t) \ , \tag{16.10}$$

and

$$P\phi(\vec{r}, t)P^{-1} = -\phi(-\vec{r}, t) \ . \tag{16.11}$$

The corresponding transformations of ψ are given by (10.5), (10.9) and (13.53). Because U is a fourth-order polynomial in ϕ, the theory is renormalizable.

The vacuum expectation value of ϕ is determined by the minimum of $U(\phi)$. As shown in Fig. 16.3, we have either

$$\langle \phi \rangle_{vac} = \rho > 0 \tag{16.12}$$

or $\langle \phi \rangle_{vac} = -\rho$. In either case, since ϕ is of $T = -1$, $CP = -1$ and $P = -1$, a nonzero expectation value of ϕ implies that the vacuum state is not an eigenstate of T (nor of CP and P). The T symmetry of the Lagrangian requires that if $\langle \phi \rangle_{vac} = \rho$ is a solution, then $\langle \phi \rangle_{vac} = -\rho$ must also be one. These two solutions transform into each other under T, but by itself neither is invariant under T. (It is also not invariant under CP and P, though it is under C and CPT.)

Because of quantum effects, the field ϕ fluctuates around its

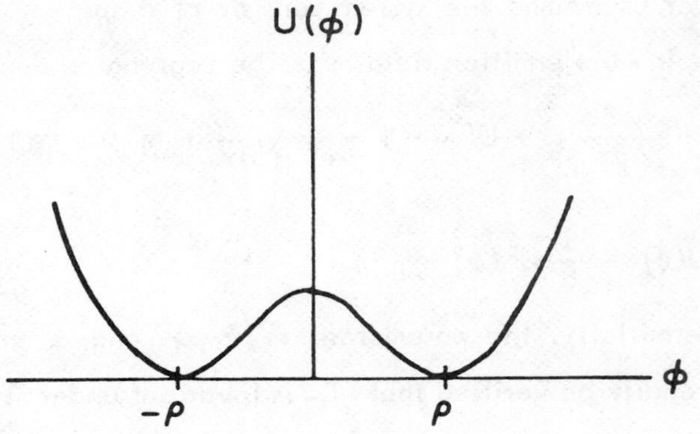

Fig. 16.3. The potential energy density $U(\phi) = \frac{1}{8}\kappa^2(\phi^2 - \rho^2)^2$.

vacuum expectation value. We may choose $<\phi>_{vac} = \rho$, and write

$$\phi = \rho + \delta\phi \quad .$$

In terms of $\delta\phi$, the potential U becomes

$$U = \tfrac{1}{2}\mu^2(\delta\phi)^2 + \tfrac{1}{2}\kappa^2\rho(\delta\phi)^3 + \frac{1}{8}\kappa^2(\delta\phi)^4 \tag{16.13}$$

where $\mu = \kappa\rho$, which is the mass of the fluctuating field $\delta\phi$.

In order to exhibit more clearly the T-violating character of the solution, we may perform a unitary transformation under which ϕ is unchanged, but

$$\psi \to e^{-i\frac{1}{2}\gamma_5\alpha}\,\psi \quad . \tag{16.14}$$

Therefore, the quadratic expressions

$$\psi^\dagger\gamma_4\psi \to \psi^\dagger e^{\frac{1}{2}i\gamma_5\alpha}\,\gamma_4\,e^{-\frac{1}{2}i\gamma_5\alpha}\psi = \psi^\dagger\gamma_4\,e^{-i\gamma_5\alpha}\psi$$

$$= \psi^\dagger\gamma_4(\cos\alpha - i\gamma_5\sin\alpha)\psi$$

and

$$i\psi^\dagger\gamma_4\gamma_5\psi \to i\psi^\dagger e^{i\frac{1}{2}\gamma_5\psi}\gamma_4\gamma_5\,e^{-i\frac{1}{2}i\gamma_5\alpha}\psi = i\psi^\dagger\gamma_4\gamma_5\,e^{-i\gamma_5\alpha}\psi$$

$$= \psi^\dagger\gamma_4(\sin\alpha + i\gamma_5\cos\alpha)\psi \quad .$$

Hence, by choosing

$$\tan\alpha = g\rho/m \quad , \tag{16.15}$$

we have

$$\psi^\dagger \gamma_4 (m + i g \rho \gamma_5) \psi \rightarrow \psi^\dagger \gamma_4 M \psi \tag{16.16}$$

where

$$M = (m^2 + g^2 \rho^2)^{\frac{1}{2}} \quad . \tag{16.17}$$

By substituting (16.14) into (16.7), we find that the Lagrangian density $\mathcal{L}$ becomes

$$-\tfrac{1}{2}\left(\frac{\partial}{\partial x_\mu}\,\delta\phi\right)^2 - U - \psi^\dagger \gamma_4\left(\gamma_\mu \frac{\partial}{\partial x_\mu} + M\right)\psi$$
$$- g\psi^\dagger \gamma_4 (\sin\alpha + i\gamma_5 \cos\alpha)\,\psi\,\delta\phi \quad . \tag{16.18}$$

Since the operator $\psi^\dagger \gamma_4 \psi$ is of $P = 1$, $C = 1$ and $T = 1$ while the operator $i\psi^\dagger \gamma_4 \gamma_5 \psi$ is of $P = -1$, $C = 1$ and $T = -1$, any exchange of the $\delta\phi$ quantum would give an interference term between these two operators that violates T, P and CP; but the product symmetry CPT remains intact. Figure 16.4 gives an example of such an interference term, whose amplitude is

$$A_- = g^2 \sin\alpha \cos\alpha \,(k^2 + \mu^2)^{-1} \quad , \tag{16.19}$$

where k denotes the 4-momentum transfer.

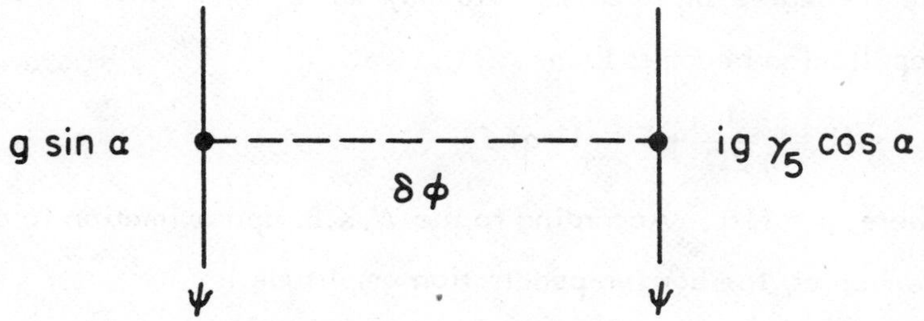

Fig. 16.4. A T-violating scattering diagram due to the exchange of the quantum of $\delta\phi$, where $\delta\phi \equiv \phi - \langle\phi\rangle_{vac}$.

The model discussed in this section illustrates the basic mechanism of a spontaneous T violation. [The same discussion applies to CP and P as well.] One assumes that the ground state of the system (called the vacuum) has a nonzero expectation value $<\phi>_{vac}$, where ϕ is a $T = -1$ phenomenological spin-0 field; thus, the vacuum is noninvariant under T, even though the Lagrangian satisfies T invariance.

The T invariance of the Lagrangian implies that the vacuum must have a double degeneracy. It is interesting to examine the barrier penetration between these two degenerate solutions: $<\phi>_{vac} = \rho$ and $-\rho$ of Fig. 16.3. By enclosing the entire system in a finite volume Ω with a periodic boundary condition, we may expand ϕ in terms of the usual Fourier series

$$\phi = q + \sum_{\vec{k} \neq 0} \frac{1}{\sqrt{\Omega}} \phi_{\vec{k}} e^{i\vec{k}\cdot\vec{r}} \tag{16.20}$$

where q is independent of $\vec{r}$. Substituting this expression into (16.7), we find

$$L = \int \mathcal{L} d^3r = \tfrac{1}{2} \dot{q}^2 \Omega - U(q)\Omega + \cdots$$

where $\cdots$ depends on $\phi_{\vec{k}}$ and the fermion field ψ. The barrier-penetration amplitude may be crudely estimated by concentrating on the q-degree of freedom. We may set $\phi_{\vec{k}} = 0$ and $\psi = 0$. The Hamiltonian becomes then

$$H = \frac{1}{2\Omega} p^2 + U(q)\Omega \tag{16.21}$$

where $p = \Omega \dot{q}$. According to the W.K.B. approximation in quantum mechanics, the barrier-penetration amplitude is

$$\sim \exp\{ -\Omega \int_{-\rho}^{\rho} [2U(q)]^{\frac{1}{2}} dq \} \tag{16.22}$$

which goes to zero exponentially as the volume of the system approaches infinity.

From (16.17) we see that the fermion mass changes from m to M when the expectation value of ϕ varies from 0 to $\langle \phi \rangle_{vac} = \rho$. Likewise the CP-violating amplitude A_- also depends on $\langle \phi \rangle_{vac}$. Hence if we follow the discussion given in Section 16.3, by applying a matter source J over a large volume Ω we may alter $\langle \phi \rangle$ inside Ω, and thereby change the mass of the particle and the symmetry-violating amplitude.

Remarks. The above example demonstrates the essence of the spontaneous symmetry-breaking mechanism. The Lagrangian is invariant under a certain group $\mathcal{G}$ of symmetry transformations. But the vacuum state is not, and that gives rise to symmetry-violating phenomena. By applying $\mathcal{G}$ onto the vacuum state, we must generate other states degenerate with the vacuum. In a realistic cosmological model, the volume Ω of the universe may be expected to be finite. In general, there would be nonzero, but very small, barrier-penetration amplitudes between these different "vacua", which can lift the degeneracy. While such effects can be safely neglected at our present stage of evolution, they may have been important at a much earlier period when Ω was still of microscopic dimensions.

In Chapter 22 we shall discuss further the application of the spontaneous symmetry-breaking mechanism to various continuous symmetry groups.

IIB. PARTICLE PHYSICS: INTERACTIONS

At present there exists a large body of supportive evidence to suggest that the theory underlying the strong interaction is quantum chromodynamics (QCD), and that the weak and electromagnetic interactions are unified via an appropriate non-Abelian gauge theory with a spontaneous symmetry-breaking mechanism. In either case, decisive experimental proof is still lacking. Hence the main motivation remains aesthetic. Nevertheless, as we shall see, the multitude of different phenomena that QCD and the unified theoretical models of weak and electromagnetic interactions have managed to explain and correlate is of sufficient scope to convince one of their essential correctness.

Chapter 17

QUARK CONFINEMENT

17.1 The Problem

In Chapter 12 we mentioned that from hadron spectroscopy there is good reason to believe that all hadrons are composites of quarks, whose flavors are up (u), down (d), strange (s), charm (c) and bottom (b). Furthermore, each flavored quark has three different colors. Their assumed masses and charges are

	u	d	s	c	b
mass	~ 0	~ 0	$\sim$ 100 MeV	$\sim$ 2 GeV	$\sim$ 5 GeV
charge/e	$\frac{2}{3}$	$-\frac{1}{3}$	$-\frac{1}{3}$	$\frac{2}{3}$	$-\frac{1}{3}$.

These assignments are quite remarkable considering that none of the quarks has been observed in its free form. A more direct way of arriving at these conclusions is through the R-value measurement in the e^+e^- collision, as we shall discuss now. Let us consider the diagrams for $e^+e^- \to \mu^+\mu^-$ and $q\bar{q}$, as shown in Fig. 17.1. The threshold energies for these two reactions are $2m_\mu$ and $2m_q$ respectively. When the center-of-mass energy E_{cm} is much larger than the threshold energy, and if we <u>neglect</u> the strong interaction between the $q\bar{q}$ pair, the ratio of the cross sections for these two

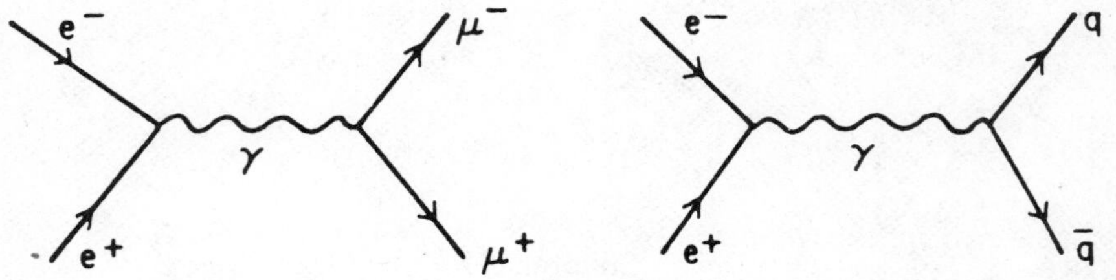

Fig. 17.1. Diagrams for $e^+e^- \rightarrow \mu^+\mu^-$, or $q\bar{q}$.

reactions is

$$\frac{\sigma(e^+e^- \rightarrow q\bar{q})}{\sigma(e^+e^- \rightarrow \mu^+\mu^-)} = 3Q_q^2 \tag{17.1}$$

where q stands for the flavor of the quark which can be $u, d, s, \cdots$, Q_q is its corresponding charge in units of e, and the factor 3 is due to the three colors of each flavor. We define

$$R \equiv \frac{\sum_q \sigma(e^+e^- \rightarrow q\bar{q} \rightarrow \text{hadrons})}{\sigma(e^+e^- \rightarrow \mu^+\mu^-)}$$

where the sum extends over all quarks whose mass m_q is $< \frac{1}{2} E_{cm}$. By using (17.1), we have

$$R \cong 3 \sum_q Q_q^2 \quad . \tag{17.2}$$

The experimental result* is consistent with the theoretical expectation, drawn schematically in Fig. 17.2. When E_{cm} is $< 2m_c$, but > 1 GeV which is assumed to be much larger than $2m_u$, $2m_d$ and

---

* See the Proceedings of the 1979 International Symposium on Lepton and Photon Interactions at High Energies, ed. T. B. W. Kirk and H. D. I. Abarbanel, Fermilab, Batavia, Illinois.

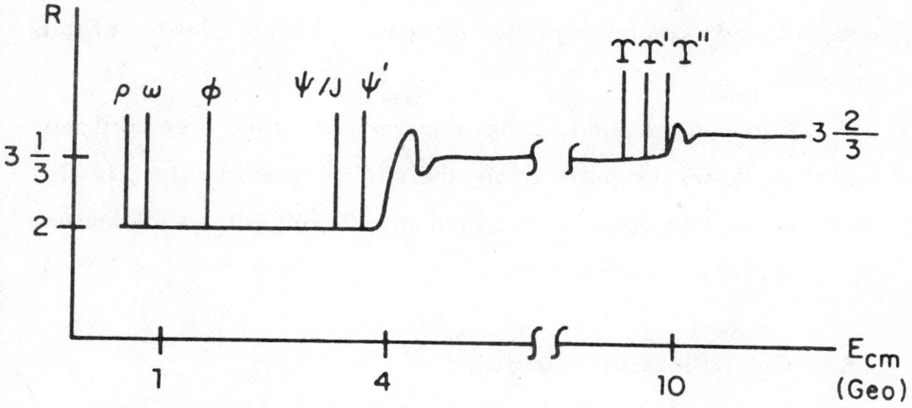

Fig. 17.2. Schematic drawing of R.

$2m_s$, we have

$$R \cong 3[(\tfrac{2}{3})^2 + (\tfrac{1}{3})^2 + (\tfrac{1}{3})^2] = 2 ,$$

since the charges for u, d, s are respectively $\frac{2}{3}$, $-\frac{1}{3}$ and $-\frac{1}{3}$. When E_{cm} is between $2m_c$ and $2m_b$, we may write $R = 2 + \Delta R$ where $\Delta R \cong 3 \cdot (\frac{2}{3})^2 = 4/3$ because the charge of the c quark is $\frac{2}{3}$. When E_{cm} is increased to above $2m_b$, then the R value should increase by another amount $\Delta R \cong 3 \cdot (\frac{1}{3})^2 = \frac{1}{3}$. The experimental confirmation of these theoretical expectations strongly supports the following conclusions:

(i) Quarks have fractional charges: $Q_q = \frac{2}{3}, -\frac{1}{3}, -\frac{1}{3}, \frac{2}{3}$ and $-\frac{1}{3}$ respectively for q = u, d, s, c and b quarks.

(ii) Each flavor variety has three colors, and the final states are symmetrical with respect to color permutations; i.e., the observed final hadron states are color singlets.

(iii) Quark "masses" are quite small (i.e., well within the energy range of our present accelerators).

(iv) Except near resonances, strong interactions between quarks and antiquarks can be neglected (at least in the calculations of total cross sections).

If indeed the quarks behave approximately like free particles and their masses are rather small, the critical question, then, is why don't we see free quarks in the final state? This is the well-known problem of quark confinement.

17.2 Color Dielectric Constant *

The details of QCD will be studied in the next chapter, so here we will only briefly mention some of its general features. QCD is the theory which describes the color SU_3-symmetric interaction. The system consists of quarks and gluons. The quarks are represented by the spin-$\frac{1}{2}$ Dirac fields ψ_q^a with q denoting the flavor u, d, s, $\cdots$ and a the color $1, 2, 3$; the gluons are represented by the vector gauge fields V_μ^ℓ where μ is the usual 4-dimensional space-time index and $\ell = 1, 2, \cdots 8$ denotes the gluon-color index. The theory is a renormalizable one, so there is no difficulty with respect to divergences in the ultraviolet region. However, there are complications in the infrared region. In order to give QCD a well-defined meaning, we may first contain the whole system within a volume of size L^3. Let g_L be the renormalized coupling constant in the long wavelength limit, momentum $k \sim L^{-1}$. It is possible to prove under rather general assumptions and valid to all orders of $(\text{coupling})^2$

$$g_L > g_\ell \qquad \text{if} \qquad L > \ell \quad . \tag{17.3}$$

The proof will be given in Section 18.6. As we shall discuss there,

---

* The discussions given in Sections 17.2 - 17.4 follow closely those given by T. D. Lee in A Festschrift for Maurice Goldhaber, edited by G. Feinberg, A. W. Sunyar and J. Weneser (New York, New York Academy of Sciences, 1980).

the above relation is closely connected to the "asymptotic freedom" property of the theory, which states that when $\ell \to 0$, g_ℓ decreases to 0. Since a pure QCD Lagrangian does not contain any mass scale, asymptotic freedom implies that when ℓ increases, g_ℓ must also increase, which leads to (17.3).

The difficulty lies in the infrared limit. When $L \to \infty$, it seems likely that g_L may $\to \infty$, or at least $>> 1$. Because the true physical system is one with $L = \infty$, we may always be in the ultra-strong coupling limit. In this chapter we shall see how this difficulty can be resolved, at least phenomenologically, by regarding the vacuum as a color dielectric medium.

Let us introduce κ_L which is called the color dielectric constant of the vacuum in a volume L^3. As a convention, we shall adopt a standard renormalized coupling constant g, defined by

$$g = g_\ell$$

when (17.4)

$$\ell = \text{some arbitrarily chosen length, say the proton radius.}$$

The constant κ_L will then be defined as

$$g_L^{\ 2} = \frac{g^2}{\kappa_L} \quad . \tag{17.5}$$

Consequently, in accordance with (17.4)

$$\kappa_\ell = 1 \qquad \text{when} \qquad \ell = \text{proton radius.} \tag{17.6}$$

Equation (17.3) now implies

$$\kappa_L < \kappa_\ell \qquad \text{if} \qquad L > \ell \quad . \tag{17.7}$$

With the convention (17.6), the above relation implies that for the vacuum in an infinite volume $L = \infty$,

$$\kappa_\infty < 1 \quad . \tag{17.8}$$

Just as in (17.3), the inequalities (17.7)-(17.8) are valid to all orders of g^2 . In the next chapter, we shall also give the lowest-order perturbation calculation. The result is (18.133), from which we obtain

$$\frac{\kappa_L}{\kappa_\ell} = \frac{1}{1 + \frac{1}{2\pi}\frac{g^2}{4\pi}(11 - \frac{2}{3}n)\ln\frac{L}{\ell} + O(g^4)} , \qquad (17.9)$$

where n is the number of quark flavor varieties (assumed to be < 17). Of course, this formula is consistent with the general inequality (17.3).

The ultra-strong coupling difficulty mentioned before corresponds to

$$\kappa_\infty = 0 , \qquad (17.10)$$

or

$$\kappa_\infty \ll 1 . \qquad (17.11)$$

In the former we call the vacuum in QCD a perfect color dia-electric medium, in the latter a nearly perfect color dia-electric medium.

17.3 A Hypothetical Problem in Classical Electromagnetism

In quantum electrodynamics, our usual convention is to set the dielectric constant of the vacuum state $\kappa_{vac} = 1$. It is then possible to prove that all physical media have their dielectric constants $\kappa \geqslant 1$. This can be seen most easily by using the familiar formula

$$\vec{D} = \vec{E} + 4\pi\vec{P} \qquad (17.12)$$

where $\vec{D}$ is the displacement vector, $\vec{E}$ the electric field, and $\vec{P}$ the polarization vector. Since under $\vec{E}$ all atoms have their polarization $\vec{P}$ in the same direction as $\vec{E}$, so as to produce a screening effect, we have $\kappa > 1$.

In this section, we shall consider a hypothetical problem. Let us imagine that in classical electromagnetism, but without the quantum theory of atoms, there could be a medium with its dielectric constant

$$\kappa \equiv \kappa_{med} \ll 1 \quad , \qquad \text{or} \qquad \cong 0 \quad , \tag{17.13}$$

i.e., this hypothetical medium is antiscreening. Now, suppose we place a small charge distribution ϵ in the medium. As we shall show, no matter how small ϵ is, the medium will crack and develop a hole surrounding the charge. Inside the hole, we have the vacuum so that $\kappa = 1$, but outside $\kappa = \kappa_{med}$, as shown in Fig. 17.3. To see this, let us assume that such a hole is formed. Because of the antiscreening nature of the medium, the induced charge on the inner surface of the hole is of the same sign as ϵ. Consequently, if we want to reduce the size of the hole, we must do work to overcome the repulsion between ϵ and the induced charge. That work is infinite if the hole is to be eliminated in toto. Hence the hole will not disappear.

This situation is completely different for a normal medium whose dielectric constant is > 1. Imagine that a similar hole were created. The corresponding induced charge would be of the opposite sign to ϵ. The hole would automatically shrink to 0, resulting in a homogeneous background medium if ϵ is sufficiently small; this gives rise to the usual Coulomb field distribution around the charge.

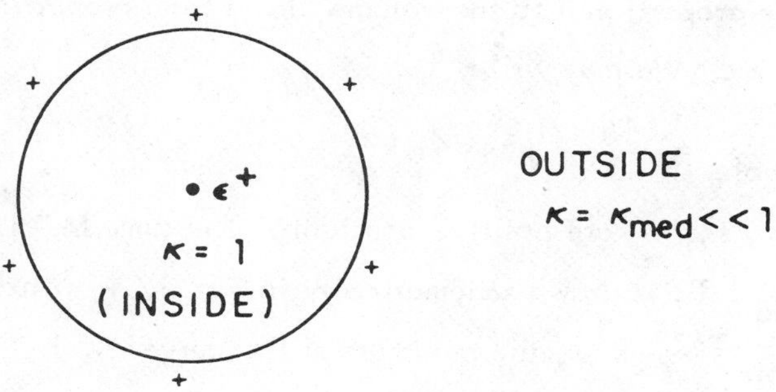

Fig. 17.3. Effect of a charge in a hypothetical dia-electric medium in classical electromagnetism.

For a dia-electric medium, however, holes (cracks) must occur whenever there are external charges. We may estimate the radius R of such a hole. Let D_{out} and E_{out} be, respectively, the normal components of $\vec{D}$ and $\vec{E}$ outside the hole when the radial coordinate $r = R+$. Similarly, let D_{in} and E_{in} be the corresponding components inside when $r = R-$. For a spherical hole we have

$$D_{in} = E_{in} = D_{out} = \epsilon / R^2$$

and

$$E_{out} = \epsilon / R^2 \kappa_{med} ,$$

where ϵ is the total charge of this small charge distribution. The electric energy inside the hole is independent of κ_{med}. The electric energy outside is given by the volume integral of $\vec{D} \cdot \vec{E}$, and is $\propto \kappa_{med}^{-1}$. It is convenient to subtract out the self-energy of the charge distribution (i.e., the energy in the absence of the medium); the change in electric energy due to the medium is

$$U_{el} = \tfrac{1}{2} \epsilon^2 (\kappa_{med}^{-1} - 1) / R . \tag{17.14}$$

In addition, there is the energy U_{hole}, needed to create such a hole. The amount U_{hole} is a function of R. When R is large, U_{hole} should be proportional to the volume plus a term proportional to the surface, etc. We may write

$$U_{hole} = \frac{4\pi}{3} R^3 p + 4\pi R^2 s + \cdots \tag{17.15}$$

where $p, s, \cdots$ are positive constants. The sum $M \equiv U_{el} + U_{hole}$, for $\kappa_{med} < 1$, is drawn schematically in Fig. 17.4. When $\kappa_{med} \to 0$ and $U_{hole} \cong \frac{4\pi}{3} R^3 p$, the minimum of the curve M is

$$M \equiv M_\epsilon \sim \frac{4}{3} \left(\frac{\epsilon^2}{2\kappa_{med}} \right)^{\frac{3}{4}} (4\pi p)^{\frac{1}{4}} . \tag{17.16}$$

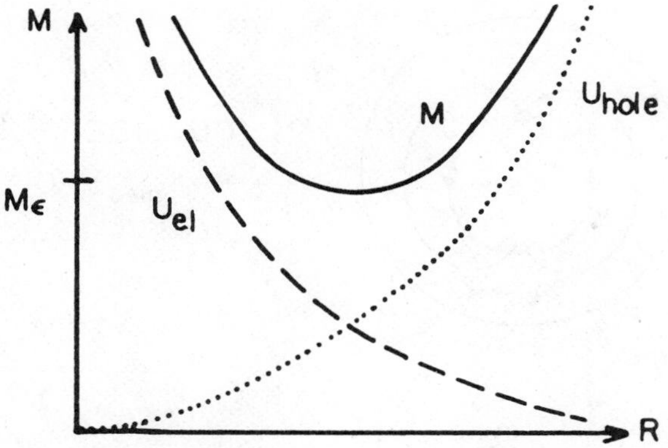

Fig. 17.4. The energy $M = U_{el} + U_{hole}$ of the system shown in Fig. 17.3, where κ_{med} is < 1.

From this we can draw the conclusion, already arrived at, that if the total charge $\epsilon \neq 0$, then $R \neq 0$. Furthermore, the energy

$$M_\epsilon \to \infty \qquad \text{when} \qquad \kappa_{med} \to 0 . \qquad (17.17)$$

It is easy to see that the same conclusion can also be reached without the approximation $U_{hole} \cong \frac{4\pi}{3} R^3 p$.

Next, we replace the single charge distribution by a dipole distribution; i.e., two small, but separate, distributions of total charge ϵ^+ and ϵ^-. It is easy to see that when κ_{med} is sufficiently small, the minimal energy state again requires the formation of a hole surrounding both charges. As before, inside the hole $\kappa = 1$, and outside $\kappa = \kappa_{med}$. When $\kappa_{med} \to 0$, it is not difficult to verify that at the surface of the hole the electric field inside should be parallel to the surface so that $\vec{D}$ is 0 outside, as shown in Fig. 17.5. Thus, U_{el} remains finite, as does the sum of energies $M_{\epsilon^+\epsilon^-} = U_{el} + U_{hole}$:

$$M_{\epsilon^+\epsilon^-} = \text{finite} \qquad \text{when} \qquad \kappa_{med} \to 0 . \qquad (17.18)$$

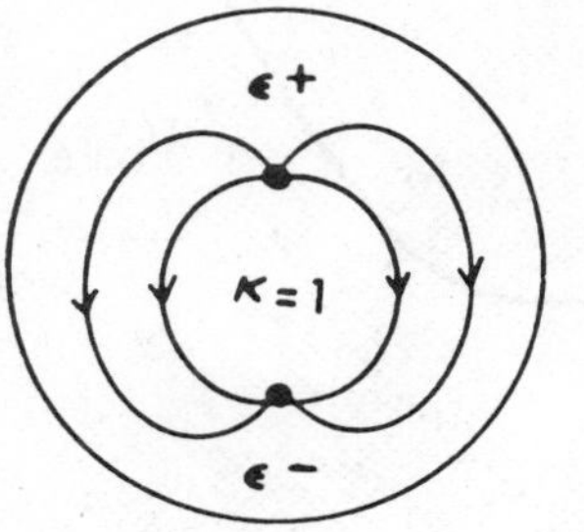

Fig. 17.5. Distribution of the electric field for a dipole placed inside the hypothetical dia-electric medium.

If we try to separate the two charges ϵ^+ and ϵ^- by an infinite distance, then because of (17.17) the work required must also be infinite. This is the analog to "quark confinement" in our hypothetical problem.

17.4 A Phenomenological Explanation

1. QCD vacuum as a perfect color dia-electric

We now return to the problem of quark confinement, discussed in Sections 17.1 and 17.2. We shall assume that the vacuum in QCD is a perfect (or nearly perfect) color dia-electric medium. By following the same argument given above, we see that whenever quarks or antiquarks are present there must be inhomogeneity in the space surrounding the particles. We may call these bags*, or domain structures, or solitons. Inside the bag, the dielectric constant κ is 1. But outside, $\kappa = \kappa_\infty$ which is 0, or $\ll 1$. If the total color is nonzero, the

---

* MIT Bag: A. Chodos, R. L. Jaffe, K. Johnson, C. B. Thorn and V. F. Weisskopf, Phys.Rev. D9, 3471 (1974).

SLAC Bag: W. A. Bardeen, M. S. Chanowitz, S. D. Drell, M. Weinstein and T.-M. Yan, Phys.Rev. D11, 1094 (1975).

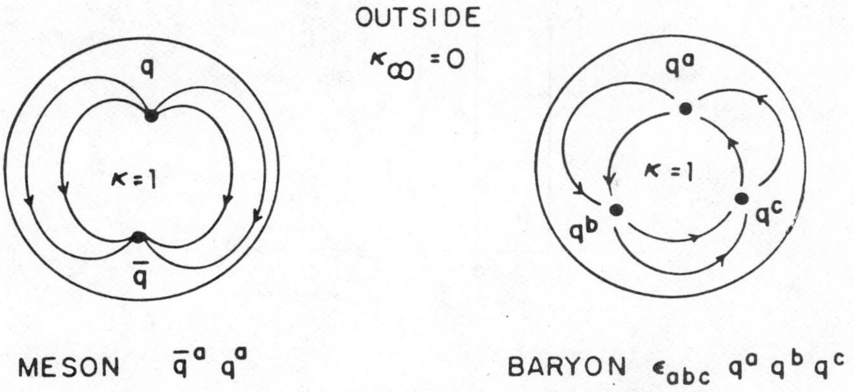

Fig. 17.6. Color electric fields inside hadrons.

mass of such a bag would be infinite when $\kappa_\infty \to 0$, in analogy with (17.17). However, just as in (17.18), if inside the bag one has a color singlet, then the bag mass remains finite when $\kappa_\infty \to 0$. Thus, by assuming all hadrons to be color singlets, we get finite masses for

mesons: $\bar{q}^a q^a$ and baryons: $\epsilon_{abc} q^a q^b q^c$

where a, b, c are color indices which can vary from 1 to 3. [See Fig. 17.6.] From this we can derive that the work required to separate the quarks to a large distance r is approximately proportional to r. The quark confinement is then "explained" by the assumption that the vacuum in QCD is a perfect (or nearly perfect) color dia-electric.

Of course, at present it is not known whether the mass M_q of a truly free quark is indeed infinite. However, the fact that no free quark has been observed so far in the final state of any high-energy collision sets a lower bound $M_q > 5$ GeV for q = u, d and s. It can be shown that this lower bound implies *

* T. D. Lee, Phys.Rev. D19, 1802 (1979).

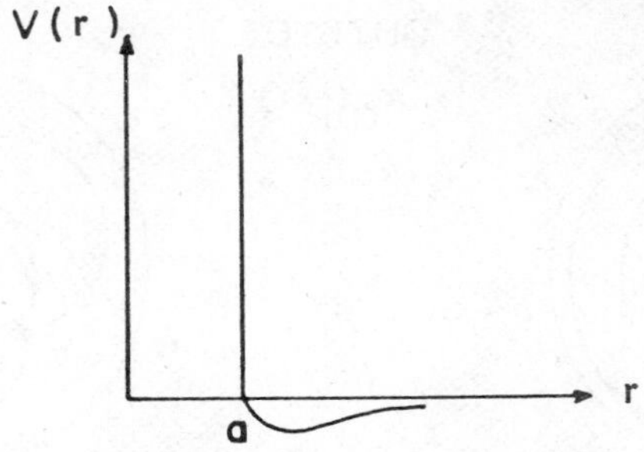

Fig. 17.7. Potential V between two He atoms.

$$\kappa_\infty < .013 \left(\frac{g^2}{4\pi}\right) \tag{17.19}$$

where g is defined by (17.4).

We may wonder in what sense we have by-passed the difficulty of ultra-strong coupling, mentioned in Section 17.2 . To understand this, let us consider other problems in physics in which interactions are also ultra-strong and yet there are no mathematical difficulties. A good example is the interaction potential $V(r)$ between two He atoms. As shown in Fig. 17.7, for $r < a$, the potential V is $\cong$ infinite; for $r > a$, V is small. Because the strong potential is repulsive, the two He atoms automatically avoid the strong-interaction region, thereby preventing any difficulty. The usual technique is to replace the strong repulsive part of V by a hard-sphere potential. Thus we need only examine a new boundary-value problem: The distance r between the two He atoms is restricted to a region $r \geqslant a$ in which the potential $V(r)$ is small, and therefore can be regarded as a perturbation.

From Fig. 17.6 we see that when $\kappa_\infty \to 0$, i.e. $g_\infty^{\ 2} = g^2/\kappa_\infty \to \infty$, the quarks are confined to a domain in which $\kappa = 1$ and the coupling = g . Thus the ultra-strong region, $g_\infty \to \infty$, exerts a repulsion against the quarks and antiquarks. Just as in the above problem of He-He interaction, the particles automatically stay away from the ultra-strong interaction region. Because of Lorentz invariance, the corresponding boundary-value problem in the present case requires the relativistic soliton solution, whose details will be discussed in Chapter 20. As we shall see, when we expand around these soliton solutions, only the coupling g inside the soliton is the relevant parameter, and that resolves the difficulty. [It is only when one attempts to make a simple-minded plane-wave expansion around a homogeneous background that one has to use g_∞ as the expansion parameter. Exactly the same kind of difficulty would appear if one were to force a similar plane-wave expansion in the aforementioned He-He problem.] As we shall also see, because QCD is asymptotically free, the quarks inside the bag behave approximately like free particles and their effective masses can be relatively small.

Since our theory is relativistically invariant, the velocity of light $c/(\kappa\mu)^{\frac{1}{2}}$ in the vacuum must remain c . Hence

$$\kappa\mu = 1 \tag{17.20}$$

where κ is the color dielectric constant and μ the corresponding "magnetic" susceptibility. When $\kappa = \kappa_\infty \to 0$ one must have $\mu = \mu_\infty \to \infty$.

With (17.20), we can consider κ to be a Lorentz-invariant quantity; it plays the same role as the ϕ-field discussed in Chapter 16. It may be convenient to introduce

$$\phi \propto 1 - \kappa \; , \tag{17.21}$$

so that inside the hadron we have $\kappa = 1$, and therefore $\phi = 0$, while outside, $\kappa = \kappa_\infty$ and $\phi = \phi_{vac}$. Thus, all hadrons can be viewed as

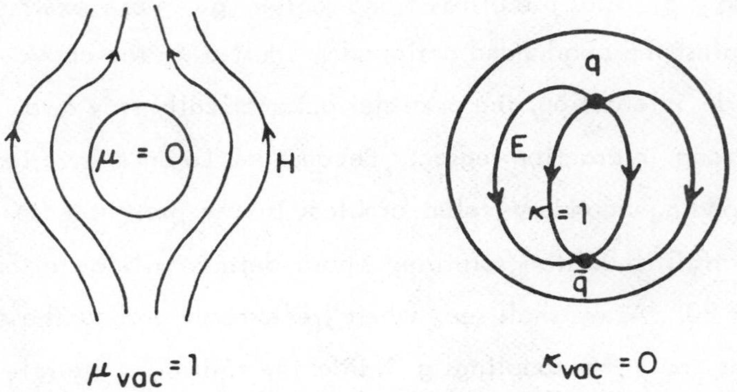

Superconductor =
Perfect Diamagnet

QCD Vacuum =
Perfect Color Dia-electric

Fig. 17.8. Superconductivity in QED vs. quark confinement in QCD.

soliton structures in the physical vacuum. Further details will be given in Chapter 20.

2. Analogy with superconductivity

As mentioned before, in QED there does not exist any dia-electric medium at zero frequency; however, there do exist dispersion laws which exhibit dia-electric natures at some nonzero frequencies. That can lead to physical effects* somewhat similar to those discussed above.

Another analog is the comparison between the superconductivity in QED and the quark confinement in our phenomenological interpretation of QCD. When we switch from QED to QCD, we replace the magnetic field $\vec{H}$ by the color electric field $\vec{E}$, the superconductor by the QCD vacuum, and the QED vacuum by the interior of the hadron;

* W. Happer and A. C. Tam, in Laser Spectroscopy III, ed. J. L. Hall and J. L. Carlsten (Berlin, Heidelberg and New York, Springer-Verlag, 1977), 333.

therefore in Fig. 17.8 the inside by the outside and the outside by the inside. In the former the magnetic field is expelled outwards from the superconductor, while in the latter the color electric field is pushed into the bag, which leads to quark (or color) confinement. This situation can be summarized as follows:

QED Superconductivity		QCD Vacuum as a Perfect Color Dielectric
$\vec{H}$	$\longleftrightarrow$	$\vec{E}$
$\mu_{inside} = 0$	$\longleftrightarrow$	$\kappa_{vacuum} = 0$
$\mu_{vacuum} = 1$	$\longleftrightarrow$	$\kappa_{inside} = 1$
inside	$\longleftrightarrow$	outside
outside	$\longleftrightarrow$	inside

3. Remarks

Quark confinement is a large-scale phenomenon. Therefore, at least on the phenomenological level, it should be understandable through a quasi-classical macroscopic theory, much like the London-Landau theory for superfluidity. We suggest that the use of the color dielectric constant of the vacuum serves such a purpose. Crucial to our explanation is the antiscreening nature of the QCD vacuum. A simple demonstration based on the lowest perturbation calculation will be given in the next chapter (Section 18.5), which may provide some physical insight into this fundamental problem.

Chapter 18

QUANTUM CHROMODYNAMICS AND GAUGE THEORIES *

Quantum chromodynamics deals with the color SU_3 symmetric interaction between quarks and gluons. It is a member of a general class of field theories, called non-Abelian gauge theories**, which we shall analyze first.

18.1 Non-Abelian Gauge Field Theories

An SU_N group is one that consists of all $N \times N$ unitary matrices u with unit determinant. In this book, we are interested in the special cases $N = 2$ and 3 for physical applications. As mentioned before, the generators of the SU_2 group $\{u\}$ are the three Pauli matrices τ^ℓ given by (3.1), and those for SU_3 are the eight Gell-Mann matrices λ^ℓ given by (12.22). In this chapter, all group indices are denoted by superscripts, instead of the subscripts used in previous chapters. It is useful to introduce a uniform notation applicable to all N. We define

$$T^\ell = \tfrac{1}{2}\tau^\ell \quad \text{for} \quad SU_2 \ , \qquad \text{and} \qquad \tfrac{1}{2}\lambda^\ell \quad \text{for} \quad SU_3 \ .$$

---

* Readers who are not interested in the mathematical details of QCD may skip this and the chapter following.

** C. N. Yang and F. Mills, Phys. Rev. 96, 191 (1954). O. Klein, in New Theories in Physics (International Institute of Intellectual Cooperation, League of Nations, 1938), pages 77-93.

For the general SU_N group, there are $M = N^2 - 1$ such T^ℓ 's, and they satisfy

$$T^\ell = T^{\ell\dagger} , \quad \text{trace } (T^\ell T^m) = \tfrac{1}{2}\delta^{\ell m}$$

and

$$[T^\ell, T^m] = i f^{\ell mn} T^n . \tag{18.1}$$

The $f^{\ell mn}$'s are the antisymmetric structure constants of the group algebra; for SU_2, $f^{\ell mn} = \epsilon^{\ell mn}$ is given by (3.4), and for SU_3 by (12.24). For an infinitesimal transformation, the matrix u can be written as

$$u = 1 - i\,\theta^\ell T^\ell \tag{18.2}$$

where the θ^ℓ 's are real and infinitesimal.

Let us consider an SU_N gauge theory consisting of a spin-$\frac{1}{2}$ fermion field ψ^a and a vector gauge field V_μ^ℓ, belonging respectively to the N and $M = N^2 - 1$ dimensional representation of the group. Hence, the superscript a varies from 1 to N, and ℓ from 1 to M. The Lagrangian density is

$$\mathcal{L} = -\tfrac{1}{4} V_{\mu\nu}^\ell V_{\mu\nu}^\ell - \psi^\dagger \gamma_4(\gamma_\mu D_\mu + m)\,\psi \tag{18.3}$$

where

$$\psi = \begin{pmatrix} \psi^1 \\ \psi^2 \\ \vdots \\ \psi^N \end{pmatrix} ,$$

with each ψ^a a 4-component Dirac spinor,

$$V_{\mu\nu}^\ell = \frac{\partial}{\partial x_\mu} V_\nu^\ell - \frac{\partial}{\partial x_\nu} V_\mu^\ell + g f^{\ell mn} V_\mu^m V_\nu^n$$

and

$$D_\mu = \frac{\partial}{\partial x_\mu} - i g T^\ell V_\mu^\ell . \tag{18.4}$$

As usual, $x_\mu = (\vec{r}, it)$, the γ_μ 's are given by (3.11) and all repeated indices are summed over. In analogy with QED, the "electric" and "magnetic" fields E_i^ℓ and B_i^ℓ are

$$E_i^\ell \equiv i V_{i4}^\ell = - \dot{V}_i^\ell - \nabla_i V_0^\ell + g f^{\ell mn} V_0^m V_i^n$$

and (18.5)

$$\epsilon_{ijk} B_k^\ell \equiv V_{ij}^\ell = \nabla_i V_j^\ell - \nabla_j V_i^\ell + g f^{\ell mn} V_i^m V_j^n$$

where $V_0^\ell = - i V_4^\ell$ and the subscripts i, j, k denote the space indices which vary from 1 to 3. By using the variational principle (2.10) and the Lagrangian density (18.3), we see that the equations of motion are

$$(\gamma_\mu D_\mu + m) \psi = 0 \qquad (18.6a)$$

and

$$\frac{\partial}{\partial x_\mu} V_{\mu\nu}^\ell + g f^{\ell mn} V_\mu^m V_{\mu\nu}^n + g I_\nu^n = 0 \qquad (18.6b)$$

where

$$I_\nu^\ell = i \psi^\dagger \gamma_4 \gamma_\nu T^\ell \psi \quad . \qquad (18.6c)$$

It is useful to introduce the $N \times N$ matrices

$$V_\mu \equiv T^\ell V_\mu^\ell \qquad \text{and} \qquad V_{\mu\nu} \equiv T^\ell V_{\mu\nu}^\ell \quad . \qquad (18.7)$$

Thus, (18.4) becomes

$$V_{\mu\nu} = \frac{\partial}{\partial x_\mu} V_\nu - \frac{\partial}{\partial x_\nu} V_\mu - i g [V_\mu, V_\nu]$$

and (18.8)

$$D_\mu = \frac{\partial}{\partial x_\mu} - i g V_\mu \quad .$$

Accordingly, the Lagrangian density (18.3) can be written as

$$\mathcal{L} = - \tfrac{1}{2} \operatorname{trace} (V_{\mu\nu} V_{\mu\nu}) - \psi^\dagger \gamma_4 (\gamma_\mu D_\mu + m) \psi \quad .$$

<u>Theorem 1.</u> The Lagrangian density is invariant under the local SU_N transformation

$$V_\nu \rightarrow V_\nu' = u V_\nu u^\dagger + \frac{i}{g} u \frac{\partial u^\dagger}{\partial x_\nu}$$

and (18.9)

$$\psi \to \psi' = u\psi$$

where $u = u(x)$ is any $N \times N$ unitary matrix function of x_μ with det $u = 1$.

Proof. From $u^\dagger u = 1$, it follows that

$$\frac{\partial u^\dagger}{\partial x_\mu} u + u^\dagger \frac{\partial u}{\partial x_\mu} = 0 \ . \qquad (18.10)$$

Hence,

$$\frac{\partial V_\nu'}{\partial x_\mu} = u \frac{\partial V_\nu}{\partial x_\mu} u^\dagger + \frac{\partial u}{\partial x_\mu} V_\nu u^\dagger + u V_\nu \frac{\partial u^\dagger}{\partial x_\mu} + \frac{i}{g} \frac{\partial u}{\partial x_\mu} \frac{\partial u^\dagger}{\partial x_\nu} + \frac{i}{g} u \frac{\partial^2 u^\dagger}{\partial x_\mu \partial x_\nu}$$

$$= u \frac{\partial V_\nu}{\partial x_\mu} u^\dagger - \left[u \frac{\partial u^\dagger}{\partial x_\mu} , u V_\nu u^\dagger \right] - \frac{i}{g} u \frac{\partial u^\dagger}{\partial x_\mu} u \frac{\partial u^\dagger}{\partial x_\nu} + \frac{i}{g} u \frac{\partial^2 u^\dagger}{\partial x_\mu \partial x_\nu} \ .$$

By substituting the above expression into

$$V_{\mu\nu}' \equiv \frac{\partial}{\partial x_\mu} V_\nu' - \frac{\partial}{\partial x_\nu} V_\mu' - ig [V_\mu' , V_\nu'] \ ,$$

and using the upper equation in (18.8), we obtain

$$V_{\mu\nu}' = u V_{\mu\nu} u^\dagger \ . \qquad (18.11)$$

From (18.9) and (18.10), we find

$$\frac{\partial}{\partial x_\mu} \psi' = u \frac{\partial \psi}{\partial x_\mu} + \frac{\partial u}{\partial x_\mu} \psi = u \left(\frac{\partial \psi}{\partial x_\mu} - \frac{\partial u^\dagger}{\partial x_\mu} u \psi \right) ,$$

and therefore

$$D_\mu' \psi' \equiv \left(\frac{\partial}{\partial x_\mu} - ig V_\mu' \right) \psi' = u D_\mu \psi \ .$$

From these it follows that $\psi'^{\dagger} \gamma_4 \gamma_\mu D'_\mu \psi' = \psi^{\dagger} \gamma_4 \gamma_\mu D_\mu \psi$ and trace $V_{\mu\nu}' V_{\mu\nu}' =$ trace $V_{\mu\nu} V_{\mu\nu}$ which establish the invariance of $\mathcal{L}$.

Remarks. QCD refers to the special case of the color-SU_3 gauge theory. In (18.3) we set $\psi = \psi_f$ and $m = m_f$ for the quark field of flavor f, and then sum over the index f.

Exercise. Show that for the infinitesimal SU_N transformation (18.2), (18.9) becomes

$$V_\nu^\ell \rightarrow V_\nu^\ell + \delta V_\nu^\ell \quad \text{and} \quad \psi \rightarrow \psi + \delta\psi$$

where

$$\delta V_\nu^\ell = f^{\ell mn} \theta^m V_\nu^n - \frac{1}{g} \frac{\partial \theta^\ell}{\partial x_\nu}$$

and

$$\delta\psi = -i T^\ell \theta^\ell \psi \ . \tag{18.12}$$

Verify directly that $\mathcal{L}$ is invariant under such an infinitesimal transformation.

Theorem 2. It is possible to choose $u(x)$ such that the transformation (18.9) can bring V_μ from any configuration $V_\mu = F_\mu(x)$ to

(i) the $V_0 = 0$, called time-axial, gauge in which the new V_μ satisfies

$$V_4(x) = i V_0(x) = 0 \quad \text{everywhere;}$$

(ii) the space-axial gauge, in which one of the spatial components, say,

$$V_1(x) = 0 \quad \text{everywhere;}$$

(iii) the Coulomb gauge, in which

$$\nabla_i V_i(x) = 0 \quad \text{everywhere.}$$

Proof. (i) From any configuration $V_\mu = F_\mu(\vec{r}, t)$, we may choose

$u^\dagger$ to be the following time-ordered function:

$$u^\dagger(\vec{r}, t) = T\, e^{-i\int_0^t g F_0(\vec{r}, t')\, dt'}$$

where $F_0 = -i F_4$ and T is the time-ordering operator defined by (5.38). Hence, $u^\dagger$ satisfies

$$\frac{\partial u^\dagger}{\partial t} = -i g F_0 u^\dagger ,$$

or

$$u F_4 u^\dagger + \frac{i}{g} u \frac{\partial u^\dagger}{\partial x_4} = 0 .$$

Thus, (18.9) transforms V_μ from the configuration $F_\mu(x)$ to the time-axial, or $V_0 = 0$, gauge.

(ii) Replacing x_4 and F_4 in the above formulas by x_1 and F_1, we establish the accessibility of the space-axial gauge.

(iii) From any given configuration $V_\mu = F_\mu(\vec{r}, t)$, we can define

$$A_i \equiv u F_i u^\dagger + \frac{i}{g} u \nabla_i u^\dagger \equiv T^\ell A_i^\ell$$

and

$$I(A_i) \equiv \int \text{trace}\,(A_i A_i)\, d^3r = \tfrac{1}{2} \int A_i^\ell A_i^\ell\, d^3r .$$

Keeping t and F_μ fixed, from the two equations above we have A_i and $I(A_i)$ both as functionals of $u(\vec{r})$. Since $I(A_i)$ is by definition $\geqslant 0$, it should have a minimum. We now vary $u(\vec{r})$ to search for this minimum. From (18.2) we see that for $\delta u = -i\theta^\ell T^\ell$

$$\delta A_i^\ell = f^{\ell mn} \theta^m A_i^n - \frac{1}{g} \nabla_i \theta^\ell .$$

Because $f^{\ell mn}$ is antisymmetric, we have

$$A_i^\ell \delta A_i^\ell = -\frac{1}{g} A_i^\ell \nabla_i \theta^\ell .$$

Hence, by using the above expression for $I(A_i)$ and through partial integration, we obtain

$$\delta I = \int A_i^\ell \delta A_i^\ell\, d^3r = \frac{1}{g} \int \theta^\ell(\vec{r})\, \nabla_i A_i^\ell(\vec{r})\, d^3r .$$

Since $\theta^{\ell}(\vec{r})$ can be an arbitrary infinitesimal function of $\vec{r}$, the minimum of I occurs when

$$\nabla_i A_i^{\ell} = 0$$

which is the Coulomb gauge.

In order to insure that in the above partial integration there is no surface term, we may enclose the system in a box of volume L^3 and impose the periodic boundary condition. In QCD, because of the infrared problem, the finite-box approach affords a certain degree of mathematical security.

Remarks. Different gauges complement each other in their applications. In order to establish unitarity, the direct route is via the Hamiltonian operator and canonical procedures. As we shall see, this can be carried out most simply in the time-axial gauge, from which we can then move on to other gauges. For practical calculations, the time-axial gauge has extra complications because there exists an infinite number of constraints on the state vectors in that gauge. In most applications, the Coulomb and the covariant gauges are the more convenient ones to use. To show Lorentz invariance, we need the covariant gauge (Section 19.6). However, by itself the covariant gauge has no Hermitian Hamiltonian, and therefore it needs other gauges for the proof of unitarity.

Exercise. The (classical) energy-momentum tensor density $T_{\mu\nu}$ is given by

$$T_{\mu\nu} = V_{\mu\sigma}^{\ell} \frac{\partial V_{\sigma}^{\ell}}{\partial x_{\nu}} + \psi^{\dagger} \gamma_4 \gamma_{\mu} \frac{\partial \psi}{\partial x_{\nu}} + L\, \delta_{\mu\nu} .$$

(i) Verify that $T_{\mu\nu}$ satisfies

$$\frac{\partial}{\partial x_{\mu}} T_{\mu\nu} = 0 .$$

Hence, the energy E and the momentum $\vec{P}$, given by

$$E = -\int T_{44}\, d^3r \quad \text{and} \quad P_k = -i \int T_{4k}\, d^3r \ ,$$

are conserved.

(ii) Prove that E and $\vec{P}$ are gauge invariant, even though $T_{\mu\nu}$ is not.

18.2 An Example

As dynamical systems, both QED and QCD belong to a class of Lagrangian theories whose generalized coordinates can be separated into two types: q_α's and ξ_n's. The Lagrangian L is a function of q_α, $\dot{q}_\alpha$, ξ_n but not of $\dot{\xi}_n$:

$$L = L(q_\alpha,\ \dot{q}_\alpha,\ \xi_n) \ . \tag{18.13}$$

In addition, L is invariant under a group of transformations whose elements u contain certain <u>arbitrary</u> functions of time (e.g., in (18.9)).

<u>Definition.</u> We call a Lagrangian system "gauge invariant" whenever the elements of its symmetry group contain arbitrary functions of time.

Clearly, the possibility of gauge invariance is not restricted only to field theories.

The Lagrangian equations of motion can be derived through the variational principle (2.10). From (18.13) we find

$$\frac{d}{dt}\frac{\partial L}{\partial \dot{q}_\alpha} - \frac{\partial L}{\partial q_\alpha} = 0$$

and

$$\frac{\partial L}{\partial \xi_n} = 0 \ . \tag{18.14}$$

In classical mechanics, usually the solution of the Lagrangian equations of motion can be uniquely fixed by the initial values of the generalized coordinates and their first time derivatives. This is no longer so for a theory with gauge invariance, because of the arbitrary time-dependent gauge transformations such as (18.9). For a quantum system, this feature generates still other complications, as we shall see.

Without gauge invariance, (18.13) by itself does not give us any problem in carrying out the quantization procedure. We may use $\partial L/\partial \xi_n = 0$ to solve ξ_n in terms of q_α and $\dot{q}_\alpha$:

$$\xi_n = \xi_n(q_\alpha , \dot{q}_\alpha)$$

which, when substituted into (18.13), gives

$$L = L(q_\alpha , \dot{q}_\alpha , \xi_n(q_\alpha , \dot{q}_\alpha)) \equiv \bar{L}(q_\alpha , \dot{q}_\alpha) \;\; .$$

The generalized momentum p_α is given by

$$p_\alpha = \frac{\partial \bar{L}}{\partial \dot{q}_\alpha} = \frac{\partial L}{\partial \dot{q}_\alpha}$$

where in the differentiation $\partial \bar{L}/\partial \dot{q}_\alpha$ we keep q_α fixed, but in $\partial L/\partial \dot{q}_\alpha$ we also keep ξ_n fixed. It is easy to see that the standard quantization procedures (1.5)-(1.9) can be carried out in a straightforward manner.

With gauge invariance, the situation is quite different. As will be shown, the time-dependence of u would in general impose a functional relation between p_α, q_α and $\partial L/\partial \xi_n$, which makes it impossible to satisfy simultaneously $\partial L/\partial \xi_n = 0$ and the quantization rule

$$[p_\alpha , q_\beta] = -i\delta_{\alpha\beta} \;\; . \tag{18.15}$$

This, then, is the underlying reason why we have to choose a specific gauge to carry out the quantization procedure.

1. A simple mechanical model

Let us consider a point particle in a three-dimensional space at position $\vec{r}$. Its Lagrangian is

$$L = \tfrac{1}{2}(\dot{\vec{r}} - \vec{\xi} \times \vec{r})^2 - V(r)$$

where $r = |\vec{r}|$, $\vec{\xi}$ is another coordinate vector, but $\dot{\vec{\xi}}$ is absent in L. It can be readily verified that this Lagrangian is invariant under the transformation

$$\begin{aligned} \vec{r} &\rightarrow \vec{r} + \vec{\alpha} \times \vec{r} \\ \text{and} \quad \vec{\xi} &\rightarrow \vec{\xi} + \vec{\alpha} \times \vec{\xi} + \dot{\vec{\alpha}} \end{aligned} \tag{18.16}$$

where $\vec{\alpha} = \vec{\alpha}(t)$ can be an arbitrary infinitesimal vector function of time t. Except for the $\frac{1}{2}(\vec{\xi} \times \vec{r})^2$ term, this would be the problem of a nonrelativistic charged particle moving in a central potential $V(r)$ and under the influence of an external magnetic field.

We may further simplify the problem by imposing the constraint that $\vec{r} = (x_1, x_2, x_3)$ lies on the (x_1, x_2) plane and $\vec{\xi} = \hat{e}\,\xi$ where $\hat{e}$ is the unit vector along the x_3-axis. The above Lagrangian then becomes

$$L = \tfrac{1}{2}(\dot{x}_1^2 + \dot{x}_2^2) - (x_1\dot{x}_2 - x_2\dot{x}_1)\,\xi + \tfrac{1}{2}\xi^2 r^2 - V(r) \tag{18.17}$$

where x_1 and x_2 are the Cartesian coordinates of $\vec{r}$ which is now a two-dimensional vector. In terms of the polar coordinates $x_1 = r\cos\theta$ and $x_2 = r\sin\theta$, L can be written as

$$L = \tfrac{1}{2}[\dot{r}^2 + r^2(\dot{\theta} - \xi)^2] - V(r) \tag{18.18}$$

and (18.16) is simply the Abelian group of transformations:

$$\begin{aligned} \theta &\rightarrow \theta + \alpha(t) \\ \text{and} \quad \xi &\rightarrow \xi + \dot{\alpha}(t) \end{aligned} \tag{18.19}$$

where $\alpha(t)$ can now be any finite function of t. The invariance group of this simple example shares with the gauge groups of QED or QCD the special feature that its elements contain arbitrary functions of t, like the u's in (18.9).

By using (18.14) and setting $\xi_n = \xi$, we find

$$\frac{\partial L}{\partial \xi} = -r^2(\dot{\theta} - \xi) = 0 \quad . \tag{18.20}$$

Since the conjugate momentum to θ is

$$p_\theta = \frac{\partial L}{\partial \dot{\theta}} = r^2(\dot{\theta} - \xi) \quad ,$$

we see that the commutation relation

$$[p_\theta, \theta] = -i \tag{18.21}$$

contradicts (18.20), confirming the necessity of choosing a gauge for the quantization procedure.

The Lagrangian equations of motion (18.14) can, however, be written down without difficulty. In this example they are, in polar coordinates

$$\ddot{r} + \frac{dV}{dr} = 0$$

and (18.22)

$$\dot{\theta} - \xi = 0 \quad .$$

2. <u>$\xi = 0$ gauge</u>

Because of the gauge transformation (18.19), any orbit $\vec{r} = \vec{r}(t)$ and $\xi = \xi(t)$ can be transformed to one in which $\xi = 0$ at all time. In this gauge, by setting $\xi = 0$ in (18.18) we find

$$L = \tfrac{1}{2}\dot{\vec{r}}^{\,2} - V(r) \quad .$$

The conjugate momentum $\vec{p}$ is given by

$$\vec{p} = \frac{\partial L}{\partial \dot{\vec{r}}} = \dot{\vec{r}} \quad ,$$

and the Hamiltonian H is $\frac{1}{2}\vec{p}^{\,2} + V(r)$. Thus, in quantum mechanics

$$\vec{p} = -i\vec{\nabla}$$

and

$$H = -\tfrac{1}{2}\nabla^2 + V(r) \quad . \tag{18.23}$$

The angular-momentum operator

$$p_\theta = -i\frac{\partial}{\partial\theta} = -i\left(x_1\frac{\partial}{\partial x_2} - x_2\frac{\partial}{\partial x_1}\right)$$

commutes with H. To be consistent with the equation of motion $r^2(\dot{\theta} - \xi) = 0$, given by (18.22), only eigenstates $|\,\rangle$ of H with

$$p_\theta\,|\,\rangle = 0 \tag{18.24}$$

will be accepted. Hence, these eigenstates are all θ-independent, and that leads to

$$H = -\frac{1}{2r}\frac{d}{dr}\left(r\frac{d}{dr}\right) + V(r) \quad . \tag{18.25}$$

Consequently, in the $\xi = 0$ gauge we reconcile (18.20) and (18.21) by keeping the quantization rule (18.15)

$$[p_1, x_1] = [p_2, x_2] = -i$$

and

$$[p_1, p_2] = [x_1, x_2] = 0$$

intact, but replacing $\partial L/\partial\xi = 0$ by the constraint (18.24) on the state vector.

3. $x_2 = 0$ gauge

From (18.19), we see that any orbit $\vec{r} = \vec{r}(t)$ and $\xi = \xi(t)$ can also be transformed to one with $x_2 = 0$ at all times, albeit there are two branches: x_1 can be > 0 or < 0, with $x_1 = 0$ being the point where the Jacobian of the transformation is zero. [This corresponds to the so-called Gribov ambiguity in the literature.] In the $x_2 = 0$ gauge, $r^2 = x_1^{\,2}$ and the Lagrangian (18.17) becomes

$$L = \tfrac{1}{2}\dot{x}_1^{\,2} + \tfrac{1}{2}\xi^2 x_1^{\,2} - V(x_1)$$

where, for definiteness, we choose the branch of positive x_1. Since the above L does not contain $\dot{\xi}$, we may follow the standard procedure to eliminate ξ through $\partial L/\partial \xi = 0$, which in the present example is simply $x_1^{\,2}\xi = 0$. Hence, the Lagrangian becomes

$$L = \tfrac{1}{2}\dot{x}_1^{\,2} - V(x_1) \quad .$$

The conjugate momentum p_1 is $\dot{x}_1$ and the classical Hamiltonian is

$$H = \tfrac{1}{2}p_1^{\,2} + V(x_1) \quad .$$

In passing over to quantum mechanics, in order that the spectrum of this Hamiltonian be identical to that of (18.25), it is important not to treat x_1 in the $x_2 = 0$ gauge as a Cartesian coordinate; in the above Hamiltonian $p_1^{\,2}$ is the operator $-\dfrac{1}{x_1}\dfrac{d}{dx_1}\left(x_1\dfrac{d}{dx_1}\right)$ and not $-\dfrac{d^2}{dx_1^{\,2}}$. Thus in the $x_2 = 0$ gauge we reconcile (18.20) and (18.21) by modifying the quantization rule (18.15), but retaining $\partial L/\partial \xi = 0$ as a bona fide operator equation.

Remarks. (i) The Lagrangian (18.18) has the same formal expression as that of a two-dimensional particle moving in a central potential $V(r)$, but observed in a rotating frame with angular velocity ξ. If ξ is regarded as a dynamical variable, then the problem acquires a gauge invariance, and it reduces to the example discussed above.

(ii) As we shall see, the $\xi = 0$ gauge is the analog of the $\vec{V}_0 = 0$ gauge in QED or QCD; the $x_2 = 0$ gauge corresponds to the usual Coulomb gauge. In this simple example, the choice of which coordinate in what gauge is Cartesian can be determined by the one who makes up the problem; it seems sensible to assume that in the $\xi = 0$ gauge the coordinates x_1 and x_2 are Cartesian, as is done in (18.23). In the non-Abelian gauge field theory, however, one is guided by the requirement of relativistic invariance. As we shall also see, that is consistent with the choice of $\vec{V}_0 = 0$ gauge as the starting point.

(iii) The group (18.19) in this simple example is Abelian. Nevertheless, there can be complications due to curvilinear coordinates. The gauge transformation in QED is not only Abelian, but also a translation in rectilinear coordinates, and that is why it is simple.

(iv) Before leaving this example, we note that if we wish we may choose a more general gauge, in which an arbitrary function $\chi(x_1, x_2, \xi) = 0$, provided that any point in the (x_1, x_2, ξ) space can be transformed onto the surface $\chi(x_1, x_2, \xi) = 0$ through the gauge transformation (18.19). In the next chapter, Section 19.3.3, when we discuss the path-integration formalism, we shall come back to this simple model. Here, we only remark that it is often useful to absorb a factor of $r^{\frac{1}{2}}$ into the state vectors so that H, given by (18.25), becomes

$$\overline{H} \equiv r^{\frac{1}{2}} H r^{-\frac{1}{2}} = -\frac{1}{2} \frac{d^2}{dr^2} + V(r) - \frac{1}{8r^2} \tag{18.26}$$

which changes the volume element from $r dr$ to dr and adds to the potential $V(r)$ a new term $-(8r^2)^{-1}$.

18.3 Quantization: $V_0 = 0$ Gauge

We now return to the non-Abelian gauge field theories discussed in Section 18.1. From Theorem 2(i) on page 410, we see that through the gauge transformation (18.9), any configuration of V_μ can be brought to the time-axial gauge; i.e.,

$$V_0^{\ell} = 0 \quad \text{everywhere.}$$

In the time-axial gauge, by setting $V_0^{\ell} = 0$ in (18.5), we find

$$E_i^{\ell} = -\dot{V}_i^{\ell}$$

and (18.3) becomes

$$\mathcal{L} = \tfrac{1}{2}(\dot{V}_i^{\ell}\dot{V}_i^{\ell} - B_i^{\ell}B_i^{\ell}) - \psi^{\dagger}\gamma_4(\gamma_{\mu}D_{\mu} + m)\psi \ . \qquad (18.27)$$

The conjugate momentum of V_i^{ℓ} is simply

$$\Pi_i^{\ell} = \dot{V}_i^{\ell} = -E_i^{\ell} \ , \qquad (18.28)$$

and that of ψ is $i\psi^{\dagger}$, as in (3.17). Thus, the Hamiltonian density $\Pi_i^{\ell}\dot{V}_i^{\ell} + i\psi^{\dagger}\dot{\psi} - \mathcal{L}$ is

$$\mathcal{H} = \tfrac{1}{2}(\Pi_i^{\ell}\Pi_i^{\ell} + B_i^{\ell}B_i^{\ell}) - gI_i^{\ell}V_i^{\ell} + \psi^{\dagger}\gamma_4(\gamma_i\nabla_i + m)\psi \ , \qquad (18.29)$$

where I_i^{ℓ} denotes the space-component of the current operator given by (18.6c). The usual canonical quantization procedure leads to

$$[V_i^{\ell}(\vec{r}, t), \Pi_j^{m}(\vec{r}', t)] = i\delta_{ij}\delta^{\ell m}\delta^3(\vec{r} - \vec{r}') \ ,$$
$$\{\psi(\vec{r}, t), \psi^{\dagger}(\vec{r}', t)\} = \delta^3(\vec{r} - \vec{r}') \ . \qquad (18.30)$$

The equal-time commutators between the V_i's and between the Π_i's are zero; likewise, the equal-time anticommutators between the ψ's and between the $\psi^{\dagger}$'s are also zero.

Remnants of the original gauge transformations (18.9) remain important. In accordance with (18.7), we denote

$$V_i = T^{\ell}V_i^{\ell}$$
and
$$\Pi_i = T^{\ell}\Pi_i^{\ell} \ . \qquad (18.31)$$

It can be readily verified that the Hamiltonian density $\mathcal{H}$ and the commutation relations are invariant under a time-independent SU_N transformation

$$V_i \rightarrow u V_i u^\dagger - \frac{i}{g} (\nabla_i u) u^\dagger ,$$

$$\Pi_i \rightarrow u \Pi_i u^\dagger \tag{18.32}$$

and

$$\psi \rightarrow u \psi$$

where $u = u(\vec{r})$ can be any $N \times N$ unitary matrix function of $\vec{r}$ with $\det u = 1$. Since the $u(\vec{r})$'s are time-independent, the invariance group $\{u(\vec{r})\}$ is generated by the r-dependent operators $\mathcal{G}^\ell$ which are conserved:

$$\mathcal{G}^\ell \equiv J^\ell + \psi^\dagger T^\ell \psi$$

where

$$J^\ell \equiv \frac{1}{g} D_i^{\ell m} \Pi_i^m \tag{18.33}$$

and

$$D_i^{\ell m} = \delta^{\ell m} \nabla_i - g f^{\ell m n} V_i^n .$$

It is straightforward to verify

$$[J^\ell(\vec{r}, t), \; J^m(\vec{r}', t)] = i f^{\ell m n} \delta^3(\vec{r} - \vec{r}') J^n(\vec{r}, t) ,$$

$$[\mathcal{G}^\ell(\vec{r}, t), \; \mathcal{G}^m(\vec{r}', t)] = i f^{\ell m n} \delta^3(\vec{r} - \vec{r}') \mathcal{G}^n(\vec{r}, t) ,$$

$$[\mathcal{G}^\ell(\vec{r}, t), \; \psi(\vec{r}', t)] = - \delta^3(\vec{r} - \vec{r}') T^\ell \psi(\vec{r}, t) , \tag{18.34}$$

$$[\mathcal{G}^\ell(\vec{r}, t), \; \Pi_i^m(\vec{r}', t)] = i f^{\ell m n} \delta^3(\vec{r} - \vec{r}') \Pi_i^n(\vec{r}, t)$$

and

$$[\mathcal{G}^\ell(\vec{r}, t), \; V_i^m(\vec{r}', t)] = i f^{\ell m n} \delta^3(\vec{r} - \vec{r}') V_i^n(\vec{r}, t) - \frac{i}{g} \delta^{\ell m} \nabla_i \delta^3(\vec{r} - \vec{r}')$$

where ∇_i is the differential operator with respect to $\vec{r}$. Consequently $\mathcal{G}^\ell$ commutes with the Hamiltonian $H = \int \mathcal{H} d^3r$ and

$$\dot{\mathcal{G}}^\ell(\vec{r}, t) = i[H, \mathcal{G}^\ell(\vec{r}, t)] = 0 .$$

By commuting H with ψ, we obtain the equation of motion for ψ. Likewise, by commuting H with V_i^ℓ and Π_i^ℓ, we derive (18.6b) for $\nu = i$ which can be 1, 2, 3, but not 4. We note that $i\,g\,\mathcal{G}^{\ell}$ is identical to the lefthand side of (18.6b) when $\nu = 4$. Thus, in order to be consistent with all the Lagrangian equations of motion, in the $V_0 = 0$ gauge we require all state vectors $|\,>$ to satisfy

$$\mathcal{G}^{\ell}\,|\,> \;=\; 0\;. \tag{18.35}$$

In the Schrödinger picture the operators $V_i^{\ell} = V_i^{\ell}(\vec{r})$ and $\Pi_i^{\ell} = \Pi_i^{\ell}(\vec{r})$ are all t-independent. The state vector in the V_i^{ℓ}-representation is the functional

$$\psi(V_i) \;\equiv\; <V_i^{\ell}\,|\,>\;. \tag{18.36}$$

In this representation,

$$\Pi_i^{\ell}(\vec{r}) \;=\; -\,i\;\frac{\delta}{\delta V_i^{\ell}(\vec{r})}\;. \tag{18.37}$$

Hence, the Hamiltonian H is

$$H \;=\; \mathcal{K} + \mathcal{V} \tag{18.38}$$

where

$$\mathcal{K} \;=\; -\tfrac{1}{2}\int \frac{\delta}{\delta V_i^{\ell}(\vec{r})}\;\frac{\delta}{\delta V_i^{\ell}(\vec{r})}\;d^3r$$

and

$$\mathcal{V} \;=\; \int\,[\tfrac{1}{2}\,B_i^{\ell}\,B_i^{\ell} - g\,I_i^{\ell}\,V_i^{\ell} + \psi^{\dagger}\,\gamma_4(\gamma_i\,\nabla_i + m)\,\psi\,]\;d^3r\;. \tag{18.39}$$

From (18.28), it follows that the operator $\mathcal{K}$ of (18.39) is simply the integral of the electric energy $\frac{1}{2}\int d^3r\,E_i^{\ell}\,E_i^{\ell}$. By using (18.33), we see that

$$g\,\mathcal{G}^{\ell} \;=\; -\,\nabla_i\,E_i^{\ell} + g\,f^{\ell mn}\,E_i^{m}\,V_i^{n} + g\,\psi^{\dagger}\,T^{\ell}\,\psi\;.$$

Hence, the constraint (18.35) implies that the state vector $|\,>$ satisfies Gauss' law, which can be written either as

$$(\nabla_i\,E_i^{\ell} - g\,\rho^{\ell})\,|\,> \;=\; 0$$

or (18.40)

$$((D_i E_i)^\ell - g\rho_F^\ell \,|\!> = 0$$

where D_i is the covariant derivative which satisfies

$$(D_i E_i)^\ell \equiv \nabla_i E_i^\ell + g f^{\ell mn} V_i^m E_i^n ,$$

and ρ^ℓ is the total "charge" density. We can decompose ρ^ℓ into two terms

$$\rho^\ell = \rho_V^\ell + \rho_F^\ell$$

with ρ_V^ℓ as the charge carried by the gauge field due to its non-Abelian nature

$$\rho_V^\ell = f^{\ell mn} E_i^m V_i^n \tag{18.41}$$

and ρ_F^ℓ as that due to the fermion field,

$$\rho_F^\ell = \psi^\dagger T^\ell \psi \ . \tag{18.42}$$

We may expand $V_i^\ell(\vec{r})$ in terms of the Fourier series:

$$V_i^\ell(\vec{r}) = \sum{}' \sqrt{2/L^3}\ [x_i^\ell(\vec{k}) \cos \vec{k}\cdot\vec{r} + y_i^\ell(\vec{k}) \sin \vec{k}\cdot\vec{r}\,]$$

where the sum $\sum'$ extends over half of the $\vec{k}$ space, and L^3 is the volume of the system. Clearly, $x_i^\ell(\vec{k})$ exists for $\vec{k} = 0$ as well as $\vec{k} \neq 0$, but $y_i^\ell(\vec{k})$ exists only for $\vec{k} \neq 0$. For each nonzero wave-number pair $\vec{k}$ and $-\vec{k}$, there are $3M$ $x_i^\ell(\vec{k})$ and $3M$ $y_i^\ell(\vec{k})$, since i runs from 1 to 3 and ℓ varies from 1 to $M = N^2 - 1$ for the SU_N group. The Hermiticity of V_i^ℓ implies that $x_i^\ell(\vec{k})$ and $y_i^\ell(\vec{k})$ are also Hermitian,

$$x_i^\ell(\vec{k})^\dagger = x_i^\ell(\vec{k}) \qquad \text{and} \qquad y_i^\ell(\vec{k})^\dagger = y_i^\ell(\vec{k}) \ .$$

In terms of these variables, $\mathcal{K}$ of (18.39) can be written as

$$\mathcal{K} = \tfrac{1}{2}\int d^3r\, E_i^\ell E_i^\ell = -\tfrac{1}{2}\sum{}' \left\{ \left[\frac{\partial}{\partial x_i^\ell(\vec{k})}\right]^2 + \left[\frac{\partial}{\partial y_i^\ell(\vec{k})}\right]^2 \right\} ,$$

and the Hamiltonian (18.38) becomes

$$H = -\frac{1}{2}\sum{}' \left\{ \left[\frac{\partial}{\partial x_i^{\ell}(\vec{k})} \right]^2 + \left[\frac{\partial}{\partial y_i^{\ell}(\vec{k})} \right]^2 \right\} + \mathcal{V} . \tag{18.43}$$

By comparing the above discussion with that given in Section 18.2 for the $\xi = 0$ gauge of the simple example, we see that ξ of the simple example is the analog of V_0; likewise, $\vec{p} = -i\vec{\nabla}$ is the analog of $\Pi_i^{\ell} = -i\,\delta/\delta V_i^{\ell}$, and $p_\theta \,|\,\rangle = 0$ is that of $\mathcal{G}^{\ell}\,|\,\rangle = 0$. Just as in the simple model, knowing the Hamiltonian operator in one gauge uniquely determines its form in any other gauge. As we shall see, through the introduction of curvilinear coordinates it is possible to eliminate the constraint (18.35), in complete analogy with the passage from (18.23) to (18.25). Furthermore, as will be discussed in Section 19.6, for the covariant gauge, this procedure does lead to a set of relativistically covariant Feynman rules.

It is useful to examine the above Hamiltonian side by side with that for the quantum soliton discussed in Section 7.6. We note that the two Hamiltonians (7.81) and (18.43) have very similar mathematical structure; both contain the Laplace operator in terms of a set of Cartesian coordinates. We recall that when we transform any set of Cartesian coordinates $x_1, x_2, \cdots$ into a set of curvilinear coordinates $q_1, q_2, \cdots,$

$$x = \begin{pmatrix} x_1 \\ x_2 \\ \vdots \end{pmatrix} \rightarrow q = \begin{pmatrix} q_1 \\ q_2 \\ \vdots \end{pmatrix} ,$$

the Lagrangian

$$L = \tfrac{1}{2}\tilde{\dot{x}}\dot{x} - U$$

becomes

$$L = \tfrac{1}{2}\tilde{\dot{q}}\, M \dot{q} - U$$

where U depends only on $x = x(q)$, and the matrix elements of M are given by

$$M_{ab} = \frac{\partial x_c}{\partial q_a}\frac{\partial x_c}{\partial q_b} \quad . \tag{18.44}$$

The conjugate momentum of q_a is $p_a = \partial L/\partial \dot{q}_a = M_{ab}\,\dot{q}_b$. Classically, the Hamiltonian is

$$H_{cl} \equiv p_a \dot{q}_a - L = \tfrac{1}{2}\tilde{p}\, M^{-1} p + U \tag{18.45}$$

where p is the column matrix whose elements are p_a.

Quantum mechanically, the Hamiltonian operator in the Cartesian coordinates $x_1, x_2, \cdots$ is

$$H = -\tfrac{1}{2}\frac{\partial^2}{\partial x_a \partial x_a} + U \quad . \tag{18.46}$$

By using (7.113), we see that in curvilinear coordinates it becomes

$$H = \frac{1}{2\mathcal{J}}\tilde{p}\, M^{-1}\mathcal{J}\, p + U = -\tfrac{1}{2}\frac{1}{\mathcal{J}}\frac{\partial}{\partial q_a}(M^{-1}_{ab}\,\mathcal{J}\frac{\partial}{\partial q_b}) + U \tag{18.47}$$

where $\mathcal{J}$ is the Jacobian given by

$$\mathcal{J} = \sqrt{\det M} \quad . \tag{18.48}$$

[See Exercise 3 below.]

Once it is known that there exists a Cartesian form (18.46) for the Hamiltonian operator, then its form in any curvilinear coordinates is uniquely determined. By setting in the classical Hamiltonian (18.45)

$$p_a = -i\frac{\partial}{\partial q_a} \quad ,$$

and with the inclusion of the Jacobian factor, we obtain the quantum operator (18.47). In Chapter 7 this procedure enabled us to derive the quantum soliton Hamiltonian (7.99). Here, the same steps also provide us a direct route to other gauges.

Exercise 1. Verify the equal-time commutators given in (18.34).

Exercise 2. Restrict the Hilbert space to state vectors that satisfy (18.35). Show that the equations of motion (18.6a)-(18.6c) can be derived by using Heisenberg's equation (1.9).

Exercise 3. The Jacobian of the transformation $x_1, x_2, \cdots \rightarrow q_1, q_2, \cdots$ is

$$\mathcal{J} = \frac{\partial(x_1, x_2, \cdots)}{\partial(q_1, q_2, \cdots)} \quad .$$

Verify that it satisfies (18.48).

18.4 Coulomb gauge

1. Coordinate transformation

Let us start from the $V_0 = 0$ gauge and consider any configuration $V_i^{\ell}(x)$. According to Theorem 2 on p. 410, there always exists a gauge-transformation matrix $u(x)$ which can bring $V_i^{\ell}(x)$ into the Coulomb gauge, and vice versa. For clarity, the gauge field in the Coulomb gauge will be denoted by $A_{\mu}^{\ell}(x)$, and that in the $V_0 = 0$ gauge by $V_i^{\ell}(x)$. Thus, we may write

$$V_i = u A_i u^{\dagger} + \frac{i}{g} u \nabla_i u^{\dagger} \tag{18.49}$$

where, as in (18.7),

$$V_i = T^{\ell} V_i^{\ell} \quad , \quad A_i = T^{\ell} A_i^{\ell} \tag{18.50}$$

and as is appropriate in the Coulomb gauge

$$\nabla_i A_i = 0 \quad .$$

From (18.9) we see that, because $V_0 = 0$, the fourth component of A_{μ} satisfies

$$0 = u A_4 u^{\dagger} + \frac{1}{g} u \dot{u}^{\dagger} \quad . \tag{18.51}$$

For SU_2, since there are three generators $\frac{1}{2}\tau^\ell$, as shown in (11.33), any 2×2 unitary matrix u with $\det u = 1$ can be characterized by three angular variables θ_1, θ_2 and θ_3, usually referred to as the group parameters. For SU_3, there are eight generators $\frac{1}{2}\lambda^\ell$. Thus, any 3×3 unitary matrix u with unit determinant depends on eight group parameters $\theta_1, \cdots, \theta_8$. In general, for SU_N we need $M = N^2 - 1$ parameters to specify the u-matrix. In a field theory, these θ_α are functions of x, and through them $u(\theta_\alpha)$ also becomes x-dependent; hence we have

$$u(x) = u(\theta_\alpha(x)) \quad . \tag{18.52}$$

At any fixed space-time point x, there are altogether $3M$ $V_i^\ell(x)$, since i varies from 1 to 3 and ℓ from 1 to M. In the Coulomb gauge there are M constraints given by

$$\nabla_i A_i^\ell = 0 \quad .$$

Hence, there are only $2M$ independent $A_i^\ell(x)$. Equation (18.49) can then be viewed as the transformation relating $3M$ Cartesian coordinate $V_i^\ell(\vec{r})$ to the curvilinear coordinates, $2M$ A_i^ℓ and M θ_α:

$$V_i^\ell(x) \to A_i^\ell(x) \quad \text{and} \quad \theta_\alpha \quad . \tag{18.53}$$

2. Rigid-body rotation

For SU_2, the three group parameters θ_α have familiar physical interpretations. We may choose them to be the Euler angles a, b and c. Thus, the u-matrix of (18.49) can be written as

$$u = e^{-\frac{i}{2}\tau_z b}\, e^{-\frac{i}{2}\tau_y a}\, e^{-\frac{i}{2}\tau_z c} \tag{18.54}$$

where τ_x, τ_y, τ_z are the usual Pauli matrices.

We recall that the origin of the Euler angles a, b, c lies in

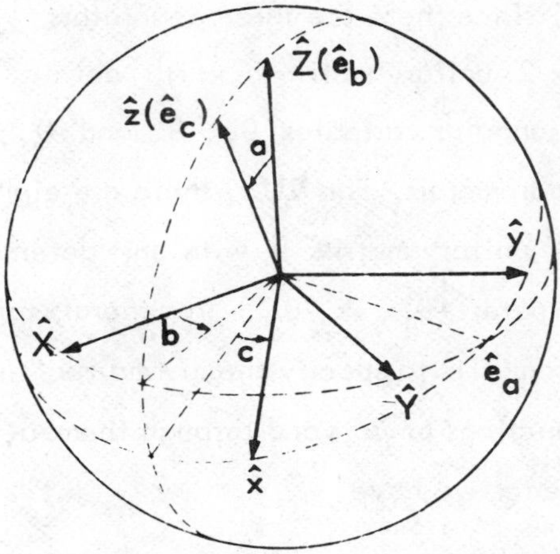

Fig. 18.1. The basis vectors of $\sum_{lab}$ are $\hat{X}$, $\hat{Y}$, $\hat{Z}$ and those of $\sum_{body}$ are $\hat{x}$, $\hat{y}$, $\hat{z}$.

the description of a rigid-body rotation. As shown in Fig. 18.1, there are two reference systems, the laboratory frame $\sum_{lab}$ and the body frame $\sum_{body}$. Each frame is defined by a basis of three orthogonal unit vectors: $\hat{X}$, $\hat{Y}$, $\hat{Z}$ for $\sum_{lab}$ and $\hat{x}$, $\hat{y}$, $\hat{z}$ for $\sum_{body}$. To go from $\sum_{lab}$ to $\sum_{body}$, we follow the "cab" sequence: First rotate an angle b along

$$\hat{Z} \equiv \hat{e}_b ,$$

which moves the y-axis from $\hat{Y}$ to $\hat{e}_a$, then an angle a along $\hat{e}_a$ which rotates the z-axis from $\hat{Z}$ to $\hat{z}$, and finally an angle c along

$$\hat{z} \equiv \hat{e}_c .$$

The inverse "bac" sequence then brings $\sum_{body}$ to $\sum_{lab}$.

Consider now a point P in space, whose coordinates in $\sum_{lab}$ and $\sum_{body}$ are, respectively, X, Y, Z and x, y, z. Let us define

$$R \equiv \tfrac{1}{2}(\tau_x X + \tau_y Y + \tau_z Z)$$

and

$$r \equiv \tfrac{1}{2}(\tau_x x + \tau_y y + \tau_z z) \quad .$$

From the definition of the Euler angles, it follows that

$$R = u r u^\dagger$$

where u is given by (18.54). [As we shall see, R plays the role of V_i in the gauge-field case, and r the role of A_i .]

We keep $\sum_{lab}$ fixed, and consider the rotation of $\sum_{body}$ by changing a , b and c . The angular-velocity vector is

$$\dot{a}\,\hat{e}_a + \dot{b}\,\hat{e}_b + \dot{c}\,\hat{e}_c \tag{18.55}$$

where, as usual, the dot denotes a time derivative. Let us refer to the components of the above vector in $\sum_{body}$ and $\sum_{lab}$ as ω^ℓ and Ω^ℓ respectively. It is useful to define

$$i\,u^\dagger \dot{u} \equiv \omega = \tfrac{1}{2}\vec{\tau}\cdot\vec{\omega}$$

and

$$i\,\dot{u}\,u^\dagger \equiv \Omega = \tfrac{1}{2}\vec{\tau}\cdot\Omega \quad . \tag{18.56}$$

Then, the matrices ω and Ω are related by

$$\Omega = u\,\omega\,u^\dagger \quad ,$$

and the components of $\vec{\omega}$ and $\vec{\Omega}$ are the aforementioned ω^ℓ and Ω^ℓ , related by

$$\Omega^\ell = U^{\ell m}\,\omega^m \tag{18.57}$$

where $U^{\ell m}$ satisfies

$$u^\dagger \tau^\ell u = U^{\ell m}\,\tau^m \quad , \tag{18.58}$$

with the components of τ^ℓ the same Pauli matrices τ_x, τ_y and τ_z used in (18.54). Equation (18.58) implies that the transformation matrix that relates $\sum_{body}$ to $\sum_{lab}$ is

$$U = (U^{\ell m}) \quad ;$$

therefore U is real and orthogonal.

The Lagrangian L of a rigid body with no external forces is a function only of ω^1, ω^2 and ω^3:

$$L = L(\omega^\ell) \quad .$$

Through (18.55), or more explicitly (18.66) below, L is also a function of a, b, c and $\dot{a}$, $\dot{b}$, $\dot{c}$. The conjugate momenta of a, b and c are given by

$$p_a = \frac{\partial \omega^\ell}{\partial \dot{a}} \frac{\partial L}{\partial \omega^\ell} \quad , \qquad p_b = \frac{\partial \omega^\ell}{\partial \dot{b}} \frac{\partial L}{\partial \omega^\ell}$$

and

$$p_c = \frac{\partial \omega^\ell}{\partial \dot{c}} \frac{\partial L}{\partial \omega^\ell} \quad . \tag{18.59}$$

It is well-known that the derivative of L with respect to ω^ℓ is the component j^ℓ of the angular-momentum vector in $\sum_{\text{body}}$:

$$j^\ell = \frac{\partial L}{\partial \omega^\ell} \quad . \tag{18.60}$$

The same vector viewed in $\sum_{\text{lab}}$ carries the component J^ℓ. Hence we have

$$\hat{x}\, j^1 + \hat{y}\, j^2 + \hat{z}\, j^3 = \hat{X}\, J^1 + \hat{Y}\, J^2 + \hat{Z}\, J^3 \equiv \vec{J} \tag{18.61}$$

or

$$J^\ell = U^{\ell m}\, j^m \quad , \tag{18.62}$$

similar to (18.57).

In the quantum theory, $\vec{J}$ is the rotation operator; the conjugate momenta to a, b, c are the derivative operators

$$p_a = -i \frac{\partial}{\partial a} \quad , \qquad p_b = -i \frac{\partial}{\partial b} \qquad \text{and} \qquad p_c = -i \frac{\partial}{\partial c} \, .$$

In $\sum_{\text{lab}}$ the components of the angular momentum satisfy the usual commutation relation

$$[J^\ell , J^m] = i \epsilon^{\ell m n} J^n \quad . \tag{18.63}$$

We shall now show that*

$$[J^\ell, j^m] = 0$$

and

$$[j^\ell, j^m] = -i\,\epsilon^{\ell mn} j^n . \qquad (18.64)$$

<u>Proof.</u> From (18.61) we have

$$j^1 = \hat{x}\cdot\vec{J} = \hat{x}^\ell J^\ell ,$$

$$j^2 = \hat{y}\cdot\vec{J} = \hat{y}^\ell J^\ell$$

and

$$j^3 = \hat{z}\cdot\vec{J} = \hat{z}^\ell J^\ell$$

where $\hat{x}^\ell$, $\hat{y}^\ell$ and $\hat{z}^\ell$ are the components of $\hat{x}$, $\hat{y}$ and $\hat{z}$ viewed in $\sum_{lab}$. Since $\hat{x}$, $\hat{y}$ and $\hat{z}$ are unit vectors fixed in the body frame, under a rotation they transform like $\vec{J}$; all rotate like vectors. Hence, the scalar products $\hat{x}\cdot\vec{J}$, $\hat{y}\cdot\vec{J}$ and $\hat{z}\cdot\vec{J}$ are invariants, and that gives the upper equation in (18.64).

The same property can also be expressed analytically; we have

$$[J^\ell, \hat{x}^m] = i\,\epsilon^{\ell mn}\hat{x}^n .$$

Together with (18.63), it gives

$$[J^\ell, \hat{x}^m J^m] = [J^\ell, \hat{x}^m] J^m + \hat{x}^m [J^\ell, J^m]$$

$$= i\,\epsilon^{\ell mn}(\hat{x}^n J^m + \hat{x}^m J^n) = 0 \qquad (18.65)$$

from which it follows that $[J^\ell, j^1] = 0$, and likewise $[J^\ell, j^2] = [J^\ell, j^3] = 0$. The commutator between j^1 and j^2 is

$$[j^1, j^2] = [\hat{x}^m J^m, \hat{y}^\ell J^\ell] = [\hat{x}^m J^m, \hat{y}^\ell] J^\ell + \hat{y}^\ell [J^\ell, \hat{x}^m J^m],$$

which because of (18.65) becomes

$$[\hat{x}^m J^m, \hat{y}^\ell] J^\ell = \hat{x}^m [J^m, \hat{y}^\ell] J^\ell$$

$$= i\hat{x}^m \epsilon^{m\ell n}\hat{y}^n J^\ell = -i\hat{z}^\ell J^\ell = -i j^3 ,$$

---

* Notice the minus sign in (18.64).

and that establishes the lower equation in (18.64). In the next section, we shall apply these elementary properties to the non-Abelian gauge field and show how the dependence on these angular variables θ_α in (18.53) can be eliminated.

Exercise 1. In terms of the Euler angles a, b, c and their time derivatives, show that the components of $\vec{\omega}$ and $\vec{\Omega}$ are given by

$$\omega^1 = \dot{a}\sin c - \dot{b}\sin a\cos c ,$$
$$\omega^2 = \dot{a}\cos c + \dot{b}\sin a\sin c ,$$
$$\omega^3 = \dot{b}\cos a + \dot{c}$$

and (18.66)

$$\Omega^1 = -\dot{a}\sin b + \dot{c}\sin a\cos b ,$$
$$\Omega^2 = \dot{a}\cos b + \dot{c}\sin a\sin b ,$$
$$\Omega^3 = \dot{b} + \dot{c}\cos a .$$

Exercise 2. The matrix $U = (U^{\ell m})$, defined by (18.58), connects any vector components r^m in the body frame to the component R^ℓ of the same vector in the laboratory frame, $R^\ell = U^{\ell m} r^m$. Show that in terms of the Euler angles a, b and c the matrix U is

$$U = \begin{pmatrix} \cos a\cos b\cos c - \sin b\sin c & -\cos a\cos b\sin c - \sin b\cos c & \sin a\cos b \\ \cos a\sin b\cos c + \cos b\sin c & -\cos a\sin b\sin c + \cos b\cos c & \sin a\sin b \\ -\sin a\cos c & \sin a\sin c & \cos a \end{pmatrix} .$$

(18.67)

Exercise 3. Prove that

$$j^1 = \sin c\, p_a - \frac{\cos c}{\sin a} p_b + \cos c \frac{\cos a}{\sin a} p_c ,$$
$$j^2 = \cos c\, p_a + \frac{\sin c}{\sin a} p_b - \sin c \frac{\cos a}{\sin a} p_c ,$$

$$j^3 = p_c \, ,$$

and

$$J^1 = -\sin b \; p_a - \cos b \; \frac{\cos a}{\sin a} \, p_b + \frac{\cos b}{\sin a} \, p_c \; , \qquad (18.68)$$

$$J^2 = \cos b \; p_a - \sin b \; \frac{\cos a}{\sin a} \, p_b + \frac{\sin b}{\sin a} \, p_c \; ,$$

$$J^3 = p_b$$

where

$$p_a = -i \, \frac{\partial}{\partial a} \, , \quad p_b = -i \, \frac{\partial}{\partial b} \quad \text{and} \quad p_c = -i \, \frac{\partial}{\partial c} \; .$$

Exercise 4. Prove the commutation relations (18.63)–(18.64) by us-ing directly the expressions of the angular–momentum operators given by Exercise 3.

3. $\underline{SU_2}$ gauge field (classical)

Let us return to the non–Abelian gauge–field theory and consider first the SU_2 group, so that in (18.53) θ_a can simply be the Euler angles a, b and c ; i.e.,

$$\theta_1(x) = a(x) \, , \qquad \theta_2(x) = b(x) \quad \text{and} \quad \theta_3(x) = c(x)$$

where, in the field–theoretic case, these angles are all functions of the space–time position x . Via (18.54), the matrix $u(a, b, c)$ also becomes x–dependent. We shall now follow the general procedure outlined in (18.45)–(18.48). Since (18.43) gives the $V_0 = 0$ gauge Hamiltonian in Cartesian coordinates, the Coulomb-gauge Hamiltonian operator can be derived by first writing down its classical expression, then setting all momenta p_a to be $-i \, \partial/\partial q_a$ and arranging these operators in the order (18.47) with the inclusion of the Jacobian (18.48). Finally, the dependence on these angular coordinates $a(x)$, $b(x)$ and $c(x)$ will be eliminated by using the constraints (18.40). The result is the quantum theory in the Coulomb gauge. In this section, we derive the classical Coulomb Hamiltonian ((18.95) below); the quantum operator will be given in the next section ((18.97) below).

In the following, the SU_2 transformation will be referred to as the "isospin" rotation, and the isovectors will be represented by arrows. The generators T^ℓ are the Pauli matrices, indicated by the isovector $\frac{1}{2}\vec{\tau}$; (18.50) becomes then

$$V_i = \tfrac{1}{2}\vec{\tau}\cdot\vec{V}_i \qquad \text{and} \qquad A_i = \tfrac{1}{2}\vec{\tau}\cdot\vec{A}_i$$

with $\vec{A}_i$ satisfying the transversality condition

$$\nabla_i \vec{A}_i = 0 \ .$$

The two matrices V_i and A_i are related through (18.49)

$$V_i = u A_i u^\dagger + \frac{i}{g} u \nabla_i u^\dagger \ .$$

The fourth component, $A_4 = i A_0$, is given by (18.51)

$$A_4 = -\frac{1}{g} \dot{u}^\dagger u = \frac{1}{g} u^\dagger \dot{u} \ ,$$

which, in terms of the matrix ω given by (18.56), is $A_4 = -i\omega/g$, or $A_0 = -\omega/g$. Hence, in isovector form, we write

$$\vec{A}_0 = -\frac{\vec{\omega}}{g} \ . \tag{18.69}$$

The table opposite summarizes some of the notations that will be used in the Coulomb gauge versus those in the time-axial (i.e., $V_0 = 0$) gauge. As in the table, let the electric and magnetic fields in the time-axial gauge be $\vec{E}_i$ and $\vec{B}_i$, those in the Coulomb gauge be $\vec{\mathcal{E}}_i$ and $\vec{\mathcal{B}}_i$. According to (18.11), the electric and magnetic fields transform homogeneously like isovectors under the gauge transformation (18.9); consequently,

$$\tfrac{1}{2}\vec{\tau}\cdot\vec{E}_i = u(\tfrac{1}{2}\vec{\tau}\cdot\vec{\mathcal{E}}_i)\, u^\dagger$$

	Time-axial gauge	Coulomb gauge
gauge field	$\vec{V}_i$ (no condition) $\vec{V}_0 = 0$	$\vec{A}_i$ $(\nabla_i \vec{A}_i = 0)$ $\vec{A}_0 = -\vec{\omega}/g$
covariant derivative	$D_i = \nabla_i + g\,\vec{V}_i \times$	$\mathcal{D}_i = \nabla_i + g\,\vec{A}_i \times$
electric field	$\vec{E}_i = -\dot{\vec{V}}_i$	$\vec{\mathcal{E}}_i = -\dot{\vec{A}}_i + \frac{1}{g}\mathcal{D}_i\vec{\omega}$ $= \vec{\mathcal{E}}_i^{tr} - \nabla_i\vec{\phi}$
magnetic field (i, j, k cyclic)	$\vec{B}_i = \nabla_j \vec{V}_k - \nabla_k \vec{V}_j$ $+ g\,\vec{V}_j \times \vec{V}_k$	$\vec{\mathcal{B}}_i = \nabla_j \vec{A}_k - \nabla_k \vec{A}_j$ $+ g\,\vec{A}_j \times \vec{A}_k$
conjugate momentum	$\vec{\Pi}_i = -\vec{E}_i$	$\vec{\Pi}_i^{tr} = -\vec{\mathcal{E}}_i^{tr}$
fermion field	ψ	ψ_c
charge density	$\vec{\rho}_V + \vec{\rho}_F$	$\vec{\sigma}_A + \vec{\sigma}_F$
where	$\vec{\rho}_F = \psi^{\dagger}\frac{1}{2}\vec{\tau}\,\psi$	$\vec{\sigma}_F = \psi_c^{\dagger}\frac{1}{2}\vec{\tau}\,\psi_c$
and	$\vec{\rho}_V = \vec{V}_i \times \vec{\Pi}_i$	$\vec{\sigma}_A = \vec{A}_i \times \vec{\Pi}_i^{tr}$
Gauss' law	$D_i\,\vec{E}_i = g\,\vec{\rho}_F$	$\mathcal{D}_i\,\vec{\mathcal{E}}_i = g\,\vec{\sigma}_F$
i.e.	$\nabla_i\,\vec{E}_i = g\,(\vec{\rho}_V + \vec{\rho}_F)$	$-\mathcal{D}_i\,\nabla_i\,\vec{\phi}$ $= g(\vec{\sigma}_A + \vec{\sigma}_F)$

Table 18.1

and

$$\tfrac{1}{2}\vec{\tau}\cdot\vec{B}_i = u(\tfrac{1}{2}\vec{\tau}\cdot\vec{\mathcal{B}}_i)\,u^\dagger \quad .$$

From (18.58) we also have

$$u^\dagger \tau^\ell u = U^{\ell m}\tau^m \quad ,$$

and therefore the components of $\vec{E}$, $\vec{\mathcal{E}}$, $\vec{B}$ and $\vec{\mathcal{B}}$ satisfy

$$E_i^\ell = U^{\ell m}\mathcal{E}_i^m \quad \text{and} \quad B_i^\ell = U^{\ell m}\mathcal{B}_i^m \quad . \tag{18.70}$$

In terms of the Euler angles, $U^{\ell m}$ is given by (18.67) and it satisfies

$$U^{\ell m}U^{\ell n} = U^{m\ell}U^{n\ell} = \delta^{mn} \quad .$$

In the language of the analog problem of rigid-body rotation, we may regard the $V_0 = 0$ gauge as the "lab" frame and the Coulomb gauge the "body" frame. The above E_i^ℓ and $\mathcal{E}_i^\ell$ then refer to the components of the same isovector, but viewed in different frames; likewise for B_i^ℓ and $\mathcal{B}_i^\ell$.

(i) For simplicity, let us assume that there is no fermion field ψ. [The generalization to include ψ will be given in (ii) below.] The Lagrangian and the Hamiltonian are both gauge invariant. By substituting (18.70) into (18.27) and (18.29), and setting $\psi = 0$, we see that

$$H = \int \mathcal{H}\,d^3r = \tfrac{1}{2}\int(\vec{\mathcal{E}}_i\cdot\vec{\mathcal{E}}_i + \vec{\mathcal{B}}_i\cdot\vec{\mathcal{B}}_i)\,d^3r$$

and

$$L = \int \mathcal{L}\,d^3r = \tfrac{1}{2}\int(\vec{\mathcal{E}}_i\cdot\vec{\mathcal{E}}_i - \vec{\mathcal{B}}_i\cdot\vec{\mathcal{B}}_i)\,d^3r \quad . \tag{18.71}$$

From (18.69) and Table 18.1, or (18.5), we see that

$$\begin{aligned}\vec{\mathcal{E}}_i &= -\dot{\vec{A}}_i - (\nabla_i + g\vec{A}_i\times)\,\vec{A}_0 \\ &= -\dot{\vec{A}}_i + \frac{1}{g}\mathcal{D}_i\vec{\omega}\end{aligned} \tag{18.72}$$

where

$$\mathcal{D}_i \equiv \nabla_i + g\vec{A}_i\times \tag{18.73}$$

is the covariant derivative in the Coulomb gauge so that

$$\mathcal{D}_i \vec{\omega} = \nabla_i \vec{\omega} + g \vec{A}_i \times \vec{\omega}$$

with the cross denoting the usual isovector product.

Our task is to express H as a function of the appropriate generalized coordinates and their conjugate momenta. In the Coulomb gauge the generalized coordinates are the transverse field A_i^ℓ and the Euler angles a, b and c. By substituting (18.72) into the Lagrangian L, given by (18.71), we observe that the dependence on $\dot{a}$, $\dot{b}$ and $\dot{c}$ in L is entirely through the $\vec{\omega}$ - dependence in the electric field. Thus, (18.59)-(18.68), derived for rigid-body rotations, are directly applicable here. According to (18.68), the conjugate momenta p_a, p_b and p_c of the Euler angles are linear functions of the components of the isovector $\vec{j} \equiv \delta L / \delta \vec{\omega}$. By differentiating L with respect to $\vec{\omega}$ and using (18.72), we find

$$\vec{j} = \frac{\delta \mathcal{E}_i^\ell}{\delta \vec{\omega}} \frac{\partial L}{\partial \mathcal{E}_i^\ell} = -\frac{1}{g} \mathcal{D}_i \vec{\mathcal{E}}_i \ . \tag{18.74}$$

In the $V_0 = 0$ gauge, the corresponding covariant derivative is, according to Table 18.1,

$$D_i \equiv \nabla_i + g \vec{V}_i \times \ .$$

From (18.28) and (18.33) it follows that

$$J^\ell = -\frac{1}{g} (D_i \vec{E}_i)^\ell \ .$$

Because $\mathcal{D}_i \vec{\mathcal{E}}_i$ and $D_i \vec{E}_i$ are covariant expressions, it can be readily verified that, similarly to (18.70)

$$J^\ell = U^{\ell m} j^m \ , \tag{18.75}$$

which is the exact analog of (18.62) in the rigid-body rotation problem. [See the Exercise below.]

To derive the conjugate momentum of $\vec{A}_i$, we can differentiate the Lagrangian L with respect to $\dot{\vec{A}}_i$, under the constraint $\nabla_i \dot{\vec{A}}_i = 0$. As we shall see, the result is *

$$-\vec{\mathcal{E}}_i^{tr} \equiv \dot{\vec{A}}_i - (\delta_{ij} - \nabla^{-2} \nabla_i \nabla_j)(\vec{A}_j \times \vec{\omega}) \tag{18.76}$$

which clearly satisfies

$$\nabla_i \vec{\mathcal{E}}_i^{tr} = 0 \quad .$$

To derive (18.76), we observe that the sum of (18.72) and (18.76) is a total gradient:

$$\vec{\mathcal{E}}_i - \vec{\mathcal{E}}_i^{tr} = \nabla_i [\frac{1}{g} \vec{\omega} + \nabla^{-2} \nabla_j (\vec{A}_j \times \vec{\omega})] \equiv - \nabla_i \vec{\phi} \, .$$

Hence, just as in (6.17)-(6.19), the electric field $\mathcal{E}_i$ can be decomposed into two terms:

$$\vec{\mathcal{E}}_i = \vec{\mathcal{E}}_i^{tr} - \nabla_i \vec{\phi} \tag{18.77}$$

where, since $\nabla_j \vec{A}_j = 0$ and $\vec{A}_0 = -\vec{\omega}/g$,

$$\begin{aligned} \vec{\phi} &= -\frac{1}{g} \vec{\omega} - \nabla^{-2} (\vec{A}_j \times \nabla_j \vec{\omega}) \\ &= \vec{A}_0 + g \nabla^{-2} (\vec{A}_j \times \nabla_j \vec{A}_0) \quad . \end{aligned} \tag{18.78}$$

---

* In matrix notation,

$$\langle \vec{r} \,|\, \nabla_i \,|\, \vec{r}\,' \rangle = \frac{\partial}{\partial r_i} \delta^3(\vec{r} - \vec{r}\,') = - \frac{\partial}{\partial r_i'} \delta^3(\vec{r} - \vec{r}\,') \, ,$$

$$\langle \vec{r} \,|\, \nabla^2 \,|\, \vec{r}\,' \rangle = \frac{\partial^2}{\partial r_i \, \partial r_i} \delta^3(\vec{r} - \vec{r}\,') \, ,$$

$$\langle \vec{r} \,|\, \nabla^{-2} \,|\, \vec{r}\,' \rangle = - \frac{1}{4\pi} \frac{1}{|\vec{r} - \vec{r}\,'|} = - \frac{1}{8\pi^3} \int \frac{1}{\vec{q}^{\,2}} e^{i\vec{q}\cdot(\vec{r} - \vec{r}\,')} d^3q \tag{18.76a}$$

so that

$$\int \langle \vec{r} \,|\, \nabla^2 \,|\, \vec{r}\,'' \rangle \langle \vec{r}\,'' \,|\, \nabla^{-2} \,|\, \vec{r}\,' \rangle \, d^3r'' = \delta^3(\vec{r} - \vec{r}\,') \quad . \tag{18.76b}$$

The compact form $F = \nabla^{-2} G$ means

$$F(\vec{r}) = \int \langle \vec{r} \,|\, \nabla^{-2} \,|\, \vec{r}\,' \rangle \, G(\vec{r}\,') \, d^3r' \quad .$$

Through partial integration, (18.71) becomes

$$H = \frac{1}{2} \int [(\vec{\mathcal{E}}_i^{tr})^2 + (\nabla_i \vec{\phi})^2 + (\vec{\mathcal{B}}_i)^2]\, d^3r$$

and (18.79)

$$L = \frac{1}{2} \int [(\vec{\mathcal{E}}_i^{tr})^2 + (\nabla_i \vec{\phi})^2 - (\vec{\mathcal{B}}_i)^2]\, d^3r \quad .$$

Keeping $\vec{A}_i$ and $\vec{\omega}$ fixed but varying $\vec{L}$ with respect to $\dot{\vec{A}}_i$, we find

$$\delta L = -\int \vec{\mathcal{E}}_i^{tr} \cdot \delta \dot{\vec{A}}_i\, d^3r \quad ,$$

where, because $\nabla_i \vec{\mathcal{E}}_i^{tr} = 0$, the integral depends only on the divergence-free part of $\delta \dot{\vec{A}}_i$. Thus, as in (6.21), the conjugate momentum of $\vec{A}_i$ is that given by (18.76):

$$\vec{\Pi}_i^{tr} = \frac{\partial L}{\partial \dot{\vec{A}}_i} = -\vec{\mathcal{E}}_i^{tr} \tag{18.80}$$

in which the constraint $\nabla_i \delta \dot{\vec{A}}_i = 0$ is automatically taken into account. Substituting (18.77) into (18.74), we find

$$\vec{j} = -\frac{1}{g} (\nabla_i + g \vec{A}_i \times) (\vec{\mathcal{E}}_i^{tr} - \nabla_i \vec{\phi})$$

$$= -\vec{A}_i \times \vec{\mathcal{E}}_i^{tr} + \frac{1}{g} \nabla_i \mathcal{D}_i \vec{\phi} \quad .$$

Because $\vec{\Pi}_i^{tr} = -\vec{\mathcal{E}}_i^{tr}$, the above expression can be written as *

$$\phi = g (\nabla_i \mathcal{D}_i)^{-1} (\vec{j} - \vec{A}_i \times \vec{\Pi}_i^{tr}) \quad . \tag{18.81}$$

---

* Equation (18.81) is the compact form of

$$\phi^{\ell}(\vec{r}) = g \int < \ell, \vec{r} \,|\, (\nabla_i \mathcal{D}_i)^{-1} \,|\, \ell', \vec{r}\,' > [j^{\ell'}(\vec{r}\,') - \epsilon^{\ell' mn} A_i^m(\vec{r}\,')\, \Pi_i^{tr}(\vec{r}\,')^n]\, d^3r'$$

where

$$< \ell, \vec{r} \,|\, \mathcal{D}_i \,|\, \ell', \vec{r}\,' > = [\delta^{\ell\ell'} \nabla_i - g\epsilon^{\ell\ell' m} A_i^m(\vec{r})]\, \delta^3(\vec{r} - \vec{r}\,') \quad ,$$

$$< \ell, \vec{r} \,|\, \nabla_i \mathcal{D}_i \,|\, \ell', \vec{r}\,' > = [\delta^{\ell\ell'} \nabla^2 - g\epsilon^{\ell\ell' m} A_i^m(\vec{r})\, \nabla_i]\, \delta^3(\vec{r} - \vec{r}\,') \quad ,$$

and $(\nabla_i \mathcal{D}_i)^{-1}$ satisfies

$$\int < \ell, \vec{r} \,|\, \nabla_i \mathcal{D}_i \,|\, \ell'', \vec{r}\,'' >< \ell'', \vec{r}\,'' \,|\, (\nabla_j \mathcal{D}_j)^{-1} \,|\, \ell', \vec{r}\,' > d^3r'' = \delta^{\ell\ell'} \delta^3(\vec{r} - \vec{r}\,') \quad .$$

The operator $(\nabla_i \mathcal{D}_i)^{-1}$ is the inverse of $\nabla_i \mathcal{D}_i$. It is useful to define λ by

$$\nabla^2 + g\lambda = \nabla_i \mathcal{D}_i = \mathcal{D}_i \nabla_i \tag{18.82}$$

where,* on account of (18.73),

$$\lambda = \nabla_i (\vec{A}_i \times) = \vec{A}_i \times \nabla_i \tag{18.83}$$

with $\times$ denoting the isospin vector product, as before. In power series of g, we have

$$(\nabla_i \mathcal{D}_i)^{-1} = (\nabla^2 + g\lambda)^{-1} = \nabla^{-2} - g\nabla^{-2}\lambda\nabla^{-2} + g^2\nabla^{-2}\lambda\nabla^{-2}\lambda\nabla^{-2} - \cdots$$

and

$$(\nabla_i \mathcal{D}_i)^{-1}(-\nabla^2)(\nabla_j \mathcal{D}_j)^{-1}$$
$$= -\nabla^{-2} + 2g\nabla^{-2}\lambda\nabla^{-2} - 3g^2\nabla^{-2}\lambda\nabla^{-2}\lambda\nabla^{-2} + \cdots \tag{18.84}$$

which will be of use later.

Without the fermion field, according to (18.33), the generators $\mathcal{G}^\ell$ and J^ℓ are equal. For the classical theory, the constraint (18.35) is simply

$$J^\ell = 0 ; \tag{18.85}$$

therefore $j^\ell = 0$, on account of (18.75). Setting $\vec{j} = 0$ in (18.81), we obtain

$$\vec{\phi} = -g(\nabla_i \mathcal{D}_i)^{-1} \vec{\sigma}_A \tag{18.86}$$

where

$$\vec{\sigma}_A = \vec{A}_i \times \vec{\Pi}_i^{tr} \tag{18.87}$$

is the "charge" carried by $\vec{A}_i$. Substituting (18.87) into the H given by (18.79), we derive the classical Coulomb gauge Hamiltonian without fermions

---

* I.e., $\langle \ell, \vec{r} \,|\, \lambda \,|\, \ell', \vec{r}\,' \rangle = \epsilon^{\ell m \ell'} A_i^m \nabla_i \delta^3(\vec{r} - \vec{r}\,')$. (18.83a)

$$H = \tfrac{1}{2}\int [\vec{\Pi}_i^{\,tr}(\vec{r})^2 + \vec{\mathcal{B}}_i(\vec{r})^2]\, d^3r + \tfrac{1}{2} g^2 \int \sigma_A^{\ell}(\vec{r}) \cdot \langle \ell, \vec{r} \,|\, (\nabla_i \mathcal{D}_i)^{-1} (-\nabla^2)(\nabla_j \mathcal{D}_j)^{-1} \,|\, \ell', \vec{r}\,' \rangle \cdot \sigma_A^{\ell'}(\vec{r}\,')\, d^3r\, d^3r' \ . \qquad (18.88)$$

(ii) Next, we extend the above discussion to include fermions. As in Table 18.1, let ψ be the fermion field in the $V_0 = 0$ gauge and ψ_c that in the Coulomb gauge; they are related by

$$\psi = u\,\psi_c$$

where u is given by (18.49). From (18.58), we have

$$\psi^\dagger \tau^\ell \gamma_4 \gamma_\mu \psi = U^{\ell m}\, \psi_c^\dagger \tau^m \gamma_4 \gamma_\mu \psi_c \ . \qquad (18.89)$$

By using (18.27)-(18.29), and taking into account

$$\psi^\dagger \gamma_4 \gamma_i (\nabla_i - i\tfrac{1}{2} g \vec{\tau} \cdot \vec{V}_i)\psi = \psi_c^\dagger \gamma_4 \gamma_i (\nabla_i - i\tfrac{1}{2} g \vec{\tau} \cdot \vec{A}_i)\psi_c \ ,$$

$$\vec{E}_i \cdot \vec{E}_i = \vec{\mathcal{E}}_i \cdot \vec{\mathcal{E}}_i \qquad \text{and} \qquad \vec{B}_i \cdot \vec{B}_i = \vec{\mathcal{B}}_i \cdot \vec{\mathcal{B}}_i \ ,$$

we see that with fermions included (18.71) is replaced by

$$H = \int \{ \tfrac{1}{2} [\vec{\mathcal{E}}_i \cdot \vec{\mathcal{E}}_i + \vec{\mathcal{B}}_i \cdot \vec{\mathcal{B}}_i] + \psi_c^\dagger \gamma_4 [\gamma_j (\nabla_j - i\tfrac{1}{2} g \vec{\tau} \cdot \vec{A}_j) + m]\psi_c \}\, d^3r$$

and (18.90)

$$L = \int \{ \tfrac{1}{2} [\vec{\mathcal{E}}_i \cdot \vec{\mathcal{E}}_i - \vec{\mathcal{B}}_i \cdot \vec{\mathcal{B}}_i] - \psi_c^\dagger \gamma_4 [\gamma_\mu (\frac{\partial}{\partial x_\mu} - i\tfrac{1}{2} g \vec{\tau} \cdot \vec{A}_\mu) + m]\psi_c \}\, d^3r \ .$$

We must convert H into a function of the generalized coordinates and their conjugate momenta. Since, as before, the dependence of L on the generalized velocities $\dot{\vec{A}}_i$, $\dot{a}$, $\dot{b}$ and $\dot{c}$ is entirely contained in $\vec{\mathcal{E}}_i \cdot \vec{\mathcal{E}}_i$, it follows that the conjugate momentum of $\vec{A}_i$ is still $\vec{\Pi}_i^{\,tr}$ given by (18.80) and those of the Euler angles a, b and c are p_a, p_b and p_c, which are related to $\vec{j}$ of (18.74) through (18.68). Thus, (18.72)-(18.78) and (18.80)-(18.84) remain intact; e.g., from (18.77), (18.80) and (18.81) we have, as before,

$$\vec{\mathcal{E}}_i = \vec{\mathcal{E}}_i^{\,tr} - \nabla_i \vec{\phi} \;, \qquad \vec{\mathcal{E}}_i^{\,tr} = -\vec{\Pi}_i^{\,tr}$$

and

$$-\nabla_i \mathcal{D}_i \vec{\phi} = g(-\vec{j} + \vec{A}_i \times \vec{\Pi}_i^{\,tr}) \quad . \tag{18.91}$$

However, (18.85) has to be replaced by

$$\mathcal{G}^\ell = J^\ell + \psi^\dagger \tfrac{1}{2}\tau^\ell \psi = 0 \;,$$

which is the classical expression for the constraint (18.35). By using (18.75) and (18.89), we can rewrite it as

$$j^m + \psi_c^\dagger \tfrac{1}{2}\tau^m \psi_c = 0 \;.$$

Substituting it into (18.91), we find that, instead of (18.86),

$$\vec{\phi} = -g(\nabla_i \mathcal{D}_i)^{-1} \vec{\sigma} \tag{18.92}$$

where, as in Table 18.1,

$$\vec{\sigma} = \vec{\sigma}_A + \vec{\sigma}_F$$

in which $\vec{\sigma}_A = \vec{A}_i \times \vec{\Pi}_i^{\,tr}$ is the charge carried by the gauge field, as before, and $\vec{\sigma}_F$ is due to the fermion,

$$\vec{\sigma}_F = \psi_c^\dagger \tfrac{1}{2}\vec{\tau}\psi_c \quad . \tag{18.93}$$

Since $\mathcal{D}_i \nabla_i = \nabla_i \mathcal{D}_i$ and

$$\mathcal{D}_i \vec{\mathcal{E}}_i = \mathcal{D}_i \vec{\mathcal{E}}_i^{\,tr} - \mathcal{D}_i \nabla_i \vec{\phi} = g\vec{A}_i \times \vec{\mathcal{E}}_i^{\,tr} - \nabla_i \mathcal{D}_i \vec{\phi} \quad ,$$

by using (18.91)-(18.93), we obtain

$$\mathcal{D}_i \vec{\mathcal{E}}_i = g\,\vec{\sigma}_F \tag{18.94}$$

which is Gauss' law.

The <u>classical Hamiltonian</u> function can be derived by

substituting (18.92) and $\vec{\mathcal{E}}_i = -\vec{\Pi}_i^{tr} - \nabla_i \phi$ into H of (18.90):

$$H = \int \{ \tfrac{1}{2} [\vec{\Pi}_i^{tr}(\vec{r})^2 + \vec{\mathcal{B}}_i(\vec{r})^2] + \psi_c^\dagger \gamma_4 [\gamma_i (\nabla_i - i\tfrac{1}{2} g \vec{\tau} \cdot \vec{A}_i) + m] \psi_c \} d^3 r$$
$$+ \tfrac{1}{2} g^2 \int \sigma^{\ell}(\vec{r}) < \ell, \vec{r} \,|\, (\nabla_i \mathcal{D}_i)^{-1} (-\nabla^2)(\nabla_j \mathcal{D}_j)^{-1} \,|\, \ell', \vec{r}' > \cdot \sigma^{\ell'}(\vec{r}')\, d^3 r\, d^3 r' \quad . \tag{18.95}$$

Exercise. By using

$$-g\vec{j} = \mathcal{D}_i \vec{\mathcal{E}}_i = (\nabla_i + g\vec{A}_i \times)\vec{\mathcal{E}}_i \quad ,$$
$$-g\vec{J} = D_i \vec{E}_i = (\nabla_i + g\vec{V}_i \times)\vec{E}_i$$

and (18.70), verify that (18.75) holds.

4. Quantum Hamiltonian

The anticommutation relations between the fermion fields in the Coulomb gauge are the usual ones,

$$\{\psi_c(\vec{r}, t), \psi_c^\dagger(\vec{r}', t)\} = \delta^3(\vec{r} - \vec{r})$$

and

$$\{\psi_c(\vec{r}, t), \psi_c(\vec{r}', t)\} = \{\psi_c^\dagger(\vec{r}, t), \psi_c^\dagger(\vec{r}', t)\} = 0 \quad .$$

Because of the transversality condition

$$\nabla_i \vec{A}_i = \nabla_i \vec{\Pi}_i^{tr} = 0 \quad ,$$

as in (6.29), we have for the gauge field

$$[A_i(\vec{r}, t)^{\ell}, \Pi_j^{tr}(\vec{r}', t)^m] = i\delta^{\ell m}(\delta_{ij} - \nabla^{-2}\nabla_i \nabla_j)\,\delta^3(\vec{r} - \vec{r}')$$

and

$$[A_i(\vec{r}, t)^{\ell}, A_j(\vec{r}', t)^m] = [\Pi_i^{tr}(\vec{r}, t)^{\ell}, \Pi_j^{tr}(\vec{r}', t)^m] = 0 \quad . \tag{18.96}$$

The equal-time commutator between ψ_c and $\vec{A}_i$ (or $\vec{\Pi}_i^{tr}$) is zero. By applying (18.47)-(18.48) onto the classical Hamiltonian (18.95), we derive the Coulomb-gauge quantum Hamiltonian:

$$
\begin{aligned}
H = \int \{ & \tfrac{1}{2} \mathscr{J}^{-1} \vec{\Pi}_i^{\,tr} \cdot \mathscr{J} \vec{\Pi}_i^{\,tr} + \tfrac{1}{2} \vec{\mathcal{B}}_i^{\,2} \\
& + \psi_c^\dagger \gamma_4 [\gamma_i (\nabla_i - i \tfrac{1}{2} g_0 \vec{\tau} \cdot \vec{A}_i) + m] \psi_c \} \, d^3 r \\
& + \tfrac{1}{2} g_0^2 \int \mathscr{J}^{-1} \sigma^{\ell}(\vec{r}) \\
& \cdot < \ell, \vec{r} \,|\, (\nabla_i \mathscr{D}_i)^{-1} (-\nabla^2) (\nabla_j \mathscr{D}_j)^{-1} \,|\, \ell', \vec{r}\,' > \\
& \cdot \mathscr{J} \sigma^{\ell'}(\vec{r}\,') \, d^3 r \, d^3 r'
\end{aligned}
\tag{18.97}
$$

where, for notational clarity, we replace the coupling g in the classical Hamiltonian by g_0, which denotes the unrenormalized coupling constant in the quantum theory. Except for this change, all other notations are the same as before; e.g., we have

$$
\vec{\Pi}_i^{\,tr} = -\vec{\mathcal{E}}_i^{\,tr} , \qquad \mathscr{D}_i = \nabla_i + g_0 \vec{A}_i \times ,
$$

$$
\vec{\sigma} = \vec{A}_i \times \vec{\Pi}_i^{\,tr} + \psi_c^\dagger \tfrac{1}{2} \vec{\tau} \psi_c = -\vec{\Pi}_i^{\,tr} \times \vec{A}_i + \psi_c^\dagger \tfrac{1}{2} \vec{\tau} \psi_c = \sigma^\dagger
$$

and

$$
\vec{\mathcal{B}}_i = \nabla_j \vec{A}_k - \nabla_k \vec{A}_j + g_0 \vec{A}_j \times \vec{A}_k
\tag{18.98}
$$

with i, j, k cyclic. As in (18.48), $\mathscr{J}$ is the Jacobian. [For the non-Abelian gauge-field theory, it is often referred to in the literature as the Faddeev–Popov determinant.] The evaluation of the Jacobian is straightforward but somewhat tedious; the result * is that in (18.97) we may write

$$
\mathscr{J} = \det \nabla_i \mathscr{D}_i ,
\tag{18.99}
$$

where $\nabla_i \mathscr{D}_i$ stands for the matrix whose elements are

$$
< \ell, \vec{r} \,|\, \nabla_i \mathscr{D}_i \,|\, \ell', \vec{r}\,' > .
$$

It is important to keep the operator ordering given in the above Hamiltonian. The generalization from SU_2 to SU_N of an arbitrary

* For details see Phys. Rev. D22, 939 (1980).

N is quite simple. We need only replace $\frac{1}{2}\tau^{\ell}$ by T^{ℓ} and $\epsilon^{\ell mn}$ by $f^{\ell mn}$ in the above discussions.

18.5 Dia-electric (Antiscreening) Nature of the Vacuum

We are now in a position to investigate the dia-electric (i.e., antiscreening) nature of the QCD vacuum, emphasized in Chapter 17. In this section we examine only the lowest-order perturbation calculation. Higher-order effects will be discussed in the next section.

1. $\underline{SU_2}$ gauge theory

Consider the special case of two external charge distributions, $\underset{\sim}{\sigma}_1(\vec{r})$ and $\underset{\sim}{\sigma}_2(\vec{r})$, located at $\vec{r}_1$ and $\vec{r}_2$ respectively, as shown in Figure 18.2. For example, these charge distributions can be formed by some heavy fermions. For notational clarity, in this section isovectors will be represented by wiggly lines and spatial vectors by arrows. Let $E(r_{12})$ denote the work done in bringing these two charge distributions from ∞ to a very large but finite r_{12}. We assume

$$r_{12} = |\vec{r}_1 - \vec{r}_2| \ggg m^{-1}, \quad \text{or} \quad d$$

where m is the fermion mass and d the extension of each of the charge distributions. In the limit $m \to \infty$, $\vec{r}_1$ and $\vec{r}_2$ can be

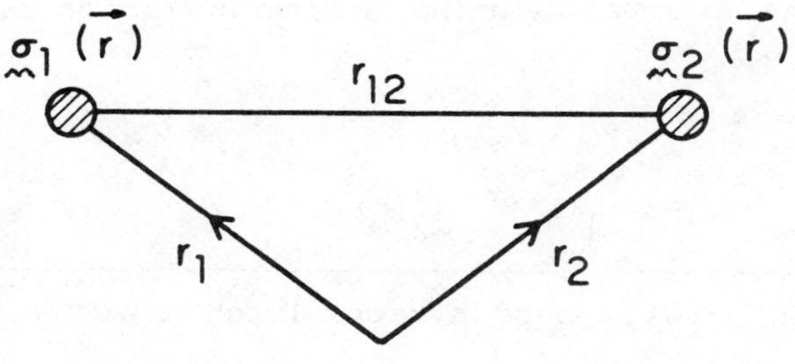

Fig. 18.2. Two external "charge" distributions, $\underset{\sim}{\sigma}_1(\vec{r})$ and $\underset{\sim}{\sigma}_2(\vec{r})$, separated by a very large distance r_{12}.

regarded as c. numbers. We shall first ignore the fermion degrees of freedom except those that are needed to characterize the charge distributions $\underset{\sim}{\sigma}_1$ and $\underset{\sim}{\sigma}_2$, e.g., their positions $\vec{r}_1$, $\vec{r}_2$ and their charges

$$\underset{\sim}{q}_1 = \int \underset{\sim}{\sigma}_1 \, d^3r \quad \text{and} \quad \underset{\sim}{q}_2 = \int \underset{\sim}{\sigma}_2 \, d^3r \ . \qquad (18.100)$$

Because of their non-Abelian nature, $\underset{\sim}{q}_1$ and $\underset{\sim}{q}_2$ are Hermitian matrices; they satisfy the commutation relations

$$[q_1^{\ell}, q_1^{m}] = i\epsilon^{\ell mn} q_1^{n} ,$$

$$[q_2^{\ell}, q_2^{m}] = i\epsilon^{\ell mn} q_2^{n}$$

and for non-overlapping distributions (i.e., r_{12} sufficiently large)

$$[q_1^{\ell}, q_2^{m}] = 0 \ .$$

Let us adopt the Coulomb gauge.* The Hamiltonian (18.97) now becomes

$$H = \int \left(\tfrac{1}{2} \mathcal{J}^{-1} \underset{\sim}{\Pi}_i^{tr} \mathcal{J} \underset{\sim}{\Pi}_i^{tr} + \tfrac{1}{2} \underset{\sim}{\mathcal{B}}_i^{2}\right) d^3r + H_{Coul} \qquad (18.101)$$

where

$$\begin{aligned} H_{Coul} = \tfrac{1}{2} g_0^2 \int \mathcal{J}^{-1} \sigma^{\ell}(\vec{r}) & \\ \cdot \langle \ell, \vec{r} \,|\, (\nabla_i \mathcal{D}_i)^{-1} (-\nabla^2)(\nabla_j \mathcal{D}_j)^{-1} \,|\, \ell', \vec{r}\,' \rangle & \\ \cdot \mathcal{J} \sigma^{\ell'}(\vec{r}\,') \, d^3r \, d^3r' \ , & \qquad (18.102) \end{aligned}$$

which plays the same role as the Coulomb interaction (6.27) in QED, and

$$\text{with} \quad \begin{aligned} \underset{\sim}{\sigma} &= \underset{\sim}{A}_i \times \underset{\sim}{\Pi}_i^{tr} + \underset{\sim}{\sigma}_{ext} \\ \underset{\sim}{\sigma}_{ext}(\vec{r}) &\equiv \underset{\sim}{\sigma}_1(\vec{r}) + \underset{\sim}{\sigma}_2(\vec{r}) \ . \end{aligned} \qquad (18.103)$$

---

* The work $E(r_{12})$ is gauge invariant. It can be written as $q_1^m (P \exp \int_1^2 -igT^{\ell} A_i^{\ell} dx_i)_{mn} q_2^n$ times a scalar function of r_{12}, where $T_{mn}^{\ell} = -i\epsilon^{\ell mn}$ and P denotes that the integration is path-ordered. In the Coulomb gauge and for static sources $A_i^{\ell} = 0$; therefore (18.105) holds for large r_{12}.

2. Perturbative calculations

To the lowest order in $g_0^{\,2}$, we may set in (18.102) $\underset{\sim}{\sigma} = \underset{\sim}{\sigma}_{ext}$, $\mathcal{J} = 1$ and $\mathcal{D}_i = \nabla_i$; the result is the familiar Coulomb's law

$$E(r_{12}) = (4\pi r_{12})^{-1} g_0^{\,2}\, \underset{\sim}{q}_1 \cdot \underset{\sim}{q}_2 + O(g_0^{\,4}) \tag{18.104}$$

where, as before, g_0 is the unrenormalized coupling. Let us define the renormalized coupling constant g by

$$E(r_{12}) = (4\pi r_{12})^{-1} g^2\, \underset{\sim}{q}_1 \cdot \underset{\sim}{q}_2 \;; \tag{18.105}$$

i.e., g^2 is directly related to the physical work done in bringing these two charges $\underset{\sim}{q}_1$ and $\underset{\sim}{q}_2$ from ∞ to r_{12}. As we shall see for QCD, g^2 is $> g_0^{\,2}$, showing the antiscreening nature of its vacuum.

Theorem 3. Without fermion loops,

$$\frac{g^2}{g_0^{\,2}} = 1 + \frac{11 g_0^{\,2}}{48\pi^2}\, C_2 \ln(\Lambda L)^2 + O(g_0^{\,4}) \tag{18.106}$$

where Λ and L^{-1} are the appropriate ultraviolet and infrared momentum cut-offs, and C_2 is the quadratic Casimir sum:

$$C_2\, \delta^{\ell m} \equiv f^{\ell n n'} f^{m n n'} = 2 \quad \text{for} \quad SU_2 \;,$$

and N for SU_N. If there are n flavors of zero-mass fermions present, then

$$\frac{g^2}{g_0^{\,2}} = 1 + (11 C_2 - 2n)\, \frac{g_0^{\,2}}{48\pi^2} \ln(\Lambda L)^2 + O(g_0^{\,4}) \,. \tag{18.107}$$

Proof.* We first expand H as a power series in g_0 :

$$H = H_0 + H_{int} \tag{18.108}$$

* Equations (18.106)-(18.107) were first obtained by D. J. Gross and F. Wilczek, Phys.Rev.Lett. 30, 1343 (1973), and H. D. Politzer, ibid. 1346. See also V. N. Gribov, lecture at the 12th Winter School of the Leningrad Nuclear Physics Institute (1977), and S. D. Drell, in A Festschrift for Maurice Goldhaber, edited by G. Feinberg, A. W. Sunyar and J. Weneser (New York, New York Academy of Sciences, 1980).

where H_0 is g_0 - independent and

$$H_{int} = g_0 H_1 + g_0^2 H_2 + g_0^3 H_3 + g_0^4 H_4 + \cdots . \qquad (18.109)$$

At any given time t, the fields $\underset{\sim}{A}_i(\vec{r},t)$ and $\underset{\sim}{\Pi}_i^{tr}(\vec{r},t)$ can be written in terms of their Fourier components, as in (6.32)-(6.33),

$$A_i(\vec{r},t)^\ell = \sum_{\vec{k}} \frac{1}{\sqrt{2\omega_k L^3}} [a_i^\ell(\vec{k})\, e^{i\vec{k}\cdot\vec{r}} + \text{h.c.}]$$

and

$$\Pi_i^{tr}(\vec{r},t)^\ell = \sum_{\vec{k}} \sqrt{\frac{\omega_k}{2L^3}}\; [-i\, a_i^\ell(\vec{k})\, e^{i\vec{k}\cdot\vec{r}} + \text{h.c.}] \quad ,$$

where L^3 is the volume

$$\omega_k = |\vec{k}| \quad ,$$

$a_i^\ell(\vec{k})$ is t - dependent and, as in (6.34), there is the transversality condition

$$k_i\, a_i^\ell(\vec{k}) = 0 \quad .$$

We may introduce a set of three righthanded unit basis vectors $\hat{e}_1(\vec{k})$, $\hat{e}_2(\vec{k})$ and $\hat{k} = \vec{k}/\omega_k$ (shown in Figure 18.3) and define, as in (6.36)-(6.37),

$$a^\ell_{\vec{k},s=\pm1} \equiv \frac{1}{\sqrt{2}} [\hat{e}_1(\vec{k}) \mp i\hat{e}_2(\vec{k})]_i\; a_i^\ell(\vec{k})$$

and

$$a^{\ell\dagger}_{\vec{k},s=\pm1} \equiv \frac{1}{\sqrt{2}} [\hat{e}_1(\vec{k}) \pm i\hat{e}_2(\vec{k})]_i\; a_i^\ell(\vec{k})^\dagger \quad .$$

From (18.96) we derive, as in (6.38) and (6.39),

$$[a^\ell_{\vec{k},s},\; a^{\ell'\dagger}_{\vec{k}',s'}] = \delta_{\vec{k}\,\vec{k}'}\, \delta_{ss'}\, \delta^{\ell\ell'} \quad ,$$

$$[a^\ell_{\vec{k},s},\; a^{\ell'}_{\vec{k}',s'}] = 0 \quad .$$

In terms of these annihilation and creation operators the zeroth-order Hamiltonian H_0 in (18.108) becomes, apart from an

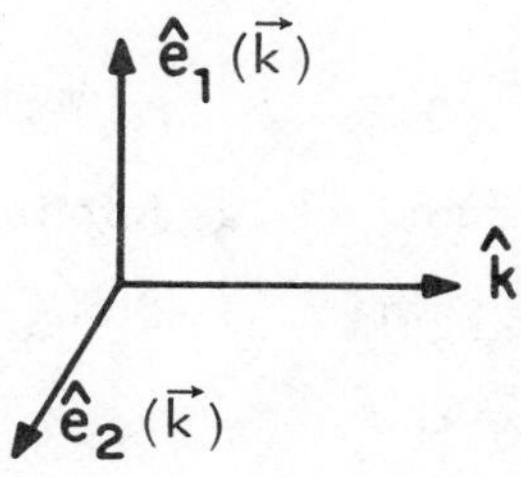

Fig. 18.3. A set of righthanded unit basis vectors.

additive constant that can be set to zero,

$$H_0 = \sum_{\vec{k},s,\ell} a^{\ell\dagger}_{\vec{k},s} a^{\ell}_{\vec{k},s} \omega_k .$$

Let the eigenstate of H_0 be $|N>$:

$$H_0 |N> = E_N^o |N> ,$$

where

$$E_N^o = \sum N^{\ell}_{\vec{k},s} \omega \tag{18.110}$$

and $N^{\ell}_{\vec{k},s}$ is the eigenvalue of the occupation-number operator $a^{\ell\dagger}_{\vec{k},s} a^{\ell}_{\vec{k},s}$ which can be 0, 1, 2, ⋯ . For the ground state $|0>$, we have

$$E_0^o = 0 .$$

From (18.101)-(18.102), we see that the interaction Hamiltonian $H_{int} = H - H_0$ contains the external charges; therefore it depends on r_{12}. With the inclusion of $H_{int}(r_{12})$, the groundstate energy shifts from $E_0^o = 0$ to $E_0(r_{12})$, given by the familiar perturbation formula

$$E_0(r_{12}) = <0|H_{int}(r_{12})|0> + \sum_{N\neq 0} \frac{|<0|H_{int}(r_{12})|N>|^2}{-E_N^o} + O(H_{int}^3) . \tag{18.111}$$

The quantity $E(r_{12})$ that we are interested in is the difference

$$E(r_{12}) = E_0(r_{12}) - E_0(\infty) = E^{(i)}(r_{12}) + E^{(ii)}(r_{12}) + \cdots \quad (18.112)$$

in which we have separated $E(r_{12})$ into terms according to the power-dependence on H_{int}, with $E^{(i)} = O(H_{int})$, $E^{(ii)} = O(H_{int}^{\ 2})$, etc.

(i) First order in H_{int}

The first-order term is determined by

$$< 0 \mid H_{int}(r_{12}) \mid 0 > \quad , \quad (18.113)$$

which can in turn be expanded as a power series in g_0 through (18.109). Since all odd g_0-power terms carry odd numbers of gauge field operators, whose vacuum-expectation values are zero, the result is an even function of g_0. [If we were in QED, then the corresponding $< 0 \mid H_{int}(r_{12}) \mid 0 >$ term would consist of only the lowest-order Coulomb energy; as we shall see, this is the crucial difference between QCD and QED. See Remark 2 on page 456.]

Because $E(r_{12})$ is proportional to $\underset{\sim}{\sigma}_1 \cdot \underset{\sim}{\sigma}_2$, we may set $H_{int} = H_{Coul}$ and, in (18.102), $\underset{\approx}{\sigma} = \underset{\sim}{\sigma}_{ext} = \underset{\sim}{\sigma}_1 + \underset{\sim}{\sigma}_2$. Since $\underset{\sim}{\sigma}_{ext}$ commutes with the Jacobian $\mathcal{J}$, we can set $\mathcal{J} = 1$. Substituting (18.84) into (18.102) we find, neglecting $O(g_0^{\ 6})$,

$$E_{Coul}(r_{12}) \equiv < 0 \mid H_{Coul}(r_{12}) \mid 0 > = -\tfrac{1}{2} g_0^{\ 2} \int \sigma^{\ell}_{ext}(\vec{r}) \cdot < \ell, \vec{r} \mid \nabla^{-2} + 3 g_0^{\ 2}\, \nabla^{-2} \lambda \nabla^{-2} \lambda \nabla^{-2} \mid \ell', \vec{r}\,' > \cdot \sigma^{\ell'}_{ext}(\vec{r}\,')\, d^3r\, d^3r' \quad (18.114)$$

where λ is given by (18.83). To $O(H_{int})$, the work done in bringing these two external charges from ∞ to r_{12} is

$$E^{(i)}(r_{12}) = E_{Coul}(r_{12}) - E_{Coul}(\infty) \quad . \quad (18.115)$$

As a reminder, we note that in terms of the Fourier component

$$\underset{\sim}{\sigma}_{\vec{p}}(a) \equiv \int \underset{\sim}{\sigma}_a(\vec{r})\, e^{-i\vec{p}\cdot\vec{r}}\, d^3r \qquad (a = 1, 2) \quad (18.116)$$

and to $O(g_0^2)$, the above expression for $E^{(i)}(r_{12})$ is simply

$$E^{(i)}(r_{12}) = g_0^2 \int \sigma_1^{\ell}(\vec{r})\, \sigma_2^{\ell}(\vec{r}\,')\, \frac{d^3r\, d^3r'}{4\pi\,|\vec{r}-\vec{r}\,'|} + O(g_0^4)$$

$$= g_0^2 \int \sigma_{-\vec{p}}^{\ell}(1)\, \sigma_{\vec{p}}^{\ell}(2)\, \frac{1}{\vec{p}^{\,2}}\, \frac{d^3p}{8\pi^3} + O(g_0^4) \quad . \qquad (18.117)$$

The $O(g_0^4)$ term is, according to (18.114),

$$-3\, g_0^4 \int \sigma_1^{\ell}(\vec{r}) < \ell, \vec{r}\, |\, \nabla^{-2} \lambda\, \nabla^{-2} \lambda\, \nabla^{-2}\, |\, \ell', \vec{r}\,' > \cdot\, \sigma_2^{\ell'}(\vec{r}\,')\, d^3r\, d^3r' \quad . \qquad (18.118)$$

From (18.83), or (18.83a), we see that it depends on the expectation value of $A_i^m(\vec{r}, t)\; A_{i'}^{m'}(\vec{r}\,', t)$. Because of the transversality condition, we have

$$< 0\, |\, a_j^m(\vec{k}) \cdot a_{j'}^{m'}(\vec{k}\,')^{\dagger}\, |\, 0 > \;=\; \delta^{mm'}\, \delta_{\vec{k}\,\vec{k}'}\, (\delta_{jj'} - \frac{k_j k_{j'}}{\vec{k}^{\,2}}) \qquad (18.119)$$

and therefore

$$< 0\, |\, A_j^m(\vec{r}, t)\; A_{j'}^{m'}(\vec{r}\,', t)\, |\, 0 > \;=\; \delta^{mm'} \int \frac{1}{2\omega_k}\, (\delta_{jj'} - \frac{k_j k_{j'}}{\vec{k}^{\,2}}) \cdot e^{i\vec{k}\cdot(\vec{r}-\vec{r}\,')}\, \frac{d^3k}{8\pi^3} \,. \quad (18.120)$$

By using (18.83), (18.76a), (18.116), (18.120) and Figure 18.4(a), we find that (18.118) is

$$-3\, g_0^4 \int \sigma_{-\vec{p}}^{\ell}(1)\, \epsilon^{\ell n m}\, \epsilon^{n \ell' m'}\, \sigma_{\vec{p}\,'}^{\ell'}(2)\, \delta^{mm'}$$

$$\cdot\, e^{-i\vec{p}\cdot\vec{r} + i\vec{p}\,'\cdot\vec{r}\,' + i(\vec{k}+\vec{q})\cdot(\vec{r}-\vec{r}\,')}$$

$$\cdot\, \frac{-1}{\vec{p}^{\,2}}\, \frac{-1}{\vec{q}^{\,2}}\, \frac{-1}{\vec{p}\,'^{\,2}}\, i p_j\, i p'_{j'}\, (\delta_{jj'} - \frac{k_j k_{j'}}{\vec{k}^{\,2}})\, \frac{1}{2\omega_k}$$

$$\cdot \left(\frac{1}{8\pi^3}\right)^4 d^3p\, d^3p'\, d^3k\, d^3q\, d^3r\, d^3r' \quad .$$

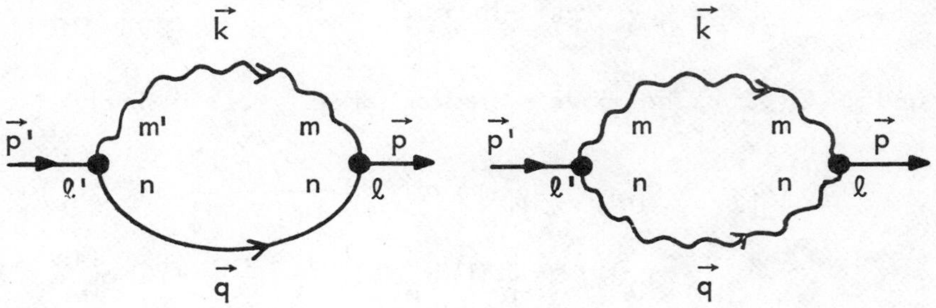

(a) (b)

Fig. 18.4. Graphic representations of radiative corrections to Coulomb interactions; (a) is first order in H_{int} and (b) second order. [These are not Feynman diagrams.]

Because of gauge invariance, the ∇_i operators in the two λ-factors in (18.118) can be written as either $i p_j\ i p'_{j'}$ or $i q_j\ i q_{j'} = i(p+k)_j \cdot i(p'+k)_{j'}$. The integration over $d^3r\, d^3r'/(2\pi)^6$ gives

$$\delta(\vec{p} - \vec{k} - \vec{q})\ \delta(\vec{p}\,' - \vec{k} - \vec{q})\ ,$$

which leads to $\vec{p} = \vec{p}\,'$ and, together with $\epsilon^{\ell nm}\ \epsilon^{n\ell' m'}\ \delta^{mm'} = -2\delta^{\ell\ell'}$, yields

$$(18.118) = \frac{3g_0^4}{(8\pi^3)^2} \int \sigma^{\ell}_{-\vec{p}}(1)\ \sigma^{\ell}_{\vec{p}}(2)\ \frac{1}{\vec{p}^{\,2}}\ \frac{1}{|\vec{k} - \vec{p}|^2}\ (1 - \cos^2\theta) \cdot \frac{1}{\omega_k} \cdot d^3p\ d^3k$$

where θ is the angle between $\vec{p}$ and $\vec{k}$. For r_{12} very large, only the $|\vec{p}| \sim O(r_{12}^{-1}) \to 0$ region is of importance; thus we may approximate

$$|\vec{k} - \vec{p}| \cong |\vec{k}| = \omega_k\ ,$$

which gives

$$(18.118) = g_0^4 \int \sigma^{\ell}_{-\vec{p}}(1)\ \sigma^{\ell}_{\vec{p}}(2)\ \frac{1}{\vec{p}^{\,2}}\ \frac{d^3p}{8\pi^3}\ \delta$$

where

$$\delta \cong \frac{3}{8\pi^3} \int \frac{1}{\omega_k^3} (1 - \cos^2 \theta)\, d^3k = \frac{1}{\pi^2} \int \frac{dk}{\omega_k} \quad .$$

Because the system is enclosed in a finite volume L^3, the k-integration starts from a lower limit L^{-1}; in order to give a mathematical meaning to ultraviolet divergence, we introduce an upper limit Λ. Thus,

$$\delta = \pi^{-2} \ln (\Lambda L) \quad . \tag{18.121}$$

With the inclusion of the above $O(g_0^4)$ term, (18.117) becomes

$$E^{(i)}(r_{12}) = (4\pi r_{12})^{-1}\, g_0^2\, \underset{\sim}{q}_1 \cdot \underset{\sim}{q}_2 (1 + g_0^2\, \delta) \quad . \tag{18.122}$$

(ii) Second order in H_{int}

From (18.111), we see that the second-order perturbation term in H_{int} is

$$E^{(ii)}(r_{12}) = \sum_{N \neq 0} \frac{|\langle 0 \,|\, H_{int}(r_{12}) \,|\, N \rangle|^2}{-E_N^{o}}$$

$$- \text{ same expression for } r_{12} = \infty , \tag{18.123}$$

in which the relevant states $|\, N \rangle$ are those of two gauge quanta

$$|\, N \rangle = a_{\vec{k},s}^{m\dagger}\, a_{\vec{q},s'}^{n\dagger} \,|\, 0 \rangle \quad , \tag{18.124}$$

and the relevant part of H_{int} is again the H_{Coul} of (18.102).

In order that $E^{(ii)}(r_{12})$ be accurate to $O(g_0^4)$, we need only $\langle N \,|\, H_{Coul}(r_{12}) \,|\, 0 \rangle$ to $O(g_0^2)$. Hence, in (18.102) we may set $\mathcal{J} \cong 1$ and $(\nabla_i \mathcal{D}_i)^{-1} \nabla^2 (\nabla_j \mathcal{D}_j)^{-1} \cong \nabla^{-2}$. Because of (18.124) and the fact that $E^{(ii)}(r_{12})$ depends on $\underset{\sim}{\sigma}_1 \cdot \underset{\sim}{\sigma}_2$, each $\langle N \,|\, H_{Coul} \,|\, 0 \rangle$ must depend linearly on σ_{ext}. By using

$\underset{\sim}{\sigma} = \underset{\sim}{\sigma}_A + \underset{\sim}{\sigma}_{ext}$, we may therefore write for the relevant part of H_{Coul}:

$$H_{Coul} \cong g_0^2 \int \sigma_A^{\ell}(\vec{r})\, \sigma_{ext}^{\ell}(\vec{r}\,')\, \langle \vec{r} \,|\, -\nabla^{-2} \,|\, \vec{r}\,' \rangle\, d^3r\, d^3r' \tag{18.125}$$

where, as in (18.87)

$$\sigma_A^{\ell} = \epsilon^{\ell\ell'\ell''} A_i^{\ell'} (\Pi_i^{tr})^{\ell''} \quad . \tag{18.126}$$

By using the Fourier expansions of $A_i^{\ell'}$ and $(\Pi_i^{tr})^{\ell''}$, we obtain the matrix element of σ_A between the two-quanta state $|\, N \rangle$ and the vacuum $|\, 0 \rangle$,

$$\langle N \,|\, \sigma_A^{\ell}(\vec{r}\,') \,|\, 0 \rangle = \frac{e^{-i(\vec{k}+\vec{q})\cdot\vec{r}\,'}}{2\sqrt{\omega_k \omega_q}\; L^3}\, \epsilon^{\ell\ell'\ell''}\, i(\omega_q - \omega_k)$$

$$\cdot \langle 0 \,|\, a^m_{\vec{k},s}\, a^n_{\vec{q},s'}\, a_i^{\dagger}(\vec{k})^{\ell'}\, a_i^{\dagger}(\vec{q})^{\ell''} \,|\, 0 \rangle, \tag{18.127}$$

in which from either $A_i^{\ell'}(\Pi_i^{tr})^{\ell''}$ or $-(\Pi_i^{tr})^{\ell'} A_i^{\ell''}$ we may extract a factor $a_i^{\dagger}(\vec{k})^{\ell'}\, a_i^{\dagger}(\vec{q})^{\ell''}$; this explains the factor $i(\omega_q - \omega_k)$. Thus, from (18.116), (18.126)-(18.127) and Fig. 18.4(b), we derive

$$\sum_{N \neq 0} |\langle 0 \,|\, H_{Coul} \,|\, N \rangle|^2 / (-E_N^0)$$

$$= \tfrac{1}{2} g_0^4 \int \left(\frac{1}{2\sqrt{\omega_k \omega_q}}\right)^2 \left(\frac{1}{8\pi^3}\right)^4 d^3p\, d^3p'\, d^3k\, d^3q\, d^3r\, d^3r'$$

$$\cdot\, \sigma^{\ell}_{-\vec{p}}\, \epsilon^{\ell mn}\, \epsilon^{\ell' mn}\, \sigma^{\ell'}_{\vec{p}\,'}\, e^{-i\vec{p}\cdot\vec{r} + i\vec{p}\,'\cdot\vec{r}\,' + i(\vec{k}+\vec{q})\cdot(\vec{r}-\vec{r}\,')}$$

$$\cdot\, \frac{1}{\vec{p}^{\,2}}\, \frac{1}{\vec{p}\,'^{2}}\, \frac{(\omega_q - \omega_k)^2}{-(\omega_q + \omega_k)} \left(\delta_{ij} - \frac{k_i k_j}{\vec{k}^2}\right)\left(\delta_{ij} - \frac{q_i q_j}{\vec{q}^2}\right) \tag{18.128}$$

in which the first $\frac{1}{2}$ factor is due to Bose statistics (i.e., interchanging $\vec{k}$ and $\vec{q}$ in (18.128) gives the same state), each of the factors in the product $\frac{1}{\vec{p}^{\,2}}\, \frac{1}{\vec{p}\,'^{2}}$ is due to $-\nabla^{-2}$ in (18.125), the denominator

$-(\omega_q + \omega_k)$ is $-E_N^o$, $\sigma^\ell_{\vec{p}} = \sigma^\ell_{\vec{p}}(1) + \sigma^\ell_{\vec{p}}(2)$ and the last two factors have the same origin as that in (18.119). The integration over $d^3r\, d^3r'/(2\pi)^6$ gives again $\delta(\vec{p} - \vec{k} - \vec{q})\, \delta(\vec{p}\,' - \vec{k} - \vec{q})$, which leads to $\vec{p} = \vec{p}\,'$. As before, for r_{12} very large, only the $|\vec{p}\,| \sim O(r_{12}^{-1}) \to 0$ region is of importance. Because $\vec{q} = \vec{p} - \vec{k}$, we have

$$\vec{q}^{\,2} = \vec{k}^2 - 2\vec{k}\cdot\vec{p} + \vec{p}^{\,2} ,$$

and since $\omega_q = |\vec{q}\,|$, $\omega_k = |\vec{k}\,|$,

$$(\omega_q - \omega_k)^2 = \vec{p}^{\,2}\cos^2\theta + O(\vec{p}^{\,3}) \cong \vec{p}^{\,2}\cos^2\theta$$

where θ is the angle between $\vec{k}$ and $\vec{p}$. Under the same approximation, we can set $(2\sqrt{\omega_k\omega_q}\,)^{-2} = (4\omega_k^{\,2})^{-1}$ and $-(\omega_q + \omega_k) = -2\omega_k$. By using $\epsilon^{\ell mn}\epsilon^{\ell' mn} = 2\delta^{\ell\ell'}$ and $(\delta_{ij} - \frac{k_i k_j}{\vec{k}^2})(\delta_{ij} - \frac{q_i q_j}{\vec{q}^{\,2}}) = 2$, and after substituting (18.128) into (18.123) we derive

$$E^{(ii)}(r_{12}) = g_0^4 \int \sigma^\ell_{-\vec{p}}(1)\, \sigma^\ell_{\vec{p}}(2)\, \frac{1}{\vec{p}^{\,2}}\, \frac{d^3p}{8\pi^3}\, \delta' \qquad (18.129)$$

where

$$\delta' = -\frac{1}{8\pi^3}\int \frac{\cos^2\theta}{2\omega_k^3}\, d^3k = -(12\pi^2)^{-1}\ln(\Lambda L) = -\frac{\delta}{12} \qquad (18.130)$$

in which the k-integration is from L^{-1} to Λ, as in (18.121) for δ. In terms of r_{12}, (18.129) can be written as

$$E^{(ii)}(r_{12}) = (4\pi r_{12})^{-1}\, g_0^4\, \underset{\sim}{q}_1 \cdot \underset{\sim}{q}_2\, \delta' . \qquad (18.131)$$

Since the $O(g_0^4)$ correction to the Coulomb interaction is the sum of (18.122) and (18.131), we find by using the definition of the renormalized g, (18.105),

$$\frac{g^2}{g_0^2} = 1 + g_0^2(\delta + \delta') = 1 + \frac{11 g_0^2}{24\pi^2}\ln(\Lambda L)^2 \qquad (18.132)$$

which is (18.106) for SU_2. We note that $g_0^2\,\delta'$ is < 0, which is a direct consequence of the second-order perturbation energy for the ground state being always negative. This, by itself, would correspond to our usual concept that vacuum polarization should always "screen" the original charge, making g less than g_0. However, due to the nonlinear nature of Gauss' law in QCD,

$$-(\nabla^2 + g_0 \vec{A}_i \times \nabla_i)\,\underset{\sim}{\phi} = g_0\,\underset{\sim}{\sigma} \quad ,$$

there is now also a g_0^4 term in the diagonal matrix element of

$$\tfrac{1}{2} \int (\nabla_i \underset{\sim}{\phi})^2 \, d^3r$$

which is always positive; this gives $g_0^2\,\delta > 0$. Since the magnitude of δ is 12 times that of δ', the net effect is to have $g^2 > g_0^2$, and that produces the antiscreening nature of the vacuum.

It is quite straightforward to extend the above considerations to SU_N. By replacing $\epsilon^{\ell mn}$ by $f^{\ell mn}$, we obtain the result given in (18.106). The quarks give an additional vacuum-polarization term which is of a screening nature. The calculation is identical * to that due to the fermion loop in QED and it will be omitted here. The result is (18.107), and that completes our proof of Theorem 3.

Remarks. 1. The decomposition of the fractional energy shift $\Delta E/E$ into $g_0^2\,\delta > 0$ and $g_0^2\,\delta' < 0$ is gauge-dependent. The sum $g_0^2(\delta + \delta')$ is gauge-independent.

2. In QED, the corresponding δ is 0 and therefore $\Delta E \propto \delta' < 0$, as expected from the sign of (18.123). This gives $\kappa > 1$ for any physical medium and also $|e_0/e| > 1$ where e_0 is the unrenormalized charge and e the renormalized one.

---

* See, for example, J. D. Bjorken and S. D. Drell, Relativistic Quantum Mechanics (New York, McGraw-Hill, 1964).

18.6 Asymptotic Freedom *

QCD is described by the color SU_3 gauge theory. From Theorem 3, we see that the renormalized coupling is a function of the unrenormalized coupling g_0, the ultraviolet momentum cut-off Λ and the volume size L^3 (which serves as an infrared cut-off parameter). Keeping g_0 and Λ fixed, we shall examine the L-dependence of the renormalized coupling, labeled g_L in this section.

The perturbation result (18.107) tells us that for two different volumes ℓ^3 and L^3, the ratio of their renormalized couplings in QCD is, neglecting $O(g_0^4)$,

$$\frac{g_\ell^2}{g_L^2} = \frac{1}{1 + \frac{1}{2\pi}\frac{g_0^2}{4\pi}\left(11 - \frac{2}{3}n\right)\ln\frac{L}{\ell}} , \tag{18.133}$$

which indicates that for quark-flavor number $n < 17$, $g_L > g_\ell$ if $L > \ell$. This conclusion will now be generalized beyond the perturbation calculation.

In the following, it will be established that under rather general assumptions (given in Remark 2 below) and valid to all orders of coupling

$$g_L > g_\ell \qquad \text{if} \qquad L > \ell \tag{18.134}$$

and

$$g_\ell \to 0 \qquad \text{if} \qquad \ell \to 0 . \tag{18.135}$$

For simplicity, we shall set all quark masses $= 0$, so that there is no natural energy scale in the theory. From dimensional considerations, g_L must depend only on g_0 and the product $L\Lambda$; i.e.,

$$g_L = G(L\Lambda,\, g_0) . \tag{18.136}$$

---

* H. D. Politzer, Phys. Rev. Lett. 30, 1346 (1973); D. Gross and F. Wilczek, ibid., 1343. G. 't Hooft, talk at the Marseilles meeting, 1972 (unpublished).

As in (18.133), let us consider two different volumes ℓ^3 and L^3, with the same g_0 and Λ. Changing L to ℓ, we also have

$$g_\ell = G(\ell\Lambda, g_0) , \qquad (18.137)$$

or its inverse $g_0 = g_0(\ell\Lambda, g_\ell)$. Eliminating g_0 between (18.136) and (18.137), we may express g_L in terms of $L\Lambda$, $\ell\Lambda$ and g_ℓ:

$$g_L = G(L\Lambda, g_0(\ell\Lambda, g_\ell)) . \qquad (18.138)$$

Since the theory is a renormalizable one, the limit $\Lambda \to \infty$ of (18.138) should exist; in this limit, Λ drops out; therefore g_L is a function of g_ℓ, ℓ and L. From dimensional considerations the limiting function must be of the form

$$g_L = g\left(\frac{\ell}{L}, g_\ell\right) \equiv \lim_{\Lambda\to\infty} G(L\Lambda, g_0(\ell\Lambda, g_\ell)) . \qquad (18.139)$$

Let us introduce

$$\lambda \equiv \ell/L . \qquad (18.140)$$

Equation (18.139) can be written as

$$g_L = g(\lambda, g_\ell) ; \qquad (18.141)$$

at $\lambda = 1$, i.e., $L = \ell$, (18.139) becomes

$$g_\ell = g(1, g_\ell) . \qquad (18.142)$$

Because $g_L = G(L\Lambda, g_0)$ is independent of ℓ, we have

$$\left(\frac{\partial g_L}{\partial \ln \ell}\right)_{g_0, \Lambda, L} = 0 . \qquad (18.143)$$

Hence, on account of (18.141), $g(\lambda, g_\ell)$ satisfies the following "renormalization group" equation*

* M. Gell-Mann and F. E. Low, Phys. Rev. 95, 1300 (1954); C. G. Callan, Phys. Rev. D2, 1541 (1970); K. Symanzik, Commun. Math. Phys. 18, 227 (1970).

$$\left(-\frac{\partial}{\partial \ln \lambda} + \beta \frac{\partial}{\partial g_\ell}\right) g(\lambda, g_\ell) = 0 \tag{18.144}$$

where

$$\beta \equiv -\frac{\partial g_\ell}{\partial \ln \ell} \quad . \tag{18.145}$$

From (18.107), we see that for QCD

$$g_\ell^{\ 2} = g_0^{\ 2}\left[1 + \frac{g_0^{\ 2}}{8\pi^2}\,(11 - \tfrac{2}{3}n)\,\ln(\Lambda\ell)\right] + O(g_0^{\ 6}) \quad .$$

To evaluate β, we substitute this expression into (18.145) and recall that in the differentiation, g_0 and Λ should be held fixed, in accordance with (18.143). Expanding the resulting β in terms of g_ℓ, instead of g_0, we derive

$$\beta = -\frac{g_\ell^{\ 3}}{16\pi^2}\,(11 - \tfrac{2}{3}n) + O(g_\ell^{\ 5}) \quad . \tag{18.146}$$

As in (18.133), the quark flavor number n will be assumed to be < 17 so that when g_ℓ is small β is < 0. [See Remark 3 below if n turns out to be $\geqslant 17$.]

The solution of (18.144) has the standard form

$$g(\lambda, g_\ell) = f(z) \tag{18.147}$$

where

$$z = \ln \lambda + \int_{g_1}^{g_\ell} \frac{dg'}{\beta(g')} \tag{18.148}$$

and g_1 is an arbitrary constant, which will be chosen to be sufficiently small so that

$$0 < g_1 \ll 1$$

and, on account of (18.146),

$$\beta(g') < 0 \qquad \text{for} \qquad 0 < g' \leqq g_1 \quad . \tag{18.149}$$

[By choosing the integration constant g_1 sufficiently small, we have made the tacit assumption that the physical region includes $g_\ell = 0+$. This is in accordance with the experimental high-energy results, as will be discussed further in Remark 2 below.]

Let $f^{-1}(z)$ be the inverse function of $f(z)$; i.e.,

$$f^{-1}(f(z)) = z \quad . \tag{18.150}$$

Thus, from (18.147) we have

$$z = f^{-1}(g(\lambda, g_\ell)) \quad ,$$

which, because of (18.142), gives

$$z = f^{-1}(g_\ell) \quad \text{at} \quad \lambda = 1 \quad .$$

By equating this expression with (18.148) at $\lambda = 1$, we derive

$$f^{-1}(g_\ell) = \int_{g_1}^{g_\ell} \frac{dg'}{\beta(g')} \quad . \tag{18.151}$$

Since $g(\lambda, g_\ell)$ is an odd function of g_ℓ , it is only necessary to consider positive values of g_ℓ . We distinguish two situations, given in Figures 18.5(a) and (b):

(a) $\beta(g_\ell) = 0$ only when $g_\ell = 0$,

(b) $\beta(g_\ell) = 0$ has more than one root. Besides $g_\ell = 0$ there is at least another root $g_\ell = \bar{g}$, chosen to be the smallest positive nonzero solution of $\beta(g_\ell) = 0$.

In Figure 18.5(a) the physical value of g_ℓ can vary from 0 to ∞ , while in Figure 18.5(b) g_ℓ can only vary between 0 and $\bar{g}$. This is because when $g_\ell = \bar{g}$ the integral in (18.151) diverges logarithmically. In either case, writing $g_\ell = x$, we have in the physical region, by using (18.151)

$$\frac{df^{-1}(x)}{dx} = \frac{1}{\beta(x)} < 0 \quad ;$$

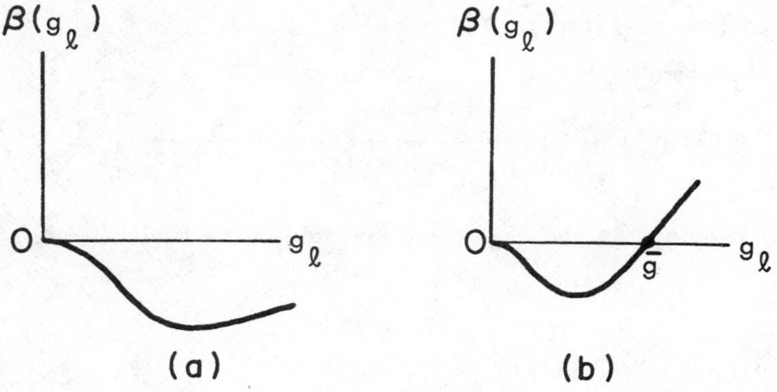

Fig. 18.5. Two possibilities for $\beta(g_\ell)$ versus g_ℓ, where g_ℓ is the renormalized coupling constant defined at a length ℓ, or momentum ℓ^{-1}, and $\beta(g_\ell) = -\partial g_\ell / \partial \ln \ell$.

hence the inverse function $f(z)$ satisfies

$$\frac{df(z)}{dz} < 0 \quad . \tag{18.152}$$

According to (18.147)-(18.148), we have, by keeping g_ℓ fixed,

$$\frac{\partial}{\partial\lambda} g(\lambda, g_\ell) = \frac{\partial f(z)}{\partial\lambda} = \frac{\partial z}{\partial\lambda} \frac{df(z)}{dz}$$

$$= \lambda^{-1} \frac{df(z)}{dz} < 0 \quad . \tag{18.153}$$

Since $\lambda = \ell / L$ and $g_L = g(\lambda, g_\ell)$, this means for $L > \ell$, $g_L > g_\ell$ and that establishes (18.134). Thus, when ℓ decreases steadily, so does g_ℓ.

For g_ℓ sufficiently small, the $O(g_\ell^5)$ term in (18.146) may be neglected. Hence, by using (18.151) we find

$$f^{-1}(g_\ell) \cong (11 - \tfrac{2}{3} n)^{-1} \, 8\pi^2 (g_\ell^{-2} - g_1^{-2}) \quad ; \tag{18.154}$$

its inverse function is, on account of the definition (18.150),

$$f(z) \cong [(8\pi^2)^{-1}(11 - \tfrac{2}{3}n)\, z + g_1^{-2}]^{-\frac{1}{2}} \quad ,$$

and therefore

$$g(\lambda, g_\ell) \cong \left[\frac{(11 - \frac{2}{3}n)^{-1}\, 8\pi^2\, g_\ell^{\,2}}{g_\ell^{\,2} \ln \lambda + (11 - \frac{2}{3}n)^{-1}\, 8\pi^2} \right]^{\frac{1}{2}} . \qquad (18.155)$$

Now keeping ℓ and g_ℓ fixed and taking the limit $L \to 0$, we derive

$$\lim_{L \to 0} g_L = 0 \quad ,$$

i.e. (18.135), called asymptotic freedom. [Cf. (23.120).]

Remarks.

1. Because QCD with zero quark mass has no natural energy scale, the volume size ℓ is equivalent to the device of introducing an energy scale ℓ^{-1}, at which we can define the renormalized coupling g_ℓ. The limit $g_\ell \to 0$ when $\ell \to 0$ means that in the sufficiently high-energy and high-momentum transfer region, quarks and gauge-quanta (called gluons) behave like free particles, thus giving rise to the name "asymptotic freedom".

2. Assume that the number of quark flavors in QCD is < 17. If $\beta(g_\ell) = 0$ has only one root at $g_\ell = 0$, as in Figure 18.5(a), then asymptotic freedom, (18.135), is established without any further assumption. However, if $\beta(g_\ell) = 0$ has more than one root, as in 18.5(b), then a priori the physical region can be either in $0 < g_\ell < \bar{g}$, or $g_\ell > \bar{g}$. Only through high-energy experiments can we know which region corresponds to our universe. As we mentioned in Section 17.1, experimental results do support the free-quark model at high energy; the physical region should lie in $0 < g_\ell < \bar{g}$. This is why in (18.148) the integration constant g_1 is chosen to satisfy (18.149), so that for sufficiently small g_ℓ the perturbation formula (18.155) becomes valid.

3. So far we have assumed the total number of quark flavors n to be < 17. At present only 5 flavors (u, d, s, c and b) are known. What happens if at some very high mass scale $M \gg 1$ GeV, n increases to $\geqslant 17$?

This means that the vacuum polarization produced by these very heavy fermion pairs would turn the QCD vacuum into para-electric instead of dia-electric (i.e., screening instead of antiscreening), at a length scale $\sim (2M)^{-1} \ll 10^{-14}$ cm. In the 10^{-14}-10^{-15} cm region, one can ignore such ultra-heavies. Consequently, in the present energy range, the application of QCD with asymptotic freedom can still be a good approximation. Of course, when we increase the energy, eventually we should reach the scale at which there are a large number of ultra-heavy quarks; then violation of asymptotic freedom would occur.

4. As discussed in Chapter 17, quark confinement can be understood phenomenologically by assuming that the QCD vacuum is a perfect color dia-electric. The result, (18.139)-(18.140), that we derive in this section supports this assumption, but does not prove it. Recently, there has been some important progress made by M. Creutz * through the application of the Monte Carlo method to a lattice formulation of QCD developed by K. Wilson.** Creutz's numerical result gives another strong support for the coexistence of color confinement and asymptotic freedom in non-Abelian gauge theories.

---

* M. Creutz, Phys.Rev. D21, 2308 (1980).

** K. G. Wilson, Phys.Rev. D10, 2445 (1974).

References.

E. S. Abers and B. W. Lee, Physics Reports 9C, 1 (1973).

W. Marciano and H. Pagels, Physics Reports 36C, 3 (1978).

See also the references at the end of Chapter 19.

Chapter 19

PATH INTEGRATION

In this chapter we develop the path-integration method, which is particularly useful for the derivation of Feynman rules in QCD. This route was also the historical one followed by Feynman when he invented his diagrams.

19.1 Cartesian Coordinates

1. One-dimensional problem

Consider a point particle moving along a straight line with x as the coordinate. Let $-id/dx$ be the momentum operator whose eigenstate is $|k\rangle$. In the x-representation, we have

$$\langle x | k \rangle = \frac{1}{\sqrt{L}} e^{ikx}$$

where, as before, we first assume that the system is enclosed in a length L with a periodic boundary condition, and then take the limit $L \to \infty$. The matrix element of any function f of the momentum operator in the coordinate space is*

$$\langle x' | f(-i\frac{d}{dx}) | x \rangle = \langle x' | k' \rangle \langle k' | f(-i\frac{d}{dx}) | k \rangle \langle k | x \rangle$$

$$= \sum_k \frac{1}{L} f(k) e^{ik(x'-x)}$$

---

* The repeated indices k and k' are summed over, as usual.

which, when $L \to \infty$ and on account of (2.29)-(2.30),

$$\to \int \frac{dk}{2\pi} \; f(k) \; e^{ik(x'-x)} \quad , \tag{19.1}$$

where the integration is from $-\infty$ to ∞. As an application of this elementary formula, we may consider a non-relativistic particle of unit mass, whose kinetic energy operator is

$$K = -\tfrac{1}{2} \frac{d^2}{dx^2} \quad .$$

The matrix element of $e^{-i(t'-t)K}$ in the coordinate space is

$$< x' \mid e^{-i(t'-t)K} \mid x > \; = \int \frac{dk}{2\pi} \; e^{-i\frac{1}{2}k^2(t'-t)+ik(x'-x)}$$

$$= \int \frac{dz}{2\pi} \; e^{-i\frac{1}{2}z^2(t'-t)+i\frac{1}{2}(x'-x)^2/(t'-t)} \tag{19.2}$$

where $z = k - (x'-x)/(t'-t)$. Since

$$\int_{-\infty}^{\infty} e^{-i\frac{1}{2}z^2\tau} \, dz = \sqrt{\frac{2\pi}{i\tau}} = \sqrt{\frac{\pi}{|\tau|}} \cdot \begin{cases} 1-i & \text{if } \tau > 0 \\ 1+i & \text{if } \tau < 0 , \end{cases} \tag{19.3}$$

we obtain

$$< x' \mid e^{-i(t'-t)K} \mid x > \; = \left(\frac{1}{2i\pi\tau}\right)^{\frac{1}{2}} e^{i\frac{1}{2}(x'-x)^2/\tau} \tag{19.4}$$

where $\tau = t' - t$. This function is called "pseudo-Gaussian" because it differs from a standard Gaussian distribution by the factor i in the exponent; its width is

$$| x' - x |_{\text{average}} = O(\sqrt{|\tau|}) \quad . \tag{19.5}$$

When $\tau \to 0$,

$$(19.4) \;\to\; \delta(x'-x) \quad .$$

For a particle moving in a potential $V(x)$, the Hamiltonian operator is

$$H = K + V(x) = -\tfrac{1}{2} \frac{d^2}{dx^2} + V(x) \quad . \tag{19.6}$$

The Green's function of the time-dependent Schrödinger equation

$$H \mid t > \; = \; i \frac{\partial}{\partial t} \mid t >$$

in the x-representation is $< x' \mid e^{-i(t' - t)H} \mid x >$; it relates the state vector from t to t' :

$$< x' \mid t' > \; = \int dx < x' \mid e^{-i(t'-t)H} \mid x > < x \mid t > \; . \qquad (19.7)$$

In terms of the eigenvectors of H

$$H \mid a > \; = \; E_a \mid a >$$

with the orthonormal condition

$$< a' \mid a > \; = \; \delta_{a'a} \; ,$$

this Green's function can also be written as

$$< x' \mid e^{-i(t'-t)H} \mid x > \; = \sum_a \psi_a(x') \, \psi_a^*(x) \; e^{-i(t'-t)\, E_a} \qquad (19.8)$$

where

$$\psi_a(x) \; = \; < x \mid a > \; .$$

2. From Hamiltonian operator to path integration

Following Feynman, we divide the time $t' - t$ into $\mathcal{N}$ intervals of ϵ each,

$$t' - t \; = \; \mathcal{N}\epsilon$$

so that when $\epsilon \to 0$, we have

$$\mathcal{N} = (t' - t)/\epsilon \; \to \; \infty \; .$$

Theorem 1. For any state vector $\mid >$,

$$<x' \mid e^{-i(t'-t)H} \mid > \; = \; \lim_{\epsilon \to 0} \int \prod_{n=1}^{\mathcal{N}} dx_n \left(\frac{1}{2 i \pi \epsilon} \right)^{\frac{1}{2}} e^{i \epsilon L_n} < x_1 \mid > \qquad (19.9)$$

where

$$L_n \; = \; \tfrac{1}{2} \dot{x}_n^{\,2} - V(\bar{x}_n) \qquad (19.10)$$

with $\dot{x}_n \equiv \dfrac{x_{n+1} - x_n}{\epsilon}$, $\bar{x}_n \equiv \dfrac{x_{n+1} + x_n}{2}$ and $x_{\mathcal{N}+1} = x'$.

<u>Proof.</u> Since $e^{-i(t'-t)H} = e^{-i\epsilon \mathcal{N} H}$, we have

$$< x' \mid e^{-i(t'-t)H} \mid x > \; = \int < x' \mid e^{-i\epsilon H} \mid x_{\mathcal{N}} > dx_{\mathcal{N}}$$
$$\cdot < x_{\mathcal{N}} \mid e^{-i\epsilon H} \mid x_{\mathcal{N}-1} > dx_{\mathcal{N}-1}$$
$$\cdots \; < x_3 \mid e^{-i\epsilon H} \mid x_2 > dx_2$$
$$\cdot < x_2 \mid e^{-i\epsilon H} \mid x > \; . \qquad (19.11)$$

When $\epsilon \to 0$,

$$e^{-i\epsilon H} = 1 - i\epsilon H + O(\epsilon^2) = 1 - i\epsilon K - i\epsilon V + O(\epsilon^2)$$
$$= e^{-i\epsilon K} e^{-i\epsilon V} + O(\epsilon^2) \; ;$$

its matrix element is, after neglecting $O(\epsilon^2)$,

$$<x_{n+1} \mid e^{-i\epsilon H} \mid x_n > \; = \int dy <x_{n+1} \mid e^{-i\epsilon K} \mid y > <y \mid e^{-i\epsilon V} \mid x_n > . \qquad (19.12)$$

By using (19.4) and

$$< y \mid e^{-i\epsilon V} \mid x_n > \; = e^{-i\epsilon V(x_n)} \, \delta(y - x_n) \; ,$$

we obtain

$$<x_{n+1} \mid e^{-i\epsilon H} \mid x_n > \; = \left(\frac{1}{2i\pi\epsilon}\right)^{\frac{1}{2}} \exp i\left[\frac{1}{2\epsilon}(x_{n+1} - x_n)^2 - \epsilon V(x_n)\right]$$
$$= \left(\frac{1}{2i\pi\epsilon}\right)^{\frac{1}{2}} e^{i\epsilon L_n} + O(\epsilon^{\frac{3}{2}}) \; .$$

Note that in L_n, given by (19.10), $\dot{x}_n$ stands for the entity $(x_{n+1} - x_n)/\epsilon$, and the argument of V is $\bar{x}_n = \frac{1}{2}(x_{n+1} + x_n)$; if we wish, we can replace $\bar{x}_n$ by x_n or x_{n+1}. Since, as indicated by (19.5), the average value of $x_{n+1} - x_n$ is $O(\epsilon^{\frac{1}{2}})$, the difference in $e^{-iV\epsilon}$ due to such a replacement is $O(\epsilon^{\frac{3}{2}})$, which can be neglected.

Substituting the above expression into (19.11) and setting $x = x_1$, we derive (19.9). Theorem 1 is proved.

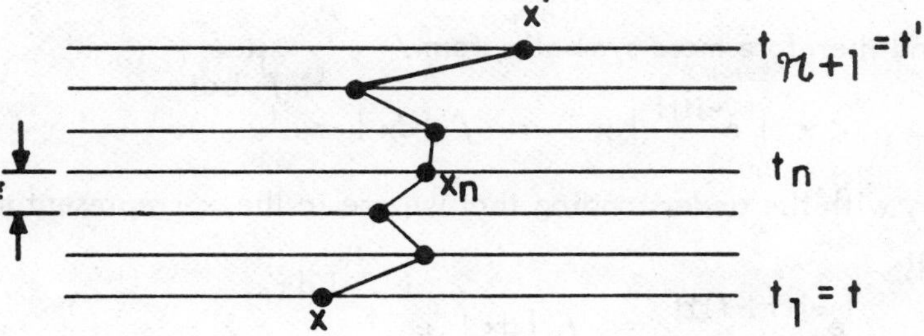

Fig. 19.1. The sequence $x = x_1, x_2, \cdots, x_{\mathcal{N}}, x_{\mathcal{N}+1} = x'$ describes a path $x(\tau)$ as τ varies from t to t'.

As shown in Figure 19.1, in the partition $t' - t = \mathcal{N}\epsilon$ we may label

$$t_n = t + (n-1)\epsilon$$

and interpret x_n as the value of x at time t_n. The sequence $x = x_1, x_2, \cdots, x_{\mathcal{N}}, x_{\mathcal{N}+1} = x'$ then describes a path $x(\tau)$ as τ varies from t to t'; the integrand in (19.9) can be interpreted as the wave amplitude due to that path. From (19.6), we see that the Lagrangian is

$$L(x, \dot{x}) = \tfrac{1}{2}\dot{x}^2 - V(x) \quad .$$

Hence, if we replace x by $\bar{x}_n \equiv \dfrac{x_{n+1} + x_n}{2}$ and $\dot{x}$ by $\dot{x}_n \equiv \dfrac{x_{n+1} - x_n}{\epsilon}$, (19.10) can be written as

$$L_n = L(\bar{x}_n, \dot{x}_n) \quad .$$

When $\epsilon \to 0$, the sum $\sum \epsilon L_n$ becomes the action integral along the path $x(\tau)$:

$$\sum_{n=1}^{\mathcal{N}} \epsilon L_n \to \int_t^{t'} L(x(\tau), \dot{x}(\tau))\, d\tau \equiv \int_t^{t'} L(\tau)\, d\tau \quad . \qquad (19.13)$$

In path-integration language, (19.9) is often written in a compact, and therefore more symbolic, form:

$$< x' \mid e^{-itH} \mid x > \; = \int [dx] \; e^{i \int_t^{t'} L dt}$$

or, with the understanding that we are in the x-representation, simply

$$e^{-i(t'-t)H} = \int [dx] \; e^{i \int_t^{t'} L d\tau} \; , \tag{19.9a}$$

so that between any two states $\mid >$ at t and $\mid t' >$ at t', we have

$$< t' \mid e^{-i(t'-t)H} \mid > \; = \int [dx] \; < t' \mid x' > \; e^{i \int_t^{t'} L d\tau} \; < x \mid >$$

where

$$[dx] \equiv \lim_{\epsilon \to 0} \left(\frac{1}{2 i \pi \epsilon} \right)^{\frac{1}{2} \mathcal{N}} \prod_{n=1}^{\mathcal{N}+1} dx_n$$

and $\int_t^{t'} L \, d\tau$ is given by (19.13). The precise meaning of (19.9a) is, of course, no more and no less than the original form (19.9).

From the Hamiltonian $H(x, p)$, it is useful to define a function $\bar{L}$ of x, $\dot{x}$ and p:

$$\bar{L}(x, \dot{x}, p) \equiv p\dot{x} - H(x, p) \; . \tag{19.14}$$

Note that $\bar{L}$ is a function of three variables x, $\dot{x}$ and p; it is not the Lagrangian function, which depends only on two variables x and $\dot{x}$. Of course, if in $\bar{L}$ we were to express p as a function of x and $\dot{x}$, then $\bar{L}$ would be related to the Lagrangian $L(x, \dot{x})$ by $\bar{L}(x, \dot{x}, p(x, \dot{x})) = L(x, \dot{x})$.

Corollary. An alternative expression of (19.9) is

$$< x' \mid e^{-i(t'-t)H} \mid > \; = \lim_{\epsilon \to 0} \int \prod_{n=1}^{\mathcal{N}} \frac{dx_n \, dk_n}{2\pi} \; e^{i \epsilon \bar{L}_n} < x_1 \mid > \tag{19.15}$$

where

$$\bar{L}_n = \bar{L}(\bar{x}_n, \dot{x}_n, k_n) \; ; \tag{19.16}$$

i.e., $\bar{L}_n$ is obtained by replacing the arguments x, $\dot{x}$ and p in the function $\bar{L}(x, \dot{x}, p)$ by

$$\bar{x}_n = \frac{x_{n+1} + x_n}{2}, \qquad \dot{x}_n = \frac{x_{n+1} - x_n}{\epsilon} \qquad \text{and} \qquad k_n .$$

<u>Proof.</u> From the first line in (19.2) we have

$$<x_{n+1} | e^{-i\epsilon K} | x_n> = \int \frac{dk_n}{2\pi} e^{-i\frac{1}{2}k_n^2\epsilon + ik_n(x_{n+1} - x_n)} .$$

By combining this expression with (19.12), we establish

$$<x_{n+1} | e^{-i\epsilon H} | x_n> = \int \frac{dk_n}{2\pi} e^{-i\frac{1}{2}k_n^2\epsilon + ik_n(x_{n+1} - x_n) - i\epsilon V(x_n)} . \tag{19.17}$$

Since in this problem $H(x, p) = \frac{1}{2}p^2 + V(x)$, the exponent is $-i\epsilon$ times

$$k_n\left(\frac{x_{n+1} - x_n}{\epsilon}\right) - H(x_n, k_n) = \bar{L}(x_n, \dot{x}_n, k_n)$$

where $\bar{L}$ is given by (19.14) and $\dot{x}_n$ stands for $(x_{n+1} - x_n)/\epsilon$, as before. We now substitute (19.17) into (19.11), setting

$$x = x_1 \quad \text{and} \quad x' = x_{\mathcal{N}+1} .$$

Then we replace x_n in $\bar{L}(x_n, \dot{x}_n, k_n)$ by $\bar{x}_n = \frac{1}{2}(x_{n+1} + x_n)$. This replacement is legitimate when $\epsilon \to 0$, because of (19.5). The resulting expression is (19.15), and that completes the proof of the corollary.

Notice that in (19.16) for $n = \mathcal{N}$, $\bar{L}_n$ contains $x_{\mathcal{N}+1} = x'$, but not $k_{\mathcal{N}+1}$. Consider now the (x, k) space. As τ varies from t to t', the sequence $(x_1, k_1), (x_2, k_2), \cdots, (x_n, k_n), \cdots, (x_{\mathcal{N}}, k_{\mathcal{N}})$ describes a path $(x(\tau), k(\tau))$ in that space (called phase space). When $\epsilon \to 0$, the sum $\sum \epsilon \bar{L}_n$ can be written as the action integral along the path in the phase space:

$$\sum_{n=1}^{\mathcal{N}} \epsilon \bar{L}_n \to \int_t^{t'} \bar{L}(x(\tau), \dot{x}(\tau), k(\tau))\, d\tau \equiv \int_t^{t'} \bar{L}(\tau)\, d\tau . \tag{19.18}$$

Equation (19.15) can also be written in a compact form:

$$e^{-i(t'-t)H} = \int [dx\, dk]\; e^{i\int_t^{t'} \bar{L}\, d\tau} \qquad (19.15a)$$

where

$$[dx\, dk] = \lim_{\epsilon \to 0} \prod_n \frac{dx_n\, dk_n}{2\pi} ,$$

with the understanding that the precise meaning of such an expression is (19.15).

Exercise 1. By direct differentiation, show that the Green's function (19.4) satisfies the free-particle Schrödinger equation

$$-\tfrac{1}{2} \frac{\partial^2}{\partial x^2} <x \,|\, e^{-i(t-t')K} \,|\, x'> = i \frac{\partial}{\partial t} <x \,|\, e^{-i(t-t')H} \,|\, x'>$$

and the boundary condition

$$\lim_{t \to t'} <x \,|\, e^{-i(t-t')H} \,|\, x'> \;\to\; \delta(x-x') \quad .$$

Exercise 2. Verify that the Green's function for the harmonic oscillator is

$$<x' \,|\, e^{-i(t'-t)H} \,|\, x> = \left(\frac{\omega}{2i\pi \sin \omega(t'-t)} \right)^{\frac{1}{2}} e^{i S(t'-t)} \qquad (19.19)$$

where

$$H = -\tfrac{1}{2} \frac{\partial^2}{\partial x^2} + \tfrac{1}{2} \omega^2 x^2$$

and

$$S(\tau) = \frac{\omega}{2 \sin \omega \tau} \left[(x'^2 + x^2) \cos \omega \tau - 2x'x \right] ,$$

which is also the classical action integral (i.e. the integral $\int L\, dt$ along the classical path).

3. N-dimensional problem

The extension to Cartesian coordinates in N dimensions is straightforward. Let

$$\vec{x} = (x_1, x_2, \cdots, x_N)$$

be the coordinate vector and

$$-i\vec{\nabla} = \left(-i\frac{\partial}{\partial x_1}, -i\frac{\partial}{\partial x_2}, \cdots, -i\frac{\partial}{\partial x_N}\right)$$

the conjugate-momentum operator. The eigenstate $|\vec{k}>$ of $-i\vec{\nabla}$ is

$$<\vec{x}\,|\,\vec{k}> = \sqrt{\frac{1}{\Omega}}\; e^{i\vec{k}\cdot\vec{x}} ,$$

where Ω is the volume of the system. The Hamiltonian is assumed to be

$$H = -\tfrac{1}{2}\nabla^2 + V(\vec{x}) \tag{19.20}$$

where

$$\nabla^2 = \frac{\partial^2}{\partial x_a\, \partial x_a}$$

is the Laplace operator in which the repeated index is summed over from 1 to N. Equations (19.9) and (19.15) can be readily generalized: For any state $|>$, we have

$$<\vec{x}'\,|\,e^{-i(t'-t)H}\,|> = \lim_{\epsilon\to 0} \int \prod_{n=1}^{\mathcal{N}} d^N x_n \left(\frac{1}{2i\pi\epsilon}\right)^{\frac{1}{2}N} e^{i\epsilon L_n} <\vec{x}_1\,|> \tag{19.21}$$

where

$$L_n = \tfrac{1}{2}\left(\frac{\vec{x}_{n+1}-\vec{x}_n}{\epsilon}\right)^2 - V\left(\frac{\vec{x}_{n+1}+\vec{x}_n}{2}\right) ,$$

and

$$<\vec{x}'\,|\,e^{-i(t'-t)H}\,|> = \lim_{\epsilon\to 0} \int \prod_{n=1}^{\mathcal{N}} \frac{d^N x_n\, d^N k_n}{(2\pi)^N}\, e^{i\epsilon\bar{L}_n} <\vec{x}_1\,|> \tag{19.22}$$

where

$$\bar{L}_n = \vec{k}_n \cdot \left(\frac{\vec{x}_{n+1}-\vec{x}_n}{\epsilon}\right) - \left[\tfrac{1}{2}\vec{k}_n^{\,2} + V\left(\frac{\vec{x}_{n+1}+\vec{x}_n}{2}\right)\right]$$

in which $t' = t + \mathcal{N}\epsilon$, $\vec{x}_n$ and $\vec{k}_n$ denote the c. number integration variables at $t_n = t + (n-1)\epsilon$, with $\vec{x}_{\mathcal{N}+1} = \vec{x}'$. [As before, $\vec{k}_{\mathcal{N}+1}$ does not enter into these expressions.] The above equations can also be written in more symbolic form, as in (19.9a) and (19.15a). We write

$$e^{-i(t'-t)H} = \int [dx]\; e^{i\int_t^{t'} L\,d\tau} \tag{19.21a}$$

where

$$[dx] = \lim_{\epsilon\to 0} \prod_n \left(\frac{1}{2i\pi\epsilon}\right)^{\frac{1}{2}N} d^N x_n ,$$

and

$$\int_t^{t'} L\, d\tau = \lim_{\epsilon\to 0} \sum \epsilon L_n ;$$

the alternative form is

$$e^{-i(t'-t)H} = \int [dx\, dk]\, e^{i\int_t^{t'} \bar{L}\, d\tau} \qquad (19.22a)$$

where

$$[dx\, dk] = \lim_{\epsilon\to 0} \prod_n \frac{d^N x_n\, d^N k_n}{(2\pi)^N} ,$$

and

$$\int_t^{t'} \bar{L}\, d\tau = \lim_{\epsilon\to 0} \sum_n \epsilon \bar{L}_n$$

with L_n and $\bar{L}_n$ given by (19.21) and (19.22).

19.2 Weyl-ordered Hamiltonian

Consider a classical nonlinear Lagrangian

$$L(q, \dot{q}) = \tfrac{1}{2} g_1(q)\, \dot{q}^2 + g_2(q)\, \dot{q} + g_3(q) \qquad (19.23)$$

where g_1, g_2 and g_3 are arbitrary real functions of q. Classically, $p \equiv \partial L/\partial\dot{q} = g_1(q)\,\dot{q} + g_2(q)$, and therefore the Hamiltonian is

$$H = H_{cl}(q, p) = \tfrac{1}{2} f_1(q)\, p^2 + f_2(q)\, p + f_3(q) \qquad (19.24)$$

where

$$f_1 = g_1^{-1} , \quad f_2 = -g_2/g_1$$

and

$$f_3 = \tfrac{1}{2}(g_2^2 - 2 g_1 g_3)/g_1 .$$

In passing to quantum mechanics, we have

$$p = p_{op} = -i \frac{\partial}{\partial q} .$$

However, knowing the classical limit does not completely determine the quantum Hamiltonian because of the different ways we may arrange the ordering of q and p.

1. From path integration to Hamiltonian operator

In the previous section, we started from a well-defined quantum-mechanical Hamiltonian operator H and showed that the Green's function $e^{-i(t'-t)H}$ can be expressed in terms of the phase-space path integration (19.15). We shall now pose the inverse question: Suppose we start from the classical nonlinear Hamiltonian $H_{cl}(q, p)$, (19.24); next we define, as in (19.14), a function $\bar{L}$ of three variables q, $\dot{q}$ and p,

$$\bar{L}(q, \dot{q}, p) \equiv p\dot{q} - H_{cl}(q, p) , \qquad (19.25)$$

and then construct the phase-space path integration by using the righthand side of (19.15). The question is: What would the corresponding operator on the lefthand side be? The answer is given by the following theorem.

Theorem 2. For any state vector $|\rangle$ that satisfies the normalization condition

$$\int dq \, |\langle q|\rangle|^2 = 1 , \qquad (19.26)$$

we have

$$\langle q'|e^{-i(t'-t)H_W}|\rangle = \lim_{\epsilon\to 0} \int \prod_{n=1}^{\mathcal{N}} \frac{dq_n \, dp_n}{2\pi} \, e^{i\epsilon \bar{L}_n} \langle q_1|\rangle \qquad (19.27)$$

where H_W is the Weyl-ordered Hamiltonian operator given by

$$H_W = \frac{1}{8}(p^2 f_1(q) + 2p f_1(q) p + f_1(q) p^2) + \tfrac{1}{2}(p f_2(q) + f_2(q) p) + f_3(q) \qquad (19.28)$$

with $p = -i\partial/\partial q$, and

$$\bar{L}_n \equiv \bar{L}(\bar{q}_n, \dot{q}_n, p_n) = p_n \dot{q}_n - H_{cl}(\bar{q}_n, p_n) \qquad (19.29)$$

in which

$$\bar{q}_n \equiv \frac{q_{n+1} + q_n}{2} \quad , \quad \dot{q}_n \equiv \frac{q_{n+1} - q_n}{\epsilon}$$

and (19.30)

$$q_{\mathcal{N}+1} = q' \quad .$$

Remarks. The righthand side of (19.27), together with (19.29)-(19.30), should be viewed as a "recipe". We first divide time $t'-t$ into $\mathcal{N}$ intervals of ϵ each, as before. From the classical function $\bar{L}(q, \dot{q}, p)$, (19.25), we obtain $\bar{L}_n$ by replacing its arguments q, $\dot{q}$ and p by $\bar{q}_n$, $\dot{q}_n$ and p_n in accordance with (19.30). Hence, we may regard q_n and p_n as the values of q and p at time $t_n = t + (n-1)\epsilon$. For arbitrary state vector $| >$, the normalization of its q-representation is determined because of (19.26). After the integration $dq_1\, dp_1 \cdots dq_{\mathcal{N}}\, dp_{\mathcal{N}}$, the righthand side of (19.27) depends only on $q_{\mathcal{N}+1} = q'$, $(t'-t)$ and $| >$. The theorem states that it is precisely the matrix element $<q' | e^{-i(t'-t)H_w} | >$ with H_w given by (19.28).

The proof of Theorem 2 will be given below, but before doing that it is useful to introduce some general properties of Weyl-ordering.

2. Weyl-ordering

Let us first discuss some simple classical polynomials of q and p: $q^n p$, $q^n p^2$, $\cdots$, $q^n p^m$, $\cdots$.

We define the corresponding Weyl-ordered quantum expressions, indicated by a subscript w, to be

$$(q^n p)_w \equiv \frac{1}{n+1} \sum_{\ell=0}^{n} q^{n-\ell}\, p\, q^{\ell}$$

$$= \frac{1}{n+1} \left[q^n p + q^{n-1} p q + \cdots + p q^n \right] \quad , \qquad (19.31)$$

$$(q^n p^2)_w \equiv \frac{2}{(n+1)(n+2)} \sum_{\ell=0}^{n} \sum_{m=0}^{n-\ell} q^{n-\ell-m}\, p\, q^{\ell}\, p\, q^{m} \quad ,$$

etc. where $p = -i\,\partial/\partial q$. In general, to derive $(q^n p^m)_W$ we first randomly order the q's and p's, with each different ordering counted once; their arithmetic mean is $(q^n p^m)_W$. The Weyl-ordering of any classical function

$$F(q, p) = \sum C_{\ell m} q^\ell p^m$$

is

$$F(q, p)_W \equiv \sum C_{\ell m} (q^\ell p^m)_W \;.$$

For our purpose, we are interested only in functions which depend quadratically on p, but arbitrarily on q, as in (19.24). It is straightforward to prove that the above forms of Weyl-ordering are identical to other definitions (see Exercises 1 and 2 on page 480):

$$(q^m p)_W \equiv (\tfrac{1}{2})^m \sum_{\ell=0}^{m} \frac{m!}{\ell!(m-\ell)!}\, q^{m-\ell}\, p\, q^\ell$$

and (19.32)

$$(q^m p^2)_W \equiv (\tfrac{1}{2})^m \sum_{\ell=0}^{m} \frac{m!}{\ell!(m-\ell)!}\, q^{m-\ell}\, p^2\, q^\ell \;;$$

or for an arbitrary function f of q,

$$[f(q)\,p]_W \equiv \tfrac{1}{2}[f(q)\,p + p\,f(q)]$$

and (19.33)

$$[f(q)\,p^2]_W \equiv \tfrac{1}{4}[f(q)\,p^2 + 2p\,f(q)\,p + p^2 f(q)] \;.$$

Note that these expressions can be readily generalized to functions that depend on higher powers of p. The generalizations of (19.32) and (19.33) are the same if we interchange q and p. By using (19.33), we see that for the classical Hamiltonian (19.24) the Weyl-ordered operator in quantum mechanics is given by (19.28).

Since with $p = -i\,\partial/\partial q$, H_W is Hermitian, its eigenfunctions, in accordance with the normalization condition (19.26), can be chosen to satisfy

$$\int \phi_{a'}^*(q)\, \phi_a(q)\, dq = \delta_{aa'}$$

where

$$H_w \, \phi_a(q) = E_a \, \phi_a(q) \quad .$$

The time-dependent Schrödinger equation * is

$$H_w \mid t > \; = \; i \, \frac{\partial}{\partial t} \mid t > \quad ,$$

or

$$H_w \, \phi(q, t) \; = \; i \, \dot{\phi}\,(q, t)$$

provided that

$$\phi(q, t) \; = \; < q \mid t > \quad .$$

Its Green's function is given by

$$< q' \mid e^{-i(t'-t) H_w} \mid q > \; = \; \sum_a \, \phi_a(q') \, \phi_a^*(q) \, e^{-i(t'-t) E_a} \quad ,$$

so that

$$\phi(q', t') \; = \; \int \, < q' \mid e^{-i(t'-t) H_w} \mid q > \, \phi(q, t) \, dq \quad .$$

From this we obtain the exact meaning of the lefthand side of (19.27).

<u>Proof of Theorem 2.</u> Let p_{op} and q_{op} be the quantum operators of p and q ; their matrix elements are

$$< q' \mid q_{op}^m \mid q > \; = \; q^m \, \delta(q' - q)$$

and, as in (19.1),

$$< q' \mid p_{op}^m \mid q > \; = \; \int \, \frac{dp}{2\pi} \, p^m \, e^{ip(q'-q)} \quad .$$

We may write (19.32) as

$$(q^m p^r)_w \; = \; (\tfrac{1}{2})^m \sum_{\ell=0}^{m} \frac{m!}{\ell!(m-\ell)!} \, q_{op}^{m-\ell} \, p_{op}^r \, q_{op}^{\ell} \quad ,$$

---

* As emphasized above, and also on p. 5, given a classical Hamiltonian such as (19.23), there can exist several different choices of quantum Hamiltonian; H_w is only one possibility. Here we are merely examining the consequences if H_w were the Hamiltonian.

where $r = 1$ or 2. Its matrix element is

$$<q'|(q^m p^r)_w|q> = (\tfrac{1}{2})^m \sum_{\ell=0}^{m} \frac{m!}{\ell!(m-\ell)!}$$

$$\cdot\, q'^{m-\ell} <q'|p_{op}^r|q> q^\ell$$

$$= \int \frac{dp}{2\pi} p^r e^{ip(q'-q)} \left(\frac{q'+q}{2}\right)^m .$$

Consequently, we obtain

$$<q'|[f(q)p^r]_w|q> = \int \frac{dp}{2\pi} p^r e^{ip(q'-q)} f\left(\frac{q'+q}{2}\right) . \tag{19.34}$$

Next, let us consider the operator $e^{-i\epsilon H_w}$. By expanding $e^{-i\epsilon H_w} = 1 - i\epsilon H_w + O(\epsilon^2)$ and using (19.28) and (19.34), we see that

$$<q_{n+1}|e^{-i\epsilon H_w}|q_n> = \delta(q_{n+1}-q_n) - i\epsilon \int \frac{dp_n}{2\pi} e^{ip_n(q_{n+1}-q_n)}$$

$$\cdot\, H_{cl}\left(\frac{q_{n+1}+q_n}{2}, p_n\right) + O(\epsilon^2) .$$

Since

$$\delta(q_{n+1}-q_n) = \int \frac{dp_n}{2\pi} e^{ip_n(q_{n+1}-q_n)} ,$$

we find

$$<q_{n+1}|e^{-i\epsilon H_w}|q_n> = \int \frac{dp_n}{2\pi} e^{i\epsilon \bar{L}_n} + O(\epsilon^2)$$

where $\bar{L}_n$ is given by (19.29). Substituting this expression into

$$<q'|e^{-i(t'-t)H_w}|> = \int <q'|e^{-i\epsilon H_w}|q_{\mathcal{N}}><q_{\mathcal{N}}|e^{-i\epsilon H_w}|q_{\mathcal{N}-1}>$$

$$\cdots <q_2|e^{-i\epsilon H_w}|q_1><q_1|> \prod_{n=1}^{\mathcal{N}} dq_n$$

and taking the limit $\epsilon \to 0$, we establish Theorem 2.

These discussions can easily be generalized to the case of several variables $q_1, q_2, \cdots$. The details will be omitted here.

Exercise 1. Show that because $[p, q] = -i$

$$\tfrac{1}{2}(q^m p + p q^m) = q^m p - \tfrac{1}{2} i m q^{m-1}$$
$$= (\tfrac{1}{2})^m \sum_0^m \frac{m!}{\ell!\,(m-\ell)!}\, q^{m-\ell}\, p\, q^{\ell}$$
$$= (q^m p)_W \quad \text{defined by (19.31).}$$

Exercise 2. By using $\sum_0^m \ell^2 = \frac{m}{6}(m+1)(2m+1)$,

$$\sum_0^m \ell^3 = [\tfrac{1}{2} m(m+1)]^2 , \quad \sum_{\ell=0}^m \frac{m!}{\ell!\,(m-\ell)!}\, \ell(\ell-1) = (\tfrac{1}{2})^{m-2} m(m-1) ,$$

show that

$$\tfrac{1}{4}(q^m p^2 + 2p\, q^m p + p^2 q^m) = q^m p^2 - i m q^{m-1} p - \tfrac{1}{4} m(m-1)\, q^{m-2}$$
$$= (\tfrac{1}{2})^m \sum_0^m \frac{m!}{\ell!\,(m-\ell)!}\, q^{m-\ell}\, p^2\, q^m$$
$$= (q^m p^2)_W \quad \text{defined by (19.31).}$$

Exercise 3. From Theorem 2, show that (19.27) is equivalent to

$$\langle q' \,|\, e^{-i(t'-t)H_W} \,|\, \rangle = \lim_{\epsilon \to 0} \int \prod_{n=1}^{\mathcal{N}} \left[\frac{g_1(\bar{q}_n)}{2\pi i \epsilon} \right]^{\frac{1}{2}} dq_n\, e^{i\epsilon L_n} \langle q_1 \,|\, \rangle$$

where

$$L_n = L(\bar{q}_n , \dot{q}_n) = \tfrac{1}{2} g_1(\bar{q}_n)\, \dot{q}_n^{\,2} + g_2(\bar{q}_n)\, \dot{q}_n + g_3(\bar{q}_n)$$

with g_i given by (19.23) and $\bar{q}_n$, $\dot{q}_n$ by (19.30).

19.3 Curvilinear Coordinates

1. Hamiltonian operator

Let us start from a problem which has a Cartesian basis, as discussed in Section 19.1. Denote the Cartesian coordinates as $x_1 , x_2 , \cdots , x_N$. Under the transformation from x_a to the curvilinear coordinates q_b ,

$$x = \begin{pmatrix} x_1 \\ x_2 \\ \vdots \\ x_N \end{pmatrix} \rightarrow q = \begin{pmatrix} q_1 \\ q_2 \\ \vdots \\ q_N \end{pmatrix} ,$$

the classical Lagrangian $L = \frac{1}{2}\tilde{\dot{x}}\,\dot{x} - V$ becomes

$$L(q, \dot{q}) = \tfrac{1}{2}\tilde{\dot{q}}\, M(q)\, \dot{q} - V(q) \tag{19.35}$$

where the matrix elements of $M(q)$ are given by $M_{ab} = \dfrac{\partial x_c}{\partial q_a}\dfrac{\partial x_c}{\partial q_b}$. [Here, the subscripts denote the components of x and q .] The classical Hamiltonian is, as in (18.45),

$$H_{cl}(q, p) = \tfrac{1}{2}\,\tilde{p}\, M^{-1}(q)\, p + V(q) \tag{19.36}$$

where $p = M\dot{q}$.

The quantum Hamiltonian H in the Cartesian coordinates is given by (19.20); the same operator in the curvilinear coordinates is, similarly to (18.47)-(18.48),

$$H(q, p) = \frac{1}{2}\,\frac{1}{\mathcal{g}}\; p_a(M^{-1}_{ab}\,\mathcal{g}\; p_b) + V(q) \tag{19.37}$$

where $p_a = -i\,\partial/\partial q_a$ and

$$\mathcal{g}\,(q) = \sqrt{\det M(q)} \quad . \tag{19.38}$$

The time-dependent Schrödinger equation is

$$H \,|\, t > \; = \; i \,\frac{\partial}{\partial t}\, |\, t > \quad ,$$

and the eigenvector of H is determined by

$$H \,|\, a > \; = \; E_a \,|\, a >$$

with the orthonormality condition $< a' \,|\, a > \; = \; \delta_{a'a}$. In the x-representation, we may define

$$\psi(x, t) \;\equiv\; < x \,|\, t >$$

and (19.39)

$$\psi_a(x) \equiv \langle x \mid a \rangle \ .$$

The $\{\psi_a(x)\}$ forms a complete orthonormal set of functions, which satisfy

$$\int \psi_{a'}^*(x)\ \psi_a(x)\ d^N x = \delta_{a'a} \ . \qquad (19.40)$$

The Green's function of the Schrödinger equation is

$$\langle x' \mid e^{-i(t'-t)H} \mid x \rangle = \sum_a \psi_a(x')\ \psi_a{}^*(x)\ e^{-i(t'-t)E_a} ,$$

which connects $\psi(x, t)$ to $\psi(x', t')$ through the relation

$$\psi(x', t') = \int \langle x' \mid e^{-i(t'-t)H} \mid x \rangle\ \psi(x, t)\ d^N x \ . \qquad (19.41)$$

Because $\mathcal{J}$ is the Jacobian of the transformation $x \to q$, we have

$$d^N x = \mathcal{J}\ d^N q \qquad (19.42)$$

where $d^N q = dq_1\ dq_2 \cdots dq_N$ and, as before, $d^N x = dx_1\ dx_2 \cdots dx_N$. In the q-representation, it is convenient to eliminate the Jacobian factor in the volume element. We introduce

$$\langle x \mid t \rangle \equiv \frac{1}{\sqrt{\mathcal{J}}} \langle q \mid t \rangle \quad \text{and} \quad \langle x \mid a \rangle \equiv \frac{1}{\sqrt{\mathcal{J}}} \langle q \mid a \rangle \ .$$

Hence,

$$\phi(q, t) \equiv \langle q \mid t \rangle = \sqrt{\mathcal{J}(q)}\ \psi(x(q), t)$$

and (19.43)

$$\phi_a(q) \equiv \langle q \mid a \rangle = \sqrt{\mathcal{J}(q)}\ \psi_a(x(q)) \ .$$

From (19.40), it follows that the ϕ_a's satisfy

$$\int \phi_{a'}^*(q)\ \phi_a(q)\ d^N q = \delta_{a'a} \ . \qquad (19.44)$$

We define the operator

$$\bar{H}(q, p) \equiv \sqrt{\mathcal{J}(q)}\ H(q, p)\ \frac{1}{\sqrt{\mathcal{J}(q)}} , \qquad (19.45)$$

so that the energy eigenfunction $\phi_a(q)$ satisfies

$$\bar{H}\,\phi_a(q) = E_a\,\phi_a(q)$$

and the time-dependent Schrödinger equation becomes

$$\bar{H}\,\phi(q, t) = i\,\dot{\phi}(q, t)\ ; \tag{19.46}$$

its Green's function is

$$\langle q' \,|\, e^{-i(t'-t)\bar{H}} \,|\, q\rangle \equiv \sum_a \phi_a(q')\,\phi_a^*(q)\,e^{-i(t'-t)E_a}\ . \tag{19.47}$$

From (19.41)-(19.43), it follows that

$$\phi(q', t') = \int \langle q' \,|\, e^{-i(t'-t)\bar{H}} \,|\, q\rangle\,\phi(q, t)\,d^N q\ . \tag{19.48}$$

Next, we want to express this Green's function in terms of path integration in the q-space.

2. Path-integration formula

As before, the time $t' - t$ is divided into $\mathcal{N}$ intervals of ϵ each. Let

$$q(n) = \begin{pmatrix} q_1(n) \\ q_2(n) \\ \vdots \\ q_N(n) \end{pmatrix}$$

denote the value of q at time $t_n = t + (n-1)\,\epsilon$. The sequence $q(1) = q$, $q(2)$, $\cdots$, $q(\mathcal{N})$, and $q(\mathcal{N}+1) = q'$ describes a path $q(\tau)$ as τ varies from t to t'. For any state vector $|\,\rangle$ at time t, (19.48) can be written in terms of the following path-integration expression:

<u>Theorem 3.</u>

$$\phi(q', t') = \langle q' \,|\, e^{-i(t'-t)\bar{H}} \,|\,\rangle$$

$$= \lim_{\epsilon\to 0} \int \prod_{n=1}^{\mathcal{N}} \left(\frac{1}{2\pi i\epsilon}\right)^{\frac{1}{2}N} d^N q(n)\ e^{i\epsilon L_{eff}(n)} \langle q(1)\,|\,\rangle \tag{19.49}$$

where

$$L_{eff}(n) = L(\bar{q}(n), \dot{q}(n)) - \frac{i}{\epsilon}\ln \mathcal{J}(\bar{q}(n)) - V_c(\bar{q}(n)) \tag{19.50}$$

in which

$$\bar{q}(n) \equiv \frac{q(n+1) + q(n)}{2} \quad , \quad \dot{q}(n) \equiv \frac{q(n+1) - q(n)}{\epsilon} \quad ,$$

the functions $L(q, \dot{q})$ and $\mathcal{J}(q)$ are given by (19.35), (19.38), and

$$V_c(q) = \frac{1}{8} \left[\frac{\partial}{\partial q_a} \left(\frac{\partial q_b}{\partial x_c} \right) \right] \left[\frac{\partial}{\partial q_b} \left(\frac{\partial q_a}{\partial x_c} \right) \right] \quad . \qquad (19.51)$$

It is interesting to note that the Jacobian term becomes an imaginary "effective" potential * with a magnitude proportional to ϵ^{-1}, which** becomes $\delta(0) = \infty$ as $\epsilon \to 0$. Furthermore, for curvilinear coordinates, there is an additional real potential term $V_c(q)$.

Proof.   From (19.33) we see that the Weyl-ordered operator H_w of (19.36) is

$$H(q, p)_w = \frac{1}{8} [p_a p_b M_{ab}^{-1} + 2 p_a M_{ab}^{-1} p_b + M_{ab}^{-1} p_a p_b] + V(q)$$

$$(19.52)$$

where, as before, $p_a = - i \partial / \partial q_a$ and

$$M_{ab}^{-1} = \frac{\partial q_a}{\partial x_c} \frac{\partial q_b}{\partial x_c} \quad .$$

By straightforward differentiation, it can be verified that, as will be given by Exercise 1 on page 487, the difference between the operators $\bar{H}$, (19.45), and H_w is $V_c(q)$ given by (19.51); i.e.

$$\bar{H}(q, p) = H(q, p)_w + V_c(q) \quad . \qquad (19.53)$$

In the partition of $t'-t$ into $\mathcal{N}\epsilon$, at each time $t_n = t + (n-1)\epsilon$, we introduce 2N integration variables $q_1(n), \cdots q_N(n), p_1(n), \cdots p_N(n)$, grouped into two $(N \times 1)$ column matrices

---

* T. D. Lee and C. N. Yang, Phys.Rev. 128, 885 (1962).

** The $\delta(t)$ function may be viewed as the limit of a positive square-well function, $= 0$ if $|t| > \frac{1}{2}\epsilon$, but $= \epsilon^{-1}$ otherwise; hence $\delta(0) = \epsilon^{-1} \to \infty$.

$$q(n) = \begin{pmatrix} q_1(n) \\ q_2(n) \\ \vdots \\ q_N(n) \end{pmatrix} \quad \text{and} \quad p(n) = \begin{pmatrix} p_1(n) \\ p_2(n) \\ \vdots \\ p_N(n) \end{pmatrix} .$$

By using Theorem 2 and (19.53), we can first cast (19.48) into a path integration in the (q, p) space: For any state vector $| >$

$$< q' | e^{-i(t'-t)\bar{H}} | > = \int < q' | e^{-i(t'-t)\bar{H}} | q > < q | > d^N q$$

$$= \lim_{\epsilon \to 0} \int \prod_{n=1}^{\mathcal{N}} \left(\frac{1}{2\pi}\right)^N d^N q(n)\, d^N p(n)\, e^{i\epsilon \bar{L}(n)} < q_1 | > \tag{19.54}$$

and

$$\bar{L}(n) = \tilde{p}(n)\, \dot{q}(n) - H_{cl}(\bar{q}(n), p(n)) - V_c(\bar{q}(n)) \tag{19.55}$$

where, as in (19.30),

$$\bar{q}(n) \equiv \frac{q(n+1) + q(n)}{2}, \quad \dot{q}(n) \equiv \frac{q(n+1) - q(n)}{\epsilon},$$

and $H_{cl}(q, p)$ and $V_c(q)$ are given by (19.36) and (19.51).

The function $\bar{L}(n)$ depends quadratically on $p(n)$. We note that the mass matrix M in (19.35) is real and symmetric. Hence, there exists a real and orthogonal matrix U which diagonalizes M:

$$\tilde{U}\, M(\bar{q}(n))\, U = \Lambda = \begin{pmatrix} \lambda_1 & 0 & 0 & \cdots & 0 \\ 0 & \lambda_2 & 0 & \cdots & 0 \\ 0 & 0 & \lambda_3 & \cdots & 0 \\ & \cdot & \cdot & \cdot & \cdot \\ 0 & 0 & 0 & \cdots & \lambda_N \end{pmatrix} .$$

It is convenient to introduce the column matrices

$$\pi \equiv \tilde{U}\, p(n) \quad \text{and} \quad v \equiv \tilde{U}\, \dot{q}(n)$$

with π_a and v_a as their a^{th} matrix elements. By substituting (19.36) into (19.55), we may write

$$\bar{L}(n) = \tilde{\pi} v - \tfrac{1}{2} \tilde{\pi} \Lambda^{-1} \pi - V - V_c$$
$$= -\tfrac{1}{2} \sum_{a=1}^{n} \left[\frac{1}{\lambda_a} (\pi_a - \lambda_a v_a)^2 - \lambda_a v_a^2 \right] - V - V_c \quad .$$

Since $d^N p(n) = d\pi_1 \, d\pi_2 \cdots d\pi_N$,

$$\sum_a \lambda_a v_a^2 = \tilde{v} \Lambda v = \tilde{\dot{q}}(n) \; M(\bar{q}(n)) \; \dot{q}(n)$$

and

$$\prod_a \sqrt{\lambda_a} = \sqrt{\det M(\bar{q}(n))} = \mathcal{J}(\bar{q}(n)) = e^{i\epsilon(-i/\epsilon) \ln \mathcal{J}(\bar{q}(n))}$$

the $p(n)$ – integration in (19.54) can be readily carried out with the aid of (19.3). The result is (19.49). Theorem 3 is proved.

In the compact notation of (19.9a) and (19.21a), (19.49) may be written as

$$e^{-i(t'-t)\bar{H}} = \int [dq] \; e^{i \int_t^{t'} L_{eff} \, d\tau} \tag{19.49a}$$

where

$$L_{eff}(q, \dot{q}) = L(q, \dot{q}) - i\,\delta(0) \ln \mathcal{J}(q) - V_c(q) \quad , \tag{19.56}$$

and

$$\delta(0) = \lim_{\epsilon \to 0} \frac{1}{\epsilon} \quad ,$$

so that between any two states $|\,\rangle$ at t, and $|\,t'\rangle$ at t' we have

$$\langle t' |\, e^{-i(t'-t)\bar{H}} \,| \,\rangle = \int [dq] \, \langle t' | q' \rangle \; e^{i \int_t^{t'} L_{eff} \, d\tau} \; \langle q | t \rangle$$

with

$$[dq] = \lim_{\epsilon \to 0} \prod_{n=1}^{\mathcal{N}} \left(\frac{1}{2\pi i \epsilon} \right)^{\frac{1}{2}N} d^N q(n)$$

and in terms of $L_{eff}(n)$ given by (19.50)

$$\int_t^{t'} L_{eff} \, d\tau = \sum_{n=1}^{\mathcal{N}} \epsilon \, L_{eff}(n) \quad .$$

From (19.38) we see that *

$$\ln \mathcal{J} = \tfrac{1}{2} \ln \det M = \tfrac{1}{2} \operatorname{trace} \ln M \quad .$$

At time τ between t and t' the matrix element of M is given by

$$\langle a \mid M(\tau) \mid b \rangle = \frac{\partial x_c(\tau)}{\partial q_a(\tau)} \frac{\partial x_c(\tau)}{\partial q_b(\tau)} \quad .$$

Let us define

$$\langle a, \tau \mid \ln \mathcal{M} \mid b, \tau' \rangle \equiv \delta(\tau - \tau') \langle a \mid \ln M(\tau) \mid b \rangle \tag{19.57}$$

so that

$$\operatorname{trace} \ln \mathcal{M} = \int d\tau \, \delta(0) \operatorname{trace} \ln M(\tau) \quad ;$$

consequently, we can express the integral of (19.56) as

$$\int_t^{t'} L_{eff} \, d\tau = \int_t^{t'} [\, L(q, \dot{q}) - V_c(q) \,] \, d\tau - i \operatorname{trace} \ln \mathcal{M}^{\frac{1}{2}} \quad . \tag{19.56a}$$

<u>Exercise 1.</u> Consider the transformation from the Cartesian coordinates $x_1, x_2, \cdots, x_N$ to the curvilinear coordinates $q_1, q_2, \cdots, q_N$. Let

$$K = -\tfrac{1}{2} \frac{\partial^2}{\partial x_a \, \partial x_a} = -\frac{1}{2\mathcal{J}} \frac{\partial}{\partial q_a} M^{-1}_{ab} \, \mathcal{J} \frac{\partial}{\partial q_b}$$

where

$$M^{-1}_{ab} = \frac{\partial q_a}{\partial x_c} \frac{\partial q_b}{\partial x_c} \quad \text{and} \quad \mathcal{J} = \sqrt{\det M} \quad .$$

By using Problem 7.3 (page 160), show that

$$\sqrt{\mathcal{J}} \, K \frac{1}{\sqrt{\mathcal{J}}} = \tfrac{1}{2} [\, \tfrac{1}{4} p_a p_b M^{-1}_{ab} + \tfrac{1}{2} p_a M^{-1}_{ab} p_b + \tfrac{1}{4} M^{-1}_{ab} p_a p_b \,] + \frac{1}{8} \left(\frac{\partial}{\partial q_a} \frac{\partial q_b}{\partial x_c} \right) \left(\frac{\partial}{\partial q_b} \frac{\partial q_a}{\partial x_c} \right) \tag{19.58}$$

---

* Denoting the eigenvalue of M as $\lambda_1, \lambda_2, \cdots$, we have $\mathcal{J} = \sqrt{\det M} = \prod_a \lambda_a^{\frac{1}{2}}$ which gives $\ln \mathcal{J} = \frac{1}{2} \sum_a \ln \lambda_a = \frac{1}{2} \operatorname{trace} \ln M$.

where $p_a = -i\,\partial/\partial q_a$.

Note that from (19.58), (19.53) follows.

<u>Exercise 2.</u> In a one-dimensional problem, let x be the Cartesian coordinate and q the curvilinear one, as in (19.35). Verify directly that (19.36), (19.37) and (19.53) hold with

$$M = \left(\frac{dx}{dq}\right)^2 \;, \qquad M^{-1} = \left(\frac{dq}{dx}\right)^2 \;, \qquad \mathcal{J} = \frac{dx}{dq}$$

and

$$V_c = \frac{1}{8}\left[\frac{d}{dq}\left(\frac{dq}{dx}\right)\right]^2 \;.$$

Apply the path-integration formalism (19.49)-(19.51) to this simple case.

3. <u>An example</u>

Consider the example discussed in Section 18.2. In the $\xi = 0$ gauge the Hamiltonian operator (18.23) can be written, in polar coordinates, as

$$H = -\frac{1}{2r}\frac{\partial}{\partial r}\left(r\frac{\partial}{\partial r}\right) - \frac{1}{2r^2}\frac{\partial^2}{\partial\theta^2} + V(r) \;.$$

The corresponding classical Lagrangian function is

$$L(r, \theta, \dot{r}, \dot{\theta}) = \tfrac{1}{2}\dot{r}^2 + \tfrac{1}{2}r^2\dot{\theta}^2 - V(r) \;. \tag{19.59}$$

The Jacobian of the transformation $x_1, x_2 \to r, \theta$, is

$$\mathcal{J} = r \;. \tag{19.60}$$

Let $\psi(x_1, x_2)$ be the wave function in Cartesian coordinates

$$\psi(x_1, x_2) \equiv \langle x_1, x_2 | \rangle \;.$$

In accordance with (19.43), we introduce

$$\psi = \frac{1}{\sqrt{r}}\,\phi \;.$$

The constraint (18.24) implies that ψ, and therefore also ϕ, is θ independent. Hence, in terms of ϕ, the time-dependent Schrödinger equation becomes

$$\bar{H}\,\phi(r,t) = i\,\dot{\phi}(r,t) \tag{19.61}$$

where $\bar{H} = \sqrt{r}\;H\;\frac{1}{\sqrt{r}}$. Because $\frac{\partial\phi}{\partial\theta} = 0$, we may write, as in (18.26),

$$\bar{H} = -\tfrac{1}{2}\,\frac{\partial^2}{\partial r^2} + V(r) + V_c(r) \tag{19.62}$$

where

$$V_c(r) = -\frac{1}{8r^2} \;. \tag{19.63}$$

Since the above Hamiltonian $\bar{H}$ is already Weyl-ordered, we may apply Theorem 2. By setting $q_n = r(n)$ and $p_n = p_r(n)$ in (19.31), we have

$$\phi(r',t') = \lim_{\epsilon\to 0}\int \prod_{n=1}^{\mathcal{N}} \frac{dr(n)\,dp_r(n)}{2\pi}\; e^{i\epsilon\bar{L}_n}\;\phi(r(1),t)$$

where

$$\begin{aligned} \bar{L}_n &= p_r(n)\,\dot{r}(n) - \tfrac{1}{2}p_r^{\,2}(n) - V(\bar{r}(n)) - V_c(\bar{r}(n)) \;, \\ \bar{r}(n) &= \tfrac{1}{2}[\,r(n+1) + r(n)\,] \;, \\ \dot{r}(n) &= [\,r(n+1) - r(n)\,]\,/\,\epsilon \;, \end{aligned} \tag{19.64}$$

and

$$r(\mathcal{N}+1) = r' \;.$$

The $p_r(n)$ integration can be readily carried out. The result is

$$\phi(r',t') = \lim_{\epsilon\to 0}\int \prod_{n=1}^{\mathcal{N}} \left(\frac{1}{2\pi i\epsilon}\right)^{\frac{1}{2}} dr(n)\; e^{i\epsilon\bar{L}_{eff}(n)}\;\phi(r(1),t) \tag{19.65}$$

where

$$\bar{L}_{eff}(n) = \tfrac{1}{2}\dot{r}^2(n) - V(\bar{r}(n)) - V_c(\bar{r}(n)) \;. \tag{19.66}$$

The same result can also be derived by using Theorem 3. By setting $N = 2$, $d^N q(n) = dr(n)\,d\theta(n)$ and

$$q(n) = \begin{pmatrix} r(n) \\ \theta(n) \end{pmatrix}$$

in (19.49), we find

$$\phi(r', \theta', t') = \lim_{\epsilon \to 0} \int \prod_{n=1}^{\mathcal{N}} \left(\frac{1}{2\pi i \epsilon}\right) dr(n)\, d\theta(n)\, e^{i\epsilon L_{eff}(n)}\, \phi(r(1), t) \tag{19.67}$$

where $r(\mathcal{N}+1) = r'$ and $\theta(\mathcal{N}+1) = \theta'$,

$$L_{eff}(n) = L(\bar{r}(n), \bar{\theta}(n), \dot{r}(n), \dot{\theta}(n)) - i\frac{1}{\epsilon} \ln \mathcal{J}(\bar{r}(n)) - V(\bar{r}(n)) - V_c(\bar{r}(n)) \tag{19.68}$$

with $\mathcal{J}$, V_c, $\bar{r}(n)$, $\dot{r}(n)$ given by (19.60) and (19.63)-(19.64),

$$\bar{\theta}(n) = \tfrac{1}{2}[\theta(n+1) + \theta(n)]$$

and

$$\dot{\theta}(n) = \frac{1}{\epsilon}[\theta(n+1) - \theta(n)] \quad .$$

Because $\phi(r(1), t)$ is independent of $\theta(1)$, we can integrate first with respect to $\theta(1)$, then $\theta(2)$, $\cdots$. The result * of these angular integrations is

$$\int \prod_{n=1}^{\mathcal{N}} \left(\frac{1}{2\pi i\, \epsilon}\right)^{\frac{1}{2}} d\theta(n)\, e^{\frac{1}{2} i\epsilon(\bar{r}(n)\,\dot{\theta}(n))^2} = \left[\prod_{n=1}^{\mathcal{N}} \bar{r}(n)\right]^{-\frac{1}{2}}$$

which cancels the Jacobian term in (19.68) and leads to the same expression (19.65), with $\phi(r', \theta', t') = \phi(r', t')$ independent of θ' .

19.4 Feynman Diagrams

The following examples illustrate how the path-integration formalism can lead naturally to Feynman diagrams. We will see the relative advantage of this approach versus the Dyson - Wick derivation given in Chapter 5.

---

* Cf. S.F. Edwards and Y.V. Gulyaev, Proc. Roy. Soc. 1A279, 229 (1964).

1. Contraction

Consider the Lagrangian of a simple harmonic oscillator

$$L = L_0(q, \dot{q}) = \tfrac{1}{2}\dot{q}^2 - \tfrac{1}{2}\omega^2 q^2 \quad ,$$

and $F(q, \dot{q})$, $G(q, \dot{q})$ are two arbitrary functions of q and $\dot{q}$. The path-integration "contraction" between

$$F(1) \equiv F(q(t_1), \dot{q}(t_1)) \quad \text{and} \quad G(2) \equiv G(q(t_2), \dot{q}(t_2)) \tag{19.69}$$

is defined to be

$$\underbrace{F(1)\,G(2)} \equiv \lim_{\substack{t' \to \infty \\ t \to -\infty}} \frac{\langle \text{vac} \,|\, \int [dq]\, F(1)\, G(2)\, e^{i\int_t^{t'} L_0 d\tau} \,|\, \text{vac} \rangle}{\langle \text{vac} \,|\, \int [dq]\, e^{i\int_t^{t'} L_0 d\tau} \,|\, \text{vac} \rangle} \tag{19.70}$$

where $|\,\text{vac}\rangle$ is the ground state (or vacuum) of the harmonic oscillator; i.e.,

$$a\,|\,\text{vac}\rangle = 0$$

where a is the standard annihilation operator defined by

$$q = \frac{1}{\sqrt{2\omega}}(a + a^\dagger)$$

and

$$p = -i\sqrt{\frac{\omega}{2}}\,(a - a^\dagger) \ . \tag{19.71}$$

In the path integration (19.70), our "recipe" is always to first replace at any time $t = t_n$

$$\begin{aligned} q(t) \quad &\text{by} \quad \tfrac{1}{2}[\, q(n+1) + q(n)\,] \equiv \bar{q}(n) \ , \\ \dot{q}(t) \quad &\text{by} \quad \frac{1}{\epsilon}[\, q(n+1) - q(n)\,] \equiv \dot{q}(n) \ , \end{aligned} \tag{19.72}$$

and then take the limit $\epsilon \to 0$, as in (19.9) or (19.49).

Theorem 4.

$$\underbrace{q(1)\,q(2)} = D_F(t_1 - t_2) \equiv \frac{i}{2\pi}\int \frac{e^{-ik_0(t_1 - t_2)}}{k_0^2 - \omega^2 + i\epsilon}\, dk_0 \ , \tag{19.73}$$

$$\underbrace{\dot{q}(1)\, q(2)} = \frac{d}{dt_1} D_F(t_1 - t_2) \tag{19.74}$$

and

$$\underbrace{\dot{q}(1)\, \dot{q}(2)} = -\frac{d^2}{dt_1^{\,2}} D_F(t_1 - t_2) = \omega^2 D_F(t_1 - t_2) + i\delta(t_1 - t_2)\,. \tag{19.75}$$

These expressions are to be contrasted with the Dyson-Wick definition of contraction given by (5.45):

$$\underline{q(1)\, q(2)} = D_F(t_1 - t_2) = \underbrace{q(1)\, q(2)} \quad \text{at all time} \tag{19.73a}$$

$$\underline{p(1)\, q(2)} = \underline{\dot{q}(1)\, q(2)} = \begin{cases} \underbrace{\dot{q}(1)\, q(2)} & \text{if } t_1 \neq t_2 \\ -\frac{i}{2} & \text{if } t_1 = t_2 \end{cases} \tag{19.74a}$$

and

$$\underline{p(1)\, p(2)} = \underline{\dot{q}(1)\, \dot{q}(2)} = \begin{cases} \underbrace{\dot{q}(1)\, \dot{q}(2)} & \text{if } t_1 \neq t_2 \\ \frac{1}{2}\,\omega & \text{if } t_1 = t_2 \end{cases}$$

$$= -\frac{d^2}{dt_1^{\,2}} D_F(t_1 - t_2) - i\,\delta(t_1 - t_2) \tag{19.75a}$$

in which the curved-line connection indicates the path-integration contraction and the straight-line connection the Dyson-Wick contraction.

<u>Proof.</u> Let

$$H_0 = \tfrac{1}{2} p^2 + \tfrac{1}{2} \omega^2 x^2 = (a^\dagger a + \tfrac{1}{2})\, \omega$$

be the Hamiltonian operator of the harmonic oscillator. The denominator of (19.70) is, on account of (19.9a),

$$\langle \text{vac} \mid e^{-i(t'-t) H_0} \mid \text{vac} \rangle = e^{-\frac{1}{2} i \omega (t'-t)} \quad ;$$

the numerator for $t_1 > t_2$ is

$$\langle \text{vac} \mid e^{-i(t'-t_1) H_0}\, F\, e^{-i(t_1 - t_2) H_0}\, G\, e^{-i(t_2 - t) H_0} \mid \text{vac} \rangle\,, \tag{19.76}$$

and the same for $t_1 < t_2$ but with t_1, F and t_2, G interchanged.

Consider first the case $F = G = q$. By using (19.7)-(19.9), we see that the path-integration formalism expresses the time-evolution of the state vector in the Schrödinger representation. Hence, from (5.15) it follows that

$$e^{itH_0}\, q\, e^{-itH_0} = q(t)$$

is the interaction representation * of the operator q at t. If we use the subscript I to denote the interaction representation, then (19.76) is

$$e^{-\frac{1}{2} i\omega(t'-t)} < vac \mid q(t_1)\, q(t_2) \mid vac >_I \quad ;$$

as a result (19.70) gives

$$\underbrace{q(1)\, q(2)} = \begin{cases} < vac \mid q(1)\, q(2) \mid vac >_I & \text{if } t_1 > t_2 \\ < vac \mid q(2)\, q(1) \mid vac >_I & \text{if } t_2 > t_1 \; . \end{cases}$$

On the other hand, the Dyson-Wick definition of contraction between F and G is given by (5.45):

$$\underbracket{F(1)\; G(2)} = \begin{cases} < vac \mid F(1)\; G(2) \mid vac >_I & \text{if } t_1 \geqslant t_2 \\ < vac \mid G(2)\; F(1) \mid vac >_I & \text{if } t_2 > t_1 \; . \end{cases} \tag{19.77}$$

By setting $F = G = q$, we find $\underbracket{q(1)\, q(2)} = \underbrace{q(1)\, q(2)}$ if $t_1 \neq t_2$. At $t_1 = t_2$, it is easy to see from their definitions that both contractions $\underbrace{q(1)\, q(2)}$ and $\underbracket{q(1)\, q(2)}$ are continuous. Thus, they are identical at all time.

In the interaction representation, we have

$$q(t) = \frac{1}{\sqrt{2\omega}} \left(a\, e^{-i\omega t} + a^{\dagger}\, e^{i\omega t} \right)$$

---

* In this simple example, the interaction representation is the same as the Heisenberg representation.

and

$$\dot{q}(t) = -i\sqrt{\frac{\omega}{2}}\,(a\,e^{-i\omega t} - a^{\dagger}\,e^{i\omega t}) = p_I(t) ,$$

where a and $a^{\dagger}$ are time-independent. Since

$$q(t)\ |\,\text{vac}> = \frac{1}{\sqrt{2\omega}}\ a^{\dagger}\,e^{i\omega t}\ |\,\text{vac}> ,$$

we derive

$$\underbrace{q(1)\,q(2)} = \underline{q(1)\,q(2)} = \frac{1}{2\omega}\,e^{\mp i\omega(t_1 - t_2)} = D_F(t_1 - t_2) \tag{19.78}$$

where in the exponent $-$ is for $t_1 > t_2$ and $+$ is for $t_1 < t_2$. The function $D_F(t_1 - t_2)$ is continuous at $t_1 = t_2$. Its integral representation in (19.73) can be readily derived by using the contour integration given in Figure 5.1, and that completes the proof of (19.73) and (19.73a).

For the rest of Theorem 4, in exactly the same way we can prove

$$\underbrace{\dot{q}(1)\,q(2)} = \underline{\dot{q}(1)\,q(2)}$$

and

$$\underbrace{\dot{q}(1)\,\dot{q}(2)} = \underline{\dot{q}(1)\,\dot{q}(2)}$$

so long as

$$t_1 \neq t_2 .$$

The question arises only when $t_1 = t_2$. From (19.72) it follows that when $t_1 = t_2$, $\underbrace{\dot{q}(1)\,q(2)}$ is the average of

$$\left(\frac{q(n+1) - q(n)}{\epsilon}\right)\left(\frac{q(n+1) + q(n)}{2}\right) = \frac{q^2(n+1.) - q^2(n)}{2\epsilon}$$

which is zero. Thus,

$$\underbrace{\dot{q}(1)\,q(2)} = 0 \qquad \text{when} \qquad t_1 = t_2 . \tag{19.79}$$

Likewise, when $t_1 = t_2$ the contraction $\underbrace{\dot{q}(1)\,\dot{q}(2)}$ is given by the average of

$$\left(\frac{q(n+1)-q(n)}{\epsilon}\right)^2 .$$

Because of (19.5), $|\, q(n+1) - q(n)\,| \sim \sqrt{\epsilon}$, this average value is $\propto \epsilon^{-1}$, as can be readily calculated by using

$$\int_{-\infty}^{\infty} \frac{x^2}{\epsilon^2}\left(\frac{1}{2\pi i\epsilon}\right)^{\frac{1}{2}} e^{i(x^2/2\epsilon)}\, dx = \frac{i}{\epsilon} .$$

Hence, from (19.70) we can verify that

$$\underbrace{\dot q(1)\, \dot q(2)} = \frac{i}{\epsilon} + \tfrac{1}{2}\omega \quad \text{when} \quad t_1 = t_2 . \tag{19.80}$$

From the D_F-function, (19.73), we see that when $t_1 = t_2$

$$\dot D_F = 0 \quad \text{and} \quad \ddot D_F = i\,\delta(0) + \tfrac{1}{2}\omega$$

which agree <u>exactly</u> with (19.79) and (19.80). This establishes (19.74) and (19.75).

On the other hand, by applying the Dyson-Wick definition (19.77) we have,* since $p = \dot q$, when $t_1 = t_2$

$$\underline{\dot q(1)\, q(2)} = \langle vac\,|\, p(1)\, q(1)\,|\, vac\rangle = -\frac{i}{2} \tag{19.81a}$$

and

$$\underline{\dot q(1)\, \dot q(2)} = \langle vac\,|\, p(1)\, p(1)\,|\, vac\rangle = \tfrac{1}{2}\omega . \tag{19.81b}$$

This leads to (19.74a) and (19.75a) and completes the proof of Theorem 4. It is important to note that in our path-integration definition (19.70) and with the "recipe" (19.72)

---

* If we modify the definition (19.77) of the Dyson-Wick contraction to

$$\underline{F(1)\, G(2)} = \begin{cases} \langle vac\,|\, F(1)\, G(2)\,|\, vac\rangle_I & \text{if } t_1 > t_2 \\ \tfrac{1}{2}\langle vac\,|\, F(1)\, G(2) + G(2)\, F(1)\,|\, vac\rangle_I & \text{if } t_1 = t_2 \\ \langle vac\,|\, G(2)\, F(1)\,|\, vac\rangle_I & \text{if } t_1 < t_2 , \end{cases}$$

then when $t_1 = t_2$, $\underline{\dot q(1)\, q(2)} = 0$ instead of $-\frac{i}{2}$, but $\underline{\dot q(1)\, \dot q(2)}$ is still $\frac{1}{2}\omega$. With this modified definition $\underbrace{\dot q(1)\, q(2)} = \underline{\dot q(1)\, q(2)}$ at all time, but (19.75a), $\underbrace{\dot q(1)\, \dot q(2)} = \underline{\dot q(1)\, \dot q(2)} + i\,\delta(t_1 - t_2)$, remains valid.

contractions of time derivatives are the same
as time derivatives of contractions, (19.82)

but in the Dyson-Wick definition this is not true.

Exercise. Let F and G be functions of q only. Show that

$$\underbrace{F(1)\,G(2)} \;=\; \underline{F(1)\,G(2)} \tag{19.83}$$

at all time.

2. Connected and disconnected diagrams

In a relativistic field theory, when we express the S-matrix as the sum of Feynman diagrams by following, say, the method developed in Chapter 5, it is not difficult to see that in general every matrix element contains disconnected as well as connected diagrams. Physically, disconnected diagrams represent independent processes, such as unrelated physical processes at different space-time regions. While the distinction is obvious graphically, it is less trivial to separate them analytically. The following example illustrates how the path-integration formalism can be of use.

Let us consider the simple problem of a free spin-0 field in the presence of a given external source. The Lagrangian is assumed to be

$$L_j(\phi, \dot{\phi}) \;\equiv\; L(\phi, \dot{\phi}) + \int j\phi\, d^3r$$

where

$$L(\phi, \dot{\phi}) \;=\; -\tfrac{1}{2} \int \left[\left(\frac{\partial \phi}{\partial x_\mu} \right)^2 + m^2 \phi^2 \right] d^3r \tag{19.84}$$

and $j = j(x)$ is an arbitrary c. number function of the space-time position $x_\mu = (\vec{r}, it)$. For $j = 0$, the Hamiltonian operator is

$$H \;=\; \int \left[\tfrac{1}{2} \Pi^2 + \tfrac{1}{2} (\vec{\nabla}\phi)^2 \right] d^3r$$

with Π and ϕ satisfying the standard canonical commutation rule

(2.5). Let $|\,\mathrm{vac}>$ be the ground state of H . We define two functionals W and S of j(x):

$$W(j) \equiv \lim_{\substack{t' \to \infty \\ t \to -\infty}} \frac{< \mathrm{vac} \,|\, \int [d\phi] \; e^{i \int_t^{t'} L_j d\tau} \,|\, \mathrm{vac} >}{< \mathrm{vac} \,|\, \int [d\phi] \; e^{i \int_t^{t'} L d\tau} \,|\, \mathrm{vac} >} \qquad (19.85)$$

and

$$e^{iS(j)} \equiv W(j) \quad . \qquad (19.86)$$

As we shall see, S(j) contains only connected graphs, and W(j) both connected and disconnected.

To give the precise meaning of the above path-integration expression, we introduce the Fourier expansion

$$\phi(\vec{r}, t) = \sqrt{\frac{2}{\Omega}} \sum{}' \left[q_{c,\vec{k}}(t) \cos(\vec{k} \cdot \vec{r}) + q_{s,\vec{k}}(t) \sin(\vec{k} \cdot \vec{r}) \right] \qquad (19.87)$$

where $q_{c,\vec{k}}(t)$ and $q_{s,\vec{k}}(t)$ are Hermitian, and the sum $\sum{}'$ extends over the half-$\vec{k}$ space, as defined on page 423. It is convenient to use one running parameter for the different subscripts:

$$\alpha \quad \text{for} \quad c, \vec{k} \quad \text{and} \quad s, \vec{k} \quad .$$

As in the previous section, we may group all these coordinates into a single column matrix q whose matrix elements are the q_α's :

$$q = \begin{pmatrix} \vdots \\ q_{c,\vec{k}} \\ q_{s,\vec{k}} \\ \vdots \end{pmatrix} \quad .$$

Equation (19.84) becomes then

$$L_j(q, \dot{q}) = L(q, \dot{q}) + \sum_\alpha j_\alpha q_\alpha$$

where

$$j_\alpha = \sqrt{\frac{2}{\Omega}} \int d^3r \; j(\vec{r}, t) \cdot \begin{cases} \cos \underset{\sim}{k} \cdot \underset{\sim}{r} & \text{if } \alpha = c, \vec{k} \\ \sin \underset{\sim}{k} \cdot \underset{\sim}{r} & \text{if } \alpha = s, \vec{k} \, , \end{cases}$$

$$L(q,\dot q) = \tfrac{1}{2}\sum_\alpha (\dot q_\alpha^{\,2} - \omega_\alpha^{\,2} q_\alpha^{\,2}) ,$$

and

$$\omega_\alpha = \sqrt{\vec k^2 + m^2} .$$

Again, we divide the time interval $t' - t \equiv \tau$ into $\mathcal{N}$ sections of ϵ each, with $q_\alpha(n)$ as the value of q_α at $t_n = t + (n-1)\epsilon$. The denominator in (19.85) is

$$< \text{vac} \,|\, \int [d\phi] \; e^{i\int_t^{t'} L d\tau} \,|\, \text{vac} >$$

$$= \lim_{\epsilon\to 0} \int dq_{\mathcal{N}+1} < \text{vac} \,|\, q(\mathcal{N}+1) > \cdot \int \prod_{n=1}^{\mathcal{N}} \prod_\alpha \left(\frac{1}{2\pi i \epsilon}\right)^{\frac{1}{2}} dq_\alpha(n)\, e^{i\epsilon L(n)} < q(1) \,|\, \text{vac} >$$

where

$$L(n) = L(\bar q(n), \dot q(n)) \tag{19.88}$$

with

$$\bar q(n) = \frac{q(n+1)+q(n)}{2} \quad \text{and} \quad \dot q(n) = \frac{q(n+1)-q(n)}{\epsilon}$$

as before; the numerator is given by the same expression, except for the replacement of L by L_j. Thus, we may regard

$$[d\phi] = \lim_{\epsilon\to 0} dq_{\mathcal{N}+1} \prod_{n=1}^{\mathcal{N}} \prod_\alpha \left(\frac{1}{2\pi i\epsilon}\right)^{\frac{1}{2}} dq_\alpha(n) ,$$

$$\int_{t_0}^{t'} L d\tau = \lim_{\epsilon\to 0} \epsilon \sum_{n=1}^{\mathcal{N}} L(\bar q(n), \dot q(n)) \tag{19.89}$$

and

$$\int_{t_0}^{t'} L_j d\tau = \lim_{\epsilon\to 0} \epsilon \sum_{n=1}^{\mathcal{N}} L_j(\bar q(n), \dot q(n)) .$$

Next, we expand the numerator as a power series in $j(x)$. The term $\exp i\int_t^{t'} L_j d\tau$ is a product of $\mathcal{N}$ factors of

$$e^{i\epsilon L_j(n)} = e^{i\epsilon L(n)} [1 + i\epsilon\, j_\alpha(t_n)\, \bar q_\alpha(n) + O(\epsilon^2)] .$$

When $\epsilon \to 0$, we keep $O(\mathcal{N}\epsilon)$ but neglect $O(\mathcal{N}\epsilon^2)$. Hence

$$e^{i\int_t^{t'} L_j d\tau} = e^{i\int_t^{t'} L d\tau} + i\int d^4x_1\, e^{i\int_{t_1}^{t'} L d\tau}\, j(1)\,\phi(1)\, e^{i\int_t^{t_1} L d\tau}$$

$$+\, i^2 \int d^4x_1\, d^4x_2\, e^{i\int_{t_2}^{t'} L d\tau}\, j(2)\,\phi(2)\, e^{i\int_{t_1}^{t_2} L d\tau}$$

$$\cdot\, j(1)\,\phi(1)\, e^{i\int_t^{t_1} L d\tau} \quad + \cdots$$

where $\int d^4x_\ell \cdots j(\ell)\,\phi(\ell)\cdots = \sum_n \epsilon \cdots \int j(\vec{r}_\ell, t_n)\,\phi(\vec{r}_\ell, t_n)\, d^3r_\ell \cdots$ with $\ell = 1,2,\cdots$. The third term on the righthand side can be written as

$$\frac{i^2}{2!} \int d^4x_1\, d^4x_2\; j(1)\, j(2)\; \phi(1)\, \phi(2)\; e^{i\int_t^{t'} L d\tau}$$

in which both the regions $t_2 > t_1$ and $t_1 > t_2$ are integrated over, and that accounts for the $1/2!$ factor. By extending the same consideration to the higher-order terms in the expansion we can write

$$e^{i\int_t^{t'} L_j d\tau} = e^{i\int_t^{t'} L d\tau} + \sum_{N=1}^{\infty} \frac{i^N}{N!} \int \prod_{\ell=1}^{N} d^4x_\ell\, j(\ell)\,\phi(\ell)\; e^{i\int_t^{t'} L d\tau} . \tag{19.90}$$

By substituting (19.90) into (19.85), we obtain the power series expansion of the functional $W(j)$ in $j(x)$. The zeroth order is 1 and the first order is 0 because $\langle vac|\, \phi(x)\, |vac\rangle = 0$. The second order is given by the limits $t' \to \infty$ and $t \to -\infty$ of

$$i\,S_2(j) \equiv \frac{i^2 \langle vac|\, \frac{1}{2}\int d^4x\, d^4y\; j(x)\,\phi(x)\, j(y)\,\phi(y)\, [d\phi]\; e^{i\int_t^{t'} L d\tau}\, |vac\rangle}{\langle vac|\, \int [d\phi]\; e^{i\int_t^{t'} L d\tau}\, |vac\rangle}$$

where S_2 is a functional of $j(x)$, with the subscript 2 indicating that it is a second-order term. Because of (19.84), we find

$$i\,S_2(j) = i^2\, \tfrac{1}{2} \int d^4x\, d^4y\; j(x)\, D_F(x-y)\; j(y) \tag{19.91}$$

in which D_F is given by (5.48)-(5.49). Graphically, we may represent

$i\,S_2$ as

$i\,S_2$ =  (19.92)

where the Feynman rules are: Each one-point vertex

 gives a factor i times $j(x)$,

each line connecting two points x and y

 gives a factor $D_F(x-y)$.

All space-time positions, such as x and y, are then to be integrated over. In addition, for each diagram there is an overall factor (= symmetry number)$^{-1}$. For example, the symmetry number* for the diagram in (19.92) is 2 because of the exchange symmetry between its two ends. It is straightforward to perform the systematic expansion of $W(j)$. The result is

$W(j)$ = 1 +  +  + 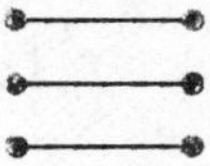 + ...

$$= 1 + i\,S_2 + \frac{(i\,S_2)^2}{2!} + \frac{(i\,S_2)^3}{3!} + \cdots \qquad (19.93)$$

where $2!\,,\ 3!\,,\ \cdots$ are the additional symmetry-number factors of these graphs. Hence we find

$$W(j) = e^{i\,S_2(j)} \quad .$$

From (19.86), we see that in this simple example $S(j) = S_2(j)$. It is clear that while $W(j)$ contains connected as well as disconnected

---

* Throughout the book, the propagator of a complex field is represented by an arrowed line, and that of a Hermitian field by a line without an arrow. The symmetry number of any diagram should be determined before the momentum assignment. We first label all points and lines (arrowed or unarrowed) of the diagram by different letters, and then consider permutations of them. The number of permutations that leaves the diagram with the letters unchanged is defined to be the symmetry number.

graphs, $S(j)$ contains only the connected graphs. As we shall show, this feature has a general validity.

3. Spin-0 field with interactions

In this section we shall extend our analysis to the more complicated case of an interacting spin-0 field. The Lagrangian is given by

$$L(\phi, \dot{\phi}) = L_0(\phi, \dot{\phi}) + L_{int}(\phi) \tag{19.94}$$

where

$$L_0 = -\tfrac{1}{2} \int \left[\left(\frac{\partial \phi}{\partial x_\mu} \right)^2 + m^2 \phi^2 \right] d^3r$$

and

$$L_{int} = \int \left[\tfrac{1}{2} \delta m^2 \phi^2 - \frac{1}{3!} g_0 \phi^3 - \frac{1}{4!} f_0 \phi^4 \right] d^3r \tag{19.95}$$

in which δm^2, g_0 and f_0 are exactly the same notations used in Section 5.6 when we discussed Feynman diagrams via Wick's theorem. Hence, according to (5.77)-(5.79) the Hamiltonian is

$$H = H_0 + H_{int}$$

where

$$H_0 = \tfrac{1}{2} \int [\Pi^2 + (\vec{\nabla}\phi)^2 + m^2 \phi^2] \, d^3r$$

and

$$H_{int} = - L_{int} \quad .$$

We first define the full Feynman propagator

$$D_F'(x-y) \equiv \lim_{\substack{t' \to \infty \\ t \to -\infty}} \frac{\langle vac | \int [d\phi] \, \phi(x) \, \phi(y) \, e^{i \int_t^{t'} L d\tau} | vac \rangle}{\langle vac | \int [d\phi] \, e^{i \int_t^{t'} L d\tau} | vac \rangle} \tag{19.96}$$

in which $| vac \rangle$ is the ground state of H_0, $[d\phi]$ and $\int_t^{t'} L d\tau$ are defined by (19.89), but with L now given by (19.94). Next, we expand $e^{i \int_t^{t'} L d\tau}$ as a power series in L_{int} :

$$e^{i \int_t^{t'} L d\tau} = e^{i \int_t^{t'} L_0 d\tau} + i \int_t^{t'} dt'' \, e^{i \int_{t''}^{t'} L_0 d\tau} \, L_{int}(t'') \, e^{i \int_t^{t''} L_0 d\tau} + \cdots .$$

Notice that L_{int} does not contain $\dot{\phi}$. By substituting this expression into (19.96) and using exactly the same argument as in the proof of (19.73) and (19.73a), we obtain

$$D_F'(x-y) = D_F(x-y) + i \int D_F(x-z)\, \delta m^2\, D_F(z-y)\, d^4z + (-ig_0)^2 \int D_F(x-z)\,[D_F(z-z')]^2\, D_F(z'-y)\, d^4z\, d^4z' + \cdots .$$

Graphically, it can be written as

$D_F'(x-y) =$ 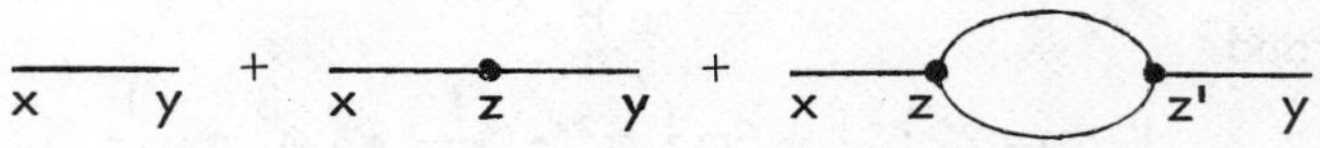

$+ \cdots$ (19.97)

in which the Feynman rules are the same as those given in Chapter 5:

a line connecting two points x and y

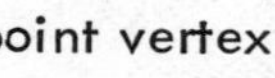

gives a factor $D_F(x-y)$,

a 2-point vertex

gives a factor $i\,\delta m^2$,

a 3-point vertex (19.98)

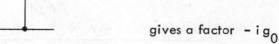

gives a factor $-ig_0$

and a 4-point vertex

gives a factor $-if_0$.

Equation (19.96) can be generalized to an N-point function

$$D_N(1,2,\cdots,N) \equiv \lim_{\substack{t'\to\infty \\ t\to-\infty}} \frac{\langle vac \,|\, \int [d\phi] \prod_1^N \phi(a)\; e^{i\int_t^{t'} L\, d\tau} \,|\, vac\rangle}{\langle vac \,|\, \int [d\phi]\; e^{i\int_t^{t'} L\, d\tau} \,|\, vac\rangle} \tag{19.99}$$

with 1, 2, ··, N denoting the space-time positions $x_1, x_2, \cdots, x_N$. For example, for $N = 4$, D_4 is given by the sum:

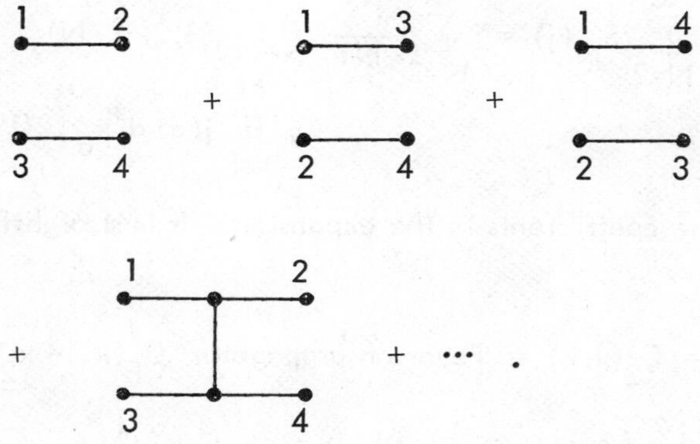

(19.100)

While $D_2(1,2) = D_F'(x-y)$ and $D_3(1,2,3)$ contain only connected diagrams, $D_{N\geq 4}$ contains disconnected as well. In order to separate the connected ones from the rest we introduce, as in (19.84),

$$L_j(\phi, \dot{\phi}) \equiv L(\phi, \dot{\phi}) + \int j\phi\, d^3r \tag{19.101}$$

where L is given by (19.94). As before, $j(x)$ is an arbitrary c. number function of space-time. Define $W(j)$ and $S(j)$ as in (19.85):

$$W(j) \equiv e^{iS(j)} \equiv \frac{\langle vac \,|\, \int [d\phi]\; e^{i\int_t^{t'} L_j\, d\tau} \,|\, vac\rangle}{\langle vac \,|\, \int [d\phi]\; e^{i\int_t^{t'} L\, d\tau} \,|\, vac\rangle}, \tag{19.102}$$

but with L_j given by (19.101) and L by (19.94). By using (19.90) and

$$< \text{vac} \mid \phi(x)\, e^{i \int_t^{t'} L\, d\tau} \mid \text{vac} > \; = \; 0 \quad ,$$

we find

$$W(j) \;=\; 1 + \sum_{N=2}^{\infty} \frac{i^N}{N!} \int D_N(1,2,\cdots,N) \prod_{a=1}^{N} j(a)\, d^4x_a \tag{19.103}$$

where D_N is given by (19.99). Let us write the functional $S(j)$ as a power series in j :

$$i\,S(j) \;=\; i \sum_{N=2}^{\infty} S_N(j) \;\equiv\; \sum_{N=2}^{\infty} \frac{i^N}{N!} \int C_N(1,2,\cdots,N) \cdot \prod_{a=1}^{N} j(a)\, d^4x_a \;, \tag{19.104}$$

with the C_N the coefficients in the expansion. It is straightforward to verify that

$$D_2(1,2) \;=\; C_2(1,2) \;=\; \text{Feynman propagator } \; D_F(x_1 - x_2) \quad ,$$

and

$$i\,S_2(j) \;=\; \frac{i^2}{2!} \int j(1)\, D_F(x_1 - x_2)\, j(2)\, d^4x_1\, d^4x_2 \quad ;$$

$$D_3(1,2,3) \;=\; C_3(1,2,3) \;=\; \text{sum of all connected 3-point diagrams}$$

and

$$i\,S_3(j) \;=\; \frac{i^3}{3!} \int C_3(1,2,3) \prod_1^3 j(a)\, d^4x_a \quad ;$$

$$D_4(1,2,3,4) \;=\; C_2(1,2)\, C_2(3,4) + C_2(1,3)\, C_3(2,4) + C_2(1,4)\, C_2(2,3) + C_4(1,2,3,4) \tag{19.105}$$

which leads to *

---

* On page 505, the coefficient $\frac{1}{8}$ outside the square bracket can be understood on the basis of the symmetry number of the disconnected diagram

We note that the permutations between 1 and 2, 3 and 4, and the pairs (1, 2) and (3, 4) give a total symmetry-number factor $2^3 = 8$.

$$\frac{i^4}{4!} \int D_4(1,\cdot\cdot 4) \prod_1^4 j(a)\, d^4x_a$$

$$= \frac{1}{8}\left[i^2 \int C_2(1,2)\, j(1)\, j(2)\, d^4x_1\, d^4x_2 \right]^2 + \frac{i^4}{4!} \int C_4(1,\cdot\cdot 4) \prod_1^4 j(a)\, d^4x_a$$

$$= \frac{1}{2!}\, (i\, S_2)^2 + i\, S_4 \ .$$

By comparing (19.100) with (19.105), we see that $C_4(1,2,3,4)$ is the sum of all connected 4-point diagrams.

Exercise. Show that $C_N(1,2,\cdot\cdot, N)$, defined by (19.104), is the sum of all connected N-point diagrams.

19.5 Fermions

To include fermions in the path-integration description, it is necessary to introduce a representation of the fermion Hilbert space as polynomials of generators $z_1, z_2, \cdots$ of a Grassmann algebra. *

1. Grassmann algebra

The generators $z_1, z_2, \cdots z_n$ of a Grassmann algebra G_n are elements that satisfy the anticommutation relations

$$\{z_i\, , z_j\} \equiv z_i z_j + z_j z_i = 0 \quad . \tag{19.106}$$

Hence,

$$z_1^2 = z_2^2 = \cdots = z_n^2 = 0 \ ,$$

and any function ϕ of z_i must be a simple polynomial of the form

$$\phi(z) = \phi_0 + \sum_i \phi_1(i)\, z_i + \sum_{i<j} \phi_2(i,j)\, z_i z_j + \sum_{i<j<k} \phi_3(i,j,k)\, z_i z_j z_k + \cdots + \phi_n(1,2,\cdot\cdot, n)\, z_1 z_2 \cdot\cdot z_n \tag{19.107}$$

* Cf. F. A. Berezin, The Method of Second Quantization (New York, Academic Press, 1966).

where the ϕ_a are ordinary c. numbers. We may choose all the $\phi_a(i_1, i_2, \cdots, i_a)$ to be totally antisymmetric functions of $i_1, i_2, \cdots, i_a$; i.e., ϕ_a changes sign under the permutation of any pair of its arguments. Equation (19.107) can also be written as

$$\phi(z) = \phi_0 + \sum_i \phi_1(i)\, z_i + \frac{1}{2!}\sum_{i,j} \phi_2(i,j)\, z_i z_j + \frac{1}{3!}\sum_{i,j,k} \phi_3(i,j,k)\, z_i z_j z_k + \cdots .$$

The Grassmann algebra G_n with n generators defines a linear space whose basis vectors are the 2^n polynomials

$$1, z_1, z_2, \cdots, z_n, z_1 z_2, z_1 z_3, \cdots, z_{n-1} z_n, \cdots, z_1 z_2 \cdots z_n ,$$

and the function $\phi(z)$ is a vector in this space.

The derivative operator $\partial/\partial z_a$ and the differential dz_a satisfy the anticommutation relations

$$\left\{ \frac{\partial}{\partial z_a}, z_b \right\} = \frac{\partial z_b}{\partial z_a} = \delta_{ab}$$

and (19.108)

$$\{dz_a, z_b\} = \{dz_a, dz_b\} = 0 .$$

Hence,

$$\frac{\partial}{\partial z_a} z_{i_1} z_{i_2} \cdots z_{i_m} = \delta_{ai_1} z_{i_2} \cdots z_{i_m} - z_{i_1} \delta_{ai_2} z_{i_3} \cdots z_{i_m} + \cdots + (-1)^{m-1} z_{i_1} \cdots z_{i_{m-1}} \delta_{ai_m} .$$

For $\phi(z)$ given by (19.107), we have

$$\frac{\partial \phi(z)}{\partial z_a} = \phi_1(a) + \sum_i \phi_2(a,i)\, z_i + \frac{1}{2!}\sum_{i,j} \phi_3(a,i,j)\, z_i z_j + \frac{1}{3!}\sum_{i,j,k} \phi_4(a,i,j,k)\, z_i z_j z_k + \cdots ,$$

$$d\phi(z) = \sum_i \phi_1(i)\, dz_i + \sum_{i,j} \phi_2(i,j)\, z_i\, dz_j + \frac{1}{2!}\sum_{i,j,k} \phi_3(i,j,k)\, z_i z_j\, dz_k + \cdots$$

and

$$\frac{\partial}{\partial z_a}\frac{\partial\phi(z)}{\partial z_b} = -\frac{\partial}{\partial z_b}\frac{\partial\phi(z)}{\partial z_a} .$$

The integration rules are *

$$\int dz_a = 0$$

and (19.109)

$$\int z_a\, dz_a = -\int dz_a\, z_a = 1 ,$$

where a can be 1, or 2, ··, or n. For $\phi(z)$ of (19.107), we have

$$\int \phi(z)\, dz_a = \phi_1(a) + \sum_i \phi_2(i,a)\, z_i + \frac{1}{2!}\sum_{i,j} \phi_3(i,j,a)\, z_i z_j + \frac{1}{3!}\sum_{i,j,k} \phi_4(i,j,k,a)\, z_i z_j z_k + \cdots ,$$

$$\int \frac{\partial\phi(z)}{\partial z_a}\, dz_a = 0$$

and if $z'_a = z_a$ + constant and $z'_i = z_i$ for all $i \neq a$

$$\int \phi(z')\, dz_a = \int \phi(z)\, dz_a .$$

Multiple integrals are simply the iterated integrals; e.g., from (19.107)

$$\int \phi(z) \prod_i dz_i = \phi_n(1,2,\cdots,n)$$

where

$$\prod_i dz_i \equiv dz_n\, dz_{n-1} \cdots dz_1 .$$

Any $\phi(z)$ can be decomposed into a sum of even and odd polynomials of z_i :

$$\phi(z) = \phi_+(z) + \phi_-(z)$$

where

$$\phi_\pm(-z) = \pm\phi_\pm(z) .$$

---

* These are really "measures." If we wish, we may regard (19.109) as definite integrals. They are not inverses of derivatives.

By using the integration rules (19.109), we find that for arbitrary $\phi(z)$ and $\phi'(z)$ the partial integration rule is

$$\int \phi_{\pm}(z) \frac{\partial \phi'(z)}{\partial z_a} dz_a = \mp \int \frac{\partial \phi_{\pm}(z)}{\partial z_a} \phi'(z) dz_a , \quad (19.110)$$

as can be readily verified by first considering the special case $\phi(z) = z_{i_1} z_{i_2} \cdots z_{i_\ell} z_a$ and $\phi'(z) = z_a z_{j_1} \cdots z_{j_m}$, and then considering the general case.

The "Fourier transform" of $\phi(z)$ is given by

$$f(\kappa) = \int \exp\left(\sum_i \kappa_i z_i\right) \phi(z) \prod_j dz_j \quad (19.111)$$

where the κ_i anticommute with each other as well as with the z_i,

$$\{\kappa_i , \kappa_j\} = \{\kappa_i , z_j\} = \{z_i , z_j\} = 0 , \quad (19.112)$$

and the running indices i and j vary from 1 to n as before. By using (19.109), (19.111) and

$$\exp\left(\sum_i \kappa_i z_i\right) = \prod_{i=1}^{n} (1 + \kappa_i z_i) ,$$

we see that

$$\kappa_a f(\kappa) = \int \exp\left(\sum_i \kappa_i z_i\right) \frac{\partial \phi(z)}{\partial z_a} \prod_j dz_j . \quad (19.113)$$

Let us define the "δ-function" to be

$$\delta(z) \equiv \int \prod_j d\kappa_j \exp\left(\sum_i \kappa_i z_i\right) = \prod_j (-z_j) . \quad (19.114)$$

We shall now prove, for arbitrary $\phi(z)$,

$$\phi(z) = \int \delta(z - z') \widetilde{\prod_i} dz'_i \, \phi(z') , \quad (19.115)$$

where the z'_i anticommute with each other as well as with z_i, and the order of product in $\widetilde{\prod_i}$ is the transpose of that in $\prod_i$; i.e., if in

(19.114)

$$\prod_j d\kappa_j = d\kappa_n \, d\kappa_{n-1} \cdots d\kappa_1 \quad , \tag{19.116}$$

then

$$\delta(z) = \prod_j (-z_j) = (-1)^n \, z_n \, z_{n-1} \cdots z_1$$

in the same order, but in (19.115)

$$\tilde{\prod_i} dz_i' = dz_1' \, dz_2' \cdots dz_n' \tag{19.117}$$

in the transposed order.

<u>Proof.</u> From the integration rule (19.109) it follows that

$$\int (z_a' - z_a) \, dz_a' \cdot \left\{ \begin{matrix} 1 \\ z_a' \end{matrix} \right. = \left\{ \begin{matrix} 1 \\ z_a \end{matrix} \right. \quad . \tag{19.118}$$

Let us examine a specific term in (19.107) by assuming

$$\phi(z) = z_{i_\ell} \, z_{i_{\ell-1}} \cdots z_{i_1} \qquad \text{with} \quad i_\ell > i_{\ell-1} \cdots > i_1 \quad .$$

Since from the definition (19.114) and with the product order given by (19.116)

$$\delta(z - z') = (z_n' - z_n)(z_{n-1}' - z_{n-1}) \cdots (z_1' - z_1) \quad , \tag{19.114a}$$

the righthand side of (19.115) is given by

$$\int (z_n' - z_n)(z_{n-1}' - z_{n-1}) \cdots (z_1' - z_1) \, dz_1' \cdots dz_n' \; z_{i_\ell}' \cdots z_{i_1}' \quad .$$

We first move the pair $(z_1' - z_1) \, dz_1'$ to the end of this product and then integrate. Next, move $(z_2' - z_2) \, dz_2$ to the end and integrate, etc. The result, because of (19.118), is

$$z_{i_\ell} \, z_{i_{\ell-1}} \cdots z_{i_1} \quad .$$

By applying the same arguments to every term in (19.107), we establish (19.115).

From this result, we can show that if $f(\kappa)$ is the Fourier transform (19.111) of $\phi(z)$, then the inverse transform is

$$\phi(z) = \int \exp\left(\sum_j z_j \kappa_j\right) f(\kappa) \widetilde{\prod_i} d\kappa_i \ . \tag{19.119}$$

Proof. Substituting (19.111) into (19.119), we can rewrite the righthand side as

$$\int \prod_{i'} dz'_{i'} \widetilde{\prod_i} d\kappa_i \exp\left[\sum_j \kappa_j (z'_j - z_j)\right] \phi(z') \ .$$

Since $\widetilde{\prod_i}$ is kept in the transposed order of $\prod_{i'}$, according to (19.116)-(19.117) we have

$$\prod_{i'} dz'_{i'} \widetilde{\prod_i} d\kappa_i = dz'_n \cdots dz'_2\, dz'_1\, d\kappa_1\, d\kappa_2 \cdots d\kappa_n$$

$$= (dz'_n d\kappa_n)(dz'_{n-1} d\kappa_{n-1}) \cdots (dz'_1 d\kappa_1) \ .$$

By using

$$\int d\kappa_i\, e^{\kappa_i (z'_i - z_i)} = (z_i - z'_i) \ , \tag{19.120}$$

and the fact that $(dz'_i d\kappa_i)$ and $(dz'_i (z_i - z'_i))$ are commuting elements, we find the righthand side of (19.119) to be

$$\int \prod_{i'} dz'_{i'} \widetilde{\prod_i} d\kappa_i \exp\left[\sum_j \kappa_j (z'_j - z_j)\right] \phi(z')$$

$$= \int dz'_n (z_n - z'_n)\, dz'_{n-1} (z_{n-1} - z'_{n-1}) \cdots dz'_1 (z_1 - z'_1)\, \phi(z')$$

$$= \int (z'_n - z_n)\, dz'_n (z'_{n-1} - z_{n-1})\, dz'_{n-1} \cdots (z'_1 - z_1)\, dz'_1\, \phi(z')$$

$$= \int \prod_j (z'_j - z_j) \widetilde{\prod_i} dz'_i\, \phi(z')$$

$$= \int \delta(z - z') \widetilde{\prod_i} dz'_i\, \phi(z') = \phi(z)$$

which is the lefthand side of (19.119).

Exercise. Any antisymmetric $n \times n$ matrix

$$a = (a_{ij}) = -\tilde{a}$$

can be converted into a block-diagonal form through an orthogonal transformation. We have

$$\tilde{S}\, a\, S = \begin{pmatrix} 0 & \lambda_1 & 0 & 0 & . & . & . & 0 & 0 \\ -\lambda_1 & 0 & 0 & 0 & . & . & . & 0 & 0 \\ 0 & 0 & 0 & \lambda_2 & . & . & . & 0 & 0 \\ 0 & 0 & -\lambda_2 & 0 & . & . & . & 0 & 0 \\ . & . & . & . & . & . & . & 0 & 0 \\ . & . & . & . & . & . & . & 0 & 0 \\ . & . & . & . & . & . & . & 0 & 0 \\ 0 & 0 & 0 & 0 & 0 & 0 & 0 & 0 & \lambda_\ell \\ 0 & 0 & 0 & 0 & 0 & 0 & 0 & -\lambda_\ell & 0 \end{pmatrix}$$

if $n = \text{even} = 2\ell$, or

$$\tilde{S}\, a\, S = \begin{pmatrix} 0 & \lambda_1 & 0 & 0 & . & . & 0 & 0 & 0 \\ -\lambda_1 & 0 & 0 & 0 & . & . & 0 & 0 & 0 \\ 0 & 0 & 0 & \lambda_2 & . & . & 0 & 0 & 0 \\ 0 & 0 & -\lambda_2 & 0 & . & . & 0 & 0 & 0 \\ . & . & . & . & . & . & 0 & 0 & 0 \\ . & . & . & . & . & . & 0 & 0 & 0 \\ 0 & 0 & 0 & 0 & 0 & 0 & 0 & \lambda_\ell & 0 \\ 0 & 0 & 0 & 0 & 0 & 0 & -\lambda_\ell & 0 & 0 \\ 0 & 0 & 0 & 0 & 0 & 0 & 0 & 0 & 0 \end{pmatrix}$$

if $n = \text{odd} = 2\ell + 1$, where $\tilde{S} S = 1$ and the a_{ij} are c. numbers. Show that the integral

$$I \equiv \int \exp\Big(\sum_{i,j} a_{ij} z_i z_j\Big)\, dz_n \cdots dz_1$$

$$= \begin{cases} 2^{n/2} \lambda_1 \cdots \lambda_n = \det |2a|^{\frac{1}{2}} & \text{if } n = 2\ell \\ 0 & \text{if } n = 2\ell + 1 . \end{cases}$$

Since $\det|a| = 0$ if $n =$ odd, this formula can also be written as

$$I = \det|2a|^{\frac{1}{2}} \quad .$$

2. Quantum mechanics

(i) Let us first consider the case of a single-fermion mode. The Hamiltonian is

$$H = \omega a^{\dagger} a$$

where ω is a constant, a and $a^{\dagger}$ are the standard annihilation and creation operators that satisfy (3.42); i.e.

$$\{a, a^{\dagger}\} = 1$$

and

$$\{a, a\} = \{a^{\dagger}, a^{\dagger}\} = 0 \quad .$$

The eigenstates of H are ket-vectors $|n>$:

$$H|n> = n\omega|n> \tag{19.121}$$

with $n = 0$ or 1.

In terms of the generator z of the Grassmann algebra G_1, we can construct the "coordinate" representation*

$$a = z \;, \quad a^{\dagger} = \frac{\partial}{\partial z} \tag{19.122}$$

and therefore

$$H = \omega \frac{\partial}{\partial z} z = \omega - \omega z \frac{\partial}{\partial z} \quad .$$

In accordance with (19.121), the states $|n>$ in the z-representation are

$$<z|0> = z \quad \text{and} \quad <z|1> = 1 \quad . \tag{19.123}$$

Likewise, the bra-vectors $<n|$ can be written as

$$<0|z> = -1 \quad \text{and} \quad <1|z> = z \;, \tag{19.124}$$

---

* If we wish, we may interchange the roles of a and $a^{\dagger}$ so that instead of (19.122), $a = \partial/\partial z$ and $a^{\dagger} = z$.

so that the usual orthonormality relation

$$< n' \mid n > \; \equiv \; \int < n' \mid z > dz < z \mid n > \; = \; \delta_{n'n} \qquad (19.125)$$

and the completeness theorem

$$\sum_{n=0,1} < z' \mid n >< n \mid z > \; = \; - z' + z \; = \; \delta(z' - z) \qquad (19.126)$$

hold. The Green's function of the time-dependent Schrödinger equation

$$H \mid t > \; = \; i \frac{\partial}{\partial t} \mid t >$$

is e^{-itH}. In the z-representation and for $H = \omega a^{\dagger} a$, we have

$$< z' \mid e^{-itH} \mid z > \; = \; \sum_{n=0,1} < z' \mid n > e^{-itn\omega} < n \mid z >$$

$$= - z' + e^{-it\omega} z . \qquad (19.127)$$

It is instructive to compute the trace of e^{-itH}. By using (19.123)-(19.124) and (19.127), we find

$$\text{trace} < z' \mid e^{-itH} \mid z >$$

$$\equiv \sum_{n} \int\int < n \mid z' > dz' < z' \mid e^{-itH} \mid z > dz < z \mid n >$$

$$= \int\int [- dz' (-z' + e^{-it\omega} z) \, dz \, z + z' \, dz' (-z' + e^{-it\omega} z) \, dz]$$

$$= 1 + e^{-it\omega}$$

which is, of course, the familiar answer. Another useful exercise is to set $t = \epsilon$ = infinitesimal in (19.127). We have, after neglecting $O(\epsilon^2)$,

$$< z' \mid e^{-i\epsilon H} \mid z > \; = \; - z' + (1 - i\epsilon\omega) z$$

which, because of (19.120), can also be written as

$$\int d\kappa \, e^{\kappa(z' - z)} \, (1 + i\epsilon\omega\kappa z)$$

where

$$\kappa^2 = \{\kappa , z\} = \{\kappa , z'\} = 0$$

and

$$- \int d\kappa \, \kappa = \int \kappa \, d\kappa = 1 .$$

Thus,

$$< z' \mid e^{-i\epsilon H} \mid z > dz \;=\; \int d\kappa \, dz \, e^{\kappa(z'-z)} \; (1 + i\epsilon\omega\kappa z) \quad , \tag{19.128}$$

where the integration refers only to $d\kappa$. This formula will be useful later on when we discuss the path integration.

Exercise. By using

$$< z \mid O \mid z' > \;\equiv\; \sum_{n,n'} < z \mid n >< n \mid O \mid n' >< n' \mid z' >$$

for any Hilbert-space operator O, show that

$$
\begin{aligned}
< z \mid z' > &= \delta(z - z') \quad , \\
< z \mid a \mid z' > &= z\,\delta(z - z') = z z' \quad , \\
< z \mid a^\dagger \mid z' > &= \frac{\partial}{\partial z}\, \delta(z - z') = -1 \quad , \\
< z \mid a^\dagger a \mid z' > &= \frac{\partial}{\partial z}\, z\,\delta(z - z') = z'
\end{aligned} \tag{19.129}
$$

and therefore

$$< z \mid a^\dagger a \mid z' > \;=\; \int < z \mid a^\dagger \mid z'' > dz'' < z'' \mid a \mid z' > \; .$$

(ii) The generalization to N fermion modes is straightforward. As in Section 3.5, each mode carries an annihilation operator a_i and its Hermitian conjugate $a_i^\dagger$. Their anticommutation relations are

$$\{a_i \, , \, a_j^\dagger\} \;=\; \delta_{ij}$$

and

$$\{a_i \, , \, a_j\} \;=\; \{a_i^\dagger \, , \, a_j^\dagger\} \;=\; 0 \quad .$$

Let the eigenvalue of the occupation number operator $a_i^\dagger a_i$ be $n_i = 0$ or 1. The Hilbert space is spanned by the 2^N basis vectors

$$
\begin{aligned}
&| 0 > , \quad a_1^\dagger \, | 0 > , \cdots , a_N^\dagger \, | 0 > , \quad a_2^\dagger a_1^\dagger \, | 0 > , \; \cdots \\
&| n > \;\equiv\; | n_1 , n_2 , \cdots , n_N > \;=\; (a_N^\dagger)^{n_N} \cdots (a_2^\dagger)^{n_2} (a_1^\dagger)^{n_1} \, | 0 > \\
&\cdots , \; | 1, 1, \cdots , 1 > \;=\; a_N^\dagger \cdots a_2^\dagger a_1^\dagger \, | 0 >
\end{aligned} \tag{19.130}
$$

where n denotes the set $\{n_i\}$ and $| 0 >$ satisfies

$$a_i \, | 0 > \; = \; 0$$

for all i.

As in (19.122), the "coordinate" representation of a_i and $a_i^\dagger$ can be expressed in terms of the generators $z_1, z_2, \cdots, z_N$ of the Grassmann algebra G_N:

$$a_i = z_i \quad \text{and} \quad a_i^\dagger = \frac{\partial}{\partial z_i} \, . \tag{19.131}$$

For the ket-vectors (19.130), we may write

$$< z \, | \, 0 > \; = \; z_1 z_2 \cdots z_N$$

and

$$\begin{aligned} < z \, | \, n > \; &\equiv \; < z \, | \, n_1, n_2, \cdots n_N > \\ &= \left(\frac{\partial}{\partial z_N} \right)^{n_N} \cdots \left(\frac{\partial}{\partial z_1} \right)^{n_1} < z \, | \, 0 > , \end{aligned} \tag{19.132}$$

as in (19.123). The corresponding bra-vectors $< n \, |$ are given by

$$< n \, | \, z > \; \equiv \; < n_1, n_2, \cdots n_N \, | \, z > \; = \; \pm \prod_i z_i^{n_i} \tag{19.133}$$

where the $\pm$ sign is chosen so that

$$\int < n' \, | \, z > \prod_i dz_i < z \, | \, n > \; = \; \delta_{nn'} \, . \tag{19.134}$$

For example, if $\prod_i dz_i = dz_N \cdots dz_2 \, dz_1$, then

$$< 0 \, | \, z > \; = \; (-1)^N ,$$

etc. Just as in (19.126), these basis vectors also satisfy the completeness theorem

$$\sum_n < z' \, | \, n >< n \, | \, z > \; = \; \widetilde{\prod_i} \, \delta(z_i' - z_i) \tag{19.135}$$

in which $\widetilde{\prod_i}$ is the transpose of $\prod_i$ in (19.134), as can be verified by multiplying (19.134) on the left by $< z' \, | \, n' >$ and summing over $n' = \{n_i'\}$. The z-representation of an arbitrary Hilbert-space state vector

$$| > \; = \; \sum C_n \, | \, n > \;,$$

where $C_n = C_{n_1 n_2 \cdots n_N}$ is the probability amplitude of finding $| >$ in $| \, n >$, is the polynomial

$$< z \, | > \; = \; \sum_n C_N < z \, | \, n >$$

$$= C_{11 \ldots 1} + C_{11 \ldots 10} z_N + \cdots$$

$$+ C_{00 \ldots 0} z_1 z_2 \cdots z_N .$$

The z-representation of its bra-vector is the complementary polynomial

$$< | \, z > \; = \; \sum_n C_n^* < n \, | \, z >$$

so that

$$\int < | \, z > \prod_i dz_i < z \, | > \; = \; \sum_n | \, C_n \, |^2 \quad .$$

As an example, let us consider the simple Hamiltonian

$$H \; = \; \sum_i \omega_i \, a_i^\dagger \, a_i \quad .$$

The coordinate representation of e^{-itH} is given by

$$< z' \, | \, e^{-itH} \, | \, z > \; = \; \sum_n < z' \, | \, n > e^{-it \sum_i n_i \omega_i} < n \, | \, z > \quad .$$

For $t = \epsilon$ = infinitesimal, by using (19.128) we find, after neglecting $O(\epsilon^2)$,

$$< z' \, | \, e^{-i\epsilon H} \, | \, z > \prod_i dz_i$$

$$= \int \prod_i d\kappa_i \, dz_i \, e^{\kappa_i (z_i' - z_i)} \cdot (1 + i\epsilon \sum_j \omega_j \, \kappa_j \, z_j) \qquad (19.136)$$

where the integration sign refers only to $d\kappa_i$, and κ_i, z_j, z_k' all anticommute with each other, as in (19.120). We note that on the righthand side the products $d\kappa_i\, dz_i\, e^{\kappa_i(z_i'-z_i)}$ and $\kappa_j z_j$ commute with each other; consequently the order of i in $\prod_i$ can be arbitrary. This introduces great convenience in its application, to be contrasted with the lefthand side in which $\prod_i dz_i$ must be kept in the same order as in (19.134). When $\epsilon \to 0$, according to (19.135) we have $\langle z' | e^{-i\epsilon H} | z \rangle \prod_i dz_i \to \widetilde{\prod_i} \delta(z_i' - z_i) \prod_i dz_i$, which gives 1 upon integration.

Exercise. Show that for

$$H = \sum_{i,j} \omega_{ij} a_i^\dagger a_j = \sum_{i,j} \omega_{ij} \frac{\partial}{\partial z_i} z_j$$

and ϵ infinitesimal, after neglecting $O(\epsilon^2)$

$$\langle z' | e^{-i\epsilon H} | z \rangle \prod_i dz_i = \int \prod_i d\kappa_i\, dz_i\, e^{\kappa_i(z_i'-z_i)} \cdot (1 + i\epsilon \sum_{j,k} \omega_{jk} \kappa_j z_k) \quad (19.137)$$

where as in (19.136) the integration sign refers only to $d\kappa_i$, and the c. number constants $\omega_{ij} = \omega_{ji}^*$. The order of the product in $\prod_i$ on the righthand side can again be arbitrary.

3. Path integration

Consider the example of a quantized spin-$\frac{1}{2}$ Dirac field $\psi(\vec{r}, t)$ moving in a given external vector potential V_μ. The Lagrangian density is

$$\mathcal{L} = -\psi^\dagger \gamma_4 \gamma_\mu (D_\mu + m) \psi \quad (19.138)$$

where

$$D_\mu = \frac{\partial}{\partial x_\mu} - i g V_\mu .$$

The corresponding Hamiltonian density is

$$\mathcal{H} = \psi^\dagger \gamma_4 (\gamma_i D_i + m) \psi + g V_0 \psi^\dagger \psi . \quad (19.139)$$

The expansion (3.32) gives

$$\psi(\vec{r},t) = \frac{1}{\sqrt{\Omega}} \sum_{\vec{p},s} \left(a_{\vec{p},s}(t)\, u_{\vec{p},s}\, e^{i\vec{p}\cdot\vec{r}} + b^{\dagger}_{\vec{p},s}(t)\, v_{\vec{p},s}\, e^{-i\vec{p}\cdot\vec{r}} \right). \tag{19.140}$$

For each pair $a_{\vec{p},s}$ and $b^{\dagger}_{\vec{p},s}$ we introduce two generators z_{2n} and z_{2n+1} of the Grassmann algebra so that *

$$a_{\vec{p},s} = z_{2n} \quad , \quad b^{\dagger}_{\vec{p},s} = z_{2n+1}$$

and

$$a^{\dagger}_{\vec{p},s} = \frac{\partial}{\partial z_{2n}} \quad , \quad b_{\vec{p},s} = \frac{\partial}{\partial z_{2n+1}} . \tag{19.141}$$

Hence, (19.140) becomes

$$\psi(\vec{r}, t) = \sum_{\alpha} \phi_{\alpha}(\vec{r})\, z_{\alpha}(t) \tag{19.142}$$

where $\phi_{\alpha}(\vec{r})$ is the appropriate c. number spinor function, either $u_{\vec{p},s}\, e^{i\vec{p}\cdot\vec{r}}/\sqrt{\Omega}$ or $v_{\vec{p},s}\, e^{-i\vec{p}\cdot\vec{r}}/\sqrt{\Omega}$, which satisfies

$$\int \phi^{\dagger}_{\alpha}(\vec{r})\, \phi_{\beta}(\vec{r})\, d^3r = \delta_{\alpha\beta} . \tag{19.143}$$

The Hamiltonian in the z-representation is

$$H = \int \mathcal{H}\, d^3r = \sum_{\alpha,\beta} H_{\alpha\beta} \frac{\partial}{\partial z_{\alpha}} z_{\beta} \tag{19.144}$$

with

$$H_{\alpha\beta} = \int \phi^{\dagger}_{\alpha} [\gamma_4(\gamma_i D_i + m) + g V_0]\, \phi_{\beta}\, d^3r .$$

The solution of the time-dependent Schrödinger equation

$$H \,|\, t > = i \frac{\partial}{\partial t} \,|\, t >$$

in the z-representation satisfies

---

* For the "positron" states, we have exchanged the roles of z_{2n+1} and $\partial/\partial z_{2n+1}$ for a temporary notational convenience; thus, representations (19.132) and (19.133) should be interchanged accordingly when dealing with these states. See the footnote on p. 512.

$$\langle z' \mid t' \rangle = \int \langle z' \mid e^{-i(t'-t)H} \mid z \rangle \prod_\alpha dz_\alpha \langle z \mid t \rangle \;.$$

We may divide $t' - t$ into $\mathcal{N}$ intervals of ϵ each. Let $z_\alpha(n)$ be the value of z_α at $t_n = t + (n-1)\epsilon$, with the set $\{z_\alpha(n)\}$ represented by $z(n)$. For $n = 1$ we have $t_n = t$ and $z(1) = z$. By applying (19.137) to the Hamiltonian (19.144), we find

$$\langle z_\alpha(n+1) \mid e^{-i\epsilon H} \mid z_\alpha(n) \rangle \prod_\alpha dz_\alpha(n)$$

$$= \int \prod_\alpha d\kappa_\alpha(n)\, dz_\alpha(n)\, e^{\kappa_\alpha(n)[z_\alpha(n+1) - z_\alpha(n)]} \cdot [1 + i\epsilon \sum_{\alpha,\beta} H_{\alpha\beta}\, \kappa_\alpha(n)\, z_\beta(n)] \;.$$

Because

$$\langle z' \mid e^{-itH} \mid z \rangle = \int \langle z' \mid e^{-i\epsilon H} \mid z(\mathcal{N}) \rangle \prod_\alpha dz_\alpha(\mathcal{N}) \cdot \langle z(\mathcal{N}) \mid e^{-i\epsilon H} \mid z(\mathcal{N}-1) \rangle \prod_\alpha dz_\alpha(\mathcal{N}-1) \cdots \prod_\alpha dz_\alpha(2) \langle z(2) \mid e^{-i\epsilon H} \mid z \rangle$$

we obtain, for any state vector $\mid\,\rangle$ at time t, the solution of the Schrödinger equation at time t' to be

$$\langle z' \mid t' \rangle = \langle z' \mid e^{-i(t'-t)H} \mid \,\rangle$$

$$= \int \langle z' \mid e^{-i(t'-t)H} \mid z \rangle \prod_\alpha dz_\alpha(z) \langle z \mid \,\rangle$$

$$= \lim_{\epsilon \to 0} \int \prod_{n=1}^{\mathcal{N}} \prod_\alpha d\kappa_\alpha(n)\, dz_\alpha(n) \cdot e^{i\epsilon L(n)} \langle z(1) \mid \,\rangle \qquad (19.145)$$

where $z_\alpha(n)$ and $\kappa_\alpha(n)$ are the values of z_α and κ_α at time $t_n = t + (n-1)\epsilon$,

$$L(n) \equiv -i \sum_\alpha \kappa_\alpha(n)\, \dot z_\alpha(n) + \sum_{\alpha,\beta} H_{\alpha\beta}\, \kappa_\alpha(n)\, z_\beta(n), \qquad (19.146)$$

$$\dot z_\alpha(n) \equiv \frac{z_\alpha(n+1) - z_\alpha(n)}{\epsilon}$$

and $z_\alpha(\mathcal{N}+1) = z'_\alpha$. It is convenient to define

$$\psi(n) \equiv \psi(\vec{r}, t_n) \equiv \sum_\alpha \phi_\alpha(\vec{r})\, z_\alpha(n) \ ,$$

$$\dot{\psi}(n) \equiv \dot{\psi}(\vec{r}, t_n) \equiv \frac{\psi(n+1) - \psi(n)}{\epsilon} = \sum_\alpha \phi_\alpha(\vec{r}) \frac{z_\alpha(n+1) - z_\alpha(n)}{\epsilon} \ ,$$

$$\psi^\dagger(n) \equiv \psi^\dagger(\vec{r}, t_n) \equiv \sum_\alpha \phi_\alpha^\dagger(\vec{r})\, \kappa_\alpha(n) \ , \tag{19.147}$$

$$\overline{\Psi}(n) \equiv \overline{\Psi}(\vec{r}, t_n) \equiv \psi^\dagger(n)\, \gamma_4 = \sum \phi_\alpha^\dagger(\vec{r})\, \gamma_4\, \kappa_\alpha(n) \ .$$

[Note that $\psi(n)$, $\dot{\psi}(n)$, $\psi^\dagger(n)$ and $\overline{\Psi}(n)$ are all polynomials of the Grassmann algebra generators $z_\alpha(n)$ and $\kappa_\alpha(n)$.] On account of (19.144) and (19.147), $L(n)$ given by (19.146) can be written as

$$\begin{aligned} L(n) &= -\int d^3r\, \overline{\Psi}(n)\, (\gamma_\mu D_\mu + m)\, \psi(n) \\ &= \int d^3r\, \psi^\dagger(n)\, [\, i\, \dot{\psi}(n) - \gamma_4(\gamma_i D_i + m + g\, \gamma_4 V_0)\, \psi(n)\,] \ . \end{aligned} \tag{19.148}$$

Equation (19.145) becomes

$$\langle \psi' |\, e^{-i(t'-t)H}\, | \rangle = \lim_{\epsilon \to 0} \int \prod_{n=1}^{\mathcal{N}} [\, d\overline{\Psi}(n)\, d\psi(n)\,]\ e^{i\epsilon L(n)} \langle \psi(1) | \rangle \tag{19.149}$$

where

$$[\, d\overline{\Psi}(n)\, d\psi(n)\,] \equiv \prod_\alpha d\kappa_\alpha(n)\, dz_\alpha(n) \ , \tag{19.150}$$

which represents a change of the basis from $z_\alpha(n)$ and $\kappa_\alpha(n)$ to $\psi(n)$ and $\psi^\dagger(n)$, with the final configuration $\psi' = \psi(\mathcal{N}+1)$. Because of the orthonormality relation (19.143) we may write

$$\begin{aligned} \prod_\alpha d\kappa_\alpha(n)\, dz_\alpha(n) &= \prod_{\vec{r}} d\psi^\dagger(\vec{r}, t_n)\, d\psi(\vec{r}, t_n) \\ &= \text{constant} \prod_{\vec{r}} d\overline{\Psi}(\vec{r}, t_n)\, d\psi(\vec{r}, t_n) \equiv [\, d\overline{\Psi}(n)\, d\psi(n)\,] \ . \end{aligned}$$

As in (19.15a) and (19.22a), (19.149) may be written in a compact form

$$e^{-i(t'-t)H} = \int [\, d\overline{\Psi}\, d\psi\,]\ e^{\, i \int_t^{t'} L(\tau)\, d\tau} \tag{19.149a}$$

where

$$[d\bar{\psi}\, d\psi] = \lim_{\epsilon\to 0} \prod_n [d\bar{\psi}(n)\, d\psi(n)] \tag{19.150a}$$

and

$$\int_t^{t'} L(\tau)\, d\tau = \epsilon \sum_n L(n) \quad .$$

Exercise. By setting $F(1) = \psi(1)$ and $G(2) = \bar{\psi}(2) = \psi^\dagger(2)\,\gamma_4$ in (19.70), but replacing $[dq]$ by $[d\bar{\psi}\, d\psi]$ and L by $-\int d^3r\, \bar{\psi}\,(\gamma_\mu \frac{\partial}{\partial x_\mu} + m)\,\psi$, show that

$$\underbrace{\psi(1)\ \bar{\psi}}(2) = S_F(x_1 - x_2)$$

given by (5.60).

19.6 QCD

We are now in a position to derive the Feynman rules in QCD. In the time-axial ($V_0 = 0$) gauge, the coordinates are the Cartesian $V_i = T^\ell\, V_i^\ell$ and ψ used in Section 18.3. By using (19.22) and (19.149), we can readily express the Green's function $e^{-i(t'-t)H}$ in terms of the path-integration formalism, where H is the operator given by (18.38)-(18.39). By dividing $t' - t$ into $\mathcal{N}$ intervals of ϵ each and setting $V_i^\ell(n) = V_i^\ell(\vec{r}, t_n)$, $\Pi_i^\ell(n) = \Pi_i^\ell(\vec{r}, t_n)$ with $t_n = t + (n-1)\,\epsilon$, we find for any state vector $|\,>$

$$< V', \psi' \,|\, e^{-i(t'-t)H} \,|\, > = \lim_{\epsilon\to 0} \int \prod_{n=1}^{\mathcal{N}} [d\Pi(n)\, dV(n)][d\bar{\psi}(n)\, d\psi(n)] \cdot e^{i\epsilon \bar{L}(n)} < V(1), \psi(1) \,|\, > \tag{19.151}$$

where

$$\bar{L}(n) = \int [\Pi_i^\ell(n)\, \dot{V}_i^\ell(n) + i\,\psi^\dagger(n)\,\dot{\psi}(n) - \mathcal{H}_{cl}(V_i(n), \Pi_i(n), \psi(n), \psi^\dagger(n))]\, d^3r \ , \tag{19.152}$$

$[d\Pi(n)\, dV(n)] = \prod_{\vec{r},i,\ell} d\Pi_i^\ell(\vec{r}, t_n)\, dV_i^\ell(\vec{r}, t_n)$, $[d\bar{\psi}(n)\, d\psi(n)]$ is given by (19.150), $\mathcal{H}_{cl}$ is the classical Hamiltonian density, (18.29),

$$\mathcal{H}_{cl}(V_i, \psi, \psi^\dagger) = \tfrac{1}{2}(\Pi_i^\ell \Pi_i^\ell + B_i^\ell B_i^\ell) + \psi^\dagger \gamma_4 (\gamma_i D_i + m)\psi \quad , \tag{19.153}$$

$\dot{V}_i^\ell(n) = [V_i^\ell(n+1) - V_i^\ell(n)]/\epsilon$ with the final configuration $V_i'^\ell = V_i^\ell(\mathcal{N}+1)$, and $\psi(n)$, $\psi^\dagger(n)$, $\dot{\psi}(n)$ are given by (19.147) with $\psi(\mathcal{N}+1) = \psi'$. By regarding the integration variables $V_i^\ell(n)$, $\Pi_i^\ell(n)$, $\psi(n)$ and $\psi^\dagger(n)$ as the values of V_i^ℓ, Π_i^ℓ, ψ and $\psi^\dagger$ at time $t_n = t + (n-1)\epsilon$, we find that the integrand in (19.151) gives the amplitude for each path in the "phase space". Just as in (19.22a) and (19.149a), the above formula can be written in a more compact form:

$$e^{-i(t'-t)H} = \int [d\Pi\, dV]\,[d\bar{\psi}\, d\psi]\; e^{i\int_t^{t'} \bar{L}(\tau)\, d\tau} \tag{19.151a}$$

where

$$\int_t^{t'} \bar{L}(\tau)\, d\tau = \epsilon \sum_n \bar{L}(n) \quad ;$$

its precise meaning is, of course, the original equation (19.151).

In the above path-integration formulation, the constraint (18.35) is automatically insured if we apply the operator $e^{-i(t'-t)H}$ only to initial states that satisfy the constraint, e.g., the vacuum state. In the derivation of Feynman diagrams discussed in Section 19.4, this would be a step that we should take in any case, as exemplified by (19.70). Consequently, it is not necessary for us to consider the constraint separately.

1. Covariant gauge

Let us denote the gauge and the quark fields in the covariant gauge by $\mathcal{A}_\mu$ and ψ_{cov}, and replace the classical Lagrangian density (18.3) by

$$\mathcal{L}_{cov}(\mathcal{A}_\mu, \dot{\mathcal{A}}_\mu, \psi_c, \dot{\psi}_c, \psi_c^\dagger) \tag{19.154}$$

$$= -\tfrac{1}{4}\mathcal{F}_{\mu\nu}^\ell \mathcal{F}_{\mu\nu}^\ell - \psi_{cov}^\dagger \gamma_4(\gamma_\mu \mathcal{D}_\mu + m)\psi_{cov} - \frac{1}{2a}\left(\frac{\partial \mathcal{A}_\mu^\ell}{\partial x_\mu}\right)^2$$

where a is a constant

$$\mathcal{F}^{\ell}_{\mu\nu} = \frac{\partial}{\partial x_\mu} \mathcal{A}^{\ell}_{\nu} - \frac{\partial}{\partial x_\nu} \mathcal{A}^{\ell}_{\mu} + g f^{\ell mn} \mathcal{A}^{m}_{\mu} \mathcal{A}^{n}_{\nu}$$

and

$$\mathcal{D}_\mu = \frac{\partial}{\partial x_\mu} - i g T^{\ell} \mathcal{A}^{\ell}_{\mu} ,$$

as in (18.4). The subscript cov denotes the covariant gauge and g is the unrenormalized coupling constant. Although the usual canonical quantization processes cannot be applied to the covariant-gauge Lagrangian density $\mathcal{L}_{cov}$ without encountering some formal difficulties,* it is possible to prove the following theorem which relates $\mathcal{L}_{cov}$, in a path-integration formalism, to the Hamiltonian H in the $V_0 = 0$ gauge.

Theorem 5. Apart from an overall constant normalization factor,

$$\langle V', \psi' | e^{-i(t'-t)H} | \rangle = \lim_{\epsilon \to 0} \left\{ \int \prod_{n=1}^{\mathcal{N}} [d \mathcal{A}_\mu (n)] \cdot [d\bar{\psi}_{cov}(n)\, d\psi_{cov}(n)]\, \mathcal{J}(n)\, e^{i\epsilon L_{cov}(n)} \right\} \cdot \langle \mathcal{A}_\mu(1), \psi_{cov}(1) | \rangle . \tag{19.155}$$

The lefthand side is identical to that of (19.151) with H given by (18.38)-(18.39). On the righthand side

$$L_{cov} = \int d^3 r\, \mathcal{L}_{cov}(\mathcal{A}_\mu(n), \dot{\mathcal{A}}_\mu(n), \psi_{cov}(n), \dot{\psi}_{cov}(n), \psi^{\dagger}_{cov}(n)), \tag{19.156}$$

in which the Lagrangian density $\mathcal{L}_{cov}$ is given by (19.154), $\mathcal{A}_\mu(n)$ is the value of $\mathcal{A}_\mu(\vec{r}, t_n)$ at time $t_n = t + (n-1)\epsilon$,

---

* If we apply the canonical rule to $\mathcal{L}_{cov}$, complications arise because $\mathcal{L}_{cov} = -\frac{1}{2a}[(\dot{\mathcal{A}}^{\ell}_0)^2 + (\nabla_i \mathcal{A}^{\ell}_i)^2] + \cdots$; therefore the conjugate momentum of A^{ℓ}_0 is $P^{\ell}_0 = \partial L/\partial \dot{\mathcal{A}}^{\ell}_0$. Consequently the Hamiltonian density $-\frac{1}{2}a(P^{\ell}_0)^2 + \frac{1}{2a}(\nabla_i \mathcal{A}^{\ell}_i)^2 + \cdots$ has no lower bound, since $-\frac{1}{2}a(P^{\ell}_0)^2 < 0$ if $a > 0$ (continued on next page)

$$\dot{\mathcal{A}}_\mu(n) = \frac{\mathcal{A}_\mu(n+1) - \mathcal{A}_\mu(n)}{\epsilon}, \quad [d\mathcal{A}_\mu(n)] = \prod_{\vec{r},\mu,\ell} d\mathcal{A}_\mu^\ell(\vec{r}, t_n),$$

$$\mathcal{J}(n) \equiv \det \left| \frac{\partial}{\partial x_\mu} \mathcal{O}_\mu \right|_{\text{at } t_n}, \tag{19.157}$$

and $\psi_{cov}(n)$, $\dot{\psi}_{cov}(n)$, $\psi_{cov}^\dagger(n)$ and $[d\bar{\psi}_{cov}(n)\, d\psi_{cov}(n)]$ are given by the same expressions (but without subscripts) as those in (19.147) and (19.150). When $n = \mathcal{N}+1$, the values of $\mathcal{A}_\mu^\ell(n)$ and $\psi_{cov}(n)$ on the righthand side of (19.155) are related to V'^ℓ_μ and ψ' on the lefthand side by

$$\mathcal{A}_i^\ell(\mathcal{N}+1) = V_i'^\ell, \quad \mathcal{A}_0^\ell(\mathcal{N}+1) = 0$$

and

$$\psi_{cov}(\mathcal{N}+1) = \psi'. \tag{19.158}$$

<u>Proof.</u> We start from the $V_0 = 0$ gauge, and introduce into the integrand of (19.151) the factor

$$1 = \int \prod_{n=1}^{\mathcal{N}} d[\theta_a(n)] \cdot |\mathcal{J}| \cdot \left\{ \exp \int \prod_{n=1}^{\mathcal{N}} d^3r \frac{-i\epsilon}{2a} [\nabla_i \mathcal{A}_i^\ell(\underset{\sim}{r}, t_n) + \frac{1}{\epsilon}(\mathcal{A}_0^\ell(\underset{\sim}{r}, t_{n+1}) - \mathcal{A}_0^\ell(\underset{\sim}{r}, t_n))]^2 \right\} \tag{19.159}$$

where $\mathcal{A}_i^\ell(n) \equiv \mathcal{A}_i^\ell(\underset{\sim}{r}, t_n)$ and $\mathcal{A}_0^\ell(n) \equiv \mathcal{A}_0^\ell(\underset{\sim}{r}, t_n)$ are given by the Hermitian matrices

$$\mathcal{A}_i(n) \equiv T^\ell \mathcal{A}_i^\ell(n) = u(n)^{-1} V_i(n)\, u(n) + \frac{i}{g} u(n)^{-1} \nabla_i u(n) \tag{19.160}$$

and

$$\mathcal{A}_0(n) \equiv T^\ell \mathcal{A}_0^\ell(n) = -\frac{i}{g\epsilon} \ln[u(n)^{-1} u(n+1)]$$

---

and $\frac{1}{2a}(\nabla_i \mathcal{A}_i^\ell)^2 < 0$ if $a < 0$. By introducing negative metric, it is possible to convert the covariant-gauge Hamiltonian into a positive-definite operator, but then it has difficulty with unitarity.

$$= -\frac{i}{g\epsilon}\, u(n)^{-1}(u(n+1) - u(n)) + \frac{i}{2g\epsilon}[u(n)^{-1}(u(n+1) - u(n))]^2 + \cdots \quad . \tag{19.161}$$

In these expressions, T^ℓ and V_i are given by (18.1) and (18.31),

$$u(n) = u(\theta_a(n)) \tag{19.162}$$

is the $N \times N$ unitary matrix function which depends on the M group parameters $\theta_a(n) = \theta_a(\underset{\sim}{r}, t_n)$ as in (18.52), and the differential $d[\theta_a(n)] = \prod_{a,\underset{\sim}{r}} d\theta_a(\underset{\sim}{r}, t_n)$. Except for a numerical factor, the Jacobian $|\bar{g}|$ is given by

$$|\bar{g}| = \det \langle \underset{\sim}{r}, t_n | \bar{g} | \underset{\sim}{r}', t_{n'} \rangle$$

$$= \det \frac{\delta}{\delta\theta_a(n)} \left[\nabla_i \mathcal{A}_i(n') + \frac{(\mathcal{A}_0(n'+1) - \mathcal{A}_0(n'))}{\epsilon} \right] . \tag{19.163}$$

In order to match the final configuration of (19.151), we set $\mathcal{A}_i^\ell(\mathcal{N}+1) = V_i^\ell(\mathcal{N}+1) = V_i'^\ell$; i.e.,

$$u(\mathcal{N}+1) = 1 \quad . \tag{19.164}$$

Furthermore, since the integrand of (19.159) is invariant under $\mathcal{A}_0^\ell(n) \to \mathcal{A}_0^\ell(n) + \text{constant}$ for all $n = 1, 2, \cdots, \mathcal{N}+1$, we may choose $\mathcal{A}_0^\ell(\mathcal{N}+1) = 0$.

We now examine the region where the summand in the exponent of (19.159) is of order 1. Because in the summand the coefficient of $[\mathcal{A}_0(n+1) - \mathcal{A}_0(n)]^2$ is proportional to ϵ^{-1}, we expect only those configurations with $\mathcal{A}_0(n+1) - \mathcal{A}_0(n)$ of order $\epsilon^{\frac{1}{2}}$ to contribute to the integral over $\theta_a(n)$. Thus since $\mathcal{A}_0(\mathcal{N}+1) = 0$, $\mathcal{A}_0(n)$ should be of order $(\mathcal{N} - n)^{\frac{1}{2}} \epsilon^{\frac{1}{2}} = O(1)$ and, from (19.161), $u(n+1) - u(n)$ is only of order ϵ.

If we perform $\int d[\Pi(n)]$ in (19.151) we find, up to a numerical factor:

$$\int d[\Pi(n)] \exp i\epsilon \sum_n \int d^3r \cdot \left\{ \Pi_i^\ell(n) [V_i^\ell(n+1) - V_i^\ell(n)]\, \epsilon^{-1} - \tfrac{1}{2}\Pi_i^\ell(n)\Pi_i^\ell(n) \right\}$$

$$= \exp i\tfrac{1}{2} \sum_n \int d^3r [V_i^\ell(n+1) - V_i^\ell(n)]^2\, \epsilon^{-1}$$

$$= \exp i \sum_n \int d^3r\, \mathrm{tr} [V_i(n+1) - V_i(n)]^2\, \epsilon^{-1} \quad . \tag{19.165}$$

Next we can use (19.160) to express this exponent in terms of $\mathcal{A}_i^\ell$ and $u^\bullet$. A particularly symmetrical form is obtained if the quantity in square brackets on the last line in (19.165) is conjugated with the matrix $[u(n)^{-1} u(n+1)]^{\frac{1}{2}} u(n+1)^{-1}$:

$$\epsilon^{-1} \mathrm{tr} [V_i(n+1) - V_i(n)]^2 = \epsilon^{-1} \mathrm{tr} \Big\{ [u(n)^{-1} u(n+1)]^{\frac{1}{2}} \mathcal{A}_i(n+1) \cdot [u(n)^{-1} u(n+1)]^{-\frac{1}{2}} + \frac{i}{g} [u(n)^{-1} u(n+1)]^{\frac{1}{2}} \nabla_i [u(n)^{-1} u(n+1)]^{-\frac{1}{2}}$$
$$- [u(n)^{-1} u(n+1)]^{-\frac{1}{2}} \mathcal{A}_i(n) [u(n)^{-1} u(n+1)]^{\frac{1}{2}} - \frac{i}{g} [u(n)^{-1} u(n+1)]^{-\frac{1}{2}} \cdot \nabla_i [u(n)^{-1} u(n+1)]^{\frac{1}{2}} \Big\}^2 \quad . \tag{19.166}$$

Using (19.161) to replace $u(n)^{-1} u(n+1)$ by a function of $\mathcal{A}_0(n)$,

$$u(n)^{-1} u(n+1) = 1 + ig\epsilon\, \mathcal{A}_0(n) + O(\epsilon^2) \quad ,$$

we can expand the righthand side of (19.166) through order ϵ :

$$\epsilon^{-1} \mathrm{tr} [V_i(n+1) - V_i(n)]^2 = \epsilon\, \mathrm{tr} \Big\{ [\mathcal{A}_i(n+1) - \mathcal{A}_i(n)]\, \epsilon^{-1} + ig[\mathcal{A}_0(n), \mathcal{A}_i(n)] + \nabla_i \mathcal{A}_0(n) \Big\}^2 + O(\epsilon^{\frac{3}{2}})$$

where we treat $\mathcal{A}_i(n+1) - \mathcal{A}_i(n)$ as of order $\epsilon^{\frac{1}{2}}$. Thus if we change integration variables from $V_i(n)$, $\psi(n)$ and $\overline{\psi}(n)$ to the unitarily equivalent set $\mathcal{A}_i(n)$,

$$\psi_{cov}(n) = u(n)^{-1} \psi(n)$$

and

$$\overline{\psi}_{cov}(n) = \overline{\psi}(n)\, u(n) \quad ,$$

the exponentials in (19.151) and (19.159) combine to give the covariant-gauge action up to terms vanishing with ϵ :

$$i\epsilon \sum_n \int d^3r \, \mathrm{tr} \Big\{ \big[(\mathcal{A}_i(n+1) - \mathcal{A}_i(n))\, \epsilon^{-1} + ig[\mathcal{A}_0(n), \mathcal{A}_i(n)] + \nabla_i \mathcal{A}_0(n) \big]^2 - [B_i(n)]^2 - \frac{1}{a} [\nabla_i \mathcal{A}_i(n) + (\mathcal{A}_0(n+1) - \mathcal{A}_0(n))\, \epsilon^{-1}]^2 \Big\}$$

$$i\epsilon \sum_n \int d^3r \, \Big\{ \bar{\psi}_{cov}(n) \Big[\frac{\gamma_4(\psi_{cov}(n+1) - \psi_{cov}(n))}{i\epsilon} + g\, \mathcal{A}_0(n)\, \gamma_4\, \psi_{cov}(n+1) + \gamma_i(\nabla_i - ig\mathcal{A}_i(n))\, \psi_{cov}(n) \Big] \Big\} .$$

Finally we must use (19.161) to change the integration variable $\theta_a(n)$ to $\mathcal{A}_0^m(n)$. The resulting Jacobian $\det[\delta\theta_a(n)/\delta\mathcal{A}_0^m(n')] \cdot |\bar{\mathcal{J}}|$, after some manipulation, can be shown to be the covariant-gauge Faddeev-Popov determinant $\mathcal{J}(n)$, (19.157), apart from a constant multiplicative factor. [For details, see Phys.Rev. D22, 939 (1980).] That completes the proof of Theorem 5.

Just as in (19.56)-(19.56a), we may exponentiate the product $\prod_n \mathcal{J}(n)$:

$$\prod_n \mathcal{J}(n) = \exp[\mathrm{trace}\ \ln \mathcal{M}^{\frac{1}{2}}] \tag{19.167}$$

where the matrix $\mathcal{M}^{\frac{1}{2}}$ is given by

$$\begin{aligned} \langle \ell, x | \mathcal{M}^{\frac{1}{2}} | \ell', x' \rangle &= \langle x | \frac{\partial}{\partial x_\mu} \mathcal{D}_\mu^{\ell\ell'} | x' \rangle \\ &= \langle x | \frac{\partial^2}{\partial x_\mu^2} \delta^{\ell\ell'} - g f^{\ell\ell' m} \frac{\partial}{\partial x_\mu} \mathcal{A}_\mu^m | x' \rangle \end{aligned} \tag{19.168}$$

and $x_\mu = (\vec{r}, i\tau)$, $x'_\mu = (\vec{r}', i\tau')$. Let us introduce a matrix G whose elements are given by

$$\begin{aligned} \langle \ell, x | G | \ell', x' \rangle &\equiv g \langle x | \left(\frac{\partial^2}{\partial x_\lambda^2} \right)^{-1} \frac{\partial}{\partial x_\mu} f^{\ell\ell' m} \mathcal{A}_\mu^m | x' \rangle \\ &= \int d^4x'' \langle x | \left(\frac{\partial^2}{\partial x_\lambda^2} \right)^{-1} | x'' \rangle \langle x'' | \frac{\partial}{\partial x_\mu} f^{\ell\ell' m} \mathcal{A}_\mu^m | x' \rangle . \end{aligned} \tag{19.169}$$

Since

$$< \ell, x \mid \mathcal{M}^{\frac{1}{2}} \mid \ell', x' > \;=\; \int d^4x'' < \ell, x \mid \frac{\partial^2}{\partial x_\lambda^{\,2}} \mid \ell'', x'' >$$

$$< \ell'', x'' \mid 1 - G \mid \ell', x' > \quad ,$$

(19.167) becomes

$$\prod_n \mathcal{J}(n) \;=\; \det \mathcal{M}^{\frac{1}{2}} \;=\; \text{constant} \cdot \det (1 - G)$$

$$= \text{constant} \cdot \exp [\, \text{trace} \ln (1 - G) \,] \tag{19.170}$$

where the constant can be absorbed into the overall normalization factor in (19.155).

As in (19.151a) we may rewrite (19.155) in a more compact form. To exhibit Lorentz invariance, it is convenient to take the limits $t' \to \infty$ and $t \to -\infty$. By using (19.170) we see that, apart from an overall constant normalization factor

$$\lim_{\substack{t' \to \infty \\ t \to -\infty}} e^{-i(t'-t)H} \;=\; \int [d\mathcal{A}\,][d\bar{\psi}_{cov}\, d\psi_{cov}\,] \cdot e^{\,i \int \mathcal{L}_{cov}\, d^4x \,+\, \text{trace} \ln (1 - G)} \tag{19.171}$$

where $\mathcal{L}_{cor}$ and G are given by (19.154) and (19.169),

$$[d\mathcal{A}\,] \;=\; \prod_{x, \mu, \ell} d\mathcal{A}_\mu^{\,\ell}(x)$$

and

$$[d\bar{\psi}_{cov}\, d\psi_{cov}\,] \;=\; \prod_x d\bar{\psi}_{cov}(x)\; d\psi_{cov}(x) \quad . \tag{19.172}$$

2. Feynman rules in covariant gauge

To derive the Feynman rules, we follow the discussions given in Section 19.3. We can either use (19.99) by considering the N-point function

$$D_N(1, 2, \cdots N) \;\equiv\; \frac{< vac \mid [d\mathcal{A}\,][d\bar{\psi}_{cov}\, d\psi_{cov}\,] \prod_{j=1}^{N} F(j)\, e^{iZ} \mid vac >}{< vac \mid [d\mathcal{A}\,][d\bar{\psi}_{cov}\, d\psi_{cov}\,]\; e^{iZ} \mid vac >} \tag{19.173}$$

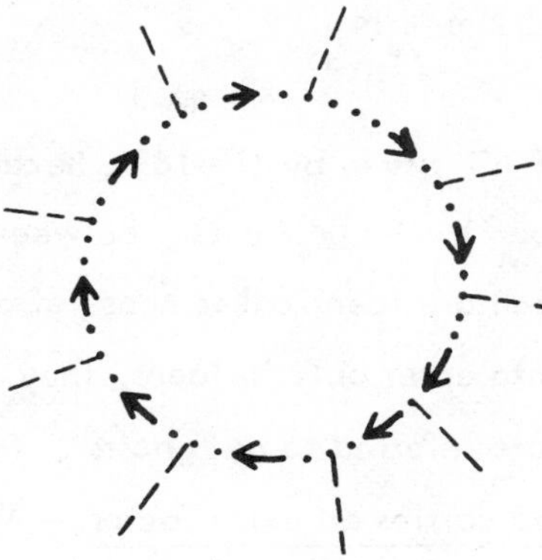

Fig. 19.2. The ghost-loop diagram due to $-\frac{1}{n}$ trace G^n in (19.174).

where

$$iZ = \int i\mathcal{L}_{cov}\, d^4x + \text{trace}\,\ln\,(1-G)$$
$$= \int i\mathcal{L}_{cov}\, d^4x - \sum_1^{\infty} \frac{1}{n}\,\text{trace}\, G^n \qquad (19.174)$$

and $F = A_\mu^\ell$, or ψ, or $\overline{\psi}$, or introduce the device (19.102)

$$W(j) = e^{iS(j)}$$
$$\equiv \frac{<\text{vac}\,|\,[dA]\,[d\overline{\psi}_{cov}\, d\psi_{cov}]\; e^{iZ + i\int j_\mu A_\mu d^4x}\,|\,\text{vac}>}{<\text{vac}\,|\,[dA]\,[d\overline{\psi}_{cov}\, d\psi_{cov}]\; e^{iZ}\,|\,\text{vac}>}\,. \qquad (19.175)$$

Both methods, of course, lead to the same Feynman rules.

In (19.174), every trace G^n term gives rise to a loop diagram* shown in Figure 19.2; there are n vertices in the loop, each stemming from the factor

---

* R. P. Feynman, Acta Physica Polonica 24, 697 (1963); L. D. Faddeev and V. N. Popov, Phys. Lett. 25B, 29 (1967).

$$\langle x \,|\, \frac{\partial}{\partial x_\mu} f^{\ell \ell' m} \mathcal{A}_\mu^m \,|\, x' \rangle \tag{19.176}$$

in the expression of G given by (19.169). Because of the additional factor $\langle x |\, (\partial^2/\partial x_\lambda{}^2)^{-1} \,| x' \rangle$ in G, between the vertices there are propagators which are identical to those of a spin-0 field. Since these "spin-0" quanta enter only in loops, they are not observables; consequently they are referred to as "ghosts". From (19.174), we see that each ghost loop carries an extra factor -1, making the ghosts behave like fermions. [The factor $\frac{1}{n}$ in (19.174) is the normal symmetry-number factor of a loop due to the cyclic permutation of its vertices.] Expression (19.176) is not symmetric with regard to x and x' because of the derivative. Correspondingly, the ghost-gluon vertex is asymmetric with respect to the two gluon lines, as shown below. This asymmetry can be taken care of by giving each ghost line an arrow before assigning its momentum, and requiring the arrow direction in each loop to be continuous, as in the case of a complex field. Every such vertex can be viewed as annihilating a ghost of momentum q and creating one of momentum p ; the latter enters explicitly in the Feynman rule given below. Hence, the ghost may be represented by a fictitious complex isovector field. In the following we represent the fermion (quark) propagator by an arrowed straight line, the gauge-field (gluon) propagator by a wavy line with no arrow, and the ghost propagator by an arrowed dashed line; the subscripts and superscripts are space-time and color indices, p , q , r denote 4-momenta with $\not p$ and p^2 standing for $\not p + i(0+)$ and $p^2 - i(0+)$.

quark propagator

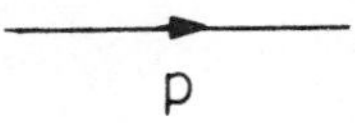

gives $\frac{i}{\not p - m}$

gluon propagator

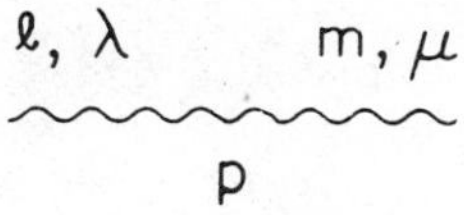

gives $\dfrac{-i\,\delta^{\ell m}}{p^2}\left[\delta_{\lambda\mu} + (\alpha - 1)\,\dfrac{p_\lambda p_\mu}{p^2}\right]$

ghost propagator

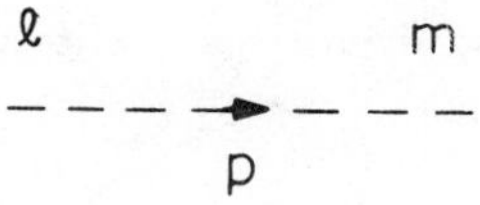

gives $\dfrac{-i\,\delta^{\ell m}}{p^2}$

quark-gluon vertex

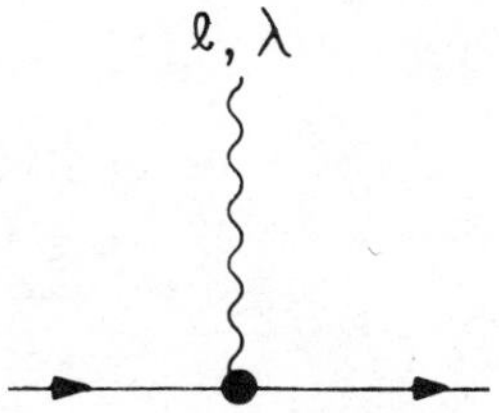

gives $-g\,\gamma_\lambda T^\ell = -\tfrac{1}{2} g\,\gamma_\lambda \begin{cases} \tau^\ell & SU_2 \\ \lambda^\ell & SU_3 \end{cases}$

gluon-gluon vertices $(p + q + r = 0$, with the direction of momentum flow indicated by the arrow)

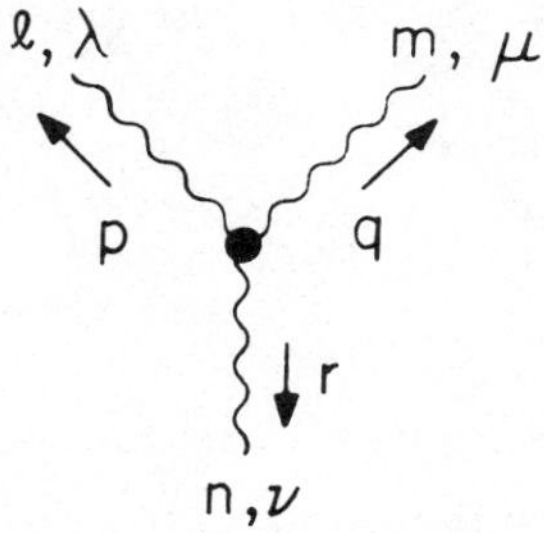

gives $g f^{\ell mn}[(p-q)_\nu \delta_{\lambda\mu} + (q-r)_\lambda \delta_{\mu\nu} + (r-p)_\mu \delta_{\nu\lambda}]$

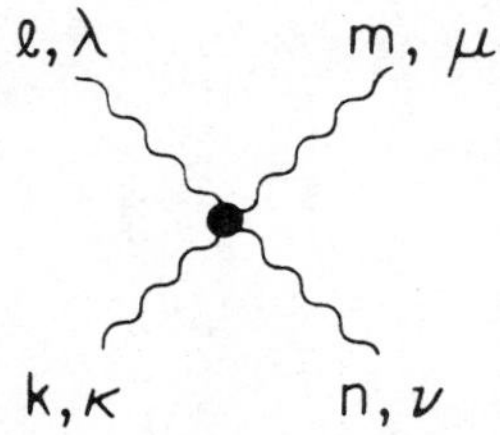

gives
$$-ig^2[f^{\ell mj} f^{nkj}(\delta_{\lambda\nu}\delta_{\mu\kappa} - \delta_{\lambda\kappa}\delta_{\mu\nu}) + f^{\ell nj} f^{kmj}(\delta_{\lambda\kappa}\delta_{\nu\mu} - \delta_{\lambda\mu}\delta_{\nu\kappa}) + f^{\ell kj} f^{mnj}(\delta_{\lambda\mu}\delta_{\kappa\nu} - \delta_{\lambda\nu}\delta_{\kappa\mu})]$$

gluon-ghost vertex

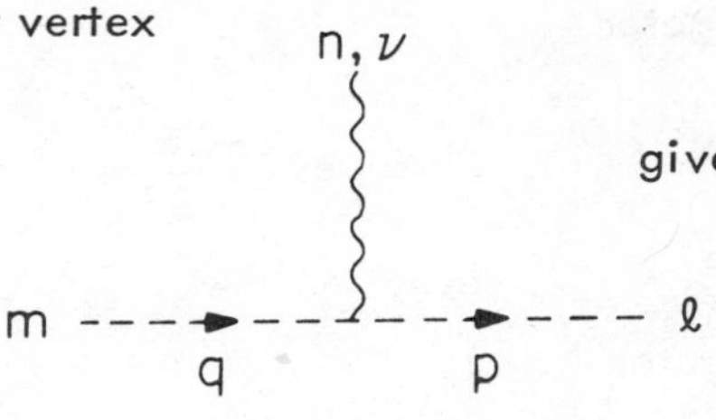

gives $g f^{\ell m n} p_\nu$.

From these Feynman rules, the Lorentz-invariance character of the theory is obvious.

3. Coulomb gauge

For the Coulomb gauge we apply Theorem 3, (19.49) - (19.51), onto the Hamiltonian (18.97). We shall omit the details, and give only the final result. As in Section 18.4, let A_μ and ψ_c denote the gauge and quark fields in the Coulomb gauge. The classical Lagrangian density $\mathcal{L}_c$ is a function of A_i^ℓ, $\dot{A}_i^\ell$, ψ_c, $\psi_c^\dagger$ and $\dot{\psi}_c$:

$$\mathcal{L}_c = -\tfrac{1}{4} F_{\mu\nu}^\ell F_{\mu\nu}^\ell - \psi_c^\dagger \gamma_4 \gamma_\mu D_\mu \psi_c \tag{19.177}$$

where

$$F_{\mu\nu}^\ell = \frac{\partial}{\partial x_\mu} A_\nu^\ell - \frac{\partial}{\partial x_\nu} A_\mu^\ell + g f^{\ell m n} A_\mu^m A_\nu^n ,$$

$$D_\mu = \frac{\partial}{\partial x_\mu} - i g T^\ell A_\mu^\ell , \tag{19.178}$$

$$\nabla_i A_i^\ell = 0$$

and A_0^ℓ is given by (18.78) and (18.92).

The Hamiltonian H in the Coulomb gauge is given by (18.97), generalized to SU_N (and with g now standing for g_0),

$$\begin{aligned} H = \int \Big\{ & \tfrac{1}{2} \mathcal{J}^{-1} (\Pi_i^{tr})^\ell \mathcal{J} (\Pi_i^{tr})^\ell + \tfrac{1}{2} \mathcal{B}_i^\ell \mathcal{B}_i^\ell \\ & + \psi_c^\dagger \gamma_4 [\gamma_i D_i + m] \psi_c \Big\} d^3 r + \tfrac{1}{2} g^2 \int \mathcal{J}^{-1} \sigma^\ell(\vec{r}) \\ & \cdot \langle \ell, \vec{r} \,|\, (\nabla_i \mathcal{D}_i)^{-1} (-\nabla^2) (\nabla_j \mathcal{D}_j)^{-1} \,|\, \ell', \vec{r}\,' \rangle \\ & \cdot \mathcal{J} \sigma^{\ell'}(\vec{r}\,') \, d^3 r \, d^3 r' \end{aligned} \tag{19.179}$$

where $(\Pi_i^{tr})^\ell$ is the conjugate momentum of A_i^ℓ,

$$\sigma^\ell = f^{\ell mn} A_i^m (\Pi_i^{tr})^n + \psi_c^\dagger T^\ell \psi_c ,$$

$$\mathcal{B}_i^\ell = \nabla_j A_k^\ell - \nabla_k A_j^\ell + g f^{\ell mn} A_j^m A_k^n$$

with i, j, k cyclic, $\nabla_i \mathcal{D}_i$ is the matrix

$$\langle \ell, \vec{r} \,|\, \nabla_i \mathcal{D}_i \,|\, \ell', \vec{r}' \rangle = (\delta^{\ell\ell'} \nabla^2 - g f^{\ell\ell' m} A_i^m(\vec{r}) \nabla_i)\, \delta^3(\vec{r} - \vec{r}')$$

and (19.180)

$$\mathcal{J} = \det \nabla_i \mathcal{D}_i(x) .$$

As in (19.45), we define

$$\overline{H} \equiv \sqrt{\mathcal{J}}\; H \frac{1}{\sqrt{\mathcal{J}}} . \qquad (19.181)$$

The time-dependent Schrödinger equation is

$$\overline{H} \,|\, t \rangle = i \frac{\partial}{\partial t} \,|\, t \rangle . \qquad (19.182)$$

In the A_i, ψ_c and $\overline{\psi}_c$ representations, if $|\, t \rangle = |\, \rangle$ at time t, then at a later time t' on account of Theorem 3 the solution of (19.182) is given by

$$\langle A_i^\ell, \psi_c, \overline{\psi}_c \,|\, e^{-i(t'-t)\overline{H}} \,|\, \rangle = \lim_{\epsilon \to 0} \int \prod_{n=1}^{\mathcal{N}} [dA_\mu(n)][d\overline{\psi}_c(n)\, d\psi_c(n)]$$
$$\cdot\, e^{i\epsilon L_{eff}(n)} \langle A_i^\ell(1), \psi_c(1), \overline{\psi}_c(1) \,|\, \rangle$$
(19.183)

where, apart from an overall normalization factor independent of n,

$$[d\overline{\psi}_c(n)\, d\psi_c(n)] = \prod_{\vec{r}} d\overline{\psi}_c(\vec{r}, t_n)\, d\psi_c(\vec{r}, t_n)$$

and

$$[dA_\mu(n)] = \prod_{\vec{r}, \ell, \mu} dA_\mu^\ell(\vec{r}, t_n)$$

in which μ can be 0, 1, 2 and 3; as before, $t_n = t + (n-1)\epsilon$ with

$\mathcal{N}\epsilon = t' - t$, and $A_i^\ell(\mathcal{N}+1)$ and $\psi_c(\mathcal{N}+1)$ denote the final configuration A_i^ℓ and ψ_c given by the lefthand side of (19.183). [Notice that $A_0(\mathcal{N}+1)$ does not appear on either side.] As in (19.50)

$$L_{eff}(n) = \int d^3r\, \mathcal{L}_c(\bar{A}_i^\ell(n),\ \dot{A}_i^\ell(n),\ \psi_c(n),\ \psi_c^\dagger(n),\ \dot{\psi}_c(n)) - \frac{i}{\epsilon}\ln \mathcal{J}(n) - V_c(n) \tag{19.184}$$

where $\mathcal{J}$ is the Jacobian (19.180) and $\mathcal{L}_c$ is the Coulomb-gauge Lagrangian density, (19.177), in which the arguments are

$$\bar{A}_i^\ell(n) = \frac{A_i^\ell(\vec{r}, t_{n+1}) + A_i^\ell(\vec{r}, t_n)}{2}, \quad \dot{A}_i^\ell(n) = \frac{A_i^\ell(\vec{r}, t_{n+1}) - A_i^\ell(\vec{r}, t_n)}{\epsilon},$$

$$\psi_c(n) = \psi_c(\vec{r}, t_n), \quad \psi_c^\dagger(n) = \psi_c^\dagger(\vec{r}, t_n)$$

and

$$\dot{\psi}_c(n) = \frac{\psi_c(n+1) - \psi_c(n)}{\epsilon}.$$

The additional potential V_c is given by*

$$\begin{aligned} V_c(A_i^\ell) = {} & \frac{1}{8} g^2 \int d^3r \langle \ell', \vec{r} | (\nabla_i \mathcal{D}_i)^{-1} \nabla_j | \ell, \vec{r} \rangle \\ & \cdot \langle m, \vec{r} | (\nabla_k \mathcal{D}_k)^{-1} \nabla_j t^{\ell'} t^{\ell} | m, \vec{r} \rangle \\ & + \frac{1}{8} g^2 \int d^3r\, d^3r' \langle \ell', \vec{r}' | \delta_{i'i} - \mathcal{D}_{i'} (\nabla_k \mathcal{D}_k)^{-1} \nabla_i | n, \vec{r} \rangle \\ & \cdot \langle \ell, \vec{r} | \delta_{ii'} - \mathcal{D}_i (\nabla_{k'} \mathcal{D}_{k'})^{-1} \nabla_{i'} | n', \vec{r}' \rangle \\ & \cdot \langle n, \vec{r} | t^\ell (\nabla_j \mathcal{D}_j)^{-1} (-\nabla^2) (\nabla_{j'} \mathcal{D}_{j'})^{-1} t^{\ell'} | n', \vec{r}' \rangle \end{aligned} \tag{19.185}$$

---

* N. H. Christ and T. D. Lee, Phys. Rev. D22, 939 (1980). The first term in (19.185) was originally obtained by J. Schwinger, Phys. Rev. 127, 324 (1962), 130, 406 (1963).

where $(t^{\ell})_{mn} = -if^{\ell mn}$ is the adjoint representation analog of the generator T^{ℓ}.

The Feynman rules can be derived by following the discussions given in Section 19.3 and considering expressions similar to (19.173) and (19.175), but in the Coulomb gauge. Because of the $\ln \mathcal{J}$ term in (19.184), on the one-loop level there are the Faddeev-Popov ghost diagrams. Due to the V_c term, there are additional two- and higher-loop diagrams.

Remarks. For interactions without derivative couplings, the Dyson-Wick derivation of Feynman rules given in Chapter 5 is perhaps the simplest. For theories with derivative couplings, such as QCD, the path-integration method is the more convenient. This is because, on account of (19.82), contractions of time derivatives are the derivatives of contractions.

For the gauge theory, the common practice in the literature has been to start from a formal path-integration representation before choosing the gauge. This approach usually encounters ill-defined expressions involving ∞/∞, resulting sometimes in mistakes such as missing the additional interaction term V_c in (19.185). Our method is to start from the time-axial gauge which has an unambiguous Hamiltonian. The Feynman rules in different gauges can then be derived by definite coordinate transformations. In order to show that each step is indeed well-defined, we have purposely avoided using the compact form of path integration; therefore, e.g., in the derivation of Theorem 5, some of the formulas have a clumsy appearance.

Problem 19.1.

(i) Write down the QED Hamiltonian in the time-axial gauge, then go over to the Coulomb gauge by following the discussions given in Section 18.4.

(ii) Derive the QED Feynman rules in the covariant and Coulomb gauges by using the method given in this chapter. Compare the result with that of Chapter 6.

Problem 19.2. The Hamiltonian of a nonrelativistic charged particle of mass μ in a magnetic field

$$\vec{B} = \vec{\nabla} \times \vec{a}$$

is

$$H = \frac{1}{2\mu}(\vec{p} - e\vec{a})^2 = -\frac{1}{2\mu}(\vec{\nabla} - ie\vec{a})^2 \quad . \tag{19.186}$$

(i) For a constant $\vec{B} /\!/ z$ axis, we may choose

$$a_x = -\tfrac{1}{2}By \quad , \qquad a_y = \tfrac{1}{2}Bx$$

and

$$a_z = 0 \quad \text{with} \quad B = |\vec{B}| \quad . \tag{19.187}$$

Prove that

$$<\vec{r}\,|\,e^{-iHt}\,|\,\vec{r}_0> = \left(\frac{\mu}{2i\pi t}\right)^{\frac{3}{2}} \frac{\omega t}{\sin\omega t}\; e^{iI} \tag{19.188}$$

where

$$I = \frac{\mu(z - z_0)^2}{2t} - \frac{eB}{2}(xy_0 - yx_0) + \frac{eB}{4}\left[(x - x_0)^2 + (y - y_0)^2\right]\cot\omega t$$

and

$$\omega = \frac{eB}{2\mu} \quad . \tag{19.189}$$

(ii) Show that

$$\text{trace}\; e^{-iHt} \equiv \int d^3r\, <\vec{r}\,|\,e^{-iHt}\,|\,\vec{r}> = \left(\frac{\mu}{2i\pi t}\right)^{\frac{3}{2}} \frac{\omega t}{\sin\omega t}\;\Omega \tag{19.190}$$

where $\Omega \to \infty$ is the volume of the system.

(iii) If instead of (19.187), we change the gauge and choose

$$a_x = a_z = 0 \qquad \text{and} \qquad a_y = Bx \ ,$$

show that

$$<\vec{r} \mid e^{-iHt} \mid \vec{r}_0> = \left(\frac{\mu}{2i\pi t}\right)^{\frac{3}{2}} \frac{\omega t}{\sin \omega t} \; e^{iI'}$$

where (19.191)

$$I' = I + \tfrac{1}{2} eB(xy - x_0 y_0) \ ,$$

and I and ω remain given by (19.189). Thus, trace e^{-iHt} is invariant under the gauge transformation.

Problem 19.3. The Lagrangian density of a charged scalar field ϕ of mass m and charge e in an external (c. number) electromagnetic 4-potential a_λ is

$$\mathcal{L} = -(\bar{D}_\lambda \phi^\dagger)(D_\lambda \phi) - m^2 \phi^\dagger \phi \qquad (19.192)$$

where

$$D_\lambda = \frac{\partial}{\partial x_\lambda} - iea_\lambda \ , \qquad \bar{D}_\lambda = \frac{\partial}{\partial x_\lambda} + iea_\lambda \qquad (19.193)$$

and the external electromagnetic field $f_{\lambda\nu}$ is related to a_λ by

$$f_{\lambda\nu} = \frac{\partial}{\partial x_\lambda} a_\nu - \frac{\partial}{\partial x_\nu} a_\lambda \ .$$

According to (19.85) and (19.96), we define the action $S(a_\lambda)$ and the full Feynman propagator $D_F'(x, x')$ to be

$$e^{iS(a_\lambda)} \equiv \frac{< vac \mid \int [d\phi^\dagger d\phi] \, e^{i\int_{-T/2}^{T/2} L d\tau} \mid vac >}{< vac \mid \int [d\phi^\dagger d\phi] \, e^{i\int_{-T/2}^{T/2} L_0 d\tau} \mid vac >}$$

and (19.194)

$$D_F'(x,x') \equiv \frac{<\mathrm{vac}\,|\int [d\phi^\dagger d\phi]\ \phi(x)\,\phi^\dagger(x')\ e^{\,i\int_{-T/2}^{T/2} L\,d\tau}\ |\,\mathrm{vac}>}{<\mathrm{vac}\,|\int [d\phi^\dagger d\phi]\ e^{\,i\int_{-T/2}^{T/2} L\,d\tau}\ |\,\mathrm{vac}>}$$

where $T \to \infty$,

$$L = \int \mathcal{L}\, d^3 r \qquad \text{and} \qquad L_0 = \int \mathcal{L}_0\, d^3 r$$

with $\mathcal{L}_0$ given by the free field Lagrangian density,

$$\mathcal{L}_0 = -\frac{\partial \phi^\dagger}{\partial x_\lambda}\frac{\partial \phi}{\partial x_\lambda} - m^2 \phi^\dagger \phi \ ,$$

and x and x' denote the space-time coordinates

$$x_\lambda = (\vec{r}, i t) \qquad \text{and} \qquad x'_\lambda = (\vec{r}\,', i t') \ .$$

When expressing the complex fields ϕ and $\phi^\dagger$ in terms of their Hermitian components ϕ_1 and ϕ_2,

$$\phi = \frac{1}{\sqrt{2}}(\phi_1 - i\phi_2) \qquad \text{and} \qquad \phi^\dagger = \frac{1}{\sqrt{2}}(\phi_1 + i\phi_2) \ ,$$

we have

$$[d\phi^\dagger d\phi] = [d\phi_1]\cdot[d\phi_2]$$

in which each factor $[d\phi_i]$ is given by the top equation in (19.89). Both S and D_F' are functionals of $a_\lambda(x)$.

(i) Show that the equation of motion is

$$(D_\lambda D_\lambda - m^2)\,\phi = 0 \ . \tag{19.195}$$

(ii) By expanding $D_F'(x,x')$ as a power series in a_λ, prove that

$$D_F'(x,x') = <x\,|\, i[D_\lambda D_\lambda - (m - i\epsilon)^2]^{-1}\,|\,x'>$$
$$= <x\,|\int_0^\infty d\ell\ e^{i\ell[D_\lambda D_\lambda - (m - i\epsilon)^2]}\,|\,x'>$$

where $\epsilon = 0+$. (19.196)

(iii) Show that in terms of Feynman graphs $S(a_\lambda)$ can be written as a sum of one-loop diagrams.

(iv) By differentiating with respect to m^2, prove that $S(a_\lambda)$ satisfies

$$-\frac{\partial S}{\partial m^2} = \int d^4x \, [\, D_F'(x,x) - D_F(x,x)\,] \tag{19.197}$$

where $D_F(x,x')$ is the free propagator given by (5.49). Thus, S is proportional to

$$\int d^4x = \Omega T$$

where Ω is the volume of the system and T is the total time duration; both $\to \infty$.

(v) The effective Lagrangian density $\mathcal{L}_{eff}$ is defined by

$$\mathcal{L}_{eff} \equiv (\Omega T)^{-1} \, S \; .$$

Prove that

$$\mathcal{L}_{eff} = -i(\Omega T)^{-1} \int_0^\infty \frac{d\ell}{\ell} \, \text{trace} \left\{ e^{i\ell[\, D_\lambda D_\lambda - (m-i\epsilon)^2\,]} - e^{i\ell[\,(\partial^2/\partial x_\lambda \partial x_\lambda) - (m-i\epsilon)^2\,]} \right\} . \tag{19.198}$$

Problem 19.4. In the above problem, assume that the external $f_{\lambda\nu}$ is a constant magnetic field. Derive that

(i) $$\text{trace } e^{i\ell D_\lambda D_\lambda} = \frac{1}{(4\pi)^2 \, i\ell} \, \frac{eB}{\sin eB\ell} \, \Omega T \tag{19.199}$$

and

(ii) $$\mathcal{L}_{eff} = -\frac{1}{(4\pi)^2} \int_0^\infty \frac{d\ell}{\ell^3} \left[\frac{eB\ell}{\sin eB\ell} - 1 \right] e^{-i\ell(m-i\epsilon)^2} . \tag{19.200}$$

Remarks. If we expand $\mathcal{L}_{eff}$ as a power series in B, the coefficient of B^2 is divergent; this is connected with the renormalization of the electromagnetic field and should be subtracted. We define the renormalized effective Lagrangian density by simply dropping this term from (19.200). The result is

$$(\mathcal{L}_{eff})_r = -\frac{1}{(4\pi)^2}\int_0^\infty \frac{d\ell}{\ell^3}\left[\frac{eB\ell}{\sin eB\ell} - 1 - \frac{1}{6}(eB\ell)^2\right] \cdot e^{-i\ell(m-i\epsilon)^2} \quad . \tag{19.201}$$

The reality of $(\mathcal{L}_{eff})_r$ can best be demonstrated by setting $\ell = -i\xi$ and rotating the integration path from $0 \to \infty$ to $0 \to -i\infty$. Thus,

$$(\mathcal{L}_{eff})_r = \frac{1}{(4\pi)^2}\int_0^\infty \frac{d\xi}{\xi^3}\left[\frac{eB\xi}{\sinh eB\xi} - 1 + \frac{1}{6}(eB\xi)^2\right] e^{-\xi m^2}$$

which is finite and equal to

$$\frac{7}{360 m^4}\left(\frac{e^2 B^2}{4\pi}\right)^2 \tag{19.202}$$

when $B \to 0$.

Problem 19.5. Assume that in Problem 19.3 the external $f_{\lambda\nu}$ consists only of a constant electric field $\vec{E} /\!/ z$ axis. Choose the gauge in which

$$a_z = -Et$$

with

$$E = |\vec{E}| \quad , \tag{19.203}$$

but all other a_λ are zero. Hence,

$$D_t = \frac{\partial}{\partial t} \equiv \partial_t \quad , \qquad D_x = \frac{\partial}{\partial x} \equiv \partial_x \quad ,$$
$$D_y = \frac{\partial}{\partial y} \equiv \partial_y \quad \text{but} \quad D_z = \frac{\partial}{\partial z} + ieEt \quad . \tag{19.204}$$

Show that

$$\text{(i)} \quad \langle x, y \,|\, e^{i\ell(\partial_x^2 + \partial_y^2)} \,|\, x_0, y_0 \rangle = \frac{1}{4\pi i \ell}\, e^{i\theta}$$

where

$$\theta = \frac{1}{4\ell}\left[(x-x_0)^2 + (y-y_0)^2\right] \quad , \tag{19.205}$$

and

$$\langle z, t \,|\, e^{i\ell(D_z^2 - \partial_t^2)} \,|\, z_0, t_0 \rangle = \frac{eE}{4\pi \sinh eE\ell}\, e^{i\phi}$$

where (19.206)

$$\phi = \frac{eE}{2}\left\{ \frac{1}{\sinh(2eE\ell)}\left[-(t^2+t_0^2)\cosh 2eE\ell + 2tt_0\right] + \tfrac{1}{2}(\coth eE\ell)\left[z - z_0 - (\tanh eE\ell)(t+t_0)\right]^2 \right\}$$

and (ii) the effective Lagrangian density (19.198) is given by

$$\mathcal{L}_{eff} = -\frac{1}{(4\pi)^2}\int_0^\infty \frac{d\ell}{\ell^3}\left[\frac{eE\ell}{\sinh eE\ell} - 1\right] e^{-i\ell(m-i\epsilon)^2},$$

or as in (19.201), the renormalized density is

$$(\mathcal{L}_{eff})_r = -\frac{1}{(4\pi)^2}\int_0^\infty \frac{d\ell}{\ell^3}\left[\frac{eE\ell}{\sinh eE\ell} - 1 + \frac{1}{6}(eE\ell)^2\right] \cdot e^{-i\ell(m-i\epsilon)^2} . \qquad (19.207)$$

(iii) Prove that trace $e^{i\ell D_\lambda D_\lambda}$, and therefore also $\mathcal{L}_{eff}$, is invariant under a gauge transformation.

(iv) Set $\ell = -i\xi$, and rotate the integration contour to path C:

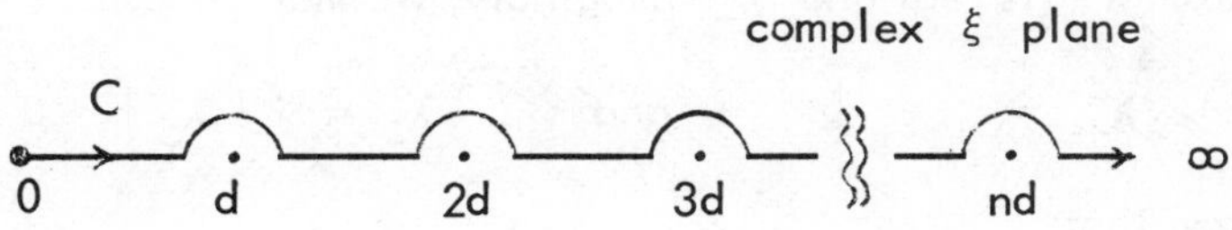

where

$$d = \pi/eE . \qquad (19.208)$$

Equation (19.207) becomes

$$(\mathcal{L}_{eff})_r = \frac{1}{(4\pi)^2}\int_C \frac{d\xi}{\xi^3}\left[\frac{eE\xi}{\sin eE\xi} - 1 - \frac{1}{6}(eE\xi)^2\right] e^{-\xi m^2} . \qquad (19.209)$$

Show that, unlike in the previous case of a constant magnetic field, $(\mathcal{L}_{eff})_r$ has now an imaginary part given by

$$i\,\mathrm{Im}\,(\mathcal{L}_{eff})_r = -i\,\frac{(eE)^2}{16\pi^3}\sum_{n=1}^{\infty}(-1)^n \frac{1}{n^2}\, e^{-\frac{n\pi m^2}{eE}} \qquad (19.210)$$

and for $eE \gg m^2$

$$\mathrm{Im}\,(\mathcal{L}_{eff})_r = \frac{(eE)^2}{192\pi} \quad .$$

(v) Discuss the physical meaning of $\mathrm{Im}\,(\mathcal{L}_{eff})_r$.

Problem 19.6. In the case of an arbitrary constant external electric and magnetic field $\vec{E}$ and $\vec{B}$,

(i) Prove that (19.198) gives*

$$\mathcal{L}_{eff} = -\frac{1}{(4\pi)^2}\int_0^\infty \frac{d\ell}{\ell^3}\left[\frac{\lambda_+\ell}{\sinh\lambda_+\ell}\,\frac{\lambda_-\ell}{\sinh\lambda_-\ell} - 1\right] e^{-i\ell(m-i\epsilon)^2} \tag{19.211}$$

where

$$\lambda_\pm = e\left[s \pm (s^2+p^2)^{\frac{1}{2}}\right]^{\frac{1}{2}} ,$$

$$s = \tfrac{1}{2}(\vec{E}^2 - \vec{B}^2)$$

and

$$p = \vec{E}\cdot\vec{B} \quad . \tag{19.212}$$

Thus, λ_+ is real and λ_- imaginary; we may choose

$$\lambda_+ = |\lambda_+| \qquad \text{and} \qquad \lambda_- = i\,|\lambda_-| \quad . \tag{19.213}$$

Show that

(ii) the renormalized effective Lagrangian is now

$$(\mathcal{L}_{eff})_r = \frac{1}{(4\pi)^2}\int_C \frac{d\xi}{\xi^3}\left[\frac{\lambda_+\xi}{\sin\lambda_+\xi}\,\frac{|\lambda_-|\,\xi}{\sinh|\lambda_-|\,\xi} - 1 + \frac{1}{6}\,g^2\,\xi^2(\lambda_+^2 + |\lambda_-|^2)\right] e^{-\xi m^2} \tag{19.214}$$

where C is the same contour as in Problem 19.5(iv) but, instead of (19.208),

$$d = \pi/\lambda_+ \quad , \tag{19.215}$$

---

* J. Schwinger, Phys.Rev. 82, 664 (1951).

(iii) the imaginary part of $(\mathcal{L}_{eff})_r$ is

$$i \, \mathrm{Im} \, (\mathcal{L}_{eff})_r = -i \frac{1}{16\pi^2} \sum_{n=1}^{\infty} (-1)^n \frac{|\lambda_+ \lambda_-|}{n \sinh \left| \frac{n \pi \lambda_-}{\lambda_+} \right|} e^{-n\pi m^2/\lambda_+} , \tag{19.216}$$

and (iv) in the special case where either $\vec{E} = 0$ ($\therefore \lambda_+ = 0$ and $\lambda_- = ieB$), or $\vec{B} = 0$ ($\therefore \lambda_- = 0$ and $\lambda_+ = eE$), $(\mathcal{L}_{eff})_r$ reduces to the results given in Problems 19.4 and 19.5.

References.

P. A. M. Dirac, Revs. Mod. Phys. 17, 195 (1945).

R. P. Feynman and A. R. Hibbs, Quantum Mechanics and Path Integrals (New York, McGraw-Hill Book Co., 1965).

F. A. Berezin, The Method of Second Quantization (New York, Academic Press, 1966).

L. D. Faddeev, Theoret. Math. Phys. 1, 3 (1969) [English translation by Consultants Bureau 1 (1969)].

E. S. Abers and B. W. Lee, "Gauge Theories", Physics Reports 9C, 1 (1973) and also other references cited therein.

Much of the presentation given in Chapters 18 and 19 is based on N. H. Christ and T. D. Lee, Phys. Rev. D22, 939 (1980).

Chapter 20

QUARK MODEL OF HADRONS

Assuming that quantum chromodynamics is the underlying theory of the strong interaction, we should in principle be able to derive the structure of all hadrons just from QCD and a given number of quark flavors. This is quite far from the present reality. The description that we shall give below will be essentially phenomenological. While such a state of affairs is certainly not satisfactory, it has not been uncommon in the evolution of physics. In this connection we may recall the relation between the well-analyzed theory of quantum electrodynamics and the extensively studied phenomena of superconductivity. There is hardly any doubt that it should be possible to start from QED plus the existence of nuclei and electrons, first to establish the crystal structure of solids, from that the electron-phonon interaction, then the B. C. S. theory *, and finally superconductivity. As yet, however, no one has succeeded through pure theoretical deduction even in the first step, proving the existence of crystal from QED. Hence, one need not be overly surprised at the inadequacy of our present theoretical deductive power with regard to QCD.

---

* J. Bardeen, L. N. Cooper and J. R. Schreiffer, Phys.Rev. 108, 1175 (1957).

20.1 Phenomenological Formulation

We start from the discussions given in Chapter 17 by assuming the QCD vacuum to be a perfect Lorentz-invariant color-dia-electric medium ($\kappa_\infty = 0$). As shown in Fig. 17.6, when the color-dielectric constant of the vacuum $\kappa_\infty \to 0$, all colored particles such as quarks and antiquarks are confined to a domain, which defines the hadron; inside it we have $\kappa = 1$ and coupling $= g^2$, which will be regarded as a relatively small number. Outside, the coupling is $g_\infty^2 = g^2/\kappa_\infty \to \infty$. The region $g_\infty \to \infty$ exerts a strong repulsion against the quarks and antiquarks. Just as in the example of the He-He atomic interaction mentioned on pages 402-403, the particles automatically stay away from this strongly repulsive region. The standard technique in the atomic problem is to replace it by a boundary condition: Restrict the interatomic distance r to $\geq$ radius a of the repulsive potential. At $r = a$ the wave amplitude is zero. In the present QCD case, because of Lorentz invariance, the corresponding boundary-value problem requires the relativistic soliton (bag) solution for the hadrons, as will be discussed in this chapter. We shall see that when we expand the QCD interaction around these soliton solutions, only the relatively small coupling g inside the hadron is the relevant parameter, and thereby the ultra-strong coupling difficulty referred to on page 395 is resolved. Because QCD is asymptotically free, the quarks inside the hadron behave approximately like free particles; furthermore, for u, d and s quarks their effective masses are relatively small.

In the following, we shall discuss how to obtain such a power-series expansion in terms of g^2 (the coupling-squared inside the hadron, not the g_∞^2 outside). For example, such an expansion for the hadron mass M is

$$M = M_0 + g^2 M_1 + g^4 M_2 + \cdots . \tag{20.1}$$

Because of (17.20), we can consider the color-dielectric constant κ to be a Lorentz-invariant quantity. It is convenient to introduce a phenomenological spin-0 field σ which is a function of κ, and to set

$$\sigma(\kappa) = \begin{cases} 0 & \text{when } \kappa = 1 \\ \sigma_{vac} & \text{when } \kappa = \kappa_\infty = 0 , \end{cases}$$

so that inside the hadron we have $\kappa = 1$ and, therefore, $\sigma = 0$. Outside, κ is $\kappa_\infty = 0$ and σ is σ_{vac}. The criterion for the function $\sigma(\kappa)$ is given below by (20.3).

The phenomenological Lagrangian density is assumed to be (not exhibiting counter terms *)

$$\mathcal{L} = -\tfrac{1}{4}\kappa V^{\ell}_{\mu\nu} V^{\ell}_{\mu\nu} - \psi^\dagger \gamma_4 (\gamma_\mu D_\mu + f\sigma + m)\psi - \tfrac{1}{2}\left(\frac{\partial \sigma}{\partial x_\mu}\right)^2 - U(\sigma) \tag{20.2}$$

where ψ is the quark field as in (18.3), $V^{\ell}_{\mu\nu}$ and D_μ are given by (18.4) in which the gauge group is the color SU_3 and g is the renormalized coupling inside the hadron, $U(\sigma)$ is a phenomenological potential function, f a new coupling constant and m the mass matrix of quarks inside the hadron.

Some explanation is needed for this seemingly complicated Lagrangian density.

1. Dielectric constant κ

The pure gauge-field Lagrangian density is, in the notation of

* The counter terms are needed for the renormalization of loop diagrams of the vector and quark fields, but not the phenomenological σ field.

(18.3), $-\frac{1}{4} V^{\ell}_{\mu\nu} V^{\ell}_{\mu\nu}$. We first discuss the case when κ is a constant.

(i) κ = constant

In this case, we can consider the scale transformation:

$$V^{\ell}_{\mu} \rightarrow \sqrt{\kappa}\, V^{\ell}_{\mu}$$

and

$$g \rightarrow g/\sqrt{\kappa} \ .$$

From (18.4), it follows that

$$V^{\ell}_{\mu\nu} \rightarrow \sqrt{\kappa}\, V^{\ell}_{\mu\nu}$$

and

$$-\frac{1}{4} V^{\ell}_{\mu\nu} V^{\ell}_{\mu\nu} \rightarrow -\frac{1}{4} \kappa V^{\ell}_{\mu\nu} V^{\ell}_{\mu\nu} \ .$$

Such a change merely brings a redefinition of g and V^{ℓ}_{μ}; therefore it has no observable consequences.

(ii) $\kappa = \kappa(x)$

If κ can vary, then relative values of κ at different space-time points do produce physical effects. As explained on page 395, our conclusion that $\kappa_{\infty} < 1$ depends on the convention (17.6). In terms of the electric and magnetic fields E^{ℓ}_{i} and B^{ℓ}_{i} given by (18.5), the Lagrangian density $-\frac{1}{4} \kappa F^{\ell}_{\mu\nu} F^{\ell}_{\mu\nu}$ can also be written as

$$\frac{1}{2} \kappa (E^{\ell}_{i} E^{\ell}_{i} - B^{\ell}_{i} B^{\ell}_{i}) = \frac{1}{2} (D^{\ell}_{i} E^{\ell}_{i} - B^{\ell}_{i} H^{\ell}_{i})$$

where

$$D^{\ell}_{i} = \kappa E^{\ell}_{i} \quad \text{and} \quad B^{\ell}_{i} = \mu H^{\ell}_{i} \ ,$$

with μ the color magnetic susceptibility. This leads to the relation

$$\kappa \mu = 1 \ ,$$

given by (17.20).

When $\kappa = \kappa(x)$, there must be a positive energy density produced by its gradient $\nabla_i \kappa$. Due to covariance, the simplest form is

$$\tfrac{1}{2} F(\kappa) \left(\frac{\partial \kappa}{\partial x_\mu}\right)^2$$

where $F > 0$ is a scalar function of κ. The function $\sigma(\kappa)$ is defined by

$$\sigma(\kappa) = \int_1^{\kappa} \sqrt{F(\kappa')}\, d\kappa' \quad . \tag{20.3}$$

For convenience, the lower limit in the above integral is chosen so that $\sigma = 0$ when $\kappa = 1$, in accordance with the convention mentioned on page 546. This leads to the $-\tfrac{1}{2}\left(\frac{\partial \sigma}{\partial x_\mu}\right)^2$ term in (20.2).

2. Energy density function $U(\sigma)$

By definition, the expectation value of $\sigma(x)$ is σ_{vac} in the vacuum state. In order to insure this, we introduce the phenomenological potential energy density function $U(\sigma)$, whose absolute minimum is at $\sigma = \sigma_{vac}$. Without any loss of generality, we may set $U(\sigma_{vac}) = 0$. In addition, as shown in Figure 20.1, we shall assume

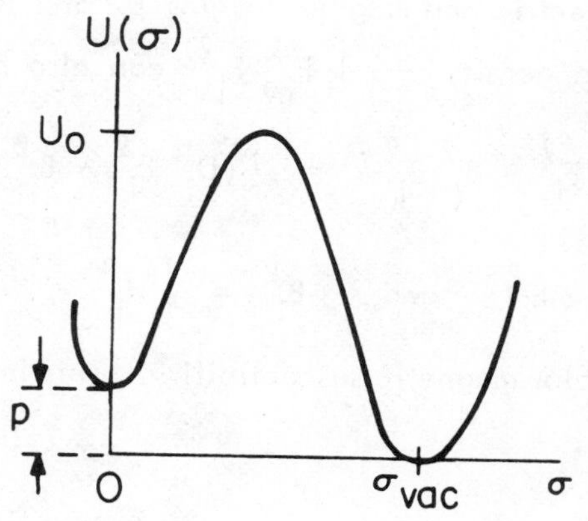

Fig. 20.1. Energy density function $U(\sigma)$ in the phenomenological Lagrangian density (20.2).

$U(\sigma)$ to have a local minimum at $\sigma = 0$ with

$$U(0) \equiv p = O(m_N^4) > 0 \qquad (20.4)$$

where m_N = nucleon mass.

Since σ is only a phenomenological field, describing the long-range collective effects of QCD, its short wavelength components do not exist in reality. We may therefore as an approximation ignore all σ loop diagrams; i.e., σ will be regarded as a classical field.

3. f - coupling

In the phenomenological Lagrangian density (20.2), there is a direct quark - σ coupling f. As explained in Chapter 17, so far as the microscopic basis of the quark-confinement problem is concerned, it suffices to have the $-\frac{1}{2}\kappa V_{\mu\nu}^{\ell} V_{\mu\nu}^{\ell}$ term with $\kappa = \kappa(x)$ and the vacuum as a perfect dia-electric (i.e. in the vacuum state $\kappa = \kappa_\infty = 0$). The origin of the f-coupling lies in the dichotomy between the two coupling constants: Inside the hadron we have the relatively small quark-gauge field coupling g; outside, the coupling is $g_\infty = g/\kappa_\infty^{\frac{1}{2}}$, which $\rightarrow \infty$ when $\kappa_\infty \rightarrow 0$. On the one hand, because of the repulsive nature of the outside region, this ultra-strong coupling g_∞ has the desirable effect of preventing the quarks from moving outside the hadron. On the other hand, it also presents a technical difficulty for the usual power series expansion. In order to overcome this obstacle, we shall devise a new expansion scheme which involves only powers of g^2, but not g^2/κ_∞. This purpose can be achieved by introducing in (20.2) the direct quark - σ coupling f together with the assumption

$$f\,\sigma_{vac} \rightarrow \infty \quad . \qquad (20.5)$$

When the quark is inside the hadron, because $\sigma = 0$, its mass is m which is assumed to be finite and relatively small; but outside, its mass becomes $(f\sigma_{vac} + m) \to \infty$. Thus, quarks do not like to go outside. Phenomenologically the f-coupling with (20.5) gives a convenient alternative formulation to confine quarks which is relativistically covariant. Because the f-coupling restricts the quarks to staying always inside the hadron, that enables us to take full advantage of the relative smallness of the quark - V_μ^ℓ coupling g in the inside region. By introducing the f-coupling we are able to expand any physical observable, say the hadron mass M, in a power series of the form (20.1). Since the quarks are already confined through the f-coupling, the parameter g^2 in the expansion is automatically the coupling inside the hadrons, and not g^2/κ_∞.

[The f-coupling plays, then, the same role here as the hard-sphere potential in the He-He problem discussed on pages 402-403 and on page 545.]

20.2 Hadrons as Solitons (Bags)

With the direct quark-σ coupling f, we can now neglect the exchange of the vector field in the zeroth-order calculation. The description of the hadron reduces to that of a simple soliton model, consisting only of the scalar σ field and the quark ψ field. The details will be given in Section 20.4. By using the soliton solution, we can then proceed to evaluate radiative corrections due to exchanges of vector mesons, in accordance with the series expansion (20.1).

Before entering into any detailed mathematical analysis, we want to show that it is possible to understand most of the features of the physical hadrons by using some simple but general arguments. Let us first consider hadrons made of light quarks u, d and s. For simplicity, we shall approximate their masses inside the hadron as zero:*

$$m_s \cong m_d \cong m_u \cong 0 \ . \tag{20.6}$$

Because σ is $\cong 0$ inside the hadron, but $\cong \sigma_{vac}$ outside, from Figure 20.1 we see that there is a volume energy

$$\frac{4\pi}{3} R^3 p \tag{20.7}$$

where R is the hadron radius and p is defined by (20.4). If we assume the function $U(\sigma)$ to be steep near $\sigma = 0$ and σ_{vac}, i.e., the local maximum U_0 in Figure 20.1 is $\gg m_N^3 \cong (1 \text{ GeV})^3$, then in order to minimize the total energy, the transition from $\sigma = 0$ to $\sigma = \sigma_{vac}$ near the hadron surface must be a rapid one. This gives rise to a surface energy

$$4\pi R^2 s \tag{20.8}$$

where s is a constant, whose exact value depends on the shape of $U(\sigma)$.

Inside the hadron, there is the kinetic energy of quarks. Pure dimensional considerations tell us that it is proportional to R^{-1} in the zero-quark-mass approximation (20.6). Let N be the number of quarks and antiquarks inside the hadron. The kinetic-energy term can be written as

$$N \frac{\xi}{R} \tag{20.9}$$

where ξ is a constant. For the low-lying hadrons states,

$$N = \begin{cases} 2 & \text{for mesons,} \\ 3 & \text{for baryons.} \end{cases} \tag{20.10}$$

If we neglect the vector exchange, i.e., set $g^2 = 0$ in (20.1), then

---

* The effect of nonzero quark masses will be discussed in Section 20.6.2. See also Exercises 1 and 2 on page 570.

by combining (20.7)-(20.9), we see that the zeroth-order hadron mass M is given by

$$M = M_0 = N \frac{\xi}{R} + \frac{4\pi}{3} R^3 p + 4\pi R^2 s \qquad (20.11)$$

which depends on three constants

$$\xi \ , \ p \ \text{ and } \ s \quad .$$

The hadron then resembles a gas bubble immersed in a medium * (i.e., the vacuum). In this analog problem, because the medium exerts a "pressure" p on the bubble, there is a volume energy $(4\pi/3)\, R^3 p$. In addition we have a "surface tension" s which accounts for the surface energy $4\pi R^2 s$, and a "thermodynamical energy" $N\xi/R$ of the gas inside the bubble. The radius R is determined by minimizing the total energy M. From $-\,dM/dR = 0$, we find

$$4\pi R^3 (2s + Rp) = N\xi \quad . \qquad (20.12)$$

The bubble is kept from collapsing by the thermodynamical energy of the gas $N\xi/R$, and it is prevented from further expanding by the surface tension s and the pressure p. In the literature, this bubble analog is sometimes referred to as a "bag". [See (20.14) below for the values of ξ in different bag models.]

In the zeroth order we neglect the vector exchange; therefore in (20.9) the energy inside the bag consists only of the kinetic energy $N\xi/R$, which is independent of the quark spin orientations. Consequently, so is the hadron mass. As we shall discuss in the next section, this leads to the approximate SU_6 symmetry.

For the low-lying hadron state, each quark is in the same

---

* Conceptually, this is merely a transfer of Dirac's idea of leptons to hadrons. See P. A. M. Dirac, Proc. Roy. Soc. 268A, 57 (1962).

relativistic $s_{\frac{1}{2}}$ orbit whose z-component angular momentum j_z can be $\frac{1}{2}$ or $-\frac{1}{2}$, which will be denoted in the following as ↑ or ↓. The details of the $s_{\frac{1}{2}}$ wave function clearly depend on the potential $U(\sigma)$ in the Lagrangian density (20.2), and will be given later.

So far, except for the general shape given by Figure 20.1, there is a great deal of arbitrariness in the function $U(\sigma)$. As already mentioned, we shall assume that $U(\sigma)$ is a steep function of σ ; i.e., its curvature $d^2U/d\sigma^2$ in units of (nucleon mass $m_N)^2$ must be large at $\sigma = 0$ and σ_{vac}. In addition, the value U_0 at its local maximum (between 0 and σ_{vac}) is $\gg m_N^4$. In Section 20.4 we shall prove that under this general assumption, the quark wave function inside the hadron, for the low-lying baryon and meson states, can always be reduced to the solutions of the following set of two dimensionless coupled differential equations:

$$\frac{d\hat{u}}{d\rho} = (-1 + \hat{u}^2 - \hat{v}^2)\,\hat{v}$$

and

$$\frac{d\hat{v}}{d\rho} + \frac{2\hat{v}}{\rho} = (1 + \hat{u}^2 - \hat{v}^2)\,\hat{u} \tag{20.13}$$

which contains no free parameter. The solution of this set of nonlinear equations is characterized by a single parameter n, varying from 0 to ∞. In the limit $n \to 0$ the solution becomes the MIT bag *, and when $n \to \infty$ the SLAC bag **. Consequently (20.13) may be regarded as the universal equations describing the whole range of all possible "bags". In these two limits, the constant ξ in (20.9) is

$$\xi = \begin{cases} 2.0428 & \text{MIT bag } (n \to 0) \\ 1 & \text{SLAC bag } (n \to \infty) \end{cases} \quad . \tag{20.14}$$

---

* A. Chodos, R. J. Jaffe, K. Johnson, C. B. Thorn and V. F. Weisskopf, Phys. Rev. D9, 3471 (1974).

** W. A. Bardeen, M. S. Chanowitz, S. D. Drell, M. Weinstein and T. M. Yan, Phys. Rev. D11, 1094 (1975).

When we take into account the exchanges of gluons, there are radiative corrections which violate SU_6 symmetry. Because in the approximation that both the quarks and gluons are massless inside the hadron, their interaction energy should also be proportional to R^{-1}, based on dimensional considerations. Furthermore, as we shall show later in Section 20.6, to the first order in g^2, the total interaction energy due to gluons depends linearly on N, the number of quarks and antiquarks in the hadron. Consequently, it is proportional to N/R, as is the kinetic energy of quarks, (20.9). Therefore, to the first order in g^2, (20.1) becomes

$$M = M_0 + g^2 M_1 = N \frac{\xi_r}{R} + \frac{4\pi}{3} R^3 p + 4\pi R^2 s \qquad (20.15)$$

where

$$\xi_r = \xi + O(g^2)$$

is a constant independent of N. The hadron radius R is determined, as before, by $-dM/dR = 0$. Equations (20.11) and (20.15) are identical in form. Hence, with the first-order g^2 corrections included, we see that the hadron masses still depend only on three constants:

$$\xi_r \, , \; p \;\text{ and }\; s \; .$$

This greatly simplifies some of our subsequent discussions. It makes the overall agreement between the theoretical results and the experimental data independent of details,* thereby giving more credence to such a phenomenological analysis.

20.3 Approximate SU_6 Symmetry

1. Mass degeneracy

The baryons and mesons are color singlets of three-quark and quark-antiquark systems. In this section, we examine their low-lying

---

* See, e.g., the mass formulas for π, ρ, N and Δ given on p. 577.

levels, which are states composed of u, d, s quarks and their antiparticles. In the zeroth approximation, we neglect the quark-mass differences inside the hadron, as well as the effect of exchanging vector gluons. Consequently, as mentioned before, in the baryon ground state, the wave functions of the three quarks can differ only in the z-component of their angular momenta. In the following we shall refer to the quark wave function as χ.

Since the phenomenological σ field is a Lorentz scalar, it is even under the charge conjugation C. Let us denote the charge-conjugate wave function of χ by

$$\overline{\chi} \equiv \gamma_2 \chi \tag{20.16}$$

where in the notation of Chapters 3 and 10, $\gamma_2 = \rho_2 \sigma_2$. [See also (20.26)-(20.27) in Section 20.4.] For the meson ground state, the quark wave function is again the same χ, and the antiquark $\overline{\chi}$; both have the same energy. The mathematical relations between χ, $\overline{\chi}$ and the solution of the universal bag equation (20.13) will be given in Section 20.4. As we shall see below, a great many of the consequences are independent of the details of χ and $\overline{\chi}$.

Because each of these wave functions describes a relativistic $s_{\frac{1}{2}}$ orbit, it is characterized by a z-component angular momentum $\uparrow$ or $\downarrow$. For simplicity, we shall call $\uparrow$ or $\downarrow$ the spin component, even though it is really the z-component of the <u>total</u> (spin plus orbital) angular momentum of the individual quark or antiquark. Each quark or antiquark has therefore $2 \cdot 3 = 6$ degrees of freedom, where 2 are due to the two spin configurations $\uparrow$ and $\downarrow$, and 3 to the three flavors u, d and s.

Let $a_\alpha^{c\dagger}$ and $b_\alpha^{c\dagger}$ be the creation operators of the quark and antiquark for the orbital wave functions χ and $\overline{\chi}$; the superscript

$c = 1, 2, 3$ denotes the color index and the subscript $\alpha = 1, 2, \cdots, 6$ the flavor-spin index. The low-lying baryon and meson state vectors are

$$T_{\alpha\beta\gamma}\, \epsilon^{c c' c''}\, a_\alpha^{c\dagger}\, a_\beta^{c'\dagger}\, a_\gamma^{c''\dagger} \mid 0 >$$

and $$M_{\alpha\beta}\, a_\alpha^{c\dagger}\, b_\beta^{c\dagger} \mid 0 > \tag{20.17}$$

where $\mid 0 >$ is the vacuum state, $\epsilon^{c c' c''}$ is given by (12.10) and $T_{\alpha\beta\gamma}$ and $M_{\alpha\beta}$ are constants, and all repeated indices are summed over, as usual. In the zeroth approximation, the energy (20.11) is independent of the flavor-spin indices α, β and γ; this leads to SU_6 symmetry.* For the baryon states, because $\epsilon^{c c' c''}$ is totally antisymmetric and $a_\alpha^{c\dagger}$ anticommute, only the symmetric part of $T_{\alpha\beta\gamma}$ contributes. Hence, without any loss of generality, we may choose $T_{\alpha\beta\gamma}$ to be a totally symmetric tensor in α, β and γ, of which there are altogether 56 independent components:

$$6 \text{ of the type } T_{111},$$

$$6 \cdot 5 = 30 \text{ of the type } T_{112}$$

and $$\frac{6 \cdot 5 \cdot 4}{3 \cdot 2 \cdot 1} = 20 \text{ of the type } T_{123}.$$

By following the discussions given in Section 12.1, we see that the $T_{\alpha\beta\gamma}$, and therefore the low-lying baryon levels, form the irreducible representation (56) of the SU_6 group. Likewise, there are $6^2 = 36$ independent components of the meson tensor $M_{\alpha\beta}$. Under the SU_6 group of transformations $M_{\alpha\beta}$ can be decomposed into

---

* F. Gürsey and L. Radicati, Phys.Rev.Lett. 13, 173 (1964); F. Gürsey, A. Pais and L. Radicati, ibid., 175; B. Sakita, Phys.Rev. 136, B1765 (1964).

$$M_{\alpha\beta} - \frac{1}{6}\,\delta_{\alpha\beta}\,M_{\gamma\gamma} \quad \text{and} \quad M_{\gamma\gamma} \,,$$

which form respectively a (35) representation and a (1) representation. In the zeroth approximation (20.11), the 56 baryon levels are all degenerate and the same holds for the 36 meson levels.

For the baryon states, the total angular momentum can be $j = \frac{3}{2}$ or $\frac{1}{2}$. Their flavor compositions are given by (12.49), from which it follows that the (56) representation can be separated into the $j = \frac{3}{2}$ decuplet (Ω^-, Ξ^*, Σ^* and Δ) and the $j = \frac{1}{2}$ octet (Ξ, Σ, Λ and N). Since there are $(2j+1)$ components for each j, the decuplet has $10 \cdot 4 = 40$ different levels and the octet $8 \cdot 2 = 16$, making a total of 56. For the mesons, $j = 0$ or 1. The (35) representation can be decomposed into the vector nonet (ϕ, ω, K^* and ρ) and the pseudoscalar octet (η, K and π); the (1) denotes the pseudoscalar singlet η', making a total of 36 approximately degenerate levels.

The zeroth-order energy is determined by (20.11)-(20.12); it depends on three parameters: ξ, s and p, of which ξ is given by (20.14). To see the physical effects of different values of s and p, let us examine the limit where either s or p equals 0.

(i) If the surface tension $s = 0$, we have the radius

$$R = \left(\frac{N\xi}{4\pi p}\right)^{\frac{1}{4}}$$

and the mass

$$M = \frac{4}{3}\,\frac{N\xi}{R} \propto N^{\frac{3}{4}} \,.$$

Hence, the ratio of the low-lying meson energy M_m vs. the baryon energy M_b is

$$\frac{M_m}{M_b} = \left(\frac{2}{3}\right)^{\frac{3}{4}} \,. \qquad (20.18)$$

(ii) If the pressure $p = 0$, we have

$$R = \left(\frac{N\xi}{8\pi s}\right)^{\frac{1}{3}}$$

and

$$M = \tfrac{3}{2}\,\frac{N\xi}{R} \propto N^{\frac{2}{3}} ,$$

and therefore

$$\frac{M_m}{M_b} = \left(\tfrac{2}{3}\right)^{\frac{2}{3}} . \tag{20.19}$$

Relations (20.18) and (20.19) are independent of ξ, and are therefore valid in both the MIT and SLAC bag limits.

2. State vectors

Consider first the baryons. In (20.17), the color superscript c in $a_\alpha^{c\dagger}$ can vary from 1 to 3 and the flavor-spin subscript α from 1 to 6; the latter will be denoted now more explicitly by $f_s = u_\uparrow$, $u_\downarrow$, $d_\uparrow$, $d_\downarrow$, $s_\uparrow$ and $s_\downarrow$ where f indicates the quark flavor and the subscript s the spin component. We write

$$| \alpha\beta\gamma > \equiv a_\alpha^{1\dagger}\, a_\beta^{2\dagger}\, a_\gamma^{3\dagger}\, | 0 > , \tag{20.20}$$

and therefore

$$| u_\uparrow u_\uparrow d_\downarrow > = a_{u_\uparrow}^{1\dagger}\, a_{u_\uparrow}^{2\dagger}\, a_{d_\downarrow}^{3\dagger}\, | 0 > , \quad \text{etc.}$$

The explicit form of baryon state vectors can be obtained quite easily by combining (12.49)-(12.50) with the top expression in (20.17).

(i) baryon decuplet

These all have total angular momentum $j = \frac{3}{2}$, and therefore their z-component j_z can be $\frac{3}{2}$, $\frac{1}{2}$, $\cdots$. From (12.50), we see that for the $j_z = \frac{3}{2}$ states:

$$| \Omega^-_{\frac{3}{2}} > \doteq | s_\uparrow s_\uparrow s_\uparrow >$$

. . . . .

$$|\Delta^+_{\frac{3}{2}}> = \frac{1}{\sqrt{3}}\left[|d_\uparrow u_\uparrow u_\uparrow> + |u_\uparrow d_\uparrow u_\uparrow> + |u_\uparrow u_\uparrow d_\uparrow>\right] \tag{20.21}$$

$$\cdots\cdots$$

$$|\Delta^-_{\frac{3}{2}}> = |d_\uparrow d_\uparrow d_\uparrow>$$

where the subscript $\frac{3}{2}$ on the lefthand sides indicates j_z. For other j_z-component states, we can use the rotational operators given by (13.47). For example, from (20.21) it follows that the $j_z = \frac{1}{2}$ states are

$$|\Omega^-_{\frac{1}{2}}> = \frac{1}{\sqrt{3}}\left[|s_\downarrow s_\uparrow s_\uparrow> + |s_\uparrow s_\downarrow s_\uparrow> + |s_\uparrow s_\uparrow s_\downarrow>\right] ,$$

$$\cdots\cdots$$

$$|\Delta^+_{\frac{1}{2}}> = \frac{1}{\sqrt{9}}\left[|d_\downarrow u_\uparrow u_\uparrow> + |d_\uparrow u_\downarrow u_\uparrow> + |d_\uparrow u_\uparrow u_\downarrow> + |u_\downarrow d_\uparrow u_\uparrow> + |u_\uparrow d_\downarrow u_\uparrow> + |u_\uparrow u_\uparrow d_\downarrow> + |u_\downarrow u_\uparrow d_\uparrow> + |u_\uparrow u_\downarrow d_\uparrow> + |u_\uparrow u_\uparrow d_\downarrow>\right],$$

$$\cdots\cdots .$$

As mentioned before, there are altogether $4 \cdot 10 = 40$ such decuplet states. The remaining $56 - 40 = 16$ members of the (56) representation are the spin-$\frac{1}{2}$ octets given below.

(ii) baryon octet

These are also given by the top expression of (20.17). By using (12.48) and their orthogonality to the decuplet states, we find, e.g.,

$$|p_{\frac{1}{2}}> = \frac{1}{\sqrt{18}}\left[-2\,|d_\downarrow u_\uparrow u_\uparrow> + |d_\uparrow u_\downarrow u_\uparrow> + |d_\uparrow u_\uparrow u_\downarrow> + |u_\downarrow d_\uparrow u_\uparrow> - 2\,|u_\uparrow d_\downarrow u_\uparrow> + |u_\uparrow d_\uparrow u_\downarrow> + |u_\downarrow u_\uparrow d_\uparrow> + |u_\uparrow u_\downarrow d_\uparrow> - 2\,|u_\uparrow u_\uparrow d_\downarrow>\right] \tag{20.22}$$

where the subscript $\frac{1}{2}$ again indicates j_z. By interchanging $u \rightleftarrows d$, we have $|n_{\frac{1}{2}}>$. Similarly, we can construct all other baryon octet state vectors.

(iii) meson states

The ⑶⑸ + ⑴ representations can be decomposed into a vector nonet (i.e., SU_3 flavor octet and singlet) and a pseudoscalar nonet. These state vectors can be constructed by using (12.39) and (20.17).

The SU_6 symmetry is broken when we include the mass differences between quarks and the radiative corrections due to gluon exchanges. These topics will be discussed in Section 20.6.

Exercise. Work out explicitly the state vectors for the 36 low-lying meson levels.

20.4 Zeroth-order Soliton Solutions

We now turn to a detailed analysis of the soliton solution. As in the previous section, we shall first neglect the quark interaction with the vector field as well as the masses of u, d and s quarks inside the hadron. The relevant part of (20.2) then consists simply of

$$\mathcal{L}_0 = -\psi^\dagger \gamma_4 (\gamma_\mu \frac{\partial}{\partial x_\mu} + f\sigma)\psi - \tfrac{1}{2}(\frac{\partial \sigma}{\partial x_\mu})^2 - U(\sigma) , \qquad (20.23)$$

which is our zeroth-order Lagrangian density. The corresponding Hamiltonian density is the normal product $:\mathcal{H}_0:$, where $\mathcal{H}_0$ is given by

$$\mathcal{H}_0 = \tfrac{1}{2}\Pi^2 + \tfrac{1}{2}(\vec{\nabla}\sigma)^2 + U(\sigma) + \psi^\dagger(-i\vec{\alpha}\cdot\vec{\nabla} + \beta f\sigma)\psi ,$$

Π is the conjugate momentum of σ, and $\vec{\alpha}$, β are the standard Dirac matrices. As mentioned before, since σ is only a phenomenological field that has no short wavelength components, we shall, as an approximation, neglect all σ loop diagrams. The remaining σ diagrams are called tree diagrams, which correspond to the classical approximation for the σ field. The quark fields are fully quantum-mechanical; i.e., they satisfy

$$\{\psi(\vec{r}, t), \psi^\dagger(\vec{r}', t)\} = \delta^3(\vec{r} - \vec{r}') \quad . \tag{20.24}$$

1. <u>Basic equations</u>

For a classical σ, the conjugate momentum Π commutes with σ; hence, the lowest-energy state always has $\Pi = 0$, and therefore σ is independent of t. From (20.23), we see that ψ satisfies

$$(-i\vec{\alpha}\cdot\vec{\nabla} + \beta f \sigma)\psi = i\dot{\psi} \quad .$$

As before, ψ stands for the column matrix whose elements are the quark field ψ_f^c of color c and flavor f. It is convenient to expand the quantum operator ψ in terms of spinors that satisfy the same equation:

$$\psi_f^c(\vec{r}, t) = \sum_n \left[(a_f^c)_n \chi_n(\vec{r})\, e^{-i\epsilon_n t} + (b_f^c)_n^\dagger\, \bar{\chi}_n(\vec{r})\, e^{i\epsilon_n t} \right] \tag{20.25}$$

where χ_n and $\bar{\chi}_n$ are the c. number spinor solutions of

$$(-i\vec{\alpha}\cdot\vec{\nabla} + \beta f \sigma)\cdot \begin{cases} \chi_n \\ \bar{\chi}_n \end{cases} = \epsilon_n \cdot \begin{cases} \chi_n \\ -\bar{\chi}_n \end{cases} \tag{20.26}$$

with ϵ_n positive. In the representation (3.10) - (3.11), $\vec{\alpha} = \rho_1 \vec{\sigma}$, $\beta = \rho_3$ and $\gamma_2 = \rho_2 \sigma_2$, we have

$$\gamma_2 \vec{\alpha}^* \gamma_2 = \vec{\alpha} \,, \qquad \gamma_2 \beta^* \gamma_2 = -\beta \,,$$

and consequently

$$\bar{\chi}_n = \gamma_2 \chi_n^* \,, \tag{20.27}$$

as can be readily verified. In order to satisfy the anticommutation relation (20.24), we have at equal time

$$\{(a_f^c)_n, (a_{f'}^{c'})_{n'}^\dagger\} = \{(b_f^c)_n, (b_{f'}^{c'})_{n'}^\dagger\} = \delta_{ff'}\,\delta^{cc'}\,\delta_{nn'} \,.$$

All other anticommutators between $(a_f^c)_n$, $(b_f^c)_n$ and their Hermitian conjugates are zero. The zeroth-order Hamiltonian $H_0 \equiv \int : \mathcal{H}_0 : d^3r$ becomes

$$H_0 = \sum_{n,f,c} \left[(a_f^c)_n^\dagger (a_f^c)_n + (b_f^c)_n^\dagger (b_f^c)_n \right] \epsilon_n + \int \left[\tfrac{1}{2}(\nabla\sigma)^2 + U(\sigma) \right] d^3r \quad . \tag{20.28}$$

Let ϵ be the smallest of ϵ_n ; χ and $\bar{\chi}$ are the corresponding χ_n and $\bar{\chi}_n$. As already discussed in the preceding section, the ground state of the baryon $(p, n, \cdots)$ will consist of three different colored quarks, all of which have, apart from spin variations, the same wave function χ ; likewise, the meson ground state consists of one quark in χ and the antiquark in $\bar{\chi} = \gamma_2 \chi^*$. Therefore, according to (20.28), the energy of the lowest hadron level is the minimum of

$$E = N\epsilon + \int \left[\tfrac{1}{2}(\vec{\nabla}\sigma)^2 + U(\sigma) \right] d^3r \tag{20.29}$$

where $N = 2$ for mesons and 3 for baryons, $\sigma(\vec{r})$ is a c. number function and ϵ is a functional of $\sigma(\vec{r})$ determined by the c. number Dirac equation (20.26),

$$(-i\vec{\alpha}\cdot\vec{\nabla} + \beta f \sigma)\chi = \epsilon\chi \tag{20.30}$$

with ϵ as the smallest positive eigenvalue, and

$$\int \chi^\dagger \chi \, d^3r = 1 \quad .$$

Since

$$\epsilon = \int \chi^\dagger(-i\vec{\alpha}\cdot\vec{\nabla} + \beta f\sigma)\chi \, d^3r \quad ,$$

by using (20.30) we see that

$$\frac{\delta\epsilon}{\delta\sigma} = f\chi^\dagger\beta\chi \quad .$$

Therefore, the minimum E occurs when $\delta E/\delta\sigma = 0$, or

$$- \nabla^2 \sigma + U'(\sigma) = - f N \chi^\dagger \beta \chi \tag{20.31}$$

where $U'(\sigma) = dU/d\sigma$. For the ground state, σ is spherically symmetric, the spin $j = \frac{1}{2}$ solution of (20.30) can be written as

$$\chi = \begin{pmatrix} u \\ i(\vec{\sigma} \cdot \vec{r}/r)\, v \end{pmatrix} \mathfrak{s} \tag{20.32}$$

where for $j_z = \frac{1}{2}$ or $-\frac{1}{2}$

$$\mathfrak{s} = \uparrow = \begin{pmatrix} 1 \\ 0 \end{pmatrix} \quad \text{or} \quad \downarrow = \begin{pmatrix} 0 \\ 1 \end{pmatrix} ,$$

$\vec{\sigma}$ is the Pauli spin matrix, $u = u(r)$ and $v = v(r)$ are the radial functions. Substitute (20.32) into (20.30) and (20.31). Because

$$(\vec{\sigma} \cdot \vec{\nabla})(\vec{\sigma} \cdot \vec{r} v/r) = \frac{dv}{dr} + \frac{2v}{r}$$

and

$$\chi^\dagger \beta \chi = u^2 - v^2 ,$$

we derive the radial equations

$$\frac{du}{dr} = (-\epsilon - f\sigma)\, v ,$$

$$\frac{dv}{dr} + \frac{2}{r} v = (\epsilon - f\sigma)\, u , \tag{20.33}$$

and

$$\frac{d^2\sigma}{dr^2} + \frac{2}{r}\frac{d\sigma}{dr} - U'(\sigma) = f N (u^2 - v^2) .$$

[See Exercises 1 and 2 on page 570 for the modification when the quark mass $m \neq 0$.]

The above set of equations is still somewhat complicated since it involves an unknown function $U(\sigma)$. As mentioned earlier, we shall assume that $U(\sigma)$ is of the general form shown in Figure 20.1, with a very steep shape; i.e., its curvatures U'' at $\sigma = 0$ and σ_{vac} are both $>> m_p^{\,2}$ and its height U_0 at its local maximum is also $>> m_N^{\,4}$. Consequently, σ can only fluctuate near 0 inside the hadron, and we may approximate $U(\sigma)$ as a quadratic function; i.e.,

$$U \cong \tfrac{1}{2} m_\sigma^{\ 2} \sigma^2$$

and its derivative $U' \cong m_\sigma^{\ 2} \sigma$ with $m_\sigma >> m_N$. Now the magnitude of $d\sigma/dr$ is $\sim \sigma/R$ where the hadron radius R is expected to be $\sim m_N^{-1}$, and therefore $R^{-1} << m_\sigma$. Hence, compared with $U'(\sigma)$ we can neglect the derivatives $d^2\sigma/dr^2$ and $r^{-1} d\sigma/dr$. As a result, inside the hadron the last equation in (20.33) becomes simply

$$\sigma \cong - \frac{Nf}{m_\sigma^{\ 2}} (u^2 - v^2) \quad . \tag{20.34}$$

Substituting this into the top two equations in (20.33), and with the scaling transformation

$$\rho \equiv \epsilon r \quad ,$$

$$\hat{u} \equiv \frac{f}{m_\sigma} \left(\frac{N}{\epsilon}\right)^{\frac{1}{2}} u$$

and

$$\hat{v} \equiv \frac{f}{m_\sigma} \left(\frac{N}{\epsilon}\right)^{\frac{1}{2}} v \tag{20.35}$$

we find that <u>inside the hadron,</u> the quark wave functions are determined by (20.13):

$$\frac{d\hat{u}}{d\rho} = (-1 + \hat{u}^2 - \hat{v}^2)\, \hat{v}$$

and

$$\frac{d\hat{v}}{d\rho} + \frac{2\hat{v}}{\rho} = (1 + \hat{u}^2 - \hat{v}^2)\, \hat{u} \quad .$$

It is quite remarkable that independently of the detailed form of $U(\sigma)$, the quark wave function inside the hadron is determined by these two equations which have no free parameters. [This situation resembles the Thomas-Fermi equation for the electron distribution inside an atom;

after a scale transformation it also reduces to a dimensionless differential equation with no parameter.] At the boundary of the hadron, $r = R$, we have zero matter density $\psi^\dagger \beta \psi = 0$. Thus, at the hadron radius $r = R \equiv \xi/\epsilon$ (i.e., $\rho = \xi$)

$$\hat{u}(\xi)^2 = \hat{v}(\xi)^2 \quad . \tag{20.36}$$

Outside the hadron radius, the quark density remains zero. Hence for $r > R$, (20.31) becomes

$$-\nabla^2 \sigma + U'(\sigma) = 0 \quad . \tag{20.37}$$

Since σ will change very rapidly from $\sigma \cong 0$ inside R to $\sigma = \sigma_{vac}$ outside R over a width $\sim m_\sigma^{-1} \ll R$, we may approximate ∇^2 by d^2/dr^2. The solution is simply the one-dimensional soliton solution (7.19):

$$r = \int^\sigma \frac{d\sigma'}{\sqrt{2\,U(\sigma')}} \quad . \tag{20.38}$$

For example, when U is a quartic function and $p = U(0)$ is neglected, we have

$$U \cong \frac{m_\sigma^2}{2\sigma_{vac}^2} (\sigma - \sigma_{vac})^2 \sigma^2$$

and the solution (20.38) for r near or $> R$ is

$$\sigma \cong \tfrac{1}{2}\left[1 + \tanh\left(\frac{m_\sigma}{2}(r - R)\right)\right] \sigma_{vac} \quad . \tag{20.39}$$

This shows explicitly that the width of the transition region is indeed $\sim m_\sigma^{-1}$.

2. Solutions

Although the fundamental equations (20.13) do not contain any explicit parameter, their solutions form a one-parameter family. From (20.32) we see that v, and therefore $\hat{v}$, is the p-wave part of the

Dirac $j = \frac{1}{2}$ spinor solution. Therefore as $r \to 0$, $v = 0$; i.e., as $\rho \to 0$, $\hat{v} = 0$. This can also be seen from (20.13); because of the $2\hat{v}/\rho$ term, as $\rho \to 0$ we must have $\hat{v} = 0$. On the other hand at the origin, $\hat{u}(0)$ can be an arbitrary constant. For a given initial value $\hat{u}(0)$, the corresponding solution can be obtained by a direct integration from $\rho = 0$ to $\rho = \xi$; at that point we have *

$$\hat{u}(\xi) = \hat{v}(\xi) \tag{20.40}$$

and therefore the boundary condition (20.36) is satisfied. Equation (20.40) determines the constant ξ. From (20.35) and the normalization condition

$$\int \chi^\dagger \chi \, d^3r = \int (u^2 + v^2)\, d^3r = 1 \quad ,$$

we find

$$\epsilon = \frac{m_\sigma}{f} \sqrt{\frac{n}{N}} \tag{20.41}$$

where

$$n = \int_0^\xi 4\pi \rho^2 (\hat{u}^2 + \hat{v}^2)\, d\rho \quad . \tag{20.42}$$

---

* From the inside solution, we arrive at the boundary condition (20.36), which admits at $\rho = \xi$ (i.e., $r = R = \xi/\epsilon$), $\hat{u}(\xi) = \pm \hat{v}(\xi)$. As we shall see, by connecting the inside with the outside solution, we can use only $\hat{u}(\xi) = \hat{v}(\xi)$, not $\hat{u}(\xi) = -\hat{v}(\xi)$. This is because when $r > R$, σ is large; the first two equations in (20.33) can be approximated by $du/dr = -f\sigma v$ and $dv/dr = -f\sigma u$. Hence, in that region the solution is

$$u \cong v \cong \exp[-\int f\sigma(r)\, dr] \quad . \tag{20.40a}$$

For the special example (20.39), we have

$$u \cong v \propto [1 - e^{m_\sigma(r-R)}]^{-(f\sigma_{vac}/m_\sigma)} \quad , \tag{20.40b}$$

which, for $r > R$, attenuates exponentially over a length $(f\sigma_{vac})^{-1} \to 0$. From (20.40a), it follows that $u = v$, and therefore $\hat{u} = \hat{v}$, at the boundary.

It turns out that the initial value $\hat{u}(0)$ can only be between 0 and a critical value $u_c = 1.7419$. When $\hat{u}(0) \to 0$, we have $n \to 0$, but as $\hat{u}(0) \to u_c-$, $n \to \infty$. For $\hat{u}(0) > u_c$, the solution has singularities and becomes unphysical. Thus, a convenient parameter to label these solutions can be either $\hat{u}(0)$ or the integral n, (20.42). Examples of the two limiting solutions, $n \to 0$ and ∞, are given in Figure 20.2.

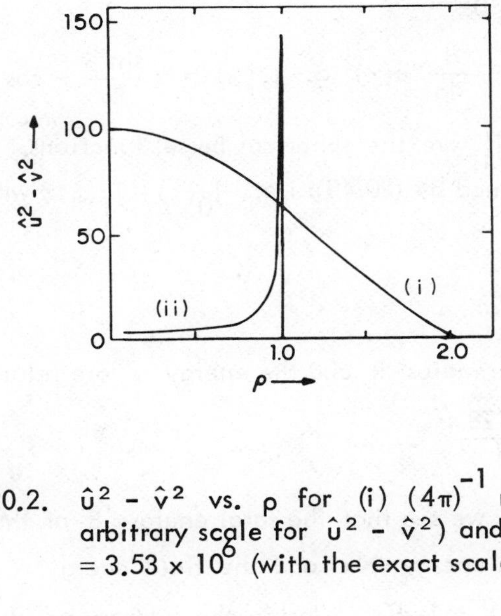

Fig. 20.2. $\hat{u}^2 - \hat{v}^2$ vs. ρ for (i) $(4\pi)^{-1} n \ll 1$ (with an arbitrary scale for $\hat{u}^2 - \hat{v}^2$) and for (ii) $(4\pi)^{-1} n = 3.53 \times 10^6$ (with the exact scale for $\hat{u}^2 - \hat{v}^2$).

3. MIT bag

The MIT bag corresponds to the limit $n \to 0$ (i.e., $\hat{u}(0) \to 0$). From (20.42) we see that, in this limit, the amplitudes $\hat{u}$ and $\hat{v}$ are uniformly small. Consequently, we may neglect the nonlinear terms in (20.13). The equations then become linearized:

$$\frac{d\hat{u}}{d\rho} = -\hat{v}$$

and

$$\frac{d\hat{v}}{d\rho} + \frac{2\hat{v}}{\rho} = \hat{u} \quad . \tag{20.43}$$

When combined, they yield

$$\left(\frac{d^2}{d\rho^2} + \frac{2}{\rho}\frac{d}{d\rho} + 1\right)\hat{u} = 0 \quad .$$

The solution is

$$\hat{u}(\rho) \propto j_0(\rho) = \frac{\sin\rho}{\rho} \quad ,$$

and

$$\hat{v}(\rho) = -\frac{d}{d\rho}\hat{u}(\rho) \propto j_1(\rho) = \left(\frac{\sin\rho}{\rho} - \cos\rho\right)\frac{1}{\rho} \quad , \tag{20.44}$$

where j_0 and j_1 are the spherical Bessel functions. The boundary $\rho = \xi$ is determined by (20.40); i.e., $j_0(\xi) = j_1(\xi)$, which gives for the ground state

$$\xi = 2.0428 \quad . \tag{20.45}$$

The hadron radius R and the energy ϵ are related by $R = \xi/\epsilon$ or

$$\epsilon = \frac{2.0428}{R} \quad . \tag{20.46}$$

By using (20.29), we see that the total energy E of the system consists of the quark energy $N\epsilon$ and the field energy $\int [\frac{1}{2}(\vec{\nabla}\sigma)^2 + U(\sigma)]\, d^3r$. Inside the hadron $r \lesssim R$, σ is $\cong 0$; near the surface $r = R + O(m_\sigma^{-1})$, σ changes rapidly from 0 to σ_{vac}, and then remains so outside. Because $U(0) = p$ and $U(\sigma_{vac}) = 0$, the integral

$$\int_{r \lesssim R} U(\sigma)\, d^3r \cong \frac{4\pi}{3} R^3 p \tag{20.47}$$

is proportional to the volume. However, the integral over the transition region

$$\int [U(\sigma) + \tfrac{1}{2}(\vec{\nabla}\sigma)^2]\, d^3r \cong \int (\vec{\nabla}\sigma)^2\, d^3r$$

$$\equiv 4\pi R^2 s \qquad (20.48)$$

is proportional to the surface area $4\pi R^2$ with the proportionality constant s defined to be the surface energy per unit area. In (20.48) the integration extends over the region $r = R + O(m_\sigma^{-1})$; the first equality is due to the virial theorem (7.44) for $D = 1$.

The total energy is then given by

$$E = N\,\frac{2.0428}{R} + \frac{4\pi}{3} R^3 p + 4\pi R^2 s \quad . \qquad (20.49)$$

4. SLAC bag

The SLAC bag corresponds to the other extreme, $n \to \infty$ (i.e., $\hat{u}(0) \to u_c = 1.7419$). From Fig. 20.2 we see that the quark wave function concentrates entirely on the surface. In this case,* ξ turns out to be 1 and $\epsilon = R^{-1}$. Instead of (20.49), the energy of the system is

$$E_{SLAC} = \frac{N}{R} + \frac{4\pi}{3} R^3 p + 4\pi R^2 s \quad . \qquad (20.50)$$

This and (20.49) confirm (20.11) with ξ given by (20.14).

Historically, the MIT bag was proposed with the surface energy $s = 0$, and the SLAC bag with the volume energy $p = 0$. The original presentations of both models are actually quite different from that given here. By regarding hadrons as solitons embedded in the QCD vacuum, we are able to give a systematic survey of all possible bag solutions; the MIT and SLAC examples represent but two extremes.

---

* For details, see Phys.Rev. D16, 1096 (1977).

Exercise 1. Suppose that inside the hadron the quark mass $m \neq 0$. Show that the first two equations of (20.33) become

$$\frac{d}{dr}\,u(r) = [-\epsilon - (f\sigma + m)]\;v(r)$$

and

$$\left(\frac{d}{dr} + \frac{2}{r}\right) v(r) = [\epsilon - (f\sigma + m)]\;u(r)\ ,$$

but the last equation remains unchanged.

Exercise 2. In the MIT bag approximation, we may set $\sigma = 0$ in the two equations above. Show that the solution is

$$u(r) \propto j_o(kr)$$

where $k = \sqrt{\epsilon^2 - m^2}$, and

$$v(r) \propto \frac{k}{\epsilon + m}\;j_1(kr)$$

with the same proportionality constant. The hadron radius R is now determined by

$$\frac{j_0(kR)}{j_1(kR)} = \frac{k}{\epsilon + m}\ .$$

20.5 Applications to the Nucleon

In the nucleon state, the quark wave function is χ given by (20.32). It depends on the spin variable $s = \uparrow$ or $\downarrow$. For clarity, we shall label the corresponding wave function χ_s . As before, we have the normalization condition

$$\int \chi_s^\dagger\, \chi_s\, d^3r = 1\ . \tag{20.51}$$

Here, as in the following, the repeated index s is not to be summed over.

1. <u>Charge radius</u>

Let $r_p{}^2$ and $r_n{}^2$ be the root-mean-squared charge radius of the proton and the neutron. It is given by the integral

$$\sum Q_f \int \chi_{\mathscr{s}}^{\dagger} \chi_{\mathscr{s}} r^2 d^3r \tag{20.52}$$

where $\mathscr{s}$ can be either $\uparrow$ or $\downarrow$, Q_f denotes the charge of the f-flavored quark in units of e, and the sum extends over the three quarks in the nucleon. From (20.32) we see that the above integral is independent of the subscript $\mathscr{s}$. For p, the total charge

$$\sum Q_f = 1 \quad ,$$

and therefore

$$r_p{}^2 = \int \chi_{\uparrow}^{\dagger} \chi_{\uparrow} r^2 d^3r = \int \chi_{\downarrow}^{\dagger} \chi_{\downarrow} r^2 d^3r \quad . \tag{20.53}$$

For n, because $\sum Q_f = 0$,

$$r_n = 0 \quad .$$

2. <u>Magnetic moment</u>

The electromagnetic current operator is

$$j_\mu = i \sum_{f,c} Q_f \psi_f^{c\dagger} \gamma_4 \gamma_\mu \psi_f^c$$

where ψ_f^c is given by (20.25). For a constant magnetic field $\vec{B}$, the electromagnetic vector potential $\vec{A}$ can be chosen to be

$$\vec{A} = \tfrac{1}{2} \vec{B} \times \vec{r} \quad ;$$

the corresponding electromagnetic-interaction Hamiltonian is

$$- e \int \vec{j} \cdot \vec{A}\, d^3r \equiv - \vec{\mu}_{op} \cdot \vec{B}$$

where μ_{op} is the magnetic-moment operator. Thus,

$$\vec{\mu}_{op} = \tfrac{1}{2} e \sum_{f,c} Q_f \int \psi_f^{c\dagger}\, \vec{r} \times \vec{\alpha}\, \psi_f^c\, d^3r \tag{20.54}$$

where, as before, the matrix $\vec{\alpha} = \rho_1 \vec{\sigma}$. Take $\vec{B}$ along the z-direction. The expectation value of the z-component $(\mu_{op})_z$ over the proton state with $j_z = \frac{1}{2}$ gives the magnetic moment of the proton. By using the $|p_{\frac{1}{2}}>$ state given by (20.22), we find for the proton

$$\mu_p \equiv < p_{\frac{1}{2}} | (\mu_{op})_z | p_{\frac{1}{2}} > = e \int \chi_\uparrow^\dagger \tfrac{1}{2} (\vec{r} \times \vec{\alpha})_z \chi_\uparrow d^3 r \,, \tag{20.55}$$

and likewise, for the neutron

$$\mu_n \equiv < n_{\frac{1}{2}} | (\mu_{op})_z | n_{\frac{1}{2}} > = \tfrac{1}{3} e \Big[\int \chi_\downarrow^\dagger \tfrac{1}{2} (\vec{r} \times \vec{\alpha})_z \chi_\downarrow d^3 r - \int \chi_\uparrow^\dagger \tfrac{1}{2} (\vec{r} \times \vec{\alpha})_z \chi_\uparrow d^3 r \Big] \,. \tag{20.56}$$

We note that in the $|p_{\frac{1}{2}}>$ state the relative probability of finding $u_\uparrow$ versus $u_\downarrow$ is 5 : 1, that of finding $d_\uparrow$ versus $d_\downarrow$ is 1 : 2 and that of finding u versus d is 2 : 1. Since the charge of u is $\frac{2}{3}$ e and that of d is $-\frac{1}{3}$ e, we see that because

$$\frac{2e}{3} \cdot 2 \cdot \frac{1}{6} - \frac{e}{3} \cdot 1 \cdot \frac{2}{3} = 0$$

the $u_\downarrow$ contribution cancels the $d_\downarrow$ contribution; that explains why in (20.55) μ_p depends only on the $\chi_\uparrow$ wave function. Similarly, one can work out the other coefficients in (20.55) and (20.56). By applying a 180° rotation along the y-axis, we see that

$$\int \chi_\downarrow^\dagger (\vec{r} \times \vec{\alpha})_z \chi_\downarrow d^3 r = - \int \chi_\uparrow^\dagger (\vec{r} \times \vec{\alpha})_z \chi_\uparrow d^3 r$$

and therefore, from (20.55)-(20.56),

$$\frac{\mu_n}{\mu_p} = -\frac{2}{3} \tag{20.57}$$

which agrees remarkably well with the experimental value - .685 .

3. g_A/g_V

As we shall discuss in the next chapter, in the β decay $n \to p + e^- + \bar{\nu}_e$, the vector and axial-vector current operators are

$$V_\mu = i \sum_c \psi_u^{c\dagger} \gamma_4 \gamma_\mu \psi_d^c$$

and (20.58)

$$A_\mu = i \sum_c \psi_u^{c\dagger} \gamma_4 \gamma_\mu \gamma_5 \psi_d^c \quad .$$

The ratio of g_A/g_V is defined to be

$$\frac{g_A}{g_V} \equiv \frac{< p_{\frac{1}{2}} \mid A_z \mid n_{\frac{1}{2}} >}{< p_{\frac{1}{2}} \mid V_0 \mid n_{\frac{1}{2}} >} \quad .$$

By using (20.22), we find

$$\frac{g_A}{g_V} = \frac{5}{3} \int \chi_\uparrow^\dagger \sigma_z \chi_\uparrow d^3r \quad . \tag{20.59}$$

As noted before, the explicit solution χ_Δ depends on a single parameter n of (20.42), which can vary between 0 and ∞. When $n \to 0$ we have the MIT limit in which χ_Δ is determined by the spherical Bessel functions given by (20.32) and (20.44). When $n \to \infty$, we have the SLAC limit in which χ_Δ is nonzero only at the surface. In the original proposals, the MIT bag has no surface tensions, and the SLAC bag is without any pressure p. Consequently, in either case there is only one constant: p for the MIT bag and s for the SLAC bag. This single constant provides the overall mass scale of the system; hence in either model there is no free parameter. By using the explicit solutions of the quark wave function χ_Δ, it is straightforward to calculate r_p, μ_p and g_A/g_V. The results are listed in the following Table; this also gives the range of variation in the soliton model of hadrons, since the MIT and SLAC bags represent two extreme limits of all such models.

Physical observable	Experimental value	MIT bag	SLAC bag
r_p	$3.86/m_N$	$4.25/m_N$	$3.21/m_N$
μ_p	$2.79/(2m_N)$	$2.36/(2m_N)$	$2.14/(2m_N)$
μ_n	$-0.685\ \mu_p$	$-\frac{2}{3}\mu_p$	$-\frac{2}{3}\mu_p$
g_A/g_V	1.25	1.09	$\frac{5}{9}$

From the value of g_A/g_V, we see that the experimental result favors the MIT-type bag model more. However, so far we are only in the zero-quark-mass approximation. From (20.59) we see that in the nonrelativistic limit $\int \chi_\uparrow^\dagger \sigma_z \chi_\uparrow d^3r \to 1$, and therefore

$$g_A/g_V \to \frac{5}{3} \quad .$$

Thus, if we take into account the effect of the quark mass, it is not difficult to raise g_A/g_V from the limiting value for the zero-mass quark to the observed value 1.25 in any of these bag models.

20.6 First-order Corrections

In the zeroth approximation, the hadron mass is given by (20.11) which is SU_6 symmetric. We shall now turn to the question of symmetry breaking; in particular we are interested in the spectroscopy of hadron levels, such as those listed below (in MeV):

π (140)	η (549)	η' (958)
ρ (770)	ω (783)	ϕ (1020)
N (940)	Δ (1232)	etc.

As noted before, the breaking of the SU_6 symmetry is due partly to the gluon exchange and partly to the fact that the s quark "mass" is heavier than those of u and d by ~ 200 MeV. The former destroys the spin - SU_2 symmetry, whereas the latter breaks the flavor - SU_3 symmetry, resulting in the familiar Gell-Mann - Okubo mass formula already discussed in Section 12.3. The first-order corrections that we shall examine include both effects.

1. <u>Gluon exchange and mass formulas</u>

For hadrons composed of u, d, $\bar{u}$ and $\bar{d}$, the zero-quark-mass approximation is a good one. * We assume

$$m_u \cong m_d \cong 0 \quad . \tag{20.60}$$

Hence, for these hadrons the first-order correction consists only of the effects due to gluon exchange, which brings in a spin - dependent interaction and thereby lifts the degeneracy between π and ρ and between N and Δ, as we shall see. For simplicity, we shall concentrate first on these four hadron states. Let g be the quark - gluon coupling inside the hadron. To first order in

$$\alpha_c = \frac{g^2}{4\pi} \quad ,$$

---

* See (20.73) and (20.74) below.

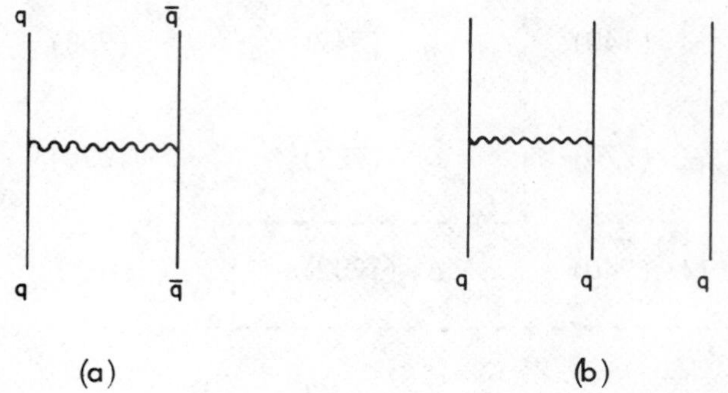

Fig. 20.3. Lowest-order gluon-exchange diagrams for (a) mesons and (b) baryons.

the diagrams are given in Figure 20.3.

As in (20.15), the mass of the hadron h can be written as

$$m_h = M = N\frac{\xi_r}{R} + \frac{4\pi}{3}R^3 p + 4\pi R^2 s \quad . \tag{20.61}$$

In the following, we shall first show that with the gluon-exchange diagrams included, ξ_r is indeed N-independent and is given by

$$\xi_r = \xi - \tfrac{2}{3}\alpha_c[I_{el} - \zeta\, I_{mag}] + O(\alpha_c^{\,2}) \tag{20.62}$$

where I_{el} and I_{mag} are positive numbers $\sim O(1)$,

$$\zeta = \begin{cases} -3 & \text{for } \pi \\ 1 & \text{for } \rho \\ -1 & \text{for } N \\ 1 & \text{for } \Delta \end{cases} \tag{20.63}$$

and ξ , determined by (20.40), is the zeroth-order value of ξ . In the limits of the MIT and SLAC bags, ξ is given by (20.14). The constants I_{el} and I_{mag} are related to the color electric and magnetic interactions, and can be calculated by using Figure 20.3.

We note that there are four parameters ξ, α_c, p and s in the mass formulas (20.61)–(20.63). Suppose that for simplicity we set the surface tension

$$s = 0 ,$$

and in addition fix ξ by choosing some particular soliton solution, say either the MIT or the SLAC bag limit. Even then, there remain two unknown parameters p and α_c. Hence, among the four masses m_π, m_ρ, m_N and m_Δ of π, ρ, N and Δ, there are maximally two mass predictions, assuming that the two constants I_{el} and I_{mag} in (20.62) have been computed by using Figure 20.3. As we shall also prove, it turns out that without actually computing these constants, we can simply derive these two mass predictions based on general properties of diagrams (a) and (b). The resulting mass formulas are

$$\left(\tfrac{2}{3}\right)^{\frac{3}{4}} \frac{m_\Delta}{m_\rho} = \begin{cases} 1 & \text{theory} \\ 1.180 & \text{experiment} \end{cases} \tag{20.64}$$

and

$$\frac{3\left(m_\rho^{\frac{4}{3}} - m_\pi^{\frac{4}{3}}\right)}{4\left(m_\Delta^{\frac{4}{3}} - m_N^{\frac{4}{3}}\right)} = \begin{cases} 1 & \text{theory} \\ 1.187 & \text{experiment} \end{cases} , \tag{20.65}$$

which are valid independently of the assumption of ξ and the numerical values of I_{el} and I_{mag}. Thus, a certain degree of success of the model can be assured without detailed computation.

<u>Proof.</u> In the zero-quark-mass approximation, on pure dimensional reasoning both diagrams in Figure 20.3 are proportional to α_c/R; its coefficient depends on N, the number of quarks and antiquarks, and the spin configuration. To derive the N-dependence, let us denote the eight Gell-Mann matrices of the ith quark (or antiquark) by λ_i^ℓ, where $\ell = 1, 2, \cdots, 8$. Both diagrams have an amplitude

proportional to the sum

$$\frac{1}{2}\sum_{i\neq j}\frac{1}{2}\lambda_i^{\ell}\cdot\frac{1}{2}\lambda_j^{\ell} = \frac{1}{8}\sum_{i\neq j}\lambda_i^{\ell}\lambda_j^{\ell} \quad ,$$

in which the first $\frac{1}{2}$ factor occurs because each pair (i, j) has been counted twice in the sum, and the other two $\frac{1}{2}$ factors are due to the Feynman rule of the quark-gluon vertex given on page 531. Because hadrons are color singlets, their state vectors $|\,\rangle$ satisfy

$$\sum_{i=1}^{N}\lambda_i^{\ell}\;|\,\rangle = 0 \quad .$$

Hence, we have for the expectation value

$$\langle\,|\,(\sum_i \lambda_i^{\ell})^2\;|\,\rangle = 0$$

and therefore, on account of (12.22),

$$\langle\,|\,\frac{1}{8}\sum_{i\neq j}\;\sum_{\ell=1}^{8}\lambda_i^{\ell}\lambda_j^{\ell}\,|\,\rangle = -\langle\,|\,\frac{1}{8}\sum_i\;\sum_{\ell=1}^{8}(\lambda_i^{\ell})^2\,|\,\rangle = -\frac{2}{3}N.$$

To obtain the spin dependence, we observe that each diagram has a spin-independent electric part I_{el} and a magnetic part I_{mag}. The latter is multiplied by a coefficient that depends on the average value of $\frac{1}{2}\sum_{i\neq j}\vec{\sigma}_i\cdot\vec{\sigma}_j$ where $\vec{\sigma}_i$ is the Pauli spin matrix of the ith quark (or antiquark).

For the mesons, π is a spin-singlet and ρ a spin-triplet. The spin-singlet π satisfies

$$\sum_i \vec{\sigma}_i\;|\,\pi\rangle = 0 \quad ,$$

and therefore the average of $(\sum_i \vec{\sigma}_i)^2$ is 0. Hence

$$\langle\pi\,|\,\frac{1}{2}\sum_{i\neq j}\vec{\sigma}_i\cdot\vec{\sigma}_j\,|\,\pi\rangle = -\langle\pi\,|\,\frac{1}{2}\sum_i \vec{\sigma}_i^{\,2}\,|\,\pi\rangle = -3 \quad .$$

For the triplet, since the matrix $\sum_{i\neq j}\vec{\sigma}_i\cdot\vec{\sigma}_j$ has zero trace, its average value must be $-\frac{1}{3}$ that of the singlet. Thus, we derive

$$\zeta \equiv \langle \, | \, \tfrac{1}{2} \sum_{i \neq j} \vec{\sigma}_i \cdot \vec{\sigma}_j \, | \, \rangle = \begin{cases} -3 & \text{for } \pi \\ 1 & \text{for } \rho \end{cases} .$$

Likewise, for the baryon states, we find

$$\zeta \equiv \langle \, | \, \tfrac{1}{2} \sum_{i \neq j} \vec{\sigma}_i \cdot \vec{\sigma}_j \, | \, \rangle = \begin{cases} -1 & \text{for } N \\ 1 & \text{for } \Delta \end{cases} .$$

Putting these factors together, we find that the color electric energy is equal to $-\frac{2}{3}\alpha_c N/R$ multiplied by a spin-independent constant I_{el}. Because it is due to an attractive force, I_{el} is positive. Similarly, the color magnetic energy can be written as $\frac{2}{3}\alpha_c \zeta N I_{mag}/R$. Since the magnetic force in a spin-singlet state is attractive, the corresponding energy for π should be negative; consequently, I_{mag} is also positive. Both I_{el} and I_{mag} are $\sim O(1)$. Equations (20.62) and (20.63) are then established.

Next, we note that since $\zeta = 1$ for both ρ and Δ, the values of ξ_r for Δ and ρ must also be the same. From (20.61), it follows that the only difference between m_ρ and m_Δ is due to $N = 2$ for ρ and $N = 3$ for Δ. In the case of zero surface tension, by minimizing the mass, (20.61), with respect to R we find it to be $\propto N^{\frac{3}{4}}$, and that leads to the mass formula (20.64).

In (20.62), ξ and $\alpha_c I_{el}$ only appear in the combination $(\xi - \frac{2}{3}\alpha_c I_{el})$. Thus, even if we treat I_{el}, I_{mag}, ξ, α_c and p all as unknown parameters, with $s = 0$ (20.61) actually depends only on three combinations $(\xi - \frac{2}{3}\alpha_c I_{el})$, $\alpha_c I_{mag}$ and p. By eliminating these three from the four hadron masses m_π, m_ρ, m_N and m_Δ, we derive (20.65). This completes the proof of (20.61)-(20.65).

The success of the bag model is therefore quite insensitive to the details of QCD, and to the specific soliton solutions. Consequently, the spectroscopy of light-quark hadrons is not the best place for the determination of α_c. [See, however, (20.72) and (20.79) below.]

2. Quark masses

The quark mass term can be written as

$$H_m = m_u \psi_u^{c\dagger} \beta \psi_u^c + m_d \psi_d^{c\dagger} \beta \psi_d^c + m_s \psi_s^{c\dagger} \beta \psi_s^c + \cdots \tag{20.66}$$

where the color superscript is summed over as before, and ... refers to the heavy-mass quarks (charm, bottom, etc.). In order to maintain isospin invariance, we assume

$$m_u \cong m_d \ .$$

In contrast, the flavor SU_3 symmetry is broken because of the mass difference between s and u (or d).

(i) vector mesons

The vector nonet consists of $\rho(770)$, $\omega(783)$, $K^*(892)$ and $\phi(1020)$. Among these, ϕ is $s\bar{s}$, the four K^* are $u\bar{s}$, $d\bar{s}$ and their charge conjugate states; ρ and ω are the isospin triplet and singlet states of $u\bar{u}$, $d\bar{d}$, $u\bar{d}$ and $d\bar{u}$. By using (20.66) and $m_u = m_d$, we derive the following mass formulas:

$$m_\rho = m_\omega ,$$

and (20.67)

$$m_\phi - m_\rho = 2(m_{K^*} - m_\rho)$$

in good agreement with the experimental data. Note that from the observed masses,

$$\frac{m_\omega - m_\rho}{m_\omega + m_\rho} \cong .0084$$

and

$$\frac{m_\phi + m_\rho - 2m_{K^*}}{m_\phi + m_\rho + 2m_{K^*}} \cong .0017 \quad .$$

The mass difference between s and u (or d) is given by

$$\tfrac{1}{2}(m_\phi - m_\rho) = 125 \text{ MeV} = (m_s - m_u) < \beta > \qquad (20.68)$$

where $< \beta >$ is the expectation value of $\psi^\dagger \beta \psi$. If we use the zeroth-order quark wave function (20.32), then

$$< \beta > = \int (u^2 - v^2)\, d^3r$$

which, e.g., in the MIT bag limit is given by

$$< \beta > = \frac{\int [j_0^2(\rho) - j_1^2(\rho)] \rho^2 d\rho}{\int [j_0^2(\rho) + j_1^2(\rho)] \rho^2 d\rho} \cong .48 \qquad (20.69)$$

where $j_0(\rho)$ and $j_1(\rho)$ are the spherical Bessel functions of order 0 and 1, and the integration is from $\rho = 0$ to $\rho = \xi = 2.0428$. Hence, we find from (20.68) and (20.69)

$$m_s - m_u \sim 260 \text{ MeV} . \qquad (20.70)$$

(ii) baryon decuplet

From (12.50), we see that the equal spacing between

$\Delta(1232)$, $\Sigma^*(1382)$, $\Xi^*(1530)$ and $\Omega^-(1672)$

can be readily explained by using the mass difference $(m_s - m_u)$. From the experimental value, we determine

$$145 \text{ MeV} \cong (m_s - m_u) < \beta > \quad . \qquad (20.71)$$

In comparison with (20.68), there is a difference between the average value $< \beta >$ for the meson state versus that for the baryon state. The % difference is

$$\cong \ (145 - 125)/(145 + 125) \ \cong \ 7.4\,\% \ ,$$

which is not unreasonable, since only the zeroth-order quark wave function is being used here.

(iii) baryon octet

Among the members of the spin-$\frac{1}{2}$ octet, N(940) has no s, Λ(1115) and Σ(1193) have one each, and Ξ(1318) two. Clearly, the ~78 MeV mass difference between Λ and Σ cannot be due to the expectation value of H_m, given by (20.66); it must be attributed to effects connected with gluon exchange. By using diagram (b) in Figure 20.3, we can calculate the color electromagnetic energy differences between two massive quarks, between one massive and one massless quark, and between two massless ones. This together with H_m can account for all the octet mass differences. We will not give the details except to remark that the actual fitting is made easier since the bulk of the mass differences

$$m_\Lambda - m_N \cong 175 \text{ MeV} \quad \text{and} \quad \tfrac{1}{2}(m_\Xi - m_N) \cong 189 \text{ MeV}$$

is not that different from the 145 MeV between the decuplets.

(iv) pseudoscalar nonet

The nine low-lying pseudoscalars π(140), K(495), η(549) and η'(958) are different because the pion has a nearly zero mass when compared with other hadrons. Also, unlike the vector-meson $\rho - \omega$ degeneracy, there is no pseudoscalar isosinglet that is near the pion mass. Why?

a. <u>Near-zero pion mass</u>

From (20.61)-(20.62), we see that the pion mass m_π is $\cong 0$, provided that the parameter ξ_r is near zero for the pion state. Phenomenologically, this is possible if the QCD fine structure constant

α_c has the value

$$\alpha_c \cong \frac{3\,\xi}{2(I_{el} + 3\,I_{mag})} \quad .$$

The constants I_{el}, I_{mag} and ξ are all positive and $O(1)$; of these, ξ can vary within a factor of 2 depending on the bag model, as shown by (20.14). If we assume these constants to be of comparable magnitude, then we have

$$\frac{\xi}{(I_{el} + 3\,I_{mag})} \approx \frac{1}{4} \quad .$$

Hence, the near-zero pion mass leads to the estimate

$$\alpha_c \approx \frac{3}{8} \quad . \tag{20.72}$$

In Chapter 24, when we discuss chiral symmetry, arguments will be given for expecting the pion mass

$$m_\pi = 0$$

when the quark mass

$$m_u = m_d = 0 \quad .$$

Furthermore, to first order in $m_u \cong m_d$ and m_s, the ratio of the pion and kaon masses is related to that of the quark masses by

$$\frac{m_u}{m_s} \cong \frac{1}{2}\left(\frac{m_\pi}{m_K}\right)^2 \cong \frac{1}{25} \quad . \tag{20.73}$$

[The derivation will be given in Section 24.6.2.]

Combining (20.70) with (20.73) we find *

---

* Cf. S. Weinberg, in A Festschrift for I. I. Rabi, ed. L. Motz (New York, New York Academy of Sciences, 1977). An estimate of the mass difference between u and d quarks is also given there.

$$m_s \sim 250 \text{ MeV}$$
and (20.74)
$$m_u \cong m_d \sim 10 \text{ MeV} \; .$$

b. $\eta - \eta'$ anomaly

The $\rho - \omega$ degeneracy is due to

$$\rho^o = \frac{(u\bar{u} - d\bar{d})}{\sqrt{2}} \quad ,$$

$$\omega^o = \frac{(u\bar{u} + d\bar{d})}{\sqrt{2}}$$

and $m_u \cong m_d$. By changing the spin configuration of ρ^o from spin-triplet to spin-singlet, we have the pseudoscalar π^o. If we were to perform the same operation on ω^o, we would expect an isosinglet pseudoscalar which should be approximately degenerate with the pion. But there is no such particle. Both η and η' are of much higher mass. This is the so-called U_1 problem. Its solution lies perhaps in part in the anomaly due to some non-perturbative effect,* and in part in the annihilation diagrams given in Figure 20.4, which have unusually large coefficients due to internal symmetry ** and exist only for singlets in flavor, color and spin.

20.7 Hadrons of Heavy Quarks

Since the dramatic discovery *** of the J/ψ (3100) level, our knowledge of the quark-lepton system has been greatly extended with

---

* G. 't Hooft, Phys. Rev. Lett. 37, 8 (1976).

** See R. Friedberg and T. D. Lee, Phys. Rev. D18, 2623 (1978) and the references mentioned therein.

*** J. J. Aubert et al., Phys. Rev. Lett. 33, 1404 (1974); J. E. Augustin et al., ibid., 1406. From these experiments we infer the existence of the charm quark c.

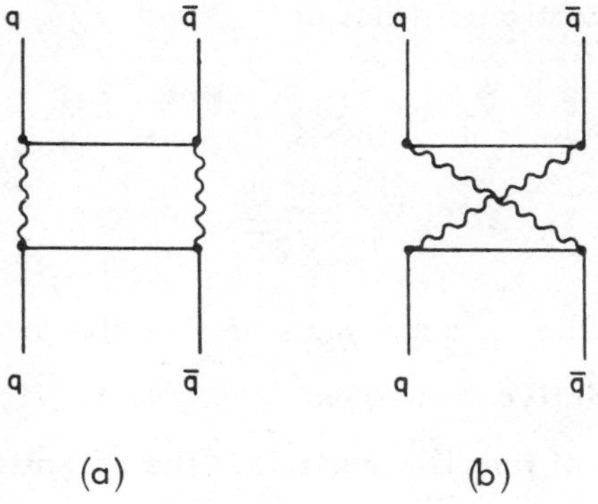

Fig. 20.4. The non-existence of a low-mass isosinglet pseudoscalar may be due, in part, to its coupling with the two-gluon state (glue-balls).

the addition * of c, b and τ. Among these, only the c quark was predicted via a compelling theoretical argument (called GIM mechanism, which will be discussed in the next chapter). In our previous discussions of the light-quark hadrons, the quarks were relativistic; therefore, we approximate the "bag" as slowly moving and the quarks as free particles inside the hadron. For the heavy-quark system, the opposite should be expected. Following the view developed by the Cornell group **, we shall regard the heavy quarks as nonrelativistic particles, and simply replace the "bag" phenomenologically by a long-range confining potential. Thus, to derive the energy levels of $c\bar{c}$ (J/ψ family) and $b\bar{b}$ (Υ family), we consider the Schrödinger equation in which the quark and its antiquark are represented

---

* The existence of the bottom quark b was inferred from the discovery of Υ by S. W. Herb et al., Phys.Rev.Lett. 39, 252 (1977). The heavy lepton was discovered by M. L. Perl et al., Phys.Rev. Lett. 35, 1489 (1975).

** E. Eichten et al., Phys.Rev.Lett. 34, 369 (1975).

by two nonrelativistic particles at $\vec{r}_1$ and $\vec{r}_2$,

$$[-\frac{1}{2m}(\vec{\nabla}_1^{\,2}+\vec{\nabla}_2^{\,2}) + U(\vec{r})]\,\psi = E\,\psi \tag{20.75}$$

where

$$\vec{r} = \vec{r}_1 - \vec{r}_2 \,, \qquad \vec{\nabla}_i = \frac{\partial}{\partial \vec{r}_i}$$

and m is either the c quark mass m_c or the b quark mass m_b. Let $\vec{p}$ be the relative momentum conjugate to $\vec{r}$, and $\vec{s}_i = \frac{1}{2}\vec{\sigma}_i$ the spin operator of the ith particle. The potential energy $U(\vec{r})$ can be written as a sum

$$U(\vec{r}) = U_c(r) + U_s(\vec{r}) \,, \tag{20.76}$$

where $U_c(r)$ is a spin-independent central potential with $r = |\vec{r}|$, and $U_s(\vec{r})$ depends on the orientation of $\vec{r}$ and the spin configuration. We can classify the spin-dependent potential $U_s(\vec{r})$ as follows:

$$\begin{aligned} U_s(\vec{r}) = {} & (\vec{r}\times\vec{p})\cdot(\vec{s}_1+\vec{s}_2)\,U_{\ell s}(r) + \vec{s}_1\cdot\vec{s}_2\,U_{ss}(r) \\ & + [\vec{s}_1\cdot\vec{s}_2 - 3(\vec{s}_1\cdot\hat{r})(\vec{s}_2\cdot\hat{r})]\,U_t(r) \end{aligned} \tag{20.77}$$

where $U_{\ell s}(r)$, $U_{ss}(r)$ and $U_t(r)$ represent respectively the angular independent factors in the spin-orbit, spin-spin and tensor interactions. It is of interest to know to what extent we can determine these potential functions from the present experimental data.

The central potential $U_c(r)$ is usually written as a sum of a "Coulomb" part $\propto \frac{-1}{r}$ and a confining part $\propto r$,

$$U_c(r) = -\frac{4}{3}\frac{\alpha_c}{r} + \frac{r}{a^2} \;. \tag{20.78}$$

In order to fit the experimental data, the parameters α_c, a and the quark masses are assumed to be *

---

* E. Eichten et al., Phys.Rev. D21, 203 (1980). The parameter $\kappa = .52$ in that paper corresponds to $4\alpha_c/3$. Here the factor $4/3$ is due to the product of the factor $2/3$ in (20.62) and the $N=2$ of (20.61). Cf. the estimate (20.72) and the alternative determination of α_c from the high-energy expression (23.123).

$$\begin{aligned} a_c &= .39 \;, \\ a &= 2.34 \ \mathrm{GeV}^{-1} \;, \\ m_c &= 1.84 \ \mathrm{GeV} \\ \text{and} \quad m_b &= 5.17 \ \mathrm{GeV} \;. \end{aligned} \tag{20.79}$$

The known levels of $c\bar{c}$ (J/ψ family) and $b\bar{b}$ (Υ family) are given in Figures 20.5 and 20.6. Let $E(nS)$ be the energy of the spin-triplet S-orbit level, n^3S_1. We see from these figures that the following relations are approximately valid:

$$\frac{E_\Upsilon(2S) - E_\Upsilon(1S)}{E_\psi(2S) - E_\psi(1S)} \cong 1$$

and

$$\frac{E_\Upsilon(3S) - E_\Upsilon(1S)}{E_\psi(3S) - E_\psi(1S)} \cong 1 \tag{20.80}$$

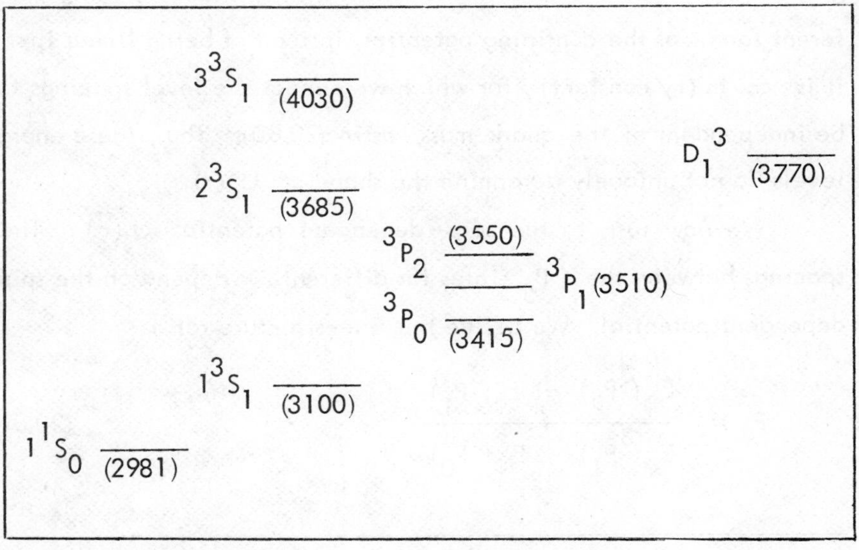

Fig. 20.5. Energy levels of the J/ψ family.

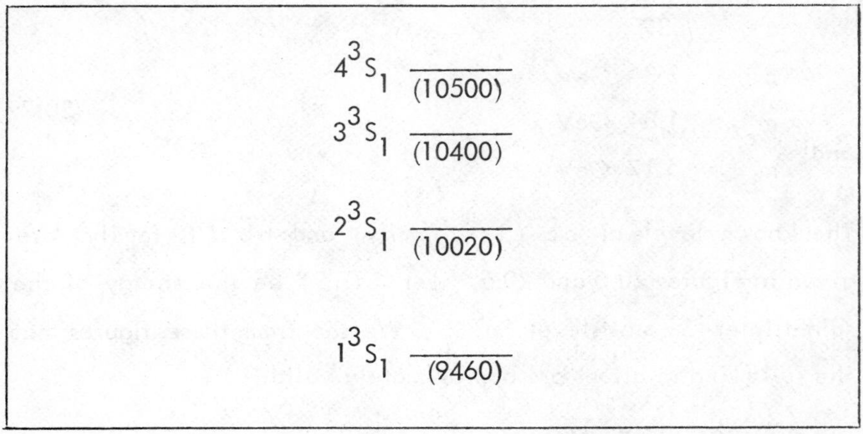

Fig. 20.6. Energy levels of the Υ family.

where the subscript ψ denotes the $c\bar{c}$ system and Υ the $b\bar{b}$ system. This near equality between these level spacings is approximately satisfied by the potential (20.78), provided that the parameters are given by (20.79). The same data can also be explained by using a different form * of the confining potential; instead of being linear in r, it is $\propto \ln(r/\text{constant})$, for which we expect the level spacings to be independent of the quark mass, as in (20.80). Thus, these energy levels do not uniquely determine the shape of $U_c(r)$.

We now turn to the spin-dependent potential $U_s(\vec{r})$. The spacings between the ${}^3P_J$ states for different J depend on the spin-dependent potential. We define the fine-structure ratio

$$F \equiv \frac{E_\psi({}^3P_2) - E_\psi({}^3P_1)}{E_\psi({}^3P_1) - E_\psi({}^3P_0)} \tag{20.81}$$

---

* C. Quigg and J. L. Rosner, Phys. Lett. 71B, 153 (1977).

which has an experimental value $\cong \frac{1}{2}$. If we assume phenomenologically the potential (20.76) to be generated by a central potential times the product of Dirac matrices $\gamma_\mu(1) \cdot \gamma_\mu(2)$ of the two particles, as in a single vector-meson exchange, then in (20.77) $U_{\ell s}$, U_{ss} and U_t would be completely determined by the central potential U_c. We find

$$U_{\ell s} = \frac{3}{2m^2 r} \frac{d}{dr} U_c ,$$

$$U_{ss} = \frac{2}{3m^2} \nabla^2 U_c \tag{20.82}$$

and

$$U_t = \frac{1}{3m^2} \left(\frac{d^2}{dr^2} - \frac{1}{r} \frac{d}{dr} \right) U_c .$$

From these equations, it can be shown that for $U_c \propto r^n$ the ratio F is

$$\frac{2}{5} \frac{(13+n)}{(5-n)} ,$$

which gives

$$F = \frac{7}{5} \qquad \text{for} \qquad n = 1 ,$$

and

$$F = \frac{4}{5} \qquad \text{for} \qquad n = -1 ;$$

both are very different from the experimental value. Consequently, the simple assumption (20.82) seems to be inconsistent with the experimental value of $F \cong \frac{1}{2}$.

In (20.78) the short-range Coulomb part, $-\frac{4}{3}\alpha_c/r$, is presumably due to a single-gluon exchange; hence there is a related spin-dependent potential given by (20.82). On the other hand, the confining potential is due to multi-gluon exchange. Thus, it is not surprising that the long-range component of the potential may deviate from (20.82); it should contain at least one part that transforms phenomenologically like a single-scalar exchange.* The relation between the

(footnote on page following)

two phenomenological potentials U_c and U_s used in the nonrelativistic Schrödinger equation is consequently expected to be different from that of a single-vector exchange.

Other heavy-quark hadrons, such as the D mesons ($c\bar{u}$, $c\bar{d}$ and their charge conjugates) and the charmed baryons have also been observed; these are given in the Appendix.

References.

R. Chand, ed., Symmetries and Quark Models (New York, Gordon and Breach, 1970).

I. M. Barbour and A. T. Davies, eds., Fundamentals of Quark Models (Edinburgh, The Scottish Universities Summer School in Physics, 1977).

G. Morpurgo, ed., Quarks and Hadronic Structure (New York, Plenum Press, 1977).

P. Hasenfratz and J. Kuti, Physics Reports 40C, 75 (1978).

C. Quigg and J. L. Rosner, Physics Reports 56C, 167 (1979).

Our discussions of the light-quark hadrons follow closely the papers of R. Friedberg and T. D. Lee, Phys. Rev. D16, 1096 (1977) and D18, 2623 (1978).

---

* If we replace the vector-vector factor $\gamma_\mu(1) \cdot \gamma_\mu(2)$ by a scalar-scalar factor $1 \cdot 1$, then instead of (20.82), we have

$$U_{\ell s} = -\frac{1}{2m^2 r}\frac{d}{dr}U_c \quad \text{and} \quad U_{ss} = U_t = 0 \ .$$

Chapter 21

WEAK INTERACTIONS

While the phenomenon of β decay was discovered near the end of the last century, the notion that the weak interaction forms a separate field of physical forces evolved rather gradually. It became clear only after the discoveries of other weak reactions, such as μ decay, μ capture processes, etc., and the observation * that all these reactions can be described by approximately the same coupling constant. Up to the present, numerous different types of weak reaction have been observed, as can be seen from the Table of Particle Properties given in the Appendix.

In order to gain a perspective on the gradual evolution of the field, we shall in this chapter give a survey of the phenomenology of the weak interaction along more or less historical lines.

All the known weak interactions can be separated phenomenologically into three classes: purely leptonic, semileptonic and nonleptonic, depending on the proportion of leptons present in the reaction. So far, their amplitudes can be adequately described by the first-order matrix elements of the effective Lagrangian

---

* O. Klein, Nature 161, 897 (1948). T. D. Lee, R. Rosenbluth and C. N. Yang, Phys. Rev. 75, 9905 (1949). G. Puppi, Nuovo Cimento 6, 194 (1949). J. Tiomno and J. A. Wheeler, Revs. Mod. Phys. 21, 153 (1949).

$$\mathcal{L}_{eff} = \mathcal{L}_{lep} + \mathcal{L}_{semilep} + \mathcal{L}_{nonlep} \tag{21.1}$$

whose specific forms will be discussed in the following.

Throughout this chapter, we assume that all neutrino masses are zero and the conservation laws of lepton numbers L_e, L_μ and L_τ are valid; these assumptions are consistent with the existing experimental information. The present limits* on the masses of ν_e, ν_μ and ν_τ are, respectively, about 35 eV, 0.56 MeV and 0.25 GeV. Since very little is known about the weak decays of b-quark hadrons, we shall restrict our main discussion to hadrons that are composites of u, d, s and c quarks. Furthermore, CP and T violations will be regarded as due to an interaction distinct from the weak. Only in Section 21.10 when we discuss the Kobayashi-Maskawa model, shall we examine the weak interaction of the b-quark, and also explore the possibility of incorporating the nonconservation of T and CP into the weak hadron currents.

21.1 Purely Leptonic Interaction

1. Phenomenological Lagrangian

In (21.1), $\mathcal{L}_{lep}$ can be written as

$$\mathcal{L}_{lep} = \frac{G}{\sqrt{2}} (j_\lambda^+ j_\lambda^- + \tfrac{1}{2} j_\lambda^o j_\lambda^o) \tag{21.2}$$

where

$$j_\lambda^- = i \sum_\ell \psi_\ell^\dagger \gamma_4 \gamma_\lambda (1 + \gamma_5) \psi_{\nu_\ell} , \tag{21.3}$$

* Throughout this chapter, all experimental values given without a direct reference are taken from Review of Particle Properties, N. Barash-Schmidt et al., Revs. Mod. Phys. 52, No. 2, Part II, (1980) and/or Proceedings of the IX International Symposium on Lepton and Photon Interactions at High Energies, Fermilab, 1979, ed. T. B. W. Kirk and H. D. I. Abarbanel.

$$j_\lambda^+ = i \sum_\ell \psi_{\nu_\ell}^\dagger \gamma_4 \gamma_\lambda (1+\gamma_5)\, \psi_\ell$$

and

$$j_\lambda^o = i \sum_\ell [\psi_{\nu_\ell}^\dagger \gamma_4 \gamma_\lambda g_\nu (1+\gamma_5)\, \psi_{\nu_\ell} + \psi_\ell^\dagger \gamma_4 \gamma_\lambda (g_v + g_a \gamma_5)\, \psi_\ell] \tag{21.4}$$

in which the sum extends over the three known charged leptons $\ell = e, \mu$ and τ, G is the Fermi constant $\sim 10^{-5}/m_p^{\,2}$, and g_ν, g_v and g_a are real dimensionless constants. The superscripts $+$, $-$ and 0 denote the change of charge Q effected by the lepton current j_λ. For example, j_λ^- converts ν_ℓ to ℓ^- and the change

$$\Delta Q \equiv Q_{final} - Q_{initial}$$

is -1; similarly, $\Delta Q = +1$ for j_λ^+, and 0 for j_λ^o.

2. Muon decay

The best-analyzed current-current matrix element is μ decay

$$\mu^- \rightarrow e^- + \nu_\mu + \bar{\nu}_e$$

and

$$\mu^+ \rightarrow e^+ + \bar{\nu}_\mu + \nu_e \ .$$

The observed muon lifetime is related to the uncorrected Fermi constant

$$G_{uncor} \equiv (G_\mu)_{uncor}$$

and the electron and muon masses m_e and m_μ by (see Problem 21.1)

$$(\text{lifetime})_\mu^{-1} = (G_\mu)^2_{uncor}\, m_\mu^5 \left(1 - \frac{8m_e^2}{m_\mu^2}\right)(192\pi^3)^{-1} , \tag{21.5}$$

which gives, from the experimental muon lifetime,

$$(G_\mu)_{uncor} = (1.4320 \pm .0002) \times 10^{-49} \text{ erg cm}^3 \tag{21.6}$$

where the subscript μ reminds us that the value is determined from μ decay and the subscript uncor indicates the absence of radiative correction. Equation (21.5) can be regarded as the definition of $(G_\mu)_{uncor}$. For a four-fermion interaction (21.2)-(21.3), the radiative correction of μ decay can be calculated; * this gives the corrected value of the Fermi constant

$$G \equiv G_\mu = [1 + \frac{\alpha}{4\pi}(\pi^2 - \frac{25}{4})] \; (G_\mu)_{uncor}$$

$$= (1.4350 \pm .0002) \times 10^{-49} \text{ erg cm}^3 , \qquad (21.7)$$

or, in the natural units, $\cong 1.029 \times 10^{-5}/m_p^{\,2}$.

In Problem 21.1, the reader is asked to verify that the normalized e distribution in the rest system of a completely polarized muon is **

$$d^2 N_e = x^2 [\, 3 - 2x \pm \cos\theta(1 - 2x)] \;\; dx \, d\cos\theta \qquad (21.8)$$

where the upper sign is for e^- and the lower for e^+, m_e/m_μ is set to be zero as an approximation, the variables x and θ are given by

$$x = (\text{momentum of e}) \,/\, \tfrac{1}{2} m_\mu$$

and

θ = angle between the muon polarization and the momentum of e ,

so that

$$\int_0^1 dx \int_0^\pi \sin\theta \, d\theta \; \frac{d^2 N_e}{dx \, d\cos\theta} = 1 .$$

Equation (21.8) has been verified experimentally to great accuracy.

---

* T. Kinoshita and A. Sirlin, Phys.Rev. 113, 1652 (1959).
S. M. Berman, Phys.Rev. 112, 267 (1958).

** T. D. Lee and C. N. Yang, Phys.Rev. 105, 1671 (1957).

Remarks. (i) The unique spectrum and angular distribution (21.8) follow from the two-component theory and the lepton-conservation law; both are assumed to be valid in the Lagrangian density (21.2). Without these assumptions the most general four-fermion interaction with no derivative coupling would contain ten complex constants. The e distribution is, then, no longer unique. For example, the normalized spectrum would be given by

$$\frac{dN_e}{dx} = 6x^2 \left[\left(2 - \frac{4}{3}\rho\right) - \left(2 - \frac{16}{9}\rho\right) x \right] \tag{21.9}$$

where ρ is called the Michel parameter* which can be any real number between 0 and 1. [Equation (21.8) corresponds to $\rho = \frac{3}{4}$.] The spectrum (21.9) is given in Figure 21.1 for several different ρ . As we can see, ρ measures the height of the end-point, at $x = 1$.

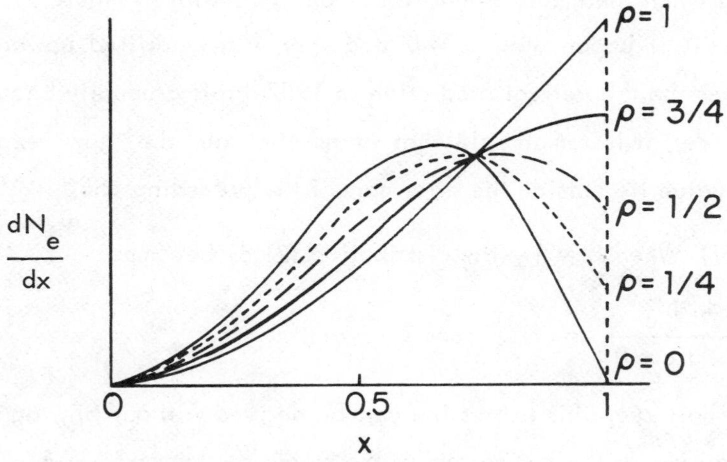

Fig. 21.1. Spectrum (21.9) for different ρ values.

---

* L. Michel, Proc.Roy.Soc. (London) A63, 514 (1950).

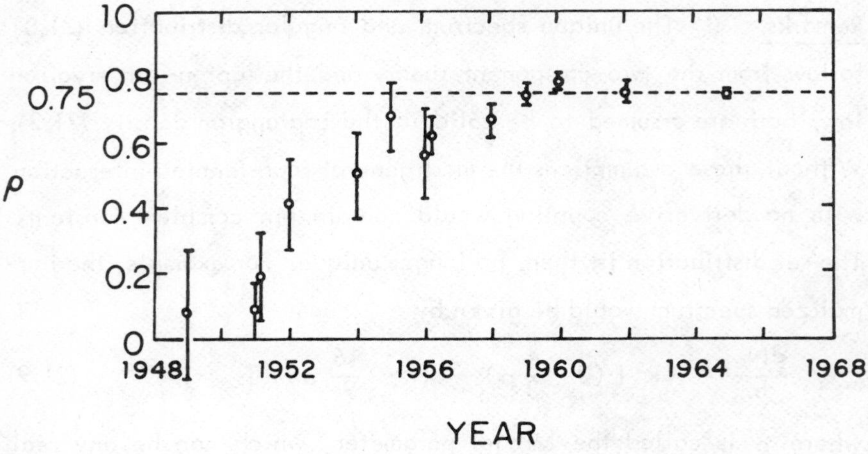

Fig. 21.2. Experimental determination of the Michel parameter ρ versus time.

It is instructive to plot the experimental value of ρ against the year when the measurement was made. As shown in Figure 21.2, historically it began with $\rho \cong 0$ and then slowly drifted upwards; only after the theoretical prediction in 1957 did it gradually become $\rho = \frac{3}{4}$. Yet, it is remarkable that at no time did the "new" experimental value lie outside the error bars of the preceding one.

(ii) When $x = 1$, the distribution (21.8) becomes

$$\frac{d^2 N_e}{dx\, d\cos\theta} = 1 \mp \cos\theta \quad .$$

As we shall see, this expression can be derived without any actual computation. In the rest system of the muon, the momentum of e, $\vec{p}_e$, has its maximum magnitude at $x = 1$. The neutrino and antineutrino momenta $\vec{p}_\nu$ and $\vec{p}_{\bar\nu}$ must therefore be parallel to each other, but antiparallel to $\vec{p}_e$. Because ν is lefthanded and $\bar\nu$ righthanded, when these two particles are moving in the same direction, they transform together like a spin-0 particle with a total momentum

$$\vec{p}_{\nu} + \vec{p}_{\bar{\nu}} = -\vec{p}_e$$

under a Lorentz transformation. Since μ and e are both of spin-$\frac{1}{2}$, as in (13.101), the angular distribution of e must be a linear function of $\cos\theta$.

In the μ^- decay, the final e^- is lefthanded. From Figure 21.3, we see that the angular momentum conservation requires $\theta = 0$ to be forbidden, but $\theta = \pi$ is allowed. Thus, the angular distribution of e^- has to be $1 - \cos\theta$; likewise for e^+, it must be $1 + \cos\theta$.

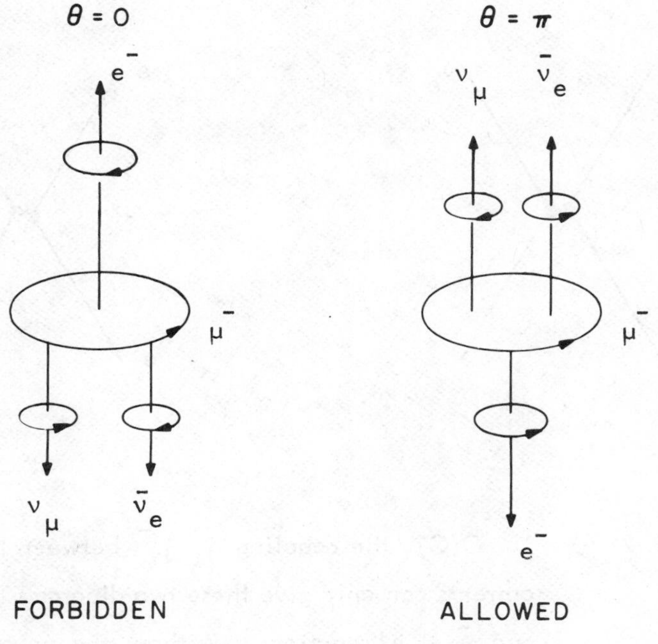

Fig. 21.3. The curled arrows indicate the directions of spin-rotation in μ^- decay.

3. <u>Neutral lepton current</u>

Next, let us consider the scattering of e^- by ν or $\bar{\nu}$. Throughout we use ν without any subscript for the generic terms ν_e, ν_μ and ν_τ, and we call $i\psi_\ell^\dagger \gamma_4 \gamma_\lambda (1+\gamma_5)\psi_{\nu_\ell}$ simply the $\bar{\ell}\nu_\ell$ current. From (21.3) and Figure 21.4, we see that in the product of the charged currents $j_\lambda^+ j_\lambda^-$, the square of the $\bar{e}\nu_e$ current allows

$$\nu_e + e^- \to \nu_e + e^- , \quad \bar{\nu}_e + e^- \to \bar{\nu}_e + e^-$$

and the cross term between the $\bar{e}\nu_e$ current and $\bar{\mu}\nu_\mu$ current gives

$$\nu_\mu + e^- \to \nu_e + \mu^- .$$

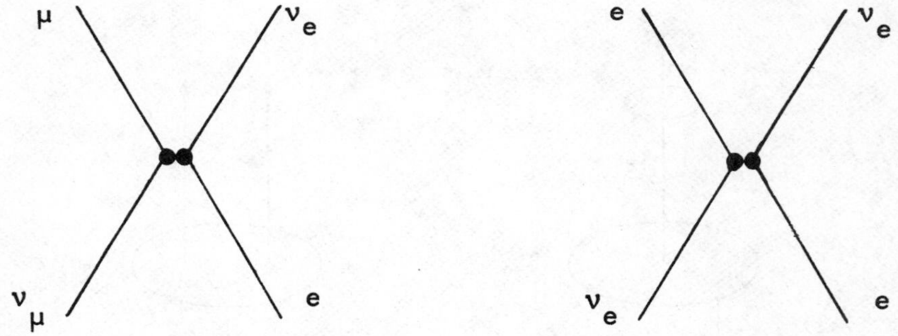

Fig. 21.4. To O(G), the coupling $j_\lambda^+ j_\lambda^-$ between the charged currents can only give these two diagrams for $\nu + e$ and $\bar{\nu} + e$ reactions. Without the neutral current, $\nu_\mu + e \to \nu_\mu + e$ and $\bar{\nu}_\mu + e \to \bar{\nu}_\mu + e$ would be forbidden.

However, to the same order in G, the elastic scattering

$$\left.\begin{matrix}\nu_\mu \\ \bar\nu_\mu\end{matrix}\right\} + e^- \rightarrow \left.\begin{matrix}\nu_\mu \\ \bar\nu_\mu\end{matrix}\right\} + e^-$$

can only be generated by the product of the neutral current $j_\lambda^o\, j_\lambda^o$.

By using (21.4) and Problem 5.4, we see that * the total cross sections of

$$\nu_\mu + e^- \rightarrow \nu_\mu + e^-$$

and

$$\bar\nu_\mu + e^- \rightarrow \bar\nu_\mu + e^-$$

are given by, after neglecting $O(m_e/E_\nu)$ as compared to 1,

$$\sigma(\nu_\mu e^-) = \frac{G^2 m_e E_\nu}{2\pi} |g_v|^2 \left(|g_v + g_a|^2 + \tfrac{1}{3}|g_v - g_a|^2\right)$$

and (21.10)

$$\sigma(\bar\nu_\mu e^-) = \frac{G^2 m_e E_\nu}{2\pi} |g_v|^2 \left(|g_v - g_a|^2 + \tfrac{1}{3}|g_v + g_a|^2\right)$$

where E_ν is the neutrino energy in the laboratory frame. From (21.7) we have

$$\frac{G^2 m_e E_\nu}{2\pi} = 4.3 \times 10^{-42} (E_\nu/\text{GeV}) \text{ cm}^2 .$$

Both $\nu_\mu e^-$ and $\bar\nu_\mu e^-$ elastic scattering events have been observed. The experimental values of their cross sections are

$$\sigma(\nu_\mu e^-) = (1.6 \pm .4) \times 10^{-42} (E_\nu/\text{GeV}) \text{ cm}^2$$

and (21.11)

$$\sigma(\bar\nu_\mu e^-) = (1.3 \pm .6) \times 10^{-42} (E_\nu/\text{GeV}) \text{ cm}^2 ,$$

---

* To derive (21.10), we first replace C_V and C_A in (5.128) by $g_\nu g_v G$ and $g_\nu g_a G$, then note that in Problem 5.4 the (center-of-mass energy)^2 is $2E_\nu m + m^2 \cong 2E_\nu m$ where m becomes m_e in the present case, and finally, integrate the variable y from 0 to 1. Further discussion will be given in Chapter 22; see (22.87) and (22.92).

which establish the existence of the weak neutral current interaction and imply that the g's are O(1).

The $\tau^{\pm}$ was discovered by Perl et al. in 1975 through the pair-production $e^{+} + e^{-} \rightarrow \tau^{+} + \tau^{-}$ and the decays

$$\tau^{\pm} \rightarrow \mu^{\pm} + \begin{cases} \bar{\nu}_{\tau} + \nu_{\mu} \\ \nu_{\tau} + \bar{\nu}_{\mu} \end{cases} ,$$

$$\tau^{\pm} \rightarrow e^{\pm} + \begin{cases} \bar{\nu}_{\tau} + \nu_{e} \\ \nu_{\tau} + \bar{\nu}_{e} \end{cases} .$$

As discussed in Section 10.4, from the high-energy neutrino reactions we know that $\nu_{\mu} + n \not\rightarrow e^{-} + p$ and, therefore, $\nu_{\mu} \neq \nu_{e}$. Likewise, it is found that

$$\nu_{\mu} + n \not\rightarrow \tau^{-} + p ,$$

and this leads to the conclusion

$$\nu_{\mu} \neq \nu_{\tau} . \tag{21.12}$$

While it seems reasonable to expect ν_e to be different from ν_τ, at present there is still a need for direct experimental evidence. If $\nu_e \neq \nu_\tau$, then there would be conservation of a third lepton number L_τ, in addition to L_e and L_μ discussed before. The assignment of L_τ is (similar to L_e and L_μ given on page 211)

	τ^{-}	τ^{+}	ν_{τ}	$\bar{\nu}_{\tau}$	other particles
L_τ =	+ 1	− 1	+ 1	− 1	0

.

21.2 Phenomenological Lagrangian for the Semileptonic Interaction

The semileptonic weak reactions can be described by the phenomenological Lagrangian

$$\mathcal{L}_{semilep} = \frac{G}{\sqrt{2}} (J_\lambda^+ j_\lambda^- + J_\lambda^- j_\lambda^+ + J_\lambda^o j_\lambda^o) \tag{21.13}$$

where G is again the Fermi constant, $j_\lambda^{\pm,o}$ denotes the same lepton current given by (21.3)-(21.4), and $J_\lambda^{\pm,o}$ the corresponding hadron current. As before, the superscripts $\pm$ and o refer to the charge change

$$\Delta Q = Q_{final} - Q_{initial} . \tag{21.14}$$

The hermiticity of the Lagrangian requires

$$(J_\lambda^\pm)^\dagger = \begin{cases} +J_\lambda^\mp & \text{if} \quad \lambda \neq 4 \\ -J_\lambda^\mp & \text{if} \quad \lambda = 4 \end{cases}$$

and

$$(J_\lambda^o)^\dagger = \begin{cases} +J_\lambda^o & \text{if} \quad \lambda \neq 4 \\ -J_\lambda^o & \text{if} \quad \lambda = 4 \end{cases} \tag{21.15}$$

where the $\pm$ sign is connected with our convention $x_4 = it$.

The current $J_\lambda^{\pm,o}$ can be decomposed into a sum of a vector- and an axial-vector part *

$$J_\lambda^{\pm,o} = V_\lambda^{\pm,o} + A_\lambda^{\pm,o} . \tag{21.16}$$

---

* R. Feynman and M. Gell-Mann, Phys.Rev. 109, 193 (1958);
R. E. Marshak and E. C. G. Sudarshan, ibid., 1860;
J. J. Sakurai, Nuovo Cimento 7, 649 (1958). In (21.16), the vector and axial-vector currents are classified by using the parity operator determined by the strong and electromagnetic interactions.

Alternatively, we may also classify $J_\lambda^{\pm,o}$ according to its selection rule with respect to the strangeness change

$$\Delta S = S_{final} - S_{initial} , \qquad (21.17)$$

and write

$$J_\lambda^{\pm,o} = [J_\lambda^{\pm,o}]_{\Delta S=0} + [J_\lambda^{\pm,o}]_{\Delta S \neq 0} . \qquad (21.18)$$

From our analysis of the $K - \bar{K}$ system we know that, in accordance with (15.49), the weak-interaction Lagrangian satisfies

$$\Delta S \neq \pm 2 .$$

Both the $\Delta S = 0$ and ± 1 components exist, as exemplified by the π and K decays that will be discussed in the next section.

21.3 $\pi_{\ell 2}$ and $K_{\ell 2}$ Decays

1. Pion decay

The two-body lepton decay

$$\pi^{\pm} \rightarrow \ell^{\pm} + \nu_\ell \;\; (\text{or } \bar{\nu}_\ell)$$

depends only on the $\Delta S = 0$ part of the charged current $J_\lambda^{\pm}$. As we shall now show, in the decomposition (21.16) the matrix element of $V_\lambda^{\pm}(x)$ between the pion and the vacuum must be zero; i.e.,

$$< vac \mid V_\lambda^{\pm}(x) \mid \pi^{\mp} > = 0 .$$

In the rest frame of π, this equation holds for $\lambda = 1, 2, 3$ because of rotational invariance and the zero-spin nature of π; it holds for $\lambda = 4$ because of space inversion and the fact that π is a pseudoscalar (under the strong and electromagnetic interactions).

According to (11.80), the x dependence of the matrix element of $A_\lambda^\pm(x)$ is known:

$$< \text{vac} \mid A_\lambda^\pm(x) \mid \pi^\mp > \;=\; < \text{vac} \mid A_\lambda^\pm(0) \mid \pi^\mp > e^{ik\cdot x}$$

where k is the 4-momentum of the pion state and, as usual,

$$k \cdot x = k_\lambda x_\lambda \quad .$$

Since π is of spin-0, we have from Lorentz invariance

$$< \text{vac} \mid A_\lambda^\pm(0) \mid \pi^\mp > \;\propto\; k_\lambda \quad ;$$

this follows because there is only one 4-vector, k, in the problem. Hence, the proportionality constant in the above expression is the only parameter that the $\pi^\pm_{\ell 2}$ decay amplitude can depend on. For reasons that will become clear, this constant will be written as the product $F_\pi \cos\theta_c$ where F_π is called the charged pion-decay constant and θ_c the Cabibbo angle. [See Section 21.5.] We may write *

$$< \text{vac} \mid A_\lambda^\pm(x) \mid \pi^\mp > \;=\; i \cos\theta_c F_\pi k_\lambda \frac{1}{\sqrt{2\omega\Omega}} e^{ik\cdot x} \qquad (21.19)$$

where ω is the pion energy ($k_4 = i\omega$) and Ω is the volume of the system.

The decay rate of $\pi^\pm_{\ell 2}$ can be readily computed by using (21.13), (21.16) and (21.19); the result is

$$\text{Rate}\,(\pi_{\ell 2}) \;=\; \frac{G^2 F_\pi^2 m_\ell^2}{8\pi m_\pi^3} \left(m_\pi^2 - m_\ell^2\right)^2 \cos^2\theta_c \, . \qquad (21.20)$$

---

* The factor $(2\omega\Omega)^{-\frac{1}{2}}$ in (21.19) is due to our convention of quantization inside a large but finite volume Ω. If we introduce a phenomenological pseudoscalar field $\phi_\pi(x)$ for the pion, then according to (2.20)

$$< \text{vac} \mid \phi_\pi(x) \mid \pi > \;=\; \frac{1}{\sqrt{2\omega\Omega}} e^{ik\cdot x} \qquad (21.19a)$$

which has the same factor.

[See Problem 21.2.] From the observed pion lifetime

$$\tau_\pi = (2.6030 \pm .0023) \times 10^{-8} \text{ sec}$$

and (21.7), we determine

$$F_\pi \cos\theta_c \cong 128 \text{ MeV} \quad . \tag{21.21}$$

The theoretical ratio of $\pi^\pm_{e2}$ and $\pi^\pm_{\mu 2}$ decay rates is

$$\frac{\text{Rate } (\pi_{e2})}{\text{Rate } (\pi_{\mu 2})} = \frac{m_e^2 (m_\pi^2 - m_e^2)^2}{m_\mu^2 (m_\pi^2 - m_\mu^2)^2} \cdot 0.965 = 1.23 \times 10^{-4} \tag{21.22}$$

where the factor 0.965 is due to a radiative correction of -3.5%. It is straightforward to show that the corresponding ratio of the phase space for these two decays is

$$\frac{\text{phase space } (\pi_{e2})}{\text{phase space } (\pi_{\mu 2})} = \frac{(m_\pi^2 - m_e^2)^2 (m_\pi^2 + m_e^2)}{(m_\pi^2 - m_\mu^2)^2 (m_\pi^2 + m_\mu^2)} \cong 3.5 \quad .$$

The smallness of the ratio 1.23×10^{-4} in (21.22) is due to the lepton current $j_\lambda^\pm$ which requires the helicity of $\bar{\nu}_\ell$ to be always $\frac{1}{2}$ (i.e., righthanded circular polarization); if the lepton mass m_ℓ were 0, it would also require the helicity of ℓ^- to be $-\frac{1}{2}$ (i.e., lefthanded circular polarization). On the other hand, from Figure 21.5 we see that in the decay

$$\pi^- \to \ell^- + \bar{\nu}_\ell \, ,$$

because of angular momentum conservation $\bar{\nu}_\ell$ and ℓ^- must have the same circular polarization. Thus, if $m_\ell = 0$, the decay $\pi_{\ell 2}$ would be forbidden. For $m_\ell \neq 0$, the probability of ℓ^- being of righthanded circular polarization in the rest frame of π is

$$1 - v_\ell = \frac{2m_\ell^2}{m_\pi^2 + m_\ell^2}$$

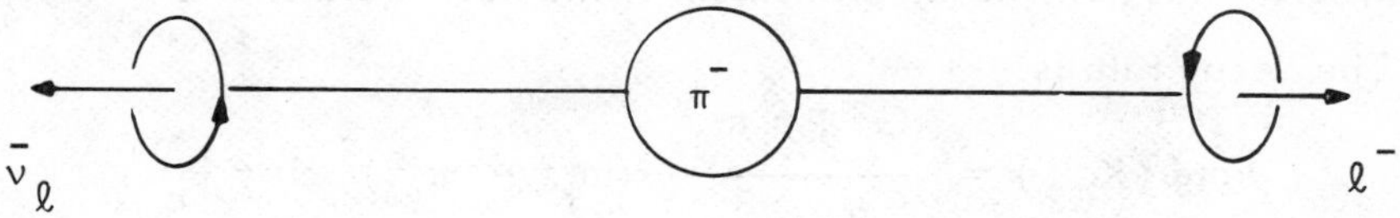

Fig. 21.5. In the rest system of π^-, angular momentum conservation requires the helicities of ℓ^- and $\bar{\nu}_\ell$ to be the same.

where v_ℓ is the velocity of ℓ^- in the rest frame of π^-. This factor $1 - v_\ell$ is much larger for μ. Multiplying the above phase-space ratio by the ratio of $1 - v_\ell$, we obtain the result 1.23×10^{-4} given by (21.22), as was first pointed out by Finkelstein and Ruderman.* The experimental verification of this small ratio supports the V − A nature of the weak interaction current. It also gives a sensitive test of the symmetry between e and μ currents; i.e., the Lagrangian is invariant under the interchange

$$e \rightleftarrows \mu \qquad \text{and} \qquad \nu_e \rightleftarrows \nu_\mu \ .$$

2. Kaon decay

Exactly the same analysis applies to the decay

$$K^\pm \to \ell^\pm + \nu_\ell \ \ (\text{or } \bar{\nu}_\ell) \ .$$

Just as in (21.19), the kaon decay amplitude depends only on a single constant, which will be written as a product $F_K \sin\theta_c$ where F_K is the kaon-decay constant and θ_c is, as before, the Cabibbo angle. We write

$$< \text{vac} \,|\, A_\lambda^\pm \,|\, K^\mp > \ = \ i \sin\theta_c \, F_K \, k_\lambda \, \frac{1}{\sqrt{2\omega\Omega}} \, e^{ik\cdot x} \tag{21.23}$$

* R. Finkelstein and M. A. Ruderman, Phys. Rev. 76, 1458 (1949).

where k is now the 4-momentum of the kaon state and $\omega = -i k_4$. The decay rate is

$$\text{Rate}(K_{\ell 2}) = \frac{G^2 F_K^2 m_\ell^2}{8\pi m_K^3} (m_K^2 - m_\ell^2)^2 \sin^2\theta_c \ .$$

From the experimental decay rate, we find

$$F_K \sin\theta_c \cong 35 \text{ MeV} \ . \qquad (21.24)$$

As we shall discuss in Section 21.5, the Cabibbo theory requires

$$F_K = F_\pi$$

and therefore

$$\tan\theta_c \cong .27 \ . \qquad (21.25)$$

However, because of the recently discovered b-quark hadron, we expect a modification of the Cabibbo theory, which will be discussed in Section 21.10. [Further analysis of the semileptonic decay amplitudes of π and K will be given in Chapter 24.]

21.4 Classical (Nuclear) β decay

Among the various weak-interaction hadron currents, the most extensively studied is the one responsible for the nuclear β decay. In these transitions we have $\Delta S = 0$ and $\Delta Q = \pm 1$. We shall now survey the main properties of the relevant current, which will simply be labeled as $J_\lambda^\pm$ with the subscript $\Delta S = 0$ omitted.*

1. Charge symmetry

As shown in (21.15), the currents J_λ^+ and J_λ^- are related through the Hermitian conjugation. The charge-symmetry property requires that they also be connected by a 180° rotation in the isospin

---

* Cf. the notation used in (21.18).

space:

$$e^{-i\pi I_2}\, J_\lambda^{\pm}\, e^{i\pi I_2} = - J_\lambda^{\mp} \quad , \tag{21.26}$$

where the operator exp $(i\pi I_2)$ is defined by (11.49)- (11.50).

2. First and second class currents

Assuming invariance under time reversal T , we have

$$T\, J_\lambda^{\pm}\, T^{-1} = - J_\lambda^{\pm} \quad . \tag{21.27}$$

Consider now the product of T and the charge - symmetry operator,

$$e^{-i\pi I_2}\, T \quad . \tag{21.28}$$

From (21.26)-(21.27), we have

$$e^{-i\pi I_2}\, T\, J_\lambda^{\pm}\, T^{-1}\, e^{i\pi I_2} = J_\lambda^{\mp} \quad . \tag{21.29}$$

Any $J_\lambda^{\pm}$ that satisfies this equality will be called the "first class" current independently of its charge-, or T- , symmetry property.

There is an alternative way * to express the same property. By using the G- parity operator given by (11.48),

$$G = C \cdot e^{i\pi I_2} \quad ,$$

we can always decompose

$$J_\lambda^{\pm} = V_\lambda^{\pm} + A_\lambda^{\pm}$$

into the following sum, without any symmetry assumption:

$$J_\lambda^{\pm} = (J_\lambda^{\pm})_1 + (J_\lambda^{\pm})_2$$

where

$$(J_\lambda^{\pm})_1 \equiv \tfrac{1}{2}(V_\lambda^{\pm} + G\, V_\lambda^{\pm}\, G^{-1}) + \tfrac{1}{2}(A_\lambda^{\pm} - G\, A_\lambda^{\pm}\, G^{-1}) \tag{21.30}$$

---

* S. Weinberg, Phys. Rev. 112, 1375 (1958).

is the first class current, and

$$(J_\lambda^\pm)_2 \equiv \tfrac{1}{2}(V_\lambda^\pm - G V_\lambda^\pm G^{-1}) + \tfrac{1}{2}(A_\lambda^\pm + G A_\lambda^\pm G^{-1})$$

is called the second class current. By using the CPT theorem, we have, in accordance with (14.19),

$$(CPT)\ V_\lambda^\pm\ (CPT)^{-1} = -(V_\lambda^\pm)^\dagger = \begin{cases} -V_\lambda^\mp & \text{if } \lambda \neq 4 \\ V_\lambda^\mp & \text{if } \lambda = 4 \end{cases}$$

and

$$(CPT)\ A_\lambda^\pm\ (CPT)^{-1} = -(A_\lambda^\pm)^\dagger = \begin{cases} -A_\lambda^\mp & \text{if } \lambda \neq 4 \\ A_\lambda^\mp & \text{if } \lambda = 4 . \end{cases}$$

Since under space inversion

$$P V_\lambda^\pm P^{-1} = \begin{cases} -V_\lambda^\pm & \text{if } \lambda \neq 4 \\ V_\lambda^\pm & \text{if } \lambda = 4 \end{cases}$$

and

$$P A_\lambda^\pm P^{-1} = \begin{cases} A_\lambda^\pm & \text{if } \lambda \neq 4 \\ -A_\lambda^\pm & \text{if } \lambda = 4 , \end{cases}$$

it follows that $(J_\lambda^\pm)_1$ satisfies (21.29); i.e.,

$$e^{-i\pi I_2}\ T\ (J_\lambda^\pm)_1\ T^{-1}\ e^{i\pi I_2} = (J_\lambda^\mp)_1 ,$$

whereas (21.31)

$$e^{-i\pi I_2}\ T\ (J_\lambda^\pm)_2\ T^{-1}\ e^{i\pi I_2} = -(J_\lambda^\mp)_2 .$$

Thus, if both T invariance and charge symmetry hold, then the second class current must be zero,

$$(J_\lambda^{\pm})_2 = 0 \quad . \tag{21.32}$$

The question whether the second class current exists in the weak interaction or not has plagued the field of β decay for nearly a decade. Only very recently has it become clear that all the present experimental information is consistent with the absence of the second class current.*

3. CVC and the isotriplet current hypothesis

The conserved vector-current structure proposed by Feynman and Gell-Mann consists of two related parts:

(i) CVC, which states that the vector current is conserved, i.e.

$$\frac{\partial V_\lambda^{\pm}}{\partial x_\lambda} = 0 \quad , \tag{21.33}$$

and

(ii) the isotriplet current hypothesis, which states that (with the inclusion of the Cabibbo modification) the three current operators

$$\frac{V_\lambda^+}{\sqrt{2}\ \cos\theta_c} \quad , \quad (J_\lambda^{el})_{I=1} \quad \text{and} \quad \frac{V_\lambda^-}{\sqrt{2}\ \cos\theta_c} \tag{21.34}$$

transform, respectively, like the $I_3 = 1$, 0 and -1 members of a single $I = 1$ triplet, where $(J_\lambda^{el})_{I=1}$ is the isovector part of the hadronic electromagnetic current operator in units of e, and θ_c is the Cabibbo angle, to be discussed in Section 21.5. Here, we may view

---

* See especially pages 549-592 of Unification of Elementary Forces and Gauge Theories (Ben Lee Memorial Conference on Parity Nonconservation, Weak Neutral Currents and Gauge Theories, Fermi National Accelerator Laboratory, October 1977), ed. D. B. Cline and F. E. Mills (London, Harwood Academic Publishers, 1978).

the factor $(\cos\theta_c)^{-1}$ simply as a normalization constant, so that (21.34) holds at zero 4-momentum transfer. As shown in (21.50), the actual value of $\cos\theta_c$ is $\cong .974$.

It is important to observe that (ii) implies (i), but not vice versa. For example, we may consider a vector current

$$K_\mu \equiv \lambda \frac{\partial}{\partial x_\nu} \psi_n^\dagger \gamma_4 \sigma_{\mu\nu} \psi_p$$

where λ is a constant, ψ_n and ψ_p are field operators for n and p, and

$$\sigma_{\mu\nu} = -\tfrac{1}{2} i (\gamma_\mu \gamma_\nu - \gamma_\nu \gamma_\mu)$$

is given by (14.15). Since $\sigma_{\mu\nu}$ is an antisymmetric tensor, the current K_μ satisfies

$$\frac{\partial K_\mu}{\partial x_\mu} = 0 .$$

However, K_μ is totally unrelated to $(J_\lambda^{el})_{I=1}$.

The isotriplet current hypothesis can also be extended to the axial-vector current, in which case, because the electromagnetic current is a pure vector, we require simply that

$$A_\lambda^+ \quad \text{and} \quad A_\lambda^- \tag{21.35}$$

be the $I_3 = 1$ and -1 members of the same $I = 1$ triplet. From these properties, (21.34)-(21.35), the charge-symmetry condition (21.26) follows.

Exercise. Prove the following statements concerning the matrix elements of $V_\lambda \equiv V_\lambda^-$ and $A_\lambda \equiv A_\lambda^-$ between the physical nucleon states:

(a) From Lorentz invariance and space-inversion symmetry (of the strong interaction) we can write

$$< n \mid V_\lambda(x) \mid p >$$
$$= i\, u_n^\dagger \gamma_4 [\gamma_\lambda f_1 + i\,(n+p)_\lambda f_2 + i\,(n-p)_\lambda f_3]\; u_p\, e^{-iq\cdot x}$$

and (21.36)

$$< n \mid A_\lambda(x) \mid p >$$
$$= i\, u_n^\dagger \gamma_4 \gamma_5 [\gamma_\lambda h_1 + i\,(n-p)_\lambda h_2 + i\,(n+p)_\lambda h_3]\; u_p\, e^{-iq\cdot x}$$

where p_λ and n_λ denote the 4-momenta of the initial p and final n, u_p and u_n the corresponding Dirac c. number spinors,

$$q_\lambda = (n-p)_\lambda \;,$$

the γ_λ are the Dirac matrices given by (3.11), and f_i, h_i ($i = 1,2,3$) are six complex functions of q^2.

(b) If the charge-symmetry condition (21.26) holds, then f_1, f_2, h_1, h_2 are real and f_3, h_3 are imaginary.

(c) If T invariance holds, then all six f_i, h_i are real.

(d) If there exists only the first-class current (i.e., (21.32) is valid, or equivalently the symmetry condition (21.29) holds), then $f_3 = h_3 = 0$ but f_1, f_2, h_1 and h_2 can remain complex.

(e) If both T invariance and the charge-symmetry condition are valid, then f_1, f_2, h_1 and h_2 are real and $f_3 = h_3 = 0$. Of course, in this case there exists only the first class current.

(f) If CVC (21.33) holds, then $f_3 = 0$ but f_1 and f_2 can remain complex (provided we ignore the mass difference between n and p).

(g) If the isotriplet vector-current hypothesis is valid, then $f_3 = 0$ and f_1, f_2 are both real; furthermore, they are related to the isovector parts of the nucleon charge form factor F_Q and the magnetic moment form factor F_M by

$$f_1 = F_Q + (\mu_p - \mu_n) F_M$$

and (21.37)

$$f_2 = (m_p + m_n)^{-1} (\mu_p - \mu_n) F_M ,$$

where F_Q and F_M are real functions of q^2, normalized to unity at $q^2 = 0$, and

$$\mu_p \cong 1.79 \quad , \quad \mu_n \cong -1.90 \tag{21.38}$$

are the anomalous magnetic moments of p and n.

(h) In general, (21.36) can also be written as

$$< n \mid V_\lambda(x) \mid p > \\ = i u_n^\dagger \gamma_4 [\gamma_\lambda g_V + q_\mu \sigma_{\mu\lambda} g_M + i q_\lambda g_S] u_p e^{-iq\cdot x}$$

and (21.39)

$$< n \mid A_\lambda(x) \mid p > \\ = i u_n^\dagger \gamma_4 \gamma_5 [\gamma_\lambda g_A + i q_\lambda g_P + i (n+p)_\lambda g_E] u_p e^{-iq\cdot x}$$

where $\sigma_{\mu\lambda}$ is given by (14.15),

$$g_V + (m_n + m_p) g_M = f_1 , \quad g_M = f_2 , \quad g_S = f_3 ,$$
$$g_A = h_1 , \quad g_P = h_2 \quad \text{and} \quad g_E = h_3 . \tag{21.40}$$

4. Experimental verification

There exist several experimental verifications of the isotriplet vector-current hypothesis, and therefore also CVC, (21.33). Historically, these were the first quantitative tests linking the electromagnetic observables with the weak-interaction observables; they provided an important impetus to our search for a unifying theory.

(i) Pion β decay

From the isotriplet vector-current hypothesis, the matrix elements of

$$\pi^{\pm} \to \pi^{o} + e^{\pm} + \nu_e \ (\bar{\nu}_e)$$

and

$$\pi^{\pm} \to \pi^{\pm} + \gamma$$

are related. It is a good approximation to neglect the dependence on the 4-momentum transfer in the pion β decay; hence its amplitude is completely determined. The theoretical branching ratio is

$$b_{th} = \frac{\text{Rate } (\pi^{\pm} \to \pi^{o} e^{\pm} \nu)}{\text{Rate } (\pi^{\pm} \to \mu^{\pm} \nu)} = 1.07 \times 10^{-8} , \qquad (21.41)$$

which is in good agreement with the experimental value * of

$$b_{exp} = (1.02 \pm .07) \times 10^{-8} .$$

(ii) β decay of B^{12} and N^{12}

Another sensitive test is provided by the $\beta^{\pm}$ spectra of the decays of B^{12} and N^{12}. As shown in Figure 21.6, B^{12}, N^{12} and the excited state of C^{12*} form an isotriplet of spin-parity 1+. The decay of these three states into the groundstate of C^{12} gives an excellent opportunity to compare the $\beta^{\mp}$ and γ matrix elements.

For such a $1+ \to 0+$ transition, when the 4-momentum transfer $q_\lambda = 0$, the $\beta^{\mp}$ decay amplitude is determined only by the axial-vector current. However, to first order in q_λ, the vector current

---

* R. Bacastow, T. Elioff, R. Larsen, C. Wiegand and T. Ypsilantis, Phys. Rev. Lett. 9, 400 (1962); P. DePommier, J. Heintze, C. Rubbia and V. Soergel, Phys. Lett. 5, 61 (1963); A. F. Dunaitsev, V. I. Petrukhin, Yu. D. Prokoshkin and V. I. Rykalin, Proceedings of the International Conference on the Fundamental Aspects of Weak Interactions (1963).

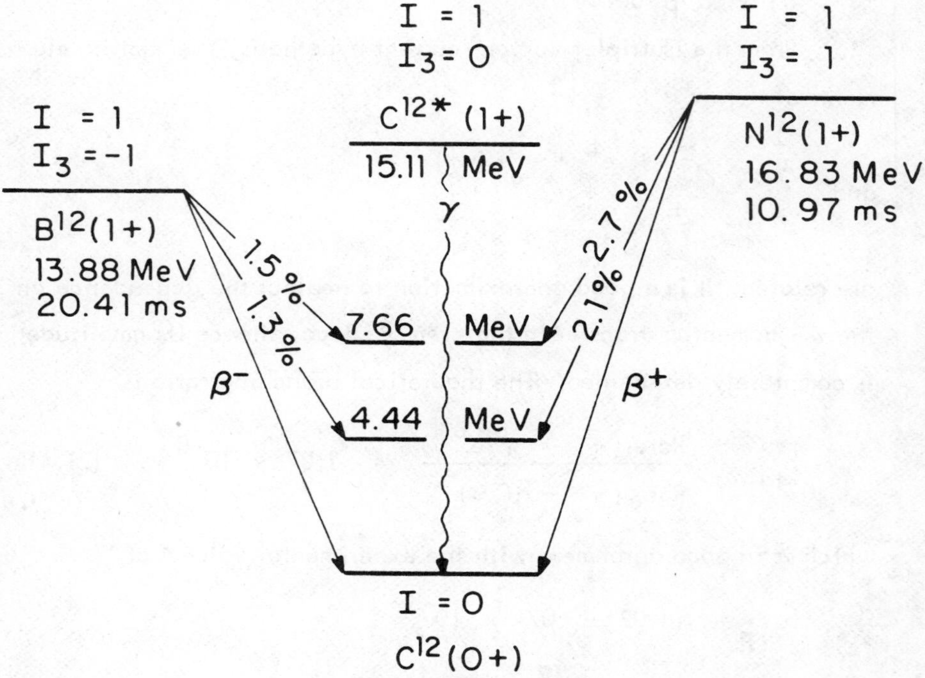

Fig. 21.6. Energy-level diagrams of $A = 12$ decays.

also contributes. Assuming the isotriplet current hypothesis and time-reversal invariance, we can set $g_S = 0$ in (21.39), i.e., no second class current. Hence, the matrix element of V_λ is

$$< n \mid V_\lambda(x) \mid p > \;=\; i\, u_n^\dagger\, \gamma_4 (\gamma_\lambda\, g_V + q_\mu\, \sigma_{\mu\lambda}\, g_M)\; u_p \quad .$$

In accordance with (21.37) and (21.40), when $q_\lambda \to 0$

$$g_M \;=\; (m_p + m_n)^{-1}\; (\mu_p - \mu_n)$$

which has an unusually large factor due to the anomalous magnetic moments of p and n , given by (21.38). Therefore, the q_λ -

dependence in this transition is dominated by the $q_\mu \sigma_{\mu\lambda} g_M$ term.

We may write the $\beta^{\mp}$ spectrum as a constant times

$$1 + a_{\mp} E$$

where E is the β energy in the laboratory frame. The isotriplet current hypothesis gives

$$a_{\mp} \cong \pm \frac{8}{3} \frac{\mu_p - \mu_n}{m_p + m_n} \left| \frac{G_V}{G_A} \right| \tag{21.42}$$

where G_V and G_A are the usual vector and axial-vector β decay constants. [See (21.45)-(21.46) below.] By using (21.42) and the experimental value $G_A/G_V \cong - 1.25$, we find

$$(a_- - a_+)_{th} \cong .84\,\%\,(\text{MeV})^{-1}$$

which is in good agreement with the experimental observation *

$$(a_- - a_+)_{exp} = .86 \pm .24\%\,(\text{MeV})^{-1} .$$

As a corollary, this agreement also supports the absence of the second class current.

5. Fermi constant in β decay

The phenomenological Lagrangian $\mathcal{L}_\beta$ for the nuclear β decay can be extracted from (21.13):

$$\mathcal{L}_\beta \equiv \frac{G}{\sqrt{2}} (J_\lambda^+ j_\lambda^- + J_\lambda^- j_\lambda^+) \tag{21.43}$$

where $j_\lambda^\pm$ contains only the $\ell = e$ part of the lepton current (21.3), and $J_\lambda^\pm$ refers to the $\Delta S = 0$ part of the hadron current in (21.18). As before, we write simply

$$J_\lambda^\pm = V_\lambda^\pm + A_\lambda^\pm ,$$

* C. S. Wu, Y. K. Lee and L. W. Mo, Phys. Rev. Lett. 39, 72 (1977).

with the subscript $\Delta S = 0$ omitted. Because the lepton current $j_\lambda^\pm$ is a direct observable, what can be measured from the hadron part are the products $G\,V_\lambda^\pm$ and $G\,A_\lambda^\pm$. Without the second class current, (21.39) becomes

$$< n \mid V_\lambda \mid p > \;=\; i\, u_n^\dagger \gamma_4 (\gamma_\lambda g_V + q_\mu \sigma_{\mu\lambda} g_M)\, u_p\, e^{-iq\cdot x}$$

and (21.44)

$$< n \mid A_\lambda \mid p > \;=\; i\, u_n^\dagger \gamma_4 (\gamma_\lambda g_A + i q_\lambda g_P)\, \gamma_5\, u_p\, e^{-iq\cdot x} \quad .$$

When the 4-momentum transfer $q_\lambda = 0$, the vector (Fermi) and the axial-vector (Gamow-Teller) coupling constants G_V and G_A are traditionally defined by

$$G_V \equiv G \cdot \lim_{q=0} g_V(q^2)$$

and (21.45)

$$G_A \equiv -G \cdot \lim_{q=0} g_A(q^2) \quad .$$

The present observed value for the ratio is

$$\frac{G_A}{G_V} = -1.253 \pm .007 \quad . \tag{21.46}$$

Historically the β decay amplitude was first analysed without the two-component neutrino theory and parity nonconservation; the factor $1/\sqrt{2}$ in (21.43) and the minus signs in (21.45)-(21.46) are reminders of this bygone period. [See (21.136) in Problem 21.4.]

As we shall see in the next section, the Cabibbo theory requires

$$\lim_{q=0} g_V(q^2) = \cos\theta_c \tag{21.47}$$

where θ_c is the same angle given by (21.25). Since according to (21.7) $G = G_\mu$, we have

$$G_V = G_\mu \cos\theta_c \quad . \tag{21.48}$$

Experimentally, G_μ can be determined by the μ decay and G_V by the nuclear β decay. Logically, it is simplest to regard (21.48) as the <u>definition</u> of $\cos\theta_c$. We note that this is the same definition provided by the normalization condition of the isotriplet vector-current hypothesis (21.34) at the zero 4-momentum transfer limit. The Cabibbo theory requires that the same θ_c can also be obtained by comparing either the $\pi_{\ell 2}$ and $K_{\ell 2}$ decay rates, or the nuclear and hyperon β decay rates. [However, as we shall discuss in Section 21.10, this particular aspect of the Cabibbo theory most likely requires some modification because of the recent discovery of the b-quark.]

In order to determine θ_c accurately from (21.48), it is necessary to examine the question of radiative correction.

First, we observe that the nucleons have strong interactions, but not the muons. From CVC we have, by integrating (21.33) and on account of Gauss' theorem,

$$\frac{\partial}{\partial t} \int V_4^{\pm}\, d^3r = -i \int \nabla_i V_i^{\pm}\, d^3r = 0$$

in which the surface term vanishes because there is no zero-mass hadron.* If we imagine that the strong interaction is turned on very slowly, the space integral $\int V_4\, d^3r$, and therefore G_V, must be unchanged. Hence, (21.48) is not affected by the strong interaction.

Next, we examine the effect of the electromagnetic interaction.** Let $(G_V)_{uncor}$ be the vector coupling in β decay without

---

* In QED, due to the zero mass of the photon, even though the electromagnetic current is conserved, there remains a substantial renormalization of the electric charge. Here, the "charge" G_V of the vector current is not renormalized.

** A. Sirlin, Nucl. Phys. B71, 21 (1974), B100, 291 (1975), Revs. Mod. Phys. 50, 573 (1978), and private communication.

the radiative correction. From the various β decay ft values (and especially that of Al^{26*}), we have

$$(G_V)_{uncor} = (1.42233 \pm .0005) \times 10^{-49} \text{ erg cm}^3 . \quad (21.49)$$

In the approximation that n and p can be represented by local field operators, a point-like four-fermion β decay interaction has a divergent radiative correction; the result is

$$(G_V)_{cor} = (G_V)_{uncor} \left[1 - \frac{\alpha}{4\pi} \left(6 \ln \frac{\lambda}{m_p} + 3 \ln \frac{m_p}{2E_m} + \frac{207}{20} - \frac{4\pi^2}{3} \right) \right]$$

where E_m is the maximum β energy and λ the ultraviolet cutoff parameter. However, if the weak interaction is mediated by the intermediate boson, then the radiative correction becomes finite. As shown by Sirlin, $\lambda = m_Z$ the mass of the neutral intermediate boson in a specific gauge model; the corrected value of θ_c as deduced from (21.48) with the appropriate radiative corrections of G_V and G_μ included is

$$\cos \theta_c = .974 \pm .002 \quad (21.50)$$

which, within the context of SU_3 symmetry as required by the Cabibbo theory, agrees reasonably well with that determined from $K_{\ell 2}$ and $\pi_{\ell 2}$, as can be seen from (21.25).

To see the importance of the radiative correction, we note that by using (21.6), (21.48) and (21.49), without the radiative correction we would have

$$(\sin \theta_c)_{uncor} \cong .12$$

which differs significantly from the above corrected value

$$\sin \theta_c = .228 \pm .01 \quad .$$

21.5 Cabibbo Theory (including the GIM modification)

The Cabibbo theory* in its original form concerned only the weak interaction of those hadrons which are now regarded as composites of u, d and s quarks. The modification due to Glashow, Iliopoulos and Maiani extended it to include the c quark.** The charged weak-hadron current operator $J_\lambda^{\pm}$ in the semileptonic Lagrangian (21.13) is assumed to be

$$J_\lambda^{+} = i\,\psi_u^{\dagger}\,\gamma_4\,\gamma_\lambda(1+\gamma_5)\,(\cos\theta_c\,\psi_d + \sin\theta_c\,\psi_s) + i\,\psi_c^{\dagger}\,\gamma_4\,\gamma_\lambda(1+\gamma_5)\,(-\sin\theta_c\,\psi_d + \cos\theta_c\,\psi_s)$$

and (21.51)

$$J_\lambda^{-} = i\,(\cos\theta_c\,\psi_d^{\dagger} + \sin\theta_c\,\psi_s^{\dagger})\,\gamma_4\,\gamma_\lambda(1+\gamma_5)\,\psi_u + i\,(-\sin\theta_c\,\psi_d^{\dagger} + \cos\theta_c\,\psi_s^{\dagger})\,\gamma_4\,\gamma_\lambda(1+\gamma_5)\,\psi_c\ ,$$

where θ_c is, as before, the Cabibbo angle. The operators ψ_u, ψ_d, ψ_s, ψ_c stand for the respective quark fields ψ_f^i (f = flavor and i = color = 1, 2, 3) but with the color superscript omitted, so that for any Dirac matrix Γ

$$\psi_u^{\dagger}\,\Gamma\,\psi_d \equiv \psi_u^{i\dagger}\,\Gamma\,\psi_d^{i}\ ,\quad \psi_s^{\dagger}\,\Gamma\,\psi_c \equiv \psi_s^{i\dagger}\,\Gamma\,\psi_c^{i}\ ,\tag{21.52}$$

etc. in which the repeated index i is summed over, as usual. The corresponding electromagnetic current operator (in units of e) is

$$J_\lambda^{el} = i\tfrac{2}{3}\,(\psi_u^{\dagger}\,\gamma_4\,\gamma_\lambda\,\psi_u + \psi_c^{\dagger}\,\gamma_4\,\gamma_\lambda\,\psi_c) - i\tfrac{1}{3}\,(\psi_d^{\dagger}\,\gamma_4\,\gamma_\lambda\,\psi_d + \psi_s^{\dagger}\,\gamma_4\,\gamma_\lambda\,\psi_s)\ .\tag{21.53}$$

---

* N. Cabibbo, Phys. Lett. 10, 513 (1963).

** S. L. Glashow, J. Iliopoulos and L. Maiani, Phys. Rev. D 2, 1258 (1970). Historically, this paper provided a strong theoretical reason for anticipating the existence of c quark hadrons several years before their actual experimental discovery. See Section 22.3.1.

[The inclusion of the b-quark will be discussed in Section 21.10.]

1. Nuclear β decay

The relevant vector and axial-vector current operators are

$$\begin{aligned} V_\lambda^+ &\equiv i\cos\theta_c\, \psi_u^\dagger \gamma_4 \gamma_\lambda \psi_d\ , \\ A_\lambda^+ &\equiv i\cos\theta_c\, \psi_u^\dagger \gamma_4 \gamma_\lambda \gamma_5 \psi_d \end{aligned} \tag{21.54}$$

and their Hermitian conjugates. The corresponding electromagnetic current operator can be separated into an isovector part

$$(J_\lambda^{el})_{I=1} = i\tfrac{1}{2}(\psi_u^\dagger \gamma_4 \gamma_\lambda \psi_u - \psi_d^\dagger \gamma_4 \gamma_\lambda \psi_d) \tag{21.55}$$

and an isoscalar part

$$\begin{aligned} (J_\lambda^{el})_{I=0} &= i\,\frac{1}{6}\,(\psi_u^\dagger \gamma_4 \gamma_\lambda \psi_u + \psi_d^\dagger \gamma_4 \gamma_\lambda \psi_d) \\ &\quad + i\,\tfrac{2}{3}\,\psi_c^\dagger \gamma_4 \gamma_\lambda \psi_c - i\,\tfrac{1}{3}\,\psi_s^\dagger \gamma_4 \gamma_\lambda \psi_s\ , \end{aligned} \tag{21.56}$$

so that their sum is

$$J_\lambda^{el} = (J_\lambda^{el})_{I=1} + (J_\lambda^{el})_{I=0}\ .$$

By using (21.51), we see that the Lagrangian $\mathcal{L}_{semilep}$, (21.13), satisfies T invariance. Assuming isospin invariance of the strong interaction, we find that $V_\lambda^\pm$ and $A_\lambda^\pm$, given by (21.54), satisfy charge-symmetry, the $|\Delta\vec{I}| = 1$ rule and the absence of the second class current condition. Furthermore, isospin symmetry implies that u and d quarks have the same mass; hence, V_λ satisfies CVC and the isotriplet current hypothesis.

2. $\pi_{\ell 2}$ and $K_{\ell 2}$ decays

The relevant current operators are

$$i\cos\theta_c\, \psi_u^\dagger \gamma_4 \gamma_\lambda \gamma_5 \psi_d \tag{21.57}$$

for $\pi^-_{\ell 2}$ decay, and

$$i \sin \theta_c \psi_u^\dagger \gamma_4 \gamma_\lambda \gamma_5 \psi_s \tag{21.58}$$

for $K^-_{\ell 2}$ decay. Thus, under the assumption of SU_3 symmetry the angle θ_c connects the decay rates of $\pi_{\ell 2}$ and $K_{\ell 2}$, which leads to (21.25).

3. Strangeness nonconserving currents

From (21.51), we see that the $\Delta S \neq 0$ hadron current is given by

$$i(\sin \theta_c \psi_u^\dagger + \cos \theta_c \psi_c^\dagger)\, \gamma_4 \gamma_\lambda (1+\gamma_5)\, \psi_s \tag{21.59}$$

and its Hermitian conjugate. Now the s quark is of strangeness $S=-1$ and charge $Q=-\frac{1}{3}$, and the u and c quarks are both of $S=0$ and $Q=\frac{2}{3}$. The above current, and therefore also $\mathcal{L}_{\text{semilep}}$, satisfy

$$|\Delta S| < 2$$

and

$$\Delta Q = \Delta S \quad . \tag{21.60}$$

In (21.59), the part that is responsible for the β decay of hyperons is

$$i \sin \theta_c \psi_u^\dagger \gamma_4 \gamma_\lambda (1+\gamma_5)\, \psi_s \quad .$$

Since the s quark is of isospin $I=0$ and the u quark of $I=\frac{1}{2}$, the hyperon β decay, and also the corresponding $\mathcal{L}_{\text{semilep}}$, satisfy

$$|\Delta \vec{I}| = \tfrac{1}{2} \quad . \tag{21.61}$$

Both (21.60) and (21.61) have been discussed in the previous chapters on symmetry violation. [See pages 243–247 and 367–370.]

4. β decay of the baryon octet

The relevant currents are

$$\begin{aligned}\mathcal{J}_\lambda^+(x) &\equiv i\psi_u^\dagger \gamma_4 \gamma_\lambda (1+\gamma_5)(\cos\theta_c \psi_d + \sin\theta_c \psi_s) \\ &\equiv \cos\theta_c [V_\lambda(x)_2^1 + A_\lambda(x)_2^1] + \sin\theta_c [V_\lambda(x)_3^1 + A_\lambda(x)_3^1] \end{aligned} \tag{21.62}$$

and its Hermitian conjugate. In the SU_3 tensor notation of Problem 12.2, we see that

$$
\begin{aligned}
V_\lambda(x)^1_2 &= i\,\psi_u(x)^\dagger\,\gamma_4\,\gamma_\lambda\,\psi_d(x) \quad , \\
A_\lambda(x)^1_2 &= i\,\psi_u(x)^\dagger\,\gamma_4\,\gamma_\lambda\,\gamma_5\,\psi_d(x) \quad , \\
V_\lambda(x)^1_3 &= i\,\psi_u(x)^\dagger\,\gamma_4\,\gamma_\lambda\,\psi_s(x) \\
\text{and} \qquad A_\lambda(x)^1_3 &= i\,\psi_u(x)^\dagger\,\gamma_4\,\gamma_\lambda\,\gamma_5\,\psi_s(x)
\end{aligned}
\tag{21.63}
$$

are members of the octet vector and octet axial-vector current operators. According to (21.55), the isovector part of the electromagnetic current is related to the same octet by

$$
(J_\lambda^{el}(x))_{I=1} = \tfrac{1}{2}[\,V_\lambda(x)^1_1 - V_\lambda(x)^2_2\,] \quad . \tag{21.64}
$$

As in (12.87), let us consider the zero 4-momentum transfer limit $q \to 0$. We may write, in expressions similar to (12.86),

$$
\lim_{q\to 0} \langle B^i_a \,|\, V_\lambda(x)^j_b \,|\, B^k_c \rangle = i(D'\, d^{ijk}_{abc} + F'\, f^{ijk}_{abc})\, u'^\dagger\, \gamma_4\, \gamma_\lambda\, u
$$

and

$$
\lim_{q\to 0} \langle B^i_a \,|\, A_\lambda(x)^j_b \,|\, B^k_c \rangle = i(D\, d^{ijk}_{abc} + F\, f^{ijk}_{abc})\, u'^\dagger\, \gamma_4\, \gamma_\lambda\, \gamma_5\, u
\tag{21.65}
$$

where u and u' are the initial and final c. number Dirac spinors, d^{ijk}_{abc} and f^{ijk}_{abc} are given by (12.82)-(12.83) and D, F, D', F' are constants. By definition, the diagonal matrix element of $(J_\lambda^{el})_{I=1}$ at $q = 0$ is proportional to the isovector part of the electric charge of the state. Thus, for example

$$
\lim_{q\to 0} \frac{\langle p \,|\, (J_\lambda^{el})_{I=1} \,|\, p \rangle}{\langle \Xi^- \,|\, (J_\lambda^{el})_{I=1} \,|\, \Xi^- \rangle} = -1 \quad . \tag{21.66}
$$

According to (12.48),

$$| p > \; = \; | B_3^{\;1} > \quad , \qquad < p \,| \; = \; < B_1^{\;3} \,| \quad ,$$

$$| \Xi^- > \; = \; | B_1^{\;3} > \quad \text{and} \quad < \Xi^- | \; = \; < B_3^{\;1} \,| \quad .$$

By using (21.64)-(21.65), we find that (21.66) can be written as

$$\frac{D'(d_{113}^{311} - d_{123}^{321}) + F'(f_{113}^{311} - f_{123}^{321})}{D'(d_{311}^{113} - d_{321}^{123}) + F'(f_{311}^{113} - f_{321}^{123})} = -1$$

which, because of (12.82)-(12.83), becomes

$$\frac{D' - F'}{D' + F'} = -1 \quad .$$

Thus, we obtain

$$D' = 0 \quad . \tag{21.67}$$

Likewise, on account of the fact that for the diagonal element

$$\lim_{q=0} < p \,| \, (J_\lambda^{el})_{I=1} \,| \, p > \; = \; \tfrac{1}{2}\, i\, u \;\; \gamma_4 \, \gamma_\lambda \, u \quad ,$$

we have

$$F' = -1 \quad . \tag{21.68}$$

Consider the β decay amplitude of

$$B' \rightarrow B + e^{\mp} + \bar{\nu}_e \;\; (\nu_e) \quad .$$

When the 4-momentum transfer $q \rightarrow 0$, we may write

$$\lim_{q \rightarrow 0} < B' \,| \, \mathcal{J}_\lambda^{\mp} \,| \, B > \; \equiv \; i\, u'^{\dagger} \, \gamma_4 \, M u \tag{21.69}$$

where M is a 4×4 matrix depending on B, B' and the subscript λ. Through (21.62)-(21.68) and Problem 12.2, M can be expressed in terms of θ_c and the two constants D and F. The results are given in the following table.

Reaction	M
$n \to p e^- \bar{\nu}_e$	$\cos\theta_c\,[\,\gamma_\lambda + (D-F)\,\gamma_\lambda\gamma_5\,]$
$\Sigma^- \to \Lambda^o e^- \bar{\nu}_e$	$\cos\theta_c\,[\,-\sqrt{\tfrac{2}{3}}\;D\,\gamma_\lambda\gamma_5\,]$
$\Sigma^+ \to \Lambda^o e^+ \nu_e$	$\cos\theta_c\,[\,-\sqrt{\tfrac{2}{3}}\;D\,\gamma_\lambda\gamma_5\,]$
$\Sigma^- \to \Sigma^o e^- \bar{\nu}_e$	$\cos\theta_c\,[\,\sqrt{2}\;\gamma_\lambda - \sqrt{2}\;F\,\gamma_\lambda\gamma_5\,]$
$\Xi^- \to \Xi^o e^- \bar{\nu}_e$	$\cos\theta_c\,[\,-\gamma_\lambda + (D+F)\;\gamma_\lambda\gamma_5\,]$
$\Lambda^o \to p e^- \bar{\nu}_e$	$\sin\theta_c\,[\,\sqrt{\tfrac{3}{2}}\;\gamma_\lambda + \sqrt{\tfrac{1}{6}}\;(D-3F)\,\gamma_\lambda\gamma_5\,]$
$\Sigma^- \to n e^- \bar{\nu}_e$	$\sin\theta_c\,[\,-\gamma_\lambda + (D+F)\;\gamma_\lambda\gamma_5\,]$
$\Xi^- \to \Lambda^o e^- \bar{\nu}_e$	$\sin\theta_c\,[\,-\sqrt{\tfrac{3}{2}}\;\gamma_\lambda + \sqrt{\tfrac{1}{6}}\;(D+3F)\,\gamma_\lambda\gamma_5\,]$
$\Xi^- \to \Sigma^o e^- \bar{\nu}_e$	$\sin\theta_c\,[\,\sqrt{\tfrac{1}{2}}\;\gamma_\lambda + \sqrt{\tfrac{1}{2}}\;(D-F)\,\gamma_\lambda\gamma_5\,]$
$\Xi^o \to \Sigma^+ e^- \bar{\nu}_e$	$\sin\theta_c\,[\,\gamma_\lambda + (D-F)\;\gamma_\lambda\gamma_5\,]$

Table 21.1. A table of M, defined by (21.69).

By comparing these theoretical expressions with the experimental results of the nuclear and various hyperon β decays, listed in the Appendix, we find the constants

$$D \cong .82\,, \qquad F \cong -.43$$

and

$$\sin\theta_c = .219 \pm .011\;. \tag{21.70}$$

The last equation agrees very well with (21.50). [See (21.133) for its modification due to the b-quark current.]

Historically, much of the experimental evidence for the validity of SU_3 flavor symmetry of the strong interaction came from the success of the Cabibbo theory for these weak decays.

5. Leptonic decay of the D mesons

The D^o and D^+ mesons are $c\bar{u}$ and $c\bar{d}$ composites, and $\bar{D}^o$ and D^- are their conjugate states. The lepton decay of the D mesons can occur via

$$c \rightarrow d + \ell^+ + \nu_\ell$$

or

$$c \rightarrow s + \ell^+ + \nu_\ell$$

with a rate ratio $\cong \tan^2 \theta_c \cong 5\%$. Because K^- and $\bar{K}^o$ are $s\bar{u}$ and $s\bar{d}$ composites, we expect

$$\frac{\text{rate } (D \rightarrow \ell^+ + \nu_\ell + \bar{K} + \cdots)}{\text{rate } (D \rightarrow \ell^+ + \nu_\ell + \cdots)} \cong 95\% \tag{21.71}$$

where $\cdots$ refers to anything, D denotes D^+ or D^o and $\bar{K}$ represents K^- and $\bar{K}^o$. At present, there are too few leptonic decay events of D mesons to test this prediction.

21.6 High-energy Neutrino Reaction

1. Kinematics

Consider the collision of a high-energy neutrino (or antineutrino) on a nucleon N :

$$\nu + N \rightarrow \ell + h \tag{21.72}$$

where the initial ν stands for ν_ℓ or $\bar{\nu}_\ell$ ($\ell = e$, μ or τ), the final ℓ can be any neutral or charged lepton and h any hadron or

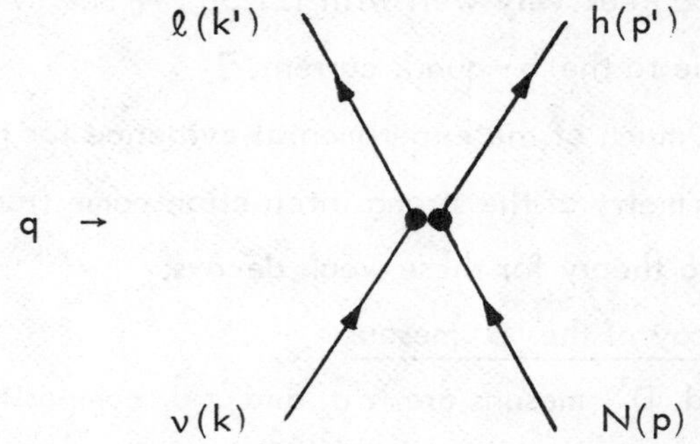

Fig. 21.7. Lowest-order Feynman diagram for $\nu + N \to \ell + h$. The Lagrangian density is given by (21.13).

complex of hadrons. Let us denote, as in Figure 21.7,

$$k = 4\text{-momentum of } \nu \text{ (or } \bar{\nu}\text{)}, \quad k' = 4\text{-momentum of } \ell,$$

$$p = 4\text{-momentum of } N, \quad p' = 4\text{-momentum of } h, \qquad (21.73)$$

and

$$s_N,\ s_h \text{ and } s_\ell = \text{helicities of } N,\ h \text{ and } \ell .$$

The 4-momentum transfer between the lepton and the hadrons is

$$q \equiv k - k' = p' - p .$$

Throughout this section, we assume that the initial nucleon is not polarized and all polarizations in the final states are summed over. Furthermore, we set

$$\text{the lepton mass} = 0 .$$

Let M be the mass of h, and m_N the nucleon mass. By taking the products of the four initial and final momenta we can construct six scalars, of which four are the initial and final masses

$$k^2 = 0, \quad k'^2 = 0, \quad p^2 = -m_N^2 \quad \text{and} \quad p'^2 = -M^2. \qquad (21.74)$$

The other two invariants may be chosen as

$$x \equiv -\frac{q^2}{2p \cdot q} \quad \text{and} \quad y \equiv \frac{q \cdot p}{k \cdot p} \quad . \tag{21.75}$$

By squaring $p + q = p'$, we have

$$2p \cdot q + q^2 = -M^2 + m_N^2 \quad ,$$

and therefore

$$x = \frac{q^2}{q^2 + M^2 - m_N^2} \quad . \tag{21.76}$$

If the final state h is a single hadron, then M is fixed; but if it refers to a specific channel of continuum states, such as $p\pi$, $n\pi K$, etc., then M is a continuous variable. Sometimes, we may be interested in summing over all possible final hadron channels; in that case the reaction is called "inclusive", and we may write (21.72) as

$$\nu + N \rightarrow \ell + \cdots \quad .$$

In the following, we shall regard M as a variable. The kinematics of the problem then depends on three independent invariants, say M, x and y, or

$$k \cdot p \ , \quad q^2 \ \text{and} \ x \ , \tag{21.77}$$

or some other equivalent choices.

In the laboratory system, the initial nucleon N is at rest. The components of the various 4-momenta given in (21.73) are

$$k = (\vec{k}_\nu \ , \ i E_\nu) \ , \quad k' = (\vec{k}_\ell \ , \ i E_\ell) \ ,$$

$$p = (0 \ , \ i m_N) \ , \quad p' = (\vec{P} \ , \ i E_h) \tag{21.78}$$

and

$$q = (\vec{q} \ , \ i q_0) \quad .$$

Since $p' = p + q$, we have

$$\vec{P} = \vec{q} \quad \text{and} \quad E_h = m_N + q_0 \quad . \tag{21.79}$$

Because $k \cdot p = -m_N E_\nu$, the three independent variables in (21.77) may also be written as

$$E_\nu \;,\; q^2 \;\text{ and }\; x \quad . \tag{21.80}$$

We can readily express other kinematic variables in terms of these three; e.g.,

$$M^2 = m_N^2 + q^2 (x^{-1} - 1) \;, \quad y = \frac{q^2}{2m_N E_\nu x} \;,$$

$$E_\ell = E_\nu - \frac{q^2}{2m_N x} \;, \quad q_0 = \frac{q^2}{2m_N x} \;,$$

$$s \equiv -(k+p)^2 = m_N^2 + 2m_N E_\nu \;, \tag{21.81}$$

$$P \equiv |\vec{P}| = \left[q^2 + \left(\frac{q^2}{2m_N x}\right)^2\right]^{\frac{1}{2}}$$

and

$$\cos\theta = 1 - \frac{q^2}{2E_\nu^2}\left(1 - \frac{q^2}{2m_N x E_\nu}\right)^{-1}$$

where $\theta = \angle(\vec{k}_\nu, \vec{k}_\ell)$. From $q^2 = (k-k')^2 = 2E_\nu E_\ell (1 - \cos\theta)$, it follows that

$$q^2 \geqslant 0 \;;$$

i.e., the 4-momentum transfer q is space-like. By using (21.76), and on account of the final hadron mass $M \geqslant m_N$, we have

$$0 \leqslant x \leqslant 1 \tag{21.82}$$

where

$$x = 1 \quad \text{when} \quad M = m_N \quad . \tag{21.83}$$

The variable $y = q \cdot p / k \cdot p$ can also be written as

$$y = \frac{q_0}{E_\nu} = \frac{E_h - m_N}{E_\nu} \tag{21.84}$$

which is the fraction of the incident neutrino energy that is transferred to the hadron in the laboratory system. Consequently, y lies within the range

$$0 \leqslant y \leqslant 1 \quad . \tag{21.85}$$

Exercise.

(i) Show that at any given E_ν , the range of q^2 is

$$0 \leqslant q^2 \leqslant \frac{(2m_N E_\nu)^2}{s} \tag{21.86}$$

where s is the (center-of-mass energy)$^2 = m_N^2 + 2m_N E_\nu$.

(ii) At fixed E_ν and q^2, show that

$$M^2 \leqslant M_0^2 \equiv s\left(1 - \frac{q^2}{2m_N E_\nu}\right) , \tag{21.87}$$

and therefore x lies within the range

$$\frac{q^2}{q^2 + M_0^2 - m_N^2} \leqslant x \leqslant 1 \quad .$$

2. Structure functions

The semileptonic Lagrangian (21.13) is a sum of terms, each of the form "current x current." Because of the two-component theory of the neutrino, the lepton current for the neutrino reaction is always of the form

$$i \psi_\ell^\dagger \gamma_4 \gamma_\lambda (1 + \gamma_5) \psi_{\nu_\ell} ,$$

or its conjugate, whether the final ℓ is a charged or neutral lepton. The differential cross section $d^2\sigma$ of reaction (21.72) depends on three variables, say E_ν, q^2 and x, as given by (21.80). The matrix elements of the lepton-current operator are readily calculable. This makes it possible for us to extract the explicit E_ν-dependence of $d^2\sigma$, as we shall see. We can then express $d^2\sigma$ in terms of three functions * which depend only on the two remaining variables q^2 and x, whatever the final hadron states may be. These functions are called structure functions.

To derive these functions, let us first consider a given final hadron channel h of a fixed M. [Later, M will be allowed to vary.] We define

$$L_{\alpha\beta} \equiv \pm \sum_{s_\ell} \langle \ell \,|\, j_\alpha \,|\, \nu \rangle \langle \ell \,|\, j_\beta \,|\, \nu \rangle^*$$

and (21.88)

$$H_{\alpha\beta} \equiv \pm \sum_{s_N, s_h} \langle h \,|\, J_\alpha \,|\, N \rangle \langle h \,|\, J_\beta \,|\, N \rangle^*$$

where j_α and J_α refer to the appropriate lepton and hadron currents, and s_N, s_h, s_ℓ are helicities defined in (21.73). In (21.88) the $\pm$ sign is due to our metric convention $x_\mu = (\vec{r}, it)$; the + sign is for $\beta \neq 4$ and the − sign for $\beta = 4$, so that under a Lorentz transformation

$$\pm \langle \ell \,|\, j_\beta \,|\, \nu \rangle^* \quad \text{and} \quad \pm \langle h \,|\, J_\beta \,|\, N \rangle^*$$

transform in the same way as

$$\langle \ell \,|\, j_\beta \,|\, \nu \rangle \quad \text{and} \quad \langle h \,|\, J_\beta \,|\, N \rangle \quad .$$

From Problem 5.3, by setting $C_V = C_A = 1$, $m_a = m_b = 0$ and replacing a_μ, b_λ by k_μ, k'_λ, we find that $L_{\alpha\beta}$ in the zero-lepton-mass limit is given by

---

* T. D. Lee and C. N. Yang, Phys. Rev. Lett. 4, 307 (1960); Phys. Rev. 126, 2239 (1962).

$$L_{\alpha\beta}(\nu_\ell) = \frac{2}{E_\nu E_\ell}(k'_\alpha k_\beta + k_\alpha k'_\beta - \delta_{\alpha\beta} k\cdot k' + \epsilon_{\alpha\beta\mu\nu} k_\mu k'_\nu)$$

if the initial neutrino is ν_ℓ, or (21.89)

$$L_{\alpha\beta}(\bar\nu_\ell) = \frac{2}{E_\nu E_\ell}(k'_\alpha k_\beta + k_\alpha k'_\beta - \delta_{\alpha\beta} k\cdot k' - \epsilon_{\alpha\beta\mu\nu} k_\mu k'_\nu)$$

if it is $\bar\nu_\ell$.

From Figure 21.7, we see that the hadron part $H_{\alpha\beta}$ depends only on two independent 4-momenta p and q, since $p' = p+q$. By considering polynomials of p and q, we can construct the following second-rank tensors:

$$\delta_{\alpha\beta}\,,\quad p_\alpha p_\beta\,,\quad \epsilon_{\alpha\beta\mu\nu} p_\mu q_\nu = \epsilon_{\alpha\beta\mu\nu} p_\mu p'_\nu$$

and (21.90)

$$p_\alpha q_\beta\,,\quad q_\alpha p_\beta\,,\quad q_\alpha q_\beta\,.$$

In the zero-lepton-mass limit, there is the conservation of lepton current $\partial j_\alpha/\partial x_\alpha = 0$, and therefore

$$q_\alpha L_{\alpha\beta} = L_{\alpha\beta} q_\beta = 0\,.$$

Consequently, of the six tensors in (21.90) only the first three are relevant; we write

$$H_{\alpha\beta} L_{\alpha\beta} = \hat{H}_{\alpha\beta} L_{\alpha\beta}$$

where

$$\hat{H}_{\alpha\beta} = \frac{1}{m_N E_h}(m_N^2 \delta_{\alpha\beta} W_1 + p_\alpha p_\beta W_2 + \tfrac{1}{2}\epsilon_{\alpha\beta\mu\nu} p_\mu p'_\nu W_3) \tag{21.91}$$

in which W_1, W_2 and W_3 are dimensionless scalar functions of q^2 and M. The factors $(E_\nu E_\ell)^{-1}$ and $(m_N E_h)^{-1}$ in (21.89) and (21.91) are due to our normalization convention (3.30). By using (5.107) we find, for a given hadron channel h of a fixed M,

$$d\sigma = (2\pi)^2 \int \frac{1}{8\pi^3} E_\ell^{\,2}\, dE_\ell\, d\cos\theta\, \left(\frac{G^2}{4}\right) L_{\alpha\beta} H_{\alpha\beta}$$
$$\delta(E_h + E_\ell - m_N - E_\nu) \qquad (21.92)$$

where $(2\pi)^2$ is due to the 2π factor in (5.107) and another such factor in

$$\int d^3k_\ell = 2\pi \int E_\ell^{\,2}\, dE_\ell\, d\cos\theta \; .$$

The factor $(G^2/4)$ is the product of $(G/\sqrt{2}\,)^2$, times $\frac{1}{2}$ due to the average of the initial nucleon spin. From (21.84), it follows that

$$y = \frac{(E_\nu - E_\ell)}{E_\nu} \; .$$

This, together with

$$E_h = (E_\ell^{\,2} + E_\nu^{\,2} - 2E_\ell E_\nu \cos\theta + M^2)^{\frac{1}{2}} \; ,$$

gives

$$dE_\ell \int d\cos\theta\, \delta(E_h + E_\ell - m_N - E_\nu) = \frac{E_h}{E_\ell}\, dy \; . \qquad (21.93)$$

Thus, by using (21.89)-(21.93) we derive, at a fixed final hadron mass M,

$$d\sigma = (2\pi q_0)^{-1} G^2 m_N E_\nu\, dy \Big\{ m_N x y^2 W_1$$
$$+ [1 - y - \tfrac{1}{2}(m_N \frac{xy^2}{q_0})]\, q_0 W_2$$
$$\pm xy(1 - \tfrac{1}{2}y)\, q_0 W_3 \Big\} \qquad (21.94)$$

where the $+$ sign is for ν_ℓ and $-$ is for $\bar{\nu}_\ell$.

Next, we consider the reaction in which we sum over the final hadron states (of any given non-coherent mixture of hadron channels h); M is now a continuous variable. From (21.81), it follows that at fixed E_ν and y

$$-dM^2 = dq^2 = (q^2/x)\, dx = 2m_N q_0\, dx \; .$$

In channel h, let J be its total angular momentum and $\rho(M^2)\, dM^2$

be the total number of states* with their (invariant mass)^2 between M^2 and $M^2 + dM^2$. Multiplying (21.94) by $(2J+1)^{-1}\rho(M^2)\,dM^2$, we find that the cross section can be written as

$$d^2\sigma = \frac{m_N E_\nu}{\pi} G^2\,dx\,dy\,[\,xy^2 F_1 + (1 - y - \frac{m_N}{2q_0} xy^2) F_2 \pm xy(1 - \tfrac{1}{2}y) F_3\,] \tag{21.95}$$

where, as in (21.94), + is for ν_ℓ and − for $\bar{\nu}_\ell$, the structure functions F_i are defined in terms of W_i by

$$F_1(q^2, x) \equiv \sum_h (2J+1)^{-1} m_N^2 \rho(M^2) W_1 ,$$

$$F_2(q^2, x) \equiv \sum_h (2J+1)^{-1} m_N q_0 \rho(M^2) W_2 \tag{21.96}$$

and

$$F_3(q^2, x) \equiv \sum_h (2J+1)^{-1} m_N q_0 \rho(M^2) W_3$$

in which the sum extends over any non-coherent mixture of different final hadron channels h of the same M. For inclusive reactions, we sum over all possible channels h. [See Problems 21.2 and 21.3 for some alternative forms of (21.94)-(21.95).]

These structure functions have been extensively measured experimentally. The fact that they depend only on two variables q^2 and x gives strong support for the local character of the lepton current; i.e., the two lepton fields

$$\psi_\ell(\vec{r}, t) \quad \text{and} \quad \psi_{\nu_\ell}(\vec{r}, t)$$

are always at the <u>same</u> space-time point in the lepton current operator $j_\lambda^\pm(\vec{r}, t)$, as shown in (21.3)-(21.4) for $j_\lambda = j_\lambda^\pm$ or $j_\lambda^{\,o}$. Further discussions will be given in Chapter 23.

---

* If h refers to a single hadron of mass m and spin J, then $\rho(M^2) = (2J+1)\,\delta(M^2 - m^2)$.

21.7 Semileptonic Neutral-current Interaction

1. $\Delta S = 0$ Rule

In the semileptonic Lagrangian (21.13), the neutral-current part is given by

$$\frac{G}{\sqrt{2}} J_\lambda^o j_\lambda^o \quad . \tag{21.97}$$

As we shall see, there exists good evidence that the neutral hadron current J_λ^o conserves the strangeness S ; i.e., unlike the charged current $J_\lambda^\pm$, it satisfies the selection rule

$$\Delta S = 0 \quad .$$

We note from the Table of Particle Properties that for the vector part of J_λ^o there is the upper limit of the branching ratio

$$\frac{\text{rate } (K^+ \to \pi^+ \nu \bar{\nu})}{\text{rate } (K^+ \to \text{all})} < 6 \times 10^{-7} \quad . \tag{21.98}$$

For the axial-vector part, we have

$$\frac{\text{rate } (K_L^o \to \mu^+ \mu^-)}{\text{rate } (K_L^o \to \text{all})} = (9.1 \pm 1.8) \times 10^{-9} \tag{21.99}$$

and

$$\frac{\text{rate } (K_L^o \to e^+ e^-)}{\text{rate } (K_L^o \to \text{all})} < 2 \times 10^{-9} \quad . \tag{21.100}$$

All the decays in the numerators are suppressed because of the $\Delta S = 0$ rule. The significance of these experimental values becomes more apparent if we compare these neutral-current rates with the ones for the corresponding charged currents:

$$\frac{\text{rate } (K^+ \to \pi^+ \nu \bar{\nu})}{\text{rate } (K^+ \to \pi^o e^+ \nu)} < 1.2 \times 10^{-5} \quad , \tag{21.101}$$

$$\frac{\text{rate}\ (K_L^o \rightarrow \mu^+\mu^-)}{\text{rate}\ (K^+ \rightarrow \mu^+\nu)} = 3.4 \times 10^{-9} \tag{21.102}$$

and

$$\frac{\text{rate}\ (K_L^o \rightarrow e^+e^-)}{\text{rate}\ (K^+ \rightarrow e^+\nu)} < 3 \times 10^{-5} \quad . \tag{21.103}$$

From the discussions given on pages 234–35, we see that K_L^o cannot decay into a μ pair via a single-photon exchange; i.e.,

$$K_L^o \not\rightarrow \text{virtual}\ \gamma \rightarrow \mu^+\mu^- \quad .$$

However, as shown in Figure 21.8,

$$K_L^o \rightarrow \text{virtual}\ \gamma\gamma \rightarrow \mu^+\mu^- \tag{21.104}$$

is allowed; therefore in terms of the fine structure constant α, we

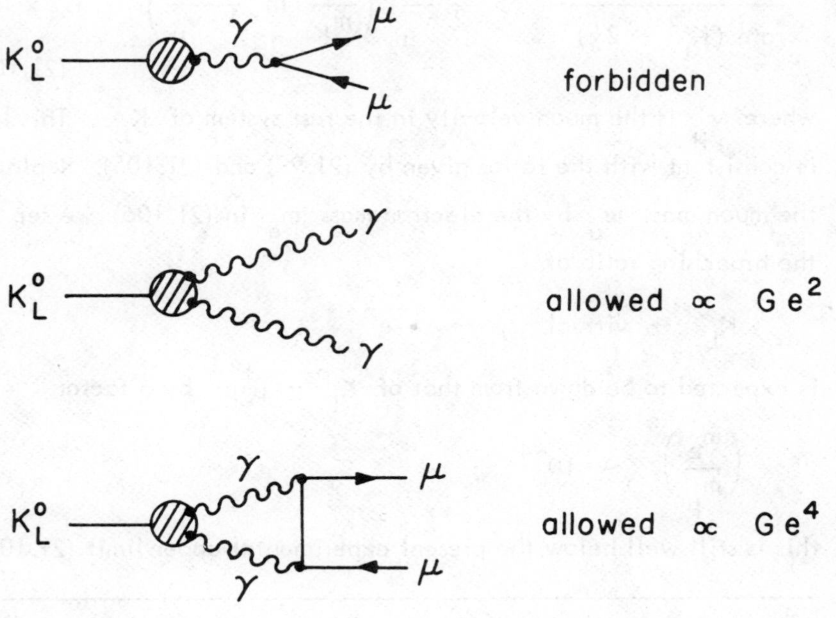

Fig. 21.8. Selection rules for some of the rare decay amplitudes of K_L^o.

may estimate

$$\frac{\text{rate}\ (K_L^o \rightarrow \mu^+\mu^-)}{\text{rate}\ (K_L^o \rightarrow \text{all})} \sim \alpha^4 \sim 10^{-8} \quad ,$$

consistent with (21.99) and (21.102). The observed 2γ decay rate of K_L^o is

$$\frac{\text{rate}\ (K_L^o \rightarrow 2\gamma)}{\text{rate}\ (K_L^o \rightarrow \text{all})} = 4.9 \times 10^{-4} \tag{21.105}$$

which, as expected from examining Figure 21.8, is of the order α^2. Through (21.104) the on-mass-shell 2γ intermediate state gives an imaginary (i.e., absorptive) part to the amplitude $K_L^o \rightarrow \mu^+\mu^-$. Hence, there is an inequality *

$$\frac{\text{rate}\ (K_L^o \rightarrow \mu^+\mu^-)}{\text{rate}\ (K_L^o \rightarrow 2\gamma)} > \frac{1}{2}\,\frac{\alpha^2}{v_\mu}\left(\frac{m_\mu}{m_K}\ \ln\ \frac{1+v_\mu}{1-v_\mu}\right)^2 \cong 1.2 \times 10^{-5} \tag{21.106}$$

where v_μ is the muon velocity in the rest system of K_L^o. This limit is consistent with the ratios given by (21.99) and (21.105). Replacing the muon mass m_μ by the electron mass m_e in (21.106), we see that the branching ratio of

$$K_L^o \rightarrow \text{virtual}\ \gamma\gamma \rightarrow e^+e^-$$

is expected to be down from that of $K_L^o \rightarrow \mu^+\mu^-$ by a factor

$$\left(\frac{m_e}{m_\mu}\right)^2 \sim 10^{-4} \quad ;$$

this is still well below the present experimental upper limit (21.100).

* M. A. B. Bég, Phys. Rev. 132, 426 (1963); A. Pais and S. B. Treiman, Phys. Rev. 176, 1974 (1968).

2. High-energy neutrino reaction

Although the experimental evidence for the weak neutral current interaction came in 1973 through the observation * of

$$\nu_\mu + N \to \nu_\mu + \cdots ,$$

the theoretical suggestion was made much earlier. ** The delay in the experimental discovery was due in part to the difficulty of separating out the neutron background from the genuine neutrino events.

The strength of the neutral-current coupling can be represented by the following ratios of total cross sections

$$R_\nu \equiv \frac{\sigma(\nu_\mu + N \to \nu_\mu + \cdots)}{\sigma(\nu_\mu + N \to \mu^- + \cdots)} = .307 \pm .008$$

and (21.107)

$$R_{\bar\nu} \equiv \frac{\sigma(\bar\nu_\mu + N \to \bar\nu_\mu + \cdots)}{\sigma(\bar\nu_\mu + N \to \mu^+ + \cdots)} = .381 \pm .025 \quad .$$

Further discussions of these reactions will be given in Chapter 23.

3. Polarized electron scattering

More recently, it was found through the scattering of polarized electrons on the deuteron,

$$e^- \text{(polarized)} + d\,\text{(unpolarized)} \to e^- + \cdots ,$$

that parity is not conserved. Consequently, there must be interactions other than the familiar parity-conserving QED, which of course dominates this reaction.

---

* F. J. Hasert et al., Phys. Lett. 46B, 121 (1973).

** See references cited on page 212.

Let $d\sigma_R$ and $d\sigma_L$ be the differential cross sections of the right-handed and lefthanded incident e^- under an otherwise identical kinematic condition. If in the semileptonic Lagrangian (21.13) the neutral-current interaction $J_\mu^o j_\mu^o$ violates parity conservation, then its interference should produce an asymmetry

$$A \equiv \frac{d\sigma_R - d\sigma_L}{d\sigma_R + d\sigma_L} \quad . \tag{21.108}$$

As shown in Figure 21.9, since the weak interaction amplitude is $\sim G$ and the electromagnetic amplitude is $\sim e^2/q^2$ where $q^2 = (4\text{-momentum transfer})^2$, we expect

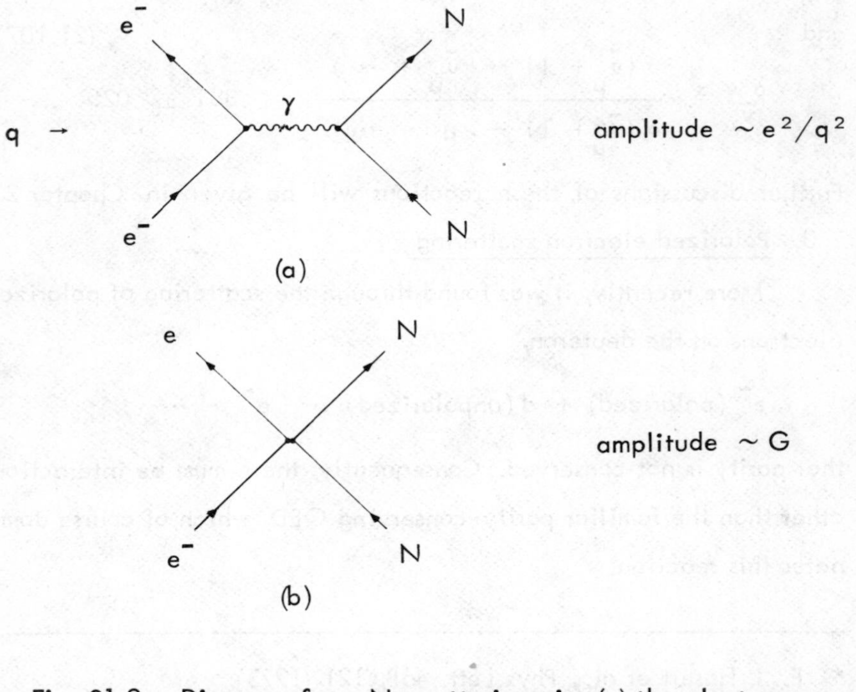

Fig. 21.9. Diagrams for eN scattering via (a) the electromagnetic and (b) the weak neutral-current interactions.

$$A \sim \frac{Gq^2}{e^2} = \frac{Gq^2}{4\pi\alpha} \sim 10^{-4} \cdot q^2/(\mathrm{GeV})^2 \quad . \tag{21.109}$$

Experimentally,* it is found that if we write

$$A = \left[a_1 + a_2 \, \frac{1-(1-y)^2}{1+(1-y)^2} \right] \cdot q^2/(\mathrm{GeV})^2 \quad , \tag{21.110}$$

then

$$\begin{aligned} a_1 &= (-.97 \pm .26) \times 10^{-4} \quad , \\ a_2 &= (.49 \pm .81) \times 10^{-4} \end{aligned} \tag{21.111}$$

and

$$y = \frac{E_{in} - E_f}{E_{in}}$$

with E_{in} and E_f the initial and final laboratory e^- energies. The agreement between the order-of-magnitude estimate (21.109) and the experimental results indicates the existence of a weak neutral-current interaction between e and hadrons. Furthermore, the kinematic dependence exhibited in (21.110) is in accordance with the general theoretical expectation, which together with other properties of $J_\lambda^{\,o}$ and $j_\lambda^{\,o}$ will be discussed in the next chapter. [See Section 22.3 and especially Problem 22.3.]

4. Atomic parity violation

Additional support for the existence of parity-violating neutral currents has come from atomic measurements by the Novosibirsk group** and the Berkeley group.*** However, in view of some early controversies**** in the experimental situation, further clarification is still

---

* C. Y. Prescott et al., Phys. Lett. 77B, 347 (1978), 84B, 524 (1979).

** L. M. Barkov and M. S. Zolotoryov, JETP Lett. 27, 357 (1978).

*** R. Conti et al., Phys. Rev. Lett. 42, 343 (1979).

**** P. E. G. Baird et al., Phys. Rev. Lett. 39, 798 (1977).
L. L. Lewis et al., ibid., 795.

needed in this important area.

21.8 Nonleptonic Interaction

Based on the intermediate boson hypothesis that will be discussed in the next section, the phenomenological nonleptonic Lagrangian is usually written as, in the limit when the boson mass $\rightarrow \infty$,

$$\mathcal{L}_{nonlep} = \frac{G}{\sqrt{2}} (J_\lambda^+ J_\lambda^- + \tfrac{1}{2} J_\lambda^o J_\lambda^o) \tag{21.112}$$

where $J_\lambda^\pm$ and J_λ^o are the same hadron currents that appear in $\mathcal{L}_{semilep}$, (21.13). As discussed before in Section 11.4.2, the nonleptonic decays of K, Λ and other strange hadrons satisfy the $|\Delta\vec{I}| = \frac{1}{2}$ rule. In accordance with (21.51), the $\Delta S = 0$ part of $J_\lambda^\pm$ contains a $|\Delta\vec{I}| = 1$ term, while the $\Delta S \neq 0$ part has a $|\Delta\vec{I}| = \frac{1}{2}$ term; their product can violate isospin by $|\Delta\vec{I}| = \frac{3}{2}$ as well as $\frac{1}{2}$. It is not clear why the $|\Delta\vec{I}| = \frac{3}{2}$ amplitude should be suppressed, as exemplified by the experimental observation

$$\frac{\text{rate } (K^\pm \rightarrow \pi^\pm + \pi^o)}{\text{rate } (K_S^o \rightarrow 2\pi)} = (1.53 \pm 0.07) \times 10^{-3} .$$

This is one of the remaining unsolved problems of the weak interaction.

The total effective weak interaction Lagrangian $\mathcal{L}_{eff}$ is given by the sum of $\mathcal{L}_{lep}$, $\mathcal{L}_{semilep}$ and $\mathcal{L}_{nonlep}$. From (21.2), (21.13) and (21.112) we see that

$$\mathcal{L}_{eff} = \frac{G}{\sqrt{2}} \left[(J_\lambda^+ + j_\lambda^+)(J_\lambda^- + j_\lambda^-) + \tfrac{1}{2}(J_\lambda^o + j_\lambda^o)(J_\lambda^o + j_\lambda^o) \right] . \tag{21.113}$$

Its lowest-order matrix elements give the observed reaction amplitudes directly; the higher-order elements are in general divergent and should be discarded, in accordance with the general idea of an effective Lagrangian.

21.9 Intermediate Boson

1. Intermediate boson hypothesis

The parallel between the weak and the electromagnetic interactions has dominated the thinking of most workers in the field since the early days of β decay. To begin with, the original Fermi theory* was made in analogy with the second order electromagnetic interaction, giving rise to the current x current form of the Lagrangian. Because of the observed short-range character of the weak interaction, the currents in the Lagrangian density of the Fermi theory are all taken at the same space-time point, in contrast to the electromagnetic case. The similarity between these two interactions was further strengthened when in the late 1940's it was discovered** that the nuclear β decay shared with μ capture and μ decay the same Fermi coupling constant G. This led to the intermediate boson hypothesis:*** All the observed weak reactions are viewed as second order processes generated through the emission and absorption of a massive charged boson, called W (with the letter W representing "weak"). Similar to the photon γ, there is a universality in the W-matter interaction; its coupling is always the same dimensionless constant. The additional assumption that this constant is identical to the electric charge e yields the estimate that the mass of W should be of the order

$$m_W \sim \sqrt{\frac{G}{\alpha}} \approx 30 \text{ GeV} \quad . \tag{21.114}$$

---

* E. Fermi, Z. Physik 88, 161 (1934).

** See the references on page 591.

*** T.D. Lee, M. Rosenbluth and C.N. Yang, Phys. Rev. 75, 9905 (1949).

From the unitarity limit which will be discussed below, we can also set an upper bound *

$$m_W < \text{about } 300 \text{ GeV} \quad . \tag{21.115}$$

2. <u>Limitation of Fermi theory</u>

There exists another compelling reason to modify Fermi theory. Even as an effective Lagrangian $\mathcal{L}_{eff}$, (21.113) cannot be adequate at very high energy. This may be demonstrated by observing that, according to (21.2)-(21.3), the reaction

$$\nu_\mu + e^- \rightarrow \nu_e + \mu^-$$

can occur, and it consists only of s-wave scattering. Its cross section can be derived by using Problem 5.4. We find

$$\sigma = \frac{4}{\pi} (Gk)^2 \tag{21.116}$$

where k is the momentum of ν_μ in the center-of-mass system. Since, from unitarity (see Problem 21.9),

$$\sigma \leqslant \frac{\pi}{2k^2} \ ; \tag{21.117}$$

(21.116) can be valid only if

$$k \leqslant \left(\frac{\pi}{2\sqrt{2}\, G}\right)^{\frac{1}{2}} \sim 300 \text{ GeV} \quad . \tag{21.118}$$

Thus, independent of the underlying mechanism of the weak interaction, this particular form of $\mathcal{L}_{eff}$ has to be modified at high momentum transfer, i.e., at small distances. For example, (21.2) may be replaced by

$$\mathcal{L}_{lep}(x) = \frac{G}{\sqrt{2}} \int j_\lambda^+(x)\, D_{\lambda\mu}(x-x')\, j_\mu^-(x')\, d^4x' + \cdots \quad . \tag{21.119}$$

To generate such "effective" nonlocality, some agents are needed to

* T. D. Lee, CERN Report 61-30 (1961).

transmit the action from x to x'. The simplest possibility is to assume the existence of an intermediate boson.

At present, the intermediate boson is still in the realm of hypothesis. For over two decades, how to unite the weak and electromagnetic interactions has been a dominant theme of particle physics, culminating in the gauge models that will be discussed in the next chapter.

21.10 Kobayashi-Maskawa Model

The discovery of b-quark hadrons and their weak decays necessitates the generalization and modification of the Cabibbo theory. Lacking detailed experimental information, we have many theoretical inventions; one of the simplest is the Kobayashi-Maskawa model.*

1. Quark and lepton generations

As discussed on page 600, we assume $\nu_e \neq \nu_\tau$. There are six lepton species which can be grouped as follows:

$$
\begin{array}{ll}
1^{st} \text{ generation} & \nu_e \,,\; e \\
2^{nd} \text{ generation} & \nu_\mu \,,\; \mu \\
3^{rd} \text{ generation} & \nu_\tau \,,\; \tau \,.
\end{array}
\tag{21.120}
$$

The Kobayashi-Maskawa model gives a similar three-generation structure for the quarks with

$$
\begin{array}{ll}
1^{st} \text{ generation} & u \,,\; d \\
2^{nd} \text{ generation} & c \,,\; s \\
3^{rd} \text{ generation} & t \,,\; b \,,
\end{array}
\tag{21.121}
$$

in which the t (top) quark is as yet only a conjecture. The present

---

* M. Kobayashi and K. Maskawa, Progr. Theor. Phys. 49, 652 (1973).

lower bound of the mass of the t quark is*

$$m_t > 17.9 \text{ GeV} \quad . \tag{21.122}$$

[Like all quark masses, m_t refers only to the mass inside a hadron.]

It is convenient to put different generation members together according to their charges. We write

$$\nu^o \equiv \begin{pmatrix} \nu_e \\ \nu_\mu \\ \nu_\tau \end{pmatrix} , \qquad \ell^- \equiv \begin{pmatrix} e \\ \mu \\ \tau \end{pmatrix}$$

$$q^{\frac{2}{3}} \equiv \begin{pmatrix} u \\ c \\ t \end{pmatrix} \quad \text{and} \quad q^{-\frac{1}{3}} \equiv \begin{pmatrix} d \\ s \\ b \end{pmatrix} \tag{21.123}$$

where the superscripts denote the electric charge in units of e. The lepton current $j_\lambda^\pm$, (21.3), becomes

$$j_\lambda^- = i\,\ell^{-\dagger}\,\gamma_4\,\gamma_\lambda(1+\gamma_5)\,\nu^o$$

and

$$j_\lambda^+ = i\,\nu^{o\dagger}\,\gamma_4\,\gamma_\lambda(1+\gamma_5)\,\ell^- \tag{21.124}$$

with ν^o and ℓ^- standing for the corresponding lepton-field operators.

2. Hadron current

The charged hadron current in the Cabibbo theory does not contain the b-quark field. The simplest generalization is to write, instead of (21.51),

$$J_\lambda^+ = i\,q^{\frac{2}{3}\dagger}\,\gamma_4\,\gamma_\lambda(1+\gamma_5)\;U\,q^{-\frac{1}{3}}$$

and

$$J_\lambda^- = i\,q^{-\frac{1}{3}\dagger}\,\gamma_4\,\gamma_\lambda(1+\gamma_5)\;U^\dagger\,q^{\frac{2}{3}} \tag{21.125}$$

where, as in the above lepton case, $q^{\frac{2}{3}}$ and $q^{-\frac{1}{3}}$ stand for the

---

* See "Mark J Collaboration", Phys.Reports 63, 337 (1980) and references mentioned therein.

corresponding quark field operators. As we shall show, for reasons of lepton-quark symmetry U is unitary.

It seems reasonable to regard the origin of quark mass differences as outside the weak interaction. Thus, so far as the symmetry of the weak interaction is concerned, we may examine the limit in which all quark masses are equal. In this case, let us consider the following transformation of the quark field operators:

$$q^{\frac{2}{3}} \rightarrow Q^{\frac{2}{3}} \equiv u^{\dagger} q^{\frac{2}{3}}$$

and

$$q^{-\frac{1}{3}} \rightarrow Q^{-\frac{1}{3}} \equiv u\, q^{-\frac{1}{3}} \quad . \qquad (21.126)$$

Because of the canonical anticommutation relations (3.21)-(3.22) satisfied by the quark fields, the matrix u must be unitary. By setting

$$u^2 = U \, ,$$

we see that (21.125) can be written as

$$J_\lambda^+ = i\, Q^{\frac{2}{3}\dagger} \gamma_4 \gamma_\lambda (1+\gamma_5)\, Q^{-\frac{1}{3}}$$

and

$$J_\lambda^- = i\, Q^{-\frac{1}{3}\dagger} \gamma_4 \gamma_\lambda (1+\gamma_5)\, Q^{\frac{2}{3}} \quad .$$

Under the interchange

$$\nu^o \rightleftarrows Q^{\frac{2}{3}} \quad \text{and} \quad \ell^- \rightleftarrows Q^{-\frac{1}{3}} \, , \qquad (21.127)$$

we have

$$j_\lambda^{\pm} \rightleftarrows J_\lambda^{\pm} \, ,$$

which gives the desired symmetry between leptons and quarks. Conversely, the requirement of such a lepton-quark symmetry (21.127) leads to (21.125) under the transformation (21.126). Because the unitarity of u implies that of U, our supposition is proved.

3. U matrix

The $N \times N$ unitary matrix U determines the weak-interaction transition amplitudes between quarks of different flavors. In the original Cabibbo theory, (21.51), $N = 2$. In the Kobayashi-Maskawa model, (21.125), $N = 3$. We shall now consider the question of how many independent parameters are needed to characterize U.

(i) $N = 2$

Any 2×2 unitary matrix depends on four real parameters, since there are $1 + 3 = 4$ generators for the $U_1 \times SU_2 = U_2$ group. To begin with, each of the quark fields u, d, c and s in the Cabibbo theory carries an arbitrary phase factor. Hence, there are altogether three arbitrary relative phases; that makes U depend only on $4 - 3 = 1$ real parameter, which can be taken to be the Cabibbo angle θ_c. We may write

$$U = \begin{pmatrix} \cos\theta_c & \sin\theta_c \\ -\sin\theta_c & \cos\theta_c \end{pmatrix} . \tag{21.128}$$

In this case, instead of (21.123), we have

$$q^{\frac{2}{3}} = \binom{u}{c} \qquad \text{and} \qquad q^{-\frac{1}{3}} = \binom{d}{s}$$

and therefore (21.125) reduces to the current (21.51) in the Cabibbo theory.

The parametrization (21.128) may also be constructed as follows: Choose first the relative phase between u and d to make the matrix element $U_{11} \equiv \cos\theta_c$ real and positive, next the relative phase between u and s to make $U_{12} \equiv \sin\theta_c$ also real and positive; likewise, the arbitrary phase of c enables us to choose U_{21} real and negative, which gives $U_{21} = -\sin\theta_c$. The unitarity of U then necessitates $U_{22} = \cos\theta_c$, and therefore det $U = 1$. Note that by this procedure, we can always set the Cabibbo angle θ_c in the first quadrant.

(ii) $N = 3$

Any 3×3 unitary matrix depends on nine real parameters, since the $U_1 \times SU_3 = U_3$ group has $1 + 8 = 9$ generators. Between the six quark fields u, d, s, c, b and t, there are five arbitrary relative phases. Hence the 3×3 matrix U depends on $9 - 5 = 4$ real parameters. Without any loss of generality, we can choose these parameters to be the four angles θ_1, θ_2, θ_3 and δ defined below:

$$U = \begin{pmatrix} c_1 & s_1c_3 & s_1s_3 \\ -s_1c_2 & c_1c_2c_3 - s_2s_3e^{i\delta} & c_1c_2s_3 + s_2c_3e^{i\delta} \\ s_1s_2 & -c_1s_2c_3 - c_2s_3e^{i\delta} & -c_1s_2s_3 + c_2c_3e^{i\delta} \end{pmatrix} \tag{21.129}$$

where

$$c_i = \cos\theta_i \quad \text{and} \quad s_i = \sin\theta_i$$

with $i = 1, 2, 3$. It can be readily verified that U is unitary and

$$\det U = e^{i\delta} \quad . \tag{21.130}$$

To begin with, without any loss of generality we may set θ_1 between 0 and π, and both θ_2 and θ_3 between 0 and 2π. It is straightforward to show that by manipulating the relative phases between (u, d), (u, s), (u, b), (c, d) and (t, d) we can choose c_1, s_1c_3, s_1s_3, s_1c_2 and s_1s_2 all to be real and positive. Hence, we can set

$$\theta_1,\ \theta_2 \text{ and } \theta_3 \text{ all in the first quadrant.} \tag{21.131}$$

Note that T invariance implies U real. Hence, the possibility of

$$\delta \neq 0 \text{ or } \pi$$

gives T violation.

4. Experimental determination

By using (21.48), (21.129) and Table 21.1, we see that $\cos\theta_1$ now replaces $\cos\theta_c$ in relating the μ decay constant G_μ to the Fermi constant G_V in the nuclear β decay. We have

$$G_V = G_\mu \cos\theta_1 \quad .$$

Thus, from (21.50), we have

$$\cos\theta_1 = .974 \pm .002 \quad . \tag{21.132}$$

For the strangeness-changing hyperon β decay, the product $\sin\theta_1 \cdot \cos\theta_3$ now replaces the factor $\sin\theta_c$ in the old Cabibbo theory. Equation (21.70) gives

$$\sin\theta_1 \cos\theta_3 = .219 \pm .011 \quad . \tag{21.133}$$

Combining this result with (21.132), we obtain

$$\theta_1 \cong 13^\circ \quad \text{and} \quad \theta_3 \leqslant 16^\circ \quad . \tag{21.134}$$

In principle, by using the appropriate weak and electromagnetic gauge theory we can determine θ_2 from the $K_L^o - K_S^o$ mass difference and the top-quark mass m_t ; we may also infer δ from the CP violating parameter $\epsilon \cong 2 \times 10^{-3}$ in K decay. Since m_t is not known and because of the theoretical uncertainties due to the strong interaction, at present only crude estimations are possible. *

* From the dimuon production cross sections by neutrinos and antineutrinos, the product $\sin\theta_1 \cos\theta_2$ can be inferred experimentally. At present the data gives $\cos\theta_2 = 1.1 \pm .1$ (J. Steinberger, private communication).

Remarks. If we regard vacuum to be a physical medium, then it is reasonable to expect that the angle θ_c in the Cabibbo theory, or the above angles θ_1, θ_2, θ_3 and δ, may all have a dynamical origin; their numerical values represent the expectation values of some long-range correlations in the vacuum state (loosely speaking, analogous to the various optical angles in a crystal). In this sense, the underlying mechanism of the CP violating angle δ in the Kobayashi-Maskawa model may share a similar origin with the corresponding angle α in Section 16.4, where we discussed the spontaneous CP symmetry breaking example.

Problem 21.1.

(i) Prove that the phenomenological μ decay Lagrangian given by (21.2) is invariant under CP and T.

(ii) Neglect radiative corrections and electron and neutrino masses. Show that by using this phenomenological Lagrangian, the muon lifetime is

$$\tau_\mu^{-1} = G_\mu^2 m_\mu^5 (192\pi^3)^{-1}$$

and that the normalized final e distribution in the rest system of a completely polarized muon is

$$d^2 N_e = x^2 [3 - 2x \pm \cos\theta (1 - 2x)] \, dx \, d\cos\theta$$

where the upper sign is for e^- and the lower for e^+,

$$x = \text{e-momentum} / \tfrac{1}{2} m_\mu ,$$

θ = angle between the muon polarization and the e-momentum.

Problem 21.2. Prove that the decay rate of $\pi_{\ell 2}$ is given by (21.20),

$$\text{Rate}\ (\pi_{\ell 2}) = \frac{G^2 F_\pi^{\ 2} m_\ell^{\ 2}}{8\pi m_\pi^{\ 3}} (m_\pi^{\ 2} - m_\ell^{\ 2})^2 \cos^2\theta_c$$

where G and $F_\pi \cos\theta_c$ are defined by (21.13) and (21.19), and the masses of $\pi^\pm$ and $\ell^\pm$ are m_π and m_ℓ.

Problem 21.3. The longitudinal polarization of a spin-$\frac{1}{2}$ particle is defined to be the expectation value of $\vec{\sigma}\cdot\hat{p}$ where $\hat{p}$ is a unit vector along the particle momentum and the components of $\vec{\sigma}$ are the Pauli spin matrices, so that

$$\vec{\sigma}\cdot\hat{p} = 2 \times \text{helicity},$$

where the helicity is defined by (3.29). By using (21.43) show that, in a nuclear β decay, the longitudinal polarization of the β particle with velocity v in the rest system of the nucleus is

$$<\vec{\sigma}\cdot\hat{p}> = \mp v \tag{21.135}$$

where the minus sign is for e^- and the plus for e^+.

Problem 21.4. In the extreme non-relativistic limit, we can regard both the initial and final nucleons as being at rest.

(i) By using (3.26)-(3.27), show that the Dirac 4×1 spinor functions u_n and u_p in (21.44) can be replaced by their upper two components χ_n and χ_p:

$$u_N \rightarrow \chi_N = \binom{1}{0} \qquad \text{for spin} \uparrow$$

$$= \binom{0}{1} \qquad \text{for spin} \downarrow$$

where $N = n$ or p.

(ii) In this limit prove that for $\beta^{\pm}$ decay the matrix elements of the Lagrangian density (21.43) at position $\vec{r}$ can be written as

$$< n e^{+} \nu \mid \mathcal{L}(\vec{r}) \mid p > \; = -\sqrt{2}\, [G_V \chi_n^{\dagger} \chi_p < 1 >_{\ell} + G_A \chi_n^{\dagger} \vec{\sigma} \chi_p \cdot < \vec{\sigma} >_{\ell}]\, \delta^3(\vec{r} - \vec{r}_N) \tag{21.136}$$

and a similar equation for $< p e^{-} \bar{\nu} \mid \mathcal{L} \mid n >$, where $\nu = \nu_e$ and $< \Gamma >_{\ell}$ denotes the matrix element of the lepton operator $< e^{+} \nu \mid \psi_{\nu}^{\dagger}(\vec{r})\, \Gamma\, \psi_e(\vec{r}) \mid vac >$ with ψ_{ν} satisfying the two component conditions, $\gamma_5 \psi_{\nu} = \psi_{\nu}$ and $\Gamma = 1$ or $\vec{\sigma}$. [Historically, the sign convention of G_A / G_V was fixed by (21.136), and that explains the minus signs in (21.45)-(21.46).]

Problem 21.5. Treat the nucleons as nonrelativistic and neglect the nuclear size as compared to the β particle's de Broglie wavelength.

(i) Show that the decay rate of a nucleus Z,

$$Z \rightarrow Z' + e^{\mp} + \bar{\nu}_e \;(\nu_e) \tag{21.137}$$

is

$$\frac{\ln 2}{t} = \frac{m_e}{2\pi^3} \left(|G_V|^2 M_F^2 + |G_A|^2 M_{GT}^2 \right) f\!\left(\frac{E_m}{m_e} \right) \tag{21.138}$$

where $E_m = \sqrt{k_m^{\,2} + m_e^{\,2}}$ is the maximum β energy, t is the half-life,

$$f(x) = \frac{1}{60} (x^2 - 1)^{\frac{1}{2}} (2x^4 - 9x^2 - 8) + \frac{x}{4} \ln(x + \sqrt{x^2 - 1}), \tag{21.139}$$

$$M_F^2 \equiv \left| \int < Z' \mid \psi_N^{\dagger} \tau_{\pm} \psi_N \mid Z > d^3 r \right|^2 ,$$
$$M_{GT}^2 \equiv \left| \int < Z' \mid \psi_N^{\dagger} \tau_{\pm} \vec{\sigma} \psi_N \mid Z > d^3 r \right|^2 , \tag{21.140}$$

$$\tau_+ = \begin{pmatrix} 0 & 1 \\ 0 & 0 \end{pmatrix} = \tfrac{1}{2}(\tau_1 + i\,\tau_2) \;,$$

$$\tau_- = \begin{pmatrix} 0 & 0 \\ 1 & 0 \end{pmatrix} = \tfrac{1}{2}(\tau_1 - i\,\tau_2) \;,$$

$$\psi_N = \begin{pmatrix} \psi_p \\ \psi_n \end{pmatrix}$$

is the quantized nucleon field operator, and τ_+ (or τ_-) is for β^- (or β^+) decay.

(ii) Prove that in this approximation the selection rule for (21.137) is

$$M_F \neq 0 \quad \text{only if} \quad P' = P \quad \text{and} \quad J' = J \qquad (21.140a)$$

and

$$M_{GT} \neq 0 \quad \text{only if} \quad P' = P \quad \text{and} \quad J' = J \,, \ \text{or} \ J \pm 1$$

but (21.140b)

$$M_{GT} = 0 \quad \text{if} \quad J' = J = 0$$

where J and P are the spin and parity of Z, and J' and P' those of Z' ; (21.140a) is called the Fermi selection rule and (21.140b) the Gamow-Teller selection rule.

(iii) Tabulate the experimental values of ft for various allowed and forbidden β decays.

<u>Problem 21.6.</u> Consider the neutrino reaction (21.72)

$$\nu + N \rightarrow \ell + h$$

where the final hadron state h is of mass M and helicity s_h .

(i) Prove that the structure functions $W_1(q^2, M)$, $W_2(q^2, M)$ and $W_3(q^2, M)$, defined by (21.91), are related to the following squared matrix elements of the hadron currents J_λ in the laboratory system

$$\alpha_+ \equiv \sum_{s_h} \tfrac{1}{2} \,|< s_h \,|\, J_x + i J_y \,|\, s_N = s_h - 1 >|^2 \quad ,$$

$$\alpha_- \equiv \sum_{s_h} \tfrac{1}{2} \,|< s_h \,|\, J_x - i J_y \,|\, s_N = s_h + 1 >|^2 \qquad (21.141)$$

and

$$q^2 \alpha_0 \equiv \sum_{s_h} |< s_h \,|\, (E_h - m) J_z + i \,|\vec{P}|\, J_4 \,|\, s_N = s_h >|^2$$

by

$$W_1 = (\alpha_+ + \alpha_-) \frac{E_h}{m_N} \quad ,$$

$$W_2 = (\alpha_+ + \alpha_- + \alpha_0)\, q^2 \frac{E_h}{m_N P^2} \qquad (21.142)$$

and

$$W_3 = 2(-\alpha_+ + \alpha_-) \frac{E_h}{P} \quad ,$$

in which $P = |\vec{P}|$ and the z-axis in the laboratory system is chosen to be parallel to $\vec{P} = \vec{q}$; all other notations are given by (21.73) and (21.78).

(ii) Show that the differential cross section $d\sigma$, (21.94), can also be written in an alternative form

$$d\sigma = (16\pi\, m_N E_\nu^2 P^2)^{-1}\, G^2 q^2 E_h\, dq^2 [(E_\nu + E_\ell)^2 - P^2]$$
$$\cdot\, [\alpha_\pm \xi + \alpha_\mp \xi^{-1} + \alpha_0] \qquad (21.143)$$

where the upper sign is for ν_ℓ and the lower sign for $\bar{\nu}_\ell$, and

$$\xi = \frac{E_\nu + E_\ell - P}{E_\nu + E_\ell + P} \quad . \qquad (21.144)$$

Problem 21.7. Next we consider the inclusive neutrino reaction

$$\nu + N \rightarrow \ell + \cdots$$

in which the final states h are being summed over. Let x and y be the dimensionless variables given by (21.75). In the limit $q^2 \to \infty$, but keeping x and y fixed, we define

$$L(x) \equiv \lim_{q^2 \to \infty} \frac{\sum_h a_-}{\sum_h (a_+ + a_- + a_0)} ,$$

$$R(x) \equiv \lim_{q^2 \to \infty} \frac{\sum_h a_+}{\sum_h (a_+ + a_- + a_0)} \tag{21.145}$$

where the sum extends over all final hadron channels h of the same mass M. Show that in the same limit, called the <u>scaling limit</u>, *

$$x(F_1 + \tfrac{1}{2}F_3)/F_2 \to L(x) ,$$

$$x(F_1 - \tfrac{1}{2}F_3)/F_2 \to R(x) ; \tag{21.146}$$

therefore, (21.95) becomes

$$d^2\sigma(\nu_\ell) = \pi^{-1} G^2 m_N E_\nu \, dx\, dy\, F_2(x) \, [\, (1-y) + y L(x) - y(1-y) R(x)]$$

and

$$d^2\sigma(\bar{\nu}_\ell) = \pi^{-1} G^2 m_N E_\nu \, dx\, dy\, F_2(x) \, [\, (1-y) + y R(x) - y(1-y) L(x)] \tag{21.147}$$

where $F_2(x)$ is the scaling limit of $F_2(q^2, x)$, i.e.,

$$F_2(x) = \lim_{q^2 \to \infty} F_2(q^2, x) .$$

<u>Problem 21.8.</u>

(i) Prove that the differential cross section for the inclusive reaction

$$e^- + N \to e^- + \cdots$$

---

* See Section 23.1 for further discussions.

can be written as

$$d^2\sigma = \frac{m_N E_e}{2\pi} \left(\frac{e^2}{q^2}\right)^2 dx\,dy\,[\,xy^2 F_1^{el}(q^2, x) + (1 - y - \frac{m_N}{2q_0} xy^2)\, F_2^{el}(q^2, x)] \tag{21.148}$$

in which we take into account only the lowest-order diagram due to photon-exchange and use the same notation as in the two previous problems, except that ν is replaced by e, J_λ by the hadronic electromagnetic current operator J_λ^{el}, and the structure functions F_1 and F_2 by F_1^{el} and F_2^{el}. (Because of parity conservation $F_3^{el} = 0$.) The mass of the electron is set $= 0$.

(ii) Show that in the scaling limit, (21.148) becomes

$$d^2\sigma = 4\pi \left(\frac{\alpha}{q^2}\right)^2 dq^2\, d\omega\, [\, \left(\frac{q^2}{s}\right)^2 F_1^{el}(x) + \left(\frac{1}{\omega} - \frac{q^2}{s}\right) F_2^{el}(x)\,] \tag{21.149}$$

where

$$\alpha = \frac{e^2}{4\pi},$$

$$\omega = x^{-1},$$

$$s = -(k+p)^2, \tag{21.150}$$

$$F_1^{el}(x) = \lim_{q^2\to\infty} F_1^{el}(q^2, x)$$

and

$$F_2^{el}(x) = \lim_{q^2\to\infty} F_2^{el}(q^2, x) \,.$$

Problem 21.9. The s-wave scattering cross section for

$$a + b \rightarrow c + d$$

is related to the S-matrix by

$$\sigma = \pi k^{-2} \, | \langle cd \,|\, S - 1 \,|\, ab \rangle |^2$$

where k is the magnitude of the momentum of a, or b, in the center-of-mass system.

Show that

(i) for the elastic s-wave scattering, $a + b \to a + b$,

$$\sigma = \pi k^{-2} \left| e^{2i\delta} - 1 \right|^2 \leqslant \frac{4\pi}{k^2} \qquad (21.151)$$

where $\exp(2i\delta)$ is the diagonal matrix element of S, and

(ii) for an inelastic scattering

$$\sigma \leqslant \frac{\pi}{k^2} \quad . \qquad (21.152)$$

(iii) Consider the process

$$e^- + \nu_\mu \to \mu^- + \nu_e$$

in the center-of-mass system. In this reaction, only the lefthanded electron, e_L^-, can interact. Prove that for an initially unpolarized e^-

$$\sigma(e^- \nu_\mu \to \mu^- \nu_e) = \tfrac{1}{2} \sigma(e_L^- \nu_\mu \to \mu^- \nu_e)$$

$$\leqslant \frac{\pi}{2k^2} \quad . \qquad (21.153)$$

References

K. Siegbahn, ed., Alpha, Beta and Gamma-ray Spectroscopy (Amsterdam, North-Holland Publishing Co., 1965).

T. D. Lee and C. S. Wu, Annual Review of Nuclear Science 15, 381 (1965), 16, 471, 511 (1966).

E. J. Konopinski, The Theory of Beta Radioactivity (Oxford, The Clarendon Press, 1966).

C. S. Wu and S. A. Moszkowski, Beta Decay (New York, Wiley-Interscience, 1966).

R. E. Marshak, Riazuddin and C. P. Ryan, Theory of Weak Interactions in Particle Physics (New York, Wiley-Interscience, 1969).

Chapter 22

WEAK AND ELECTROMAGNETIC GAUGE THEORY

22.1 Nambu-Goldstone and Higgs Mechanisms

As remarked in Section 21.9, the desire to unify the weak and the electromagnetic forces has been shared by many physicists since the early days of the weak interaction. The obvious path of assuming a non-Abelian gauge theory encountered several serious difficulties: All gauge field quanta are, at least to begin with, of zero mass. Yet the intermediate boson, if it exists, must be of a very heavy mass, perhaps in the range of 30 - 300 GeV as estimated by (21.114)-(21.115). Furthermore, the symmetry properties of the weak and the electromagnetic interactions appear to be quite different. Up to the 1960's, it was not clear how these disparities could be reconciled. The resolution came gradually. Of particular importance in this evolution were the concepts of "spontaneous symmetry breaking" (Section 16.4) and the Nambu-Goldstone * and Higgs mechanisms ** , which we shall discuss in this section.

---

* Y. Nambu, Phys.Rev.Letters 4, 380 (1960); J. Goldstone, Nuovo Cimento 19, 154 (1961); Y. Nambu and G. Jona-Lasinio, Phys. Rev. 122, 345 (1961), 124, 246 (1961).

** P. W. Higgs, Phys.Lett. 12, 132 (1964), Phys.Rev.Lett. 13, 508 (1964), and Phys.Rev. 145, 1156 (1966); F. Englert and R. Brout, Phys.Rev. Lett. 13, 321 (1964); G. S. Guralnik, C. R. Hagen and T. W. B. Kibble, Phys.Rev.Lett. 13, 585 (1964); T. W. B. Kibble, Phys.Rev. 155, 1554 (1967).

1. <u>An example</u>

Let us consider the example of an SO_3 gauge theory.* The group elements in $\{u\}$ are 3×3 real orthogonal matrices of unit determinant:

$$u^\dagger u = 1 \quad , \quad u^* = u$$

and

$$\det u = 1 \quad .$$

There are three generators

$$T^x \; , \; T^y \text{ and } T^z \tag{22.1}$$

which satisfy the commutation relations

$$[T^\ell \; , \; T^m] = i \, \epsilon^{\ell m n} \, T^n \tag{22.2}$$

with each of the superscripts = x, y or z, and $\epsilon^{\ell mn}$ given by (12.10). The SO_3 transformations will be referred to as "isospin" rotations. To construct the gauge theory, we follow the procedures and notations of Section 18.1. Corresponding to each T^ℓ, there is a gauge field $V_\mu^{\ \ell}$; altogether there are three spin-1 fields $V_\mu^{\ x}$, $V_\mu^{\ y}$ and $V_\mu^{\ z}$, forming an "isospin" vector $\vec{V}_\mu$. In addition, let us assume there is a spin-0 matter field $\vec{\phi}$, which is also an isovector. The Lagrangian density is

$$\mathcal{L} = -\tfrac{1}{4} \vec{V}_{\mu\nu}^{\ 2} - \tfrac{1}{2} (D_\mu \vec{\phi})^2 - U(\phi) \tag{22.3}$$

where

$$\vec{V}_{\mu\nu} = \frac{\partial}{\partial x_\mu} \vec{V}_\nu - \frac{\partial}{\partial x_\nu} \vec{V}_\mu + g(\vec{V}_\mu \times \vec{V}_\nu) \quad ,$$

$\vec{\phi}$ is Hermitian,

$$\phi = |\vec{\phi}| \quad ,$$

---

* H. Georgi and S. Glashow, Phys.Rev.Lett. <u>13</u>, 168 (1964). [See, e.g., pages 264–65 for a discussion of the $\overline{SO_3}$ group.]

$$D_\mu \vec{\phi} = \left(\frac{\partial}{\partial x_\mu} + g \vec{V}_\mu \times \right) \vec{\phi} \tag{22.4}$$

and, as in (16.8), in order to have a spontaneous symmetry breaking

$$U(\phi) = \frac{\mu^2}{8\rho^2} (\phi^2 - \rho^2)^2 \quad . \tag{22.5}$$

The constants μ^2, ρ^2 and g are all real so that $\mathcal{L}$ is a Hermitian. Furthermore, we choose both μ and $\rho > 0$. It is straightforward to show that, similarly to (18.12), $\mathcal{L}$ is invariant under the infinitesimal local gauge transformation

$$\vec{V}_\mu \rightarrow \vec{V}_\mu + \delta \vec{V}_\mu$$

and

$$\vec{\phi} \rightarrow \vec{\phi} + \delta \vec{\phi} \quad ,$$

where

$$\begin{aligned} \delta \vec{V}_\mu &= \vec{\theta} \times \vec{V}_\mu - \frac{1}{g} \frac{\partial \vec{\theta}}{\partial x_\mu} \quad , \\ \delta \vec{\phi} &= \vec{\theta} \times \vec{\phi} \end{aligned} \tag{22.6}$$

and $\vec{\theta} = \vec{\theta}(x)$ is an arbitrary infinitesimal isovector function of space and time. The quantization procedure can be carried out in the standard way by following the steps outlined in Chapter 18.

As in (16.12), the vacuum expectation value of $\vec{\phi}$ is determined by the minimum of $U(\phi)$. From (22.5), or Fig. 22.1, we see that the magnitude of the vacuum expectation value of $\vec{\phi}$ is

$$| \langle \vec{\phi} \rangle_{vac} | = \rho > 0 \quad , \tag{22.7}$$

which is independent of the direction of $\langle \vec{\phi} \rangle_{vac}$. Here, because SO_3 is a continuous group, the vacuum degeneracy is a continuous one. [In contrast, the vacuum discussed in Section 16.4 has only a discrete two-fold degeneracy.]

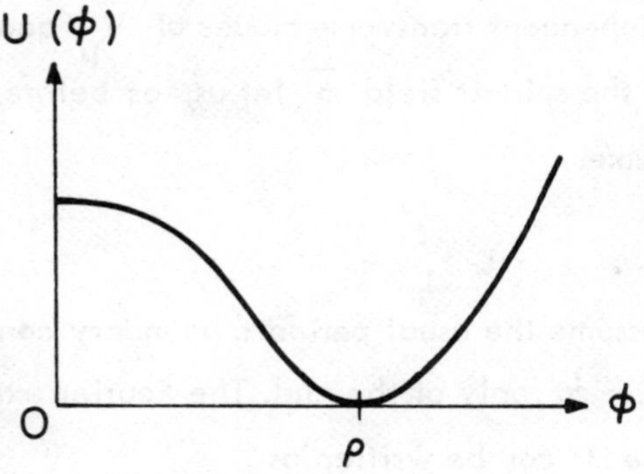

Fig. 22.1. A schematic drawing of $U(\phi) = \frac{\mu^2}{8\rho^2}(\phi^2 - \rho^2)^2$ where $\phi = |\vec{\phi}|$.

2. Limit $g = 0$

In this limit, the vector field is decoupled from the scalar field, and the Lagrangian density (22.3) becomes

$$\mathcal{L} \rightarrow \mathcal{L}_0 = \mathcal{L}_V + \mathcal{L}_\phi \tag{22.8}$$

where

$$\mathcal{L}_V = -\frac{1}{4}\left(\frac{\partial \vec{V}_\nu}{\partial x_\mu} - \frac{\partial \vec{V}_\mu}{\partial x_\nu}\right)^2 , \tag{22.9}$$

$$\mathcal{L}_\phi = -\frac{1}{2}\left(\frac{\partial \vec{\phi}}{\partial x_\mu}\right)^2 - U(\phi) . \tag{22.10}$$

By comparing (22.9) with (6.3), we see that each isospin component of $\vec{V}_\mu$ can be quantized in the same way as the free electromagnetic field. For example, in the Coulomb gauge, we have the transversality condition

$$\nabla_i \vec{V}_i = 0 . \tag{22.11}$$

As in (6.32)-(6.39), for a given momentum and isospin direction, there are only two independent transverse modes of $\vec{V}_\mu$ quanta.

To discuss the spin-0 field $\vec{\phi}$ let us, as before, introduce a finite cubic volume

$$\Omega = L^3$$

for the system, assume the usual periodic boundary condition, and take the limit $\Omega \to \infty$ only at the end. The Fourier expansion of $\vec{\phi}(x)$ at any given time t can be written as

$$\vec{\phi}(x) = \vec{Q}(t) + \sum_{k \neq 0} \frac{\vec{q}_k(t)}{\sqrt{\Omega}} \exp(i k_j x_j) \tag{22.12}$$

where x is the four-dimensional spatial coordinate

$$x_\mu = (x_1, x_2, x_3, it) ,$$

but k stands for the 3-vector k_j which satisfies, as in (2.13),

$$k_j = \frac{2\pi \ell_j}{L}$$

with

$$\ell_j = 0, \pm 1, \pm 2, \cdots .$$

In (22.12) the repeated index j is summed over from 1 to 3, as usual. On account of (22.7), we anticipate $\langle \vec{\phi} \rangle_{vac}$ to be $O(\Omega^0)$, not $O(\Omega^{-\frac{1}{2}})$. Hence, in the above Fourier expansion it is more convenient to separate out the $k = 0$ component and write it as Q, instead of $q_0/\sqrt{\Omega}$ as in (2.11).

From (22.10), it follows that the conjugate momentum of $\vec{\phi}(x)$ is

$$\vec{\Pi} = \frac{\partial \mathcal{L}_\phi}{\partial \dot{\vec{\phi}}} = \dot{\vec{\phi}} , \tag{22.13}$$

which leads to the Hamiltonian density

$$\mathcal{H}_\phi = \tfrac{1}{2} \vec{\Pi}^2 + \tfrac{1}{2} (\nabla_i \vec{\phi})^2 + U(\phi) . \tag{22.14}$$

The Fourier expansion of $\Pi(x)$ is

$$\Pi(x) = \frac{\vec{P}(t)}{\Omega} + \sum_{k\neq 0} \frac{\vec{p}_{-k}(t)}{\sqrt{\Omega}} \exp(i k_j x_j) \tag{22.15}$$

in which the $k = 0$ component is written as P/Ω, instead of $p_0/\sqrt{\Omega}$ as in (2.12). This difference in notation can be appreciated by substituting (22.12) into the Lagrangian. We find

$$L_\phi = \int \mathcal{L}_\phi \, d^3r = \tfrac{1}{2}\Omega \dot{\vec{Q}}^2 + \cdots$$

where, on account of (22.10), the $\cdots$ terms are all independent of $\dot{\vec{Q}}$. The conjugate momentum of $\vec{Q}$ is therefore

$$\vec{P} = \frac{\partial L_\phi}{\partial \dot{\vec{Q}}} = \Omega \dot{\vec{Q}} ,$$

and the Hamiltonian is

$$H_\phi \equiv \int \mathcal{H}_\phi \, d^3r = \frac{\vec{P}^2}{2\Omega} + \cdots \tag{22.16}$$

where $\cdots$ represents the $\vec{P}$-independent terms. The same expression must also result if we directly substitute the Fourier expansion of $\vec{\Pi}(x)$ into the Hamiltonian density; this then explains the first term on the righthand side of (22.15).

To carry out the quantization, we apply the standard canonical commutation relations. As in (2.23)-(2.24), we have

$$[P^\ell(t), Q^m(t)] = -i\delta^{\ell m} ,$$

$$[p_k^\ell(t), q_{k'}^m(t)] = -i\delta^{\ell m}\delta_{kk'}$$

and all other equal-time commutators between these coordinates and momenta are zero.

From (22.15) we see that the "inertia" associated with the $k = 0$ mode tends to infinity as $\Omega \to \infty$. Hence, although $\vec{P}$ and $\vec{Q}$ do not commute, we can regard $\vec{Q}$ as a classical variable in the infinite-volume limit. Equation (22.7) can then be written as

$$< \vec{\phi} >_{vac} = < \vec{Q} >_{vac} = \rho \hat{e} \tag{22.17}$$

where $\hat{e}$ is an arbitrary time-independent unit vector in the isospin space, since its time derivative is proportional to Ω^{-1}, which is zero as $\Omega \to \infty$. Without any loss of generality, we may call

$$\hat{e} \equiv \hat{z} \tag{22.18}$$

which is a unit vector along the z-axis in the isospin space.

Next, as on page 386 we expand $\phi(x)$ around its vacuum expectation value, and write

$$\vec{\phi}(x) = \rho \hat{z} + \delta\vec{\phi}(x) \quad . \tag{22.19}$$

In terms of $\delta\vec{\phi}(x)$, (22.5) and (22.14) become

$$U(\phi) = \tfrac{1}{2}\mu^2(\delta\phi^z)^2 + \cdots$$

and

$$\mathcal{H}_\phi = \tfrac{1}{2}(\vec{\Pi})^2 + \tfrac{1}{2}(\nabla_i\, \delta\vec{\phi})^2 + \tfrac{1}{2}\mu^2(\delta\phi^z)^2 + \cdots \tag{22.20}$$

where the $\cdots$ terms are all cubic or quartic in $\delta\vec{\phi}$. Ignoring these higher-order terms, by examining the Hamiltonian in its quadratic form we see that the mass of the quantum due to oscillations in $\delta\phi^z$ is very different from those due to $\delta\phi^x$ and $\delta\phi^y$. We have

$$\text{mass }(\delta\phi^z) = \mu \neq 0$$

and

$$\text{mass }(\delta\phi^x) = \text{mass }(\delta\phi^y) = 0 \quad . \tag{22.21}$$

This is, of course, a direct consequence of $U(\phi)$ being invariant under an isospin rotation. As shown in Fig. 22.2, with the spontaneous symmetry breaking (22.17), a variation $\delta\vec{\phi} \perp \hat{e}$ is a rotation in $\vec{\phi}$, whereas a $\delta\vec{\phi} \,/\!/\, \hat{e}$ is not.

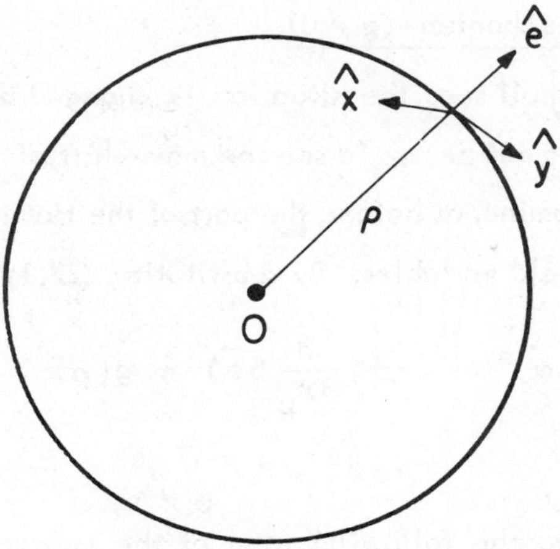

Fig. 22.2. In the $\vec{\phi}$ space, the vacuum expectation value $\langle \vec{\phi} \rangle_{vac} = \rho \hat{e}$ selects a direction $\hat{e}$. Since $U(\phi)$ is unchanged under oscillations perpendicular to $\hat{e}$, this gives rise to the zero-mass Goldstone bosons (if the vector coupling $g = 0$).

If the parameter ρ were zero, $U(\phi)$ would be proportional to ϕ^4 and there would be three spin-0 zero-mass bosons (at least in the approximation of neglecting radiative corrections). In that case, the vacuum would be invariant under any SO_3 rotation. However, because of the spontaneous symmetry breaking, $\rho \neq 0$ and a direction $\hat{e}$ is selected in the isospin space. The vacuum remains invariant under a two-dimensional rotation around $\hat{e}$. Consequently, there are still two spin-0 zero-mass fields; their quanta are called Goldstone bosons. *

In the limit $g = 0$, the theory consists of one massive spin-0 field $\delta\phi^z$, two zero-mass Goldstone boson fields $\delta\phi^x$ and $\delta\phi^y$, and three zero-mass spin-1 fields V_μ^x, V_μ^y and V_μ^z.

---

* See Section 24.3 for further discussion.

3. Higgs mechanism $(g \neq 0)$

As we shall see, the situation is changed drastically once the coupling g is not zero. To see the mass-shift of these particles, we need only examine, as before, the part of the Hamiltonian that is quadratic in the field variables. By substituting (22.19) into (22.4), we obtain

$$-\tfrac{1}{2}(D_\mu \vec{\phi})^2 = -\tfrac{1}{2}\left(\frac{\partial}{\partial x_\mu}\delta\vec{\phi}\right)^2 + g(\rho\hat{z}\times\vec{V}_\mu)\cdot\frac{\partial}{\partial x_\mu}\delta\vec{\phi} - \tfrac{1}{2}g^2(\rho\hat{z}\times\vec{V}_\mu)^2 + \cdots , \qquad (22.22)$$

which leads to the following form of the Lagrangian density (22.3)

$$\mathcal{L} = -\tfrac{1}{4}\left(\frac{\partial}{\partial x_\mu}\vec{V}_\nu - \frac{\partial}{\partial x_\nu}\vec{V}_\mu\right)^2 - \tfrac{1}{2}\left(\frac{\partial}{\partial x_\mu}\delta\vec{\phi}\right)^2 - \tfrac{1}{2}g^2\rho^2[(V_\mu^x)^2 + (V_\mu^y)^2] - \tfrac{1}{2}\mu^2(\delta\phi^z)^2 + g\rho\left[V_\mu^x\frac{\partial}{\partial x_\mu}\delta\phi^y - V_\mu^y\frac{\partial}{\partial x_\mu}\delta\phi^x\right] + \cdots \qquad (22.23)$$

where the $\cdots$ terms are all cubic or quartic in $\vec{V}_\mu$ and $\delta\vec{\phi}$. By defining

$$W_\mu^x \equiv V_\mu^x - \frac{1}{g\rho}\frac{\partial}{\partial x_\mu}\delta\phi^y$$

and

$$W_\mu^y \equiv V_\mu^y + \frac{1}{g\rho}\frac{\partial}{\partial x_\mu}\delta\phi^x , \qquad (22.24)$$

we can rewrite (22.23) as

$$\mathcal{L} = \sum_{\ell=x,y}\left[-\tfrac{1}{4}\left(\frac{\partial}{\partial x_\mu}W_\nu^\ell - \frac{\partial}{\partial x_\nu}W_\mu^\ell\right)^2 - \tfrac{1}{2}m^2(W_\mu^\ell)^2\right] - \tfrac{1}{4}\left(\frac{\partial}{\partial x_\mu}V_\nu^z - \frac{\partial}{\partial x_\nu}V_\mu^z\right)^2 - \tfrac{1}{2}\left(\frac{\partial}{\partial x_\mu}\delta\phi^z\right)^2 - \tfrac{1}{2}\mu^2(\delta\phi^z)^2 + \cdots \qquad (22.25)$$

where

$$m = g\rho \quad . \tag{22.26}$$

By comparing the terms inside the sum with those in (4.1), we see that the theory now consists of only one zero-mass field, V_μ^z. The rest are all massive; the spin-1 fields W_μ^x and W_μ^y are of mass m and the spin-0 field $\delta\phi^z$ is of mass μ.

When $g=0$, as noted before, for a given momentum each of the three zero-mass spin-1 fields V_μ^x, V_μ^y and V_μ^z has only two transverse modes; in addition there are three spin-0 fields, two massless and one massive. Altogether, there are

$$2 \times 3 + 3 = 9$$

different quantum modes for a given momentum, among which

$$2 \times 3 + 2 = 8$$

are of zero mass. For $g \neq 0$, there are two massive spin-0 fields W_μ^x and W_μ^y; each has three quantum modes for a given momentum. These together with the zero-mass spin-1 field V_μ^z and the massive spin-0 field $\delta\phi^z$ give, again, a total of

$$3 \times 2 + 2 + 1 = 9$$

different quantum modes for a given momentum. The longitudinal modes of W_μ^x and W_μ^y are generated by $\partial\delta\phi^y/\partial x_\mu$ and $\partial\delta\phi^x/\partial x_\mu$, as shown in (22.24). The conversion of "massless" gauge fields and Goldstone bosons into massive spin-1 mesons is the "Higgs mechanism." In the above example, the massive vector fields are prototypes of the intermediate boson, the zero-mass gauge field V_μ^z simulates the photon, and the remaining massive spin-0 field is called the Higgs boson.

4. Unitary gauge

An alternative derivation of the above result is to take advantage of the gauge invariance of the Lagrangian density (22.3). Through the continuous application of the gauge transformation (22.6), we can, at any space-time point, rotate $\vec{\phi}(x)$ so that it is parallel to the z-axis in the isospin space. * The result is

$$\vec{\phi}(x) = \hat{z}\, \phi^z(x) \tag{22.27}$$

everywhere. The gauge thus chosen is called the unitary gauge. Because of (22.7) we can write

$$\phi^z(x) = \rho + \delta\phi^z \quad . \tag{22.28}$$

Substituting (22.27)-(22.28) into (22.3)-(22.4), we obtain

$$D_\mu \vec{\phi} = \hat{z} \frac{\partial}{\partial x_\mu} \delta\phi^z + g(\vec{V}_\mu \times \hat{z})(\rho + \delta\phi^z)$$

and, therefore,

$$\mathcal{L} = -\tfrac{1}{4}\left(\frac{\partial \vec{V}_\mu}{\partial x_\nu} - \frac{\partial \vec{V}_\nu}{\partial x_\mu}\right)^2 - \tfrac{1}{2} m^2 (V_\mu^x V_\mu^x + V_\mu^y V_\mu^y)$$
$$-\tfrac{1}{2}\left(\frac{\partial}{\partial x_\mu} \delta\phi^z\right)^2 - \tfrac{1}{2}\mu^2 (\delta\phi^z)^2 + \cdots \tag{22.29}$$

where m is given by (22.26), and the ⋯ terms are all cubic or quartic in $\vec{V}_\mu$ and $\delta\phi^z$. It is straightforward to carry out the canonical quantization in the unitary gauge. From (22.29), we see that V_μ^x and V_μ^y are massive, but V_μ^z is massless. In the unitary gauge,

$$\phi^x = \phi^y = 0$$

and $\delta\phi^z$ is massive.

---

* This is possible only for states without topological solitons. [See page 127.]

Remarks. In a gauge theory, the gauge fields V_μ^ℓ transform according to (18.9); thus, they must belong to the same representation as that of the generators T^ℓ of the group. On the other hand, the spin-0 field in the Higgs mechanism does not have to. Such an example will be discussed in the next section.

Exercise 1. In the unitary gauge (22.27)-(22.28), define

$$\chi \equiv \delta\phi^z , \quad A_\mu \equiv V_\mu^z ,$$

$$W_\mu \equiv \frac{1}{\sqrt{2}}(V_\mu^x - i V_\mu^y) ,$$

$$\bar{W}_\mu \equiv \frac{1}{\sqrt{2}}(V_\mu^x + i V_\mu^y) ,$$

$$F_{\mu\nu} \equiv \frac{\partial}{\partial x_\mu} A_\nu - \frac{\partial}{\partial x_\nu} A_\mu , \tag{22.30}$$

$$W_{\mu\nu} \equiv \left(\frac{\partial}{\partial x_\mu} - igA_\mu\right) W_\nu - \left(\frac{\partial}{\partial x_\nu} - igA_\nu\right) W_\mu$$

and

$$\bar{W}_{\mu\nu} \equiv \left(\frac{\partial}{\partial x_\mu} + igA_\mu\right) \bar{W}_\nu - \left(\frac{\partial}{\partial x_\nu} + igA_\nu\right) \bar{W}_\mu .$$

Hence,

$$\bar{W}_\mu = \begin{cases} W_\mu^\dagger & \text{if } \mu \neq 4 \\ -W_4^\dagger & \text{if } \mu = 4 . \end{cases} \tag{22.31}$$

Show that the Lagrangian density (22.3) can be written as

$$\begin{aligned} \mathcal{L} = & -\tfrac{1}{4} F_{\mu\nu}^2 - \tfrac{1}{2} \bar{W}_{\mu\nu} W_{\mu\nu} - m^2 \bar{W}_\mu W_\mu \\ & -\tfrac{1}{2}\left(\frac{\partial \chi}{\partial x_\mu}\right)^2 - \tfrac{1}{2}\mu^2\chi^2 - igF_{\mu\nu}\bar{W}_\mu W_\nu \\ & -\frac{g^2}{2}[(\bar{W}_\mu W_\mu)^2 - (\bar{W}_\mu W_\nu)^2] \\ & - \bar{W}_\mu W_\mu (2mg\chi + g^2\chi^2) - \frac{\mu^2}{2\rho}\chi^3\left(1 + \frac{\chi}{4\rho}\right) \end{aligned} \tag{22.32}$$

where m is given by (22.26).

Exercise 2. If we regard A_μ in the above problem as the electromagnetic field and the coupling g as the electric charge e, show that the electromagnetic current is

$$j_\mu = -ie\,[\bar{W}_\nu W_{\mu\nu} - \bar{W}_{\mu\nu} W_\nu + \frac{\partial}{\partial x_\nu}(\bar{W}_\mu W_\nu - \bar{W}_\nu W_\mu)] \tag{22.33}$$

which satisfies the current conservation law

$$\frac{\partial j_\mu}{\partial x_\mu} = 0 .$$

Therefore, the quanta associated with W_μ and $\bar{W}_\mu$ are charged and will be called $W^\pm$, with the superscript denoting the charge.

Exercise 3. (i) Carry out the quantization process for the Lagrangian density (22.32) by following the method given in Chapters 2, 4 and 6. Show that the operator W_μ annihilates W^+ and creates W^-.

(ii) Prove that the magnetic moment vector for a W^+ at rest is

$$\vec{M} = \frac{e}{m}\vec{S} \tag{22.34}$$

where $\vec{S}$ is its spin-vector. Since it is customary to write

$$\vec{M} = (1+\kappa)\frac{e}{2m}\vec{S}$$

with κ as the anomalous moment, the W-boson has an "anomalous moment"

$$\kappa = 1 . \tag{22.35}$$

(iii) The quadrupole moment $\mathcal{Q}$ of the W-boson is defined to be the integral

$$\int (3z^2 - r^2)\,\rho\, d^3r$$

for a W^+ at rest with $S_z = +1$, where $i\rho$ is the expectation value of the fourth component of the current operator, j_4. Show that

$$\mathcal{Q} = -\frac{e}{m^2} . \tag{22.36}$$

22.2 Standard Model *

We shall now discuss the simplest realistic gauge model ** that attempts to unify the weak and electromagnetic interactions. At present, because the intermediate boson is only a theoretical invention, the high-energy behavior of the standard model that we shall discuss in this section is still far from being proved experimentally. Nevertheless, the remarkable agreement of the model with all available data already establishes it as a good low-energy phenomenological theory.

1. Gauge group

To arrive at the appropriate gauge group for the unification, we first start from the known lepton currents for the (e^-, ν_e) system:

$$j_\lambda^- = i\,\psi_e^\dagger\,\gamma_4\,\gamma_\lambda(1+\gamma_5)\,\psi_\nu$$

and

$$j_\lambda^+ = i\,\psi_\nu^\dagger\,\gamma_4\,\gamma_\lambda(1+\gamma_5)\,\psi_e \tag{22.37}$$

for the weak interaction, and

$$j_\lambda^{el} = -i\,\psi_e^\dagger\,\gamma_4\,\gamma_\lambda\,\psi_e \tag{22.38}$$

---

* S. Weinberg, Phys.Rev.Lett. 19, 1264 (1967).

** See also S. L. Glashow, Nucl.Phys. 22, 579 (1961); A. Salam and J. C. Ward, Nuovo Cimento 11, 568 (1959), Phys.Lett 13, 168 (1964). Cf. in addition the articles by A. Salam, Proceedings of the Eighth Nobel Symposium, ed. N. Svartholm (New York, Wiley-Interscience, 1968), and M. Veltman, Proceedings of the Sixth International Symposium on Electron and Photon Interactions at High Energies, eds. H. Rollnik and W. Pfeil (Amsterdam, North-Holland Publishing Co., 1974).

for the electromagnetic interaction, where the subscript ν stands for ν_e and the minus sign in (22.38) is on account of the negative charge of e^-. The structures of these currents have been well determined experimentally; their support ranges from the extensive β decay measurements to the numerous QED tests. Let us consider the "charges" of these currents:

$$Q^- \equiv -i \int j_4^- d^3r = \int \psi_e^\dagger (1+\gamma_5)\, \psi_\nu\, d^3r \;,$$

$$Q^+ \equiv -i \int j_4^+ d^3r = \int \psi_\nu^\dagger (1+\gamma_5)\, \psi_e\, d^3r \qquad (22.39)$$

and

$$Q \equiv -i \int j_4^{el} d^3r = -\int \psi_e^\dagger\, \psi_e\, d^3r$$

in which the Q without superscript denotes the electric charge. Clearly, these charges are all physical observables; the same must therefore also be true for their commutators. The algebra satisfied by these observables can be determined by repeatedly commuting these charges and their commutators, as we shall see.

It is convenient to define

$$T^x \equiv \tfrac{1}{4}(Q^- + Q^+) \;,$$

$$T^y = \frac{i}{4}(Q^- - Q^+) \;, \qquad (22.40)$$

and to combine ψ_e and ψ_ν into a single column matrix

$$\psi = \begin{pmatrix} \psi_\nu \\ \psi_e \end{pmatrix} . \qquad (22.41)$$

Thus, T^x, T^y and Q can also be written as

$$T^x = \tfrac{1}{4} \int \psi^\dagger (1+\gamma_5)\, \tau^x\, \psi\, d^3r \;,$$

$$T^y = \tfrac{1}{4} \int \psi^\dagger (1+\gamma_5)\, \tau^y\, \psi\, d^3r \qquad (22.42)$$

and

$$Q = \frac{1}{2} \int \psi^{\dagger}(\tau^{z} - 1)\, \psi\, d^3r$$

where τ^x, τ^y and τ^z are the Pauli matrices τ_1, τ_2 and τ_3 given by (3.1). Because of (3.2) and $\gamma_5{}^2 = 1$, we have

$$[\tfrac{1}{4}(1+\gamma_5)\tau^i,\ \tfrac{1}{4}(1+\gamma_5)\tau^j] = i\,\tfrac{1}{4}(1+\gamma_5)\,\epsilon^{ijk}\,\tau^k$$

where i, j, k can be x, or y, or z. By using (3.24a) we see that the commutator of T^x and T^y leads to

$$T^z \equiv -i[\,T^x, T^y] = \tfrac{1}{4}\int \psi^{\dagger}(1+\gamma_5)\,\tau^z\,\psi\, d^3r\ . \qquad (22.43)$$

Likewise, we can verify

$$[\,T^i,\ T^j] = i\,\epsilon^{ijk}\,T^k\ . \qquad (22.44)$$

Just as with an isospin vector, we can regard T^i as the component of

$$\vec{T} = \tfrac{1}{4}\int \psi^{\dagger}(1+\gamma_5)\,\vec{\tau}\,\psi\, d^3r\ . \qquad (22.45)$$

From (22.42) - (22.43) we see that T^z commutes with Q, but T^x and T^y do not. Let T' be the difference between Q and T^z:

$$\begin{aligned} T' &\equiv Q - T^z \\ &= \int \psi^{\dagger}\,[\tfrac{1}{4}(1-\gamma_5)\,\tau^z - \tfrac{1}{2}]\ \psi\, d^3r\ . \end{aligned} \qquad (22.46)$$

Since

$$(1+\gamma_5)(1-\gamma_5) = 0\ ,$$

it follows that T' commutes with all the T_i 's; i.e.,

$$[\,T',\ \vec{T}] = 0\ . \qquad (22.47)$$

Thus, from the three observed charges Q^-, Q^+, Q and their commutators we obtain these four Hermitian operators T' and T^i. Because of (22.45) and (22.47), they generate a group that is the direct product

$$U_1 \times SU_2 \tag{22.48}$$

with T' as the generator of U_1 and $\vec{T}$ that of SU_2.

As in (3.101), we decompose ψ_e into a lefthanded component

$$\psi_L = \tfrac{1}{2}(1 + \gamma_5)\,\psi_e \tag{22.49}$$

and a righthanded component

$$\psi_R = \tfrac{1}{2}(1 + \gamma_5)\,\psi_e \equiv R \ . \tag{22.50}$$

The former will be combined with the two-component neutrino field ψ_ν, which satisfies

$$\gamma_5\,\psi_\nu = \psi_\nu$$

to form

$$L \equiv \begin{pmatrix} \psi_\nu \\ \psi_L \end{pmatrix} = \tfrac{1}{2}(1 + \gamma_5)\,\psi \tag{22.51}$$

where ψ is given by (22.41).

Let ν_e, e_L^- and e_R^- be the one-particle states generated by applying $\psi_\nu^\dagger$, $\psi_L^\dagger$ and $\psi_R^\dagger$ to the vacuum. These states are all eigenstates of T, T^z, the electric charge Q and, therefore, also $T' = Q - T^z$. We have

	T	T^z	Q	T'	
ν_e	$\frac{1}{2}$	$\frac{1}{2}$	0	$-\frac{1}{2}$	
e_L^-	$\frac{1}{2}$	$-\frac{1}{2}$	-1	$-\frac{1}{2}$	(22.52)
e_R^-	0	0	-1	-1 .	

Hence, on account of (22.50) – (22.51), we see that

$$\begin{aligned} &L \quad \text{is of} \quad T = \tfrac{1}{2} \ , \quad T' = -\tfrac{1}{2} \\ \text{and} \quad & \\ &R \quad \text{is of} \quad T = 0 \ , \quad T' = -1 \ . \end{aligned} \tag{22.53}$$

2. Lagrangian density

Once the gauge group is selected, the gauge fields are determined. Corresponding to each generator there must be a gauge field. For the group $SU_2 \times U_1$, there should be four:

$$\begin{array}{lll} \text{generator} & \vec{T} \ , & T' \\ \text{gauge field} & \vec{B}_\mu \ , & C_\mu \ . \end{array}$$

The pure gauge-field part of the Lagrangian is

$$-\tfrac{1}{4} \vec{B}_{\mu\nu}^{\ 2} - \tfrac{1}{4} C_{\mu\nu}^{\ 2}$$

where, as in (18.4),

$$\begin{aligned} \vec{B}_{\mu\nu} &= \frac{\partial}{\partial x_\mu} \vec{B}_\nu - \frac{\partial}{\partial x_\nu} \vec{B}_\mu + g \vec{B}_\mu \times \vec{B}_\nu \\ \text{and} \quad & \\ C_{\mu\nu} &= \frac{\partial}{\partial x_\mu} C_\nu - \frac{\partial}{\partial x_\nu} C_\mu \ . \end{aligned} \tag{22.54}$$

By themselves, the quanta of these four fields would all be of zero mass; at a given momentum each quantum would have only two transverse polarization degrees of freedom.

As in Section 22.1.3, we introduce the Higgs mechanism to give masses to three of the gauge fields, leaving only one massless, which will then be the photon. This means that there must be three would-be Goldstone bosons, so that they can be incorporated into the three massive vector particles as their appropriate longitudinal modes. In addition, there should also be at least one physical spin-0 Higgs particle. Thus, our system must contain a minimum of four spin-0

Hermitian fields, which is equivalent to two spin-0 complex fields. The simplest way is to put these two complex fields, ϕ_1 and ϕ_2, into a single two-dimensional representation under the $SU_2 \times U_1$ transformations. We write

$$\phi = \begin{pmatrix} \phi_1 \\ \phi_2 \end{pmatrix} \tag{22.55}$$

and assume it to be of the representation

$$T = \tfrac{1}{2} \qquad \text{and} \qquad T' = \tfrac{1}{2} \tag{22.56}$$

so that, because of (22.53),

$$L^\dagger R \phi \qquad \text{is of} \qquad T = T' = 0 \; . \tag{22.57}$$

The Lagrangian density of this system of leptons, gauge fields and ϕ is

$$\begin{aligned} \mathcal{L} = & -\tfrac{1}{4} \vec{B}_{\mu\nu}^{\,2} - \tfrac{1}{4} C_{\mu\nu}^{\,2} - R^\dagger \gamma_4 \gamma_\mu D_\mu R - L^\dagger \gamma_4 \gamma_\mu D_\mu L \\ & - (\overline{D}_\mu \phi^\dagger) D_\mu \phi - U(|\phi|) - f(L^\dagger \gamma_4 R\phi + R^\dagger \gamma_4 L \phi^\dagger) \end{aligned} \tag{22.58}$$

where $|\phi|^2 = \phi^\dagger \phi$,

$$\begin{aligned} D_\mu L &= \Big[\frac{\partial}{\partial x_\mu} - i g \tfrac{1}{2} \vec{\tau} \cdot \vec{B}_\mu - i(-\tfrac{1}{2}) g' C_\mu\Big] L \; , \\ D_\mu R &= \Big[\frac{\partial}{\partial x_\mu} - i(-1) g' C_\mu\Big] R \; , \\ D_\mu \phi &= \Big[\frac{\partial}{\partial x_\mu} - i g \tfrac{1}{2} \vec{\tau} \cdot \vec{B}_\mu - i(\tfrac{1}{2}) g' C_\mu\Big] \phi \end{aligned} \tag{22.59}$$

and

$$\overline{D}_\mu \phi^\dagger = \Big[\frac{\partial}{\partial x_\mu} + i g \tfrac{1}{2} \vec{\tau} \cdot \vec{B}_\mu + i(\tfrac{1}{2}) g' C_\mu\Big] \phi^\dagger$$

in which the numerical value inside the parenthesis denotes the T' of the field.

Exercise. Prove that the above Lagrangian density is invariant under the local $SU_2 \times U_1$ transformation:

$$B_\mu \to u B_\mu u^\dagger + \frac{i}{g} u \frac{\partial u^\dagger}{\partial x_\mu} ,$$

$$C_\mu \to C_\mu + \frac{1}{g'} \frac{\partial \alpha}{\partial x_\mu} ,$$

$$L \to \exp(i \alpha T_L') \, u L , \tag{22.60}$$

$$R \to \exp(i \alpha T_R') \, R$$

and

$$\phi \to \exp(i \alpha T_\phi') \, u \phi$$

where $\alpha(x)$ is an arbitrary real function, $u(x)$ is any 2×2 unitary matrix function with $\det u = 1$,

$$B_\mu \equiv \tfrac{1}{2} \vec{\tau} \cdot \vec{B}_\mu ,$$

$$T_L' = -\tfrac{1}{2} , \quad T_R' = -1 \tag{22.61}$$

and

$$T_\phi' = \tfrac{1}{2} .$$

3. Spontaneous symmetry breaking

Similarly to (22.5), we assume

$$U(|\phi|) = \frac{\mu^2}{4\rho^2} (\phi^\dagger \phi - \rho^2)^2 ; \tag{22.62}$$

its shape is again of the form given by Figure 22.1. The minimum of U is at $|\phi| = \rho$. Just as in (22.7), this means

$$| \langle vac \,|\, \phi \,|\, vac \rangle | = \rho . \tag{22.63}$$

To see the physical content of the theory, it is simplest to adopt the unitary gauge. Let us consider first the classical theory in that gauge. As in (22.27), at any space-time point we may choose the gauge transformation $u(x)$ so that in the last equation of (22.60) the final $\phi(x)$ is of the form

$$\phi = \begin{pmatrix} 0 \\ \phi_2 \end{pmatrix} ,$$

then by using the appropriate $\alpha(x)$ we can always set ϕ_2 real. Similarly to (22.28) and the first equation in (22.30), we may write

$$\phi_2(x) = \rho + \chi(x)$$

where $\chi(x)$ is a real field; hence, in the unitary gauge

$$\phi = \begin{pmatrix} 0 \\ \rho + \chi \end{pmatrix} . \tag{22.64}$$

By substituting it into the third equation of (22.59), we obtain

$$D_\mu \phi = \begin{pmatrix} 0 \\ \dfrac{\partial \chi}{\partial x_\mu} \end{pmatrix} - i \tfrac{1}{2} (\rho + \chi) \begin{pmatrix} g B_\mu^x - i g B_\mu^y \\ -g B_\mu^z + g' C_\mu \end{pmatrix} . \tag{22.65}$$

It is convenient to define

$$W_\mu \equiv \frac{1}{\sqrt{2}} (B_\mu^x - i B_\mu^y) ,$$

$$\bar{W}_\mu \equiv \frac{1}{\sqrt{2}} (B_\mu^x + i B_\mu^y) ,$$

$$Z_\mu \equiv \cos \theta_w B_\mu^z - \sin \theta_w C_\mu , \tag{22.66}$$

$$A_\mu \equiv \sin \theta_w B_\mu^z + \cos \theta_w C_\mu$$

and

$$\tan \theta_w \equiv \frac{g'}{g} ,$$

so that (22.65) becomes

$$D_\mu \phi = \begin{pmatrix} 0 \\ \\ \dfrac{\partial \chi}{\partial x_\mu} \end{pmatrix} - \frac{i}{\sqrt{2}} (\rho + \chi) g \begin{pmatrix} W_\mu \\ \\ \dfrac{-Z_\mu}{\sqrt{2} \cos \theta_W} \end{pmatrix} . \quad (22.67)$$

The Lagrangian density (22.58) is a fourth-order polynomial of χ, W_μ, $\bar{W}_\mu$, Z_μ, A_μ, ψ_e and ψ_ν. To find the masses of the physical particles we need only examine its quadratic part $\mathcal{L}_{quad}$ which is given by

$$\mathcal{L}_{quad} = \mathcal{L}_0 + \mathcal{L}_m , \quad (22.68)$$

with

$$\mathcal{L}_0 = -\tfrac{1}{2}\left(\frac{\partial}{\partial x_\mu} \chi\right)^2 - \tfrac{1}{2}\left(\frac{\partial}{\partial x_\mu} \bar{W}_\lambda - \frac{\partial}{\partial x_\lambda} \bar{W}_\mu\right)\left(\frac{\partial}{\partial x_\mu} W_\lambda - \frac{\partial}{\partial x_\mu} W_\lambda\right)$$

$$- \tfrac{1}{4}\left(\frac{\partial}{\partial x_\mu} Z_\lambda - \frac{\partial}{\partial x_\lambda} Z_\mu\right)^2 - \tfrac{1}{4}\left(\frac{\partial}{\partial x_\mu} A_\lambda - \frac{\partial}{\partial x_\lambda} A_\mu\right)^2$$

$$- \psi_\nu^\dagger \gamma_4 \gamma_\mu \frac{\partial}{\partial x_\mu} \psi_\nu - \psi_e^\dagger \gamma_4 \gamma_\mu \frac{\partial}{\partial x_\mu} \psi_e$$

and, as we shall see, (22.69)

$$\mathcal{L}_m = - m_W^2 \bar{W}_\mu W_\mu - \tfrac{1}{2} m_Z^2 Z_\mu Z_\mu - m_e \psi_e^\dagger \gamma_4 \psi_e - \tfrac{1}{2} \mu^2 \chi^2$$

where

$$m_W = \frac{g\rho}{\sqrt{2}} ,$$

$$m_Z = \frac{m_W}{\cos \theta_W} , \quad (22.70)$$

$$m_e = f\rho$$

and μ is defined in (22.62). Since $\mathcal{L}_m$ does not contain A_μ and ψ_ν, we have

$$m_A = 0$$
and
$$m_\nu = 0 \ ; \tag{22.71}$$

i.e., photon and neutrino are of zero mass.

4. Lepton and gauge field coupling

From (22.59), we see that

$$-L^\dagger \gamma_4 \gamma_\mu D_\mu L = -L^\dagger \gamma_4 \gamma_\mu \frac{\partial}{\partial x_\mu} L + i\tfrac{1}{2} L^\dagger \gamma_4 \gamma_\mu \cdot \begin{pmatrix} g B_\mu^z - g' C_\mu & g B_\mu^x - i g B_\mu^y \\ g B_\mu^x + i g B_\mu^y & -g B_\mu^z - g' C_\mu \end{pmatrix} L$$

and
$$-R^\dagger \gamma_4 \gamma_\mu D_\mu R = -R^\dagger \gamma_4 \gamma_\mu \frac{\partial}{\partial x_\mu} R - i g' C_\mu R^\dagger \gamma_4 \gamma_\mu R . \tag{22.72}$$

By using (22.66), we can express $\vec{B}_\mu$ and C_μ in terms of the normal modes W_μ, $\bar{W}_\mu$, Z_μ and A_μ:

$$B_\mu^x = \frac{1}{\sqrt{2}} (W_\mu + \bar{W}_\mu) ,$$

$$B_\mu^y = \frac{i}{\sqrt{2}} (W_\mu - \bar{W}_\mu) ,$$

$$B_\mu^z = \sin\theta_w A_\mu + \cos\theta_w Z_\mu \tag{22.73}$$

and
$$C_\mu = \cos\theta_w A_\mu - \sin\theta_w Z_\mu .$$

The Lagrangian density (22.58) can be written as

$$\mathcal{L} = \mathcal{L}_{quad} + \mathcal{L}_{\ell W} + \mathcal{L}_{\ell Z} + \mathcal{L}_{\ell A} + \cdots \tag{22.74}$$

where $\mathcal{L}_{quad}$ is the quadratic part given by (22.68), $\mathcal{L}_{\ell W}$, $\mathcal{L}_{\ell Z}$

and $\mathcal{L}_{\ell A}$ denote the interactions between the lepton fields ψ_e, ψ_ν and the vector-meson fields W_μ, Z_μ and A_μ. All other interaction terms are relegated to $\cdots$. By substituting (22.49)-(22.51) and (22.73) into (22.72), we can verify that

$$\mathcal{L}_{\ell W} = i \frac{g}{2\sqrt{2}} W_\lambda \psi_\nu^\dagger \gamma_4 \gamma_\lambda (1 + \gamma_5) \psi_e + \text{h.c.} \ ,$$

$$\mathcal{L}_{\ell Z} = i \tfrac{1}{4} \sqrt{g^2 + g'^2} \ Z_\lambda \left\{ \psi_\nu^\dagger \gamma_4 \gamma_\lambda (1 + \gamma_5) \psi_\nu \right.$$
$$\left. - \psi_e^\dagger \gamma_4 \gamma_\lambda [(1 - 4\sin^2 \theta_w) + \gamma_5] \psi_e \right\} \qquad (22.75)$$

and

$$\mathcal{L}_{\ell A} = -i \, g \sin \theta_w A_\lambda \psi_e^\dagger \gamma_4 \gamma_\lambda \psi_e \ .$$

Since A_μ is the photon field, the electric charge e is related to g, g' and θ_w by

$$e = g \sin \theta_w = g' \cos \theta_w > 0 \ , \qquad (22.76)$$

as shown in Figure 22.3.

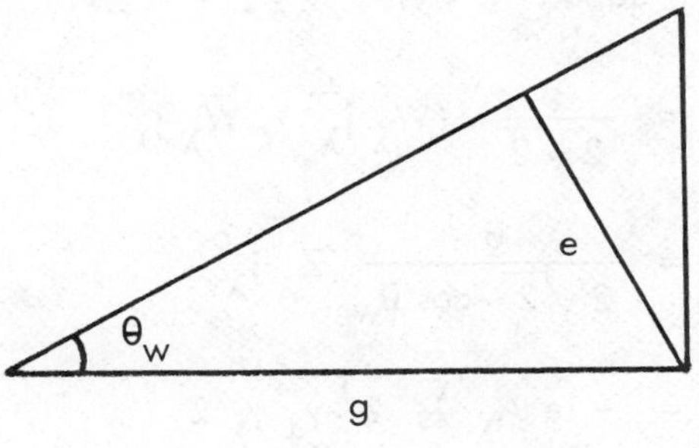

Fig. 22.3.   Relations between the Weinberg angle θ_w and the coupling constants e, g and g'.
[See (22.66) and (22.76).]

So far, we have considered only the (e^-, ν_e) system. To include other leptons (ℓ^-, ν_ℓ), we replace ψ_ν by ψ_{ν_ℓ}, ψ_e by ψ_ℓ, and for any 4×4 Dirac matrix Γ

$\psi_\nu^\dagger \Gamma \psi_\nu$ by

$$\sum_\ell \nu_\ell^\dagger \Gamma \nu_\ell \equiv \sum_\ell \psi_{\nu_\ell}^\dagger \Gamma \psi_{\nu_\ell} ,$$

$\psi_\nu^\dagger \Gamma \psi_e$ by

$$\sum_\ell \nu_\ell^\dagger \Gamma \ell \equiv \sum_\ell \psi_{\nu_\ell}^\dagger \Gamma \psi_\ell , \tag{22.77}$$

and

$\psi_e^\dagger \Gamma \psi_e$ by

$$\sum_\ell \ell^\dagger \Gamma \ell \equiv \sum_\ell \psi_\ell^\dagger \Gamma \psi_\ell$$

where the sums extend over $\ell = e$, μ and τ. Thus, (22.74) and (22.75) become

$$\mathcal{L} = \mathcal{L}_{quad} + \mathcal{L}_{\ell W} + \mathcal{L}_{\ell Z} + \mathcal{L}_{\ell A} + \cdots$$

where

$$\mathcal{L}_{\ell W} = \frac{g}{2\sqrt{2}} (W_\lambda j_\lambda^+ + \bar{W}_\lambda j_\lambda^-) ,$$

$$\mathcal{L}_{\ell Z} = \frac{g}{2\sqrt{2}\cos\theta_W} Z_\lambda j_\lambda^o \tag{22.78}$$

and

$$\mathcal{L}_{\ell A} = - ie A_\lambda \sum_\ell \ell^\dagger \gamma_4 \gamma_\lambda \ell ,$$

with

$$j_\lambda^+ = i \sum_\ell \nu_\ell^\dagger \gamma_4 \gamma_\lambda (1 + \gamma_5) \ell ,$$

$$j_\lambda^- = -\sum_\ell \ell^\dagger \gamma_4 \gamma_\lambda (1+\gamma_5) \nu_\ell \ , \tag{22.79}$$

$$j_\lambda^o = \frac{i}{\sqrt{2}} \sum_\ell [\nu_\ell^\dagger \gamma_4 \gamma_\lambda (1+\gamma_5) \nu_\ell - \ell^\dagger \gamma_4 \gamma_\lambda (1 - 4\sin^2\theta_w + \gamma_5) \ell] \ .$$

[See (22.114) for the complete form of $\mathcal{L}$, including quarks.]

5. Second order processes

The phenomenological Lagrangian densities

$$\frac{G}{\sqrt{2}}\ j_\lambda^+ j_\lambda^- \qquad \text{and} \qquad \frac{G}{2\sqrt{2}}\ j_\lambda^o j_\lambda^o$$

in (21.2) can be derived by using the second order diagrams, such as those in Figure 22.4. From Problem 4.2 and by following the arguments given in Sections 5.3 and 19.1, we can show that the propagator of a massive spin-1 boson is

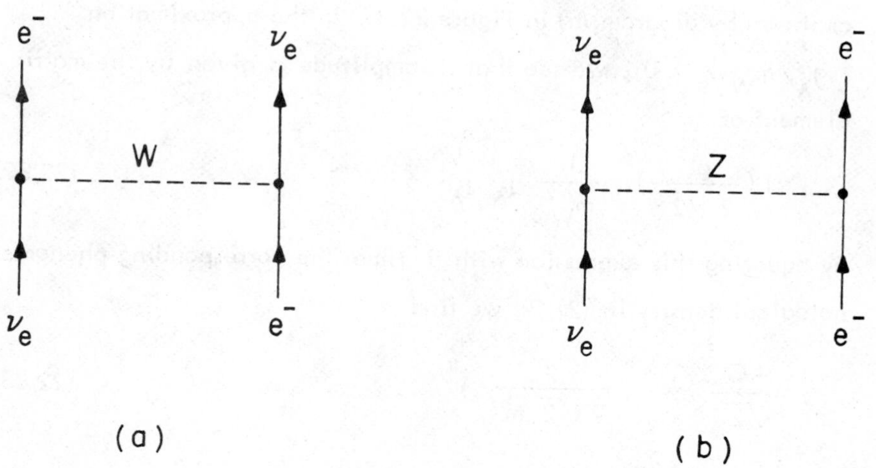

Fig. 22.4. $e\nu_e$ - scattering diagrams: (a) due to the exchange of $W^\pm$ and (b) due to the exchange of Z^o.

$$- i [\delta_{\lambda\mu} + (\frac{q_\lambda q_\mu}{m^2})] \quad (q^2 + m^2 - i\epsilon)^{-1} \tag{22.80}$$

where $\epsilon = 0+$. [See Exercise 1 on page 687.] For the low-energy phenomena discussed in Chapter 21, we have

$$| q_\lambda | \ll m = m_W \text{ or } m_Z \quad .$$

Thus, the boson propagators in Figure 22.4 become simply

$$- i \, \delta_{\lambda\mu} / m_W^2 \quad \text{for} \quad (a)$$

and

$$- i \, \delta_{\lambda\mu} / m_Z^2 \quad \text{for} \quad (b) \quad .$$

The second order amplitude of a $W^\pm$ exchange diagram can be obtained by contracting $\overline{W}_\lambda(x)$ and $W_\mu(y)$ in the product $i \mathcal{L}_{\ell W}(x) \cdot i \mathcal{L}_{\ell W}(y)$. Such a second order process can lead to, e.g., the scattering

$$\nu_e + e^- \rightarrow \nu_e + e^- \quad , \tag{22.81}$$

as shown by diagram (a) in Figure 22.4. In the approximation $| q_\lambda / m_W | \cong 0$, we see that its amplitude is given by the matrix element of

$$i \, (\frac{g}{2\sqrt{2}})^2 \, \frac{1}{m_W^2} \, j_\lambda^+ \, j_\lambda^- \quad . \tag{22.82}$$

By equating this expression with i times the corresponding phenomenological density in (21.2), we find

$$\frac{G}{\sqrt{2}} = \left(\frac{g}{2\sqrt{2}\, m_W} \right)^2 \quad . \tag{22.83}$$

Because of (22.76), this leads to

$$m_W^2 = \frac{\sqrt{2}\, e^2}{8 G \sin^2 \theta_w} \geqslant \frac{\sqrt{2}\, e^2}{8G} = (37.3 \text{ GeV})^2 \quad . \tag{22.84}$$

From (22.70), we also have

$$m_Z^2 = \frac{m_W^2}{\cos^2\theta_w} = \frac{e^2}{\sqrt{2}\, G \sin^2 2\theta_w} \tag{22.85}$$

$$\geqslant \frac{e^2}{\sqrt{2}\, G} = (74.6\ \text{GeV})^2 \quad .$$

Similarly, the amplitude of the corresponding Z^o exchange diagram (b) in Figure 22.4 is given by the matrix element of

$$\tfrac{1}{2}\, i \left(\frac{g}{2\sqrt{2}}\right)^2 \left(\frac{1}{m_Z \cos\theta_w}\right)^2 j_\lambda^o\, j_\lambda^o \quad , \tag{22.86}$$

which reduces to i times $G\, j_\lambda^o j_\lambda^o / 2\sqrt{2}$ in (21.2) provided that the constants g_v, g_a and g_ν in (21.4) are given by

$$g_v = -\frac{1}{\sqrt{2}}\,(1 - 4\sin^2\theta_w) \quad ,$$

$$g_a = -\frac{1}{\sqrt{2}} \qquad \text{and} \qquad g_\nu = \frac{1}{\sqrt{2}} \quad . \tag{22.87}$$

For $\nu_e e^-$ scattering, on account of (22.79), (22.83) and (22.85), the relevant parts of (22.82) and (22.86) are

$$M_a = -i\,\frac{G}{\sqrt{2}}\, \psi_\nu^\dagger \gamma_4 \gamma_\lambda (1+\gamma_5)\, \psi_e \cdot \psi_e^\dagger \gamma_4 \gamma_\lambda (1+\gamma_5)\, \psi_\nu$$

and

$$M_b = i\,\frac{G}{2\sqrt{2}}\, \psi_\nu^\dagger \gamma_4 \gamma_\lambda (1+\gamma_5)\, \psi_\nu$$

$$\cdot\, \psi_e^\dagger \gamma_4 \gamma_\lambda [\,(1 - 4\sin^2\theta_w) + \gamma_5\,]\, \psi_e \,.$$

By using the Fierz identity (Exercise 2 on page 687)

$$\psi_e^\dagger \gamma_4 \gamma_\lambda (1+\gamma_5)\, \psi_\nu \cdot \psi_\nu^\dagger \gamma_4 \gamma_\lambda (1+\gamma_5)\, \psi_e$$

$$= \psi_\nu^\dagger \gamma_4 \gamma_\lambda (1+\gamma_5)\, \psi_\nu \cdot \psi_e^\dagger \gamma_4 \gamma_\lambda (1+\gamma_5)\, \psi_e \quad , \tag{22.88}$$

we find that the total $\nu_e e^-$ scattering amplitude is

$$M_a + M_b = -i \frac{G}{\sqrt{2}} \psi_\nu^\dagger \gamma_4 \gamma_\lambda (1+\gamma_5) \psi_\nu$$

$$\cdot \psi_e^\dagger \gamma_4 \gamma_\lambda (\tfrac{1}{2} + 2 \sin^2 \theta_w + \tfrac{1}{2} \gamma_5) \psi_e \quad . \qquad (22.89)$$

Next, we consider

$$\nu_\mu + e^- \rightarrow \nu_\mu + e^-$$

and

$$\bar{\nu}_\mu + e^- \rightarrow \bar{\nu}_\mu + e^- \quad .$$

These reactions are determined only by the Z^o exchange diagram; their cross sections can be readily derived by substituting (22.87) into (21.10). From the experimental values (21.11), we find

$$\sin^2 \theta_w = .225 \begin{matrix} + .06 \\ - .05 \end{matrix} \quad . \qquad (22.90)$$

Remarks.

1. The experimental value of θ_w is quite near 30°. If we assume

$$\sin^2 \theta_w \cong \tfrac{1}{4} \qquad (22.91)$$

then $g_v \cong 0$, and

$$j_\lambda^o \cong \frac{i}{\sqrt{2}} \sum_\ell [\nu_\ell^\dagger \gamma_4 \gamma_\lambda (1+\gamma_5) \nu_\ell - \ell^\dagger \gamma_4 \gamma_\lambda \gamma_5 \ell] \quad . \qquad (22.92)$$

The neutral current of the charged lepton $\ell^\pm$ is then purely axial.

2. The sign of the Fermi constant G is positive according to (22.83). Experimentally, by accurately measuring the scattering amplitude of, say,

$$e + e \rightarrow e + e \quad ,$$

we can deduce the interference between the photon-exchange and the Z^o exchange diagrams, and from that the sign of G. [See (22.126).]

Exercise 1. Use the path-integration contraction (19.70) for the definition of the vector-meson propagator of mass m

$$\lambda \bullet\!-\!-\!-\!-\!\bullet \mu \qquad .$$
$$q$$

Show that it is given by (22.80). [See also (24.142)-(24.144).]

Exercise 2. Show that for any 4×1 c. number column matrices ψ_e and ψ_ν , the Fierz identity (22.88) holds.

22.3 Extension to Hadrons

Because hadrons are composites, their current operators are expressed in terms of the quark field ψ_q. The connection between ψ_q and the observed hadron is not as direct as that between the lepton field ψ_ℓ and the physical lepton. This complicates the extension of the above results from leptons to hadrons. In order to gain a proper perspective, we shall retrace the steps that led to the present formulation.

Before 1950, one might have linked leptons with hadrons through the naive parallel relation between the lepton doublet (ν_e, e) and the nucleon doublet (p, n). After the discoveries of the strange particles and the approximate validity of SU_3 symmetry, there was the Cabibbo theory which suggested that we should replace (p, n) by (u, d_c), where u stands for the up-quark field ψ_u and d_c for the Cabibbo-rotated field

$$\psi_{d_c} = \cos\theta_c\, \psi_d + \sin\theta_c\, \psi_s \quad , \tag{22.93}$$

as in (21.51), with

θ_c = Cabibbo angle.

If we employ only the (u, d_c) system for the hadrons, by applying the same arguments given in Section 22.2.1 we would find, in place of the leptonic weak charges (22.39), the hadronic weak charges *

$$Q^- \sim \int d_c^\dagger (1+\gamma_5)\, u\, d^3r$$

and (22.94)

$$Q^+ \sim \int u^\dagger (1+\gamma_5)\, d_c\, d^3r \quad .$$

Their commutator is

$$Q^o \equiv [\,Q^-, Q^+\,] \sim 2\int [\,d_c^\dagger(1+\gamma_5)\,d_c - u^\dagger(1+\gamma_5)\,u\,]\, d^3r \qquad (22.95)$$

where, as well as in the following, the quark field ψ_q is represented simply by q. Because

$$d_c^\dagger(1+\gamma_5)\,d_c = \cos^2\theta_c\, d^\dagger(1+\gamma_5)\,d + \sin^2\theta_c\, s^\dagger(1+\gamma_5)\,s$$
$$+ \cos\theta_c \sin\theta_c\, [\,d^\dagger(1+\gamma_5)\,s + s^\dagger(1+\gamma_5)\,d\,] \; ,$$

the neutral "charge" Q^o can change the strangeness quantum number S by 0 or ± 1. Since Q^o plays the same role as the generator T^z of (22.43), this would lead to a neutral hadron current J_λ^o satisfying the same selection rule:

$$\Delta S = 0 \quad \text{and} \quad \pm 1 \; .$$

The second order amplitude $G J_\lambda^o J_\lambda^o$ would then contain a $\Delta S = \pm 2$ component, in violation of experimental fact. [See page 352.] Furthermore, there would be a hadron-lepton neutral current interaction $G J_\lambda^o j_\lambda^o$ which could give an unsuppressed amplitude for

$$K_L^o \to \mu^+ + \mu^- \quad \text{or} \quad e^+ + e^- \; .$$

---

* Throughout this chapter we do not explicitly exhibit the color superscripts. [See (21.52).]

This would contradict the very small branching ratios $< 10^{-8}$ for these decays. [See page 634.]

1. GIM mechanism

In order to overcome this difficulty, Glashow, Iliopoulos and Maiani* proposed the existence of the charm quark. The charged hadronic weak currents are given by (21.51); i.e.,

$$J_\lambda^- = i\, d_c^\dagger \gamma_4 \gamma_\lambda (1+\gamma_5)\, u + i\, s_c^\dagger \gamma_4 \gamma_\lambda (1+\gamma_5)\, c$$

and

$$J_\lambda^+ = i\, u^\dagger \gamma_4 \gamma_\lambda (1+\gamma_5)\, d_c + i\, c^\dagger \gamma_4 \gamma_\lambda (1+\gamma_5)\, s_c \;, \qquad (22.96)$$

where c stands for the charm-quark field ψ_c, and s_c for

$$\psi_{s_c} = -\sin\theta_c\, \psi_d + \cos\theta_c\, \psi_s \;. \qquad (22.97)$$

The hadronic electromagnetic current operator is

$$J_\lambda^{el} = i\tfrac{2}{3}\,(u^\dagger \gamma_4\gamma_\lambda u + c^\dagger \gamma_4\gamma_\lambda c) - i\tfrac{1}{3}\,(d^\dagger \gamma_4\gamma_\lambda d + s^\dagger \gamma_4\gamma_\lambda s) \qquad (22.98)$$

as in (21.53). Instead of (22.94), the hadronic weak charges are given by

$$Q^- \equiv -i \int J_4^- \, d^3r = \int [\, d_c^\dagger (1+\gamma_5)\, u + s_c^\dagger (1+\gamma_5)\, c \,]\, d^3r$$

$$Q^+ \equiv -i \int J_4^+ \, d^3r = \int [\, u^\dagger (1+\gamma_5)\, d_c + c^\dagger (1+\gamma_5)\, s_c \,]\, d^3r \,. \qquad (22.99)$$

The hadronic electric charge is

$$Q \equiv -i \int J_4^{el}\, d^3r = \int [\, \tfrac{2}{3}(u^\dagger u + c^\dagger c) - \tfrac{1}{3}(d^\dagger d + s^\dagger s)\,]\, d^3r \,. \qquad (22.100)$$

From (22.93) and (22.97), it follows that for any 4×4 Dirac matrix Γ

$$d_c^\dagger \Gamma d_c + s_c^\dagger \Gamma s_c = d^\dagger \Gamma d + s^\dagger \Gamma s \;. \qquad (22.101)$$

Thus, the commutator between Q^- and Q^+ is now

---

* Loc. cit., page 619.

$$Q^o \equiv [\, Q^- , Q^+]$$
$$= 2 \int [\, d^\dagger(1+\gamma_5)\, d + s^\dagger(1+\gamma_5)\, s - u^\dagger(1+\gamma_5)\, u - c^\dagger(1+\gamma_5)\, d\,]\, d^3r ,$$

which conserves the strangeness.

As in (22.40), (22.43) and (22.46), we introduce

$$T^x \equiv \tfrac{1}{4}(Q^- + Q^+) ,$$
$$T^y \equiv \frac{i}{4}(Q^- - Q^+) ,$$
$$T^z \equiv -i[\, T^x , T^y] \qquad (22.102)$$

and

$$T' \equiv Q - T^z .$$

Hence,

$$T^z = -Q^o/8$$
$$= \tfrac{1}{4}\int [\, u^\dagger(1+\gamma_5)\, u + c^\dagger(1+\gamma_5)\, c - d^\dagger(1+\gamma_5)\, d - s^\dagger(1+\gamma_5)\, s\,]\, d^3r . \qquad (22.103)$$

Let us define

$$\psi_1 \equiv \begin{pmatrix} u \\ d_c \end{pmatrix}$$

and

$$\psi_2 \equiv \begin{pmatrix} c \\ s_c \end{pmatrix} . \qquad (22.104)$$

Then, as in (22.45), we have

$$T^i = \sum_h \int \psi_h^\dagger \tfrac{1}{4}(1+\gamma_5)\, \tau^i\, \psi_h\, d^3r \qquad (22.105)$$

where the τ^i are the Pauli matrices and the sum extends over $h = 1$ and 2. By using (22.100) and (22.102)–(22.103), we see that

$$T' = \sum_h \int \psi_h^\dagger [\, \tfrac{1}{4}(1-\gamma_5)\, \tau^z + \tfrac{1}{6}\,]\; \psi_h . \qquad (22.106)$$

These generators T^i and T' satisfy the commutation relations (22.44) and (22.47). Thus, we have the same $SU_2 \times U_1$ gauge group as before. In analogy to (22.51), we introduce

$$L_h \equiv \tfrac{1}{2}(1+\gamma_5)\,\psi_h \tag{22.107}$$

so that for $h = 1$

$$L_1 = \tfrac{1}{2}(1+\gamma_5)\begin{pmatrix} u \\ d_c \end{pmatrix}$$

and for $h = 2$

$$L_2 = \tfrac{1}{2}(1+\gamma_5)\begin{pmatrix} c \\ s_c \end{pmatrix} .$$

As in (22.50), we define

$$R_q \equiv \tfrac{1}{2}(1-\gamma_5)\,q \tag{22.108}$$

where $q = u, d, s$ and c .

Let q_L and q_R denote the one-particle states generated by applying the creation operators $\frac{1}{2}(1+\gamma_5)\,q^\dagger$ and $\frac{1}{2}(1-\gamma_5)\,q^\dagger$ to the vacuum. These states are all eigenstates of T, T^z, Q and therefore also T' . We have

	T	T^z	Q	T'
u_L or c_L	$\frac{1}{2}$	$\frac{1}{2}$	$\frac{2}{3}$	$\frac{1}{6}$
d_L or s_L	$\frac{1}{2}$	$-\frac{1}{2}$	$-\frac{1}{3}$	$\frac{1}{6}$
u_R or c_R	0	0	$\frac{2}{3}$	$\frac{2}{3}$
d_R or s_R	0	0	$-\frac{1}{3}$	$-\frac{1}{3}$.

(22.109)

Since the quarks have strong interactions, these states are also eigenstates of the isospin I, its z-component I^z, the baryon number N, the strangeness S, the charm number C and a new hypercharge Y, defined by *

$$Y = N + S + C .$$

It is also of interest to list the strong interaction quantum numbers of these states:

	I	I^z	N	S	C	Y
u	$\frac{1}{2}$	$\frac{1}{2}$	$\frac{1}{3}$	0	0	$\frac{1}{3}$
d	$\frac{1}{2}$	$-\frac{1}{2}$	$\frac{1}{3}$	0	0	$\frac{1}{3}$
s	0	0	$\frac{1}{3}$	-1	0	$-\frac{2}{3}$
c	0	0	$\frac{1}{3}$	0	-1	$-\frac{2}{3}$.

Because the strong interaction is parity conserving, these quantum numbers depend only on the quark states q, not on their helicity subscript L or R. From (21.121) we see that quarks of the same generation have the same Y.

The generalization to b and t quarks in the context of the Kobayashi and Maskawa model (Section 21.10) is straightforward and will be omitted here.

* Cf. (11.89).

2. Lagrangian density

In addition to the quark fields L_h and R_q given by (22.107)-(22.108), there are also the lepton fields L_ℓ, R_ℓ, the gauge fields $\vec{B}_\mu$, C_μ, and the Higgs field ϕ. In accordance with the notations used in (22.77), we generalize (22.50) and (22.51) by defining

$$L_\ell \equiv \tfrac{1}{2}(1+\gamma_5)\begin{pmatrix}\nu_\ell \\ \ell\end{pmatrix}$$

and (22.110)

$$R_\ell \equiv \tfrac{1}{2}(1+\gamma_5)\ \ell$$

where $\ell = e$, μ and τ. As before, to each generator T^i or T', there is a gauge field B_μ^i or C_μ. The Higgs field ϕ remains given by (22.55). It is useful to introduce its conjugate field $\bar{\phi}$:

$$\bar{\phi} \equiv \exp\left(i\,\frac{\pi}{2}\,\tau^y\right)\cdot \text{ transpose of } \phi^\dagger \quad . \tag{22.111}$$

Thus, for

$$\phi = \begin{pmatrix}\phi_1 \\ \phi_2\end{pmatrix} \quad ; \quad \phi^\dagger = (\phi_1^\dagger\ \phi_2^\dagger)$$

and $\bar{\phi}$ is

$$\bar{\phi} = \begin{pmatrix}\phi_2^\dagger \\ -\phi_1^\dagger\end{pmatrix} \quad . \tag{22.112}$$

From (11.38)-(11.39), we see that $\bar{\phi}$, like ϕ, belongs to the $T=\frac{1}{2}$ representation; however,

$$T'(\bar{\phi}) = -T'(\phi) \quad .$$

Under the global $SU_2 \times U_1$ transformation, each of these fields $\vec{B}_\mu$, C_μ, L_ℓ, $\cdots$ belongs to an irreducible (T, T') representation. From (22.53), (22.56) and (22.109), we find

	$\vec{B}_\mu$	C_μ	ϕ	$\bar{\phi}$	L_ℓ	R_ℓ
T	1	0	$\frac{1}{2}$	$\frac{1}{2}$	$\frac{1}{2}$	0
T'	0	0	$\frac{1}{2}$	$-\frac{1}{2}$	$-\frac{1}{2}$	-1

and (22.113)

	L_h	R_u	R_d	R_s	R_c
T	$\frac{1}{2}$	0	0	0	0
T'	$\frac{1}{6}$	$\frac{2}{3}$	$-\frac{1}{3}$	$-\frac{1}{3}$	$\frac{2}{3}$.

For this system, the complete Lagrangian density is (excluding QCD)

$$\mathcal{L} = -\tfrac{1}{4}\vec{B}_{\mu\nu}^{\,2} - \tfrac{1}{4}C_{\mu\nu}^{\,2} - R_\ell^\dagger \gamma_4\gamma_\mu D_\mu R_\ell - L_\ell^\dagger \gamma_4\gamma_\mu D_\mu L_\ell$$
$$- R_q^\dagger \gamma_4\gamma_\mu D_\mu R_q - L_h^\dagger \gamma_4\gamma_\mu D_\mu L_h - (\bar{D}_\mu \phi^\dagger) D_\mu \phi$$
$$- U(|\phi|) + \mathcal{L}_{\ell\phi} + \mathcal{L}_{q\phi} \qquad (22.114)$$

where $\vec{B}_{\mu\nu}$ and $C_{\mu\nu}$ are given by (22.54), and $\bar{D}_\mu \phi^\dagger$ and the covariant derivative D_μ by (22.59); i.e.,

$$D_\mu = \frac{\partial}{\partial x_\mu} - ig\vec{\tau}\cdot\vec{B}_\mu T - ig' C_\mu T' \qquad (22.115)$$

with T and T' listed in (22.113). In (22.114) the repeated indices are summed over; as before, $\ell = e, \mu$ and τ, $h = 1$ and 2, and $q = u, d, s$ and c. The potential $U(|\phi|)$ is given by (22.62), so that in the unitary gauge, (22.64) remains valid; i.e.,

$$\phi = \begin{pmatrix} 0 \\ \rho + X \end{pmatrix} \tag{22.116}$$

where X is a Hermitian field and ρ is a real and positive constant. Thus, from (22.112) it follows that

$$\bar{\phi} = \begin{pmatrix} \rho + X \\ 0 \end{pmatrix} \quad . \tag{22.117}$$

By using (22.58) and the last equation in (22.70), we see that the lepton-Higgs interaction $\mathcal{L}_{\ell\phi}$ is

$$\mathcal{L}_{\ell\phi} = -\sum_{\ell} \frac{m_\ell}{\rho} (L_\ell^\dagger \gamma_4 R_\ell \phi + \text{h.c.}) \tag{22.118}$$

where m_ℓ is the mass of the charged lepton $\ell^\pm$. [The neutrinos are all assumed to be of zero mass.] Since none of the quarks is truly massless, the quark-Higgs interaction $\mathcal{L}_{q\phi}$ is somewhat more complicated. Let m_q be the mass * of quark q. We write

$$\begin{aligned} \mathcal{L}_{q\phi} = &- \frac{m_u}{\rho} [L_1^\dagger \gamma_4 R_u \bar{\phi} + \text{h.c.}] \\ &- \frac{m_d}{\rho} [(\cos\theta_c L_1 - \sin\theta_c L_2)^\dagger \gamma_4 R_d \phi + \text{h.c.}] \\ &- \frac{m_s}{\rho} [(\sin\theta_c L_1 + \cos\theta_c L_2)^\dagger \gamma_4 R_s \phi + \text{h.c.}] \\ &- \frac{m_c}{\rho} [L_2^\dagger \gamma_4 R_c \bar{\phi} + \text{h.c.}] \quad . \end{aligned} \tag{22.119}$$

From (22.113) we see that $\mathcal{L}_{q\phi}$ is invariant under the T' rotation. Under the $\vec{T}$ rotation, ϕ, $\bar{\phi}$, L_1 and L_2 transform in the same way

---

* Here, as in Chapter 20, m_q refers to the quark mass inside the hadron.

and therefore $\mathcal{L}_{q\phi}$ is also invariant. It is straightforward to verify that because of (22.107)-(22.108), (22.110) and (22.116)-(22.117)

$$\mathcal{L}_{\ell\phi} = - \sum_{\ell} m_{\ell} \, \ell^{\dagger} \gamma_4 \, \ell + \cdots$$

and (22.120)

$$\mathcal{L}_{q\phi} = - \sum_{q} m_{q} \, q^{\dagger} \gamma_4 \, q + \cdots$$

where $\cdots$ are cubic in the Higgs field χ and the fermion fields ℓ and q.

As in (22.60), the Lagrangian density (22.114) is invariant under the local gauge transformation

$$B_{\mu} \equiv \tfrac{1}{2}\vec{\tau}\cdot\vec{B}_{\mu} \rightarrow u\, B_{\mu}\, u^{\dagger} + \frac{i}{g}\, u\, \frac{\partial u^{\dagger}}{\partial x_{\mu}} ,$$

$$C_{\mu} \rightarrow C_{\mu} + \frac{1}{g'} \frac{\partial \alpha}{\partial x_{\mu}} ,$$

$$L_{\ell} \rightarrow e^{i\alpha T'} u\, L_{\ell} , \qquad R_{\ell} \rightarrow e^{i\alpha T'} R_{\ell} , \tag{22.121}$$

$$R_{h} \rightarrow e^{i\alpha T'} u\, R_{h} , \qquad R_{q} \rightarrow e^{i\alpha T'} R_{q}$$

and

$$\phi \rightarrow e^{i\alpha T'} u\, \phi$$

where the T' of the various fields are listed in (22.113). The normal modes W_{μ}, Z_{μ} and A_{μ} remain given by (22.66), with their masses by (22.70)-(22.71).

3. Quark and gauge field coupling

The interaction terms $\mathcal{L}_{\ell W}$, $\mathcal{L}_{\ell Z}$ and $\mathcal{L}_{\ell A}$ between the leptons and the gauge fields are given by (22.78)-(22.79). Likewise, we can derive the corresponding interaction terms $\mathcal{L}_{qW}$, $\mathcal{L}_{qZ}$ and $\mathcal{L}_{qA}$ between the quarks and the gauge fields. We observe that if we uniformly shift the electric charge of each quark by an amount $-\frac{2}{3}$, the Q of u and d_c (or c and d_s) would be identical to those of ν_e and e (or ν_μ and μ). Such a shift does not alter T ; it changes only the T' of each quark by the same amount $-\frac{2}{3}$. Correspondingly, there is a change in the Lagrangian density

$$i\,\tfrac{2}{3}\,g'\,C_\lambda\,[\sum_h L_h^\dagger \gamma_4\gamma_\lambda L_h + \sum_q R_q^\dagger \gamma_4\gamma_\lambda R_q]$$
$$= i\,\tfrac{2}{3}\,g'(\cos\theta_w\,A_\lambda - \sin\theta_w\,Z_\lambda)\sum_q q^\dagger \gamma_4\gamma_\lambda q \qquad (22.122)$$

due to this shift. Hence by changing, in (22.78)-(22.79),

$$(\nu_e\,,\ e) \rightarrow (u\,,\ d_c)$$

and

$$(\nu_\mu\,,\ \mu) \rightarrow (c\,,\ s_c)$$

and by including (22.122), we find

$$\mathcal{L}_{qW} = \frac{g}{2\sqrt{2}}\,(W_\lambda J_\lambda^+ + \bar{W}_\lambda J_\lambda^-)\ ,$$

$$\mathcal{L}_{qZ} = \frac{g}{2\sqrt{2}\cos\theta_w}\,Z_\lambda J_\lambda^o \qquad (22.123)$$

and $$\mathcal{L}_{qA} = e\,A_\lambda J_\lambda^{el}$$

where $g = e/\sin\theta_w$, and as in (21.51) and (21.53)

$$J_\lambda^+ = i u^\dagger \gamma_4\gamma_\lambda(1+\gamma_5)\,d_c + i c^\dagger \gamma_4\gamma_\lambda(1+\gamma_5)\,s_c\ ,$$

$$J_\lambda^- = i d_c^\dagger \gamma_4\gamma_\lambda(1+\gamma_5)\,u + i s_c^\dagger \gamma_4\gamma_\lambda(1+\gamma_5)\,c$$

and $$J_\lambda^{el} = i\,\tfrac{2}{3}\,(u^\dagger\gamma_4\gamma_\lambda u + c^\dagger\gamma_4\gamma_\lambda c) - i\,\tfrac{1}{3}\,(d^\dagger\gamma_4\gamma_\lambda d + s^\dagger\gamma_4\gamma_\lambda s)\ ; \qquad (22.124)$$

in addition,

$$J_\lambda^{o} = \frac{i}{\sqrt{2}} [u^\dagger \gamma_4 \gamma_\lambda (1 - \frac{8}{3} \sin^2\theta_w + \gamma_5) u + c^\dagger \gamma_4 \gamma_\lambda (1 - \frac{8}{3} \sin^2\theta_w + \gamma_5) c - d^\dagger \gamma_4 \gamma_\lambda (1 - \frac{4}{3} \sin^2\theta_w + \gamma_5) d - s^\dagger \gamma_4 \gamma_\lambda (1 - \frac{4}{3} \sin^2\theta_w + \gamma_5) s] .$$

In deriving J_λ^{el} and J_λ^{o}, we have used the identities $g' \cos\theta_w = e$ and $g' = g \tan\theta_w$.

4. Second order processes

At small 4-momentum transfer, the second order $W^\pm$-exchange and Z^o-exchange diagrams lead to the following effective Lagrangian

$$\mathcal{L}_{eff} = \frac{G}{\sqrt{2}} [(J_\lambda^{+} + j_\lambda^{+})(J_\lambda^{-} + j_\lambda^{-}) + \tfrac{1}{2}(J_\lambda^{o} + j_\lambda^{o})(J_\lambda^{o} + j_\lambda^{o})] , \tag{22.125}$$

similarly to (21.113).

A sensitive test of the specific forms of J_λ^{o} and j_λ^{o} is the parity violation in the electron-deuteron scattering experiment discussed in Section 21.7.3. It can be shown * that the asymmetry parameter A of (21.110) is given by (see Problem 22.3)

$$\frac{A}{k^2} = - \frac{9G}{5\sqrt{2}\, e^2} \left[(1 - \frac{20}{9} \sin^2\theta_w) + (1 - 4\sin^2\theta_w) \frac{1-(1-y)^2}{1+(1-y)^2} \right] \tag{22.126}$$

where $k^2 = $ (4-momentum transfer)^2. From the experimental values given by (21.111), we determine that

$$\sin^2\theta_w = .224 \pm .02 , \tag{22.127}$$

in agreement with (22.90).

* R. N. Cahn and F. J. Gilman, Phys.Rev. D17, 1313 (1978).

Problem 22.1. Let

$$\mathcal{L}_{\ell Z} + \mathcal{L}_{qZ} \equiv e Z_\lambda \mathcal{J}_\lambda^o$$

be the sum of lepton - Z and quark - Z interactions given by (22.78)-(22.79) and (22.123)-(22.124).

(i) Show that the weak neutral charge coupled to Z,

$$Q^Z \equiv -i \int \mathcal{J}_4^o \, d^3r ,$$

is related to T^z and the electric charge Q by

$$Q^Z = \frac{e}{\sin\theta_w \cos\theta_w} (T^z - Q \sin^2\theta_w) . \qquad (22.128)$$

(ii) Establish the following table for the various left and right-handed leptons and quarks:

	$Q^Z \sin\theta_w \cos\theta_w / e$
ν_ℓ	$\frac{1}{2}$
ℓ_L	$-\frac{1}{2} + \sin^2\theta_w$
ℓ_R	$\sin^2\theta_w$
u_L or c_L	$\frac{1}{2} - \frac{2}{3}\sin^2\theta_w$
u_R or c_R	$-\frac{2}{3}\sin^2\theta_w$
d_L or s_L	$-\frac{1}{2} + \frac{1}{3}\sin^2\theta_w$
d_R or s_R	$\frac{1}{3}\sin^2\theta_w$

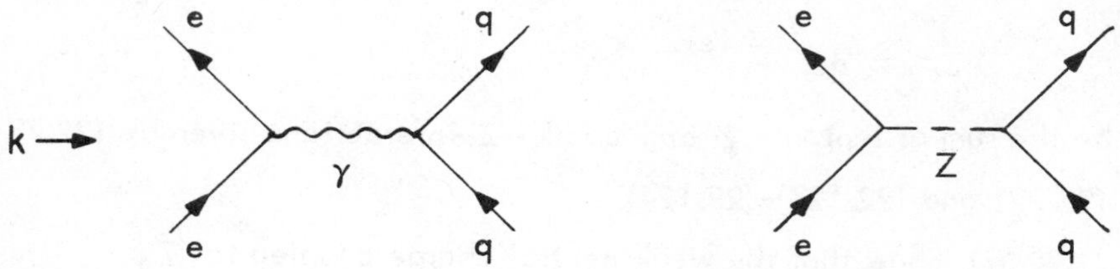

Fig. 22.5. Feynman diagrams for $e + q \to e + q$ via γ and Z.

Problem 22.2. Consider the elastic scattering between an electron and a quark:

$$e + q \to e + q \quad .$$

The Feynman diagrams are given in Figure 22.5. At high energy and in the center-of-mass system, we can neglect all masses. Hence, the helicities of these particles, left (L) or right (R), are unchanged during the collision.

(i) Prove that the differential cross section $d\sigma(e_L q_L)$ of a lefthanded electron and a lefthanded quark is proportional to

$$\left| \frac{Q(e)\,Q(q)}{k^2} + \frac{Q^Z(e_L)\,Q^Z(q_L)}{k^2 + m_Z^2} \right|^2$$

where k is the 4-momentum transfer, Q^Z refers to the weak neutral charge and Q to the electric charge.

(ii) By using Problem 5.4, show that for different helicities the differential cross sections satisfy

$$d\sigma(e_L q_R) \propto \left| \frac{Q(e)\,Q(q)}{k^2} + \frac{Q^Z(e_L)\,Q^Z(q_R)}{k^2 + m_Z^2} \right|^2 (1 - y)^2 \ ,$$

$$d\sigma(e_R q_R) \propto \left| \frac{Q(e)\,Q(q)}{k^2} + \frac{Q^Z(e_R)\,Q^Z(q_R)}{k^2 + m_Z^2} \right|^2$$

and

$$d\sigma(e_R q_L) \propto \left| \frac{Q(e)\,Q(q)}{k^2} + \frac{Q^Z(e_R)\,Q^Z(q_L)}{k^2 + m_Z^2} \right|^2 (1 - y)^2$$

with the same proportionality. As in (21.110)-(21.111), the parameter y is related to the initial and final laboratory energies of the electron, E_{in} and E_f, by

$$y = \frac{E_{in} - E_f}{E_{in}} .$$

Problem 22.3. In the scattering of a polarized electron on an unpolarized deuteron, the asymmetry parameter A is defined by (21.108):

$$A = \frac{d\sigma(e_R) - d\sigma(e_L)}{d\sigma(e_R) + d\sigma(e_L)} .$$

Neglecting the strong interaction, show that to the first order in $\frac{k^2}{m_Z^2}$, the asymmetry parameter A is given by (22.126); i.e.,

$$A = -\frac{9}{10}\,\frac{k^2}{m_Z^2 \sin^2 2\theta_W}\left[1 - \frac{20}{9}\sin^2\theta_W + \frac{1-(1-y)^2}{1+(1-y)^2}\,(1 - 4\sin^2\theta_W)\right].$$

Reference.

Proceedings of the IX International Symposium on Lepton and Photon Interactions at High Energies, Fermilab, 1979, ed. T.B.W. Kirk and H.D.I. Abarbanel.

Chapter 23

QUARK-PARTON MODEL AND HIGH-ENERGY PROCESSES

23.1 Scaling Approximation

Let us first consider the inclusive lepton reactions

$$\ell^- + N \rightarrow \ell^- + \cdots \tag{23.1}$$

and

$$\nu_\ell \ (\text{or } \bar{\nu}_\ell) + N \rightarrow \ell^{\mp} + \cdots \tag{23.2}$$

where N denotes the nucleon n or p, ℓ can be e, μ or τ and $\cdots$ indicates that all possible final hadron channels are being summed over. The Feynman diagrams of these reactions are given in Figure 23.1. As in (21.73), we set

$$\begin{aligned} k, k' &= \text{4-momenta of the initial and final leptons,} \\ p &= \text{4-momentum of the initial nucleon,} \\ p' &= \text{4-momentum of the final hadron complex,} \end{aligned} \tag{23.3}$$

and

$$q = k - k' = p' - p$$

is the 4-momentum transfer. Throughout this chapter, we adopt the notations used in Section 21.6.

For the ℓ^--reaction (23.1), the ratio of the amplitude of the Z-exchange diagram (a') to that of the γ-exchange diagram (a) is

$$\approx \frac{q^2}{q^2 + m_Z^2} .$$

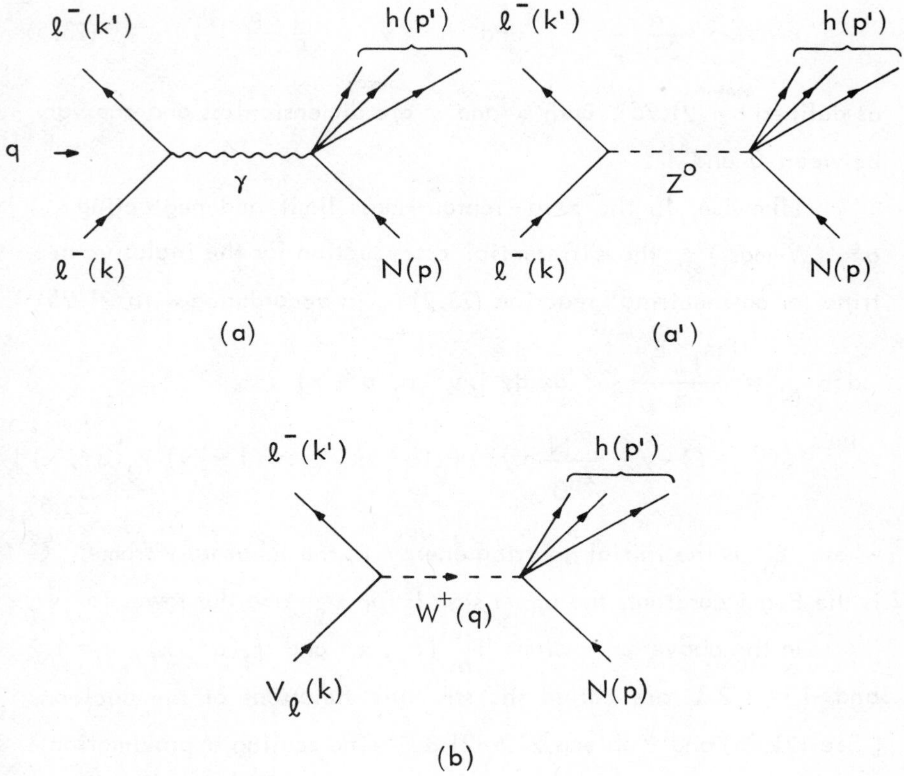

Fig. 23.1. Diagrams for lepton reactions (23.1)-(23.2).

where m_Z is the Z° mass. Hence, for $q^2 \ll m_Z^2$, the Z-exchange diagram can be neglected. According to (21.148), in the zero-lepton-mass limit the inclusive cross section is given by

$$d^2\sigma_{\ell N} = \frac{m_N E_\ell}{2\pi} \left(\frac{e^2}{q^2}\right)^2 dx\,dy \left[xy^2 F_1^{el}(q^2, x) + \left(1 - y - \frac{m_N}{2q_0} xy^2\right) F_2^{el}(q^2, x) \right] \quad (23.4)$$

where m_N is the nucleon mass, E_ℓ the initial ℓ^- energy in the laboratory frame,

$$x = -\frac{q^2}{2q \cdot p} \qquad \text{and} \qquad y = \frac{q \cdot p}{k \cdot p} \,, \tag{23.5}$$

as defined by (21.75). Both x and y are dimensionless and can vary between 0 and 1 .

Likewise, in the zero-lepton-mass limit and neglecting $q^2/(\text{W-mass})^2$, the differential cross section for the inclusive neutrino (or antineutrino) reaction (23.2) is, in accordance with (21.95),

$$d^2\sigma_{\nu N} = \frac{m_N E_\nu}{\pi} G^2 \, dx \, dy \, \Big[xy^2 F_1(q^2, x) + \Big(1 - y - \frac{m_N}{2q_0} xy^2\Big) F_2(q^2, x) \pm xy\big(1 - \tfrac{1}{2}y\big) F_3(q^2, x) \Big] \tag{23.6}$$

where E_ν is the initial neutrino energy in the laboratory frame, G is the Fermi constant, the upper sign is for ν_ℓ and the lower for $\bar{\nu}_\ell$.

In the above expressions $F_a^{el}(q^2, x)$ and $F_i(q^2, x)$, $a = 1,2$ and $i = 1,2,3$ are called the structure functions of the nucleon. [See (21.96) and Problems 21.6–21.8.] The scaling approximation * assumes that at any given x when q^2 is $\gg m_N^2$, we can regard these structure functions as approximately q^2-independent; i.e.,

$$F_a^{el}(q^2, x) \cong F_a^{el}(x)$$

and

$$F_i(q^2, x) \cong F_i(x) \,. \tag{23.7}$$

Since F_a^{el} and F_i can be measured directly, the validity of this approximation may be readily checked. On pages 847–8 the experimental values of some of these structure functions are plotted against q^2 for different x values. As we can see, the scaling approximation holds remarkably well for q^2 in the range from a few $(\text{GeV})^2$

---

* J. D. Bjorken, Phys. Rev. 179, 1547 (1969).

to about 20 $(\mathrm{GeV})^2$; beyond that there is a systematic departure.

Such a departure is expected from QCD, on account of the radiative correction due to gluons. By following arguments similar to those given in Section 8.7, we can show that the correction term is of the order

$$\frac{g^2}{4\pi} \ln q^2 \tag{23.8}$$

where g is the quark-gluon coupling inside the hadron. According to (20.72) and (20.79), $\alpha_c = g^2/4\pi$ is not a large number. Because $\ln q^2$ is a slowly varying function of q^2, (23.8) is almost a constant, at least for a limited range of q^2 ; this then explains the approximate validity * of the scaling assumption.

For $q^2 \gg m_N{}^2$, we have from (21.81) $m_N/q_0 \ll 1$; hence, in the scaling approximation (23.4) and (23.6) become

$$d^2\sigma_{\ell N} = \frac{m_N E_\ell}{2\pi} \left(\frac{e^2}{q^2}\right)^2 dx\, dy\, [\, xy^2 F_1^{el}(x) + (1-y) F_2^{el}(x)\,] \tag{23.9}$$

and

$$d^2\sigma_{\nu N} = \frac{m_N E_\nu}{\pi} G^2\, dx\, dy\, [\, xy^2 F_1(x) + (1-y) F_2(x) \pm xy(1-\tfrac{1}{2}y) F_3(x)\,] \tag{23.10}$$

where + is for ν_ℓ and - for $\bar{\nu}_\ell$, as before.

Recall that in order to characterize the kinematics of the inclusive neutrino reaction (23.2), we need three independent variables, which can be

$$E_\nu\,,\quad x \quad \text{and} \quad y\quad .$$

Since the ranges of x and y are both from 0 to 1, independent

---

* Further mention of this is made on pages 722–3.

of E_ν, the scaling approximation (23.10) implies that the total inclusive neutrino (or antineutrino) cross section must be proportional to E_ν. This agrees quite well with the experimental results given on page 846 of the Appendix, with the proportionality constants equal to

$$(.63 \pm .02) \times 10^{-38} \ \mathrm{cm}^2/\mathrm{GeV} \tag{23.11}$$

for ν_μ, and

$$(.30 \pm .01) \times 10^{-38} \ \mathrm{cm}^2/\mathrm{GeV} \tag{23.12}$$

for $\bar{\nu}_\mu$; both values are for a single target nucleon, averaged over its isospin. [See also the discussions given in Section 8.5.]

As in (21.146), we may define

$$\begin{aligned} L(x) &\equiv x(F_1 + \tfrac{1}{2} F_3)/F_2 \\ \text{and} \qquad & \\ R(x) &\equiv x(F_1 - \tfrac{1}{2} F_3)/F_2 \ . \end{aligned} \tag{23.13}$$

Equation (23.10) becomes then

$$d^2\sigma_{\nu N} = \frac{m_N E_\nu}{\pi} G^2 \, dx \, dy \, F_2 [1 - y + yL - y(1-y) R] \tag{23.14}$$

for the neutrino reaction, and

$$d^2\sigma_{\bar{\nu} N} = \frac{m_N E_\nu}{\pi} G^2 \, dx \, dy \, F_2 [1 - y + yR - y(1-y) L] \tag{23.15}$$

for the antineutrino reaction.

Exercise. Consider the region $q^2 = O(m_W^2)$ and $O(m_Z^2)$. Assuming that the scaling approximation is valid for the nucleon, derive the expressions of $d^2\sigma_{\ell N}$ and $d^2\sigma_{\nu N}$ by using Figure 23.1 and including the full Z and W propagators.

23.2 Quark-parton Model

We shall now examine the physical interpretation of the structure function $F_a^{el}(x)$ and $F_i(x)$, within the context of the scaling approximation.

1. Parton picture

Let us represent (23.1) or (23.2) as

$$\ell(k) + N(p) \rightarrow \ell(k') + \cdots \qquad (23.16)$$

where $\ell(k)$ and $\ell(k')$ stand for the initial and final leptons; each can be either neutral or charged. The labels k, p and k' still denote the appropriate 4-momenta given by (23.3). In the quark model, the nucleon is a composite of three u or d quarks, plus quark pairs and gluons (i.e., quanta of the color gauge fields):

$$\begin{aligned} \text{proton} &= \text{u u d} + \text{quark pairs} + \text{gluons}, \\ \text{neutron} &= \text{d d u} + \text{quark pairs} + \text{gluons} \end{aligned} \qquad (23.17)$$

in which the first three, uud in p or ddu in n, are called valence quarks, the rest are referred to as the sea. Unlike the quarks, the gluons do not directly interact with the electromagnetic and the weak intermediate boson fields. Because the gluons are flavorless, in the sea for every quark of flavor f there must also be an antiquark of flavor $\bar{f}$.

It is sometimes convenient to stay in the center-of-mass system. In this system when the total energy is very high so is the 3-momentum $\vec{p}$ of the initial nucleon. When

$$\vec{p} \rightarrow \infty , \qquad (23.18)$$

we may neglect the quark-gluon interaction during the collision, and therefore the quarks and gluons can be viewed simply as " free " particles, called partons by Feynman.* [This is similar in spirit to the impulse approximation used in, say, high-energy atomic collisions.] Reaction (23.16) becomes then a superposition of "elastic" processes, each of the form

$$\ell(k) + \text{quark}(P) \rightarrow \ell(k') + \text{quark}(P') \qquad (23.19)$$

where "quark" stands for either a quark or an antiquark, with P and P' as its initial and final 4-momenta. Since different final quark channels do not interfere, the inclusive cross section of the inelastic ℓ-nucleon reaction (23.16) can be written as a sum of the "elastic" ℓ-quark cross sections. In this approximation, and neglecting the mass difference between the initial and final quarks (in case they are of different flavors), we have

$$P^2 = P'^2 .$$

Since

$$P' - P = k - k' = q ,$$

it follows that

$$q^2 + 2q \cdot P = 0$$

which, together with the first equation in (23.5), leads to

$$x = \frac{q \cdot P}{q \cdot p} . \qquad (23.20)$$

Let us choose the z-axis parallel to $\vec{p}$. The 4-momenta p and P can be written as

$$p = (0, 0, p_z, i p_0)$$

* R. P. Feynman, Phys.Rev.Lett. 23, 1415 (1969); Proceedings of the Third International Conference on High-energy Collisions at Stony Brook (New York Gordon and Breach, 1969); J. D. Bjorken and E. A. Paschos, Phys.Rev. 185, 1975 (1969).

and

$$P = (P_x, P_y, P_z, i P_0)$$

where

$$p_0 = \sqrt{p_z^2 + m_N^2}$$

and (23.21)

$$P_0 = \sqrt{P_\perp^2 + P_z^2 + m_q^2}$$

with

$$P_\perp^2 = P_x^2 + P_y^2$$

and m_q the quark mass inside the nucleon. The limit (23.18) becomes

$$p_z \to \infty ; \tag{23.22}$$

in this coordinate system, called the infinite momentum frame, P_z must also $\to \infty$, since the quark is a constituent of the nucleon. However, the quark momentum in the x or y direction remains finite:

$$P_\perp = O(R^{-1})$$

where R is the radius of the nucleon bag. Thus, we have from (23.21)-(23.22)

$$p_0 = p_z + O(m_N^2/p_z)$$

and

$$P_0 = P_z + O(m_q^2/P_z \, , \, P_\perp^2/P_z) \quad ,$$

which means that the 4-momenta P and p become parallel when $\vec{p} \to \infty$. Combining this conclusion with (23.20), we see that in the infinite momentum frame

$$P_\mu = x \, p_\mu \quad . \tag{23.23}$$

The parameter x now acquires a different physical meaning.

To begin with, in the inclusive ℓ - nucleon process (23.16)

$$x \equiv - \frac{q^2}{2q \cdot p}$$

is a parameter that is determined by the specific kinematic configuration of the reaction. On the other hand, according to (23.23), in the frame where the nucleon has an infinite momentum, x is also the fraction of the nuclear momentum carried by the quark. The interplay between these two definitions of x leads to a new interpretation of the structure functions, as we shall see.

2. <u>Quark - distribution functions</u>

In the infinite momentum frame of the nucleon, let $u(x)\,dx$ be the probability of finding a u-quark carrying the fraction between x and $x + dx$ of the total nuclear momentum. Likewise, $\bar{u}(x)$, $d(x)$, $\bar{d}(x)$, $s(x)$, ··· are the corresponding probabilities of the $\bar{u}$ - quark, d - quark, $\bar{d}$ - quark, s - quark, ··· . These functions are called quark-distribution functions.

In the proton, there are two valence u - quarks and one valence d - quark; hence,

$$\int [u(x) - \bar{u}(x)]\, dx = 2$$

and

$$\int [d(x) - \bar{d}(x)]\, dx = 1 \quad . \tag{23.24}$$

For the neutron the roles of u and d are switched.

Since there is no valence s - quark in the nucleon, we have

$$\int [s(x) - \bar{s}(x)]\, dx = 0 \quad . \tag{23.25}$$

Likewise, we find

$$\int [c(x) - \bar{c}(x)]\, dx = 0$$

and a similar relation for the b - and $\bar{b}$ - quarks. All x - integrations are from 0 to 1. In the next section, we shall discuss how these quark-distribution functions can be related to the structure functions $F_a^{el}(x)$ and $F_i(x)$.

23.3 Deep Inelastic e-nucleon Scattering

In the literature, the high-energy inclusive inelastic scattering between e and N is often referred to as deep inelastic scattering. In accordance with the quark-parton picture, we shall now resolve this process in terms of the elastic scattering between an electron and a quark of flavor f :

$$e(k) + \text{quark}(P) \rightarrow e(k') + \text{quark}(P') \tag{23.26}$$

in which k, P, k' and P' denote the appropriate 4-momenta, as in (23.19). The (center-of-mass energy)^2 of (23.26) is

$$s^2 \equiv -(k+P)^2 = -2k \cdot P + m_q^2 + m_e^2$$

where, as before, m_q is the quark mass inside the nucleon and m_e the electron mass. At very high energy, $s \rightarrow \infty$, we can set

$$\frac{m_q^2}{2k \cdot P} = 0 \qquad \text{and} \qquad \frac{m_e^2}{2k \cdot P} = 0 \ . \tag{23.27}$$

By using Problem 6.2 and considering only the γ-exchange diagram (i.e., neglecting the Z-exchange diagram for simplicity), we find that the differential cross section of (23.26) is

$$d\sigma_{ef} = \frac{dq^2}{4\pi}\left(\frac{e_f e}{q^2}\right)^2 \left[1 + \left(\frac{q^2}{2k \cdot P}\right) + \tfrac{1}{2}\left(\frac{q^2}{2k \cdot P}\right)^2\right] \tag{23.28}$$

where f denotes the flavor of the quark and e_f its electric charge. From (23.5) and $P = xp$, we see that

$$xy = -\frac{q^2}{2k \cdot p} = -\frac{xq^2}{2k \cdot P}$$

where, as in (23.16), p refers to the 4-momentum of the initial nucleon. Therefore,

$$y = -\frac{q^2}{2k\cdot P}$$

and (23.29)

$$\frac{dq^2}{dy} = -2k\cdot P = -2xk\cdot p \quad .$$

The differential cross section (23.28) can then be written as

$$d\sigma_{ef} = \frac{dy}{2\pi}\left(\frac{e_f e}{q^2}\right)^2(-xk\cdot p)\left(1-y+\tfrac{1}{2}y^2\right) \quad . \qquad (23.30)$$

Let f(x) be the distribution function of the quark (or antiquark) of flavor f in the nucleon, with

$$f = u, d, s, c, \bar{u}, \bar{d}, \ldots \quad . \qquad (23.31)$$

Multiplying (23.30) by f(x) dx and summing over the different flavors, we obtain the deep inelastic eN cross section:

$$d^2\sigma_{eN} = \sum_f f(x)\,dx\,d\sigma_{ef}$$

$$= \frac{1}{2\pi}\left(\frac{e^2}{q^2}\right)^2(-k\cdot p)\,dx\,dy\sum_f (e_f/e)^2\,x\,f(x)\left[1-y+\tfrac{1}{2}y^2\right] .$$

(23.32)

A comparison between (23.9) and (23.32) relates the nuclear structure functions and the quark-distribution functions. We find

$$2x\,F_1^{el}(x) = F_2^{el}(x)$$

and (23.33)

$$F_2^{el}(x) = \sum_f (e_f/e)^2\,x\,f(x) \quad .$$

Thus, for the ep collision the corresponding $F_2^{el}(x)$ is

$$(F_2^{el})_p = x\left\{\frac{4}{9}\left[u(x)+\bar{u}(x)+c(x)+\bar{c}(x)\right] + \frac{1}{9}\left[d(x)+\bar{d}(x)+s(x)+\bar{s}(x)\right]\right\} \qquad (23.34)$$

where, for simplicity, we neglect the b and $\bar{b}$ distributions. Likewise, for the en collision the corresponding $F_2^{el}(x)$ is

$$(F_2^{el})_n = x\left\{\frac{4}{9}[d(x) + \bar{d}(x) + c(x) + \bar{c}(x)] \right.$$
$$\left. \frac{1}{9}[u(x) + \bar{u}(x) + s(x) + \bar{s}(x)]\right\} \qquad (23.35)$$

in which, as well as in the following, $u(x)$, $\bar{u}(x)$, $d(x)$, $\bar{d}(x)$, $\cdots$ always refer to the distribution functions in the proton state. [Hence, e.g., $d(x)$ in (23.35) equals the distribution function of the u-quark in the neutron.] The experimental values of the sum and difference of $(F_2^{el})_p$ and $(F_2^{el})_n$ are given on page 847 (but referred to there as F_2^{ep} and F_2^{en}).

We emphasize that x, in accordance with its definition (23.5), is a scalar; so are the structure functions and the quark distribution functions. They do not change under any Lorentz transformation. However, only in the infinite momentum frame do these distribution functions $u(x)$, $d(x)$, $\cdots$ acquire the additional probabilistic interpretation given on page 710.

23.4 High-energy Neutrino Reaction

In the same way, we can resolve the high-energy inelastic neutrino reactions

$$\nu(k) + N(p) \rightarrow \ell(k') + \cdots$$
and
$$\bar{\nu}(k) + N(p) \rightarrow \ell(k') + \cdots, \qquad (23.36)$$

into incoherent mixtures of elastic scatterings between ν (or $\bar{\nu}$) and free quarks. In (23.36), the final ℓ can be any lepton, neutral or charged, and ν stands for any neutrino ν_e, ν_μ or ν_τ. As in (23.32), the differential cross section of (23.36) can be written as

$$d^2\sigma_{\nu N} = \sum_f f(x)\, dx\, d\sigma_{\nu f}$$

and (23.37)

$$d^2\sigma_{\bar{\nu}N} = \sum_f f(x)\, dx\, d\sigma_{\bar{\nu}f}$$

where $d\sigma_{\nu f}$ and $d\sigma_{\bar{\nu}f}$ refer to the elastic cross sections of ν_ℓ and $\bar{\nu}_\ell$ on a quark of flavor f, and the sum extends over all f given by (23.31).

1. Neutrino-quark cross sections

To derive $d\sigma_{\nu f}$ and $d\sigma_{\bar{\nu}f}$, let us consider, e.g.,

$$\nu_\ell(k) + d(P) \rightarrow \ell^-(k') + u(P')$$

and (23.38)

$$\bar{\nu}_\ell(k) + u(P) \rightarrow \ell^+(k') + d(P') \quad .$$

From Problem 5.4 and (22.124), we see that these differential cross sections can be obtained by setting in (5.128)

$$C_V = C_A = G \cos\theta_c \quad ,$$

where G is the Fermi constant and θ_c the Cabibbo angle. On account of (23.27), at high energy the results are

$$d\sigma^-_{\nu d} = \frac{2}{\pi}(-k\cdot P)\, dy\, (G\cos\theta_c)^2$$

and (23.39)

$$d\sigma^+_{\bar{\nu}u} = \frac{2}{\pi}(-k\cdot P)\, dy\, (G\cos\theta_c)^2\, (1-y)^2$$

where the superscript $\mp$ indicates the final lepton charge.

Because $P = xp$, we have

$$-k\cdot P = -xk\cdot p = xm_N E_\nu$$

where E_ν is the incident neutrino energy in the rest frame of the nucleon. By applying CP, or CPT, to (23.38)-(23.39), we find that the differential cross sections for

$$\nu_\ell(k) + \bar{u}(P) \rightarrow \ell^-(k') + \bar{d}(P')$$

and

$$\bar{\nu}_\ell(k) + \bar{d}(P) \rightarrow \ell^+(k') + \bar{u}(P')$$

are

$$d\sigma^{-}_{\nu\bar{u}} = \frac{2}{\pi} m_N E_\nu \times dy(G\cos\theta_c)^2 (1-y)^2$$

and

$$d\sigma^{+}_{\bar{\nu}\bar{d}} = \frac{2}{\pi} m_N E_\nu \times dy(G\cos\theta_c)^2 \quad . \tag{23.40}$$

In a similar way we can derive $d\sigma^{-}_{\nu f}$ and $d\sigma^{+}_{\bar{\nu}f}$ for quarks of other flavors. The results are given in the following two tables:

Reaction	$\frac{d\sigma^{-}_{\nu f}}{dy}$ in units of $\frac{2}{\pi} G^2 m_N E_\nu$
$\nu_\ell + d \rightarrow \ell^- + u$	$\times \cos^2\theta_c$
$\nu_\ell + d \rightarrow \ell^- + c$	$\times \sin^2\theta_c$
$\nu_\ell + s \rightarrow \ell^- + u$	$\times \sin^2\theta_c$
$\nu_\ell + s \rightarrow \ell^- + c$	$\times \cos^2\theta_c$
$\nu_\ell + \bar{u} \rightarrow \ell^- + \bar{d}$	$\times \cos^2\theta_c (1-y)^2$
$\nu_\ell + \bar{u} \rightarrow \ell^- + \bar{s}$	$\times \sin^2\theta_c (1-y)^2$
$\nu_\ell + \bar{c} \rightarrow \ell^- + \bar{d}$	$\times \sin^2\theta_c (1-y)^2$
$\nu_\ell + \bar{c} \rightarrow \ell^- + \bar{s}$	$\times \cos^2\theta_c (1-y)^2$

Table 23.1

Reaction	$\frac{d\sigma^{+}_{\bar{\nu}f}}{dy}$ in units of $\frac{2}{\pi} G^2 m_N E_\nu$
$\bar{\nu}_\ell + u \rightarrow \ell^+ + d$	$\times \cos^2 \theta_c (1-y)^2$
$\bar{\nu}_\ell + u \rightarrow \ell^+ + s$	$\times \sin^2 \theta_c (1-y)^2$
$\bar{\nu}_\ell + c \rightarrow \ell^+ + d$	$\times \sin^2 \theta_c (1-y)^2$
$\bar{\nu}_\ell + c \rightarrow \ell^+ + s$	$\times \cos^2 \theta_c (1-y)^2$
$\bar{\nu}_\ell + \bar{d} \rightarrow \ell^+ + \bar{u}$	$\times \cos^2 \theta_c$
$\bar{\nu}_\ell + \bar{d} \rightarrow \ell^+ + \bar{c}$	$\times \sin^2 \theta_c$
$\bar{\nu}_\ell + \bar{s} \rightarrow \ell^+ + \bar{u}$	$\times \sin^2 \theta_c$
$\bar{\nu}_\ell + \bar{s} \rightarrow \ell^+ + \bar{c}$	$\times \cos^2 \theta_c$

Table 23.2

The differential cross sections of these two tables are related to each other through CP, or CPT. We have

$$d\sigma^{-}_{\nu f} = d\sigma^{+}_{\bar{\nu}\bar{f}} \tag{23.41}$$

where, as before, the flavor f can be u, $\bar{u}$, d, $\bar{d}$, $\cdots$.

Next, we consider the neutral current cross sections $d\sigma^{o}_{\nu f}$ and $d\sigma^{o}_{\bar{\nu}f}$; the neutrality is indicated by the superscript o. The reactions are

$$\nu_\ell(k) + f(P) \rightarrow \nu_\ell(k') + f(P')$$

and
$$\bar{\nu}_\ell(k) + f(P) \rightarrow \bar{\nu}_\ell(k') + f(P') \ . \qquad (23.42)$$

From page 698 we see that the neutral current $J_\lambda^{\,o}$ conserves flavor; hence, the initial and final quarks in (22.42) are of the same f. Table 23.3 lists $d\sigma^{\,o}_{\nu f}$. The corresponding $d\sigma^{\,o}_{\bar{\nu}\bar{f}}$ can be obtained by using

$$d\sigma^{\,o}_{\nu f} = d\sigma^{\,o}_{\bar{\nu}\bar{f}} \quad ,$$

which follows from CP or CPT invariance.

Reaction	$\dfrac{d\sigma^{\,o}_{\nu f}}{dy}$ in units of $\dfrac{2}{\pi} G^2 m_N E_\nu$
$\nu_\ell + u \rightarrow \nu_\ell + u$	$\frac{x}{4}[(1-\frac{4}{3}\sin^2\theta_w)^2 + \frac{16}{9}\sin^4\theta_w(1-y)^2]$
$\nu_\ell + d \rightarrow \nu_\ell + d$	$\frac{x}{4}[(1-\frac{2}{3}\sin^2\theta_w)^2 + \frac{4}{9}\sin^4\theta_w(1-y)^2]$
$\nu_\ell + s \rightarrow \nu_\ell + s$	$\frac{x}{4}[(1-\frac{2}{3}\sin^2\theta_w)^2 + \frac{4}{9}\sin^4\theta_w(1-y)^2]$
$\nu_\ell + c \rightarrow \nu_\ell + c$	$\frac{x}{4}[(1-\frac{4}{3}\sin^2\theta_w)^2 + \frac{16}{9}\sin^4\theta_w(1-y)^2]$
$\bar{\nu}_\ell + u \rightarrow \bar{\nu}_\ell + u$	$\frac{x}{4}[\frac{16}{9}\sin^4\theta_w + (1-\frac{4}{3}\sin^2\theta_w)^2(1-y)^2]$
$\bar{\nu}_\ell + d \rightarrow \bar{\nu}_\ell + d$	$\frac{x}{4}[\frac{4}{9}\sin^4\theta_w + (1-\frac{2}{3}\sin^2\theta_w)^2(1-y)^2]$
$\bar{\nu}_\ell + s \rightarrow \bar{\nu}_\ell + s$	$\frac{x}{4}[\frac{4}{9}\sin^4\theta_w + (1-\frac{2}{3}\sin^2\theta_w)^2(1-y)^2]$
$\bar{\nu}_\ell + c \rightarrow \bar{\nu}_\ell + c$	$\frac{x}{4}[\frac{16}{9}\sin^4\theta_w + (1-\frac{4}{3}\sin^2\theta_w)^2(1-y)^2]$

Table 23.3

2. Neutrino-nucleon cross sections

By first substituting these neutrino-quark cross sections into (23.37) and then comparing the result with (23.10) and (23.14)-(23.15), we see that *

$$2x\, F_1(x) = F_2(x)$$

and

$$L(x) + R(x) = 1 \quad . \tag{23.43}$$

These, and a similar relation

$$2x\, F_1^{el}(x) = F_2^{el}(x)$$

given by (23.38), are the consequences of the spin-$\frac{1}{2}$ nature of quarks in the parton model. From Tables 23.1-3, it is strightforward to derive the nucleon structure functions in terms of the quark distribution functions.

(i) Charged current reactions

For the reaction

$$\nu_\ell + p \rightarrow \ell^- + \cdots \quad ,$$

the products $F_2\, L$ and $F_2\, R$ are

$$(F_2\, L)_p^- = 2x\, [\, d(x) + s(x)\,]$$

and

$$(F_2\, R)_p^- = 2x\, [\, \bar{u}(x) + \bar{c}(x)\,] \quad , \tag{23.44}$$

and for

$$\bar{\nu}_\ell + p \rightarrow \ell^+ + \cdots$$

they are

$$(F_2\, L)_p^+ = 2x\, [\, u(x) + c(x)\,]$$

and

$$(F_2\, R)_p^+ = 2x\, [\, \bar{d}(x) + \bar{s}(x)\,] \tag{23.45}$$

---

* C. G. Callan and D. J. Gross, Phys.Rev.Lett. 22, 156 (1969).

where, as before, the superscript $\pm$ indicates the final lepton charge, which of course equals the change in the hadron charge. Likewise, for the $\nu_\ell n$ collision

$$(F_2 L)_n^- = 2x [u(x) + s(x)]$$
and
$$(F_2 R)_n^- = 2x [\bar{d}(x) + \bar{c}(x)] , \tag{23.46}$$

and for the $\bar{\nu}_\ell n$ collision

$$(F_2 L)_n^+ = 2x [d(x) + c(x)] ,$$
$$(F_2 R)_n^+ = 2x [\bar{u}(x) + \bar{s}(x)] . \tag{23.47}$$

Throughout, the quark distributions $u(x)$, $\bar{u}(x)$, $d(x)$, $\cdots$ refer to those in the proton state, as in (23.34)-(23.35).

In terms of these structure functions, because $L + R = 1$, the differential cross sections (23.14)-(23.15) can be written as

$$d^2\sigma_{\nu N}^\alpha = \frac{m_N E_\nu}{\pi} G^2 \, dx \, dy \, [(F_2 L)_N^\alpha + (1-y)^2 (F_2 R)_N^\alpha]$$
and
$$d^2\sigma_{\bar{\nu} N}^\alpha = \frac{m_N E_\nu}{\pi} G^2 \, dx \, dy \, [(F_2 R)_N^\alpha + (1-y)^2 (F_2 L)_N^\alpha] \tag{23.48}$$

where

$$N = p \text{ or } n \qquad \text{and} \qquad \alpha = \pm , \text{ or } 0 .$$

The case $\alpha = 0$ is for the neutral current reactions which will be discussed below.

(ii) Neutral current reactions

For the reactions

$$\nu_\ell + p \rightarrow \nu_\ell + \cdots$$
and
$$\bar{\nu}_\ell + p \rightarrow \bar{\nu}_\ell + \cdots ,$$

the structure functions are

$$(F_2 L)_p^o = \frac{x}{2}\left\{(1-\frac{4}{3}\sin^2\theta_w)^2[u(x)+c(x)] + (1-\frac{2}{3}\sin^2\theta_w)^2[d(x)+s(x)] + \frac{16}{9}\sin^4\theta_w[\bar{u}(x)+\bar{c}(x)] + \frac{4}{9}\sin^4\theta_w[\bar{d}(x)+\bar{s}(x)]\right\} \tag{23.49}$$

and

$$(F_2 R)_p^o = \frac{x}{2}\left\{\frac{16}{9}\sin^4\theta_w[u(x)+c(x)] + \frac{4}{9}\sin^4\theta_w[d(x)+s(x)] + (1-\frac{4}{3}\sin^2\theta_w)^2[\bar{u}(x)+\bar{c}(x)] + (1-\frac{2}{3}\sin^2\theta_w)^2[\bar{d}(x)+\bar{s}(x)]\right\}; \tag{23.50}$$

for $\nu_\ell n$ and $\bar{\nu}_\ell n$ neutral current reactions, we have

$$(F_2 L)_n^o = \frac{x}{2}\left\{(1-\frac{4}{3}\sin^2\theta_w)^2[d(x)+c(x)] + (1-\frac{2}{3}\sin^2\theta_w)^2[u(x)+s(x)] + \frac{16}{9}\sin^4\theta_w[\bar{d}(x)+\bar{c}(x)] + \frac{4}{9}\sin^4\theta_w[\bar{u}(x)+\bar{s}(x)]\right\} \tag{23.51}$$

and

$$(F_2 R)_n^o = \frac{x}{2}\left\{\frac{16}{9}\sin^4\theta_w[d(x)+c(x)] + \frac{4}{9}\sin^4\theta_w[u(x)+s(x)] + (1-\frac{4}{3}\sin^2\theta_w)^2[\bar{d}(x)+\bar{c}(x)] + (1-\frac{2}{3}\sin^2\theta_w)^2[\bar{u}(x)+\bar{s}(x)]\right\}. \tag{23.52}$$

The corresponding differential cross sections can be obtained by substituting (23.49)-(23.52) into (23.48).

Exercise 1. Prove the following sum rules:

(i) $$\frac{1}{2}\int \frac{dx}{x}\,[(F_2)_n^- - (F_2)_p^-] = \frac{1}{2}\int \frac{dx}{x}\,[(F_2)_p^+ - (F_2)_n^+]$$

$$= \int dx\,[(u-\bar{u}) - (d-\bar{d})] = 1\;, \qquad \text{(Adler)}$$

(ii) $$\frac{1}{4}\int dx\,[(F_3)_n^- + (F_3)_n^+ + (F_3)_p^- + (F_3)_p^+]$$

$$= \int dx\,[(u-\bar{u}) + (d-\bar{d})] = 3 \qquad \text{(Gross-Llewellyn Smith)}$$

and

(iii) $$\int x^n\,dx\,[6(F_2^{el})_p - 6(F_2^{el})_n + x(F_3)_p^- - x(F_3)_n^-]$$

$$= \int x^n\,dx\,[(F_2^{el})_p - (F_2^{el})_n + x(F_3)_n^+ - x(F_3)_p^+] = 0$$

(Llewellyn Smith)

where n is arbitrary.

As before, the integrations are all from $x = 0$ to $x = 1$, and $u, \bar{u}, d, \bar{d}, \cdots$ are the quark-distribution functions in the proton.

Exercise 2. Consider lepton collisions with isoscalar nuclei. The structure functions are:

$$F_a^{el} = \frac{1}{2}[(F_a^{el})_p + (F_a^{el})_n]$$

where $a = 1$ or 2, and

$$F_i^{\alpha} = \frac{1}{2}[(F_i^{\alpha})_p + (F_i^{el})_n]$$

where $i = 1, 2$ or 3 and $\alpha = +, -$ or 0, as before. Show that in the quark-parton model

(i) $$F_2^{el} = \frac{5x}{18}\,[q + \bar{q} - \frac{3}{5}\,(s + \bar{s} - c - \bar{c})]\;,$$

$$F_2^{\mp} = x\,[q + \bar{q} \pm (s - \bar{s} - c + \bar{c})]$$

and (23.53)

$$F_3^{\mp} = q - \bar{q} \pm (s + \bar{s} - c - \bar{c})$$

with

$$q \equiv u + d + c + s$$

and (23.54)

$$\bar{q} \equiv \bar{u} + \bar{d} + \bar{c} + \bar{s} \ ,$$

(ii) $\int dx\, F_2^{-}(x) = \int dx\, F_2^{+}(x)$,

and therefore at $y = 0$, if we integrate (23.48) over x,

$$\frac{d\sigma_{\nu}^{-}}{dy} = \frac{d\sigma_{\bar{\nu}}^{+}}{dy} \qquad (23.55)$$

(Eq. (23.55) is also valid if we replace the superscript $\mp$ by 0, i.e., replace the charged by the neutral current), and

(iii) in the approximation of zero sea quark,

$$\bar{q} = 0 \ ,$$

the ratio of the total inclusive cross section of

$$\nu_{\ell} + N \rightarrow \ell^{-} + \cdots$$

to that of

$$\bar{\nu}_{\ell} + N \rightarrow \ell^{+} + \cdots$$

is

$$3 : 1 \ . \qquad (23.56)$$

3. Experimental results

High-energy e, μ, ν and $\bar{\nu}$ experiments have been extensively performed. As noted before, in the range of q^2 = (4-momentum transfer)^2 from a few (GeV)^2 to ~ 20 (GeV)^2, the scaling approximation and the quark-parton model hold remarkably well. Even at larger q^2, between 20 to 200 (GeV)^2, the deviation remains relatively small. A good measure is to study the percentage variation of the q^2 - dependence of the structure functions at a fixed x. As we can see from the figures on pages 847-8, these variations are only $\approx$ 15% or less, depending on x ; they can be accounted for by including the QCD radiative corrections.* A detailed discussion lies outside the scope of this book.

Within the context of the quark-parton model, it is possible to determine the quark-distribution functions from experiment. We list some of the results.**

(i) The importance of the sea quarks can best be represented by the experimental value

$$\frac{\int \bar{q}(x)\, dx}{\int [\, q(x) + \bar{q}(x)\,]\, dx} = .15 \pm .03 \quad , \qquad (23.57)$$

that of the strange quarks by

$$\frac{\int \bar{s}(x)\, dx}{\int [\, q(x) + \bar{q}(x)\,]\, dx} = .025 \pm .01 \qquad (23.58)$$

where $q(x)$ and $\bar{q}(x)$ are defined by (23.54). At present there is no accurate determination of c and $\bar{c}$, except that they are extremely small; this is reasonable considering their large masses.

---

* See the review article by A. H. Mueller, "Perturbative QCD at High Energies," to be published as a Physics Report.

** J. G. H. de Groot et al., Zeits. für Physik (Particles and Fields), 1, 143 (1979). Proceedings of the IX International Symposium on Lepton and Photon Interactions at High Energies, Fermilab, 1979, ed. T. B. W. Kirk and H. D. I. Abarbanel. CDHS Collaboration 2, 92 (1979), Neutrino '79, University of Bergen and NORDITA.

(ii) Phenomenologically, the experimental results of $q(x)$ and $\bar{q}(x)$ have been fitted to simple expressions:

$$\begin{aligned} q(x) - \bar{q}(x) &\propto \sqrt{x}\,(1-x)^n \;, \\ \bar{q}(x) + \bar{s}(x) &\propto (1-x)^m \end{aligned} \tag{23.59}$$

with $n = 3.5 \pm .5$ and $m = 6.5 \pm .5$. The former is due to the valence quarks and the latter to the sea quarks.

(iii) The Weinberg angle θ_W can be determined by the neutral current cross sections. From inclusive ν_μ and $\bar{\nu}_\mu$ reactions, one finds

$$\sin^2 \theta_W = .228 \pm .018 \tag{23.60}$$

in good agreement with other measurements. [See (22.90) and (22.127).]

(iv) In the infinite-momentum frame of the nucleon, the fractional momentum carried by the quarks and antiquarks can be measured by the integral of $x[\,q(x) + \bar{q}(x)\,]$. From (23.53) we see that

$$F_2(x) \equiv \tfrac{1}{2}[\,F_2^-(x) + F_2^+(x)\,] = x[\,q(x) + \bar{q}(x)\,] \quad .$$

The value from the neutrino and antineutrino experiments is

$$\int dx\, F_2(x) \cong .45 \quad . \tag{23.61}$$

This means that approximately $\frac{1}{2}$ of the nucleon momentum is carried by the gluons. The identical conclusion can be reached by using the electron and muon results.

23.5 KLN Theorem *

At high energy, perturbative QCD should be applicable, since it is an "asymptotically free" theory. On the other hand, in the perturbation series gluons are all massless and so, approximately, are the quarks. As we know, none of these zero-mass particles can be actually observed due, presumably, to the nonperturbative color confinement mechanism. Hence, there are inherent difficulties in comparing the perturbative QCD predictions with experimental results. This is why so far we have restricted our discussions to inclusive reactions in which all final channels are summed over.

There is still another problem. As discussed in Section 8.7, the presence of zero-mass particles can lead to "mass singularities." This is connected with the fact that the system has a high degree of degeneracy. For states that consist only of parallel moving massless particles, the same total momentum implies the same total energy. In such a theory, the perturbation expansion of the on-mass-shell transition amplitude in general carries infinities. As we shall see, such divergences can be removed provided we average the relevant transition probability over an appropriate ensemble of degenerate states. Actually, the occurrence of such singularities and their cancellations are consequences of an elementary theorem in quantum mechanics which can be established without any explicit use of Feynman graphs or the detailed form of the Hamiltonian. The well-known problem of the elimination of infrared divergence in QED is one other example of such a case. [See Section 23.6.1.]

---

* T. Kinoshita, J.Math.Phys. 3, 650 (1962). T. D. Lee and M. Nauenberg, Phys.Rev. 133, $\overline{\text{B}}$1549 (1964).

Consider a field theory whose total Hamiltonian is

$$H = H_0 + g H_1 \tag{23.62}$$

where H_0 is the free-particle Hamiltonian, H_1 is the perturbation, and g is an expansion parameter. Since we are interested only in the singularities that appear in the power series in g due to degenerate states, nonperturbative effects such as bound states (even if they exist) will be ignored. Hence, as noted on page 70, in the case of relativistic field theories we may set the spectra of H and H_0 to be the same. The eigenstates of H_0 will be denoted by $|\, a >$, with

$$H_0 \,|\, a > \;=\; E_a \,|\, a > \tag{23.63}$$

and

$$< a \,|\, a > \;=\; 1 \quad .$$

The S matrix can be expressed in terms of the $U(t, t_0)$ matrix given by (5.20)-(5.21):

$$S = U(\infty, -\infty) \quad . \tag{23.64}$$

It is convenient to define

$$U_- \equiv U(0, -\infty)$$

and

$$U_+ \equiv U(0, +\infty) \quad . \tag{23.65}$$

From (6.58)-(6.61), we see that either U_- or U_+ can diagonalize H:

$$U_-^\dagger H U_- = U_+^\dagger H U_+ = E \tag{23.66}$$

where E is a diagonal matrix whose diagonal matrix elements are the same E_a given by (23.63).

Two complete sets of eigenstates of H can be constructed:

$$|\, a^{in} > \;\equiv\; U_- \,|\, a >$$

and

$$|\, a^{f} > \;\equiv\; U_+ \,|\, a > \quad . \tag{23.67}$$

The S matrix is the unitary transformation connecting these two sets. We have

$$S = U_+^\dagger U_- \tag{23.68}$$

and its matrix elements

$$S_{ab} \equiv \langle a | S | b \rangle = \langle a^f | b^{in} \rangle .$$

Hence, the superscripts in and f denote the initial and final states of a reaction in the representation that the total Hamiltonian is diagonal. The corresponding transition probability is given by

$$| S_{ab} |^2 = \sum_{i,j} [(U_+)^*_{ia} (U_+)_{ja}] [(U_-)_{ib} (U_-)^*_{jb}] \tag{23.69}$$

where i and j go over the complete set of vectors that satisfy (23.63).

1. First order perturbation

For clarity, we assume that the problem contains a parameter μ; the degeneracy in the total Hamiltonian appears when $\mu \to 0$. For $\mu \neq 0$, the $(j, a)^{th}$ matrix element of $U_\mp$ can be expanded into the familiar power series:

$$(U_\mp)_{ja} = \delta_{ja} + g \frac{1 - \delta_{ja}}{E_a - E_j \pm i\epsilon} (H_1)_{ja} + O(g^2) \tag{23.70}$$

where $\epsilon = 0+$ and δ_{ja} is the matrix element of the unit matrix. The $\pm i\epsilon$ in the denominator gives rise to the outgoing and incoming waves in the eigenstates $| a^{in} \rangle$ and $| a^f \rangle$, as noted on pages 114 - 115. When $\mu \to 0$, the state a may become degenerate with other states which lie within a certain subset $D(E_a)$. From (23.70) we see that for a state j belonging to this subset, the O(g) term in the expansion of $(U_\mp)_{ja}$ can become singular. It is simple to see that in the same limit $\mu \to 0$, similar divergences also appear in the higher-order terms in the series. Since according to (23.68)

$$S_{ab} = \sum_j (U_+)^*_{ja} (U_-)_{jb} \quad ,$$

the power series expansion of S is also afflicted with infinities of the same origin. We may call these divergences "degeneracy singularities." In the case that the degeneracy is due to zero-mass particles, they will be referred to as "mass singularities."

We now introduce the sums

$$T^-(E_a)_{ij} \equiv \sum_{D(E_a)} (U_-)_{ia} (U_-)^*_{ja}$$

and

$$T^+(E_a)_{ij} \equiv \sum_{D(E_a)} (U_+)_{ia} (U_+)^*_{ja} \tag{23.71}$$

where the summations extend over all states a in the same degenerate subset $D(E_a)$. It is important that the domain of $D(E_a)$ in the Hilbert space be held fixed when the parameter $\mu \to 0$. For example, we can think of $D(E_a)$ as consisting of all states within the energy interval $E_a - \delta E$ and $E_a + \delta E$; in the limit $\mu \to 0$, the width δ should be held fixed, but otherwise can be arbitrarily small. As an illustration, we may consider the case of infrared divergence in QED: the parameter μ can be the photon mass and $D(E_a)$ can consist of any photon whose frequency is $< \omega$, where

$$\omega = \text{resolution of the detection equipment,} \tag{23.72}$$

which is clearly independent of the fictitious photon mass $\mu \to 0$. Furthermore, as will be discussed in Section 23.6.1, in the power-series expansion to every finite order in the coupling constant each of the relevant states in $D(E_a)$ contains only a finite number of infrared photons.

A substitution of (23.70) into (23.71) gives

$$T^{\mp}(E_a)_{ij} = \sum_{D(E_a)} \left[\delta_{ia}\delta_{ja} + g\,\frac{\delta_{ia}(1-\delta_{ja})}{E_a - E_j \mp i\epsilon}\,(H_1)^*_{ja} \right.$$

$$\left. + g\,\frac{\delta_{ja}(1-\delta_{ia})}{E_a - E_i \pm i\epsilon}\,(H_1)_{ia} \right] + O(g^2). \quad (23.73)$$

If both states i and j are outside the degenerate set $D(E_a)$, the three terms in the above square bracket are all 0 ; if one of the states, say i, is within $D(E_a)$ but the other, j , is not, then $\delta_{ja} = 0$, and the sum of these three terms becomes

$$g\,\frac{\delta_{ia}}{E_a - E_j}\,(H_1)^*_{ja} = \text{finite} \quad .$$

In either case, there is no singularity. [It is, of course, understood that the matrix elements of H_1 are free of degeneracy singularities.]

The interesting situation is when both i and j lie within the degenerate set; in this case, after the summation over $D(E_a)$, the terms inside the square bracket can be written as

$$\delta_{ij} + g(1-\delta_{ij}) \left[\frac{1}{E_i - E_j \mp i\epsilon}\,(H_1)^*_{ji} + \frac{1}{E_j - E_i \pm i\epsilon}\,(H_1)_{ij} \right] = \delta_{ij}$$

which is also <u>not</u> singular. Thus, although the $O(g)$ element of the S matrix has mass singularities, to the same order,

$$T^-(E_a) \equiv \text{matrix } T^-(E_a)_{ij}$$

and

$$T^+(E_a) \equiv \text{matrix } T^+(E_a)_{ij}$$

do not. From (23.69), we see that the same conclusion also applies to the transition probability * for $a \to b$

---

* Although $E_a = E_b$, $D(E_a)$ and $D(E_b)$ can be different because of the generalization (23.93).

$$P(a,b) \equiv \sum_{D(E_a)} \sum_{D(E_b)} |S_{ab}|^2 = \sum_{i,j} T^{+}(E_a)_{ji}\, T^{-}(E_b)_{ij}$$

$$= \text{trace}\, [T^{+}(E_a)\, T^{-}(E_b)] \quad . \tag{23.74}$$

In the next section we shall show that our analysis can be readily extended to higher order terms in g. The cancellation of degeneracy singularities remains true to every order, provided we deal with $T^{-}(E_a)$, $T^{+}(E_a)$ and $P(a,b)$, instead of U_-, U_+ and S.

2. <u>General case</u>

From (23.62), (23.66) and the unitarity of $U_{\mp}$, it follows that

$$(H_0 + g H_1) U = U E \tag{23.75}$$

where, as well as in the following,

$$U = U_- \text{ or } U_+ \tag{23.76}$$

and E is the diagonal matrix whose diagonal elements are the eigenvalues of $H = H_0 + g H_1$. Let us define

$$\Delta \equiv H_0 - E \ . \tag{23.77}$$

Equation (23.75) can be written as

$$[U, E] = (g H_1 + \Delta) U \quad . \tag{23.78}$$

In the representation where H_0 is diagonal, Δ is also; its diagonal elements denote the shift of energy levels from H_0 to H. As mentioned before, we are particularly interested in theories in which H_0 and H have the same spectra, i.e.,

$$\Delta = 0 \ . \tag{23.79}$$

For example, in perturbative QCD with zero-mass quarks, both H_0

and H have identical spectra made of the same variety of massless particles. [Actually, the theorem that we shall prove is also applicable to theories in which $\Delta \neq 0$. See the exercise at the end of this section.]

On account of (23.79), we can write (23.78) as

$$[U, E] = g H_1 U . \tag{23.80}$$

By substituting the power-series expansion

$$U = \sum_{n=0}^{\infty} g^{\ell} U_{\ell} \tag{23.81}$$

into (23.80) and equating the coefficients of g^{ℓ} on both sides, we find

$$[U_{\ell}, E] = H_1 U_{\ell-1} . \tag{23.82}$$

In addition, the unitarity

$$U^{\dagger} U = U U^{\dagger} = 1$$

gives for $\ell = 0$

$$U_0 = 1 , \tag{23.83}$$

and for $\ell \geq 1$

$$\sum_{m=0}^{\ell} U_m^{\dagger} U_{\ell-m} = \sum_{m=0}^{\ell} U_m U_{\ell-m}^{\dagger} = 0 . \tag{23.84}$$

Equation (23.82) enables us to express the matrix elements of U_{ℓ} in terms of those of $U_{\ell-1}$. For the off-diagonal $(i, a)^{th}$ element, we have

$$(U_{\ell})_{ia} = \frac{1}{E_a - E_i \pm i\epsilon} \sum_b (H_1)_{ib} (U_{\ell-1})_{ba} \tag{23.85}$$

where, as in (23.70), $\epsilon = 0+$, the upper sign is for $U = U_-$ and the lower for $U = U_+$. By taking the diagonal element of (23.82), we obtain for any given a

$$\sum_b (H_1)_{ab} (U_{\ell-1})_{ba} = 0 \ , \tag{23.86}$$

which is the consequence of $\Delta = 0$. The corresponding element $(U_\ell)_{aa}$ can be obtained from the unitarity condition (23.84). Alternatively, we may also use (23.85); when $i = a$, both the numerator and the denominator are zero, but by taking the so-called adiabatic limit * we can arrange their ratio to give the correct $(U_\ell)_{aa}$.

Whenever there are degenerate states, say

$$E_i = E_a \qquad \text{but} \qquad i \neq a \ ,$$

the denominator in (23.85) becomes zero; that gives rise to degeneracy singularities in $(U_\ell)_{ia}$. Likewise, the power-series expansion of the S matrix also has degeneracy singularities, as noted before. Our purpose is to show that, in contrast, the power-series expansions of $T^-(E_a)_{ij}$ and $T^+(E_a)_{ij}$ are free of such singularities. We can combine the two equations in (23.71):

$$T(E_a)_{ij} = \sum_{D(E_a)} U_{ia} U_{ja}^* \ , \tag{23.87}$$

so that when $U = U_-$ or U_+,

$$T(E_a) = T^-(E_a) \quad \text{or} \quad T^+(E_a)$$

where, as before,

$$T(E_a) = \text{matrix } T(E_a)_{ij} \ . \tag{23.88}$$

---

* See the footnote on page 114.

Let us examine the power-series expansion of $T(E_a)$,

$$T(E_a) = \sum_{\ell=0}^{\infty} g^{\ell} T_{\ell}(E_a) \quad . \tag{23.89}$$

The matrix elements of $T_{\ell}(E_a)$ are related to those of U_m by

$$[T_{\ell}(E_a)]_{ij} = \sum_{m=0}^{\ell} \sum_{D(E_a)} (U_{\ell-m})_{ia} (U_m^*)_{ja} \quad . \tag{23.90}$$

Theorem. $[T_{\ell}(E_a)]_{ij}$ is free of degeneracy singularities.

Proof. For $\ell = 0$ the theorem is obviously true because of (23.83). In the preceding section, we have also established the validity of the theorem for $\ell = 1$. Assume that when $\ell = n - 1 \geq 0$ the theorem holds. To analyze the matrix element $[T_{\ell}(E_a)]_{ij}$ for $\ell = n$, we shall consider separately the following three situations:

(i) The state i lies outside the subset of states $D(E_a)$ that are degenerate with a, but the state j is arbitrary. In this case, from (23.83) it follows that

$$(U_0)_{ia} = 0 \quad .$$

By substituting (23.85) into (23.90) and setting $\ell = n$, we find

$$[T_n(E_a)]_{ij} = (E_a - E_i)^{-1} \sum_b (H_1)_{ib} [T_{n-1}(E_a)]_{bj}$$

where $E_a - E_i \neq 0$ and, as before, b runs over all states. Hence, if $[T_{n-1}(E_a)]_{ij}$ is free of degeneracy singularities, so is $[T_n(E_a)]_{ij}$.

(ii) The state j lies outside $D(E_a)$, but i is arbitrary. Because of the hermiticity relation

$$[T_n(E_a)]_{ij} = [T_n(E_a)]^*_{ji} \quad ,$$

this case becomes the same as the preceding.

(iii) Both i and j are inside $D(E_a)$. From the unitarity condition (23.84) and by setting $\ell = n$, we can write (23.90) as

$$[T_n(E_a)]_{ij} = - \sum_{m=0}^{\ell} \sum_b{}' (U_{n-m})_{ib} (U_m^*)_{jb}$$

where the summation $\sum_b{}'$ now extends only over those b not in $D(E_a)$. Therefore, the righthand side can be written as a sum of $[T_n(E_b)]_{ij}$ where i and j are nondegenerate with b. Case (iii) then also reduces to (i). The theorem is proved by induction on n.

Remarks.

(i) Since $U_{\mp}$ is unitary, its matrix elements cannot have a magnitude bigger than 1. However, the coefficients of its power-series expansion may carry divergences due to degeneracy singularities, e.g., as shown by (23.70). From the unitarity condition

$$U_-^{\dagger} U_- = 1$$

or

$$U_+^{\dagger} U_+ = 1 ,$$

the expansion of $U_-^{\dagger} U_-$ or $U_+^{\dagger} U_+$ in powers of g gives 1 in the zeroth order and 0 in all others, each of which is of course nonsingular; therefore, both expansions are free of degeneracy singularities. The $T^-(E_a)$ and $T^+(E_a)$, defined by (23.71), refer only to what appear to be the dangerous parts in these expansions; naturally, all the apparent singular terms must cancel each other, and that is the origin of the KLN theorem. It states that there is no degeneracy singularity in the power-series expansion of $T^{\mp}(E_a)$. The same conclusion also applies to the power-series expansion of P(a, b), defined by (23.74):

$$P(a,b) = \sum_{D(E_a)} \sum_{D(E_b)} |S_{ab}|^2 = \sum_{\ell=0}^{\infty} g^{\ell} P_{\ell}(a,b) . \quad (23.91)$$

(ii) So far we have restricted our discussions to theories in which the unperturbed Hamiltonian H_0 and the total Hamiltonian H have the same spectra. Because the KLN theorem is the consequence of unitarity, we expect it to remain valid when this restriction is removed. The exercise given below extends the theorem to the general case.

(iii) Up to now, the set $D(E_a)$ refers to all states that are degenerate with E_a ; hence, the theorem can be readily applied to total transition probabilities when states of equal energy are by definition summed over. As we shall see, the set $D(E_a)$ in most applications can be substantially reduced. A simple technique is to consider a different problem in which H_1 is changed into a truncated Hamiltonian H_1' , where

$$(H_1')_{ij} = \begin{cases} (H_1)_{ij} & \text{if } i \text{ and } j \text{ are in } S \,, \\ 0 & \text{otherwise} \end{cases} \tag{23.92}$$

and the set S can be arbitrarily chosen. The only requirement is that H_1' remain Hermitian. Applying the theorem to this new problem, we find that the relevant degenerate set becomes the intersection

$$\mathcal{D}(E_a) = D(E_a) \wedge S \,, \tag{23.93}$$

which can be much smaller than the original $D(E_a)$. In this way, the theorem can also be extended to partial transition rates. [See Section 23.6.2.]

<u>Exercise.</u> Let $H = H_0 + g\,H_1$ and

$$U^\dagger H U = E = \text{diagonal}.$$

Hence, as in (23.78), we have

$$[U, E] = (g H_1 + \Delta) U ,$$

but now assuming

$$\Delta = H_0 - E \neq 0 .$$

Consider the power-series expansions

$$\Delta = \sum_1^\infty g^\ell \Delta_\ell ,$$

$$U = \sum_0^\infty g^\ell U_\ell$$

and

$$T(E_a) = \sum_0^\infty g^\ell T_\ell(E_a) ,$$

where $T(E_a)$ is given by (23.87)-(23.88). Show that if all Δ_ℓ are free of degeneracy singularities, then so are $T_\ell(E_a)$.

23.6 Applications to QED

1. Infrared divergence

Electrodynamics contains degeneracies because photons have zero mass. This is the well-known infrared divergence problem already mentioned. Consider the collision between an electron and an external potential V :

$$e(\vec{p}) \xrightarrow{V} e(\vec{p}\,') \tag{23.94}$$

where $\vec{p}$ and $\vec{p}\,'$ are the initial and final electron momenta. For a static V,

$$|\vec{p}| = |\vec{p}\,'| .$$

In Figure 23.2, diagram (i) is for (23.94), and (ii) gives the lowest-order radiative correction. As we shall see, diagram (ii) is infrared

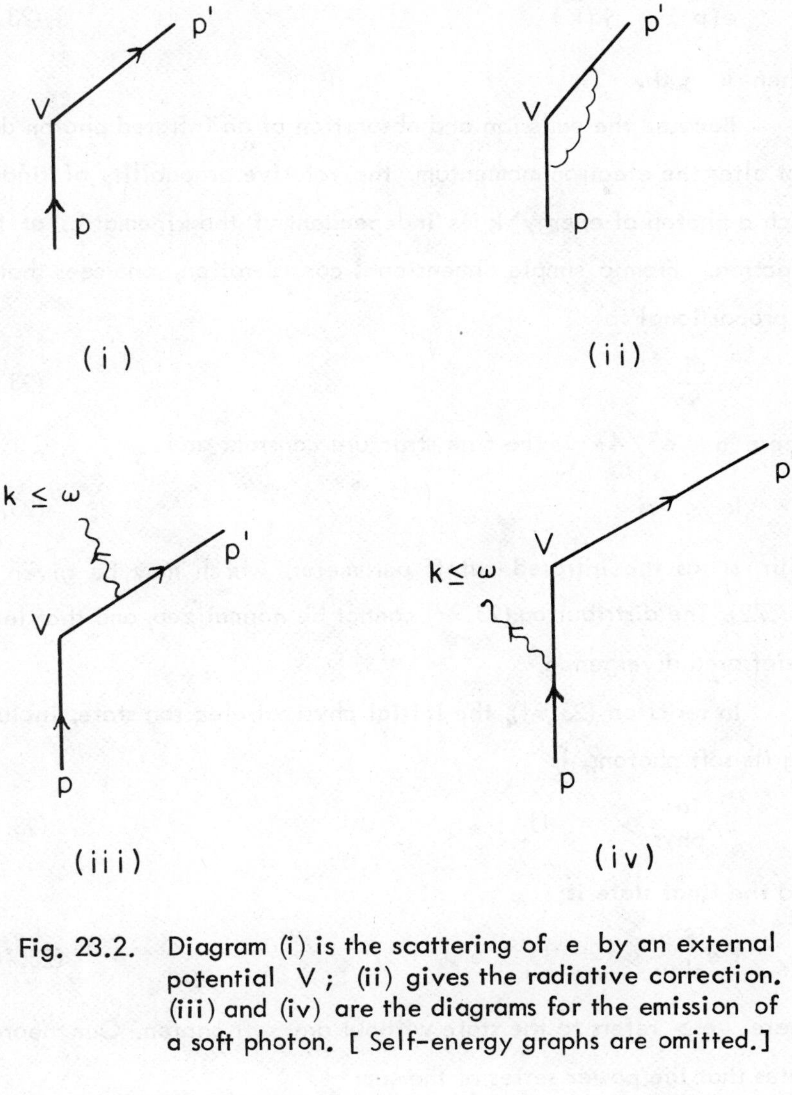

Fig. 23.2. Diagram (i) is the scattering of e by an external potential V ; (ii) gives the radiative correction. (iii) and (iv) are the diagrams for the emission of a soft photon. [Self-energy graphs are omitted.]

divergent. The initial state $e(\vec{p})$ is degenerate with the state

$$e(\vec{p}) + \gamma(\vec{k}) \tag{23.95}$$

when the photon momentum $\vec{k} \to 0$; likewise, the final state $e(\vec{p}\,')$ is also degenerate with the state

$$e(\vec{p}\,') + \gamma(\vec{k}) \tag{23.96}$$

when $\vec{k} \to 0$.

Because the emission and absorption of an infrared photon does not alter the electron momentum, the relative probability of finding such a photon of energy k is independent of the kinematics of the electron. From a simple dimensional consideration, one sees that it is proportional to

$$\alpha \frac{dk}{k} \tag{23.97}$$

where $\alpha = e^2/4\pi$ is the fine structure constant and

$$k \lessgtr \omega \tag{23.98}$$

with ω as the infrared cutoff parameter, which may be given by (23.72). The distribution (23.97) cannot be normalized, and that leads to infrared divergence.

In reaction (23.94), the initial physical electron state, including its soft photons, is

$$| e^{in}_{phys} > = U_- | e > \tag{23.99}$$

and the final state is

$$| e^{f}_{phys} > = U_+ | e > \tag{23.100}$$

where $| e >$ refers to the state without any soft photon. Our theorem states that the power series of the sum

$$\sum_{D(E_a)} U_{ia} U_{ja}^* \tag{23.101}$$

is free of infrared divergence, where

$$U = U_- \quad \text{or} \quad U_+$$

and the sum extends over states (23.95), or (23.96), with k satisfying (23.98). On the other hand, in Figure 23.2 let i and ii be the Feynman amplitudes of the top two diagrams. The square

$$|\, i + ii \,|^2$$

is the probability of observing the reaction (23.94) without any soft photon. This probability is of course zero. However, if we force a power-series expansion, then on account of (23.97) the same probability becomes proportional to

$$1 - c\alpha \int_0^\omega \frac{dk}{k} + O(\alpha^2)$$

where c is a positive numerical constant, and that explains why $|\, i + ii \,|^2$ is infrared divergent.

In the case of soft photons, U_- and U_+ of (23.99) - (23.100) can be obtained explicitly by using the Bloch-Nordsieck approximation,* in which the electron currents $e j_\mu$ can be regarded as a static classical distribution. Let A_μ represent the electromagnetic field which consists only of soft photons ($k \lesssim \omega$). The interaction Hamiltonian is given by

$$e \int j_\mu A_\mu d^3 r \quad .$$

For each given static classical distribution j_μ, the entire Hamiltonian for the soft photons can be diagonalized by a unitary matrix U(j). Its explicit form can be readily derived by using the unitary matrix in Problem 5.1(i).

Consider a problem in which the electron has an initial current distribution j_μ which becomes j_μ' after the collision with an external

---

* F. Bloch and H. Nordsieck, Phys.Rev. 52, 54 (1937).

potential. Let q be the momentum transfer given to the external potential and V(q) the corresponding matrix element. The S matrix (to first order in V) is then given by

$$S = U^\dagger(j') \; V(q) \; U(j) \quad . \tag{23.102}$$

In (23.99)–(23.100),

$$U_- = U(j) \text{ and } U_+ = U(j') \quad .$$

In this approximation, the Hilbert space consists only of soft photons; hence, (23.101) becomes simply the unitarity of U(j) :

$$U(j) \; U^\dagger(j) = 1$$

for arbitrary j_μ. Because the initial and final soft photon amplitudes are represented by independent factors, U(j) and U(j'), in (23.102), it is not difficult to show that the power series expansions of

$$\sum_{D(E_a)} | S_{ab} |^2$$

and

$$\sum_{D(E_b)} | S_{ab} |^2 \tag{23.103}$$

are separately free of infrared divergence.

Thus, in Figure 23.2, if we also include diagrams iii and iv, then the sum

$$| i + ii |^2 + | iii + iv |^2$$

is free from infrared divergence. The calculation is straightforward and the result is *

$$d\sigma_1 = d\sigma_0 \left\{ 1 + \frac{\alpha}{\pi} \left[2\left(\ln \frac{q^2}{m_e^2} - 1\right) \ln \frac{\omega}{E} + \frac{3}{2} \ln \frac{q^2}{m_e^2} + O(1) \right] \right\} \tag{23.104}$$

---

* J. Schwinger, Phys. Rev. 76, 790 (1949).

where q^2 is the square of the momentum transfer, ω is the infrared cutoff given by (23.72), m_e is the electron mass,

$$E = (\vec{p}^{\,2} + m_e^{\,2})^{\frac{1}{2}} = (\vec{p}'^{\,2} + m_e^{\,2})^{\frac{1}{2}} \tag{23.105}$$

and $d\sigma_0$ is the differential cross section for the collision (23.94) without the radiative correction. The $O(1)$ term is finite when $\omega \to 0$ and $m_e \to 0$.

2. Mass singularities and jets

Another application is to examine QED in the limit when $m_e \to 0$. In this case, the state of an electron with a three-momentum $\vec{p}$ is degenerate with the state consisting of an electron with momentum $\vec{p} - \vec{k}$ and a photon with momentum $\vec{k}$, provided $\vec{k} /\!/ \vec{p}$. Here, the magnitude of $\vec{k}$ can be comparable to $\vec{p}$; these photons will be referred to as "hard," in contrast to the soft ones in the infrared region. It has been shown in Section 8.7 that this leads to mass singularities.

Consider again the scattering (23.94) of an electron by a fixed external potential V. In Figure 23.3, diagram (i) denotes such a collision without the presence of any hard photon (i.e., (i) here represents the totality of diagrams given previously in Figure 23.2). The corresponding differential cross section is given by (23.104), which diverges as $m_e \to 0$, in accordance with the expected behavior of the mass singularity.

In order to remove the mass singularity in the differential cross section, we must include the states that are degenerate with either the initial or the final electron. For clarity, let us first assume that the external potential V is not there. In that case, we would have a closed system of photons, electrons and positrons. The states that

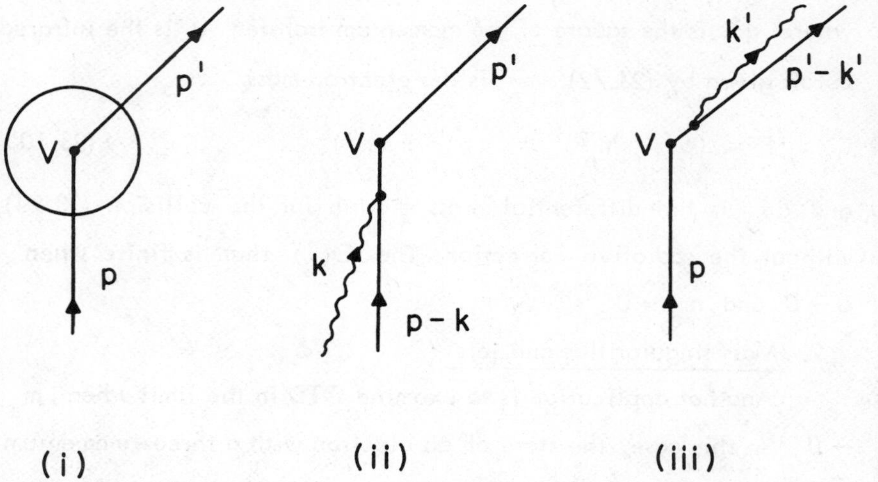

Fig. 23.3. Diagrams for the scattering of jet $(\vec{p})$ to jet $(\vec{p}\,')$ by an external potential: (i) gives the elastic scattering of a single electron with the radiative correction and soft photon effects included, (ii) corresponds to an inelastic scattering with the absorption of a hard photon by the initial electron and (iii) illustrates the emission of a single hard photon in the final state.

are degenerate with, and have the same total momentum as, a single electron of momentum $\vec{p}$ must contain, besides infrared particles, only hard particles ($e^{\mp}$ and γ) all moving parallel to $\vec{p}$. We call such an ensemble of parallel moving particles a <u>jet</u>. Each jet is characterized by a total momentum and a narrow cone of a fixed half-angle δ, which can be arbitrarily small. All particles in the jet have momenta lying within this cone. For δ sufficiently small and for zero-mass particles, the total energy of the jet is simply the magnitude of its total momentum.

The existence of V gives rise to the scattering of particles. The total momentum of the $e^{\mp}$, γ system is no longer conserved. Consider again, e.g., the state of a single electron of momentum $\vec{p}$; it is also degenerate with a state consisting of an electron of momentum $\vec{q}$ and a photon of momentum $\vec{k}$, at a wide angle to $\vec{q}$, provided

$$|\vec{p}| = |\vec{k}| + |\vec{q}| .$$

The presence of V now allows

$$\vec{p} \neq \vec{k} + \vec{q} .$$

By following the argument given in Section 8.7, we know that these states are irrelevant to the mass singularity in (23.104). In order to eliminate these extraneous degenerate states, we may adopt the strategy outlined in (23.92)-(23.93). Let us modify the QED amplitude for the emission and absorption

$$e(\vec{p}) \rightleftarrows e(\vec{p} - \vec{k}) + \gamma(\vec{k}) \tag{23.106}$$

of a <u>hard</u> photon (i.e., a photon whose energy satisfies

$$k = |\vec{k}| > \omega \tag{23.107}$$

where ω is given by (23.72)). We replace the usual QED Hamiltonian H_1 by a truncated Hamiltonian H_1', so that for the hard-photon transition (23.106),

$$H_1' = \begin{cases} H_1 & \text{if the angle between } \vec{k} \text{ and } \vec{p} \lesssim \delta , \\ 0 & \text{otherwise,} \end{cases} \tag{23.108}$$

where, as before, δ can be an arbitrarily small, but fixed, angle. For soft photons, H_1' remains the same as H_1. Such a replacement clearly does not alter the mass singularity term in (23.104). By applying the KLN theorem to the truncated Hamiltonian, we see that if reaction (23.94) is replaced by

$$\text{jet}\ (\vec{p}) \xrightarrow{V} \text{jet}\ (\vec{p}\,') \ , \qquad (23.109)$$

the differential cross section should be free from mass singularities.

To first order in α, we must consider the other two diagrams in Figure 23.3. In diagram (ii), the final state remains the single particle state $e(\vec{p}\,')$, but the initial state $e(\vec{p})$ is replaced by that of an electron of momentum $\vec{p}-\vec{k}$ plus a hard photon of momentum $\vec{k}$, with the jet condition satisfied; i.e.,

$$\text{all momenta lie within a cone of half-angle } \delta \ , \qquad (23.110)$$

where the cone axis is along the original direction of the electron momentum. Likewise, in diagram (iii), the initial state remains $e(\vec{p})$, but the final state now consists of a jet in which there are an electron of momentum $\vec{p}\,' - \vec{k}\,'$ and a hard photon of momentum $\vec{k}\,'$.

It can be readily verified that, after summing over the final and initial sets of degenerate states in these jets and neglecting terms which remain finite as $m_e \to 0$, the differential cross sections $d\sigma_2$ and $d\sigma_3$ for diagrams (ii) and (iii) are given by

$$d\sigma_2 = d\sigma_3 = d\sigma_0 \frac{\alpha}{\pi} \left[2 \ln \frac{E\delta}{m_e} \right] \left[\ln \frac{E}{\omega} - \frac{3}{4} \right] , \qquad (23.111)$$

where $E = |\vec{p}| = |\vec{p}\,'|$. From (23.104) and (23.111) we see that the sum $d\sigma_1 + d\sigma_2 + d\sigma_3$ contains no mass singularity, in agreement with our general theorem.

<u>Remarks.</u> If m_e were zero, then it would be physically impossible to have a single electron in motion without its jet of hard photons and e^+e^- pairs. Thus, we must replace both the initial and the final electron states by their appropriate jets.

3. Radiative correction to μ decay

We shall now examine the zero electron mass limit in the radiative correction to μ decay. The muon mass m_μ will be kept fixed in our discussion. The zeroth-order electron spectrum in $\mu^\pm$ decay has been analyzed in Problem 21.1. If one takes into account the first order radiative correction, with the inclusion of infrared photons, then the final e distribution in the rest system of a completely polarized muon is * (for $m_e \to 0$)

$$d^2 N_{e^\mp} \propto x^2 \left\{ 3 - 2x + \frac{\alpha}{2\pi} f(x) \pm \cos\theta \left[1 - 2x + \frac{\alpha}{2\pi} g(x) \right] \right\} dx\, d\cos\theta \tag{23.112}$$

where

$$f(x) = \left[\left(3 + 4 \ln \frac{1-x}{x}\right)(3-2x) + \frac{1-x}{3x^2}(5 + 17x - 34x^2) \right] \ln \frac{m_\mu}{m_e} + O(1),$$

$$g(x) = \left[\left(3 + 4 \ln \frac{1-x}{x}\right)(1-2x) + \frac{1-x}{3x^2}(1 + x + 34x^2) \right] \ln \frac{m_\mu}{m_e} + O(1) \tag{23.113}$$

and, as before,

θ = angle between the electron momentum $\vec{p}$ and the muon spin ,

$$x = \frac{2p}{m_\mu} \quad \text{and} \quad p = |\vec{p}| \ .$$

The $O(1)$ term is finite, but $\ln(m_\mu/m_e) \to \infty$ when $m_e \to 0$. Hence, $d^2 N_e$ exhibits the typical mass-singularity behavior.

---

* T. Kinoshita and A. Sirlin, Phys.Rev. 113, 1652 (1959); S. Berman, Phys.Rev. 112, 267 (1958). Historically, the investigation of mass singularities was stimulated by the Kinoshita–Sirlin paper on μ decay.

Let us consider, say, the μ^- decay

$$\mu^- \to e^-(\vec{p}) + \nu_\mu + \bar{\nu}_e \quad . \tag{23.114}$$

By following the same argument given in the preceding section, we see that in order to remove the mass singularity we need only replace (23.114) by

$$\mu^- \to \text{jet}\,(\vec{p}) + \nu_\mu + \bar{\nu}_e \quad . \tag{23.115}$$

To $O(\alpha)$, it means that we must also consider

$$\mu^- \to e^-(\vec{p} - \vec{k}) + \gamma(\vec{k}) + \nu_\mu + \bar{\nu}_e \tag{23.116}$$

where the final momenta $\vec{k}$ and $\vec{p} - \vec{k}$ satisfy the jet condition (23.110). According to the KLN theorem, the sum of the partial decay rates of (23.114) and (23.116) should be free of any mass singularity in m_e. The result, for a sufficiently small opening angle δ of the jet, is simply to replace in (23.113)

$$\ln \frac{m_\mu}{m_e} \quad \text{by} \quad \ln \frac{m_\mu}{p\delta} \quad , \tag{23.117}$$

which is finite when $m_e \to 0$.

For the same reason, the radiative correction to the total decay rate of the muon must also be free of mass singularity. This is of course the case, as can be seen from (21.7).

23.7 Jets in QCD

By applying the KLN theorem to quantum chromodynamics, we can likewise eliminate the mass singularities in its perturbation series. The procedure is identical to that used in QED. As in the preceding section, we replace any fast-moving quark or gluon in a reaction by a jet of particles. Each jet is again characterized by a small cone in the momentum space; inside the cone there is an ensemble of parallel moving massless particles. Through the use of hadron jets instead of specific hadrons, we can test the predictions of perturbative QCD against experiment. Since the differential cross sections of the jets are free from mass singularities, we can extend the quark-parton model to processes beyond the inclusive reactions discussed in Sections 23.3 and 23.4.

1. 2 - jet cross section

As a concrete example, let us consider the e^+e^- annihilation of hadrons at high energy. To first order in the fine structure constant $\alpha = e^2/4\pi$ but arbitrary order in QCD coupling, this reaction must be via the exchange of γ and Z^o; the final states will be decomposed into different numbers of jets, as shown in Figure 23.4. The simplest is the 2 - jet event:

$$e^+ + e^- \rightarrow \gamma \text{ (or } Z^o) \rightarrow 2 \text{ hadron - jets.} \tag{23.118}$$

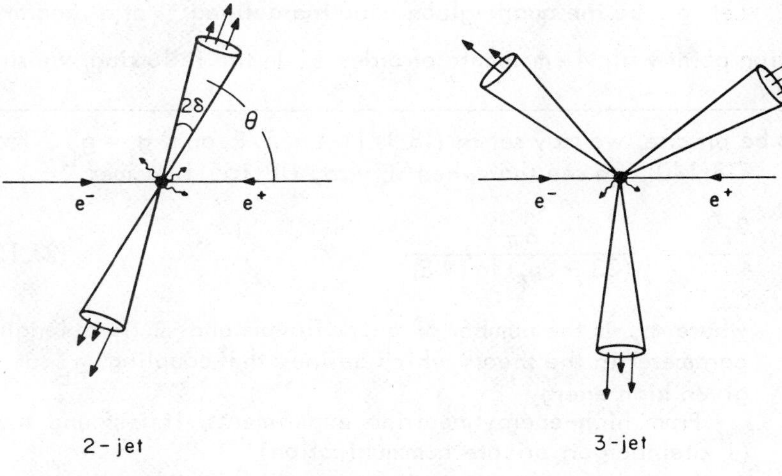

Fig. 23.4. $e^+ + e^- \rightarrow 2$ jets and 3 jets. [Short wavy lines denote soft particles.]

Let us choose the c.m. (center-of-mass) system, and consider the partial cross section

$$d\Omega \frac{d\sigma}{d\Omega} (E, \theta, \omega, \delta) \tag{23.119}$$

in which the total energy of the reaction is $2E$, and the total energy in the 2 jets is more than

$$2E - \omega , \tag{23.120}$$

where ω serves as an infrared cutoff parameter with

$$\epsilon \equiv \frac{\omega}{2E} \ll 1 . \tag{23.121}$$

In the momentum space, each jet consists of particles whose momenta lie within a narrow cone of half-angle

$$\delta \ll 1 ; \tag{23.122}$$

the directions of the two cones are opposite each other with their axis at an angle θ to the e^+e^- beam direction and within a small solid angle $d\Omega$.

Let g_E be the quark-gluon coupling defined * at a renormalization point with 4-momenta of order E. In the following, we shall

---

* To be precise, we may set in (18.141) $L = 1/E$ and $g_L = g_E$. From (18.140), we see that when $E \to \infty$, (18.155) becomes

$$\frac{g_E^2}{4\pi} = \frac{6\pi}{(33 - 2n_f)\ln(\ell E)} \tag{23.123}$$

where n_f is the number of quark flavors and ℓ is a length parameter in the theory which defines the coupling g_E at a given high energy E.

From high-energy neutrino experiments, it is found that (J. Steinberger, private communication)

$$\ell^{-1} = (.20 \pm .05) \text{ GeV} . \tag{23.124}$$

(continued on next page)

neglect the Z^o - exchange amplitude, and consider only the lowest-order diagrams $\alpha = e^2/4\pi$.

To the zeroth order in g_E , reaction (23.118) reduces to

$$e^- + e^+ \rightarrow q + \bar{q} \quad , \tag{23.125}$$

given by diagram a in Figure 23.5; its corresponding differential cross section is, after neglecting the mass of the electron,

$$\left(\frac{d\sigma}{d\Omega}\right)_0 = \frac{\alpha^2}{16E^2}(1+\cos^2\theta)\sum_f 3(e_f/e)^2 \tag{23.126}$$

where e_f is the electric charge of the f - flavored quark and θ is now simply the angle between the e^- - momentum $\vec{\ell}$ and the quark momentum $\vec{p}$ in the c.m. system. [See Problem 6.1 and (17.1) for the derivation.]

To $O(g_E{}^2)$, there is the QCD radiative correction diagram b in Figure 23.5. Its amplitude is divergent on account of the mass singularity. If we keep the quark mass zero, but give the gluon a fictitious mass μ , then the square of the sum of the amplitudes of diagrams a and b gives*

$$\left(\frac{d\sigma}{d\Omega}\right)_{a+b} = \left(\frac{d\sigma}{d\Omega}\right)_0 \left\{1 - \frac{g_E^2}{3\pi^2}\left[2\left(\ln\frac{2E}{\mu}\right)^2 - 3\ln\frac{2E}{\mu} + O(1)\right]\right\} \tag{23.127}$$

where the $O(1)$ term remains finite when $\mu \rightarrow 0$. As expected, in this limit (23.127) diverges.

We may compare (23.123) with our previous value $\alpha_c = .39$ given by (20.79). By setting $n_f = 5$ because of u , d , s , c and b quarks, and $(4\pi)^{-1} g_E{}^2 = .39$, we find the relevant E is $\cong$ 1.6 GeV, which is quite reasonable since the average mass of these five quarks is, according to (20.74) and (20.79), about 1.45 GeV.

* G. Sterman and S. Weinberg, Phys.Rev.Lett. 39, 1436 (1977).

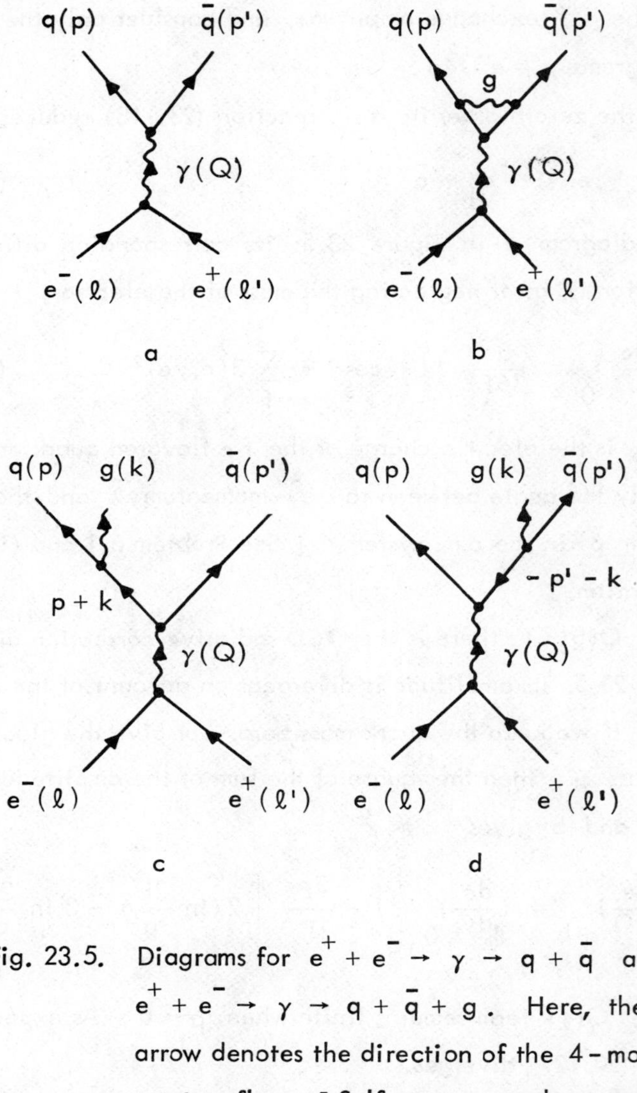

Fig. 23.5. Diagrams for $e^+ + e^- \rightarrow \gamma \rightarrow q + \bar{q}$ and $e^+ + e^- \rightarrow \gamma \rightarrow q + \bar{q} + g$. Here, the arrow denotes the direction of the 4-momentum flow. [Self-energy graphs are omitted.]

On the other hand, to the same order in $g_E{}^2$, the 2-jet process also includes

$$e^-(\ell) + e^+(\ell') \rightarrow q(p) + \bar{q}(p') + g(k) , \qquad (23.128)$$

provided that conditions (23.120)-(23.122) are satisfied. In (23.128), g is the gluon, and ℓ , ℓ', p , p' and k denote the 4-momenta of the particles. In the c.m. system, the components of these momenta are

$$\begin{aligned} \ell &= (\vec{\ell}\,, i\,E) = (\hat{\ell}\,, i)\,E\,, \\ \ell' &= (-\vec{\ell}, i\,E) = (-\hat{\ell}\,, i)\,E\,, \\ p &= (\vec{p}\,, i\,p_0) = (\hat{p}\,, i)\,p_0\,, \\ p' &= (\vec{p}\,', i\,p_0') = (\hat{p}', i)\,p_0'\,, \\ k &= (\vec{k}\,, i\,k_0) = (\hat{k}\,, i)\,k_0 \end{aligned} \tag{23.129}$$

and

$$Q \equiv \ell + \ell' = p + p' + k = (0, i2E)\ .$$

Because of (23.120)-(23.122), in order that the three-particle final state in this reaction can qualify as a 2-jet event, we must have <u>either</u> one of the following two conditions satisfied:

(i) Two of the final momenta are within

$$\text{an angle} \leqslant 2\delta \ll 1\ , \tag{23.130}$$

irrespective of whether they are soft or hard.

(ii) One of the final particles is soft, i.e., whose energy is less than

$$\omega = 2\epsilon E \ll E\ ; \tag{23.131}$$

in addition, the angle between the momenta of the other two particles is

$$\text{between}\ \ \pi - \delta\ \ \text{and}\ \ \pi\ . \tag{23.132}$$

Let $(d\sigma/d\Omega)_i$ and $(d\sigma/d\Omega)_{ii}$ denote the respective differential cross sections of (23.128) in these two regions, (i) and (ii). The 2-jet differential cross section, to $O(g_E{}^2)$, is given by the sum

$$\left(\frac{d\sigma}{d\Omega}\right)_{2\text{-jet}} \equiv \left(\frac{d\sigma}{d\Omega}\right)_{a+b} + \left(\frac{d\sigma}{d\Omega}\right)_{i} + \left(\frac{d\sigma}{d\Omega}\right)_{ii}\ , \tag{23.133}$$

which, according to the KLN theorem, should remain finite when the fictitious mass $\mu = 0$. As we shall show in the next section (derived on page 764), the result is*

$$\left(\frac{d\sigma}{d\Omega}\right)_{2\text{-jet}} = \left(\frac{d\sigma}{d\Omega}\right)_0 \left\{1 - \frac{g_E^2}{3\pi^2}\left[(3 + 4\ln 2\epsilon)\ln\delta + O(1)\right]\right\}, \tag{23.134}$$

and is indeed free from mass singularity. In the above expression, the $O(1)$ term remains finite, $\rightarrow (2\pi^2 - 15)/6$, when the physical parameters δ and $\epsilon \rightarrow 0$.

Because in (23.134) the factor inside the curly bracket is independent of θ, the angular distribution of the 2-jet is given simply by the lowest-order perturbation formula (23.126). Experimentally, there is a good verification ** of this

$$1 + \cos^2\theta$$

distribution, giving support to the general notion of the quark-parton model and QCD.

2. <u>3-jet events</u>

In a similar way, we see that to derive the distribution of 3-jet events we need examine only its lowest-order perturbation formula. In (23.128) any final configuration of $q\bar{q}g$ that lies outside the regions (i) and (ii) stated on the preceding page is considered a 3-jet event. Let us now calculate its differential cross section. In the amplitude sum of diagrams c and d in Figure 23.5, the matrix

---

* G. Sterman and S. Weinberg, loc. cit. Cf. also P. M. Stevenson, Phys. Lett. 78B, 451 (1978).

** See page 38 of the Proceedings of the IX International Symposium on Lepton and Photon Interactions at High Energies, Fermilab, 1979, ed. T. B. W. Kirk and H. D. I. Abarbanel.

element of the lepton current operator is

$$< 0 \mid j_\mu \mid e^+ e^- > \;=\; -ie\, v^\dagger_{\vec{\ell}',s'}\, \gamma_4\, \gamma_\mu\, u_{\vec{\ell},s} \quad , \tag{23.135}$$

and that of the hadron current is, on account of the rules given on pages 530-31,

$$< q\bar{q}g \mid J_\mu \mid 0 > \;=\; \tfrac{1}{2} e_f\, g_E$$
$$\cdot\, u^\dagger_{\vec{p},s}\, \gamma_4 \left[\gamma_\nu V^\ell_\nu \lambda^\ell \frac{1}{\not{p}+\not{k}} \gamma_\mu - \gamma_\mu \frac{1}{\not{p}'+\not{k}} \gamma_\nu V^\ell_\nu \lambda^\ell \right] v_{\vec{p}',s'} \tag{23.136}$$

where, as before, $e > 0$ is the electric charge, e_f is the charge of the f-flavored quark, g_E is the QCD coupling at energy E, V^ℓ_μ is the amplitude of the gluon, λ^ℓ and γ_μ are given by (12.22) and (3.12), and $u_{\vec{k},s}$ and $v_{\vec{k}',s'}$ (with $\vec{k} = \vec{p}$ or $\vec{\ell}$) are the spinor solutions of (3.26)-(3.27) when mass = 0. In the c.m. system, the e^+-momentum $\vec{\ell}'$ is related to that of e^- by

$$\vec{\ell}' \;=\; -\vec{\ell} \quad .$$

As in (5.124), we introduce $L_{\mu\nu}$, obtained by squaring the lepton matrix element and averaging over the initial spins:

$$L_{\mu\nu} \;\equiv\; \pm \tfrac{1}{4} \sum_{s,s'} < 0 \mid j_\mu \mid e^+e^- > < 0 \mid j_\nu \mid e^+e^- >^* \tag{23.137}$$

where * denotes the complex conjugation, + is for $\nu \neq 4$ and - for $\nu = 4$. From (5.125) we see that

$$L_{\mu\nu} \;=\; \frac{e^2}{4E^2} \left(\ell_\mu \ell'_\nu + \ell'_\mu \ell_\nu - \delta_{\mu\nu}\, \ell \cdot \ell' \right) \tag{23.138}$$

where, as before, 2E is the total c.m. energy. Likewise, we define $H_{\mu\nu}$ by similarly squaring the hadron matrix element (with the same sign convention) and summing over all final polarizations, colors and flavors:

$$H_{\mu\nu} \equiv \pm \sum_{\substack{s,s',t\\ c,f}} <q\bar{q}g|J_\mu|0><q\bar{q}g|J_\nu|0>^* \qquad (23.139)$$

where c and f are the color and flavor indices, and t the polarization vector of the gluon. For a gluon of momentum $\vec{k}$,

$$t = \text{unit vectors } \hat{e}_1 \text{ and } \hat{e}_2 \qquad (23.140)$$

given by Figure 6.1.

To sum over c, we note that, * on account of (6.32) - (6.34) and (12.23),

$$\begin{aligned} \text{trace}\,(V_\mu^\ell \lambda^\ell V_\nu^m \lambda^m) &= \frac{1}{2k_0} t_\mu t_\nu \delta^{\ell m}\ \text{trace}\,(\lambda^\ell \lambda^m) \\ &= \frac{8}{k_0} t_\mu t_\nu \;. \qquad (23.141) \end{aligned}$$

Thus, by substituting (23.136) into (23.139) and using (23.141)

$$\gamma_\nu t_\nu = i\not{t} \;,$$

$$(\not{p} + \not{k})^2 = -(p+k)^2 = -2p\cdot k \;,$$

$$\frac{1}{\not{p}+\not{k}} = -\frac{1}{2p\cdot k}(\not{p}+\not{k})$$

and the exercise on page 36, we find

$$\begin{aligned} H_{\mu\nu} = -\frac{1}{8p_0p_0'k_0} \sum_{f,t} (e_f g_E)^2 \\ \cdot\ \text{trace}\Big\{ \not{p}\Big[\not{t}\frac{1}{p\cdot k}(\not{p}+\not{k})\gamma_\mu - \gamma_\mu \frac{1}{p'\cdot k}(\not{p}'+\not{k})\not{t}\Big] \\ \cdot\ \not{p}'\Big[\gamma_\nu \frac{1}{p\cdot k}(\not{p}+\not{k})\not{t} - \not{t}\frac{1}{p'\cdot k}(\not{p}'+\not{k})\gamma_\nu\Big]\Big\}. \end{aligned} \qquad (23.142)$$

---

* In (23.141), the trace refers only to the color-degrees of freedom. Also, as in (5.125), we set the volume of the system = 1 for convenience.

Because of (5.111), the differential cross section of (23.128) is

$$d\sigma = \pi \left(\frac{1}{8\pi^3}\right)^2 \int |\mathcal{M}(p, p', k)|^2 \, d^3p \, d^3p' \, \delta(p_0 + p_0' + k_0 - 2E) \tag{23.143}$$

where, on account of

$$\frac{1}{Q^2} = -\frac{1}{4E^2}$$

in the photon propagator,

$$|\mathcal{M}(p, p', k)|^2 = \left(\frac{1}{4E^2}\right)^2 L_{\mu\nu} H_{\mu\nu} \quad . \tag{23.144}$$

The evaluation of the trace in (23.142) is straightforward, and it is worked out in Problems 23.1 and 23.2. By using (23.177), derived in these problems, we obtain

$$\begin{aligned} |\mathcal{M}(p, p', k)|^2 = & - \frac{1}{16E^6 k_0 p_0 p_0'} \sum_f (e\, e_f\, g_E)^2 \\ & \cdot \left\{ \frac{p \cdot p'}{(p \cdot k)(p' \cdot k)} \left[2(p \cdot \ell)(p' \cdot \ell') + 2(p \cdot \ell')(p' \cdot \ell) \right. \right. \\ & \qquad \left. - 2(k \cdot \ell)(k \cdot \ell') + (Q \cdot \ell)(Q \cdot k) \right] \\ & + \frac{1}{p \cdot k} \left[(Q \cdot \ell)(Q \cdot p') - 2(p \cdot \ell)(p \cdot \ell') - 2(p' \cdot \ell)(p' \cdot \ell') \right] \\ & \left. + \frac{1}{p' \cdot k} \left[(Q \cdot \ell)(Q \cdot p) - 2(p \cdot \ell)(p \cdot \ell') - 2(p' \cdot \ell)(p' \cdot \ell') \right] \right\} \end{aligned} \tag{23.145}$$

where ℓ, ℓ', p, p', k and Q are given by (23.129).

In order to convert the above $d\sigma$ into that of the 3-jet events, we must exclude the regions (i) and (ii), the details of which we shall now examine.

3. Dalitz plot

In the c.m. system, let us denote angles

$$\theta = \angle(\vec{p}, \vec{\ell}) \, , \quad \theta' = \angle(\vec{p}', \vec{\ell})$$

and

$$\alpha = \angle(\vec{p}, \vec{p}') \quad , \tag{23.146}$$

and introduce the dimensionless parameters

$$x \equiv \frac{p_0}{E} \,, \quad y \equiv \frac{p_0'}{E}$$

and

$$z \equiv \frac{k_0}{E} \quad . \tag{23.147}$$

The conservation of energy gives

$$x + y + z = 2 \quad . \tag{23.148}$$

By following the discussions given in Section 15.1, we see that each final configuration of $q\bar{q}g$ defines a point in the Dalitz plot, which we may take to be the shaded triangle in graph (b) of Figure 15.3 with the height of the larger triangle $= 2$. [See also Figure 23.7.] Because

$$\vec{k} + \vec{p} + \vec{p}' = 0 \quad ,$$

the angle α is related to x, y and z by

$$z^2 = x^2 + y^2 + 2xy \cos\alpha \quad . \tag{23.149}$$

It is convenient to consider a unit sphere; from its center o, we draw the unit vectors $\hat{\ell}$, $\hat{p}$ and $\hat{p}'$, which defines the spherical triangle $\ell p p'$, illustrated in Figure 23.6. If we regard the great circle that passes through p and p' as the equator, then the north pole n defines a unit vector

$$\hat{n} \equiv \overrightarrow{on} \,/\!/\, \hat{p} \times \hat{p}' \quad . \tag{23.150}$$

Let P be an arbitrarily chosen point on the equator with

$$\phi = \widehat{Pp} \quad .$$

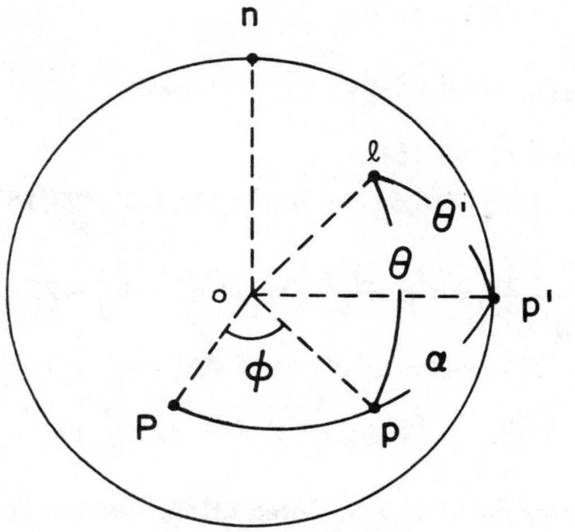

Fig. 23.6. The radial vectors $\overrightarrow{o\ell}$, $\overrightarrow{op}$ and $\overrightarrow{op}'$ equal to $\hat{\ell}$, $\hat{p}$ and $\hat{p}'$.

We shall now consider the coordinate <u>transformation</u>

from $\hat{p}$, $\hat{p}'$ to $\hat{n}$, α and ϕ.

As we shall show,

$$d^2\hat{p}\, d^2\hat{p}' = \sin\alpha\, d\alpha\, d\phi\, d^2\hat{n} \quad . \tag{23.151}$$

This can be derived by considering in Figure 23.6 four infinitesimal rotations: we rotate along $\overrightarrow{on}$ first an angle $d\alpha$ (keeping P and p fixed), then an angle $d\phi$ (keeping P and p' fixed). Next, we keep the arc lengths of the sides of the spherical triangle npp' fixed, and rotate an angle $d\xi$ along $\overrightarrow{op}$, then an angle $d\eta$ along $\overrightarrow{op}'$. The points n, p and p' each move two displacements, which form the sides of a parallelogram; their areas are

$$d^2\hat{n} = \sin\alpha\, d\xi\, d\eta \quad ,$$

$$d^2\hat{p} = \sin\alpha \, d\eta \, d\phi$$

and

$$d^2\hat{p}' = \sin\alpha \, d\xi \, d\alpha$$

and that leads to (23.151).

As in (15.5)-(15.6), by using (23.147)-(23.151) we see that

$$\int \frac{|\mathcal{M}|^2}{k_0 p_0 p_0'} d^3p \, d^3p' \, \delta(p_0 + p_0' + k_0 - 2E)$$
$$= \int d\phi \, d^2\hat{n} \, dp_0 \, dp_0' \, |\mathcal{M}|^2$$
$$= 2\pi E^2 \int d^2\hat{n} \, dx \, dy \, |\mathcal{M}|^2 \quad . \tag{23.152}$$

To carry out the $d^2\hat{n}$ integration, we can consider the spherical triangle npp' fixed, but average $|\mathcal{M}|^2$ over all possible directions of $\hat{\ell}$. By using the averages

$$(\hat{p}\cdot\hat{\ell})^2_{av} = (\hat{p}'\cdot\hat{\ell})^2_{av} = (\hat{k}\cdot\hat{\ell})^2_{av} = \frac{1}{3} \quad ,$$

$$[(\hat{p}\cdot\hat{\ell})(\hat{p}'\cdot\hat{\ell})]_{av} = \frac{1}{3}\hat{p}\cdot\hat{p}' = \frac{1}{3}\cos\alpha \quad ,$$

we find * that after the angular integration the differential cross section for

$$e^+ + e^- \to q + \bar{q} + g$$

is

$$d\sigma = \frac{2}{3E^2}\alpha^2 \frac{g_E^2}{4\pi} dx \, dy \, \frac{x^2+y^2}{(1-x)(1-y)} \sum_f (e_f/e)^2 \tag{23.153}$$

where x and y are defined by (23.147).

4. 3-jet cross section

To the lowest order in g_E, reaction

$$e^+ + e^- \to \text{jet}\,(p_1) + \text{jet}\,(p_2) + \text{jet}\,(p_3) \tag{23.154}$$

---

* J. Ellis, M. K. Gaillard and G. Ross, Nucl. Phys. B111, 253 (1976); T. A. De Grand, Y. J. Ng and S.-H. H. Tye, Phys. Rev. D16, 3251 (1977).

is related to

$$e^+ + e^- \rightarrow q(p) + \bar{q}(p') + g(k)$$

by identifying each of the particles q, $\bar{q}$ and g as a jet. Thus the 4-momentum p_i of the ith jet can be either p, or p', or k, and the 3-jet differential cross section can be derived from (23.143) by symmetrization:

$$d\sigma_{3\text{-jet}} = \frac{1}{3}\pi\left(\frac{1}{8\pi^3}\right)^2 \int \Big\{ |\mathcal{M}(p_1, p_2, p_3)|^2 + |\mathcal{M}(p_2, p_3, p_1)|^2 + |\mathcal{M}(p_3, p_1, p_2)|^2 \Big\} \cdot d^3p_1\, d^3p_2\, d^3p_3\, \delta^4(p_1 + p_2 + p_3 - Q) \qquad (23.155)$$

where Q is defined in (23.129), $|\mathcal{M}(p_1, p_2, p_3)|^2$ is given by (23.144) and is already symmetric in the first two of its variables; i.e.

$$|\mathcal{M}(p_1, p_2, p_3)|^2 = |\mathcal{M}(p_2, p_1, p_3)|^2 .$$

Let us define

$$x_1 \equiv \frac{|\vec{p}_1|}{E}, \qquad x_2 = \frac{|\vec{p}_2|}{E}$$

and

$$x_3 = \frac{|\vec{p}_3|}{E}, \qquad (23.156)$$

then, as in (23.148), we have

$$x_1 + x_2 + x_3 = 2 . \qquad (23.157)$$

Since the 3-jet distribution is, by definition, symmetric with respect to p_1, p_2 and p_3, we need only consider in the Dalitz plot the domain, say,

$$x_1 \leqslant x_2 \leqslant x_3 . \qquad (23.158)$$

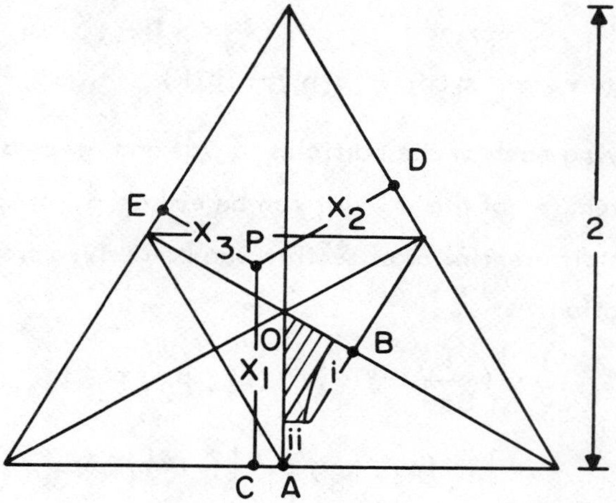

Fig. 23.7. Dalitz plot. The shaded region is for 3-jet events; i and ii are for 2-jet events. [See (23.167) and (23.172) for their boundaries.]

In addition, they must be outside the regions (i) and (ii) discussed on page 751. This allowed region, called $R_{3\text{-jet}}$, is illustrated by the shaded domain in Figure 23.7. Experimentally, every 3-jet event gives a point in $R_{3\text{-jet}}$. The distribution in the Dalitz plot can be obtained by carrying out the angular integrations, as in the previous section. By symmetrizing (23.153) with respect to the three jets, and multiplying by 6 to account for the restriction (23.158), we find the differential and total cross sections for (23.154) to be

$$d\sigma_{3\text{-jet}} = \frac{4}{3E^2}\,\alpha^2\,\frac{g_E^{\,2}}{4\pi}\;P(x_1, x_2, x_3)\,dx_1\,dx_2 \sum_f (e_f/e)^2$$

and

$$\sigma_{3\text{-jet}} = \int_{R_{3\text{-jet}}} d\sigma_{3\text{-jet}}\,, \tag{23.159}$$

where

$$P(x_1, x_2, x_3) = \frac{x_1^{\,2} + x_2^{\,2}}{(1-x_1)(1-x_2)} + \frac{x_2^{\,2} + x_3^{\,2}}{(1-x_2)(1-x_3)} + \frac{x_3^{\,2} + x_1^{\,2}}{(1-x_3)(1-x_1)}\,. \tag{23.160}$$

Along the line AB in Figure 23.7, we have $x_3 = 1$ and, therefore, the second and third terms on the righthand side in (23.160) are singular. Because of the excluded regions (i) and (ii), the 3-jet cross section is, of course, finite in $R_{3\text{-jet}}$. To examine the boundary of (i) and (ii), it is only necessary to consider the neighborhood of AB.

We recall that in the c.m. system, the sum of the three 3-momentum vectors $\vec{p}_1$, $\vec{p}_2$ and $\vec{p}_3$ is zero, and because of the choice (23.158), their relative orientations are of the general form illustrated in Figure 23.8 with their angles

$$\theta_{ij} \equiv \angle\,(\vec{p}_i\,,\ \vec{p}_j)$$

satisfying

$$\theta_{12} + \theta_{23} + \theta_{31} = 2\pi$$

and

$$\sin\tfrac{1}{2}\,\theta_{ij} = \left[\frac{1 - x_k}{x_i\, x_j}\right]^{\frac{1}{2}}\ , \qquad (23.161)$$

where (i, j, k) is a permutation of $(1, 2, 3)$.

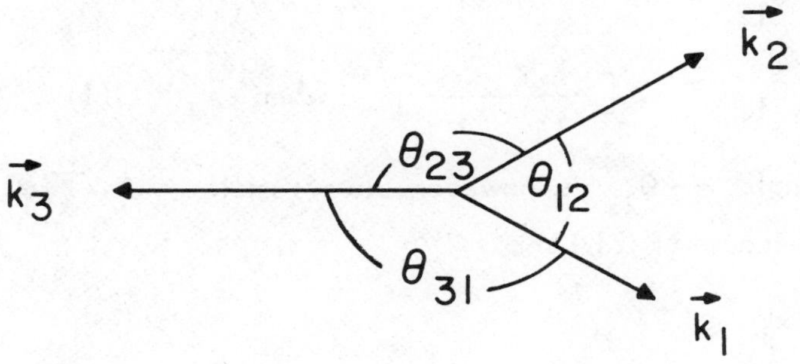

Fig. 23.8. The angle between $\vec{k}_i$ and $\vec{k}_j$ is θ_{ij}.

In Figure 23.7, near AB we have

$$\zeta \equiv 1 - x_3 \ll 1 \ ; \tag{23.162}$$

therefore,

$$x_2 = 2 - x_1 - x_3 = 1 - x_1 + \zeta \cong 1 - x_1$$

and (23.163)

$$\sin \tfrac{1}{2} \theta_{12} \cong \left[\frac{\zeta}{x_1(1-x_1)} \right]^{\frac{1}{2}} .$$

Likewise, by neglecting higher powers in ζ, we find

$$\pi - \theta_{23} \cong 2 \left[\frac{\zeta(x_1 - \zeta)}{1 - x_1} \right]^{\frac{1}{2}}$$

$$\cong \begin{cases} 2 \left[\dfrac{\zeta x_1}{1 - x_1} \right]^{\frac{1}{2}} & \text{if } x_1 = O(1) \\ 2 \left[\zeta(x_1 - \zeta) \right]^{\frac{1}{2}} & \text{if } x_1 = O(\zeta) . \end{cases} \tag{23.164}$$

Along ζ = constant $\ll 1$ and within the triangle OAB, x_1 ranges from

$$2\zeta \text{ to } \tfrac{1}{2}(1 + \zeta) \cong \tfrac{1}{2} \ ;$$

hence, θ_{12} decreases from

$$\theta_{12} \cong \frac{\pi}{2} \quad \text{when} \quad x_1 = 2\zeta \tag{23.165}$$

to

$$\theta_{12} \cong 2 \left[\frac{\zeta}{x_1(1-x_1)} \right]^{\frac{1}{2}} \quad \text{when} \quad x_1 = O(1) . \tag{23.166}$$

The angle $\pi - \theta_{23}$ is, however, always small, $\leqslant 2\sqrt{\zeta}$. In region (ii), because of (23.131),

$$x_1 \leqslant 2\epsilon \ . \tag{23.167}$$

The condition (23.132) becomes

$$0 \leqslant \pi - \theta_{23} \leqslant \delta \tag{23.168}$$

which, on account of the last expression in (23.164), is

$$\zeta(x_1 - \zeta) \leqslant \frac{\delta^2}{4} \quad ,$$

which is identical to

$$\left(\frac{x_1}{2} - \zeta\right)^2 \geqslant \frac{1}{4}(x_1^2 - \delta^2) \quad . \tag{23.169}$$

If we regard ϵ and δ both as infinitesimals and choose

$$2\epsilon < \delta \quad , \tag{23.170}$$

then in the region $x_1 \leqslant 2\epsilon$, (23.169) is automatically satisfied; therefore, so is (23.168). In region (i), in view of (23.130), we require

$$\theta_{12} \leqslant 2\delta \tag{23.171}$$

which gives for its curved boundary

$$\zeta = x_1(1 - x_1)\,\delta^2 \quad . \tag{23.172}$$

By using (23.159)-(23.160) and integrating $d\sigma_{3\text{-jet}}$ over the entire region $R_{3\text{-jet}}$ we find, to $O(g_E^2)$,

$$\sigma_{3\text{-jet}} = \frac{1}{E^2}\,\alpha^2\left(\frac{g_E^2}{3\pi}\right)[\,(3+4\ln 2\epsilon)\ln\delta + O(1)]\sum_f (e_f/e)^2 \tag{23.173}$$

where the $O(1)$ term is finite when $\epsilon \to 0$ and $\delta \to 0$. [See Problem 23.3.] To the same order in g_E^2,

$$\sigma_{total} \equiv \sigma_{2\text{-jet}} + \sigma_{3\text{-jet}}$$

is the <u>total</u> cross section of

$$e^+ + e^- \rightarrow \text{hadrons}, \tag{23.174}$$

which is free from mass singularity in accordance with the KLN theorem; σ_{total} is also independent of δ and ϵ since in (23.174) we have integrated over all angular and energy ranges. Thus, the above derivation of (23.173) also provides a verification of (23.134), given earlier for $d\sigma_{2\text{-jet}}$.

5. <u>Correlations between jets</u>

In passing from the differential cross section (23.153) for $q\bar{q}\,g$ to that for 3-jet, (23.159), information about the difference between quarks and gluons is lost. This is because in the actual experimental observation, each jet is a cone of multiparticle flux. Due to the soft particles which may lie outside the jets, the quantum numbers of each jet have large fluctuations. Therefore, by examining a single jet we will not be able to tell whether it is a quark-jet or a gluon-jet. As we shall see, through correlation measurement, in principle it should be possible to retrieve this missing information.

Consider, for example, the charge $Q(i)$ in units of e of the ith jet. Due to symmetry under particle - antiparticle conjugation, the average of $Q(i)$ over a large ensemble of jet - events is clearly zero; i.e.,

$$< Q(i) > \;=\; 0 \quad .$$

Next, let us examine the two-body charge correlation

$$< Q(1)\, Q(2) > \quad . \tag{23.175}$$

Its value averaged over an ensemble of 2-jet events

$$e^+ + e^- \rightarrow \text{jet (1)} + \text{jet (2)}$$

is given by

$$< Q(1)\, Q(2) >_{2\text{-jet}} \;=\; -\,\frac{\sum\limits_f (e_f/e)^4}{\sum\limits_f (e_f/e)^2} \tag{23.176}$$

which, to the lowest order in g_E^2 , has the value

$$-\frac{1}{9}\ \frac{16+1+1}{4+1+1} = -\frac{1}{3}$$

for $m_s < E < m_c$, and

$$-\frac{1}{9}\ \frac{16+1+1+16}{4+1+1+4} = -\frac{17}{45}$$

for $m_c < E < m_b$.

A more interesting problem is to study the same two-body correlation function, but averaged over an ensemble of 3-jet events

$$e^+ + e^- \rightarrow \text{jet }(1) + \text{jet }(2) + \text{jet }(3) \quad .$$

As in (23.158), we differentiate these three jets by arranging their energies in the order

$$x_3 \geqslant x_2 \geqslant x_1 \quad .$$

Consequently, there are three different two-body correlation functions $< Q(1)\,Q(2) >$, $< Q(2)\,Q(3) >$ and $< Q(3)\,Q(1) >$ that can be measured. Let us consider, e.g., $< Q(1)\,Q(2) >$ at a fixed point in the shaded region $R_{3\text{-jet}}$ of the Dalitz plot in Figure 23.7. When one of the jets, 1 or 2 , is due to the gluon, then since the gluon has zero charge, $Q(1)\,Q(2)$ is zero. Hence, to the lowest order in g_E^2 we have for the ensemble average

$$< Q(1)\,Q(2) >_{3\text{-jet}}$$

$$= \; < Q(1)\,Q(2) >_{2\text{-jet}} \frac{x_1^{\,2} + x_2^{\,2}}{(1-x_1)(1-x_2)\,P(x_1\,,\,x_2\,,\,x_3)} \tag{23.177}$$

where $P(x_1\,,\,x_2\,,\,x_3)$ is given by (23.160). Through permutations of (1, 2, 3), the averages of $Q(2)\,Q(3)$ and $Q(3)\,Q(1)$ can be similarly derived. Together, they satisfy

$$< Q(1)\,Q(2) + Q(2)\,Q(3) + Q(3)\,Q(1) >_{3\text{-jet}}$$
$$= < Q(1)\,Q(2) >_{2\text{-jet}} \, . \qquad (23.178)$$

This in turn insures that, to the next order in $g_E{}^2$, the expression for $< Q(1)\,Q(2) >_{2\text{-jet}}$, (23.176), has no mass singularity.

However, to $O(g_E{}^4)$, there is a mass-singular term due to the virtual process

$$e^+ e^- \rightarrow q(p) + \bar{q}(p') + g \text{ (off-shell but soft)} \qquad (23.179)$$

followed by

$$g \rightarrow q(k) + \bar{q}(k') \qquad (23.180)$$

with $\vec{k}$ lying inside one of the 2-jet cones (defined by $q(p)$ and $q(p')$ in (23.179)) and $\vec{k}'$ inside the other. The combined process, (23.179)-(23.180), is therefore a 2-jet event. The net result is that (23.176) should be multiplied by a mass-singular correction factor of the form

$$1 + \delta^4 \; O\left(\left(\frac{g_E{}^2}{2\pi}\right)^2 \; \ln \frac{E\delta^2}{\mu} \right) \qquad (23.181)$$

where μ should be 0 in the perturbative QCD series expansion. The δ^4 - factor is due purely to the probability of having the two momenta $\vec{k}$ and $\vec{k}'$ in (23.180) falling within the two opposite jet-cones. For practical applications we expect μ to be of the order of a typical meson mass, ~ a few hundred MeV. Hence, for sufficiently small δ, the $g_E{}^4$ term in (23.181) is not expected to play any important role. Furthermore, by varying δ it should be possible to extract a δ^4 - dependent term directly from the experimental data, and thereby one may gain some insight into the nonperturbative aspect of QCD.

Likewise, there is also a similar $g_E^4 \; \delta^4$ - correction to the correlation formula (23.177) for 3-jet. Similar considerations can, of course, be extended to other additive quantum numbers. The determinations of the coefficients of such mass-singular δ^4 - dependent terms are quite interesting from the theoretical point of view, because it gives us a new class of physical parameters that can be measured experimentally, but is divergent in a perturbation series expansion.

Remarks

At present, neither the spin nor any other quantum number of the gluon has been determined experimentally. In order to verify its spin-1 character, we must establish the distribution (23.160) for the 3-jet events. In order to determine its other quantum numbers, it is necessary to measure correlations between jets, such as that given by (23.177).

Problem 23.1. The $H_{\mu\nu}$, defined by (23.142), is a quadratic function of the polarization t_μ, given by (23.140). We may write

$$H_{\mu\nu} = T_{\mu\nu,\,\alpha\beta} \, t_\alpha t_\beta$$

where $T_{\mu\nu,\,\alpha\beta}$ is symmetric with respect to α and β. Show that, because of (6.51),

$$k_\alpha T_{\mu\nu,\,\alpha\beta} = 0$$

and therefore the polarization sum can be converted into

$$\sum_t H_{\mu\nu} = T_{\mu\nu,\,\alpha\alpha} \tag{23.182}$$

in which the repeated index α is summed over from 1 to 4 .

Problem 23.2. In (23.138), we may write

$$L_{\mu\nu} = \frac{e^2}{4E^2} (\ell , \ell')_{\mu\nu}$$

where

$$(\ell , \ell')_{\mu\nu} \equiv \ell_\mu \ell'_\nu + \ell'_\mu \ell_\nu - \delta_{\mu\nu} \ell \cdot \ell' . \qquad (23.183)$$

Prove that by carrying out the trace and the gluon polarization sum, (23.142) becomes

$$H_{\mu\nu} = - \frac{2}{p_0 p'_0 k_0} \sum_f (e_f g_E)^2$$

$$\cdot \left\{ \frac{p \cdot p'}{(p \cdot k)(p' \cdot k)} [2(p , p')_{\mu\nu} + (k , p + p')_{\mu\nu}] \right.$$

$$+ \frac{1}{p \cdot k} [(p' , p + k)_{\mu\nu} - (p , p)_{\mu\nu}]$$

$$\left. + \frac{1}{p' \cdot k} [(p , p' + k)_{\mu\nu} - (p' , p')_{\mu\nu}] \right\} . \qquad (23.184)$$

Problem 23.3. Integrate $d\sigma_{3\text{-jet}}$, (23.159), over the shaded region $R_{3\text{-jet}}$ in Figure 23.7 and show that $\sigma_{3\text{-jet}}$ is given by (23.173).

References

C. H. Llewellyn Smith, "Neutrino Reactions at Accelerator Energies," Physics Reports 3C, 261 (1972).

J. Kogut and L. Susskind, "The Parton Picture of Elementary Particles," Physics Reports 8C, 75 (1973).

The Mark J Collaboration, "Physics with High Energy Electron-positron Colliding Beams with the Mark J Detector," Physics Reports 63, 337 (1980).

Chapter 24

CHIRAL SYMMETRY

The quark model provides us a simple picture of the internal structure of hadrons. It also gives us an effective way to describe their dynamics at high energy. Corrections due to QCD are in most cases relatively small; these are often computable and can serve as tests of the general correctness of our theoretical basis. Much of the success of the model lies in the circumstance that to a reasonably good approximation we can regard quarks as free or weakly interacting particles (except for the confining mechanism). However, this does not hold for hadron reactions at low energy. In terms of quarks, these would involve the dynamics of many bodies, and that presents problems which may appear difficult.

Fortunately, we are helped by the fact that pions have a much lighter mass than any other hadrons. Most of the low-energy hadronic reactions are therefore dominated by pion exchanges. The underlying reason is the small masses of u and d quarks inside hadrons, which makes it possible to derive the approximate chiral symmetry and PCAC. Although historically these subjects were developed relatively independently of the quark model, as we shall see they fit harmoniously with our overall picture and form one of the cornerstones of our understanding of the strong interaction.

24.1 Current Algebra *

In this chapter, we restrict ourselves to the discussion of low-energy theorems concerning hadrons made of u, d and s quarks. Let $\psi_u^c(x)$, $\psi_d^c(x)$ and $\psi_s^c(x)$ be their respective field operators and the superscript $c = 1, 2$ and 3 be the color index. We define

$$q(x) = \begin{pmatrix} \psi_u^c(x) \\ \psi_d^c(x) \\ \psi_s^c(x) \end{pmatrix} , \tag{24.1}$$

and for any color-independent matrix Γ

$$q^\dagger \Gamma q \equiv \sum_c q^\dagger \Gamma q \tag{24.2}$$

with the color index c summed over. From Chapters 21 and 22, we see that the weak and electromagnetic interactions of these hadrons are closely connected with the eight vector current operators

$$v_\mu^\ell(x) \equiv i q^\dagger(x) \gamma_4 \gamma_\mu \tfrac{1}{2} \lambda^\ell q(x) \tag{24.3}$$

and the eight axial-vector current operators

$$a_\mu^\ell(x) \equiv i q^\dagger(x) \gamma_4 \gamma_\mu \gamma_5 \tfrac{1}{2} \lambda^\ell q(x) \tag{24.4}$$

where, as usual, γ_μ are the Dirac matrices and λ^ℓ the Gell-Mann matrices, which satisfy (12.22).

1. Naive commutators

Consider two space-time points

$$x_\mu = (\vec{r}, it) \quad \text{and} \quad x_\mu' = (\vec{r}', it') . \tag{24.5}$$

---

* M. Gell-Mann, Phys. Rev. 125, 1067 (1962); Physics 1, 63 (1964).

At equal time

$$t = t' ,$$

we may apply (3.24a) and obtain

$$\left[v_4^{\ell}(x),\ v_\mu^{m}(x')\right] = \left[a_4^{\ell}(x),\ a_\mu^{m}(x')\right] = -f^{\ell mn}\, v_\mu^{n}(x)\, \delta^3(\vec{r}-\vec{r}\,')$$

and (24.6)

$$\left[v_4^{\ell}(x),\ a_\mu^{m}(x')\right] = \left[a_4^{\ell}(x),\ v_\mu^{m}(x')\right] = -f^{\ell mn}\, a_\mu^{n}(x)\, \delta^3(\vec{r}-\vec{r}\,') ,$$

where $f^{\ell mn}$ is given* by (12.24). These expressions will be referred to as naive commutators because, as we shall see in the next section, except for $\mu = 4$, modifications must be introduced. However, it will turn out that for practically all applications these modifications can simply be ignored. Accepting this for the moment, we may integrate (24.6) and derive the equal - time commutation relations between these local currents and their charges,

$$Q_v^{\ell}(t) \equiv -i \int v_4^{\ell}(x)\, d^3r$$

and (24.7)

$$Q_a^{\ell}(t) \equiv -i \int a_4^{\ell}(x)\, d^3r .$$

We find

$$\left[Q_v^{\ell}(t),\ v_\mu^{m}(x)\right] = \left[Q_a^{\ell}(t),\ a_\mu^{m}(x)\right] = i\, f^{\ell mn}\, v_\mu^{n}(x)$$

---

* The λ^{ℓ} and $f^{\ell mn}$ here are identical to the λ_{ℓ} and $f_{\ell mn}$ used in Chapter 12.

and (24.8)

$$\left[Q_V^{\ell}(t)\,,\; a_\mu^{m}(x)\right] = \left[Q_a^{\ell}(t)\,,\; v_\mu^{m}(x)\right] = i f^{\ell mn} a_\mu^{n}(x) \;.$$

By setting $\mu = 4$ in (24.8) and repeating the integration, we have, at any given time t,

$$\left[Q_V^{\ell}(t)\,,\; Q_V^{\ell}(t)\right] = \left[Q_a^{\ell}(t)\,,\; Q_a^{\ell}(t)\right] = i f^{\ell mn}\, Q_V^{n}(t)$$

and (24.9)

$$\left[Q_V^{\ell}(t)\,,\; Q_a^{m}(t)\right] = \left[Q_a^{\ell}(t)\,,\; Q_V^{m}(t)\right] = i f^{\ell mn}\, Q_a^{n}(t) \;.$$

The group structure can be most easily exhibited by forming the combination

$$Q_\pm^{\ell} \equiv \tfrac{1}{2}(Q_V^{\ell} \pm Q_a^{\ell}) \tag{24.10}$$

which then satisfies,

$$\left[Q_+^{\ell}(t)\,,\; Q_+^{m}(t)\right] = i f^{\ell mn}\, Q_+^{n}(t) \;,$$

$$\left[Q_-^{\ell}(t)\,,\; Q_-^{m}(t)\right] = i f^{\ell mn}\, Q_-^{n}(t) \tag{24.11}$$

and

$$\left[Q_+^{\ell}(t)\,,\; Q_-^{m}(t)\right] = 0 \;.$$

Thus, Q_+^{ℓ} and Q_-^{ℓ} ($\ell = 1, 2, \cdots, 8$) separately are the generators of an SU_3 group; since they commute with each other, the group $\{g\}$ generated by these sixteen operators is

$$\{g\} = SU_3 \times SU_3 \;. \tag{24.12}$$

The currents v_μ^ℓ and a_μ^ℓ form an $(8, 8)$ representation of the group.

It is useful to define the parity operator $P(t)$ by

$$P(t)\, q(\vec{r}, t)\, P(t)^\dagger \equiv \eta\, \gamma_4\, q(-\vec{r}, t) \tag{24.13}$$

where q is the quark field and, as in (10.9), η is a phase factor. From (10.60), we see that

$$P(t)\, Q_v^\ell(t)\, P(t)^\dagger = Q_v^\ell(t)$$

and, similarly, (24.14)

$$P(t)\, Q_a^\ell(t)\, P(t)^\dagger = -\, Q_a^\ell(t) \ .$$

Consequently, $P(t)$ connects the generators $Q_+^\ell(t)$ and $Q_-^\ell(t)$ of the $SU_3 \times SU_3$ group:

$$P(t)\, Q_+^\ell(t)\, P(t)^\dagger = Q_-^\ell(t)$$

and

$$P(t)\, Q_-^\ell(t)\, P(t)^\dagger = Q_+^\ell(t) \ .$$

The group

$$\{ g\, ,\ gP \} \tag{24.15}$$

is called the chiral $SU_3 \times SU_3$ group.

So far, we have not studied the time-dependence of these operators. Their equal-time commutators are independent of the Hamiltonian of the system.

2. Goto–Imamura–Schwinger modification *

In deriving (24.6), we make use of (3.24a) which relies on formal manipulations of some rather singular quantities involving the product of four field operators at the same space-time point. We shall now demonstrate the inadequacy of this formal approach.

It is convenient to combine v_μ^ℓ and a_μ^ℓ to form a set of 16 current operators J_μ^α :

$$\{ J_\mu^\alpha \} \equiv \{ v_\mu^\ell , a_\mu^m \} , \tag{24.16}$$

where

$$\alpha = 1, 2, \cdots , 16$$

but (24.17)

$$\ell \text{ or } m = 1, 2, \cdots , 8 .$$

The set of 16 charges Q_v^ℓ and Q_a^ℓ can be similarly put together as

$$\{ Q^\alpha \} = \{ Q_v^\ell , Q_a^m \} \tag{24.18}$$

where

$$Q^\alpha (t) = -i \int J_4^\alpha (x) \, d^3 r ; \tag{24.19}$$

their equal-time commutation relations (24.9) can be written as

$$[Q^\alpha (t) , Q^\beta (t)] = C^{\alpha\beta\gamma} Q^\gamma (t) \tag{24.20}$$

where

$$C^{\alpha\beta\gamma} = - C^{\beta\alpha\gamma} = - C^{\alpha\gamma\beta} \tag{24.21}$$

is the antisymmetric structure constant of the chiral $SU_3 \times SU_3$ group (24.15). Likewise, (24.8) becomes

$$[Q^\alpha (t) , J_\mu^\beta (x)] = i \, C^{\alpha\beta\gamma} J_\mu^\gamma (x) . \tag{24.22}$$

---

* T. Goto and T. Imamura, Prog. Theor. Phys. 14, 396 (1955). J. Schwinger, Phys. Rev. Lett. 3, 296 (1959). Readers who are not interested in the subtlety of local field operators are encouraged to skip this section.

Since (24.20) expresses the algebra of the group generators and (24.22) gives the representation of $J_\mu^\alpha(x)$ under the group transformations, neither validity is under question. What we would like to examine is the correctness of the local current - current commutation relations (24.6) which, as we shall show, will now be replaced by

$$[J_4^\alpha(\vec{r}, t) , J_\mu^\beta(\vec{r}', t)] = - C^{\alpha\beta\gamma} J_\mu^\gamma(x) \delta^3(\vec{r} - \vec{r}') + \text{s.t.} \tag{24.23}$$

where s. t. stands for the Schwinger term. If s. t. were zero, then the above expression would reduce to the naive commutation relations given by (24.6). A relatively simple argument that establishes the necessity of the Schwinger term is to take the vacuum expectation of (24.23). In particular, we note that when $\alpha = \beta$ and without s. t. we would have

$$< \text{vac} \,|[J_4^\alpha(\vec{r}, t) , J_\mu^\alpha(\vec{r}, t')] |\, \text{vac} > = 0 \tag{24.24}$$

because of the antisymmetric property of $C^{\alpha\beta\gamma}$. On the other hand, the following theorem demonstrates that this vacuum expectation value cannot be zero for $\mu = i \neq 4$.

Theorem.

$$< \text{vac} \,|\, [J_4^\alpha(\vec{r}, t) , J_i^\alpha(\vec{r}', t)] \,|\, \text{vac} > \neq 0 \tag{24.25}$$

where $i = 1, 2$ and 3. [Thus, the Schwinger term is not zero.]

Proof. Let us define

$$K_{\mu\nu}^{\alpha\beta} \equiv < \text{vac} \,|[J_\mu^\alpha(x) , J_\nu^\beta(x')] |\, \text{vac} > \tag{24.26}$$

and decompose it into a sum over all eigenstates $| >$ of the entire physical system in the Hilbert space. By using (11.74), we have

$$K_{\mu\nu}^{\alpha\beta}(x-x') = \sum [< \mathrm{vac} \mid J_\mu^{\alpha}(0) \mid > < \mid J_\nu^{\beta}(0) \mid \mathrm{vac} > e^{ik\cdot(x-x')}$$

$$- < \mathrm{vac} \mid J_\nu^{\beta}(0) \mid > < \mid J_\mu^{\alpha}(0) \mid \mathrm{vac} > e^{ik\cdot(x'-x)}] \tag{24.27}$$

where

$$k_\mu = (\vec{k}, i\, k_0)$$

denotes the 4-momentum of the state $| >$, x and x' are given by (24.5) and, as usual,

$$k\cdot(x-x') = k_\mu (x-x')_\mu \quad .$$

For clarity, we write

$$| > = | n, \vec{k} >$$

$$= | \cdots, h_n, j_n, m_n, \vec{k} > \tag{24.28}$$

where n denotes all other quantum numbers besides $\vec{k}$ that are needed to specify the state, e.g. its invariant mass m_n , spin j_n, helicity h_n , electric charge Q_n , $\cdots$. In our metric,

$$m_n^2 = -k_\mu^2 = k_0^2 - \vec{k}^2 \quad .$$

In (24.27), the sum extends over all possible eigenstates, including those that are related by Lorentz transformations. More explicitly, we may write for the sum in (24.27)

$$\sum \cdots = \int dM \sum_n \delta(M-m_n) \frac{1}{8\pi^3} \int d^3k \cdots . \tag{24.29}$$

Recall that under a Lorentz transformation parallel to the direction of $\vec{k}$, we have

$$\vec{k} \rightarrow \vec{k}' \quad , \qquad k_0 \rightarrow k_0' \quad , \qquad m_n \rightarrow m_n \quad ,$$

$$j_n \rightarrow j_n \quad , \qquad h_n \rightarrow h_n' = \pm h_n$$

(where the $\pm$ sign depends on the velocity of the state versus that of the transformation), and

$$k_0^{\frac{1}{2}} \mid \cdots, h_n, j_n, m_n, \vec{k} >$$

$$\rightarrow k_0'^{\frac{1}{2}} \mid \cdots, h_n', j_n, m_n, \vec{k}' > \tag{24.30}$$

so that *

$$d^3k \mid \cdots, h_n, j_n, m_n, \vec{k} > < \cdots, h_n, j_n, m_n, \vec{k} \mid \tag{24.31}$$

---

* A remark concerning our normalization convention is perhaps necessary. Let us consider the simple example of a free spin-0 field discussed in Chapter 2. In a finite volume Ω the one-particle state is, according to (2.38),

$$\mid \vec{k}, \Omega > = a_{\vec{k}}^{\dagger} \mid vac >$$

which satisfies

$$< \vec{k}, \Omega \mid \vec{q}, \Omega > = \delta_{\vec{k}, \vec{q}}$$

and therefore

$$\sum_{\vec{k}} < \vec{k}, \Omega \mid \vec{q}, \Omega > = 1 .$$

We define

$$\mid \vec{k} > \equiv \Omega^{\frac{1}{2}} a_k^{\dagger} \mid vac > , \tag{24.31a}$$

so that when $\Omega \rightarrow \infty$, the states $\mid k >$ satisfy the orthonormal relations

$$< \vec{k} \mid \vec{q} > = 8\pi^3 \delta^3(\vec{k} - \vec{q}) \tag{24.31b}$$

and

$$\int \frac{d^3k}{8\pi^3} < \vec{k} \mid \vec{q} > = 1 . \tag{24.31c}$$

Under the Lorentz transformation, we have $\mid \vec{k} > \rightarrow \mid \vec{k}' >$ and $\mid \vec{q} > \rightarrow \mid \vec{q}' >$. Since $\frac{d^3k}{k_0} = \frac{d^3k'}{k_0'}$, in order that

$$\int \frac{d^3k'}{8\pi^3} < \vec{k}' \mid \vec{q}' > = 1$$

we must have $k_0^{\frac{1}{2}} \mid \vec{k} > \rightarrow k_0'^{\frac{1}{2}} \mid \vec{k}' >$ which explains (24.30)-(24.31).

is an invariant. The $\vec{k}$ - integration in (24.29) can be most conveniently performed by transforming all states to their respective rest frames. Under a 3 - dimensional rotation the space-components of $J_\mu^\alpha(0)$ transform like a 3 - vector and its time-component like a scalar. Consequently in the rest frame, from rotational invariance we derive

$$< vac \mid J_\mu^\alpha(0) \mid n, \vec{k} = 0 > \; = \; 0$$

$$\text{if the spin} \begin{cases} j_n > 1 \; , \text{ or} \\ j_n = 1 \text{ and } \mu = 4 \; , \text{ or} \\ j_n = 0 \text{ and } \mu = i \neq 4 \; . \end{cases} \qquad (24.32)$$

For a fixed n, by using (24.30), (24.32) and summing over the different helicities we find, in any frame,

$$\sum_{h_n} < vac \mid J_\mu^\alpha(0) \mid n, \vec{k} > < n, \vec{k} \mid J_\nu^\beta(0) \mid vac >$$

$$= \begin{cases} 0 & \text{if } j_n > 1 \\ 3\sigma_1^{\alpha\beta}(n)\left(\delta_{\mu\nu} + \dfrac{k_\mu k_\nu}{m_n^{\;2}}\right) \dfrac{m_n}{k_0} & \text{if } j_n = 1 \\ \sigma_0^{\alpha\beta}(n) \dfrac{k_\mu k_\nu}{m_n^{\;2}} \dfrac{m_n}{k_0} & \text{if } j_n = 0 \end{cases} \qquad (24.33)$$

where, because J_i^α $(i = 1, 2, 3)$ is Hermitian but J_4^α is anti-Hermitian (and k_i real but k_4 imaginary), we have for $j_n = 0$

$$\sigma_0^{\alpha\beta}(n) = \delta^{\alpha\beta} < vac \mid J_4^\alpha(0) \mid n, \vec{k} = 0 > < vac \mid J_4^\beta(0) \mid n, \vec{k} = 0 >^* \qquad (24.34)$$

and for $j_n = 1$

$$\sigma_1^{\alpha\beta}(n) = \tfrac{1}{3}\, \delta^{\alpha\beta} \sum_{i=1}^{3} < vac \mid J_i^\alpha \mid n, \vec{k} = 0 > < vac \mid J_i^\beta \mid n, \vec{k} = 0 >^* , \qquad (24.35)$$

in which, as usual, $*$ denotes the complex conjugation.

The simplest way to establish (24.33) is first to verify its validity in the frame $\vec{k}=0$, and then transform it to other reference frames. Note that in order to simplify the formulas, we assume Q^{α} to be good quantum numbers and that accounts for the factor $\delta^{\alpha\beta}$ in (24.34) and (24.35). Substituting (24.33) into (24.27) and using (24.29), we obtain

$$i K_{\mu\nu}^{\alpha\beta}(x-x') = \int dM^2 \left[\delta_{\mu\nu} - \frac{1}{M^2} \frac{\partial^2}{\partial x_\mu \, \partial x_\nu} \right] \sigma_1^{\alpha\beta}(M)\, D_M(x-x')$$

$$+ \int dM^2 \left[- \frac{1}{M^2} \frac{\partial^2}{\partial x_\mu \, \partial x_\nu} \right] \sigma_0^{\alpha\beta}(M)\, D_M(x-x') \tag{24.36}$$

where the spectral functions $\sigma_0^{\alpha\beta}(M)$ and $\sigma_1^{\alpha\beta}(M)$ are related to those in (24.34)-(24.35) by

$$\sigma_0^{\alpha\beta}(M) = \sum_n \delta(M-m_n)\, \sigma_0^{\alpha\beta}(n)$$

and

$$\sigma_1^{\alpha\beta}(M) = \sum_n \delta(M-m_n)\, \sigma_1^{\alpha\beta}(n) \tag{24.37}$$

in which, because of (24.32), these sums extend only over states with $j_n = 0$ and 1 respectively,

$$D_M(x-x') = \frac{1}{8\pi^3} \int d^3k \; e^{i\vec{k}\cdot(\vec{r}-\vec{r}\,')} \; \omega^{-1} \sin \omega(t-t')$$

and

$$\omega = \sqrt{\vec{k}^{\,2} + M^2} \;,$$

as in Problem 2.1. When $t \to t'$,

$$D_M(x-x') \to 0$$

but

$$\frac{\partial}{\partial t} D_M(x-x') \to \delta^3(\vec{r}-\vec{r}\,') \;;$$

therefore in (24.26), for $\mu = 4$ and $\nu = i \neq 4$, $K_{4i}^{\alpha\beta}(x-x')$ becomes

$$< \text{vac} \,|\, [J_4^\alpha(\vec{r}, t) \;,\; J_i^\beta(\vec{r}', t)] \,|\, \text{vac} >$$

$$= \int \frac{dM^2}{M^2} [\sigma_1^{\alpha\beta}(M) + \sigma_0^{\alpha\beta}(M)] \, \nabla_i \, \delta^3(\vec{r} - \vec{r}') . \qquad (24.38)$$

Regarding $\sigma_0^{\alpha\beta}(M)$ and $\sigma_1^{\alpha\beta}(M)$ as matrices, we see that, from (24.34)-(24.35) and (24.37), they are both positive definite. In particular, if we set $\alpha = \beta$, then since

$$\int \frac{dM^2}{M^2} [\sigma_1^{\alpha\alpha}(M) + \sigma_0^{\alpha\alpha}(M)] > 0 \;,$$

(24.25) follows. Thus, we establish the theorem. By comparing (24.24) with (24.38), we see that the Schwinger term is not zero; furthermore, it is manifestly noncovariant.

(i) Equation (24.38) can also be written as

$$< \text{vac} \,|\, [J_4^\alpha(\vec{r}, t) \,,\; J_i^\beta(\vec{r}', t)] \,|\, \text{vac} > = \lambda \, \delta^{\alpha\beta} \, \nabla_i \, \delta^3(\vec{r} - \vec{r}') \qquad (24.39)$$

where

$$\lambda = \frac{1}{16} \sum_{\alpha=1}^{16} \int \frac{dM^2}{M^2} [\sigma_1^{\alpha\alpha}(M) + \sigma_0^{\alpha\alpha}(M)] > 0 \;.$$

(ii) To be consistent with the Theorem, we replace (24.6) by

$$\begin{aligned}
[v_4^\ell(\vec{r}, t), \; a_4^m(\vec{r}', t)] &= [a_4^\ell(\vec{r}, t), \; v_4^m(\vec{r}', t)] \\
&= - f^{\ell mn} \; a_4^n(\vec{r}, t) \; \delta^3(\vec{r} - \vec{r}') \;, \\
[v_4^\ell(\vec{r}, t), \; v_4^m(\vec{r}', t)] &= [a_4^\ell(\vec{r}, t), \; a_4^m(\vec{r}', t)] \\
&= - f^{\ell mn} \; v_4^n(\vec{r}, t) \; \delta^3(\vec{r} - \vec{r}') \;, \qquad (24.40) \\
[v_4^\ell(\vec{r}, t), \; a_i^m(\vec{r}', t)] &= [a_4^\ell(\vec{r}, t), \; v_i^m(\vec{r}', t)] \\
&= - f^{\ell mn} \; a_i^n(\vec{r}, t) \; \delta^3(\vec{r} - \vec{r}')
\end{aligned}$$

and

$$[v_4^\ell(\vec{r},t),\ v_i^m(\vec{r}',t)] = [a_4^\ell(\vec{r},t),\ a_i^m(\vec{r}',t)]$$
$$= -f^{\ell mn}\ v_i^n(\vec{r},t)\ \delta^3(\vec{r}-\vec{r}') + \text{s.t.}$$

where the simplest form * of the Schwinger term is

$$\text{s. t.} = \lambda\ \delta^{\ell m}\ \nabla_i\ \delta^3(\vec{r}-\vec{r}') \qquad (24.41)$$

with λ = a constant. By integrating these equations with respect to d^3r, we see that the Schwinger term makes no contributions. Consequently, (24.8)-(24.11) remain valid.

24.2 CVC and CAC

From the Cabibbo theory (Section 21.5), we know that because of the conservation of the electromagnetic current and the flavor-SU_3 symmetry there is the CVC condition; i.e., the eight vector-current operators (24.3) satisfy

$$\frac{\partial}{\partial x_\mu}\ v_\mu^\ell = 0 \quad . \qquad (24.42)$$

Hence, the space integrals of v_4^ℓ are constants of motion; i.e.,

$$\dot{Q}_V^\ell = 0 \quad . \qquad (24.43)$$

These eight vector charges are the generators of the flavor-SU_3 group. Each physical hadron state belongs to an irreducible representation of the SU_3 group, as discussed in Chapter 12.

The same conclusion can also be arrived at by using the QCD Lagrangian. Setting the gauge group in (18.3) to be the color - SU_3 and ψ the quark field, we see that if we assume different quarks all have the same mass m, flavor - SU_3 symmetry follows and so does CVC.

---

* T. D. Lee, S. Weinberg and B. Zumino, Phys. Rev. Lett. 18, 1029 (1967).

In our discussions on hadrons in Chapter 20, we mentioned that a fairly good approximation for the strong interaction is to regard the u, d and s quarks all as of zero mass. By setting

$$m = 0 \tag{24.44}$$

in (18.3), we see that in addition to CVC, the axial-vector current operators are also conserved; i.e.,

$$\frac{\partial}{\partial x_\mu} a_\mu^\ell = 0 \quad , \tag{24.45}$$

and therefore

$$\dot{Q}_a^\ell = 0 \tag{24.46}$$

where a_μ^ℓ and Q_a^ℓ are given by (24.4) and (24.7). The chiral $SU_3 \times SU_3$ group given by (24.15) is generated by Q_v^ℓ and Q_a^ℓ, together with the parity operator P. Under the same approximation of the strong interaction, each group element is a constant of motion.

Consider now any physical hadron state, say the proton $|p>$ at rest. It is an eigenstate of the strong interaction Hamiltonian H_{st}:

$$H_{st} \,|\, p> \;=\; m_p \,|\, p> \quad . \tag{24.47}$$

It is also an eigenstate of the parity P. For convenience, we choose the arbitrary phase factor η in (24.13) such that

$$P \,|\, p> \;=\; |\, p> \quad . \tag{24.48}$$

By operating the vector charge Q_v^ℓ on $|p>$, we obtain a linear superposition of the baryon octet states given by (12.48). Within the approximation of flavor-SU_3 symmetry, these states are all degenerate with $|p>$ and have the same parity.

Let us extend the same reasoning to the axial charge Q_a^ℓ. On account of (24.46) we have

$$[\,Q_a^{\ell}\,,\ H_{st}\,] = 0\ . \tag{24.49}$$

Equations (24.47)-(24.49) give

$$H_{st}\,Q_a^{\ell}\,|\,p> \;=\; m_p\,Q_a^{\ell}\,|\,p>\ . \tag{24.50}$$

Furthermore, because of (24.14), the parity of $Q_a^{\ell}\,|\,p>$ must differ from that of $|\,p>$ by a minus sign. In contrast to (24.48), we have

$$P\,Q_a^{\ell}\,|\,p> \;=\; -\,Q_a^{\ell}\,|\,p>\ . \tag{24.51}$$

What are these $Q_a^{\ell}\,|\,p>$ states that are approximately degenerate with $|\,p>$, but of opposite parity? Clearly, they cannot be any of the known single hadron states, since none exist in nature. Thus, we have to identify $Q_a^{\ell}\,|\,p>$ as one of the continuum states such as $|\,p\,\pi>$, $|\,n\,\pi>$, ... The degeneracy implies that in the approximation that the axial current is conserved, the low-lying pseudoscalar mesons should be viewed as zero-mass particles, which as we shall see fits neatly into the Nambu - Goldstone mechanism described in Section 22.1.2. The assumption of axial-current conservation must be used in conjunction with the approximation of zero-mass pseudoscalar mesons. * Together they will be referred to as the two conditions of CAC. It is important that we should distinguish these unusual features of CAC from those (more conventional ones) of CVC.

The kaon is much heavier than the pion (i.e., s is much heavier than u and d). A better approximation is to set only the u and d quark masses to be zero. In this case, the superscript ℓ in (24.45)

---

* In the literature, sometimes one refers to the assumption of axial-current conservation and the approximation of zero-mass pions and/or kaons as PCAC (partially conserved axial current). In order to avoid unnecessary confusion we shall restrict the term PCAC only to cases when violations to the axial-current conservation law (24.45) become important. See Section 24.6.

goes only from 1 to 3 ; the symmetry group is

$$\text{chiral } SU_2 \times SU_2 \ , \tag{24.52}$$

and only the pions are Goldstone bosons, as will be discussed in Sections 24.4 and 24.5.

24.3 Goldstone Theorem *

To appreciate the generality of the Nambu–Goldstone mechanism, let us consider a relativistic quantum system of, among others, L spin-0 Hermitian fields $\phi_\ell(x)$, with

$$\ell = 1, 2, .. L \ .$$

The theory is assumed to be invariant under the continuous transformations

$$\phi_\ell(x) \rightarrow \phi_\ell(x) + \delta\phi_\ell(x)$$

with

$$\delta\phi_\ell(x) = i\,\theta^\alpha\, T^\alpha_{\ell\ell'}\,\phi_{\ell'}(x) \tag{24.53}$$

where, similarly to (18.12), θ^α are real x-independent infinitesimals, and

$$T^\alpha = \text{matrix } (T^\alpha_{\ell\ell'}) = T^{\alpha\dagger}$$

is the $L \times L$ representation of the αth generator Q^α of the symmetry group $\{g\}$ of the theory,

$$\alpha = 1, 2, .. N \ .$$

Furthermore, we assume, as in (24.19), the operator Q^α is related to the <u>conserved</u> current J^α_μ by

$$Q^\alpha = -i \int J^\alpha_4\, d^3r \ .$$

---

* See the references cited on page 658, and also J. Goldstone, A. Salam and S. Weinberg, Phys.Rev. 127, 965 (1962).

Because

$$\frac{\partial J_\mu^\alpha}{\partial x_\mu} = 0 \quad , \tag{24.54}$$

therefore

$$\dot{Q}_\alpha = 0 \quad .$$

In accordance with (24.53), there are the commutation relations

$$[\, Q^\alpha \, , \, \phi_\ell(x) \,] = i \, T^\alpha_{\ell \ell'} \, \phi_{\ell'}(x) \quad . \tag{24.55}$$

To take into account the necessary ingredient of the spontaneous symmetry-breaking mechanism we assume, as in (22.7) and (22.63), that the vacuum expectation value

$$< \text{vac} \mid \phi_\ell \mid \text{vac} > \; \equiv \; \rho_\ell \tag{24.56}$$

satisfies

$$T^\alpha_{\ell \ell'} \, \rho_{\ell'} \neq 0 \tag{24.57}$$

at least for some α and ℓ .

<u>Theorem.</u>* If (24.57) holds, then there must exist a massless particle which has the same quantum numbers as those of T^α .

<u>Proof.</u> Let us decompose

$$< \text{vac} \mid J_\mu^\alpha(x) \; \phi_\ell(0) \mid \text{vac} > \tag{24.58}$$

into a sum over the complete set of eigenstates $\mid n \, , \vec{k} >$ given by (24.28), with $\vec{k}$ as the momentum of the state and n all its other quantum numbers. By using (11.74) and (24.29), we obtain

---

* See, however, remarks (i) and (ii) at the end of this section.

$$(24.58) = \sum_n \int dM\, \delta(M - m_n) \int \frac{1}{8\pi^3} d^3k < vac \mid J_\mu^\alpha(0) \mid n, \vec{k} >$$

$$\cdot < n, \vec{k} \mid \phi_\ell(0) \mid vac > e^{ik\cdot x} . \qquad (24.59)$$

For states of zero 3-momentum, because of angular-momentum conservation we have

$$< n, \vec{k} = 0 \mid \phi_\ell(0) \mid vac > \neq 0 \qquad \text{only if}$$

the spin $j_n = 0$.

Using (24.30) we can transform these zero-spin states to any Lorentz frame. In addition, from the arguments given on page 235, it follows that

$$< vac \mid J_\mu^\alpha(0) \mid n, \vec{k} > \; \propto \; k_\mu .$$

Combining these results with (24.32), we may write

$$< vac \mid J_\mu^\alpha(0) \mid n, \vec{k} > < n, \vec{k} \mid \phi_\ell(0) \mid vac > = \frac{k_\mu}{k_0} c_\ell^{\alpha}(n) \qquad (24.60)$$

where

$$c_\ell^{\alpha}(n) = \begin{cases} -i < vac \mid J_4^\alpha(0) \mid n, \vec{k} = 0 > < n, \vec{k} = 0 \mid \phi_\ell(0) \mid vac > \\ \qquad\qquad\qquad\qquad \text{if } j_n = 0 \\ 0 \qquad\qquad\qquad\qquad \text{if } j_n \neq 0 \end{cases} \qquad (24.61)$$

and

$$k_0 = -i k_4 = \sqrt{\vec{k}^{\,2} + m_n^{\,2}}$$

with m_n = the rest mass of the state $\mid n, \vec{k} >$. Substituting (24.60) into (24.59) and changing the variable from M to

$$\omega = \sqrt{\vec{k}^{\,2} + M^2} ,$$

we obtain

$$(24.58) = \int_0^\infty d\omega \frac{1}{8\pi^3} \int d^3k\, e^{ik\cdot x} \sum_n \frac{k_\mu}{M} \delta(M - m_n)\, c_\ell^{\alpha}(n) . \qquad (24.62)$$

Define

$$C_\ell^\alpha(M) \equiv \frac{1}{M} \sum_n \delta(M - m_n)\, c_\ell^\alpha(n) \ . \tag{24.63}$$

Equation (24.62) may be written as

$$< \text{vac} \mid J_\mu^\alpha(x)\, \phi_\ell(0) \mid \text{vac} > \; = \int \frac{d^4k}{8\pi^3}\, \theta(k_0)\, e^{ik\cdot x}\, k_\mu\, C_\ell^\alpha(M) \tag{24.64}$$

where

$$\theta(k_0) = \begin{cases} 1 & \text{if } k_0 \geqslant 0 \\ 0 & \text{if } k_0 < 0 \ , \end{cases}$$

M is a function of k_μ given by

$$M^2 = -k_\mu^2 = k_0^2 - \vec{k}^2 \tag{24.65}$$

and the integration of $d^4k = dk_0\, d^3k$ ranges over the entire k-space.* By differentiating (24.64) with respect to x_μ and applying the current conservation law (24.54), we derive, on account of (24.65),

$$M^2\, C_\ell^\alpha(M) = 0 \ . \tag{24.66}$$

Setting $\mu = 4$, $x_4 = it = 0$ in (24.64) and integrating both sides over the 3-space, we find

$$< \text{vac} \mid Q^\alpha\, \phi_\ell(0) \mid \text{vac} > \; = \int_{-\infty}^{\infty} k_0\, \theta(k_0)\, C_\ell^\alpha(M)\, dk_0 \ ;$$

its complex conjugate is

$$< \text{vac} \mid \phi_\ell(0)\, Q^\alpha \mid \text{vac} > \; = \int_{-\infty}^{\infty} k_0\, \theta(k_0)\, C_\ell^\alpha(M)^*\, dk_0 \ .$$

Because of (24.55) and (24.56), their difference leads to

$$T_{\ell\ell'}^\alpha\, \rho_{\ell'} = 2 \int_{-\infty}^{\infty} k_0\, \theta(k_0)\, \text{Im}\, C_\ell^\alpha(M)\, dk_0$$

---

* Hence, $\int_{-\infty}^{\infty} dk_0\, \theta(k_0) \cdots$ in (24.64) is the same as $\int_0^{\infty} d\omega \cdots$ in (24.62).

where Im denotes the imaginary part. The lefthand side of the above equation is, by assumption, not zero for some α and ℓ. Since according to (24.66)

$$C_\ell^\alpha(M) = 0 \quad \text{when} \quad M \neq 0 ,$$

we conclude that for these α and ℓ,

$$C_\ell^\alpha(M) = \lambda_\ell^\alpha \, \delta(M^2) \neq 0 . \tag{24.67}$$

Recalling the definition (24.63) for $C_\ell^\alpha(M)$, we establish that there must exist states $|\, n, \vec{k} >$ with their rest mass $m_n = 0$; furthermore, for these states the corresponding $c_\ell^\alpha(n) \neq 0$. Consequently, in view of (24.61), we must have

$$j_n = 0 ,$$

$$< n, \vec{k} \,|\, \phi_\ell \,|\, vac > \neq 0$$

and

$$< vac \,|\, Q^\alpha \,|\, n, \vec{k} > \neq 0 . \tag{24.68}$$

The last equation enables us to determine the symmetry properties of these states, and that completes the proof of the theorem. These zero-mass particles are called Goldstone bosons.

Remarks.

(i) Substituting (24.67) into (24.64) we have

$$< vac \,|\, J_\mu^\alpha(x) \; \phi_\ell(0) \,|\, vac > = -i \lambda_\ell^\alpha \frac{\partial}{\partial x_\mu} F_D(x)$$

where, in the four-dimensional space $(D = 4)$, $F_D(x)$ is

$$F_4(x) = \frac{1}{8\pi^3} \int d^4k \; e^{ik\cdot x} \; \delta(k^2) \; \theta(k_0) .$$

In the two-dimensional space $(D = 2)$, the corresponding function is

$$F_2(x) = \frac{1}{2\pi} \int d^2k \; e^{ik\cdot x} \; \delta(k^2) \; \theta(k_0) , \tag{24.69}$$

which is however singular, as can be seen by performing the k_0 integration: Since

$$k \cdot x = k_1 x_1 - k_0 x_0 ,$$
$$\delta(k^2) = \delta(k_0^2 - k_1^2)$$

and

$$dk_0 = \frac{1}{2k_0} dk_0^2 ,$$

we have

$$\int d^2k \, e^{ik \cdot x} \, \delta(k^2) \, \theta(k_0)$$
$$= \int_{-\infty}^{\infty} dk_1 \frac{1}{2|k_1|} \exp[ik_1 x_1 - i|k_1| x_0]$$

which diverges over the integration region that contains the origin $k_1 = 0$. In order to have a well-defined theory, we may require the assumption (24.57) of the Goldstone theorem to be invalid. Thus,

$$< vac \mid \delta\phi_\ell(x) \mid vac > = 0$$

for all ℓ where $\delta\phi_\ell$ is defined by (24.53). This implies that there is no Goldstone boson * in two dimensions.

(ii) In Section 22.1.3, we have shown that in a gauge theory, by introducing a suitable set of Higgs fields it is possible to have the Goldstone bosons appear only in the unphysical sector which can be gauged away. As a result, the physical spectrum does not contain any zero-mass spin-0 particles. Through the Higgs mechanism we can convert the "massless" gauge fields and Goldstone bosons into physical massive vector particles. [Details have been given on pages 666-68.]

---

* S. Coleman, Commun. Math. Phys. 31, 259 (1973).

24.4 Goldberger-Treiman Relation

In this and the following sections, we shall examine some of the applications of chiral $SU_2 \times SU_2$ symmetry. As in Section 24.1, both the vector and the axial currents, $\vec{v}_\lambda$ and $\vec{a}_\lambda$, are assumed to be conserved, where

$$\vec{v}_\lambda = i\, q^\dagger(x)\, \gamma_4\, \gamma_\lambda\, \tfrac{1}{2}\vec{\tau}\, q(x)$$

and (24.70)

$$\vec{a}_\lambda = i\, q^\dagger(x)\, \gamma_4\, \gamma_\lambda\, \gamma_5\, \tfrac{1}{2}\vec{\tau}\, q(x)$$

in which

the components of $\vec{\tau}$ = Pauli matrices,

and the quark field $q(x)$ is given by the 2×1 column matrix

$$q(x) = \begin{pmatrix} \psi_u^{\,c}(x) \\ \psi_d^{\,c}(x) \end{pmatrix} \qquad (24.71)$$

instead of (24.1). In accordance with the discussions of the previous sections, the pion is regarded as a zero-mass Goldstone boson. Thus, while both the vector and the axial charges

$$\vec{Q}_v = -i \int \vec{v}_4\, d^3r$$

and

$$\vec{Q}_a = -i \int \vec{a}_4\, d^3r$$

commute with the Hamiltonian, the vacuum state $|\, vac >$ is only an eigenstate of $\vec{Q}_v$, not $\vec{Q}_a$. We have

$$\vec{Q}_v \,|\, vac > \; = \; 0$$

but (24.72)

$$\vec{Q}_a \,|\, vac > \; \neq \; 0 \quad .$$

The former is the result of CVC, and the latter that of the axial-current conservation,

$$\frac{\partial \vec{a}_\lambda}{\partial x_\lambda} = 0 \quad , \tag{24.73}$$

plus the approximation of massless pions. [(24.72) follows from the arguments given on page 783, provided $|\, p >$ is replaced by $|\, \text{vac} >$.]

1. π decay amplitude

The pion decay has been studied in Section 21.3. Its amplitude is given by (21.19):

$$< \text{vac} \,|\, A_\lambda^{\pm}(x) \,|\, \pi^{\mp} > \;=\; i \cos\theta_c \, F_\pi \, q_\lambda \, \frac{1}{\sqrt{2\omega\Omega}} \, e^{iq\cdot x} \tag{24.74}$$

where q_λ is the 4 – momentum of the pion with

$$q_4 = i\omega \quad ,$$

Ω is the volume of the system, θ_c is the Cabibbo angle given by (21.50), and from (21.21)

$$F_\pi \cong 130 \text{ MeV} \quad .$$

In terms of the components a_λ^{ℓ} ($\ell = 1, 2$ and 3) of the current operator (24.70), we write

$$A_\lambda^{+} = (a_\lambda^{1} + i\, a_\lambda^{2}) \cos\theta_c$$

and

$$A_\lambda^{-} = (a_\lambda^{1} - i\, a_\lambda^{2}) \cos\theta_c \quad . \tag{24.75}$$

Hence, the above matrix element (24.74) can be rewritten as

$$< \text{vac} \,|\, a_\lambda^{\ell}(x) \,|\, \pi^{m} > \;=\; i\,\delta^{\ell m} \, f_\pi \, q_\lambda \, \frac{1}{\sqrt{2\omega\Omega}} \, e^{iq\cdot x} \tag{24.76}$$

where $\delta^{\ell m}$ is the Kronecker symbol,

$$f_\pi = \frac{F_\pi}{\sqrt{2}} \cong 93 \text{ MeV} \tag{24.77}$$

is called the pion-decay constant, and $|\pi^m>$ are related to the charged and neutral pion states by

$$|\pi^1> = \frac{1}{\sqrt{2}}(|\pi^+> + |\pi^->) ,$$

$$|\pi^2> = -\frac{i}{\sqrt{2}}(|\pi^+> - |\pi^->)$$

and

$$|\pi^3> = |\pi^o> .$$

To represent the matrix element (24.74) or (24.76) graphically, a convenient method is first to think of the intermediate boson theory and write down all the Feynman diagrams for $\pi_{\ell 2}$ decay. Such diagrams have to go through the sequence

$$\pi^{\pm} \rightarrow W_\lambda^{\pm} \rightarrow \ell^{\pm} + \nu_\ell \ (\bar{\nu}_\ell) .$$

Their sum is given by the upper graph in Figure 24.1, in which the external pion line carries the usual factor

$$\frac{1}{\sqrt{2\omega\Omega}}$$

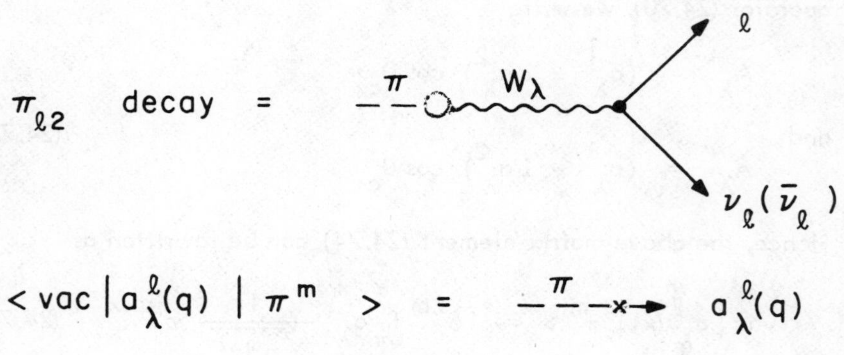

Fig. 24.1. Diagrams for $\pi_{\ell 2}$ decay (via W) and for the corresponding matrix element (24.79) of a_λ^ℓ .

and the $\pi - W_\lambda$ vertex a factor *

$$i \cos \theta_c \, F_\pi \, q_\lambda \quad . \tag{24.78}$$

We then remove the W-propagator and the W-lepton vertex from the diagram. The remainder is the matrix element (24.74). In a similar way, we can also express graphically the alternate form (24.76). Its Fourier transform

$$< vac \,|\, a_\lambda^{\ell}(q) \,|\, \pi^m > \;\equiv\; \int e^{-iq\cdot x} < vac \,|\, a_\lambda^{\ell}(x) \,|\, \pi^m > d^4x \tag{24.79}$$

is represented by the lower graph in Figure 24.1 where the vertex carries the factor

$$i f_\pi q_\lambda \delta^{\ell m} \tag{24.80}$$

in accordance with (24.76).

Note that by differentiating (24.76) we have

$$< vac \,|\, \frac{\partial a_\lambda^{\ell}}{\partial x_\lambda} \,|\, \pi^m > \;\propto\; q^2 \quad .$$

The lefthand side is zero because of the axial-current conservation assumption and the righthand side also vanishes due to the zero-pion-mass approximation. Consequently, (24.74) and (24.76) are consistent with the two conditions of CAC.

2. <u>β decay of the nucleon</u>

From general symmetry considerations, we know that the axial-current matrix element of $A_\lambda \equiv A_\lambda^-$ between the nucleon states can be written in the form (21.44):

$$< n \,|\, A_\lambda(x) \,|\, p > \; = i\, u_n^\dagger \gamma_4 (\gamma_\lambda g_A - i\, q_\lambda g_P) \gamma_5 u_p \, e^{iq\cdot x} \tag{24.81}$$

---

* According to (22.123), the hadron - W vertex is $(2\sqrt{2})^{-1} \cdot g(W_\lambda A_\lambda^+ + h.c.) + \cdots$. Here, because we are only interested in $A_\lambda^\pm$, the irrelevant factor $(2\sqrt{2})^{-1} g$ is dropped.

where u_p and u_n are the initial and final c. number spinors for the nucleons, p_λ and n_λ are the corresponding initial and final 4-momenta,

$$q_\lambda = (p - n)_\lambda$$

is the 4-momentum transfer,

$$g_P = g_P(q^2)$$

is called the (induced) pseudoscalar form factor and

$$g_A = g_A(q^2)$$

the axial form factor. To give a graphic representation of this matrix element we can follow steps similar to those taken in the previous section. We first think of the sum of all Feynman diagrams for the proton β decay via the sequence

$$\bar{n} p \rightarrow W_\lambda^+ \rightarrow e^+ \nu_e \quad . \tag{24.82}$$

By removing the W-propagator and the W-lepton vertex from each of these diagrams, we are left with the graphs representing the matrix element $< n \mid A_\lambda(x) \mid p >$.

The diagrams for (24.82) can be grouped into two classes:

(i) those where we can completely detach the W_λ lepton part from the diagram by cutting open a single pion line, and

(ii) all others.

The former must go through the sequence

$$\bar{n} p \rightleftarrows \pi^+ \rightleftarrows W_\lambda^+ \rightleftarrows e^+ \nu_e \ ,$$

which is a subclass of (24.82); each of these diagrams carries a pion propagator factor

$$\frac{-i}{q^2} \tag{24.83}$$

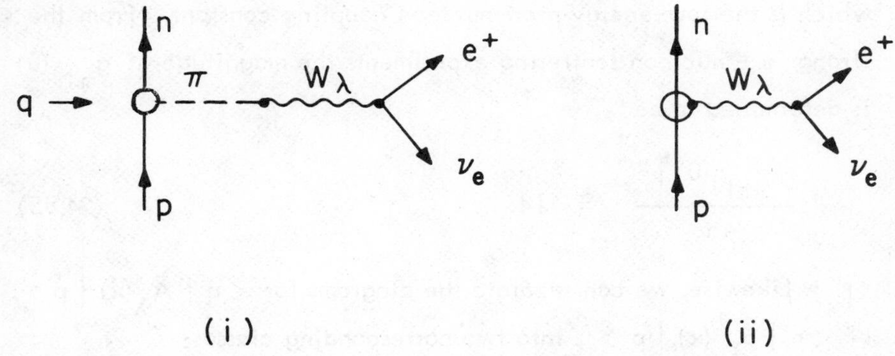

Fig. 24.2. Two classes of diagrams for the β decay of the proton: (i) those that carry the pion pole factor $-i/q^2$ and (ii) those that do not.

and therefore dominates the decay amplitude when $q^2 \to 0$ (but q_λ can be nonzero). The sum of all diagrams in this class is represented by (i) in Figure 24.2, in which because of Lorentz invariance the pion-nucleon vertex can always be written as *

$$\sqrt{2}\; u_n^{\dagger} \gamma_4 \gamma_5 u_p\, g_{\pi N}(q^2) \quad . \tag{24.84}$$

The factor $\sqrt{2}$ is due to the usual normalization convention ** of the nuclear isospin matrix element for the emission and absorption of a charged pion. When $q^2 \to 0$,

$$g_{\pi N}(q^2) \to g_{\pi N}(0)$$

---

* The pion-nucleon vertex must be a pseudoscalar. Because the external nucleon lines are on the mass shell, scalar products like $p \cdot q$, $n \cdot q$, etc. are all linear functions of q^2. Furthermore, invariants such as $\not{p}$ and $\not{n}$ can all be converted into constants. [See (24.90).] This explains the general form (24.84).

** The same $\sqrt{2}$ factor also appears when we compare f_π with F_π. [See (24.77).]

which is the low-energy pion-nucleon coupling constant. From the strong π - nucleon scattering experiments the magnitude of $g_{\pi N}(0)$ is determined to be *

$$\frac{|g_{\pi N}(0)|^2}{4\pi} \cong 14 \quad . \tag{24.85}$$

Likewise, we can separate the diagrams for $\langle n | A_\lambda(x) | p \rangle$, of $\langle n | a_\lambda^{\ell}(x) | p \rangle$, into two corresponding classes:

(i) those that contain the pion pole (24.83), and
(ii) those that do not.

This is illustrated in Figure 24.3. By multiplying together the three factors (24.78), (24.83) and (24.84), we see that the amplitude of (i)

$\langle n | A_\lambda(x) | p \rangle$ =

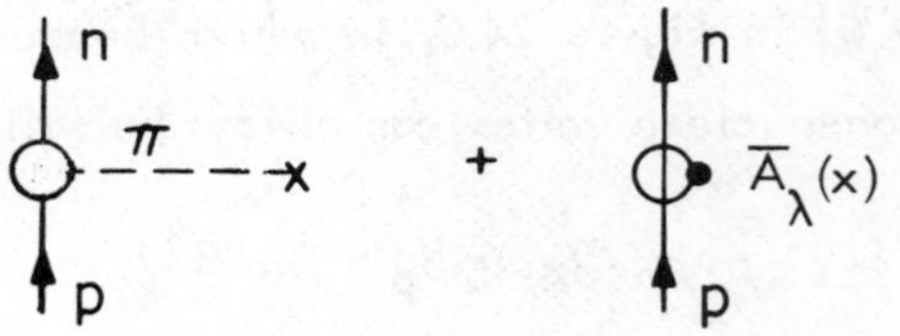

= (i) + (ii)

Fig. 24.3. Decomposition of diagrams for $\langle n | A_\lambda(x) | p \rangle$ into (i) those that carry the pion pole factor and (ii) those that do not; the latter is defined to be $\langle n | \bar{A}_\lambda(x) | p \rangle$. [(i) and (ii) here are obtained by removing the W propagator and the W lepton vertex from the corresponding diagrams in Figure 24.2.]

---

* See, e.g., A. J. Bearden, Phys.Rev.Lett. 4, 240 (1960).

is

$$(i) = \sqrt{2}\ \cos\theta_c \frac{q_\lambda}{q^2} F_\pi g_{\pi N}(q^2)\, u_n^\dagger \gamma_4 \gamma_5 u_p\, e^{iq\cdot x} \quad .$$

By defining

$$(ii) \equiv \langle n \mid \bar{A}_\lambda(x) \mid p \rangle \tag{24.86}$$

and using (24.77), we have

$$\langle n \mid A_\lambda(x) \mid p \rangle = 2\cos\theta_c \frac{q_\lambda}{q^2} f_\pi g_{\pi N}(q^2)\, u_n^\dagger \gamma_4 \gamma_5 u_p\, e^{iq\cdot x} + \langle n \mid \bar{A}_\lambda(x) \mid p \rangle \quad . \tag{24.87}$$

When $q^2 \to 0$, it becomes

$$\lim_{q^2\to 0} \langle n \mid A_\lambda(0) \mid p \rangle \to 2\cos\theta_c \frac{q_\lambda}{q^2} f_\pi g_{\pi N}(0)\, u_n^\dagger \gamma_4 \gamma_5 u_p + O(1) . \tag{24.88}$$

Comparing this expression with (24.81), we find that when $q^2 \to 0$,

$$g_P(q^2) \to \frac{2}{q^2} \cos\theta_c\ f_\pi\, g_{\pi N}(0) + O(1)$$

and (24.89)

$$g_A(q^2) \to g_A(0) = O(1)$$

where the $O(1)$ terms are all finite.

So far we have only used the zero-pion-mass aspect of CAC. Next, we apply the condition of axial-current conservation. By differentiating (24.81) with respect to x_λ and because of

$$\not{p}\, u_p = m_N u_p$$

and (24.90)

$$\not{n}\, u_n = m_N u_n \ ,$$

we derive

$$2m_N\, g_A(q^2) + q^2 g_P(q^2) = 0 \tag{24.91}$$

where m_N is the nucleon mass. Taking the limit $q^2 \to 0$ and using

(24.89), we obtain

$$|\, g_A(0)\, | \;=\; \cos\theta_c \left| \frac{f_\pi g_{\pi N}(0)}{m_N} \right| . \tag{24.92}$$

From CVC and the Cabibbo theory, the corresponding vector form factor at $q^2 = 0$ is given by (21.47):

$$g_V(0) \;=\; \cos\theta_c \;.$$

The ratio

$$\left| \frac{g_A(0)}{g_V(0)} \right| \;=\; \frac{|\, f_\pi g_{\pi N}\, |}{m_N} \;\cong\; 1.31 \tag{24.93}$$

is the Goldberger–Treiman relation, which agrees quite well with the experimental value

$$1.253 \;\pm\; .007$$

given by (21.46).

Remarks. While the underlying theory of strong interaction is QCD, for the low-energy meson-baryon scattering processes it is not easy to extract useful Feynman diagrams directly from QCD because of the quark-confinement mechanism. In this chapter we are only interested in results that follow from the assumption of chiral symmetry. Hence, the details of the Hamiltonian are irrelevant; the only requirement is the validity of the symmetry conditions. In our discussions of the π-nucleon Feynman diagrams, it is more convenient to think of a phenomenological strong-interaction Hamiltonian

$$H \;=\; H_0 + H_{int} \;. \tag{24.94}$$

As usual, the free Hamiltonian H_0 is assumed to have the physical masses of the hadrons, and both H_0 and H_{int} are supposed to be chiral symmetric. The scattering matrix S is given by (5.23):

$$S = \lim_{\substack{t' \to -\infty \\ t \to \infty}} U(t, t')$$

where $U(t, t')$ is the Green's function of the Schrödinger equation (5.17). Let $|\alpha>$ denote the free-particle eigenstate of H_0,

$$H_0 \, |\alpha> \; = \; E_\alpha \, |\alpha> \quad . \tag{24.95}$$

According to (6.58)-(6.59) the corresponding eigenstate of H can be either the in (initial) state

$$|\alpha^{in}> \; \equiv \; U(0, -\infty) \, |\alpha> \tag{24.96}$$

or the f (final) state

$$|\alpha^{f}> \; \equiv \; U^{\dagger}(\infty, 0) \, |\alpha> \; , \tag{24.97}$$

so that

$$H \, |\alpha^{in}> \; = \; E_\alpha \, |\alpha^{in}>$$
and
$$H \, |\alpha^{f}> \; = \; E_\alpha \, |\alpha^{f}> \; . \tag{24.98}$$

We may choose $|\alpha>$ to consist only of plane waves of mesons and baryons. Then asymptotically $|\alpha^{in}>$ consists of plane waves plus outgoing waves and $|\alpha^{f}>$ plane waves plus incoming waves. [See pages 114-115.]

These three sets of states

$$\{ \, |\alpha> \} \quad , \quad \{ \, |\alpha^{in}> \} \quad \text{and} \quad \{ \, |\alpha^{f}> \}$$

are all complete and can be set to be orthonormal. The S-matrix is the unitary transformation matrix connecting the in and f states:

$$<\beta \, | \, S \, | \, \alpha> \; = \; <\beta^{f} \, | \alpha^{in}> \; , \tag{24.99}$$

as given by (6.62). The states $|vac>$, $|\pi>$, $|p>$ and $|n>$ used in (24.74)-(24.88) all refer to the physical states; i.e., eigenstates of the total Hamiltonian H. Consequently, in the notations of (24.96)-(24.97), they should be written as

$$| \text{vac}^{in} > \quad \text{or} \quad | \text{vac}^{f} > ,$$

$$| \pi^{in} > \quad \text{or} \quad | \pi^{f} > ,$$

$$| p^{in} > \quad \text{or} \quad | p^{f} > , \quad \text{etc.}$$

Because the vacuum and these single particle states are all stable, the in-states $| a^{in} >$ differ from the corresponding f-states $| a^{f} >$ only by a phase factor. It is immaterial which set we use. However, for the multiparticle states that we shall discuss in the following section, care is necessary and we shall label the superscripts in and f explicitly.

Because the strong-interaction Hamiltonian H is assumed to be chiral symmetric, both the vector current $\vec{v}_\lambda$ and the axial current $\vec{a}_\lambda$ are conserved. In writing down the current-conservation law, such as (24.73), we have implicitly assumed that these operators are in the Heisenberg representation. Hence, the operators *

$$O(t) = v_\lambda^{\ell}(x) , \quad a_\lambda^{\ell}(x) , \quad Q_v^{\ell}(t) , \quad Q_a^{\ell}(t) , \cdots$$

all satisfy Heisenberg's equation (1.9)

$$[H, O(t)] = -i \dot{O}(t) .$$

The corresponding operators $O_I(t)$ in the interaction representation are related to $O(t)$ by

$$O(t) = U^\dagger(t, 0) \, O_I(t) \, U(t, 0) . \tag{24.100}$$

* In the notations used in Section 5.1, an operator O(t) in the Heisenberg representation would be labeled $O_H(t)$. Here, we omit the subscript H.

Exercise. Using (1.9), (5.20) and (24.100), verify that $O_I(t)$ satisfies the equation of motion (5.6) in the interaction representation.

3. Generalization

Let us consider the matrix element of $a_\lambda^\ell(x)$ between any two multiparticle states α and β. The x-dependence of these matrix elements is known. We have

$$<\beta^f \mid a_\lambda^\ell(x) \mid \alpha^{in}> \;\propto\; e^{iq\cdot x} \tag{24.101}$$

where $a_\lambda^\ell(x)$ is in the Heisenberg representation and

$$q_\lambda = (\alpha - \beta)_\lambda$$

is the difference between α_λ, the 4-momentum of the initial state, and β_λ, that of the final state. The Fourier transform of the above matrix element is defined to be

$$<\beta^f \mid a_\lambda^\ell(q) \mid \alpha^{in}> \;\equiv\; \int e^{-iq\cdot x} <\beta^f \mid a_\lambda^\ell(x) \mid \alpha^{in}> d^4x \quad . \tag{24.102}$$

In terms of the corresponding operator $a_\lambda^\ell(x)_I$ in the interaction representation and the free-particle eigenstates of H_0 we can write, on account of (24.96)-(24.97) and (24.100),

$$<\beta^f \mid a_\lambda^\ell(x) \mid \alpha^{in}> \;=\; <\beta \mid U(\infty, t)\; a_\lambda^\ell(x)_I\; U(t, -\infty) \mid \alpha> \tag{24.103}$$

where the components of x are

$$x_\mu = (\vec{r}, it) \quad .$$

By using the Wick theorem and following the steps outlined in Chapter 5, we can express the righthand side of (24.103) in terms of Feynman diagrams. Just as in Figure 24.3, these diagrams can be separated into two classes: (i) those that carry the pion pole factor

$$\frac{-i}{q^2}$$

and (ii) those that do not. Since a pion emission of momentum q is identical to a pion absorption of momentum $-q$, the amplitude for

$$\alpha \rightarrow \beta + \pi^{\ell}(q) \tag{24.104}$$

is equal to

$$\alpha + \pi^{\ell}(-q) \rightarrow \beta \tag{24.105}$$

and will be denoted by

$$< \pi^{\ell}(q)\ \beta^{f} \mid \alpha^{in} > \ = \ < \beta^{f} \mid \pi^{\ell}(-q)\ \alpha^{in} > \quad . \tag{24.106}$$

Similarly to (24.87), by using the product (24.80) times the pion propagator, we can write

$$< \beta^{f} \mid a_{\lambda}^{\ell}(q) \mid \alpha^{in} > \ = \ < \pi^{\ell}(q)\ \beta^{f} \mid \alpha^{in} > f_{\pi} \frac{q_{\lambda}}{q^{2}} + < \beta^{f} \mid \bar{a}_{\lambda}^{\ell}(q) \mid \alpha^{in} > \tag{24.107}$$

where the first term on the righthand side represents the sum of (i) and the second term that of (ii). Because

$$q_{\lambda} < \beta^{f} \mid a_{\lambda}^{\ell}(q) \mid \alpha^{in} > \ = -i \int e^{-iq\cdot x} < \beta^{f} \mid \frac{\partial a_{\lambda}^{\ell}(x)}{\partial x_{\lambda}} \mid \alpha^{in} > d^{4}x \ = \ 0 \ ,$$

through the multiplication of (24.107) by q_{λ} we obtain the generalization of (24.91):

$$q_{\lambda} < \beta^{f} \mid \bar{a}_{\lambda}^{\ell}(q) \mid \alpha^{in} > \ = \ - f_{\pi} < \pi^{\ell}(q)\ \beta^{f} \mid \alpha^{in} > \ = \ - f_{\pi} < \beta^{f} \mid \pi^{\ell}(-q)\ \alpha^{in} > . \tag{24.108}$$

The diagram-decomposition* of (24.107) into (i) + (ii) is

---

* Quite often, one would factor out a phase $e^{i\phi} \equiv < vac \mid U(\infty, -\infty) \mid vac >$ from the diagram representation of the matrix element $< \beta^{f} \mid a_{\lambda}^{\ell}(q) \mid \alpha^{in} >$. However, because $e^{i\phi}$ is a common multiplicative factor present in every term in (24.107), it is immaterial whether we do or do not take it out explicitly.

$$< \beta^{f} \mid a_{\lambda}^{\ell}(q) \mid \alpha^{in} >$$

= 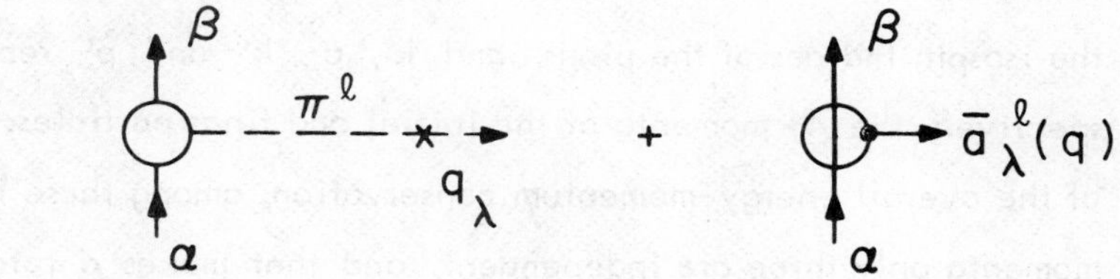

= (i) + (ii)

Fig. 24.4. Decomposition of $< \beta^{f} \mid a_{\lambda}^{\ell}(q) \mid \alpha^{in} >$ into (i) those diagrams that carry a pion pole factor $-i/q^2$ and (ii) those that do not.

illustrated in the above figure, where the rules are:

the pion propagator

- - - - - - - (q) gives $\frac{-i}{q^2}$

the cross vertex

- - -x——> (q_λ) gives $i f_\pi q_\lambda$ (24.109)

and

the dot vertex

•——> $\bar{a}_{\lambda}^{\ell}(q)$ gives the matrix element of $\bar{a}_{\lambda}^{\ell}(q)$.

24.5 Low-energy πN Scattering

Let us consider the low-energy π-nucleon scattering

$$\pi^{\ell}(k) + N(p) \rightarrow \pi^{m}(k') + N'(p') \tag{24.110}$$

where N and N' can be p or n, the superscripts ℓ and m are the isospin indices of the pions, and k, p, k' and p' represent respectively the 4-momenta of the initial and final particles. Because of the overall energy-momentum conservation, among these four 4-momenta only three are independent, and that makes a total of $\frac{1}{2}\cdot 4\cdot 3 = 6$ independent Lorentz scalars, of which we have

$$p^2 = p'^2 = -m_N^2 \tag{24.111}$$

and

$$k^2 = k'^2 = -m_\pi^2 \tag{24.112}$$

where m_N and m_π are the physical nucleon and pion masses. Hence, there are only two independent scalar variables in the problem; these can be chosen to be

$$\nu \equiv k\cdot p$$

and

$$\nu' \equiv k\cdot p' \ . \tag{24.113}$$

1. Matrix element

The scattering matrix element can be written as *

$$< \pi^{m}(k')\, N'(p') \,|\, S \,|\, \pi^{\ell}(k)\, N(p) >$$

$$\equiv \frac{1}{2\sqrt{k_0 k_0'}}\,(2\pi)^4\,\delta^4(k+p-k'-p')\,M \tag{24.114}$$

where the amplitude M is, apart from a covariant factor dependent

---

* Cf. (5.106) and (5.108). The factor $(2\sqrt{k_0 k_0'})^{-1}$ is due to the initial and final pions. For convenience, we have set the volume of the system = unity.

on the nucleon spinors, an invariant function of ν and ν' (and, of course, also of k^2, k'^2, p^2 and p'^2). In the limit of chiral $SU_2 \times SU_2$ symmetry, we have instead of (24.112),

$$k^2 = k'^2 = 0$$

which means that in evaluating M we should keep the variables ν and ν' (also p^2 and p'^2) fixed and let the pion mass approach 0; i.e.,

$$M = M(\nu, \nu') \equiv \lim_{\substack{k^2=0 \\ k'^2=0}} M(\nu, \nu'; k^2, k'^2) \quad .$$

To derive $M(\nu, \nu')$ in the low-energy region, let us consider the matrix element

$$< f \mid T(a_\lambda^{\ell}(x)\; b(x')) \mid in > \tag{24.115}$$

where $b(x')$ can be any local operator, say $a_\mu^{m}(x')$ or its derivative, $\mid in >$ and $\mid f >$ can be any initial and final states given by (24.98), the components of x and x' are, as before,

$$x_\mu = (\vec{r}, it) \quad \text{and} \quad x_\mu' = (\vec{r}', it') \quad ,$$

and T denotes the time-ordered product. In accordance with (5.38), we write

$$T(a_\lambda^{\ell}(x)\; b(x')) = \begin{cases} a_\lambda^{\ell}(x)\; b(x') & \text{if} \quad t \geqslant t' \\ b(x')\; a_\lambda^{\ell}(x) & \text{if} \quad t < t' \quad , \end{cases}$$

which is a step function at $t = t'$. For $\lambda = 4$ the discontinuity of this step function is the commutator

$$[\, a_4^{\ell}(\vec{r}, t) \; , \; b(\vec{r}', t)] \quad .$$

Hence, by differentiating (24.115) with respect to x_λ, we obtain

$$\frac{\partial}{\partial x_\lambda} < f \mid T(a_\lambda^{\ell}(x)\; b(x')) \mid in >$$

$$= < f \mid T (\frac{\partial a_\lambda^{\ell}(x)}{\partial x_\lambda} \; b(x')) \mid in >$$

$$+ < f \mid [a_0^{\ell}(\vec{r}, t), \; b(\vec{r}', t)] \mid in > \; \delta(t-t') \qquad (24.116)$$

where

$$a_4^{\ell}(x) \;=\; i \, a_0^{\ell}(x) \; .$$

We now set

$$b(x') \;=\; a_\mu^m(x') \quad , \quad \mid in > \;=\; \mid N(p)^{in} >$$

$$\text{and} \qquad \mid f > \;=\; \mid N'(p')^f > \; .$$

Because of the current algebra, (24.40), and CAC, (24.73), equation (24.116) becomes

$$\frac{\partial}{\partial x_\lambda} < N'(p')^f \mid T(a_\lambda^{\ell}(x) \; a_\mu^m(x')) \mid N(p)^{in} >$$

$$= i \, \epsilon^{\ell m n} < N'(p')^f \mid v_\mu^n(x) \mid N(p)^{in} > \; \delta^4(x-x')$$

$$+ \; s.\, t. \qquad (24.117)$$

where s. t. denotes the Schwinger term which, as shown by (24.39) and (24.41), is manifestly noncovariant under a Lorentz transformation. In (24.116) and (24.117) the current operators are all in the Heisenberg representation. By using (24.96), (24.97) and (24.100), we can write

$$< N'(p')^f \mid T(a_\lambda^{\ell}(x) \; a_\mu^m(x')) \mid N(p)^{in} >$$

$$= \; < N'(p') \mid T(U(\infty, -\infty) \; a_\lambda^{\ell}(x)_I \; a_\mu^m(x')_I) \mid N(p) > \qquad (24.118)$$

where the subscript I denotes the interaction representation and the T product means that we should substitute the series expansion (5.70) for $U(\infty, -\infty)$ into (24.118) and then arrange the products of

$$H_{int}(t_1), \cdots, H_{int}(t_n), \quad a_\lambda^{\ell}(x)_I \quad \text{and} \quad a_\mu^{m}(x')_I$$

in time-ordered sequences. By using the Wick theorem, we can convert the righthand side into Feynman diagrams; in general, they will consist of the usual covariant diagrams (given in Figure 24.5) plus some manifestly noncovariant terms.* Delete these noncovariant terms and define the remaining part of the righthand side of (24.118) to be

$$< N'(p')^f \,|\, T(a_\lambda^{\ell}(x)\; a_\mu^{m}(x')) \,|\, N(p)^{in} >_{cov} \;. \qquad (24.119)$$

By equating the covariant parts on both sides of (24.117), we derive **

$$\frac{\partial}{\partial x_\lambda} < N'(p')^f \,|\, T(a_\lambda^{\ell}(x)\; a_\mu^{m}(x')) \,|\, N(p)^{in} >_{cov}$$
$$= i\,\epsilon^{\ell m n} < N'(p')^f \,|\, v_\mu^{n}(x) \,|\, N(p)^{in} > \delta^4(x-x') \;. \qquad (24.120)$$

As in (24.102), we introduce the Fourier transforms

$$< N'(p')^f \,|\, v_\mu^{n}(p-p') \,|\, N(p)^{in} >$$
$$\equiv \int e^{-i(p-p')\cdot x} < N'(p')^f \,|\, v_\mu^{n}(x) \,|\, N(p)^{in} > d^4x \qquad (24.121)$$

and

$$< N'(p')^f \,|\, T(a_\lambda^{\ell}(-k)\; a_\mu^{m}(k')) \,|\, N(p)^{in} >$$
$$\equiv \int e^{i(k\cdot x - k'\cdot x')} < N'(p')^f \,|\, T(a_\lambda^{\ell}(x)\; a_\mu^{m}(x')) \,|\, N(p)^{in} >_{cov} \cdot d^4x\, d^4x' \;. \qquad (24.122)$$

---

* See the Remark at the end of this section for further discussion.

** Equation (24.117) implies that, in addition to (24.120), the noncovariant terms on both sides must also equal each other. Hence, the Schwinger term is exactly cancelled by that due to the noncovariant part in (24.118), i.e., the difference between (24.118) and (24.119).

$$< N'(p')^f \,|\, T(a_\lambda^\ell(-k) \;\; a_\mu^m(k')) \,|\, N(p)^{in} >$$

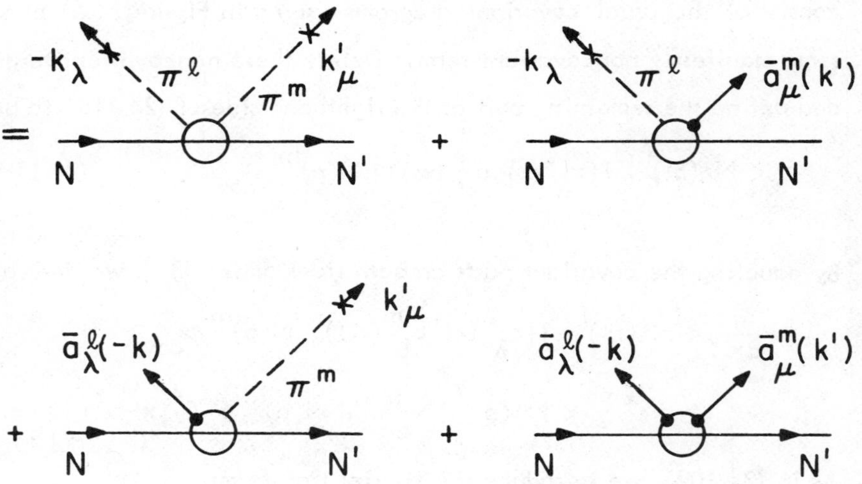

Fig. 24.5. Graphic representation of (24.122) and (24.127).

We multiply both sides of (24.120) by a factor

$$i k'_\mu \int e^{i(k \cdot x - k' \cdot x')} \, d^4x \, d^4x'$$

and then integrate. Since $k - k' = p' - p$, the result is

$$k'_\mu k_\lambda < N'(p')^f \,|\, T(a_\lambda^\ell(-k) \;\; a_\mu^m(k')) \,|\, N(p)^{in} >$$
$$= -k'_\mu \, \epsilon^{\ell mn} < N'(p')^f \,|\, v_\mu^n(p - p') \,|\, N(p)^{in} > \quad . \qquad (24.123)$$

2. Reduction

The Feynman diagrams for (24.122) can be grouped into three types according to factors of pion propagators: those that carry

(i) the product of two pion pole factors

$$\frac{-i}{k^2} \cdot \frac{-i}{k'^2} \quad ,$$

(ii) only one pion pole factor

$$\frac{-i}{k^2} \quad \text{or} \quad \frac{-i}{k'^2}$$

and (iii) none. These are illustrated by Figure 24.5 in which by using the rules (24.109), we see that the first diagram gives

$$(i) = f_\pi^2 \frac{-k_\lambda k'_\mu}{k^2 k'^2} < \pi^m(k') N'(p')^f \mid \pi^\ell(k) N(p)^{in} > , \tag{24.124}$$

the sum of the second and third diagrams is

$$(ii) = f_\pi \frac{-k_\lambda}{k^2} < N'(p')^f \mid \bar{a}_\mu^m(k') \mid \pi^\ell(k) N(p)^{in} > + f_\pi \frac{k'_\mu}{k'^2} < \pi^m(k') N'(p')^f \mid \bar{a}_\lambda^\ell(-k) \mid N(p)^{in} > , \tag{24.125}$$

and the fourth diagram gives

$$(iii) = < N'(p')^f \mid T(\bar{a}_\lambda^\ell(-k)\ \bar{a}_\mu^m(k')) \mid N(p)^{in} > . \tag{24.126}$$

The matrix element (24.122) is the sum of all these three terms:

$$< N'(p')^f \mid T(a_\lambda^\ell(-k)\ a_\mu^m(k')) \mid N(p)^{in} > \ = (i) + (ii) + (iii) . \tag{24.127}$$

From (24.108), it follows that

$$\begin{aligned} - f_\pi < \pi^m(k') N'(p')^f \mid \pi^\ell(k) N(p)^{in} > \\ = k'_\mu < N'(p)^f \mid \bar{a}_\mu^m(k') \mid \pi^\ell(k) N(p)^{in} > \\ = - k_\lambda < \pi^m(k') N'(p')^f \mid \bar{a}_\lambda^\ell(-k) \mid N(p)^{in} > . \end{aligned} \tag{24.128}$$

Multiplying (24.124) and (24.125) by $k_\lambda k'_\mu$ and using (24.128), we see that

$$k_\lambda k'_\mu [(i) + (ii)] = f_\pi^2 < \pi^m(k')\, N'(p')^f \mid \pi^\ell(k)\, N(p)^{in} > .$$

Now, substitute (24.127) into (24.123). By using the above expression, we derive, for the covariant amplitude M defined * in (24.114)

$$\begin{aligned} f_\pi^2 M &= f_\pi^2 < \pi^m(k')\, N'(p')^f \mid \pi^\ell(k)\, N(p)^{in} > \\ &= -k'_\mu\, \epsilon^{\ell mn} < N'(p')^f \mid v_\mu^n(p-p') \mid N(p)^{in} > \\ &\quad - k_\lambda k'_\mu < N'(p')^f \mid T(\bar{a}_\lambda^\ell(-k)\, \bar{a}_\mu^m(k')) \mid N(p)^{in} > . \end{aligned} \tag{24.129}$$

3. s-wave scattering length

At very low energy, we regard the components k_λ and k'_μ (in the rest system of the nucleon) as infinitesimals:

$$k_\lambda = O(\epsilon) \quad \text{and} \quad k'_\mu = O(\epsilon)$$

where $\epsilon \to 0$. The scattering process is dominated by the s-wave. As k_λ and $k'_\mu \to 0$, we have

$$p' \to p = (0, i\, m_N)$$

and (24.130)

$$\epsilon^{\ell mn} < N'(p')^f \mid v_\mu^n(p-p') \mid N(p)^{in} > \to \begin{cases} 0 & \text{if } \mu \neq 4 \\ < i\, \epsilon^{\ell mn}\, \tfrac{1}{2}\tau^n > & \text{if } \mu = 4 \end{cases}$$

where τ^n is the nuclear isospin matrix and $< >$ denotes its appropriate matrix element. Because the nucleons can be regarded as fixed particles, when $\epsilon \to 0$ the initial and final pion energies become the same,

$$k'_4 = k_4 = i\, k_0 , \tag{24.131}$$

* As in M, the factors of external pion amplitudes, $(2k_0)^{-\frac{1}{2}}$ and $(2k'_0)^{-\frac{1}{2}}$, are also not included in (24.124)-(24.129).

and (24.129) reduces to

$$f_\pi^2 M = k_0 < \epsilon^{\ell mn} \tfrac{1}{2}\tau^n > + O(\epsilon^2) \qquad (24.132)$$

where the $O(\epsilon^2)$ term can be neglected. In deriving this, we note that the second term on the righthand side of (24.129) is proportional to $k_\lambda k'_\mu$ and * therefore $O(\epsilon^2)$.

Next, we have to evaluate the isospin matrix element. Let $| I, I_z >$ be the isospin part of the $\pi - N$ wave function where, as in Chapter 11, I is the total isospin quantum number and I_z its z-component. For $I_z = -\frac{1}{2}$, these states are (see (11.108))

$$| \tfrac{1}{2}, -\tfrac{1}{2} > = \sqrt{\tfrac{1}{3}}\, | n\pi^o > - \sqrt{\tfrac{2}{3}}\, | p\pi^- >$$

and

$$| \tfrac{3}{2}, -\tfrac{1}{2} > = \sqrt{\tfrac{2}{3}}\, | n\pi^o > + \sqrt{\tfrac{1}{3}}\, | p\pi^- > .$$

Furthermore, by using the relations between $| \pi^- >$, $| \pi^o >$ and $| \pi^\ell >$ given on page 792, we find

$$< p\pi^- | \epsilon^{\ell mn} \tfrac{1}{2}\tau^n | p\pi^- > = i\,\epsilon^{12n} < p | \tfrac{1}{2}\tau^n | p > = \tfrac{1}{2} i ,$$

$$< n\pi^o | \epsilon^{\ell mn} \tfrac{1}{2}\tau^n | n\pi^o > = \epsilon^{33n} < n | \tfrac{1}{2}\tau^n | n > = 0 .$$

Because $< \epsilon^{\ell mn} \tau^n >$ refers to the matrix element of an isoscalar **, we have

---

* Because the pions are pseudoscalars, the s-wave scattering amplitude does not carry a nucleon pole factor. Otherwise, there would be a factor $[(p+k)^2 + m_N^2]^{-1} = (2p\cdot k)^{-1}$ which might change the estimate.

** Properly written, this isoscalar is $a^{m\dagger} a^\ell \epsilon^{\ell mn} \tau^n$ where $a^{m\dagger}$ and a^ℓ denote the appropriate pion creation and annihilation operators.

$$< \tfrac{3}{2}, -\tfrac{1}{2} \mid \epsilon^{\ell mn} \tfrac{1}{2} \tau^n \mid \tfrac{1}{2}, -\tfrac{1}{2} > \; = \; 0 \; ;$$

therefore, it follows that

$$< \tfrac{3}{2}, -\tfrac{1}{2} \mid \epsilon^{\ell mn} \tfrac{1}{2} \tau^n \mid \tfrac{3}{2}, -\tfrac{1}{2} > \; = \; -\frac{i}{2}$$

and

$$< \tfrac{1}{2}, -\tfrac{1}{2} \mid \epsilon^{\ell mn} \tfrac{1}{2} \tau^n \mid \tfrac{1}{2}, -\tfrac{1}{2} > \; = \; i \tag{24.133}$$

in which the righthand sides are clearly independent of I_z. By using (24.132)-(24.133), we find that for $I = \frac{3}{2}$ the amplitude M is *

$$M_{\frac{3}{2}} = -\tfrac{1}{2} i \frac{k_0}{f_\pi^2} \tag{24.134}$$

and for $I = \frac{1}{2}$ it is

$$M_{\frac{1}{2}} = i \frac{k_0}{f_\pi^2} \tag{24.135}$$

where the $O(\epsilon^2)$ term is neglected. It is convenient to absorb the factor

$$\frac{1}{2\sqrt{k_0 k_0'}} = \frac{1}{2k_0}$$

in (24.114) into the scattering amplitude and define

$$\mathcal{M}_I \equiv \frac{1}{2\sqrt{k_0 k_0'}} M_I \; .$$

Equations (24.134) and (24.135) can then be written as

$$\mathcal{M}_I = \frac{i}{2f_\pi^2} \cdot \begin{cases} -\tfrac{1}{2} & \text{for } I = \tfrac{3}{2} \\ 1 & \text{for } I = \tfrac{1}{2} \; . \end{cases} \tag{24.136}$$

Let us return to the physical πN scattering (24.110) in which the pion mass is

$$m_\pi \neq 0 \; .$$

---

* In terms of invariants, $k_0 = -k \cdot p / m_N$.

Suppose that the πN system is in an eigenstate of total isospin I. The scattering cross section is, according to (5.94),

$$\sigma_I = 2\pi \int \frac{d^3 k'}{8\pi^3 v} \delta(k_0' - k_0) \, |\mathcal{M}_I|^2$$

where v is the pion velocity and

$$k_0 = \sqrt{\vec{k}^2 + m_\pi^2} \quad .$$

When $\vec{k} \to 0$, we have

$$\sigma_I = \frac{m_\pi^2}{\pi} |\mathcal{M}_I|^2 \equiv 4\pi a_I^2$$

where a_I is the scattering length. Adopting the usual convention that the scattering length is positive for a repulsive potential (i.e., the phase shift is negative), we have

$$a_I = i \frac{m_\pi}{2\pi} \mathcal{M}_I \tag{24.137}$$

where the factor i is due to the fact that $\mathcal{M}_I$ has the same phase as $-i H_{int}$. From (24.136) we see that *

$$a_{\frac{1}{2}} = - \frac{m_\pi}{4\pi f_\pi^2} = -.18/m_\pi$$

and (24.138)

$$a_{\frac{3}{2}} = \frac{m_\pi}{8\pi f_\pi^2} = .09/m_\pi$$

which agree quite well with the experimental values **

$$(a_{\frac{1}{2}})_{exp} = -(.171 \pm .005)/m_\pi$$

---

* S. Weinberg, Phys.Rev.Lett. 17, 616 (1966). Y. Tomozawa, Nuovo Cimento 46A, 707 (1966).

** S. W. Barnes, H. Winick, K. Miyake and K. Kinsey, Phys.Rev. 117, 238 (1960).

and

$$(a_{\frac{3}{2}})_{exp} = (.088 \pm .004)/m_\pi \ .$$

Remarks. Since (24.117) is an identity, the equality holds separately for both the covariant and the noncovariant parts of both sides. As remarked before, the Schwinger term on its righthand side must then be cancelled by a corresponding term on its lefthand side, which is due to the difference between (24.118) and (24.119). To illustrate why a matrix element such as

$$< f \mid T(a_\lambda^{\ell}(x)\ a_\mu^{m}(0)) \mid in > \qquad (24.139)$$

should carry a noncovariant component, we may consider an elementary example. In (24.139), let the initial and final states be the vacuum and the local axial operator be simply a canonical massive axial-vector field. To make matters even simpler, we assume the axial-vector field to be a free massive spin-1 field (in which case there is no difference between a vector and an axial-vector field). We may then use the Lagrangian density (4.1) and set $a_\mu^m(x)$ to be the field $A_\mu(x)$ in Chapter 4. The matrix element of interest becomes the Dyson-Wick contraction

$$\underbrace{A_\mu(x)\ A_\nu(0)} = < vac \mid T(A_\mu(x)\ A_\nu(0)) \mid vac> \ . \qquad (24.140)$$

From (4.7), (4.11) and (4.12) it follows that, in the notations of Chapter 4,

$$\vec{A}(x) = \sum_{\vec{k}} \frac{1}{\sqrt{2\omega\Omega}} \left\{ \left[\hat{k}\, a_L(\vec{k})\, \frac{\omega}{m} + \sum_{T=1,2} \hat{e}_T\, a_T(\vec{k}) \right] e^{i\vec{k}\cdot\vec{r} - i\omega t} + \text{h.c.} \right\}$$

and

$$A_4(x) = \sum_{\vec{k}} \frac{1}{\sqrt{2\omega\Omega}} \left\{ i\, a_L(\vec{k})\, \frac{|\vec{k}|}{m}\, e^{i\vec{k}\cdot\vec{r} - i\omega t} - \text{h.c.} \right\} .$$

Since

$$T(A_\mu(x)\ A_\nu(0)) \equiv \begin{cases} A_\mu(x)\ A_\nu(0) & \text{if } t \geqslant 0 \\ A_\nu(0)\ A_\mu(x) & \text{if } t < 0\ , \end{cases} \tag{24.141}$$

we have

$$\underbrace{A_\mu(x)\ A_\nu(0)} = \sum_{\vec{k}} \frac{1}{2\omega\Omega}\, e^{i\vec{k}\cdot\vec{r}}$$

$$\cdot \begin{cases} (\delta_{\mu\nu} + q_\mu q_\nu m^{-2})\ e^{-i\omega t} & \text{if } t \geqslant 0 \\ (\delta_{\mu\nu} + q_\mu^* q_\nu^* m^{-2})\ e^{i\omega t} & \text{if } t < 0 \end{cases} \tag{24.142}$$

where

$$q_4 = i\omega = i\sqrt{\vec{k}^{\,2} + m^2}\ ,$$

$$q_j = k_j \qquad (j = 1, 2, 3)$$

and q_μ^* is the complex conjugate of q_μ. Upon converting the summand in (24.142) into a Feynman-type integral, we find

$$\underbrace{A_\mu(x)\ A_\nu(0)} = -i \sum_{\vec{k}} \frac{1}{2\pi\Omega} \int_{-\infty}^{\infty} dk_0$$

$$\cdot \left[\frac{\delta_{\mu\nu} + m^{-2} k_\mu k_\nu}{k^2 + (m - i\epsilon)^2} - \frac{1}{m^2}\, \delta_{4\mu}\, \delta_{4\nu} \right] e^{ik\cdot x}$$

where $k_4 = i k_0$. Taking the limit $\Omega \to \infty$, we obtain

$$\underbrace{A_\mu(x)\ A_\nu(0)} = \overbrace{A_\mu(x)\ A_\nu(0)} + \frac{i}{m^2}\, \delta_{4\mu}\, \delta_{4\nu}\, \delta^4(x) \tag{24.143}$$

where

$$\overbrace{A_\mu(x)\ A_\nu(0)} = -i\, \frac{1}{(2\pi)^4} \int \frac{\delta_{\mu\nu} + m^{-2} k_\mu k_\nu}{k^2 + (m - i\epsilon)^2}\, d^4k \tag{24.144}$$

is the covariant path-integration contraction given by (22.80). Note that (24.143) remains valid if we alter the definition of the T product (24.141) to

$$T(A_\mu(x)\, A_\nu(0)) = \begin{cases} A_\mu(x)\, A_\nu(0) & \text{if} \quad t > 0 \\ A_\nu(0)\, A_\mu(x) & \text{if} \quad t \leqslant 0 \quad . \end{cases}$$

Thus, we see that the Dyson-Wick contraction (24.143), typical of matrix elements of the form (24.139), indeed carries a manifestly non-covariant term which is zero if $x \neq 0$, similar to the Schwinger term.

24.6 PCAC

In this section we consider the effects of nonzero quark masses and, consequently, nonzero pion and kaon masses.

1. Chiral symmetry breaking

We discuss first the breaking of the chiral $SU_2 \times SU_2$ symmetry. Assume that the up and down quark masses are equal but nonzero; i.e.,

$$m_u = m_d \neq 0 \tag{24.145}$$

in the QCD Lagrangian. [See, e.g., (18.3).] Thus, the isospin SU_2 symmetry remains intact. By using the equation of motion (18.6a), we find the axial-current conservation law (24.73) is replaced by

$$\frac{\partial}{\partial x_\lambda} \vec{a}_\lambda = i m_u q^\dagger \gamma_4 \gamma_5 \vec{\tau} q \tag{24.146}$$

where q is the quark field defined by (24.71).

Consider now the π decay amplitude formula (21.19),

$$< vac \mid A_\lambda^{\pm}(x) \mid \pi^{\mp} > = i \cos\theta_c F_\pi k_\lambda \frac{1}{\sqrt{2\omega\Omega}} e^{ik\cdot x} \quad .$$

We recall that it was orignally derived in Chapter 21 for the physical pion state with the pion 4-momentum k satisfying

$$k_4 = i\omega = i\sqrt{\vec{k}^{\,2} + m_\pi^{\,2}}$$

and

$$-k^2 = m_\pi^{\ 2} = (\text{physical pion mass})^2 \quad . \tag{24.147}$$

Taking the divergence of $A_\lambda^{\pm}(x)$, we have

$$< \text{vac} \left| \frac{\partial A_\lambda^{\pm}(x)}{\partial x_\lambda} \right| \pi^{\mp} > \;=\; \cos\theta_c \, F_\pi m_\pi^{\ 2} \frac{1}{\sqrt{2\omega\Omega}} \, e^{ik\cdot x} \; .$$

Converting $A_\lambda^{\pm}$ into $\vec{a}_\lambda$ by using (24.75), we find

$$< \text{vac} \left| \frac{\partial a_\lambda^{\ell}(x)}{\partial x_\lambda} \right| \pi^{\ell'} > \;=\; \delta^{\ell\ell'} f_\pi m_\pi^{\ 2} \frac{1}{\sqrt{2\omega\Omega}} \, e^{ik\cdot x} \quad . \tag{24.148}$$

On the phenomenological level, we may introduce an isovector pion field operator $\vec{\phi}_\pi(x)$. The necessary conditions are that it should be a local operator and have a nonzero matrix element between the vacuum and a physical single pion state; in addition, its normalization is

$$< \text{vac} \mid \phi^{\ell}(x) \mid \pi^{\ell'} > \;=\; \delta^{\ell\ell'} \frac{1}{\sqrt{2\omega\Omega}} \, e^{ik\cdot x} \; . \tag{24.149}$$

We may relax our requirement by regarding these necessary conditions as sufficient conditions, and call any such local operator an <u>interpolating</u> field for the pion. With this understanding, (24.148) and (24.149) yield

$$\frac{\partial}{\partial x_\lambda} \vec{a}_\lambda(x) \;=\; f_\pi m_\pi \vec{\phi}_\pi(x) \quad , \tag{24.150}$$

which will be referred to as the basic equation for PCAC (partially conserved axial current *). Combining (24.146) and (24.150), we see that the pion interpolating field is also given by

---

* M. Gell-Mann and M. Lévy, Nuovo Cimento 16, 705 (1960). J. Bernstein, S. Fubini, M. Gell-Mann and W. Thirring, Nuovo Cimento 17, 757 (1960). Chou Kuang-chao, Soviet Physics 12, 492 (1961) [JETP 39, 703 (1963)] .

$$\vec{\phi}_\pi = i \frac{m_u}{f_\pi m_\pi^2} q^\dagger \gamma_4 \gamma_5 \vec{\tau} q \quad .$$

The generalization to chiral SU_3 symmetry breaking is straightforward. We introduce the mass matrix

$$m = \begin{pmatrix} m_u & 0 & 0 \\ 0 & m_u & 0 \\ 0 & 0 & m_s \end{pmatrix} , \tag{24.151}$$

in which u and d have the same mass but s has a different one. Thus, the flavor SU_3 symmetry is also broken, although the isospin symmetry is intact. Instead of (24.146), we now have

$$\frac{\partial}{\partial x_\lambda} a_\lambda^\ell = i \tfrac{1}{2} q^\dagger \gamma_4 \gamma_5 (m \lambda^\ell + \lambda^\ell m) q \tag{24.152}$$

where $q(x)$ and $a_\lambda^\ell(x)$ are given by (24.1) and (24.4).

For $\ell = 1, 2$ and 3, (24.148) remains valid, consequently so is (24.150). For $\ell = 4, 5, 6$ and 7, we use the expressions for K-decay given on pages 605–6; by following the same arguments leading from the π-decay amplitude (21.19) to (24.148), we find that the K-decay amplitude (21.23) gives

$$< vac \mid \frac{\partial a_\lambda^\ell(x)}{\partial x_\lambda} \mid K^{\ell'} > = \delta^{\ell \ell'} f_\pi m_K^2 \frac{1}{\sqrt{2\omega\Omega}} e^{ik \cdot x} \tag{24.153}$$

where

$$\ell \text{ and } \ell' = 4, 5, 6, 7,$$

$$k_4 = i\omega = \sqrt{\vec{k}^2 + m_K^2} , \tag{24.154}$$

$$-k^2 = m_K^2$$

with

$$m_K = \text{physical kaon mass} \quad .$$

Because of the Cabibbo theory, the constant in (24.153) is the same pion decay constant $f_\pi \cong 93$ MeV. Similarly to (24.150) we may introduce the kaon interpolating field $\phi_K^\ell(x)$ which satisfies the PCAC equation

$$\frac{\partial}{\partial x_\lambda} a_\lambda^\ell(x) = f_\pi m_K^2 \phi_K^\ell(x) \tag{24.155}$$

and the normalization condition

$$< 0 \mid \phi_K^\ell(x) \mid K^{\ell'} > = \delta^{\ell\ell'} \frac{1}{\sqrt{2\omega\Omega}} e^{ik\cdot x}$$

with ℓ, ℓ', ω and k given by (24.154).

2. Relations between quark and pseudoscalar masses

By treating the quark mass matrix m, (24.151), as a perturbation, we can study its effect on the physical pion and kaon masses. When $m = 0$, both the pion and the kaon become zero-mass Goldstone bosons. In this section, we shall calculate m_π^2 and m_K^2 to the first order in m. We begin our analysis by setting in (24.116)

$$b(x') = \frac{\partial a_\mu^{\ell'}(x')}{\partial x'_\mu}$$

and

$$\mid in > = \mid f > = \mid vac > .$$

This gives

$$\frac{\partial}{\partial x_\lambda} < vac \mid T(a_\lambda^\ell(x) \frac{\partial a_\mu^{\ell'}(x')}{\partial x'_\mu}) \mid vac >$$
$$= < vac \mid T(\frac{\partial a_\lambda^\ell(x)}{\partial x_\lambda} \frac{\partial a_\mu^{\ell'}(x')}{\partial x'_\mu}) \mid vac >$$
$$+ < vac \mid [a_0^\ell(x), \frac{\partial a_\mu^{\ell'}(x')}{\partial x'_\mu}] \mid vac > \delta(t-t') . \tag{24.156}$$

Let us first examine the pion sector by considering the case

ℓ and $\ell' = 1, 2, 3$.

Substituting (24.150) into the lefthand side of (24.156) and (24.146) into the righthand side, we find

$$f_\pi m_\pi^2 \frac{\partial}{\partial x_\lambda} < vac \,|\, T(a_\lambda^\ell(x) \;\; \phi_\pi^{\ell'}(x')) \,|\, vac >$$

$$= i m_u < vac \,|\, \left[a_0^\ell(\vec{r}, t), \;\; q^\dagger(\vec{r}', t) \, \gamma_4 \gamma_5 \tau^{\ell'} \; q(\vec{r}', t) \right] \,|\, vac >$$

$$\cdot \; \delta(t - t') + O(m_u^2) \qquad (24.157)$$

where to $O(m_u)$, we may evaluate the vacuum expectations on both sides in the limit of perfect chiral symmetry. Since

$$[\tfrac{1}{2} \gamma_5 \tau^\ell, \;\; \gamma_4 \gamma_5 \tau^{\ell'}] \;=\; -\tfrac{1}{2} \gamma_4 (\tau^\ell \tau^{\ell'} + \tau^{\ell'} \tau^\ell)$$

$$= \; - \gamma_4 \, \delta^{\ell \ell'} \;\; ,$$

the righthand side of (24.157) is, on account of (3.24a) and (24.70)-(24.71),

$$- i m_u < vac \,|\, \psi_u^{c\dagger} \gamma_4 \psi_u^c + \psi_d^c \gamma_4 \psi_d^c \,|\, vac > \delta^4(x - x') \; \delta^{\ell \ell'}$$

$$= - 2i \, m_u \, \rho \, \delta^{\ell \ell'} \qquad (24.158)$$

where

$$\rho \;=\; < vac \,|\, \psi_u^{c\dagger} \gamma_4 \psi_u^c \,|\, vac > \delta^4(x - x')$$

$$= \; < vac \,|\, \psi_d^{c\dagger} \gamma_4 \psi_d^c \,|\, vac > \delta^4(x - x') \;\; . \qquad (24.159)$$

Because of isospin conservation, we have

$$< vac \,|\, T(a_\lambda^\ell(x) \; \phi_\pi^{\ell'}(x')) \,|\, vac > \; = \; 0 \qquad \text{if} \quad \ell \neq \ell' .$$

Hence by defining, in the limit of isospin symmetry,

$$- i \sigma \delta^{\ell \ell'} \;\equiv\; \frac{\partial}{\partial x_\lambda} < vac \,|\, T(a_\lambda^\ell(x) \; \phi_\pi^{\ell'}(x')) \,|\, vac > \qquad (24.160)$$

and neglecting $O(m_u^2)$, we can rewrite (24.157) as

$$f_\pi m_\pi^2 \, \sigma \;=\; 2 m_u \, \rho \;\; . \qquad (24.161)$$

In exactly the same way, we can generalize our discussions to the kaon sector. For

$$\ell \quad \text{and} \quad \ell' = 4, 5, 6, 7, \tag{24.162}$$

and on account of (24.152) and (24.155), (24.157) is replaced by

$$f_\pi m_K^2 \frac{\partial}{\partial x_\lambda} < vac \,|\, T(a_\lambda^{\ell}(x)\ \phi_K^{\ell'}(x'))\,|\ vac >$$
$$= i\tfrac{1}{2} < vac \,|\, [a_0^{\ell}(\vec{r}, t),\ q^\dagger(\vec{r}', t)\ \gamma_4 \gamma_5 (m\lambda^{\ell'} + \lambda^{\ell'} m)\, q(\vec{r}', t)] \cdot |\, vac >$$
$$+ \ O(m^2) \tag{24.163}$$

where, according to (24.151),

$$m = \tfrac{1}{3}(2m_u + m_s) + \frac{1}{\sqrt{3}}(m_u - m_s)\,\lambda^8 \ . \tag{24.164}$$

From (12.23)-(12.24), (24.162) and (24.164), it follows that

$$\tfrac{1}{2}(m\lambda^{\ell} + \lambda^{\ell} m) = \tfrac{1}{2}(m_u + m_s)\,\lambda^{\ell}$$

and

$$[\tfrac{1}{2}\gamma_5 \lambda^{\ell},\ \gamma_4 \gamma_5 \lambda^{\ell'}]$$
$$= \begin{cases} -\gamma_4 \begin{pmatrix} 1 & 0 & 0 \\ 0 & 0 & 0 \\ 0 & 0 & 1 \end{pmatrix}, & \text{if } \ell = \ell' = 4 \text{ or } 5 \\[2ex] -\gamma_4 \begin{pmatrix} 0 & 0 & 0 \\ 0 & 1 & 0 \\ 0 & 0 & 1 \end{pmatrix}, & \text{if } \ell = \ell' = 6 \text{ or } 7 \\[2ex] \text{matrix with zero diagonal elements, otherwise.} \end{cases}$$

In the limit of flavor - SU_3 symmetry, we have

$$< vac \mid \psi_u^{c\dagger} \gamma_4 \psi_u^c \mid vac > \;=\; < vac \mid \psi_d^{c\dagger} \gamma_4 \psi_d^c \mid vac >$$
$$= \; < vac \mid \psi_s^{c\dagger} \gamma_4 \psi_s^c \mid vac >$$

and

$$< vac \mid \psi_u^{c\dagger} \gamma_4 \psi_d^c \mid vac > \;=\; < vac \mid \psi_d^{c\dagger} \gamma_4 \psi_s^c \mid vac >$$
$$= \; < vac \mid \psi_s^{c\dagger} \gamma_4 \psi_u^c \mid vac >$$
$$= \; 0 \; ,$$

and therefore the righthand side of (24.163) is

$$- i \, (m_u + m_s) \, \rho \; \delta^{\ell \ell'}$$

where ρ is given by (24.159); in the same limit we also have

$$- i \, \sigma \, \delta^{\ell \ell'} \;=\; \frac{\partial}{\partial x_\lambda} < vac \mid T(a_\lambda^{\ell}(x) \; \phi_K^{\ell'}(x')) \mid vac >$$

where σ is given by (24.160). Thus, neglecting $O(m^2)$, (24.163) gives

$$f_\pi \, m_K^2 \, \sigma \;=\; (m_u + m_s) \, \rho \; . \tag{24.165}$$

Combining with (24.161), we derive

$$\left(\frac{m_\pi}{m_K} \right)^2 \;=\; \frac{2 m_u}{m_u + m_s} \tag{24.166}$$

or

$$\tfrac{1}{2} \left(\frac{m_\pi}{m_K} \right)^2 \;\cong\; \frac{m_u}{m_s} \tag{24.167}$$

on account of $m_s \gg m_u$.

The above relation has been used in Section 20.6.2 to determine the quark masses. [See (20.73)-(20.74).]

Problem 24.1. Show that the low-energy s-wave $\pi\pi$ scattering lengths are

$$a_I = \frac{m_\pi}{32\pi f_\pi^2} \begin{cases} -7 & \text{for } I = 0 \\ 2 & \text{for } I = 2 \end{cases}$$

where I is the total isospin.

[See S. Weinberg, loc. cit., p. 813.]

References

S. L. Adler and R. F. Dashen, Current Algebra (New York, Benjamin Inc., 1968).

J. Bernstein, Elementary Particles and Their Currents (New York, W. H. Freeman and Co., 1968).

M. Weinstein, Chiral Symmetry, Springer Tracts in Modern Physics, Vol. 60 (Berlin, Heidelberg and New York, Springer-Verlag, 1971).

B. W. Lee, Chiral Dynamics (New York, Gordon and Breach Science Publishers, 1972).

S. B. Treiman, R. Jackiw and D. J. Gross, Lectures on Current Algebra and Its Applications (Princeton, Princeton University Press, 1972).

Chapter 25

OUTLOOK

At the present it appears that we have arrived at a closed system of physical laws, with QCD for the strong interaction and a unifying gauge theory for the weak and electromagnetic interactions,
plus of course the Einstein theory of gravitation. However, we are
still in urgent need of most of the crucial proofs of the theory. The
spin and other quantum numbers of the gluon have not yet been determined experimentally.* The intermediate boson remains a theoretical hypothesis, and the efforts to detect gravitational waves have so
far yielded null results. Of the four mediating fields of our basic interactions, only the photon has a solid experimental foundation.
Nevertheless, we can be encouraged to feel that verification will
come in due time. But even then, is there any basis for us to believe
that we have arrived at our final goal—the understanding of all the
fundamental laws of nature?

The standard model of the unifying gauge theory alone needs
~ 20 parameters: e, G, θ_w, various masses for the three generations of leptons and quarks and the four weak decay angles θ_1, θ_2,
θ_3 and δ. [If neutrino oscillations are observed, more parameters
may be necessary.] Therefore, while we may have achieved a rather

---

* See Section 23.6.5.

effective description of physical processes up to about 100 GeV, the theory we have should more appropriately be viewed as essentially phemenological. After all, who has ever heard of a fundamental theory that requires twenty-some parameters? The following are some possible directions that may change our present frame of thinking.

1. Size of leptons and quarks

Suppose that neither the lepton nor the quark is elementary. Instead of being point particles, they all have a size. This size cannot be too large. Consider, for example, the electron; its size ℓ has to be less than $\sim 10^{-16}$ cm due to tests of QED. Now, $\ell \neq 0$ implies that there should be a nonzero moment of inertia, and through that the electron can acquire an angular momentum in a collision, resulting in a high-spin $j = 3/2, 5/2, \cdots$ heavy electron. The slope of its mass vs. j should be

$$\sim \frac{1}{\ell} \gtrsim 100 \text{ GeV} .$$

On the basis of known interactions, we may predict the minimal phenomena that such a heavy electron, called E, would exhibit. We expect among its dominant decay modes

$$\begin{aligned} E &\to e + \gamma \\ &\to e + \text{lepton pairs} + \gamma\text{'s} . \end{aligned}$$

Experimentally, if E can be produced we will see high-energy jets of pure γ and lepton showers, which will make the observation quite spectacular. If such high-spin leptons and quarks are discovered, then we will have to make a fresh start in formulating our theories and search for new building blocks to construct the universe.

2. Possibility of vacuum engineering

Based on either the spontaneous symmetry-breaking mechanism or the quark-confinement phenomena, we believe our vacuum, though Lorentz invariant, to be quite complicated. Like any other physical medium, it can carry long-range-order parameters and it may also undergo phase transitions. Hitherto, in any high-energy experiment the higher the energy, the smaller is the spatial region that we are able to explore. Consequently, we have avoided the opportunity to study coherent phenomena which may be connected with the vacuum. To explore such possibilities, we have to distribute high energy or high matter density over a large spatial volume. The experimental method to alter the properties of the vacuum may be called vacuum engineering. An effective way may well be to use relativistic heavy ions. In order to overcome the short-distance strong nuclear repulsive forces generated by ω^o and other vector mesons, we need an energy $\gg 1$ GeV per nucleon in the center-of-mass frame. If we can create vacuum excitations or vacuum phase transitions, then any of the constants in our present theory, θ_w, θ_c, m_u, m_d, $\cdots$, can be subject to change. If indeed we are able to alter the vacuum, then we may encounter some new phenomena, totally unexpected.

3. Improvement on conventional quantum mechanics

At present, the correctness of quantum mechanics is viewed almost universally as beyond question. Yet, how sure are we? How solid is its foundation? In quantum mechanics, the system that is being observed satisfies the standard time-dependent Schrödinger equation

$$H \psi_{system} = i \dot{\psi}_{system} \tag{25.1}$$

in which the observer is not included as part of its dynamical variables. The effect of the observation on the system is described by the well-known rules of probability amplitude decomposition of ψ_{system}. Here, an argument will be given which indicates that this cannot be the final form of a complete theory. We note that the converse effect of the system on the observer is not at all well described by our usual formulation of quantum mechanics. To be sure, such an effect must be there, as is obvious if we consider the example of a physicist who inadvertently drinks a cup of the liquid helium superfluid that he happens to be observing. However, while the effect of the observer on the system is quantum-mechanical, the converse effect is invariably expressed in classical terms. Yet, by reciprocity such an asymmetry must be viewed only as an approximation.

We may wonder how to express the effect on the observer in terms of quantum-mechanical probability amplitude and not classical probability. At present, there even seems to be a lack of vocabulary adequate to express such a phenomenon. A way out of this dilemma could be to consider a state vector in a larger Hilbert space

$$\psi = \psi_{observer} \otimes \psi_{system} ,$$

which tentatively may be assumed to satisfy the usual quantum-mechanical law of motion. By taking into account the quantum-mechanical effect of ψ_{system} on $\psi_{observer}$ and then reiterating the effect of $\psi_{observer}$ back onto ψ_{system}, we may derive an improved equation for ψ_{system} which carries a correction to (25.1). The result of this corrected equation can then be expressed in terms of the usual language which ignores the non-classical effect of the system on the observer. By carrying out successive approximations in this direction, hopefully we may arrive at a more fundamental equation for the system which may be experimentally tested.

Of course, we are not able to predict the future. The history of particle physics has been full of unexpected discoveries, which in turn led into unforeseen directions. In its evolution we have witnessed many examples that showed both the wisdom and the folly of physicists. It seems more than likely that our present understanding is also transitory and that our basic concepts and theories will undergo further major change. Indeed, long ago it was said: *

道可道非常道
名可名非常名

The principle that can be stated
Cannot be the absolute principle.

The name that can be given
Cannot be the permanent name.

* Laotse, Dao Dé Jing (~ 550 B.C.)

APPENDIX *

Tables of Particle Properties

and

Plots of Cross Sections and Structure Functions

N. Barash-Schmidt, C. Bricman, R. L. Crawford, C. Dionisi, C. P. Horne, R. L. Kelly, M. J. Losty, M. Mazzucato, L. Montanet, A. Rittenberg, M. Roos, T. Shimada, T. G. Trippe, C. G. Wohl, C. P. Yost

[For references see Revs. Mod. Phys. 52, No. 2, Part II, April 1980.]

* I am grateful to the Particle Data Group of the Lawrence Berkeley Laboratory (Dr. Robert L. Kelly, Director) for their kind permission to reproduce this material.

Stable Particle Table

For additional parameters, see Addendum to this table.

Quantities in italics have changed by more than one (old) standard deviation since April 1978.

Particle	$I^G(J^P)C_n$ [a]	Mass (MeV) / Mass^2 (GeV)^2	Mean life (sec) / $c\tau$ (cm)	Partial decay mode: Mode	Fraction [b]	p or Pmax [c] (MeV/c)
				PHOTON		
γ	$0,1(1^-)-$	$0(<6\times10^{-22})$	——	stable		
				LEPTONS		
ν_e	$J=\frac{1}{2}$	$0(<0.00006)$	stable $(>3\times10^{8}m_{\nu_e}(\text{MeV}))$	stable		
e	$J=\frac{1}{2}$	0.5110034 $\pm.0000014$	stable $(>5\times10^{21}\text{y})$	stable		
ν_μ	$J=\frac{1}{2}$	$0(<0.57)$	stable $(>2.6\times10^{4}m_{\nu_\mu}(\text{MeV}))$	stable		
μ	$J=\frac{1}{2}$	105.65946	2.197120×10^{-6}	$\mu^- \xrightarrow{d}$ $e^-\bar{\nu}\nu$	$(98.6$ $)\%$	53
		$\pm.00024$	$\pm.000077$	$e^-\bar{\nu}\nu\gamma$	[e]$(1.4^{\pm0.4})\%$	53
		$m^2=0.01116392$	$c\tau=6.5868\times10^{4}$	$e^-\gamma\gamma$	$(<4)\times10^{-6}$	53
		$m_\mu-m_{\pi^\pm}=-33.9074$		$e^-e^+e^-$	$(<1.9)\times10^{-9}$	53
		$\pm.0012$		$e^-\gamma$	$(<1.9)\times10^{-10}$	53
				$e^-\nu_e\bar{\nu}_\mu$	$(<25)\%$	53
τ	$J=\frac{1}{2}$[f]	*1784*	$<2.3\times10^{-12}$	$\tau^- \xrightarrow{d}$ $\mu^-\bar{\nu}\nu$	$(17.9\ \pm1.5)\%$	889
		±4		$e^-\bar{\nu}\nu$	$(17.0\ \pm1.1)\%$	892
		$m^2=$ *3.18*	$c\tau<0.07$	hadron$^-$ neutrals	$(33\ \pm10)\%$	
				†[$\pi^-\nu$	$(8.2\ \pm2.6)\%]$	887
				†[$\rho^-\nu$	$(22\ \pm4)\%]$	723
				†[K^- neutrals	(small)]	
				$e^-\gamma+\mu^-\gamma$	$(<12)\%$	892
				3(hadron$^\pm$) neutrals	$(35\ \pm11)\%$	
				†[$\pi^-\rho^0\nu$	$(4.2\ \pm1.3)\%]$	715
				†[$\pi^-\pi^-\pi^+\nu$ (incl. $\pi\rho\nu$)	$(7\ \pm5)\%]$	864
				†[$\pi^-\pi^-\pi^+(\geq0\pi^0)\nu$	$(18\ \pm7)\%]$	864
				($\geq$3chgd.) neutrals	$(32\ \pm5)\%$	
				†[e^- chgd.parts. + μ^- chgd.parts.	$(<4)\%]$	
				NONSTRANGE MESONS[a]		
$\pi^\pm$	$1^-(0^-)$	139.5669	2.6030×10^{-8}	$\pi^+ \xrightarrow{d}$ $\mu^+\nu$	$100\ \%$	30
		$\pm.0012$	$\pm.0023$	$e^+\nu$	$(1.267\pm0.023)\times10^{-4}$	70
		$m^2=0.0194789$	$c\tau=780.4$	$\mu^+\nu\gamma$	[e]$(1.24\pm0.25)\times10^{-4}$	30
			$(\tau^+-\tau^-)/\bar{\tau}=$	$e^+\nu\pi^0$	$(1.02\pm0.07)\times10^{-8}$	5
			$(0.05\pm0.07)\%$	$e^+\nu\gamma$	[e]$($*5.6 ±0.7*$)\times10^{-8}$	70
			(test of CPT)	$e^+\nu e^+e^-$	$(<5)\times10^{-9}$	70
π^0	$1^-(0^-)+$	134.9626	0.828×10^{-16}	$\gamma\gamma$	$(98.85\pm0.05)\%$	67
		$\pm.0039$	$\pm.057$ S=1.8*	γe^+e^-	$(1.15\pm0.05)\%$	67
		$m^2=0.0182149$	$c\tau=2.5\times10^{-6}$	$\gamma\gamma\gamma$	$(<1.5)\times10^{-6}$	67
		$m_{\pi^\pm}-m_{\pi^0}=4.6043$		$e^+e^-e^+e^-$	[g]$(3.32)\times10^{-5}$	67
		$\pm.0037$		$\gamma\gamma\gamma\gamma$	$(<4)\times10^{-5}$	67
				e^+e^-	$(2.2\ ^{+2.4}_{-1.1})\times10^{-7}$	67
η	$0^+(0^-)+$	548.8	$\Gamma=(0.85\pm0.12)$keV	Neutral decays: $\gamma\gamma$	$(38.0\ \pm1.0)\%$ S=1.2*	274
		±0.6	Neutral decays	$\pi^0\gamma\gamma$	[h]$(3.1\ \pm1.1)\%$ S=1.2*	258
		S=1.4*	$(71.0\pm0.7)\%$	$3\pi^0$	$(29.9\ \pm1.1)\%$ S=1.1*	180
		$m^2=0.3012$	S=1.1*	Charged decays: $\pi^+\pi^-\pi^0$	$(23.6\ \pm0.6)\%$ S=1.1*	175
				$\pi^+\pi^-\gamma$	$(4.89\pm0.13)\%$ S=1.1*	236
				$e^+e^-\gamma$	$(0.50\pm0.12)\%$	274
				$e^+e^-\pi^0$	$(<4)\times10^{-5}$	258
				$\pi^+\pi^-$	$(<0.15)\%$	236
			Charged decays	$e^+e^-\pi^+\pi^-$	$(0.1\ \pm0.1)\%$	236
			$(29.0\pm0.7)\%$	$\pi^+\pi^-\pi^0\gamma$	$(<6)\times10^{-4}$	175
			S=1.1*	$\pi^+\pi^-\gamma\gamma$	$(<0.2)\%$	236
				$\mu^+\mu^-$	$(2.2\ \pm0.8)\times10^{-5}$	253
				$\mu^+\mu^-\gamma$	$(1.5\ \pm0.8)\times10^{-4}$	253
				$\mu^+\mu^-\pi^0$	$(<5)\times10^{-4}$	211
				e^+e^-	$(<3)\times10^{-4}$	274

Stable Particle Table *(cont'd)*

Particle	$I^G(J^P)C_n$ [a]	Mass (MeV) Mass^2 (GeV)^2	Mean life (sec) $c\tau$ (cm)	Partial decay mode: Mode	Fraction [b]	p or p_{max} [c] (MeV/c)
				STRANGE MESONS [e]		
				$K^+ \rightarrow$ [d]		
$K^\pm$	$\frac{1}{2}(0^-)$	493.669	1.2371×10^{-8}	$\mu^+\nu$	(63.50±0.16)%	236
		±0.015	±.0026 S=1.9*	$\pi^+\pi^0$	(21.16±0.15)%	205
		m^2=0.24371	$c\tau$=370.9	$\pi^+\pi^+\pi^-$	(5.59±0.03)% S=1.1*	125
			$(\tau^+-\tau^-)/\bar{\tau}$=	$\pi^+\pi^0\pi^0$	(1.73±0.05)% S=1.3*	133
			(.11±.09)%	$\mu^+\nu\pi^0$	(3.20±0.09)% S=1.7*	215
			(test of CPT)	$e^+\nu\pi^0$	(4.82±0.05)% S=1.1*	228
			S=1.2*	$\mu^+\nu\gamma$	[e] (5.8 ± 3.5)$\times10^{-3}$	236
				$e^+\nu\pi^0\pi^0$	($1.8^{+2.4}_{-0.6}$)$\times10^{-5}$	207
	$m_{K^\pm}-m_{K^0}$=−4.01			$e^+\nu\pi^+\pi^-$	(3.90±0.15)$\times10^{-5}$	203
		±0.13		$e^-\bar{\nu}\pi^+\pi^+$	(<5)$\times10^{-7}$	203
		S=1.1*		$\mu^+\nu\pi^+\pi^-$	(0.9 ±0.4)$\times10^{-5}$	151
				$\mu^-\bar{\nu}\pi^+\pi^+$	(<3.0)$\times10^{-6}$	151
				$e^+\nu$	(1.54±0.09)$\times10^{-5}$	247
				$e^+\nu\gamma$ (SD+) [i]	(1.52±0.23)$\times10^{-5}$	247
				$e^+\nu\gamma$ (SD−) [i]	(<1.0)$\times10^{-4}$	247
				$\pi^+\pi^0\gamma$	[j,e] (2.75±0.16)$\times10^{-4}$	205
				$\pi^+\pi^+\pi^-\gamma$	[e] (1.0 ±0.4)$\times10^{-4}$	125
				$\mu^+\nu\pi^0\gamma$	[e] (<6)$\times10^{-5}$	215
				$e^+\nu\pi^0\gamma$	[e] (3.7 ±1.4)$\times10^{-4}$	228
				$e^+e^-\pi^+$	(2.6 ±0.5)$\times10^{-7}$	227
				$e^+e^+\pi^-$	(<1)$\times10^{-8}$	227
				$\mu^+\mu^-\pi^+$	(<2.4)$\times10^{-6}$	172
				$\pi^+\gamma\gamma$	[e] (<3.5)$\times10^{-5}$	227
				$\pi^+\gamma\gamma\gamma$	[e] (<3.0)$\times10^{-4}$	227
				$\pi^+\nu\bar{\nu}$	(<0.6)$\times10^{-6}$	227
				$\pi^+\gamma$	(<4)$\times10^{-6}$	227
				$e^+\mu^\pm\pi^\mp$	(<7)$\times10^{-9}$	214
				$e^-\mu^+\pi^+$	(<5)$\times10^{-9}$	214
				$e^+\nu\nu\bar{\nu}$	(<6)$\times10^{-5}$	247
				$\mu^+\nu\nu\bar{\nu}$	(<6)$\times10^{-6}$	236
				$\mu^+\nu e^+e^-$	(11 ±3)$\times10^{-7}$	236
				$\mu^-\nu e^+e^+$	(<2.0)$\times10^{-8}$	236
				$e^+\nu e^+e^-$	(2^{+2}_{-1})$\times10^{-7}$	247
K^0 $\bar{K}^0$	$\frac{1}{2}(0^-)$	497.67	50% K_{Short}, 50% K_{Long}			
		±0.13				
		S=1.1*				
		m^2=0.24768				
K^0_S	$\frac{1}{2}(0^-)$		0.8923×10^{-10}	$\pi^+\pi^-$	(68.61 ±0.24)% S=1.1*	206
			±.0022	$\pi^0\pi^0$	(31.39)%	209
			$c\tau$=2.675	$\mu^+\mu^-$	(<3.2)$\times10^{-7}$	225
				e^+e^-	(<3.4)$\times10^{-4}$	249
				$\pi^+\pi^-\gamma$	[e] (1.85±0.10)$\times10^{-3}$	206
				$\gamma\gamma$	(<0.4)$\times10^{-3}$	249
K^0_L	$\frac{1}{2}(0^-)$		5.183×10^{-8}	$\pi^0\pi^0\pi^0$	(21.5 ±0.7)% S=1.3*	139
			±.040	$\pi^+\pi^-\pi^0$	(12.39±0.18)% S=1.2*	133
			$c\tau$=1554	$\pi^\pm\mu^\mp\nu$	(27.0 ±0.5)% S=1.1*	216
				$\pi^\pm e^\mp\nu$ (incl. $\pi e\nu\gamma$)	(38.8 ±0.5)% S=1.1*	229
				$\pi e\nu\gamma$	[e] (1.3 ±0.8)%	229
	$m_{K_L}-m_{K_S}$= $0.5349\times10^{10}\hbar$ sec^{-1}			$\pi^+\pi^-$	[k] (0.203±0.005)%	206
	±0.0022			$\pi^0\pi^0$	[k] (0.094±0.018)% S=1.5*	209
				$\pi^+\pi^-\gamma$	[e] (6.0 ±2.0)$\times10^{-5}$	206
				$\pi^0\gamma\gamma$	(<2.4)$\times10^{-4}$	231
				$\gamma\gamma$	(4.9 ±0.5)$\times10^{-4}$	249
				$e\mu$	(<2.0)$\times10^{-9}$	238
				$\mu^+\mu^-$	(9.1 ±1.9)$\times10^{-9}$	225
				$\mu^+\mu^-\gamma$	(<7.8)$\times10^{-6}$	225
				$\mu^+\mu^-\pi^0$	(<5.7)$\times10^{-5}$	177
				e^+e^-	(<2.0)$\times10^{-9}$	249
				$e^+e^-\gamma$	(<2.8)$\times10^{-5}$	249
				$\pi^+\pi^-e^+e^-$	(<8.8)$\times10^{-6}$	206
				$\pi^0\pi^\pm e^\mp\nu$	(<2.2)$\times10^{-3}$	207

Stable Particle Table *(cont'd)*

Particle	$I^G(J^P)C_n$ [a]	Mass (MeV) / Mass^2 (GeV)^2	Mean life (sec) / $c\tau$ (cm)	Partial decay mode: Mode	Fraction [b]	p or p_{max} [c] (MeV/c)
			CHARMED MESONS[a]			
$D^{\pm}$	$\frac{1}{2}(0^-)$ [f]	1868.3 [l]	$(2.5^{+3.5}_{-1.5})\times10^{-13}$	$D^+ \xrightarrow{d}$ K^-anything	(10 ±7)%	
		±0.9		†[$K^-\pi^+\pi^+$(incl. $K^*\pi$)	(3.9 ±1.0)%]	845
		m^2= 3.491	$c\tau$=0.007	†[$\overline{K}^*(892)^0\pi^+$	(seen)]	456
		$m_{D^\pm}-m_{D^0}$= 5.0		†[$K^-K^+\pi^+$	(<0.6)%]	743
		±0.8		$\overline{K}^0$anything	(39 ±29)%	
				†[$\overline{K}^0\pi^+$	(1.5 ±0.6)%]	862
				$e^{\pm}$anything	[m](*8.2 ±1.2*)%	
				$\pi^+\pi^+\pi^-$	(<0.31)%	908
				K^+anything	(6 ±6)%	
				$K^+\pi^+\pi^-$	(<0.20)%	845
D^0 $\overline{D}^0$	$\frac{1}{2}(0^-)$ [f]	1863.1 [l]	$(3.5^{+3.5}_{-1.7})\times10^{-13}$	$D^0 \xrightarrow{d}$ $K^{\mp}$anything	(35 ±10)%	
		±0.9		†[$K^-\pi^+$	(1.8 ±0.5)%]	860
		m^2= 3.471	$c\tau$=0.01	†[$K^-\pi^+\pi^0$	(12 ±6)%]	843
				†[$K^-\pi^+\pi^+\pi^-$	(3.5 ±0.9)%]	812
		$\frac{\Gamma(D^0\to\overline{D}^0\to K^+\pi^-)}{\Gamma(D^0\to K\pi)}<0.16$		$\overline{K}^0$anything + K^0any	(57 ±26)%	
				†[$\overline{K}^0\pi^0 + K^0\pi^0$	(<6)%]	859
				†[$\overline{K}^0\pi^+\pi^- + K^0\pi^+\pi^-$	(4.4 ±1.1)%]	841
				$e^{\pm}$anything	[m](*8.2 ±1.2*)%	
				$\pi^+\pi^-$	(5.9 ±3.2)$\times10^{-4}$	921
				K^+K^-	(2.0 ±0.8)$\times10^{-3}$	790
			NONSTRANGE BARYONS[a]			
p	$\frac{1}{2}(\frac{1}{2}^+)$	938.2796	stable ($>10^{30}$y)	stable		
		±0.0027				
		m^2= 0.880369		$\lvert q_p\rvert-\lvert q_e\rvert < 10^{-21}\lvert q_e\rvert$ [n]		
n	$\frac{1}{2}(\frac{1}{2}^+)$	939.5731	917±14	$pe^-\overline{\nu}$	100 %	1
		±0.0027	$c\tau=2.75\times10^{13}$	$p\nu\overline{\nu}$ (chg.noncons.)	(<3)$\times10^{-19}$	1
		m^2= 0.882798				
		m_p-m_n=-1.29343				
		±0.00004		$\lvert q_n\rvert < 10^{-21}\lvert q_e\rvert$ [n]		
			STRANGENESS −1 BARYONS[a]			
Λ	$0(\frac{1}{2}^+)$	1115.60	2.632×10^{-10}	$p\pi^-$	(64.2 ± 0.5)%	100
		±0.05	±.020 S=1.6*	$n\pi^0$	(35.8 ± 0.5)%	104
		S=1.2*	$c\tau$=7.89	$pe^-\nu$	(8.07±0.28)$\times10^{-4}$	163
		m^2= 1.2446		$p\mu^-\nu$	(1.57±0.35)$\times10^{-4}$	131
		$m_\Lambda-m_{\Sigma^0}$= −76.86		$p\pi^-\gamma$	[e](0.85±0.14)$\times10^{-3}$	100
		±0.08				
Σ^+	$1(\frac{1}{2}^+)$	1189.36	0.800×10^{-10}	$p\pi^0$	(51.64 ±0.30)%	189
		±0.06	±.004	$n\pi^+$	(48.36 ±0.30)%	185
		S=1.8*	$c\tau$=2.40	$p\gamma$	(1.24±0.18)$\times10^{-3}$ S=1.4*	225
		m^2= 1.4146		$n\pi^+\gamma$	[e](0.93±0.10)$\times10^{-3}$	185
				$\Lambda e^+\nu$	(2.02±0.47)$\times10^{-5}$	71
		$m_{\Sigma^+}-m_{\Sigma^-}$= −7.98	$\frac{\Gamma(\Sigma^+\to \ell^+ n\nu)}{\Gamma(\Sigma^-\to \ell^- n\nu)}<.04$ ←{	$n\mu^+\nu$	(<3.0)$\times10^{-5}$	202
		±.08		$ne^+\nu$	(<0.5)$\times10^{-5}$	224
		S=1.2*		pe^+e^-	(<7)$\times10^{-6}$	225
Σ^0	$1(\frac{1}{2}^+)$ [p]	1192.46	5.8×10^{-20}	$\Lambda\gamma$	100 %	74
		±0.08	±1.3	Λe^+e^-	[g](5.45)$\times10^{-3}$	74
		m^2= 1.4220	$c\tau=1.7\times10^{-9}$	$\Lambda\gamma\gamma$	(<3)%	74
Σ^-	$1(\frac{1}{2}^+)$	1197.34	1.482×10^{-10}	$n\pi^-$	100 %	193
		±0.05	±.011 S=1.3*	$ne^-\nu$	(1.08±0.04)$\times10^{-3}$	230
		m^2= 1.4336	$c\tau$=4.44	$n\mu^-\nu$	(0.45±0.04)$\times10^{-3}$	210
				$\Lambda e^-\nu$	(0.61±0.05)$\times10^{-4}$	79
		$m_{\Sigma^0}-m_{\Sigma^-}$= −4.88		$n\pi^-\gamma$	[e](4.6 ±0.6)$\times10^{-4}$	193
		±.06				

Stable Particle Table *(cont'd)*

Particle	$I^G(J^P)C_n$ [a]	Mass (MeV) / Mass^2 (GeV)^2	Mean life (sec) / $c\tau$ (cm)	Partial decay mode: Mode	Fraction [b]	p or p_{max} [c] (MeV/c)
				STRANGENESS −2 BARYONS[a]		
Ξ^0	$\frac{1}{2}(\frac{1}{2}^+)$[q]	1314.9	2.90×10^{-10}	$\Lambda\pi^0$	100 %	135
		±0.6	±.10	$\Lambda\gamma$	(0.5 ±0.5)%	184
		m^2= 1.7290	$c\tau$=8.69	$\Sigma^0\gamma$	(<7)%	117
				$p\pi^-$	(<3.6)$\times10^{-5}$	299
				$pe^-\nu$	(<1.3)$\times10^{-3}$	323
				$\Sigma^+e^-\nu$	(<1.1)$\times10^{-3}$	120
				$\Sigma^-e^+\nu$	(<0.9)$\times10^{-3}$	112
		$m_{\Xi^0}-m_{\Xi^-}$=−6.4		$\Sigma^+\mu^-\nu$	(<1.1)$\times10^{-3}$	64
		±.6		$\Sigma^-\mu^+\nu$	(<0.9)$\times10^{-3}$	49
				$p\mu^-\nu$	(<1.3)$\times10^{-3}$	309
Ξ^-	$\frac{1}{2}(\frac{1}{2}^+)$[q]	1321.32	1.641×10^{-10}	$\Lambda\pi^-$	100 %	139
		±0.13	±.016	$\Lambda e^-\nu$	(*2.8 ±1.2*)$\times10^{-4}$	190
		m^2= 1.7459	$c\tau$=4.92	$\Sigma^0e^-\nu$	(<5)$\times10^{-4}$	123
				$\Lambda\mu^-\nu$	(3.1 ±1.2)$\times10^{-4}$	163
				$\Sigma^0\mu^-\nu$	(<8)$\times10^{-4}$	70
				$n\pi^-$	(<1.1)$\times10^{-3}$	303
				$ne^-\nu$	(<3.2)$\times10^{-3}$	327
				$n\mu^-\nu$	(<1.5)%	313
				$\Sigma^-\gamma$	(<1.2)$\times10^{-3}$	118
				$p\pi^-\pi^-$	(<4)$\times10^{-4}$	223
				$p\pi^-e^-\nu$	(<4)$\times10^{-4}$	304
				$p\pi^-\mu^-\nu$	(<4)$\times10^{-4}$	250
				$\Xi^0e^-\nu$	(<2.3)$\times10^{-3}$	6
				STRANGENESS −3 BARYON[a]		
Ω^-	$0(\frac{3}{2}^+)$[q]	1672.22	0.82×10^{-10}	ΛK^-	(68.6 ±1.3)%	211
		±.31	±.03	$\Xi^0\pi^-$	(23.4 ±1.3)%	293
		m^2= 2.7963	$c\tau$=2.5	$\Xi^-\pi^0$	(8.0 ±0.8)%	290
				$\Xi^0e^-\nu$	(~1)%	319
				$\Xi(1530)^0\pi^-$	(~2)$\times10^{-3}$	15
				$\Lambda\pi^-$	(<1.3)$\times10^{-3}$	449
				$\Xi^-\gamma$	(<3.1)$\times10^{-3}$	314
				NONSTRANGE CHARMED BARYON[a]		
Λ_c^+	$0(\frac{1}{2}^+)$[r]	2273	$\sim7\times10^{-13}$	$\Lambda\pi^+\pi^+\pi^-$	(seen)	798
		±6	$c\tau$~0.02	$pK^-\pi^+$	(2.2 ±1.0)%	814
		S=1.6*		$pK^*(892)^0$	(seen)	567
		m^2= 5.17		$\Delta(1232)^{++}K^-$	(seen)	700

ADDENDUM TO

Stable Particle Table

e	**Magnetic moment** 1.001 159 652 41 ±.000 000 000 20 $\frac{e\hbar}{2m_ec}$	
μ	1.001 165 924 ±.000 000 009 $\frac{e\hbar}{2m_\mu c}$	**μ Decay parameters** [s] ρ = 0.752±0.003 η = − 0.12 ±0.21 ξ = 0.972±0.013 δ = 0.755±0.009 h = 1.00±0.13 $\lvert g_A/g_V\rvert=0.86^{+0.33}_{-0.11}$ ϕ = 180°±15°

η	**Mode**	**Left-right asymmetry**	**Sextant asymmetry**	**Quadrant asymmetry**
	$\pi^+\pi^-\pi^0$	(0.12±.17)%	(0.19±0.16)%	(−0.17±0.17)%
	$\pi^+\pi^-\gamma$	(0.88±.40)%		β=0.047±0.062

	Mode	**Partial rate (sec^{-1})**	
$K^\pm$	$\mu\nu$	$(51.33\pm0.17)\times10^6$	S=1.2*
	$\pi\pi^0$	$(17.10\pm0.13)\times10^6$	S=1.1*
	$\pi\pi^+\pi^-$	$(4.52\pm0.02)\times10^6$	S=1.1*
	$\pi\pi^0\pi^0$	$(1.40\pm0.04)\times10^6$	S=1.3*
	$\mu\pi^0\nu$	$(2.58\pm0.07)\times10^6$	S=1.7*
	$e\pi^0\nu$	$(3.90\pm0.04)\times10^6$	S=1.1*
K^0_S	$\pi^+\pi^-$	k$(0.7689\pm.0033)\times10^{10}$	
	$\pi^0\pi^0$	k$(0.3517\pm.0029)\times10^{10}$	S=1.1*
K^0_L	$\pi^0\pi^0\pi^0$	$(4.14\pm0.15)\times10^6$	S=1.3*
	$\pi^+\pi^-\pi^0$	$(2.39\pm0.04)\times10^6$	S=1.2*
	$\pi\mu\nu$	$(5.21\pm0.10)\times10^6$	S=1.1*
	$\pi e\nu$	$(7.49\pm0.11)\times10^6$	S=1.1*
	$\pi^+\pi^-$	k$(3.91\pm0.10)\times10^4$	
	$\pi^0\pi^0$	k$(1.81\pm0.35)\times10^4$	S=1.5*

Slope parameters for $K\to3\pi$ [t]

$K^+\to\pi^+\pi^+\pi^-$ g=−0.215±.004 S=1.4*
$K^-\to\pi^-\pi^-\pi^+$ g=−0.217±.007 S=2.5*
$K^\pm\to\pi^0\pi^0\pi^\pm$ g= *0.607±.030* S=1.3*
$K^0_L\to\pi^+\pi^-\pi^0$ g= 0.670±.014 S=1.6*

See Data Card Listings for quadratic coefficients.

K^+_{l3}: λ^e_+= 0.029±.004; λ^μ_+= 0.026±.008 S=1.5*; λ^μ_0= −0.003±.007 S=1.5*

K^0_{l3}: λ^e_+= 0.0301±.0016 S=1.2*; λ^μ_+= 0.034 ±.006 S=2.5*; λ^μ_0= 0.020 ±.007 S=2.5*

See Data Card Listings for ξ, f_s, and f_t.

CP violation parameters [u,k]

$\lvert\eta_{+-}\rvert=(2.274\pm.022)\times10^{-3}$ $\lvert\eta_{00}\rvert=(2.33\pm.08)\times10^{-3}$ S=1.1*
$\phi_{+-}=(44.6\pm1.2)°$ $\phi_{00}=(54\pm5)°$
$\lvert\eta_{+-0}\rvert^2<0.12$ $\lvert\eta_{000}\rvert^2<0.28$ $\delta=(0.330\pm.012)\times10^{-2}$

$\Delta S = -\Delta Q$

Re x=0.009±.020 S=1.4*   Im x = −0.004±.026 S=1.1*

	Magnetic moment ($e\hbar/2m_pc$)	Decay parameters[v]						
			Measured		Derived			
			α	ϕ(degree)	γ	Δ(degree)	g_A/g_V	g_V/g_A
p	2.7928456 ±.0000011							
n [w]	−1.91304184 ±.00000088	$pe^-\nu$					−1.254±0.007 δ=(180.11±0.17)°	
Λ [w]	−0.614 ±.005	$p\pi^-$	0.642±0.013	(−6.5±3.5)°	0.76	$(7.7^{+4.0}_{-4.1})°$		
		$n\pi^0$	0.646±0.044					
		$pe\nu$					−0.62±0.05 S=1.2*	
Σ^+	*2.33* *±.13*	$p\pi^0$	−0.979±0.016	(36±34)°	0.17	(187±6)°		
		$n\pi^+$	+0.068±0.013	(167±20)°	−0.97	$(-72^{+132}_{-11})°$		
		$p\gamma$	$-1.03^{+0.52}_{-0.42}$	S=1.1*				
Σ^-	−1.41 ±.25	$n\pi^-$	−0.068±0.008	(10±15)°	0.98	$(249^{+12}_{-115})°$		
		$ne^-\nu$					±(0.385±0.070) S=2.3*	
		$\Lambda e^-\nu$						0.10±0.22 S=1.5*
Ξ^0	−1.20 ±.06	$\Lambda\pi^0$	−0.47±0.05 S=1.3*	(21±12)°	0.84	$(216^{+13}_{-19})°$		
Ξ^-	−1.85 ±.75	$\Lambda\pi^-$	−0.403±0.017	(2±6)° S=1.1*	0.92	(185±13)°		
Ω^-		ΛK^-	*−0.26±0.33* S=1.5*					

Meson Table

Quantities in italics are new or have changed by more than one (old) standard deviation since April 1978.

Name (G\I: 0, 1; −: ω/φ, π; +: η, ρ)	$I^G(J^P)C_n$ (underline: estab.)	Mass M (MeV)	Full Width Γ (MeV)	M^2 $\pm\Gamma M$[(a)] $(GeV)^2$	Partial decay mode: Mode	Fraction (%) [Upper limits are 1σ (%)]	p or p_{max}[(b)] (MeV/c)
NONSTRANGE MESONS							
$\pi^\pm$	$1^-(0^-)+$	139.57	0.0	0.019479	See Stable Particle Table		
π^0		134.96	7.95 eV ±.55 eV	0.018215			
η	$0^+(0^-)+$	548.8 ±0.6	0.85 keV ±.12 keV	0.301 ±.000	Neutral	71.0 See Stable	
					Charged	29.0 Particle Table	
$\rho(770)$	$1^+(1^-)-$	776¶ ±3§	158¶ ±5§	0.602 ±.123	$\pi\pi$	≈ 100	362
					$\pi\gamma$	0.024 ±.007	375
					e^+e^-	0.0043±.0005 (d)	388
					$\mu^+\mu^-$	0.0067±.0012 (d)	373
					$\eta\gamma$	seen¶	194
M and Γ from neutral mode.					For upper limits, see footnote (e)		
$\omega(783)$	$0^-(1^-)-$	782.4 ±0.2 S=1.1*	10.1 ±.3	0.612 ±.008	$\pi^+\pi^-\pi^0$	89.8±0.5	327
					$\pi^+\pi^-$	1.4±0.2	365
					$\pi^0\gamma$	8.8±0.5	380
					e^+e^-	0.0076±.0017 S=1.9*	391
					$\eta\gamma$	seen¶	199
					For upper limits, see footnote (f)		
$\eta'(958)$	$0^+(0^-)+$¶	957.57 ±0.25	0.28 ±0.10	0.917 ±.0003	$\eta\pi\pi$	65.6±1.6	231
					$\rho^0\gamma$	29.8±1.6	164
					$\omega\gamma$	2.7±0.5	159
					$\gamma\gamma$	1.9±0.2	479
					For upper limits, see footnote (g)		
$\delta(980)$	$1^-(0^+)+$	981(h) ±3	52(h) ±8	0.962 ±.051	$\eta\pi$	seen	319
					$K\bar{K}$	seen¶	
$S^*(980)$	$0^+(0^+)+$	∿ 980(c)§ ±10§	40(c)§ ±10§	0.960 ±.039	$K\bar{K}$	seen¶	
					$\pi\pi$	seen	470
See note on $\pi\pi$ and $K\bar{K}$ S-wave¶.							
$\phi(1020)$	$0^-(1^-)-$	1019.6 ±0.1 S=1.3*	4.1 ±.2	1.040 ±.004	K^+K^-	48.6±1.2 S=1.3*	127
					K_LK_S	35.2±1.2 S=1.5*	111
					$\pi^+\pi^-\pi^0$ (incl. $\rho\pi$)	14.7±0.7 S=1.2*	462
					$\eta\gamma$	1.5±0.2	362
					$\pi^0\gamma$	0.14±0.05	501
					e^+e^-	.031±.001 S=1.1*	510
					$\mu^+\mu^-$	.025±.003	499
					For upper limits, see footnote (i)		
A_1(1100-1300)	$1^-(1^+)+$	1100¶ to 1300	∿ 300¶	1.44 ±.36	$\rho\pi$	*dominant*	329
					$\pi(\pi\pi)_{S\text{-wave}}$	*seen*	558
B(1235)	$1^+(1^+)-$	1231 ±10§	129 ±10§	1.52 ±.16	$\omega\pi$	only mode seen	348
					[D/S amplitude ratio = .29±.05]		
					For upper limits, see footnote (j)		
f(1270)	$0^+(2^+)+$	1273 ±5§	178 ±20§	1.62 ±.23	$\pi\pi$	*83.1±1.9* S = 1.4*	621
					$2\pi^+2\pi^-$	2.9±0.3 S = 1.1*	558
					$K\bar{K}$	2.8±0.3 S = 1.3*	397
					$\pi^+\pi^-2\pi^0$	seen	561
					For upper limits, see footnote (ℓ)		

Meson Table *(cont'd)*

Name (G\I: 0, 1; −: ω/φ, π; +: η, ρ)	$I^G(J^P)C_n$ ⊢ estab.	Mass M (MeV)	Full Width Γ (MeV)	M^2 $\pm\Gamma M^{(a)}$ $(GeV)^2$	Partial decay mode: Mode	Fraction (%) [Upper limits are 1σ (%)]	p or p_{max}(b) (MeV/c)
D(1285)	$0^+(1^+)+$	1284§ ±10§	27§ ±10§	1.65 ±.03	$K\bar{K}\pi$	*10±2*	303
					$\eta\pi\pi$	*49±6*	483
					†[$\delta\pi$	*36±7*]	239
					4π (prob. $\rho\pi\pi$)	*41±13*	564
ε(1300)	$0^+(0^+)+$	∿ 1300	200-400		$\pi\pi$	*∿ 90*	635
					$K\bar{K}$	*∿ 10*	423
See note on ππ and $K\bar{K}$ S wave¶.							
A_2(1310)	$1^-(2^+)+$	1317§ ±5§	102§ ±5§	1.73 ±.13	$\rho\pi$	70.0±2.2	414
					$\eta\pi$	14.6±1.1	534
					$\omega\pi\pi$	10.6±2.5	360
					$\bar{K}K$	4.8±0.5	434
					$\eta'\pi$	<1	285
					$\pi\gamma$	0.45±0.11	651
E(1420)	$0^+(1^+)+$	1418§ ±10§	50§ ±10§	2.01 ±.07	$K\bar{K}\pi$ (prob. $K^*\bar{K}$ + $\bar{K}^*K$)	seen	423
					$\eta\pi\pi$	possibly seen	565
					†[$\delta\pi$	possibly seen]	350
f'(1515)	$0^+(2^+)+$	1516§ ±12§	67§ ±10§	2.30 ±.10	$K\bar{K}$	dominant	572
					$\pi\pi$	seen	745
					For upper limits, see footnote (k)		
ρ'(1600)	$1^+(1^-)-$	∿ 1600¶	∿ 300¶	2.56 ±.48	4π (incl. $\rho\pi^+\pi^-$)	∿ 85	738
					$\pi\pi$	∿ 15	788
A_3(1660)	$1^-(2^-)+$	*1660*§ *±10*§	*200*§ *±50*§	2.76 ±.33	$f\pi$	*∿ 60*	320
					$\rho\pi$	*∿ 30*	640
					$\pi(\pi\pi)_{S\text{-wave}}$	*∿ 10*	802
ω(1670)	$0^-(3^-)-$	1666 ±5	166§ ±15§	2.78 ±.28	$\rho\pi$	seen	644
					3π	possibly seen	805
					5π	seen	739
					†[$\omega\pi\pi$ (prob. $B\pi$)	seen]	614
g(1700)¶	$1^+(3^-)-$	1700§ ±20§	200§ ±20§	2.89 ±.34	2π	24.0±1.3	838
					4π (incl. $\pi\pi\rho$, $\rho\rho$, $A_2\pi$, $\omega\pi$)	*72.1±1.6*	792
					$K\bar{K}\pi$ (incl. $K^*\bar{K}$)	*2.4±0.7*	631
					$K\bar{K}$	*1.5±0.3*	689
J^P, M and Γ from the 2π and $K\bar{K}$ modes.							
S(1935)¶		1936§ ±5§		3.74	$N\bar{N}$	seen	236
Not a well established resonance.¶							
h(2040)	$0^+(4^+)+$	2040§ ±20§	150§ ±50§	4.16 ±.31	$\pi\pi$	seen	1010
					$K\bar{K}$	seen	890

Meson Table *(cont'd)*

G \ I	0	1
−	ω/φ	π
+	η	ρ

Name	$I^G(J^P)C_n$ ⊢ estab.	Mass M (MeV)	Full Width Γ (MeV)	M^2 $\pm\Gamma M$ [a] $(GeV)^2$	Partial decay mode: Mode	Fraction (%) [Upper limits are 1σ (%)]	p or p_{max} [b] (MeV/c)
J/ψ(3100)	$\underline{0^-(1^-)-}$	3097±1	0.063±0.009	9.598 ±.000	e^+e^-	7±1	1549
					$\mu^+\mu^-$	7±1	1545
					hadrons	86±2	
					†[all stables		
					$2(\pi^+\pi^-)\pi^0$	3.7±0.5	1496
					$3(\pi^+\pi^-)\pi^0$	2.9±0.7	1433
					$\pi^+\pi^-\pi^0K^+K^-$	1.2±0.3	1368
					$\pi^+\pi^-K^+K^-$	*0.72±0.23*	1407
					$4(\pi^+\pi^-)\pi^0$	0.9±0.3	1345
					$p\bar{p}\pi^+\pi^-$	*0.55±0.06*	1107
					$2(\pi^+\pi^-)$	0.4±0.1	1517
					$3(\pi^+\pi^-)$	0.4±0.2	1466
					$\Xi\bar{\Xi}$	*0.32±0.08*	818
					$2(\pi^+\pi^-)K^+K^-$	0.31±0.13	1320
					$K^0_SK^\pm\pi^\mp$	0.26±0.07	1440
					$p\bar{p}\eta$	0.23±0.04	948
					$p\bar{n}\pi^-$ or $\bar{p}n\pi^+$	*0.21±0.02*	1174
					$p\bar{p}$	0.22±0.02	1232
					$n\bar{n}$	*0.18±0.09*	1231
					$p\bar{p}\pi^+\pi^-\pi^0$	*0.16±0.06* (n)	1033
					$\Sigma^0\bar{\Sigma}^0$	*0.13±0.04*	988
					$\Lambda\bar{\Lambda}$	0.11±0.02	1074
					$p\bar{p}\pi^0$	0.11±0.01]	1176
					†[with resonances		
					$\rho\pi$	1.2±0.1	1448
					$\omega 2\pi^+2\pi^-$	*0.85±0.34*	1392
					ρA_2	*0.84±0.45*	1124
					$\omega\pi\pi$	*0.68±0.19*	1435
					$K^{*0}(892)\bar{K}^{*0}(1430)$	0.67±0.26	1007
					$K\bar{K}^* + \bar{K}K^*$	0.61±0.08	1373
					$B^\pm(1235)\pi^\mp$	*0.29±0.07*	1300
					ωf	*0.23±0.08* S=1.2*	1144
					$\phi\pi^+\pi^-$	0.21±0.09	1365
					$\eta' p\bar{p}$	*0.18±0.06*	596
					$\phi K\bar{K}$	0.18±0.08	1176
					$\omega p\bar{p}$	*0.16±0.03*	768
					$\omega K\bar{K}$	*0.16±0.10*	1265
					$\phi\eta$	0.10±0.06]	1320
					†[radiative decays		
					$\gamma\eta'$	0.25±0.06	1400
					γf	0.15±0.05]	1287
					For smaller branching ratios and upper limits see listing.¶		
χ(3415)	$\underline{0^+(0^+)+}$	3414±4		11.655	$2(\pi^+\pi^-)$ (incl. $\pi\pi\rho$)	4.6±0.9	1678
					$\pi^+\pi^-K^+K^-$ (incl. $\pi K\bar{K}^*$)	3.7±0.9	1580
					$\gamma J/\psi(3100)$	2.7±1.0 (m)	302
					$3(\pi^+\pi^-)$	1.9±0.7	1632
					$\pi^+\pi^-$	0.9±0.2	1701
					K^+K^-	0.9±0.2	1634
					$p\bar{p}\pi^+\pi^-$	0.6±0.2	1319
P_c or χ(3510)	$\underline{0^+(A)+}$	3507±4		12.299	$\gamma J/\psi(3100)$	31.5±5.2 S = 1.3*	386
					$3(\pi^+\pi^-)$	2.7±1.1	1681
					$2(\pi^+\pi^-)$ (incl. $\pi\pi\rho$)	2.0±0.6	1726
					$\pi^+\pi^-K^+K^-$ (incl. $\pi K\bar{K}^*$)	1.1±0.4	1630
					$\pi^+\pi^-p\bar{p}$	0.17±0.11	1379
$J^P = 1^+$ preferred.							
χ(3550)	$\underline{0^+(N)+}$	3551±5		12.610	$\gamma J/\psi(3100)$	15.4±2.4	425
					$2(\pi^+\pi^-)$ (incl. $\pi\pi\rho$)	2.4±0.6	1748
					$\pi^+\pi^-K^+K^-$ (incl. $\pi K\bar{K}^*$)	2.1±0.6	1654
					$3(\pi^+\pi^-)$	1.3±0.8	1704
					$\pi^+\pi^-$ and K^+K^-	0.27±0.11	
					$\pi^+\pi^-p\bar{p}$	0.37±0.14	1407
$J^P = 2^+$ preferred.							

Meson Table *(cont'd)*

Name	$I^G(J^P)C_n$ estab.	Mass M (MeV)	Full Width Γ (MeV)	M^2 $\pm\Gamma M$[a] $(GeV)^2$	Partial decay mode: Mode	Fraction (%) [Upper limits are 1σ (%)]	p or Pmax[b] (MeV/c)
ψ(3685)	$0^-(1^-)-$	3685±1 S=1.1*	0.215±0.040	13.579 ±.001	e^+e^-	0.9±0.1	1842
					$\mu^+\mu^-$	0.8±0.2	1839
					hadrons	98.1±0.3	
					+[$J/\psi\ \pi^+\pi^-$	33±2]	476
	$m_{\psi(3685)} - m_{\psi(3100)}$	= 588.2±0.9 S=1.2*			+[$J/\psi\ \pi^0\pi^0$	17±2]	480
					+[$J/\psi\ \eta$	3.7±0.4]	194
					+[$2(\pi^+\pi^-)\pi^0$	0.4±0.2]	1798
					+[$\pi^+\pi^-K^+K^-$	0.16±0.04]	1725
					+[$p\bar{p}\pi^+\pi^-$	*0.08±0.02*]	1490
					+[$2(\pi^+\pi^-)$	*0.05±0.01*]	1816
					+[$\gamma\ \chi(3415)$	7±2]	261
					+[$\gamma\ \chi(3510)$	7±2]	174
					+[$\gamma\ \chi(3550)$	7±2]	132
For smaller branching ratios and upper limits see Listings.¶							
ψ(3770)	$(1^-)-$	3768 ±3	25 ±3	14.198 ±.094	e^+e^-	0.0013±0.0002	1884
					$D\bar{D}$	dominant	243
	$m_{\psi(3770)} - m_{\psi(3685)}$	= *82.5±3.7* S=2.2*					
ψ(4030)	$(1^-)-$	*4030*§*±5*§	*52±10*	*16.241 ±0.210*	e^+e^-	*0.0014±0.0004*	2015
					hadrons	*dominant*	
ψ(4160)	$(1^-)-$	*4159±20*	*78±20*	*17.297 ±0.324*	e^+e^-	*0.0010±0.0004*	2079
					hadrons	*dominant*	
ψ(4415)	$(1^-)-$	4415±6	43±20§	19.492 ±.190	e^+e^-	0.0010±0.0003	2207
					hadrons	dominant	
ϒ(9460)	$(1^-)-$	*9458±6*	*∿ 0.060*	*89.454 ±0.0006*	$\mu^+\mu^-$	*2.2±2.0*	4728
					e^+e^-	*2.5±2.1*	4729
ϒ(10020)	$(1^-)-$	*10016±14*	*< 12*	*100.320*	$\mu^+\mu^-$	*seen*	5007
					e^+e^-	*seen*	5008
	$m_{\Upsilon(10020)} - m_{\Upsilon(9460)}$	= *559±7*					
Additional structure at m = *10410±30* is seen¶.							
STRANGE MESONS							
K^+	$1/2(0^-)$	493.67		0.244	See Stable Particle Table		
K^0		497.67		0.248			
$K^*(892)$	$1/2(1^-)$	891.8 ±0.4 S=1.1*	50.3 ±0.8	0.795 ±.045	Kπ	≈ 100	288
					Kππ	< 0.07	216
					Kγ	0.15±0.07	309
M and Γ from charged mode; $m^0 - m^\pm$ = *6.7±1.2* MeV.							
$Q_1(1280)$	$1/2(1^+)$	∿ 1280	∿ 120	1.64 ±.15	Kππ	dominant	501
					+[Kρ	large]	62
					+[$K^*\pi$	possibly seen]	307
					Kω	possibly seen	
$Q_2(1400)$	$1/2(1^+)$	*∿ 1400*	*∿ 150*	*1.96 ±.21*	Kππ	*dominant*	576
					+[$K^*\pi$	*large*]	399
					+[Kρ	*possibly seen*]	286
$K^*(1430)$	$1/2(2^+)$	1434§ ±5§	100§ ±10§	2.06 ±.14	Kπ	49.1±1.6	623
					$K^*\pi$	27.0±2.2	424
					$K^*\pi\pi$	11.2±2.5	374
					Kρ	6.6±1.5 S=1.1*	327
					Kω	3.7±1.6	320
					Kη	2.5±2.6	492

Meson Table *(cont'd)*

Name	$I^G(J^P)C_n$ estab.	Mass M (MeV)	Full Width Γ (MeV)	M^2 $\pm\Gamma M$[a] $(GeV)^2$	Partial decay mode: Mode	Fraction (%) [Upper limits are 1σ (%)]	p or p_{max}[b] (MeV/c)
κ(1500)	1/2(0^+)	∼ *1500*	∼ *250*	2.25 ±.36	Kπ	seen	661
See note on Kπ S wave¶.							
L region	1/2	1600 to 2000			Kππ	seen	
Not a well established resonance¶.							
K^*(1780)¶	1/2(3^-)	1785 ±6	126 ±20§	3.19 ±.22	Kππ	large	798
					+[Kρ	large]	619
					+[$K^*\pi$	large]	660
					Kπ	19±5§	817
CHARMED, NONSTRANGE MESONS							
D^+	1/2(0^-)	1868.3		3.491	See Stable Particle Table		
D^0		1863.1		3.471			
D^{*+}(2010)	1/2(1^-)	2008.6 ±1.0	< 2.0	4.034	$D^0\pi^+$	64±11	40
					$D^+\pi^0$	*28±9*	37
	$m_{D^{*+}} - m_{D^0}$ = 145.3 ± 0.4 MeV				$D^+\gamma$	*8±7*	135
D^{*0}(2010)	1/2(1^-)	2006.0 ±1.5	< 5	4.024	$D^0\pi^0$	55±15	45
					$D^0\gamma$	45±15	138

→

Contents of Meson Data Card Listings

Non-strange (S = 0, C = 0)						Strange (|S| = 1, C = 0)	
entry	$I^G(J^P)C_n$	entry	$I^G(J^P)C_n$	entry	$I^G(J^P)C_n$	entry	I (J^P)
π	$1^-(0^-)+$	A_2(1310)	$1^-(2^+)+$	→ e^+e^-(1100-3100)		K	1/2(0^-)
η	$0^+(0^-)+$	E (1420)	$0^+(1^+)+$	→ X (2830)		K^*(892)	1/2(1^-)
ρ (770)	$1^+(1^-)-$	→ X (1410-1440)		→ U (2980)		Q_1(1280)	1/2(1^+)
ω (783)	$0^-(1^-)-$	f′(1515)	$0^+(2^+)+$	J/ψ (3100)	$0^-(1^-)-$	Q_2(1400)	1/2(1^+)
→ M (940-953)		→ F_1(1540)	1 (A)	χ (3415)	$0^+(0^+)+$	→ K′(1400)	1/2(0^-)
η′ (958)	$0^+(0^-)+$	ρ′(1600)	$1^-(2^-)-$	→ χ (3455)		K^*(1430)	1/2(2^+)
δ (980)	$1^-(0^+)+$	A_3(1660)	$1^+(1^-)-$	P_c or χ(3510)	0^+(A)+	κ (1500)	1/2(0^+)
S^* (980)	$0^+(0^+)+$	ω (1670)	$0^-(3^-)-$	χ (3550)	0^+(N)+	→ L (1580)	1/2(2^-)
→ H (990)		g (1700)	$1^+(3^-)-$	→ χ (3590)		→ K^*(1650)	1/2(1^-)
φ (1020)	$0^-(1^-)-$	→ X (1690)		ψ (3685)	$0^-(1^-)-$	→ K_N(1700)	1/2
→ M (1033-1040)		→ A_4(1900)	1^-	ψ (3770)	$(1^-)-$	L region	1/2(A)
→ η_N(1080)	0^+(N)+	→ A_2(1900)	$1^-(4^+)+$	ψ (4030)	$(1^-)-$	K^*(1780)	1/2(3^-)
→ M (1150-1170)		S (1935)		ψ (4160)	$(1^-)-$	→ K^*(2200)	
A_1(1100-1300)	$1^-(1^+)+$	h (2040)	$0^+(4^+)+$	ψ (4415)	$(1^-)-$	→ I (2600)	
		→ T0(2150)	$0^+(2^+)+$	T (9460)	$(1^-)-$	Charmed (|C| = 1)	
B (1235)	$1^+(1^+)-$	→ T1(2190)	1	T (10020)	$(1^-)-$		
→ ρ′(1250)	$1^+(1^-)-$	→ X (2200)		T (10400)	$(1^-)-$	D (1870)	1/2(0^-)
f (1270)	$0^+(2^+)+$	→ U0(2350)	0			D^*(2010)	1/2(1^-)
→ η (1275)	$0^+(0^-)+$	→ U1(2400)	1			→ F (2030)	
D (1285)	$0^+(1^+)+$	→ $\bar{N}N$(1400-3600)				→ F^*(2140)	
ε (1300)	$0^+(0^+)+$	→ X (1900-3600)				→ Exotics	

Baryon Table

The following short list gives the status of all the Baryon States in the Data Card Listings. In addition to the status, the name, the nominal mass, and the quantum numbers (where known) are shown. States with three- or four-star status are included in the main Baryon Table; the others have been omitted because the evidence for the existence of the effect and/or for its interpretation as a resonance is open to considerable question.

N(939) P11 ****	Δ(1232) P33 ****	Λ(1115) P01 ****	Σ(1193) P11 ****	Ξ(1317) P11 ****
N(1470) P11 ****	Δ(1550) P31 **	Λ(1330) Dead	Σ(1385) P13 ****	Ξ(1530) P13 ****
N(1520) D13 ****	Δ(1650) S31 ****	Λ(1405) S01 ****	Σ(1480) *	Ξ(1630) **
N(1535) S11 ****	Δ(1670) D33 ****	Λ(1520) D03 ****	Σ(1560) **	Ξ(1680) S11 **
N(1540) P13 *	Δ(1690) P33 ***	Λ(1600) P01 **	Σ(1580) D13 **	Ξ(1820) 13 ***
N(1650) S11 ****	Δ(1890) F35 ****	Λ(1670) S01 ****	Σ(1620) S11 **	Ξ(1940) **
N(1670) D15 ****	Δ(1900) S31 **	Λ(1690) D03 ****	Σ(1660) P11 ***	Ξ(2030) 1 ***
N(1688) F15 ****	Δ(1910) P31 ****	Λ(1800) S01 ***	Σ(1670) D13 ****	Ξ(2120) *
N(1700) D13 ****	Δ(1950) F37 ****	Λ(1800) P01 **	Σ(1670) **	Ξ(2250) *
N(1710) P11 ****	Δ(1960) P33 **	Λ(1800) G09 *	Σ(1690) **	Ξ(2370) 1 **
N(1810) P13 ****	Δ(1960) D35 ***	Λ(1800) *	Σ(1750) S11 ***	Ξ(2500) **
N(1990) F17 ***	Δ(2160) ***	Λ(1815) F05 ****	Σ(1765) D15 ****	
N(2000) F15 **	Δ(2300) H39 *	Λ(1830) D05 ****	Σ(1770) P11 *	Ω(1672) P03 ****
N(2040) D13 **	Δ(2420) H311***	Λ(1860) P03 ***	Σ(1840) P13 *	
N(2100) S11 *	Δ(2500) G39 *	Λ(2010) *	Σ(1880) P11 **	Λ_c(2260) ***
N(2100) D15 **	Δ(2750) I313*	Λ(2020) F07 *	Σ(1915) F15 ****	
N(2190) G17 ****	Δ(2850) ***	Λ(2100) G07 ****	Σ(1940) D13 ***	Σ_c(2430) **
N(2200) G19 ****	Δ(2950) K315*	Λ(2110) F05 ***	Σ(2000) S11 *	
N(2220) H19 ****	Δ(3230) ***	Λ(2325) D03 *	Σ(2030) F17 ****	
N(2600) I111***		Λ(2350) ****	Σ(2070) F15 *	Dibaryons
N(2700) K113*		Λ(2585) ***	Σ(2080) P13 **	S = 0 *
N(2800) G19 *	Z0(1780) P01 *		Σ(2100) G17 *	S = -1 **
N(3030) ***	Z0(1865) D03 *		Σ(2250) ****	S = -2 *
N(3245) *	Z1(1900) P13 *		Σ(2455) ***	
N(3690) *	Z1(2150) *		Σ(2620) ***	
N(3755) *	Z1(2500) *		Σ(3000) **	
			Σ(3170) *	

**** Good, clear, and unmistakable.
*** Good, but in need of clarification cr not absolutely certain.
** Needs confirmation.
* Weak.

[See notes on N's and Δ's, Z*'s, Λ's and Σ's, Ξ*'s, and dibaryons at the beginning of those sections in the Baryon Data Card Listings; also see notes on individual resonances in the Baryon Data Card Listings.]

Particle[a]	I (J^P)[a] — estab.	π or K beam[b] p_{beam} (GeV/c) $\sigma = 4\pi\lambda\!\!\!^{-2}$ (mb)	Mass M[c] (MeV)	Full Width Γ[c] (MeV)	$M^2 \pm \Gamma M$[b] (GeV^2)	Partial decay mode[f]: Mode	Fraction[c] %	p or p_{max}[d] (MeV/c)
S=0 I=1/2 NUCLEON RESONANCES (N)								
p n	$1/2(1/2^+)$		938.3 939.6		0.880 0.883	See Stable Particle Table		
N(1470)	$1/2(1/2^+)P'_{11}$	p = 0.66 σ = 27.8	1400 to 1480	120 to 350 (200)	2.16 ±0.29	Nπ	50-65	420
						Nη	~18	*d*
						Nππ	~25	368
						[Nε	~ 7][e]	*d*
						[Δπ	~23][e]	177
						[Nρ	~ 7][e]	*d*
N(1520)	$1/2(3/2^-)D'_{13}$	p = 0.74 σ = 23.5	1510 to 1530	100 to 140 (125)	2.31 ±0.19	Nπ	~55	456
						Nππ	~45	410
						[Nε	< 5][e]	*d*
						[Nρ	~19][e]	*d*
						[Δπ	~23][e]	228
						Nη	< 1	*d*
N(1535)	$1/2(1/2^-)S'_{11}$	p = 0.76 σ = 22.5	1520 to 1560	100 to 250 (150)	2.36 ±0.23	Nπ	~40	467
						Nη	~55	182
						Nππ	~ 5	422
						[Nρ	~ 3][e]	*d*
						[Nε	~ 2][e]	*d*
						[Δπ	~ 1][e]	243
N(1650)	$1/2(1/2^-)S''_{11}$	p = 1.05 σ = 14.3	1620 to 1680	100 to 200 (150)	2.72 ±0.25	Nπ	~60	547
						Nππ	~30	511
						[Nε	<10][e]	*d*
						[Nρ	7-21][e]	*d*
						[Δπ	4-15][e]	344
						ΛK	~10	161
						ΣK	2-7	*d*
						Nη	~ 1	346

Baryon Table *(cont'd)*

Particle[a]	I (J^P)[a] estab.	π or K beam[b] p_{beam} (GeV/c) $\sigma = 4\pi\lambda\!\!\!^{-2}$ (mb)	Mass M[c] (MeV)	Full Width Γ[c] (MeV)	M^2 $\pm\Gamma M$[b] (GeV^2)	Partial decay mode[f] Mode	Fraction[c] %	p or p_{max}[d] (MeV/c)
N(1670)	$1/2(5/2^-)D'_{15}$	p = 1.00	1660 to	120 to	2.79	Nπ	~40	560
		σ = 15.6	1690	180	±0.26	Nππ	~60	525
				(155)		[Δπ	~50][e]	360
						[Nρ	~ 5][e]	d
						ΛK	< 0.3	200
						Nη	< 0.5	368
N(1688)	$1/2(5/2^+)F'_{15}$	p = 1.03	1670 to	110 to	2.85	Nπ	~60	572
		σ = 14.9	1690	140	±0.22	Nππ	~40	538
				(130)		[Nε	~22][e]	340
						[Nρ	~13][e]	d
						[Δπ	~18][e]	375
						Nη	< 0.3	388
N(1700)	$1/2(3/2^-)D''_{13}$	p = 1.05	1670 to	70 to	2.89	Nπ	~10	580
		σ = 14.3	1730	120[g]	±0.20	Nππ	~90	547
				(120)		[Nε	<40][e]	355
						[Nρ	< 5][e]	d
						[Δπ	15-40][e]	385
						ΛK	< 1	250
						Nη	~ 4	400
N(1710)	$1/2(1/2^+)P''_{11}$	p = 1.20	1680 to	100 to	2.92	Nπ	~20	587
		σ = 12.2	1740	140[h]	±0.21	Nππ	>50	554
				(120)		[Nε	15-40][e]	d
						[Nρ	40-65][e]	d
						[Δπ	10-20][e]	393
						ΛK	< 5	264
						ΣK	~10	138
						Nη	2-20[i]	410
N(1810)	$1/2(3/2^+)P''_{13}$	p = 1.26	1690 to	150 to	3.28	Nπ	~17	652
		σ = 11.5	1800	250	±0.36	Nππ	~70	624
				(200)		[Nε	~20][e]	468
						[Nρ	45-70][e]	297
						[Δπ	~20][e]	471
						ΛK	1-4	386
						ΣK	~ 2	307
						Nη	< 5	503
N(1990)	$1/2(7/2^+)F_{17}$	p = 1.62	1950 to	100 to	3.96	Nπ	~5	772
		σ = 8.35	2050	400	±0.50	Nη	~3	655
				(250)		ΛK	seen	562
						ΣK	seen	506
N(2190)	$1/2(7/2^-)G_{17}$	p = 2.07	2120 to	<400	4.80	Nπ	~15	888
		σ = 6.21	2180	(250)	±0.55	Nη	~ 2	790
						ΛK	< 1	712
N(2200)	$1/2(9/2^-)G_{19}$	p = 2.10	2130 to	200 to	4.84	Nπ	~10	894
		σ = 6.12	2270	350	±0.55	Nη	~ 2	810
				(250)				
N(2220)	$1/2(9/2^+)H_{19}$	p = 2.14	2150 to	~300	4.93	Nπ	~20	905
		σ = 5.97	2300	(300)	±0.67	Nη	~ 1	811
N(2600)	$1/2(11/2^-)I_{111}$	p = 3.26	2580 to	>300	6.76	Nπ	~ 5	1014
		σ = 3.67	2700	(400)	±1.04			
N(3030)	1/2(?)	p = 4.41	~3030	~400	9.18	Nπ	(J +1/2) x	1366
		σ = 2.62		(400)	±1.21		<0.1[k]	

Baryon Table *(cont'd)*

Particle[a]	I (J^P)[a] —— estab.	π or K beam[b] p_{beam} (GeV/c) $\sigma = 4\pi\lambda\!\!\!^{-2}$ (mb)	Mass M[c] (MeV)	Full Width Γ[c] (MeV)	M^2 $\pm\Gamma M$[b] (GeV^2)	Partial decay mode[f] Mode	Fraction[c] %	p or p_{max}[d] (MeV/c)
			S=0 I=3/2 DELTA RESONANCES (Δ)					
Δ(1232)	$3/2(3/2^+)P'_{33}$ (estab.)	p = 0.30 σ = 94.3	1230 to 1234	110 to 120 (115)	1.52 ±0.14	Nπ $N\pi^+\pi^-$	~99.4 ~ 0	227 80
	Δ(++) Pole position:[l] $M - i\Gamma/2 = (1211.0 \pm 0.8) - i(49.9 \pm 0.6)$ Δ(0) Pole position:[l] $M - i\Gamma/2 = (1210.5 \pm 1.0) - i(52.9 \pm 1.0)$							
Δ(1650)	$3/2(1/2^-)S'_{31}$ (estab.)	p = 0.96 σ = 16.4	1600 to 1650	120 to 160 (140)	2.72 ±0.23	Nπ Nππ [Nρ [Δπ	~32 ~65 <50][e] ~40][e]	547 511 *d* 344
Δ(1670)	$3/2(3/2^-)D_{33}$ (estab.)	p = 1.00 σ = 15.6	1630 to 1740	190 to 300 (200)	2.79 ±0.33	Nπ Nππ [Nρ [Δπ	~15 ~85 ~40][e] <50][e]	560 525 *d* 361
Δ(1690)	$3/2(3/2^+)P''_{33}$ (estab.)	p = 1.03 σ = 14.9	1500 to 1900[m]	150 to 350 (250)	2.86 ±0.42	Nπ Nππ [Nρ [Δπ	~20 ~80 <10][e] 30-45][e]	573 540 *d* 377
Δ(1890)	$3/2(5/2^+)F_{35}$ (estab.)	p = 1.42 σ = 9.88	1890 to 1930	250 to 400 (250)	3.57 ±0.47	Nπ Nππ [Nρ [Δπ ΣK	~15 ~80 ~60][e] 10-30][e] < 3	704 677 403 531 400
Δ(1910)	$3/2(1/2^+)P''_{31}$ (estab.)	p = 1.46 σ = 9.54	1850 to 1950	200 to 330 (220)	3.65 ±0.42	Nπ Nππ [Nρ [Δπ ΣK	20-25 >40 <40][e] small][e] 2-20	716 691 429 545 420
Δ(1950)	$3/2(7/2^+)F_{37}$ (estab.)	p = 1.54 σ = 8.90	1910 to 1950	200 to 340 (240)	3.80 ±0.47	Nπ Nππ [Nρ [Δπ ΣK	~40 >30 ~20][e] ~30][e] < 1	741 716 471 574 460
Δ(1960)	$3/2(5/2^-)D_{35}$ (estab.)	p = 1.56 σ = 8.75	1890 to 1940	150 to 300 (200)	3.84 ±0.39	Nπ ΣK	4-12 <10	748 469
Δ(2160)[n]	$3/2(\ ?^-)$	p = 2.00 σ = 6.46	2150 to 2280	200 to 440 (300)	4.67 ±0.65	Nπ	(J +1/2)x = 0.2 - 1.2[k]	870
Δ(2420)	$3/2(11/2^+)H_{311}$ (estab.)	p = 2.64 σ = 4.68	2380 to 2450	300 to 500 (300)	5.86 ±0.73	Nπ	~10	1023
Δ(2850)	$3/2(\ ?^+)$	p = 3.85 σ = 3.05	2800 to 2900	~400 (400)	8.12 ±1.14	Nπ	(J +1/2)x ~0.25[k]	1266
Δ(3230)	3/2(?)	p = 5.08 σ = 2.25	3200 to 3350	~440 (440)	10.43 ±1.42	Nπ	(J +1/2)x ~0.05[k]	1475

Z*	Evidence for states with strangeness +1 is inconclusive. See the Baryon Data Card Listings for data and discussion.

Baryon Table *(cont'd)*

Particle[a]	I (J^P)[a] — estab.	π or K beam[b] p_{beam} (GeV/c) $\sigma = 4\pi\lambda\!\!\!^{-2}$ (mb)	Mass M[c] (MeV)	Full Width Γ[c] (MeV)	M^2 $\pm\Gamma M$[b] (GeV^2)	Partial decay mode[f]: Mode	Fraction[c] %	p or[d] p_{max} (MeV/c)
S=-1 I=0 LAMBDA RESONANCES (Λ)								
Λ	$0(1/2^+)$		1115.6		1.245	See Stable Particle Table		
Λ(1405)	$0(1/2^-)S'_{01}$	Below K^-p threshold	1405 ±5[o]	40 ± 10[o] (40)	1.97 ±0.06	Σπ	100	142
Λ(1520)	$0(3/2^-)D'_{03}$	p = 0.389 σ = 84.5	1519.5 ±1.5[o]	15.5 ± 1.5[o] (16)	2.31 ±0.02	$N\bar{K}$	46 ± 1	234
						Σπ	42 ± 1	258
						Λππ	10 ± 1	250
						Σππ	0.9 ± 0.1	140
Λ(1670)	$0(1/2^-)S''_{01}$	p = 0.74 σ = 28.5	1660 to 1680	20 to 60 (40)	2.79 ±0.07	$N\bar{K}$	15-25	410
						Λη	15-35	64
						Σπ	20-60	393
Λ(1690)	$0(3/2^-)D''_{03}$	p = 0.78 σ = 26.1	1690 ±10[o]	50 to 70 (60)	2.86 ±0.10	$N\bar{K}$	20-30	429
						Σπ	20-40	409
						Λππ	~25	415
						Σππ	~20	352
Λ(1800)	$0(1/2^-)S'''_{01}$	p = 1.16 σ = 14.2	1700 to 1850	200 to 400 (300)	3.50 ±0.56	$N\bar{K}$	25-40	525
						Σπ	seen	488
						Σ(1385)π	seen	346
						$N\bar{K}^*(892)$	seen	*d*
Λ(1815)	$0(5/2^+)F'_{05}$	p = 1.05 σ = 16.7	1820 ±5[o]	70 to 90 (80)	3.29 ±0.15	$N\bar{K}$	55-65	542
						Σπ	5-15	508
						Σ(1385)π	5-10	362
Λ(1830)	$0(5/2^-)D_{05}$	p = 1.09 σ = 15.8	1810 to 1830	60 to 110 (95)	3.35 ±0.17	$N\bar{K}$	<10	554
						Σπ	35-75	519
						Σ(1385)π	>15	375
Λ(1860)	$0(3/2^+)P'_{03}$	p = 1.14 σ = 14.7	1850 to 1920	60 to 200 (100)	3.46 ±0.19	$N\bar{K}$	15-40	576
						Σπ	3-10	534
						Σ(1385)π	seen	396
						$N\bar{K}^*(892)$	seen	162
Λ(2100)	$0(7/2^-)G_{07}$	p = 1.68 σ = 8.68	2080 to 2120	100 to 300 (250)	4.41 ±0.53	$N\bar{K}$	~30	748
						Σπ	~ 5	699
						Λη	< 3	617
						ΞK	< 3	483
						Λω	< 8	443
						$N\bar{K}^*(892)$	10-20	514
Λ(2110)	$0(5/2^+)F''_{05}$	p = 1.70 σ = 8.48	2080 to 2140	150 to 250 (200)	4.45 ±0.42	$N\bar{K}$	5-25	756
						Σπ	<40	709
						$N\bar{K}^*(892)$	20-60	524
						Λω	seen	454
						Σ(1385)π	seen	589
Λ(2350)	$0(9/2^+)$ —	p = 2.29 σ = 5.85	2340 to 2420	100 to 250 (120)	5.52 ±0.28	$N\bar{K}$	~12	913
						Σπ	~10	865
Λ(2585)	0(?) —	p = 2.91 σ = 4.37	~2585	~300 (300)	6.68 ±0.78	$N\bar{K}$	(J+1/2)x ~1.0[k]	1058
S=-1 I=1 SIGMA RESONANCES (Σ)								
Σ	$1(1/2^+)$		(+)1189.4 (0)1192.5 (-)1197.3		1.415 1.422 1.434	See Stable Particle Table		
Σ(1385)	$1(3/2^+)P'_{13}$	Below K^-p threshold	(+)1382.3±0.4 S = 1.6[p]	(+)35±2 S = 2.2[p]	1.92 ±0.05	Λπ	88 ± 2	208
						Σπ	12 ± 2	117

Baryon Table *(cont'd)*

Particle[a]	I (J^P)[a] —— estab.	π or K beam[b] p_{beam} (GeV/c) $\sigma = 4\pi\lambda\!\!\!^{-2}$ (mb)	Mass M[c] (MeV)	Full Width Γ[c] (MeV)	M^2 $\pm\Gamma M$[b] (GeV^2)	Partial decay mode[f] Mode	Fraction[c] %	p or p_{max}[d] (MeV/c)
Σ(1660)[q]	$1(1/2^+)P'_{11}$	p = 0.72 σ = 30.1	1580 to 1690	30 to 200 (100)	2.76 ±0.17	$N\bar{K}$ Σπ Λπ	<30 seen seen	402 383 440
Σ(1670)	$1(3/2^-)D''_{13}$	p = 0.74 σ = 28.5	1675 ±10[o]	40 to 60 (50)	2.79 ±0.08	$N\bar{K}$ Σπ Λπ	5-15 20-60 < 20	410 387 447
Σ(1750)	$1(1/2^-)S''_{11}$	p = 0.91 σ = 20.7	1730 to 1820	50 to 160 (75)	3.06 ±0.13	$N\bar{K}$ Λπ Σπ Ση	10-40 5-20 < 8 15-55	483 507 450 54
Σ(1765)	$1(5/2^-)D_{15}$	p = 0.94 σ = 19.6	1774 ±7[o]	105 to 135 (120)	3.12 ±0.21	$N\bar{K}$ Λπ Λ(1520)π Σ(1385)π Σπ	~41 ~14 ~19 ~ 9 ~ 1	496 518 187 315 461
Σ(1915)	$1(5/2^+)F'_{15}$	p = 1.25 σ = 13.0	1905 to 1930	70 to 160 (100)	3.67 ±0.19	$N\bar{K}$ Λπ Σπ Σ(1385)π	5-15 10-20 seen < 5	612 619 568 437
Σ(1940)[q]	$1(3/2^-)D'''_{13}$	p = 1.32 σ = 12.0	1900 to 1950	150 to 300 (220)	3.76 ±0.43	$N\bar{K}$ Λπ Σπ Λ(1520)π Δ(1232)$\bar{K}$ $N\bar{K}^*$(892) Σ(1385)π	<20 seen seen seen seen seen seen	678 680 589 370 410 320 461
Σ(2030)	$1(7/2^+)F_{17}$	p = 1.52 σ = 9.93	2020 to 2040	120 to 200 (180)	4.12 ±0.37	$N\bar{K}$ Λπ Σπ ΞK Λ(1520)π Σ(1385)π Δ(1232)$\bar{K}$ $N\bar{K}^*$(892)	~20 ~20 5-10 < 2 10-20 5-15 10-20 < 5	700 700 652 412 429 530 498 438
Σ(2250)[q]	1(?)[r]	p = 2.04 σ = 6.76	2200 to 2300	50 to 150 (100)	5.06 ±0.22	$N\bar{K}$ Λπ Σπ	< 10 seen seen	849 841 801
Σ(2455)	1(?)	p = 2.57 σ = 5.09	~2455	~120 (120)	6.03 ±0.29	$N\bar{K}$	(J+1/2)x ~0.2[k]	979
Σ(2620)	1(?)	p = 2.95 σ = 4.30	~2600	~200 (200)	6.86 ±0.52	$N\bar{K}$	(J+1/2)x ~0.3[k]	1064

S=−2 I=1/2 CASCADE RESONANCES (Ξ)

Particle	I (J^P)		Mass M (MeV)	Full Width Γ (MeV)	M^2 $\pm\Gamma M$ (GeV^2)	Mode	Fraction %	p or p_{max} (MeV/c)
Ξ	$1/2(1/2^+)$		(0)1314.9 (−)1321.3		1.729 1.746	See Stable Particle Table		
Ξ(1530)	$1/2(3/2^+)P_{13}$		(0)1531.8±0.3 S = 1.3[p] (−)1535.0±0.6	(0)9.1±0.5 (−)10.1±1.9 (10)	2.34 ±0.02	Ξπ	100	144
Ξ(1820)	1/2(3/2)		1823 ±6[o]	20^{+15}_{-10}[o] (20)	3.31 ±0.04	Λ$\bar{K}$ Ξ(1530)π Σ$\bar{K}$ Ξπ	~45 ~45 ~10 small	396 234 306 413
Ξ(2030)	1/2(?)		2024 ±6[o]	16^{+15}_{-5}[o] (16)	4.12 ±0.03	Σ$\bar{K}$ Λ$\bar{K}$ Ξπ Ξ(1530)π	~80 ~20 small small	524 587 573 418

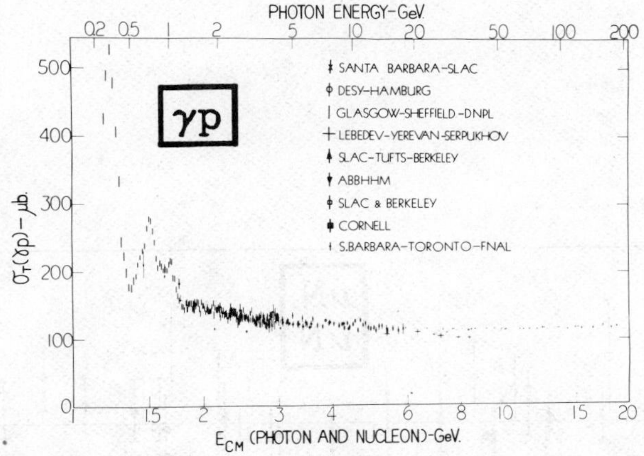

γp total cross section versus photon energy (top scale) and photon-plus-nucleon total center-of-mass energy (lower scale).

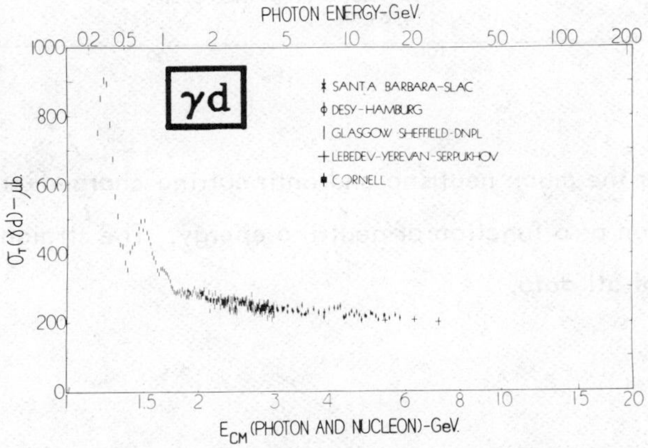

γd total cross section versus photon energy (top scale) and photon-plus-single-nucleon total center-of-mass energy (lower scale).

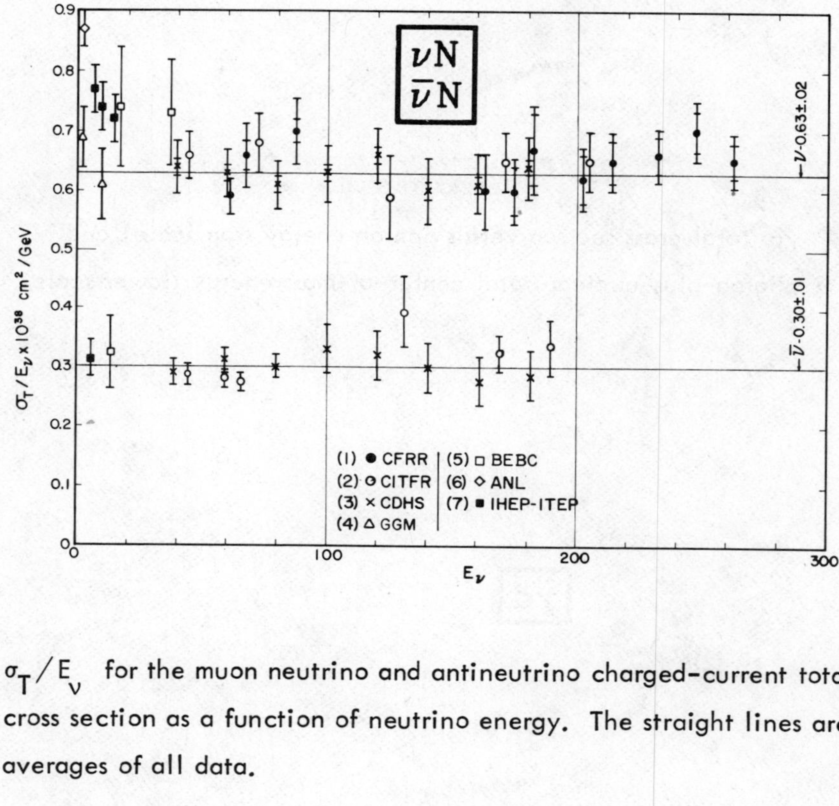

σ_T / E_ν for the muon neutrino and antineutrino charged-current total cross section as a function of neutrino energy. The straight lines are averages of all data.

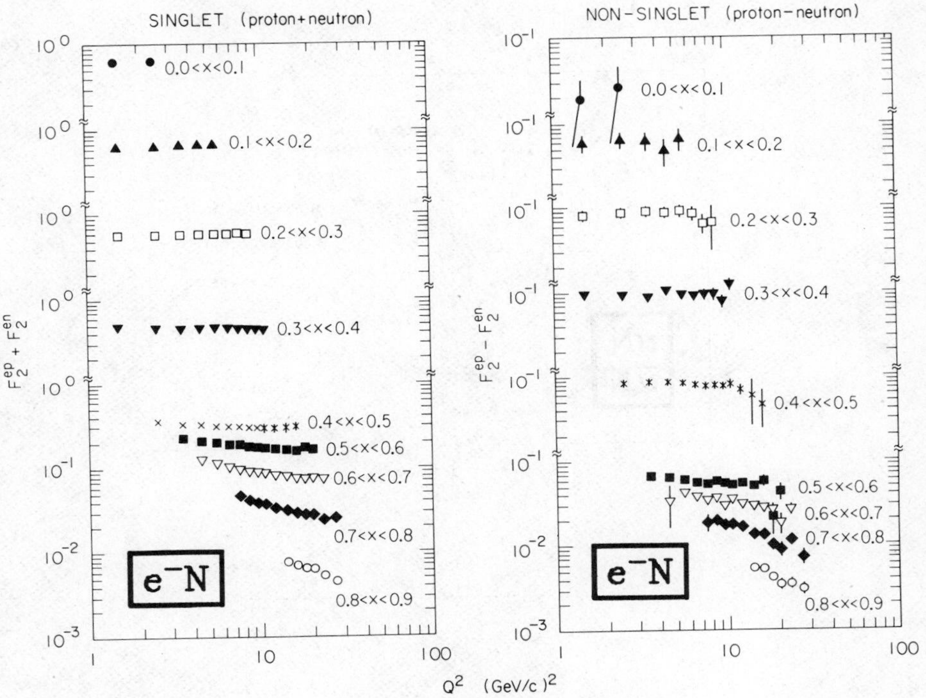

F_2 structure functions derived from inelastic electron – nucleon data taken at SLAC with recoil mass > 2 GeV and four-momentum transfer squared $q^2 > 1$ (GeV/c)^2. $R \equiv \sigma_L/\sigma_T = 0.21$ was assumed.

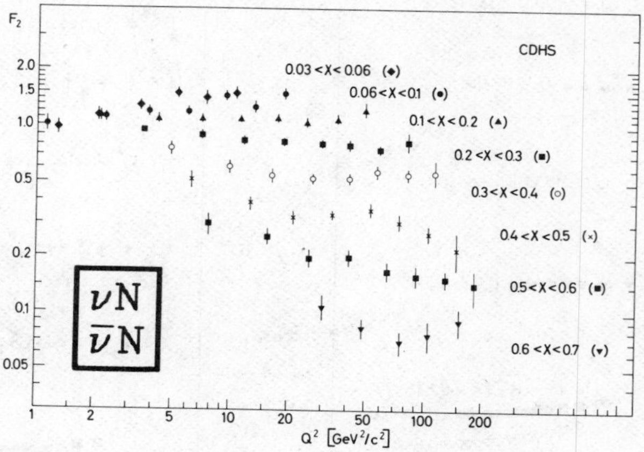

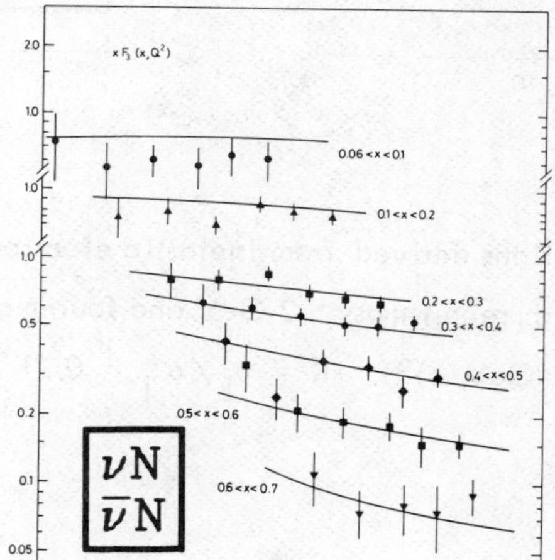

Nucleon structure functions as measured by the CDHS collaboration in high energy (30–200 GeV) charged – current neutrino– and anti-neutrino–nucleon scattering.

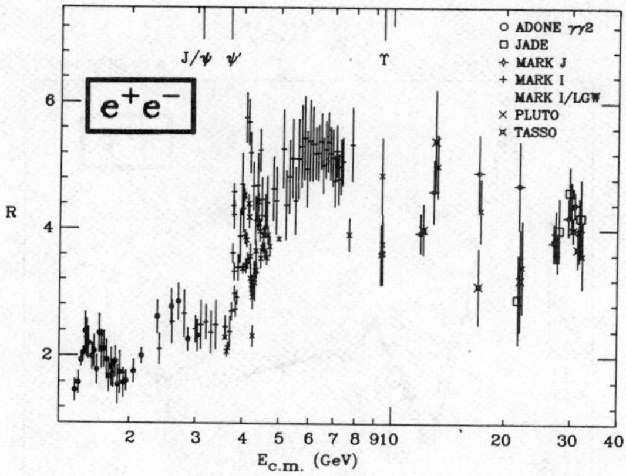

Measurements of $R \equiv \sigma(e^+e^- \to \text{hadrons})/\sigma(e^+e^- \to \mu^+\mu^-)$, where the annihilation proceeds via one photon. [The numerator includes $q\bar{q}$ and $\tau^+\tau^-$ production.]

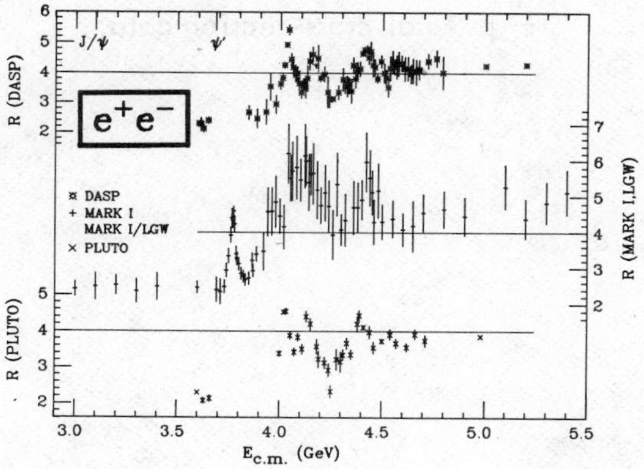

An expanded view of R measurements around charm threshold.

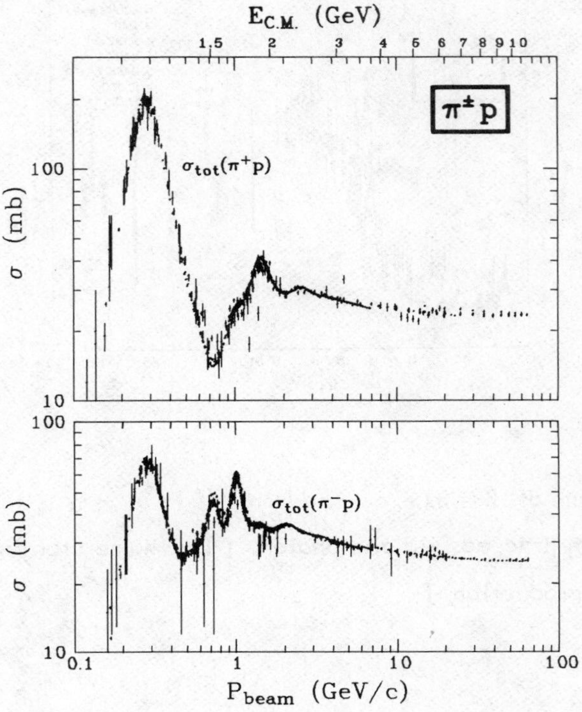

$\pi^{\pm}$ p total cross-section data.

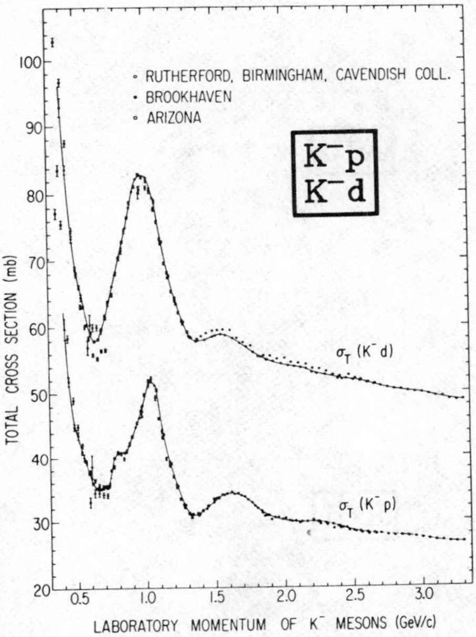

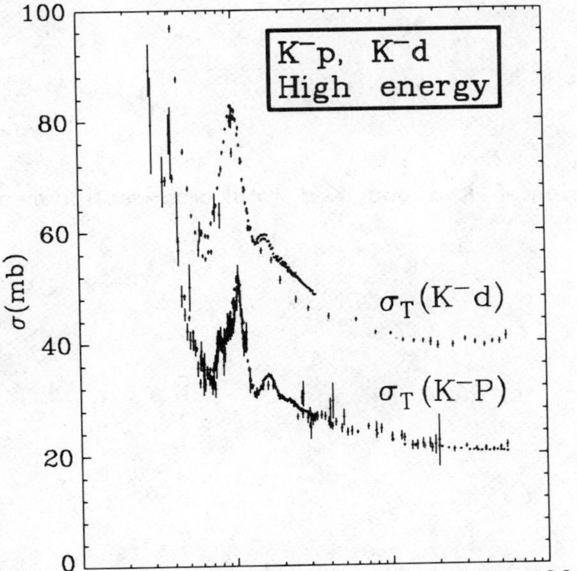

K^-p and K^-d total cross-section data.

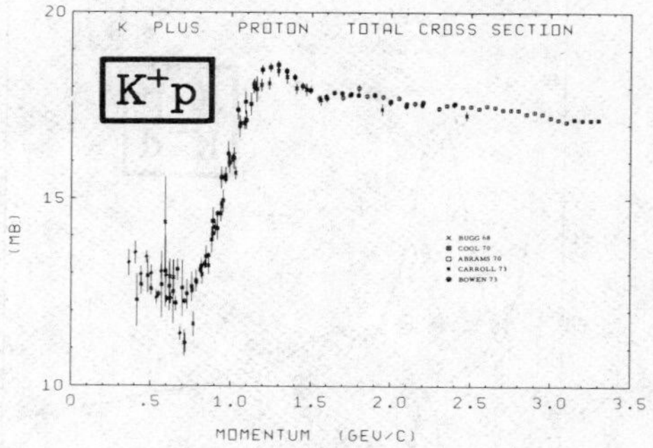

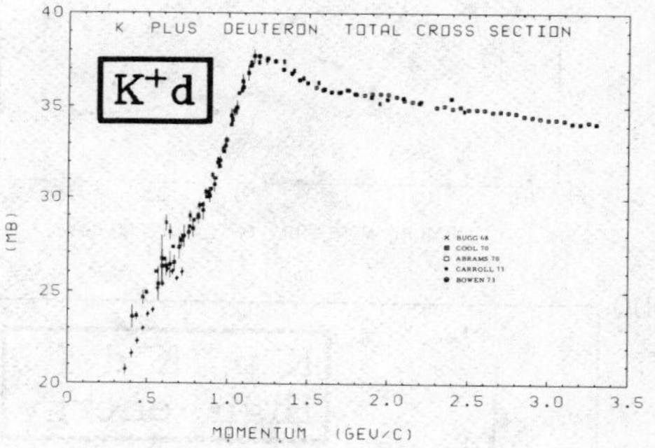

Compilation of K^+p and K^+d total cross-section measurements.

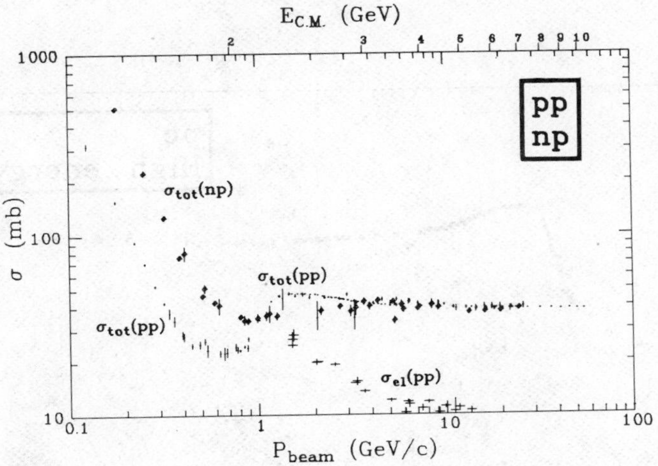

pp and np cross sections.

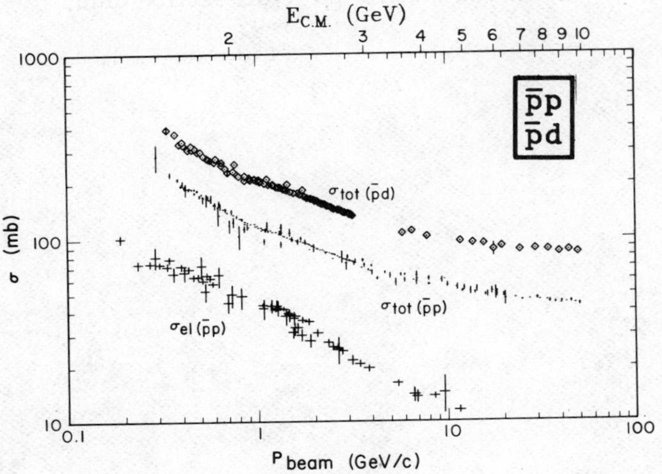

$\bar{p}p$ and $\bar{p}d$ cross sections.

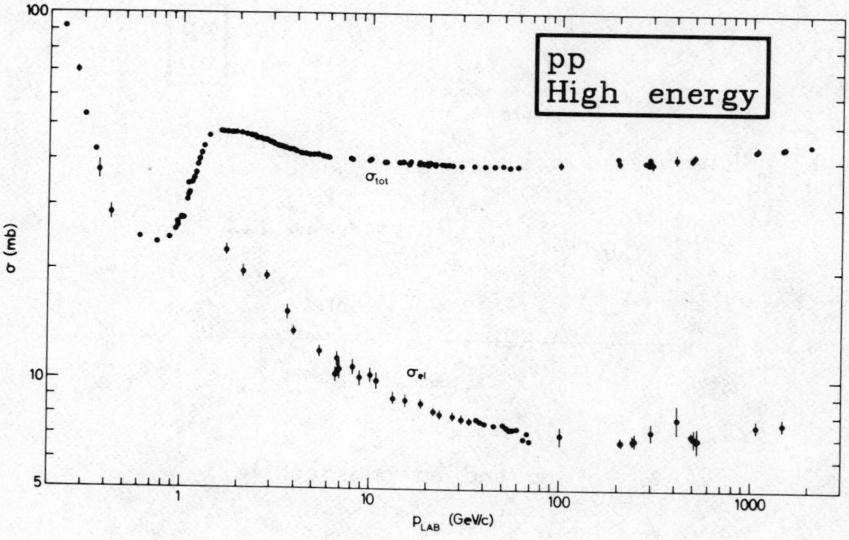

Total and elastic pp cross-section data.

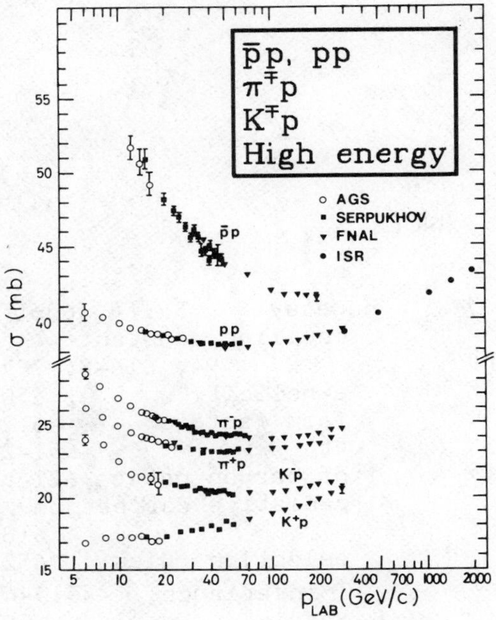

$\bar{p}p$, pp, π^-p, π^+p, K^-p and K^+p total cross sections.

$\bar{p}n$, np, pn, K^-n and K^+n total cross sections.

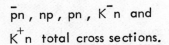

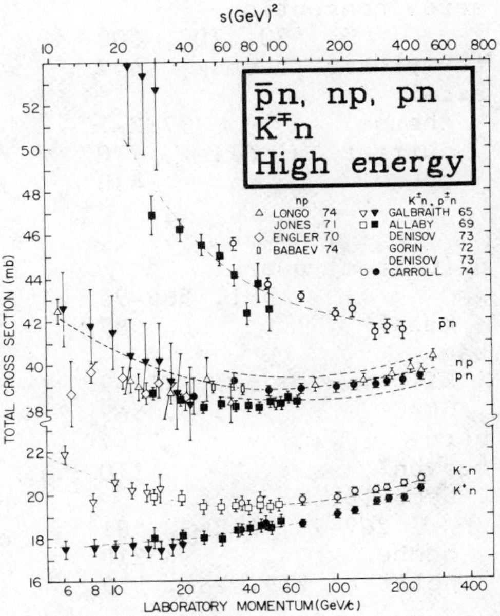

INDEX